Jihād

Jihād

From Qur'ān to bin Laden

Richard Bonney

Foreword by
Sheikh Dr Zaki Badawi

palgrave
macmillan

First published 2004 by
PALGRAVE MACMILLAN
Houndmills, Basingstoke, Hampshire RG21 6XS and
175 Fifth Avenue, New York, N.Y. 10010
Companies and representatives throughout the world

PALGRAVE MACMILLAN is the global academic imprint of the Palgrave Macmillan division of St. Martin's Press, LLC and of Palgrave Macmillan Ltd.
Macmillan® is a registered trademark in the United States, United Kingdom and other countries. Palgrave is a registered trademark in the European Union and other countries.

ISBN 1–4039–3372–3 hardback

This book is printed on paper suitable for recycling and made from fully managed and sustained forest sources.

A catalogue record for this book is available from the British Library.

Library of Congress Cataloging-in-Publication Data
Bonney, Richard, 1947–
Jihad : from Qur'an to Bin Laden / Richard Bonney.
p. cm.
Includes bibliographical references and index.
ISBN 1–4039–3372–3 (cloth)
1. Jihad. 2. Islam and politics. 3. Religion and politics. 4. Terrorism—Religious aspects—Islam. I. Title.

BP182.B66 2004
297.7'2—dc22

2004051152

10 9 8 7 6 5 4 3 2 1
13 12 11 10 09 08 07 06 05 04

Printed and bound in Great Britain by
Antony Rowe Ltd, Chippenham and Eastbourne

Contents

Foreword

Sheikh Dr Zaki Badawi

Dr Badawi is Chair of the Imams and Mosques Council, UK; Co-Founder of the Three Faiths Forum; and Vice-Chair of the World Congress of Faiths.

The great majority of Muslims world-wide were horrified by the events of 11 September 2001. At the time, I stated:

> The atrocity of September 11 is a violation of Islamic law and ethics. Neither the people who were killed or injured, nor the properties that were destroyed, qualified as legitimate targets in any system of law, especially Islamic law… Taking revenge on the innocent as sanctioned by tribalism is abhorrent to Islam as it is abhorrent to ethical principle…

Though many Muslims condemned the atrocity, and though no religious leader of any standing condoned what happened, the fact is that the image of Islām has been damaged in several ways by the events of 9/11. Firstly, it has increased fear and suspicion of Muslims in the West and led to a rise in prejudice and stereotyping. Secondly, it has isolated states with a Muslim majority population, who have had to satisfy the demands of the United States by going that much further than others in the so-called 'war on terrorism'. Thirdly, two states have been attacked and occupied by the United States and its allies in wars of occupation of questionable legitimacy. The common perception is that the United States seems to be using the 'war on terrorism' as an excuse for a state of permanent war and as justification for the new doctrine, alien in international law, of the pre-emptive strike. Finally, the reputation of Islām itself as a peaceful and tolerant religion has been damaged. Because Osama bin Laden and other terrorist leaders use an historically inaccurate and distorted view of the Islamic concept of just war (*jihād*) to justify their actions, Islām itself has been depicted by its enemies and estranged friends as condoning unethical, unlimited and almost unthinkable acts of violence and terrorism, which it does not.

This is why this new reappraisal of the evolution of the concept of *jihād* in Islamic history by Professor Richard Bonney is particularly timely and welcome. It has needed someone who is both sympathetic to the mainstream Muslim position yet who stands outside the world of Islam itself to explain the nature of the problem both to Muslims themselves and to non-Muslims, particularly in the West. Richard Bonney does not believe in any inevitability of a 'clash of civilizations', though he is fully aware that the present 'war on terrorism' may slide into something leading to the dreaded clash of civilizations, alienating the Islamic world. Instead, he argues cogently in this book – refreshingly for Muslims and non-Muslims alike – that violent Islamist *jihādists* of the present and recent generations are a minority aberration who have created a fictional world of conflict to suit their own interests in seeking power and in an attempt to provide a unifying ideology which seeks to mobilize and radicalize various disunited but real political, economic and social discontents in the Muslim world. The threat posed by the radical or revolutionary Islamists is directed equally at the regimes in majority Muslim countries as at the United States and its allies.

Whereas Orientalists in the West have sought to depict *jihād* as a state of permanent war, in which its proponents will not rest until they have has converted everyone else to Islām or to accept a position of inferiority under an Islamic state, Richard Bonney correctly depicts the *jihād* as two concepts which coexist: one is the Muslim's struggle against his or her own lower nature, the struggle within the self (*jihād al-nafs*); the other, more political concept, is the Muslim view of the 'just war'. As he observes, this has changed and developed over time. At first, in the early centuries of Islam when its borders were not settled, it has to be admitted that it was a warlike concept. But this view changed, once an Islamic world had been established and had stabilized its frontiers: then the world of Islām accepted that it did in fact (and should also in theory) live in harmony with the world outside or beyond Islām. Richard Bonney argues that to use concepts of *jihād* from the early centuries of Islām's development to define the modern Islamic understanding of just war is clearly anachronistic as well as damaging to the reputation of Islām itself. Mainstream Muslims can only welcome this reappraisal of the significance of *jihād* in their history and hope that it is read as widely in the Islamic world as it undoubtedly will be in the West. Indeed, it is mandatory reading for all who seek to avert any 'clash of civilizations' and to isolate and defeat radical elements who seek to subvert the rules of ethics and justice to pursue their own wild and unworkable political ambitions.

Sheikh Dr Zaki Badawi
The Muslim College
London
16 September 2003

Author's Preface

> The academics say, so consequently do the politicians, that… violence and terrorism actually goes back deep into the roots of Islām, into its religious roots. The call of *jihād*… with which the Qur'ān is full, the division of the world into tribes, the [abode of Islām] *Dār al-Islām* and [the abode of war] *Dār al-Ḥarb*, and the dream of world domination, are deemed to be the roots of Islām. This is why the terminology is carefully tailored to fit this pattern. If Pakistan makes a bomb, a nuclear bomb, it is christened as an Islamic bomb. The bomb which was dropped on Hiroshima was not a Christian bomb, and the bomb which was made by Israel not is not a Jewish bomb, the bomb made by India is not a Hindu bomb but if Pakistan succeeds in making a bomb, it is an Islamic bomb…
>
> Khurram Murad, 1998[1]

> Rachid Ghannouchi [Rāshid al-Ghannūshī] may have something very sensible to say, but how many people can read him? In other words, he is not accessible to everyone.
>
> M. Nejatullah Siddiqi, 1998[2]

'In September 1970, the Popular Front for the Liberation of Palestine hijacked four aeroplanes; three of them were taken to Jordan and one to Cairo. On 13 September, the three in Jordan were blown up in front of the assembled world media. This was the starting point of international terrorism appearing before a worldwide audience.'[3] Ever since this time, violent Islamist movements have sought to capture media attention and have had no difficulty in achieving success. That very success has led to a potentially fatal confusion of issues and terms in the mind of the politicians, opinion-makers and the general public in the West. It is a confusion of terms made most manifest in the widespread use of the expressions 'Islamic terrorism/Islamic terrorist'.

In this study, which is intended to be read both in the West and in the Islamic world, a careful (and it is hoped, consistent) use of terms is employed. *There is no such thing, in our view, as Islamic terrorism. There is terrorism perpetrated by violent Islamists*, that is to say, by those who are acting in a political cause but who seek to motivate people, gaining support and recruits thereby, by using the terminology of the faith of Islām, and in particular the ambiguous but key concept of *jihād*, the subject of this book.

Violent Islamists have to be distinguished from two other groups of Muslims. One group of Muslims, the minority, are those peaceful Islamists who view the interrelationship between their faith and politics as the central organizational principle of their political activity. They may be considered radical. They may, abusively, be called 'terrorists' even if they have publicly eschewed violence. In such cases, the authoritarian regime in power is extending the definition of 'terrorism' in an unwarranted way. The history of the FIS in Algeria (see Chapter 11) – which won the first round of the national elections in 1991, but was then declared an illegal organization by the Algerian military – is a case in point. Similarly, Rāshid al-Ghannūshī's political party in Tunisia has been outlawed by the government and he has been forced to live in exile. That is one reason why his voice has not been heard, as Nejatullah Siddiqi states in the second of our opening quotations. (The other reason that he is not heard in the West is because he writes in Arabic, a point to which we will return.) We consider that dialogue with peaceful Islamist leaders and peaceful Islamist movements is both necessary and potentially fruitful. There is every interest in seeking to include them in such a dialogue on the agenda of 'enlightened moderation in Islām', both in the positive interests of humanity and also to preclude any later abandonment by them of the principle of non-violence and subsequent inclusion into violent Islamist movements.

A second group of Muslims, the great majority of them, are not Islamists at all, that is to say, they may or may not recognize the interrelationship between their faith and politics but they certainly do not make this relationship the central organizational principle of their political activity. They do not support Islamist parties, whether these are peaceful political movements or militant organizations seeking to overthrow the existing political and social system. It is this mainstream body of Muslims to whom the agenda of 'enlightened moderation' is addressed (Conclusion) and with whom, for the foreseeable future, the main dialogue between the West and Islām will take place.

This book is intended as a helpful contribution to such a process of dialogue, since it is clear that a 'false consciousness', a misunderstanding of the nature of Islamic history, has potentially devastating consequences in perpetuating myths and misconceptions of 'the other'. Such myths and misconceptions can exist among Muslims quite as much as among non-Muslims: there is no alternative to an objective account of the historical context, causation, achievements and consequences of *jihād* in history. This book does not claim to address all the issues completely, because in the existing state of scholarship this is an impossible task. It does, however, mark a considerable advance in an area of widespread concern and controversy, especially in the West.

The book will no doubt be scorned by some within the academic community: it makes no use of Arabic sources in Arabic; it is a work of synthesis, reliant on the detailed research of others; finally, perhaps worst of all, it unashamedly makes no excuse for using internet resources.

On the first point, the use of Arabic sources has two answers, from almost the first and last words of this book. The second quotation to this preface was the remark that 'Rachid Ghannouchi [Rāshid al-Ghannūshī] may have something very sensible to say, but how many people can read him? In other words, he is not accessible to everyone.' The last comments in the Conclusion concern the Muslim community at present 'hiding their light under a bushel' and not projecting effectively to the West the very real and positive progress that has taken place within the Islamic community on issues of concern to the West. The reverence for Arabic is understandable because of the nature of the Islamic revelation (see Chapter 1). It is, however, an obstacle to effective communication in the twenty-first century. The languages of the West, above all English, have to be recognized as the medium of communication for the debate on the history and future of Islām. This is simply practical politics, not an issue of principle: if Muslims fail to get their point of view across so that it is understood in the West, the misunderstanding will continue; moreover, the role of interpretation is surrendered to scholars in the West, some of whom try hard to write without bias and consult their Muslim friends in order to do so; others, regrettably, do not.

As to the second point, that this book is a work of synthesis, the answer is a simple one: how else could a book with this range be written? There is a vast literature in the languages accessible to the author, not all of which has been encompassed. The richness and diversity of Islamic history is so great that no one scholar can hope to do it justice. The aim of the book is a more modest one. It is to try to inform people in the West about the richness and diversity of Islamic history; to explain that history does not 'determine' the present, let alone the future; and, if possible, to stimulate constructive debate and discussion as well as further research in areas where the author is only too well aware that he has only 'scratched the surface' of the problems. But an overview is needed to explain why the issues are so important.

The third issue, the use of internet resources, is in the author's view a non-issue. There are historians and social scientists who may believe that such resources are somehow 'not for them' or even 'beneath them' and that real knowledge is confined to books. No student of *jihād* can afford to take this view. For *jihād* is out and about, and very loud, in cyberspace (see Chapter 11). Muslims and non-Muslims ignore what is being said on *jihād*, and about the faith of Islām, whether accurate or inaccurate, in cyberspace at their peril. The violent Islamists have made it their medium *par excellence*. To understand them, you have to consult their statements. It is time that the Muslim mainstream takes its 'public diplomacy' more seriously and projects itself more effectively using the same medium for communication (see Conclusion).

The intentions of the author in writing this book are entirely constructive. There is no intention to show disrespect of any person's faith, be it Islām, Judaism, Christianity or any other. Dates have been given in two calendars (Muslim Era [ME]/Common Era [CE]) up to the modern period (in the chapters before Chapter 11). It is not customary in British academic discourse for the name 'Muḥammad' and the term 'Prophet' to be given the additional designation of PBUH, 'Peace and Blessings be Upon Him', but the author is happy for Muslims to read and understand the text in that way.

This is a work of history, not theology. There is no wish here to undermine the duties or purpose of Islamic scholars who are the experts in theology. It is obvious, however, that in the history of a religious idea there is necessarily some overlap between the two disciplines of history and theology: the sources for the early history of the idea are in essence the same as for Islamic theology. The method of citation used for a *ḥadīth* here arises from the nature of the material displayed in the MSA–USC *ḥadīth* database <www.usc.edu/dept/MSA/reference/searchhadith.html>. Thus, for example, the entry in the database for Al-Bukhārī, volume 1, book 2, number 25 (<www.usc.edu/dept/MSA/fundamentals/hadithsunnah/bukhari/002.sbt.html#001.002.025>) would be cited as Al-Bukhārī 1/2/25 in the endnotes. Where there is only a book and number given, as for Abu-Dāwūd, Book 14, Number 2510, then the reference would be cited as Abu-Dāwūd 14/2510 in the endnotes.

This book is a *jihād* in itself, not only to increase understanding, especially in the West, of the varieties of *jihād* in history, but also to facilitate greater understanding of mainstream Islām, whether of the Sunnī or Shī'a traditions. This discussion takes place principally in the Conclusion. The earlier chapters analyse different ideas of *jihād* and their application within specific historical contexts. In Chapter 2, the political actions of the Four Rightly Guided Caliphs are considered: some actions may have been more successful than others but this does not imply that the Caliphs were other than Rightly Guided. Chapter 9 on the Shī'a depiction of *Jihād* and Martyrdom (*shahādah*) tries to develop certain themes which appear to emanate from that specific tradition. It does not seek to stereotype all Shī'a as holding the views thus described, any more than Chapter 6 seeks to imply that all Wahhābis are desecrators of monuments associated with traditions other than their own. The author advocates 'Muslim ecumenism' (see Chapter 6 and the Conclusion), not the continuance of damaging sectarian divisions. It is, however, important for such different traditions within Islām to be understood by non-Muslims, just as those who are not Christians need to understand something of the differences between Orthodox, Catholics, Protestants, Anglicans and so on if they are to gain an understanding of the Christian tradition.

There are several Muslims who have encouraged the author to write this book, and who should be thanked for their many kindnesses in lending materials and exchanging views. If they are unnamed, it is to ensure that they are not included in any criticism that this book may receive. Thanks are also due to my family who put up with my *jihād* to complete *jihād*, and to the patient Commissioning Editor of Palgrave, Luciana O'Flaherty, who encouraged me to end the first stage of a *jihād* which, almost by definition, cannot end but is ongoing. As Churchill said of the battle for Egypt in World War II, 'it is not the end. It is not even the beginning of the end. But it is, perhaps, the end of the beginning.'

Professor Richard Bonney
Director, Centre for the History of Religious and Political Pluralism
Director, Institute for the Study of Indo-Pakistan Relations (INPAREL)
University of Leicester
15 July 2004

Glossary

adab	propriety; good conduct
'adālah	justice
'ādāt	norms; habits
'ahd	pledge; covenant
aḥkām	legal rulings
ahl	family
ahl al-dhimmah	non-Muslim citizens of Islamic states
'alā	salaries
al-aḥkām al-nihā'iyyah	final legal rulings
al-arbāb al-arḍiyyah	earthly lords
al-bāṭil	falsehood
al-ḥarām	that which is unlawful
aḥadīth	plural of *ḥadīth*
al-Ḥujjah	proof
al-'Ījābīyyah	positiveness of Islamic faith
al-Ikhwān al-Muslimūn	Muslim Brotherhood
al-ithm	sinful act
al-Jamā'at al-Islāmīyyah	the Islamic Group
al-Jamā'at al-Jihād	the Group of *Jihād*
al-jamā'at-i jāhilīyyah	party of pagans (Mawdūdī's name for Jinnah's Muslim League)
al-jihād al-Islāmī	Islamic fighting/Islamic equivalent of 'holy war'
al-jihāz al-sirrī	'secret apparatus' (section of the Muslim Brotherhood)
al-kāffa	the masses
al-khulafā'-ar-Rāshidūn	the (four) rightly-guided caliphs
al-khurūj	revolt
al-Mahdī	the Awaited One or Saviour
al-maqāṣid	objectives
al-māriq	dissenter
al-Mu'minūn	true believers
al-nafs al-ammārah	the soul that enjoins evil
al-nafs al-lawwāmah	the self-accusing, or reproachful, or admonishing, soul
al-nafs al-muṭma'innah	the satisfied, or tranquil, soul
al-Nakbah	disaster

al-Nāṣir	the one who gives victory
al-niẓām al-khāṣṣ	special organization (section of the Muslim Brotherhood)
al-qadar	predestination
al-qā'idah al-ṣulbah	strong foundation
al-salaf al-ṣāliḥ	Righteous Predecessors
al-salām al-'ālamī	universal peace
al-shumūlīyyah	comprehensiveness of Islām
al-silm al-kādhib	superficial peace
al-sukūt	quietism
al-talī'at al-Islāmīyyah	vanguard of Islām
al-tamkīn	empowerment
al-tasākun al-bārid	cold cohabitation
al-Tatār	Mongols
al-tawāzun	balance or equilibrium
al-thabāt	constancy
al-'ulūhīyyah	God's divinity
al-wāqi'īyyah	realism or pragmatism of Islām
amān	safety; grant of safe conduct
amīr	ruler
Amīr al-Mu'minīn	Commander of the Faithful
amr bi'l-ma'rūf	commanding that which is good
amthāl	parables
anṣār	helpers of the Prophet, those believers at Medina who helped him after his exile from Mecca; also the name for the forces of the Mahdī
Aq Taghliqs	White Mountaineers (zealous Muslims in exile in Khoqand)
arkān-i-dīn	religious obligations
ashrāf	descendants of the Prophet
'atā'	payment for military service
a'wān	armed helpers
awliyā'	people in authority; saints
awṣiyā'	trustees
āyah (sing.)	verse
a'yān	provincial notables
āyat al-sayf	verse of the sword
āzādī	freedom/independence
bāb	gate
Bābur	lion
barakah	blessing/grace
bātin	inner meaning

bay'ah	oaths of allegiance/oaths of fealty
bayān	policy statement
bedel-i 'askerī	military payment-in-lieu
bedel-i naqdī	cash payment
beylerbeylik	province
bid'ah	innovation
cihād	Turkish term for *jihād*
cizye or haraç	poll tax (Turkish term)
daghābāz	treacherous
dā'ī	religio-political missionary
dajjāl (sing.)	false Messiah
dajjālūn (pl.)	wily deceivers
dār	abode
Dār al-Ḥarb	House or Abode of War
Dār al-Īmān	House of Faith
Dār al-Islām	House or Abode of Islām
Dār al-Kufr	House of Disbelief
Dār al-Ṣulḥ	Abode of Truce
dār al-ẓarb	city of the mint, capital
ḍarūrah	necessity
da'wah	missionary work/persuasion
dawr al-ṣatr	period of concealment
defterdar	head of the treasury
devlet-i Islām	Muslim state
dhikr	remembrance of God
dhimmī	non-Muslim subject of Islamic state
dīn	religion
dīn-i ilāhī	religion of God
dīwān	army rolls, register
elimān	title to rule
eretz Israel	the land of Israel
faḍl	surplus revenue
farāghat	space
farā'iḍ	obligatory duty
farḍ	religious duty
farḍ al-kifāyah	collective obligation
farḍ 'ayn	greatest obligation
farḍ taklīf	personal responsibility
fasād	corruption
fatāwā (pl.)	religious edicts/(according to some views: legal opinions)

fatwā (sing.)	religious edict/(according to some views: legal opinion)
fay'	booty
fedayeen (or fidā'iyūn)	those who sacrifice themselves
fetvā	Turkish term for *fatwā*
fī sabīl Allāh	in the path of God
fidā'ī (sing.)	redemption or self-sacrifice
fidā'īs (pl.)	devotees
fiqh	Islamic jurisprudence
fitnah	sedition
fuqahā'	jurisprudents
futūḥāt	conquests; 'carrying the message abroad'
futuuwah	spiritual chivalry
ghanīmah	spoils of war
ghärbiy	eastern region of Xinjiang Uīghur Autonomous region
ghayba	concealment of the hidden *imām*; first occultation
ġhāza	Ottoman term for *jihād*
ghazawāt/gazavāt	Chechen term for sanctified violence = *gazavāt*
ġhāzīs	holy warriors for Islām
ġhazw	the practice of collecting booty by conducting raids on traders' caravans, on rival tribes or on peaceful and poorly-defended communities
ghazwah (sing.)	military campaign
hac	Turkish term for *Ḥajj*
hadāra	appointment
hadd	maximum punishment
ḥadīth	tradition
ḥadīth qudsī	holy tradition (originally from a divine saying)
hāfiḍ	guardian/memorizer of the Qur'ān
hāīrat al-quds	holy enclosure
Ḥajj	major pilgrimage
ḥākim	governor
ḥākimīyyah	divine governance
ḥākimīyyat Allāh	sovereignty of God
Hamas	Ḥarakat al-Muqāwamah al-Islāmīyyah = Islamic Resistance Movement; the word '*ḥamās*' also means courage and bravery
hamidiyyah	light cavalry regiments
haq baat	the truth
ḥaqīqah	reality

hijrah	exile/detachment from the world of heresy, to establish and strengthen a community of believers outside it, in the path of the Prophet Muḥammad
ḥikmah	wisdom
Ḥizbu'llah (*ḥizb Allāh*)	party of God
hudnah	truce
ḥudūd	fixed penalties; fixed punishment for crimes
iane-i 'askerī	military assistance
'ibādah/'ibādāt	act of worship/acts of worship
iḥsān	excellence/spiritual excellence
iḥyā'	revival
ijmā'	consensus (of scholars)
ijtihād	independent reasoning
ikhtilāf	disagreement
ikhtiyār	act of choosing
Ikhwān	Wahhābi agents of enforcement; Muslim Brothers
Ikhwān al-Qassām	Brethren of al-Qassām
ilḥād	heresy
'illat al-qatl wa al-qitāl	the cause of killing and fighting
'ilm	knowledge
imām	head of state, leader
imām jā'ir	tyrant ruler
imām qā'im	imām inaugurating the resurrection
īmān	faith
imāra	emirate
intifāḍah	uprising; term for the uprising of the Palestinians
irādah istishhādiyyah	martyrological will
ishārāh	allusion/spiritual allusion (pl. ishārāt)
iṣlaḥ	reform
Islāḥat Fermani	Reform Charter of 1272/1856
'iṣmah	inerrancy; infallibility; divine protection
isnād	chain of transmission
istighāthah	calling for help
istiḥsān	theory of 'just preference'
isti'rāḍ	parade
ithbāt	affirmation
jahanbani	world-wide rule
jahangiri	world subduing (Persian term)
jāhil ghāll	an ignorant person who brought about evil
jāhilī	neo-pagan
jāhilīyyah	ignorance, barbarism; falsehood

janubiy	southern region of Xinjiang Uīghur Autonomous region
jelali	bandit
jihād	holy war, struggle
jihād al-qalb	*jihād* of the heart
Jihād-e-Akbar	the greater *jihād*
Jihād-e-Ashgar	the smaller *jihād*
jihād waṭanī	patriotic struggle
jizya	tax
kadhdhāb	liar
kāfir	unbeliever
kalām	scholastic theology
kalimah	declarations of faith
Kanuni (or Qānūnī)	law-giver
Kanunname	sultanic laws
karāmah	honour/miracle
kararname	verdict
khalīfah	head of the Muslim state
khānqāh	Dervish Lodge
khilāf	conflict; divergence
khilāfah khāṣṣah	special or specific caliphate
khuṭbah	sermon
khwāja	Muslim leader
kuffār	unbelievers
kufr	unbelief; ingratitude to God
kurh	hatred
madhāhib (pl.)	schools (hence 'schools of law')
madhhab (sing.)	school (hence 'school of law')
madrasahs	schools or colleges
mafsadah	adverse effects
Mahdī	Awaited One
Mahdīyyah	the Mahdist state
maḥzar	declaration
majlis al-shūrā	consultation committee
majūsī	Magi
marabout	member of a Ṣūfī brotherhood
marātib	levels; ranks
marḥaliyyah	graduation in following the Islamic vision
ma'rūf	that which is good
masjid	mosque
maṣlaḥah	benefit
ma'ṣūm	infallible

mawlā	master/client
millet-i sadika	faithful nation
mu'adhdhin	one who calls the faithful to prayer
mu'āmalāt	social relations
muballigh	propagator of the faith
mubtadi' ḍall muḍill	a misguided and misguiding innovator
mubtadi'ah	radical heretics
müceddid	renewer of the religion
mudjāhidūn/mujāhidīn	warriors for God; 'freedom fighters'; fighters
muftī	religious scholar who issues religious edicts
muhājirūn	emigrants
muḥaribūn	enemies
mujāhadah	striving; struggle
mujāhadat al-nafs	to do battle with the ego
mujtahid	a scholar capable of independent reasoning
munāfiq	hypocrite
munkar	evil; that which is wrong
muqātilah	warriors
murīd	disciple or aspirant
murshid-i kāmil	perfect spiritual master
murtadd	apostate
muṣālaḥah	peace treaty or truce
mushādah	contemplation of God
musḥaf	written book = the Holy Qur'ān
mushrikūn	polytheist
musliḥūn	reformers, or those who bring about *ṣalāḥ*
muslimūn	Muslims
mutaṭawwi'	volunteer (*muṭawwiīn*)
muwāda'ah	peace treaty or truce
Nabī 'Isā	Prophet Jesus
nafs	lower self; soul; ego
naskh	abrogation
naṣṣ	scriptural text (by extension: explicit designation)
nāṭiq	speaking
nihā'iyyah	finality
Pādshāh	ruler of the empire
Pesdaran	revolutionary guards
peshmerga	Shī'a resistance
Q.	(abbreviation for Qur'ān used in this book)
qā'im yawm al-qiyāmah	Lord of the day of resurrection
qāḍī	judge
qadar	decree; destiny

Qiblah	the direction of Mecca, particularly the Ka'ba and therefore the direction for prayer for all Muslims
qitāl	fighting
qiyāmah	Resurrection or the Last Day
Qizilbāsh	'red heads' (troops of Shāh Ismā'īl)
qulunsuwah	turban cap
Qur'ān	Holy book of Islām
Rasūl	Messenger
Rātib	collection of prayers by the Mahdī
ribāṭ	a post of the Muslim army at the frontiers of the enemy (in practice may be a fortified monastery)
ribāṭ al-khayl	the place for the tethering of horses (hence, by inference, 'holding in readiness mounted troops')
riddah	return, as in 'return' to apostasy
rifq	gentleness (hence, civility)
risālah	letter/message (hence, treatise)
rubūbiyyah	lordship
rūḥāniyyah	spirituality
sābiqah	precedence
ṣadaqah	alms; charity
ṣadd	blocking; preventing something or somebody
Ṣaḥābah	Companions of the Prophet
sahib-kiran	world conqueror
Ṣaḥīfāt al-Madīnah	the constitution of Medina or the Medina Agreement; the so-called first written constitution in the world
ṣaḥīḥ	soundness, particularly soundness of tradition
salām	peace
ṣalāḥ	righteousness
salafī	one who follows the salaf, the Companions of the Prophet and the pious Muslims of the first three generations of Islām. Three sub-groups may be identified: traditionalism; reformism; and the political and literalist *salafīyyah*
ṣalāt	prayer
ṣalāt al-khawf	prayer of fear
sāliḥāt	good works
ṣāmit	silent
Sanad ḍa'īf	weak chain of narrators
ṣawm	fasting
sayf al-ḥaqq	sword of God
Sayyid al-shuhadā'	chief (or lord) among martyrs

Sepay-e Padaran	Guardians of the Islamic Revolution Corps (Iran)
ṣeri'at	Turkish term for *sharīʻah*
ṣeyhülislām	Chief Muftī
shafāʻah	intercession
shahādah	oral profession of faith; testimony; martyrdom
shāhid	martyr; witness
sharīʻah	Islamic law
shawkah	authority
Shīʻa	(minority) independent faith community within Islām
shimaliy	northern region of Xinjiang Uīghur Autonomous region
shirk	polytheism
shūrā	electoral council; consultation
sīrah	practice; life; achievement
sirr	secret
siyar	international law
Ṣūfī	individual aligned with the Islamic mystical tradition, and member of a separate *ṭarīqah*
sujūd	prostration
Sunnah	Prophet's tradition
Sunnī	(majority) independent faith community within Islām
sūrah	chapter of the Qur'ān
sūrat al-Anfāl	chapter on the spoils of war (in the Qur'ān)
tablīgh	proselytism
tābūt	ark of the covenant
tafsīr	interpretation of the Qur'ān
taḥrif	alteration; corruption
tajdīd	revival/movement for the renewal of Islām
takfīr	the charge of unbelief levelled against other Muslims who do not conform
talibés	disciples or soldiers
taʻlīm	authoritative instruction or teaching
Tanzīmāt	reorganization or restructuring (edict of 1254/1839)
taqīyah	dissimulation
taqlīd	the following or emulation of a particular authority (as opposed to independent reasoning)
ṭarīqah	path or way
ṭarīqah Muḥammadiyyah	Muḥammadan path

taṣawwuf	Islamic mysticism; the Ṣūfī path of 'Purification of the Self'
tawassul	calling for help
tawbah	repentance
tawḥīd	monotheism; one God, the Creator, the Provident, the Law-giver
ta'wīl	hermeneutics
ta'ziyah	martyrdom enactment
thiqah	trustworthy
timars	independent fiefs
Tungan	Hui or Chinese-speaking Muslims
tūz	obligation of salt
'udwān	act of aggression
'ulamā'	scholars (pl.; sing.: 'alīm)
ūlī al-amr	those who hold authority
ummah	community of believers, Muslims
ummah mujāhidah	combative community
'umrah	lesser or minor pilgrimage
'urf	custom
vilāyat-i faqīh	(see also *wilāyat al-faqīh*) = Chief Jurisprudent or Juriconsult
wājib	mandatory; obligatory
wājib sharī'	religious legal obligation
walāyah/wilāyah	sainthood/authority
walī (sing.)	saint
walīs (pl.)	saints; holy men
waqf	endowment
waṣaṭiyyah	middle-roadness
waṣī	executor
wāzīr	imam's representative; vizier
wilāyat	legal competence
wilāyat al-faqīh	Chief Jurisprudent or Juriconsult
wird	litany
wurathā'	inheritors
Yāsā	Mongol legal code; Mongol order/decree
ẓāhir	outward; outward meaning
zakāt	charity, alms-giving
ẓann	speculation; conjecture
zindīq	atheist
ẓulm	injustice, oppression

Introduction

> It is the prosecution's case, in a nutshell, that the arrangement they became concerned with was for the purposes of supporting and promoting militant Islamic extremism which specifically embraces the use or threat of terrorism as a means of advancement or influence. That is a form of violent *jihād* often referred to as a holy war against the perceived enemies of Islam.

The above quotation from Mark Ellison, QC for the prosecution, Leicester Crown Court, 5 February 2003, in the case against two Leicester-resident Algerians accused of aiding al-Qaeda, was reported in *The Times* the following day.[1] The two accused were convicted and sentenced. American critics of the British multicultural city were quick to rush in and assert that instead of Leicester being 'a model city of sorts' in issues of racial integration and multiculturalism 'things had gone horribly wrong on the multicultural front'. Leicester 'may indeed be a model for some new kind of social pattern, in the United States as well as Europe – but not one [of which] the city fathers are going to be too proud'. Thus the conviction of two very recent immigrants, from a nationality which is uncharacteristic of migrants to Leicester, was taken to condemn or at least question the loyalty and quiet citizenship of 30,885 Muslims living in Leicester.[2]

It is difficult to imagine a more telling piece of evidence revealing the post-11 September 2001 paranoia in the United States with regard to the alleged clash of civilizations, and how to deal with Islām as a world faith and to relate to Muslims in our society. It just happened that this incident took place in the author's home city and imputed an accusation against one particular community among other diverse communities with which the author happens to be in contact. The name 'al-Qaeda' and the term '*jihād*' had been uttered in court. That was enough. Fear and incomprehension are only a short step away from rejection of difference within our society and the stereotyping of the vast majority of peaceful Muslims as closet *jihādī* terrorists. Clearly, for anyone who seeks better understanding of, and a more fruitful relationship with, the Muslim communities which live within the British or American multicultural city the central concept of *jihād* in

Islām has to be addressed. For if we cannot begin to understand our immediate neighbours, how can we hope to make sense of the problems of the Middle East or the Indian subcontinent?

Enter a bookshop in the United States or the United Kingdom, and examine the shelves on contemporary history or current affairs, and what do you find? The array of titles portraying an inevitable conflict of civilizations, between 'the West' and Islām, or depicting Muslim intolerance, fanaticism and violence is truly staggering. Titles such as *Islam Unveiled*, *Preachers of Hate*, *The Two Faces of Islam*, *Onward Muslim Soldiers* abound.[3] Never have there been so many publications in English on the contemporary Islamic world. To the extent that these books serve to increase public understanding and awareness of the issues at stake between 'the West and Islām', and within the Islamic world itself, since 9/11 we should be grateful. But do these books actually achieve this purpose?

Regrettably they do not. Their purpose is to 'sell copy'. Public alarm in the West at the phenomenon of suicide bombings has created an atmosphere of distrust against both Muslims and the faith of Islām as such. On the whole, the alarmist publications are written by journalists with an eye to a good storyline. They know how to fuel public alarm and succeed in doing so. Their characterization of Muslims and the faith of Islām is cast in apocalyptic terms, because apocalypticism 'sells copy'. For every radical Islamist 'cleric' (the credentials of such individuals to speak for the faith is in any case often open to question) who can be quoted in such books there may be dozens of mainstream Muslims who reject what is claimed on behalf of their faith. But their views do not count. The silent Muslim mainstream is a majority, but it is a majority that is shouted down by the violent Islamists on the one hand and those who do their publicity for them, the apocalyptic journalists in the West.

These journalists may not be Islamophobic themselves; but by using language such as 'Islamofascism'[4] they certainly create or perpetuate stereotypes which lend themselves to Islamophobia. These writings would not be quite so dangerous but for their effect on public opinion and because of the apparent credulity of some government advisers who are looking around desperately for a 'quick fix' to what is perceived as the problem of the age. It was to such advisers that Samuel Huntington's rather slight publication in 1993 originally appealed; and, events have subsequently demonstrated, apocalyptic prophesies become dangerously self-fulfilling if they are accepted at face value without objective analysis of the source material on which these views are allegedly based. More dangerous still, such views pander to the prejudices of the neo-conservative right and the Israel 'right or wrong' lobbies in the USA and may have already had the effect of hardening unrealistic and potentially self-defeating political standpoints. For it is a matter of report, not conjecture, that there has been a rapid decline of the US image in the Muslim world since 9/11, and that this decline is so severe that greater levels of expenditure on 'public diplomacy' will not succeed in reversing it without a change of policy.[5] Even Prime Minister Tony Blair, in his press statement on 20 November 2003 following the horrific attack on the British consulate and the HSBC bank in Istanbul, talked of 'the wretched, backward, philosophy of these terrorists', a philosophy of hate which had to be confronted by the West's commitment to a philosophy of tolerance and freedom.[6] Such comments do not suggest that the government advisers have studied in depth the nature of the ideological challenge from extreme Islamist terrorists and the reasons why the philosophy of hate appears to be gaining in its appeal in parts of the Middle East, particularly in Palestine.

As Douglas E. Streusand remarked in 1997, non-Muslims should not 'assert that *jihād* always means violence or that all Muslims believe in *jihād* as warfare'. But the historical enquiry underlying this book has also served to confirm his words that 'the discord over the meaning of *jihād* permits deliberate deception... A Muslim can honestly dismiss *jihād* as warfare, but he cannot deny the existence of this concept...'[7] In the aftermath of 11 September 2001, neo-conservative criticism even of moderate Muslims has been in the ascendancy. In this respect, we might take an assault on the brief talk given by Dr Zaki Badawi soon after the event, when he cited the Qur'ān [henceforth 'Q'], 5:32. 'He failed to mention... the very next verse [Q.5:33]... Nor did he mention the other warlike and intolerant verses from the Qur'ān such as *sūrahs* 9:5, 9:29, 4:89 and 8.39', argue Caroline Cox and John Marx. These critics continue: 'Dr Badawi cannot be unaware of these and other provocative verses given his lifetime experience as an Islamic scholar. Surely it is incumbent on him to explain to the British public just what they mean and which is the true voice of Islām.'[8] Are there no difficult passages in other faith traditions? What of the difficult texts in the Hebrew Scriptures where genocide appears to be advocated (for example, the destruction of the Amalekites in 1 Samuel 15:2–3, 20: see the prologue in this volume) and even Jesus' statement, in a passage dear to Calvin, that he had 'come not to send peace on earth... but a sword' (Matthew 10:34)? Neither Rabbis nor Archbishops are usually required to give a complete exegesis of their scriptures in one soundbite, nor should this be expected of Muslim leaders.

The meaning of the term 'al-Islām' is to surrender to Allāh, 'self-surrender to God' or 'submission of the whole self to God'.[9] Although in most respects we can categorize him as a fundamentalist, Ḥasan al-Bannā''s essay on 'Peace in Islām' published in 1948 contains these words:[10]

> Islām is uncompromisingly a law of peace and a religion of mercy. Only he who is ignorant of its teachings, hostile to its system, or is arrogant enough not to accept clear evidence, will dispute this fact. The word Islām is itself derived from the word peace (i.e. *salām*). And Muslim is the best description of those who believe in this religion. (Q.22:78)

Professor Hassan Hanafī notes that the word *salām* appears in the Qur'ān 129 times, while *al-Salām*, 'the Peace' or 'the One on whom all salvation rests', is one of the 99 divine names (Q.59:23). Peace in the individual precedes peace in society; peace in the soul precedes peace in the world (Q.49:14). No nation shall be discouraged from pursuing peace (Q.47:35), but there is no peace without safety and security, the distinguishing features of Paradise (Q.5:16; Q.15:46; Q.50:34). Equality and justice bring in their wake peace, while inequality and injustice are causes of war. The inclination towards peace must be respected, while specific guarantees help to bring about a secure peace.[11] For Hanafī, in the Islamic Revelation, there are five Universal Intentions, which serve as the positive foundation of law: the preservation of human life as an absolute and primary value; the protection of human reason; the struggle for true knowledge; the affirmation of human honour and the dignity of the person; and finally, the protection of individual and national wealth.[12]

If a simple answer is to be given to those who, like Caroline Cox and John Marks, question the basic peaceful credentials of Islām as a world faith, then it is to be found chiefly in two places in the Qur'ān. The first is in the requirement that there shall be no compulsion in religion (Q.2:256). This verse is considered by mainstream Muslims to 'abrogate', that is to supersede, all the aggressive or warlike verses in the Qur'ān.[13] The second is contained in the verse 'enjoining right and forbidding wrong' (Q.7:157). This is to be viewed as the decisive force within the faith, which determines other responses such as *jihād*. *Jihād* does not abrogate the commandment to enjoin right and forbid wrong; instead maxims which are enjoined within the context of *jihād* have to conform to this ethical principle.[14] Such is a modern, mainstream, understanding of the central, peaceful, purpose of Islām. It goes without saying that, just as in Christian history, by no means all Christians in the past have conducted themselves according to what would be considered a modern understanding, so by no means all Muslims have done so. And just as within Christianity today there are many Christian 'exclusivists', so there are Muslim exclusivists.

Nevertheless, for Cox and Marks, *jihād* is one of the key problems of Islām:[15]

> *Jihād* can be interpreted spiritually as a struggle to lead a holy life. But it can be extended to mean an obligation – imposed by Allāh on all Muslims – to strive unceasingly to convert or to subjugate non-Muslims. *Jihād* in this latter sense is without limit of time or space and continues until the whole world accepts Islām or submits to the Islamic state.

The idea of an Islamic doctrine of permanent war may please the opponents of Islām,[16] and appears to be asserted by Islamist theoreticians such as Sayyid Quṭb,[17] but is in fact erroneous. As one recent historian has argued,

> even if it may have been a major force in the ideological matrix of medieval western Asian and [western and] eastern European frontier regions, the 'championing of one's faith' could never function as the sole concern of historical actors in that stage or as a single-minded zeal.[18]

In any case there has been, and remains, both a practical and a philosophical opportunity for accommodation. 'The people were one community (*ummah*); then God sent forth the Prophets, good tidings to bear and warning, and He sent down with them the Book with the truth, that He might decide [between] the people touching their differences' (Q.2:213). Here is a Qur'ānic conception of religious pluralism: mankind is united under One God; the teaching of the prophets brought about the particularity of religions; and divine revelations ('the Book') have a key role in resolving the differences that touch communities of faith.[19] Islām makes no distinction between one or other of the previous revelations, prophets and apostles (Q.2:135; Q.2:285; Q.3:84).[20] Not committing mischief, and doing good on Earth, become the highest implementation of faith (Q.11:88; Q.38:28).[21] Dialogue and conflict resolution ('coming to terms') are imperative, and are not optional extras (Q.3:64; Q.21:108; Q.29:46).[22] A draft memorandum of understanding between Jews and Muslims in recent times contains the following observation:[23]

> The holy Qur'ān revealed to Muḥammad, peace be upon him, the prophet of Islām, calls for the respect and honour of every human being regardless of race or creed. Moreover, the Qur'ān states that special respect and feeling of brotherhood are due to all believers in the faith of the one God. Thus, Jews, who worship the same God as the Muslims, are primary recipients of these feelings of brotherhood.

Such may be the highest ideal of Islām. The ideal is far from practical realization in key areas of the world under territorial dispute, primarily the Middle East. There is no doubt that this conflict has brought in its wake virulent anti-semitism among some sections of Arab opinion. One Arab-language columnist, on 2 May 2002, referred back to Hitler's campaign of extermination: 'if only you had done it, brother'.[24] Other references to Hitler abound among the 'preachers of hate' that radicalized Wahhābism and the protracted conflict of the *intifāḍah* have spawned. One Hamas activist claimed, in August 2003,

> When we compare the Zionists to the Nazis, we insult the Nazis – despite the abhorrent terror they carried out, which we cannot but condemn. The crimes perpetrated by the Nazis against humanity, with all their atrocities, are no more than a tiny particle compared to the Zionists' terror against the Palestinian people.[25]

Holocaust denial, myths of Jewish conspiracies – in particular, belief in the fabricated *Protocols of the Elders of Zion*[26] is widespread – and naked anti-semitism (against Jews)[27] are rife in extremist circles; and, unpardonably, extreme violence is even indoctrinated to young children.[28] Since 2003, one of the Palestinian authority's textbooks has propagated a doctrine of violent *jihād* to children in the eleventh grade.[29] It should be noted, however, that at least one prominent Islamic voice has been raised against the use of children in the *intifāḍah*:[30]

> the Prophet did not hide his need for fighters in the battle of Badr, which was the first battle of Islām. There were three times as many infidels as Muslims [in this battle] and it was possible to use the youths from a distance as archers. Nevertheless, the Prophet did not allow them to fight. Moreover, Islām defended the souls of non-Muslim children when it forbade the killing of the enemy's women and children. Today, on the other hand, we see in the *intifāḍah*, children who are less than the age of maturity, thrown unarmed and undefended to be targets for the [Israelis] who are armed from head to toe so that they can hit these children as they wish. The Prophet even forbade the use of animals as targets. So what is there left to say about the Palestinian people who have turned their children into targets?

In the viewpoint of others, the divine grace of Islām is, unforgivably, said to require the extreme suffering of the Jewish people for its fulfilment, otherwise called 'the honour of reaping as great a harvest as possible of Israeli lives…'[31] 'By means of *jihād* Allāh tortures [the Infidels] with killing.'[32] Whatever the faults of the Israeli government and its unwillingness or inability to settle with the Palestinians, there can be no excuse for such horrific reinterpretations or reformulations of Islām which have no real basis in the faith. Though the chief political adviser to President Mubārāk of Egypt wrote a significant criticism of anti-semitic propaganda in January 2003, in which he counselled against conspiracy theories and recognized that the *Protocols of the Elders of Zion* was a fabrication,[33] subsequent statements such as that of Prime Minister Mahathir Mohammed of Malaysia in October 2003, that 'the Jews rule this world by proxy' potentially fuel Muslim anti-semitism. However, what was less noticed by the critics of Mahathir's statement was his reliance on another widespread historical myth:[34]

> Remember Ṣalāḥ al-Dīn and the way he fought against the so-called Crusaders, King Richard of England in particular. Remember the considerateness of the Prophet to the enemies of Islām. We must do the same. It is winning the struggle that is important, not angry retaliation, not revenge.

Here we have the myth of Saladin, born a Kurd but Arabized by mythology and given an heroic status, who allegedly thwarted the first Western assault on the Islamic world. Patience will be needed, so the argument runs, before Israel is destroyed in the same way as was the Latin kingdom of Jerusalem. Based on an interpretation of a passage in the Qur'ān by Shaykh Aḥmad Yassīn, its leader (who was assassinated in 2004), Hamas believes that Israel will cease to exist by the year 1448/2027.[35] 'International Zionism' is seen as an ideology employed by the imperialism of the outside world to mask its 'Crusaderism': the ambition of the old Christian enemy to subvert Islām and destroy its followers. Parts of the world not traditionally associated with crusading are now considered to be theatres of the same war, while new 'Crusader states', above all the United States, have arisen. Professor Jonathan Riley-Smith, the distinguished historian of the Crusades, comments that 'so many share' Osama bin Laden's 'historical vision' – we would call it historical myth – 'that one is tempted to call it mainstream'.[36] A lack of historical vision and a myopic world view of what is misleadingly referred to as 'the West' is a weakness in the Muslim world; 'the West' can be criticized for a similar lack of historical curiosity about the Muslim world. It is this collective amnesia which produces on either side the 'false consciousness' of a clash of civilizations.

This book is not, nor should it be, a history of the crusades against the Muslim world; but two brief, but very important, comments on the Crusades are necessary to place this historical experience in context. The first concerns the start and end dates of the Crusading movement. As a demonstrable historical phenomenon, there is reasonable agreement that the Crusades, defined in Riley-Smith's terms as 'Christian penitential war-pilgrimages authorized by the Popes and fought by volunteers, who were privileged in various ways',[37] began by 456/1063–64 or by 487/1095 at the latest[38] and had ceased on any large scale by 987/1580, and certainly well before 1214/1800. The Enlightenment of the eighteenth century was extremely hostile to the Crusading movement, one Enlightenment writer calling it 'the most signal and durable monument of human folly that has yet appeared in any age or nation'.[39] A second comment concerns the object of the Crusading movement. Here, the consensus of scholars is that they were not specifically anti-Islamic. In the words of Riley-Smith,[40]

> they manifested themselves in many different theatres of war against many different enemies: Muslims of course, but also Pagan Wends, Balts and Lithuanians, Shamanist Mongols, Orthodox Russians and Greeks, Cathar and Hussite heretics and even Catholic political opponents of the Papacy.

Professor Norman Housley concludes that religious warfare in the West acted[41]

> as a bridge between the twelfth and thirteenth centuries, when crusading was at its height of popularity, and the Wars of Religion. These patterns are

remarkable in themselves. Furthermore, they show that, given the way their ancestors had acted, the escalation of religious violence amongst Catholics and Protestants in Early Modern Europe becomes less bizarre, without becoming any less exceptional or indeed horrific. The conviction that human armies could literally fight God's war was not to be easily uprooted from the European consciousness.

We are thus able to conclude, from the chronology established below in Chapter 2, that the doctrine of *jihād* preceded the ideology of the Crusades in Christianity, and that it has also lasted longer, notably in the struggle against the colonial powers in Africa and Asia after the eighteenth century. Its essential purpose is, however, little different. For Henry the Navigator of Portugal, and his biographer Zurāra, 'the Infidels are our enemies by nature'. 'And for what glory will they be able to praise me on the day when I am made knight', Henry the Navigator mused when he reflected on the capture of Ceuta in 817/1415, 'if my sword has not been dipped to the hilt in the blood of the Infidels?'[42] Unlike the Muslim world, the concept of European nationhood was decisively fashioned by the attention paid by the Christian Church to the Hebrew Scriptures (the Christian 'Old Testament'), which provided the concept of a 'Holy People' divinely chosen to endure the rigours of a confusing, but on the whole divinely-determined, history. Exodus 32:26–8, in which Moses recruits the sons of Levi to carry out a ruthless programme of execution in the name of God, is the *locus classicus* of Christian sanctified violence. But Deuteronomy 20:10–14, the terms of surrender which the Israelites were allowed to offer to the inhabitants of any town which they besieged, received this commentary from the Salamanca jurist Francisco de Vitoria: 'in wars against the infidel... peace can never be hoped for on any terms; therefore the only remedy is to eliminate all of them who are capable of bearing arms against us, given that they are all guilty'. In his *On the Law of War* (938/1532), Francisco de Vitoria pronounced that all captured infidel combatants should be killed and their women and children enslaved.[43]

Is this Christian tradition of religious violence so very different from the warlike interpretation of the doctrine of *jihād* founded on the distinction made by the classical Islamic jurists between the House or Abode of Islām (*Dār al-Islām*) and the House or Abode of War (*Dār al-Ḥarb*), with at best an indeterminate area of negotiation known as the Abode of Truce (*Dār al-Ṣulḥ*)?[44] We should also note that there is no Qur'ānic sanction for the theological division of the world into the *Dār al-Islām* and the *Dār al-Ḥarb*. According to the Qur'ān, the world is divided between believers or the House of Faith (*Dār al-Īmān*) and non-believers or the House of Disbelief (*Dār al-Kufr*). The Qur'ān repeatedly states that the believers together constitute one people and the disbelievers together constitute another people, as in 'the believers are brethren of one another' (Q.49:10) and 'those who disbelieve are friends of one another' (Q.8:72).

Differences of belief are thus seen as part of God's plan. The abolition of such differences is not the purpose of the Qur'ān, nor was the Prophet sent for that reason (see Conclusion). In Qamaruddin Khan's words, nowhere does the Qur'ān demand[45]

> that the Muslims should remain permanently at war with the non-believers. The verses (for instance Q.4:89 and Q.9:5) which seem to give the impression of perpetual war between the world of Islām and the world of *Kufr*, are decidedly topical and circumstantial in their import, and cannot be taken as permanent injunctions of God… [The Qur'ān] enjoins the incessant struggle until the whole world has been submitted to the message of Muḥammad. But the struggle is to be done by *da'wah* (persuasion and preaching). Resort to force is allowed only as a defensive or self-protective measure…

Jihād 'in the way of Allāh' may thus mean a peaceful struggle by persuasion and preaching, the summons to which the Qur'ān refers (Q.14:26); though, as Khan notes, this was not how the medieval Islamic jurists tended to regard it.[46]

Any new interpretation such as the one proposed in this book, which enters the contested waters of the various meanings of *jihād*, and which attempts to depict the mainstream Islamic view against fundamentalist variants, is open to immediate objections. The subject is so complex and so wide-ranging in its ramifications it might be safer not to attempt an analysis at all. What right does a non-Muslim have to pronounce on such matters, when the Muslim community itself is divided? Is there some hidden agenda, perhaps a Christian desire to demonstrate the superiority of his faith, determining the judgements made?

No doubt some who disagree with the interpretation contained in this book will wish to project one or other of such views, to which clear answers can be given in advance. Firstly, the subject is indeed complex (there are differences between the Sunnī and Shī'a schools and within the Sunnī traditions):[47] but the importance of the term '*jihād*' and the widespread condemnation of it (and the confusion with the term 'terrorism') in the West after the attacks on 11 September 2001 mean that discussion and clarification of the term are not optional extras, but mandatory for any understanding of the relationship between the Muslim world and the West.

Secondly, it is true that the Muslim community is itself divided on the matter. It is a struggle which those who wish their Muslim friends well hope will be won by those with a mainstream viewpoint. If a non-Muslim offers suggestions then the purpose is clear: it is to emphasize the traditional appreciation of Islām as a peace-loving religion. There is indeed a problem of Muslims who are terrorists. But the use of the term 'Islamic terrorist' is objectionable. There are also Christians who are terrorists. But they are never referred to as 'Christian terrorists'. That is, in essence, what the sectarian killers in Northern Ireland for the last 40 years or so have been, but the term has never been used because there is a clear understanding that Christianity is a religion of peace. We need to return to the proper understanding of Islām as a religion of peace.[48] This means that there has to be a clear understanding of the different meanings of 'Islamic' and the 'rule–faith' distinction: while Islamic rule may be sought by all legitimate means, there can be no forcing of consciences to create new converts to Islām.[49] Only the conscience of the apostate is forced: he is *de jure* dead from the moment of his abjuration and has to choose between Islām and the sword.[50] The Islamist theoretician Sayyid Quṭb referred to Islām as 'an original genuine system with its own unique bases and an integral comprehensive plan, not mere adjustments to current outstanding conditions'.[51] This helps explain the radical Islamists' drive to assert the supremacy of the 'Islamic' over other political systems.

On the third point, whether there is some hidden agenda, perhaps a Christian desire to demonstrate the superiority of his faith, which might determine the judgements made in this book, this needs to be repudiated at the outset. Such a book does indeed exist, though the agenda is not hidden but explicit. It is clear from the text of John MacArthur's *Terrorism, Jihād and the Bible: A Response to the Terrorist Attacks* (2001), which has been banned in Pakistan's North-West Frontier Province for using inflammatory language against Islām,[52] that the purpose is avowedly to demonstrate the supposed inferiority of Islām to Christianity. MacArthur comments: 'within the first hundred years after Muḥammad, the Arab world was unified to a remarkable degree, as that part of the world succumbed to the power of Islam, mostly by the edge of the sword'. Islām teaches and many Muslims believe that war is a legitimate means of converting non-believers. '"Convert or die" has always been the most persuasive tool in the Islamic missionary's arsenal.' It is true that not all Muslims are terrorists, support terrorism or rejoice when terrorism strikes at their enemies, 'but nonetheless violence against infidels and the concept of *jihād* is fundamental to Islām and an inescapable part of Islamic history'.[53] But to attack Islām for its failure to accept the Christian doctrine of original sin is perverse. In his essay on 'Peace in Islām', Ḥasan al-Bannā' (the twentieth-century Islamist who has already been quoted above) stated categorically: 'Islām has supported its theoretical consideration and practical plans with the spreading of the best of human sentiments in the hearts and souls. These feelings of love of the good for mankind and the attitude of altruism [extend] even [to] the time of need' (Q.59:9; Q.2:195; Q.18:30; Q.16:90). In other words, though the entire philosophical premise about man's nature is different in Islām from Christianity it does not follow that such an optimistic view has provided its followers with a religion of hatred for the other.

Quoting Ibn Sīrīn's 'golden rule' ('be wary from whom you take your religion') as reported by Muslim, Muḥammad al-A'zamī argues that 'only a devout Muslim has the legitimate prerogative to write on Islām… and its related subjects. Some may consider this biased', he argues, 'but then who is not?' Non-followers, he contends, 'cannot claim neutrality, for their writings swerve [*sic*] depending on whether Islām's tenets agree or disagree with their personal beliefs, so any attempts at interpretation from Christians, Jews, atheists or non-practising Muslims must be unequivocally discarded'.[54] Such arguments preclude any serious inter-faith dialogue. They are contradicted by no less an authority than Wilfrid Cantwell Smith, who argues that it is possible for an outsider to state 'the meaning of a faith in, say, modern terms more successfully than a believer'.[55] Moreover, the argument is deficient in that it fails to address the assault on contemporary Islām arising from neo-conservative critics in the United States.[56] For unless these criticisms are addressed in their own terms, mainstream Islām will not find its voice listened to.

In any case, given the 'dominant position that discussion of *jihād* occupies in modern Muslim apologetics', the subject has to be considered by informed outsiders, whatever their origin. For Rudolph Peters, 'of all Islamic institutions, *jihād* is certainly the one which offers the most admirable resources for studies on the inexhaustible and complex theme of the relationship between Islām and Western colonialism',[57] or, we might add in the aftermath of the Second Gulf War (2003), between the Muslim world and Western neo-colonialism. Writing shortly before the events of 11 September 2001, Ḥilmī M. Zawātī commented that 'the classical sources of Islamic legal theory maintain that all kinds of warfare are outlawed except the *jihād*, which is an exceptional war waged by Muslims to defend the freedom of religious belief for all humanity, and constitutes a deterrent against aggression, injustice and corruption'. The ideas expressed by Hugo Grotius in *The Law of Peace and War* (*De jure belli ac pacis*, 1034/1625) were taken from the Spanish jurists Francisco de Vitoria and Francisco Suárez who in turn had derived their ideas from Islamic law (as they themselves acknowledged).[58] 'Although it would be hard to dispute the fact that the idea of just war existed before Islām', Zawātī contends that

> this notion has been developed and refined by Muslim jurists. It becomes evident… that *jihād*, in the form of armed struggle, must be just in its causes, defensive in its initiative, decent in its conduct and peaceful in its conclusion. Hence, as a defensive war, *jihād* can be exercised individually or collectively by contemporary Muslim States, since such a type of war is definitely sanctioned by the norms of international law, particularly the United Nations Charter.[59]

In contrast, Daniel Pipes castigates the majority view of senior American scholars of Islām[60] and instead provides a verdict on *jihād* which is entirely negative:[61]

> Despite *jihād*'s record as a leading source of conflict for 14 centuries, causing untold human suffering, academic and Islamic apologists claim it permits only defensive fighting, or even that it is entirely non-violent…

> It would be wonderful were *jihād* to evolve into nothing more aggressive than controlling one's anger, but that will not happen simply by wishing away a gruesome reality. To the contrary, the pretence of a benign *jihād* obstructs serious efforts at self-criticism and reinterpretation…

Writing in 1997, Douglas E. Streusand comments: 'in the Qur'ān and in later Muslim usage, *jihād* is commonly followed by the expression *fī sabīl Allāh*, "in the path of God"'.[62] The description of warfare against the enemies of the Muslim community as *jihād fī sabīl Allāh* 'sacralized an activity that otherwise might have appeared as no more than the tribal warfare endemic in pre-Islamic Arabia.'[63]

Daniel Pipes concludes that the violent tendency in Islām has been 'mainly associated' with the thinker Ibn Taymīyah (661/1268–728/1328: see below, Chapter 4) and holds 'that born Muslims who fail to live up to the requirements of their faith are themselves to be considered unbelievers, and so legitimate targets of *jihād*…' The second variant, usually associated with the Ṣūfī, or Muslim mystical tradition, was the doctrine customarily translated as 'greater *jihād*' but perhaps more usefully termed 'higher *jihād*'. This Ṣūfī variant invokes allegorical modes of interpretation to turn *jihād*'s literal meaning of armed conflict upside-down, calling instead for a withdrawal from the world to struggle against one's baser instincts in pursuit of numinous awareness and spiritual depth.[64] Politicians appeal to the greater *jihād* as a matter of course, as for example did President Musharraf of Pakistan in his celebrated speech on 12 January 2002 against terrorism.[65] Such talk of a non-military *jihād* has not been confined to President Musharraf. A number of modern scholars contend that *jihād* encompasses all forms of political and social action to establish justice in order to accomplish Islām's social and political agenda. 'There is no doubt that the Qur'ān wanted Muslims to establish a political order on earth for the sake of creating an egalitarian and just moral-social order. *Jihād* is the instrument for doing so.'[66] In this spirit, President Habib Bourguiba of Tunisia used the term *jihād* to describe the struggle for economic development in Tunisia, much as Lyndon Johnson spoke of a 'War on Poverty'. In this context, *jihād* no more implies violence than do the terms 'war' or 'crusade' on poverty in today's English. Bourguiba clearly did not advocate violence to improve education and development in Tunisia.[67] The Muslims in India proclaim on their website: 'we have kept our religion within our mosques and houses and are fighting a *jihād* against ignorance, illiteracy, poverty and diseases and not against the state'.[68]

Contrary to a frequent projection in the West, in its original sense *jihād* does not mean 'war', let alone 'holy war'.[69] It means 'struggle' (*jahd*), exertion, striving; in the juridico-religious sense, it signifies the exertion of one's power to the utmost of one's capacity in the cause of Allāh: it is thus the opposite of being inert, the antonym to the word *qu'ūd* (sitting) in the Qur'ān (Q.4:95). Whereas war may be fought for territorial ambitions, as Liaquat 'Ali Khan, the first prime minister of the newly independent Pakistan remarked, a struggle can be pacific ('...*jihād* really means to strive for justice and truth whereas war means to fight others for territorial ambitions...', he contended).[70] It may comprise a campaign for justice and truth, or (as in one passage in the Qur'ān), 'that you believe in Allāh and His Messenger and that you strive hard and fight in the Cause of Allāh with your wealth and your lives...' (Q.61:11).[71] Dr Amīr 'Alī comments:

> if we translate the words 'holy war' back into Arabic, we find *harbun muqaddasatu*, or for 'the holy war', *al-ḥarbu al-muqaddasatu.* We challenge any researcher or scholar to find the meaning of *jihād* as holy war in the Qur'ān or authentic *ḥadīth* collections or in early Islamic literature.[72]

Professor Sohail H. Hāshmī refutes the idea of an offensive *jihād* condoned by the Qur'ān, considering that 'the *jihād* tradition is parallel to the "just war" tradition in the West, acknowledging that violence is evil, but may be justifiable in certain circumstances'. Only two verses, the 'verses of the sword' suggest an interpretation of 'conquering people with the ultimate goal of converting them to Islām'. Hāshmī considers that 'scholars after Muḥammad's death concentrated on [these verses] because they wanted to justify the expansion of Islām as an empire (an expansionist *jihād*), whereas the Qur'ān overwhelmingly speaks of defensive war only'. For revivalist writers, un-Islamic regimes include those ruling in most Muslim countries. The immediate goal of the revivalist *jihād* is to replace hypocritical leaders with true Muslims. Only when this long and painstaking internal struggle has succeeded in re-establishing an authentically Islamic base can the external *jihād* resume. Thus, Hāshmī argues, *jihād* 'is today largely synonymous with Islamic revolution in the works of most Muslim activists'.[73] The practical difficulty that Osama bin Laden and his followers encounter is that they seek to undertake both the internal and the external *jihād* simultaneously. The challenge to authoritarian regimes in majority Muslim lands is nevertheless real enough. In 2000, Thomas Scheffler published an important article on 'West–Eastern Cultures of Fear: Violence and Terrorism in Islam'.[74] Scheffler is in no doubt: the causes of political violence in the region lie for the most part with the authoritarian regimes in most Muslim countries ('the continued existence of authoritarian structures in the region is due in no small part to Western participation', he further comments). 'The most obvious way to reduce the dangerous potential [for violence]', he argues, 'is not to focus exclusively on terrorism, but to encourage democratization of the region.' Nevertheless, as the Arab press has vociferously commented with regard to the comments of Condoleezza Rice, the US National Security Adviser in October 2002, democracy cannot be imposed from outside.[75] The much-vaunted 'Greater Middle East Initiative' is doomed from the outset if it proves insensitive to the cultural and religious mindset of the region.[76]

It has recently been stated that '*jihād* can never be a war for the sake of war, a war of instrumental reasoning and worldly glory', while, in contrast, 'terrorism holds nothing inviolable and is therefore the offspring of the same nihilism which is the antithesis of faith'.[77] Nonetheless, as Jonathan Riley-Smith has argued, Osáma bin Laden's 'historical vision, although extreme and in western terms a fantasy, is not that of an isolated eccentric; nor, in the context of Islamist thought, is his terminology "archaic"'.[78] There is widespread, but still minority, support for the Islamist viewpoints on *jihād* which has to be confronted. As Professor Khalid Masud has observed, 'it is... essential for Muslims to begin rethinking *jihād* in the light of the modern developments of warfare... *Jihād* should be revived as a doctrine of peace and security against [the] prevailing concept of violence and aggression.'[79]

Such rethinking has to take place against a clear understanding of both the texts and the historical context, which is the task of this book. Bin Laden appeals to the Qur'ān and certain preferred post-classical writers such as Ibn Taymīyah to justify what he asserts as the 'true' understanding of *jihād*. But as John Kelsay argues, it is bin Laden who is the innovator: 'bin Laden's *jihād* is new, not so much in the sense of "up to date", as in the sense of a departure from tradition, an innovation.'[80] Bin Laden quotes the 'verse of the sword' (Q.9:5) as justification for his particular interpretation of an offensive, transnational, or even global *jihād* 'in order to establish truth and abolish falsehood'.[81] 'Falsehood' (*al-bātil*) is reduced to shorthand to refer to 'illegitimate' Muslim rulers who collaborate with Western (that is, Christian) powers. The aggressive *jihād* is proclaimed as an act of Islamic self-defence. Most Muslims understand that a process of derailment has occurred; and that however popular bin Laden may be in some circles, his is not the way forward, but a dead end. This mainstream Muslim majority nevertheless lacks a vision for how to rectify the situation. It is hoped that this book will provide, in some measure, an analysis of how the process of derailment has occurred and some suggestions as to the way forward for a modern, enlightened Islām, as well as for a progressive (rather than oppressive and neo-imperialist) Western world which is prepared to collaborate with it.

Prologue
The War of Annihilation (*Ḥerem*) in the Hebrew Scriptures

In the introduction we have already had cause to remark that Islām is not unique as a great world religion in having 'difficult texts', which require some further commentary to be explicable. Both the Torah (the Tawrāt for Islām) and the New Testament (the Injīl for Islām) have difficult passages too. Why in a book about the history of Islām should we concern ourselves with the Hebrew Scriptures, when it might be considered that these had been abrogated by the Divine Revelation given to the Prophet? The reason is that the doctrine of abrogation (which will be considered in more detail in Chapter 1) does not work in this way. The general principle in Islām is that there is no time limit for the validity of previous prophets and their revelations. A prophet sent with an earlier version of the law (*sharīʿa*) is not rendered irrelevant because another, final Prophet, was sent after him. There are clear verses in the Qur'ān in which believers are said to have faith in God, in his Angels, in His books and in His messengers without making any distinction between them (Q.2:136; Q.2:285; Q.3:84). The presumption was for the compatibility of the laws revealed to the various prophets. All remained valid unless changed or replaced by abrogation.[1]

Paradoxically, there is sufficient concentration on the theme of *jihād* in the Qur'ān, with some verses having themselves been subject to abrogation, to suppose that the concept of the war of annihilation (*ḥerem*) as it appears in the Hebrew Scriptures was indeed abrogated by the subsequent revelation to the Prophet. However, there remains every reason to consider this concept at the outset, since either directly or indirectly it may have influenced the Islamic tradition of *jihād*. The most likely hypothesis is that it influenced it indirectly by leading Islām to repudiate some of its more extreme formulations.

The concept of *ḥerem* has received full treatment from Philip Stern,[2] so that it is only necessary here to recount some of the more important biblical passages where it appears and to summarize some of the author's conclusions. The Book of Deuteronomy emphasizes that the people of Israel are a 'holy people' (Deuteronomy 7:6) who reject idolatry and worship the monotheistic god

Yahweh (YHWH). In chapters 7 and 20 of Deuteronomy a 'religiously motivated xenophobia'[3] is evident: Israel owed its possession of the land to divine favour. In the 'promised land' there is no place for idolatry. What seems at first sight to be an all-destructive war (*ḥerem*) serves to remove abomination (that is, the worship of other gods and idolatry) and to create holiness.[4] The idea of a consecration to the deity through destruction can apply both to groups of Israelites as well as to specifically-designated foreign nations. In Deuteronomy 13:16–18 a holocaust is prescribed for the city which has sinned against God and therefore must atone for its sins:[5]

> You shall utterly smite the dwellers of the city by the sword, devoting it and all in it and its cattle by the sword; and all its booty you shall gather in the middle of its square, and then you shall burn in flames the city and its booty, and it shall become a ruin, never to be rebuilt.

Elsewhere in Deuteronomy (7:1 and 20:1), the Hittites, Amorites, Canaanites, Perizzites, Hivites and Jesubites (and in 7:1, the Girgashites as well) were described as foreign tribes who were subject to all-destructive war (*ḥerem*). Philip Stern argues that 'in the Northern Kingdom in the eighth–seventh centuries [before the Common Era], the memory, if not the practice of [*ḥerem*] was alive and well'. There were short-term purposes in describing the existence of a previous practice of warfare. Only if Yahweh willed it, could it once more be put into operation.[6] Moreover, there is some ambivalence in the use of the verb 'destroy', illustrated by its use in Deuteronomy 4:26–27: what seems to be suggested is the destruction of the people as an entity subsisting on the land, their expulsion but not their wholesale physical annihilation (cf. Deuteronomy 7:1).[7]

Elsewhere in the Hebrew Scriptures there is less ambiguity. Joshua 6 is a spectacular chapter, which recalls the first conquest of a city on the soil of the land west of the Jordan river, the 'promised land'. Here, the creation of the holy community was possible only as a result of the destruction of the walled city. The seventh day involved not rest, for this was holy war *par excellence*, but a special, sevenfold, heightening of activity as a result of which the forces resisting the new order were crushed.[8] Thereafter, killing without a remnant or without a living body left alive became commonplace in the history of the conquest in which Yahweh fought for Israel. The failure to observe the law to the letter led to extremely serious consequences: Achan, a peasant, took some of the plunder from Ai and infuriated Yahweh. A tenfold increase in the size of the army was necessary as a consequence, and then the soldiers only succeeded because of Yahweh's direct intervention.[9]

A further key passage for an understanding of *ḥerem* is 1 Samuel 15:1–3, where Samuel delivers the order that the Amalekites of whatever age and all their livestock are to be slaughtered. This is the oldest source for 'the war against

Amelek from generation to generation'.[10] As Philip Stern comments, it is 'as a uniquely dangerous and perpetual challenger to the divine order of YHWH that the Amalekite nation appears in the Bible'. He suggests that Amalek 'filled the unenviable role of "chaos" and so served as the object' of *ḥerem*.[11] Shortly after the deliverance of the people of Israel from Egypt, Amalek attacked them at their most vulnerable moment in the wilderness, before they were accustomed to fighting (Exodus 7:8–15; Deuteronomy 25:17–19). Fatally for his own reign, and the future of his dynasty, King Saul wanted to spare the life of Agag of Amalek, but in so doing he defied Yahweh and the commandment to pursue *ḥerem*. Such an act of disobedience, while less than that of Achan the peasant, was sufficient to lead Saul to a crushing defeat in his last battle at Gilboa, for him to lose his sceptre and for his dynasty to be supplanted by that of David.[12]

Philip Stern concludes that the biblical narratives favour a view of *ḥerem* as 'an *ad hoc* activity, brought about by the most elemental circumstances of a people's struggle for life and land. This *ad hoc* activity had its source in a broader ancient world view…'[13] In particular, it seems 'that a mentality in which warfare in general was seen as a battle against the forces of chaos was widespread in the ancient Near East from long before the advent of Israel'. Nothing could have been more palpable, Stern argues, than 'the human longing to live in a liveable environment'. The people of Israel had to ensure their survival through an exclusivist relationship with Yahweh, which carried with it the requirement of stringent anti-idolatry laws, for Yahweh had made it clear that he would accept no sacrifices to other gods.[14] Similarly (as there would be within Islām), there was an anti-iconic tendency in the early history of the religion of Israel: the ark of the covenant had images of the cherubim but none of YHWH himself.[15]

Monotheism in the Hebrew Scriptures was therefore not just about what one believed, but much more about how one practised one's faith. Yahweh was the source of all order. An internal plague of worshipping other gods, that might spread from place to place, bringing disorder and disaster in its wake, as in Sodom and Gomorrah, was to be averted at all costs. The mythicization of the enemy helped to justify the massacre of large populations. Absolute obedience was owed to Yahweh: it was through faith in him alone that the practice of *ḥerem* took place. The chosen means might be destructive (though just how destructive might be a matter of dispute),[16] but the objective was to create a holier, as well as a safer, world. There were social benefits to be gained by the people of Israel from eliminating a predatory people like the Amalekites, while their economic organization was insufficiently advanced for them to be able to absorb vanquished soldiers as slaves, even had the Israelites had the will to do so.[17]

The people of Israel were called to follow 'the paths of Yahweh', but the gold and silver idols of the world of disorder were never far from their minds. The struggle for land was equally a struggle to create a 'sacred space', one which was consecrated by the presence of God. In this respect, there is an important philological link between *ḥerem* and the Arabic *ḥaram* in its basic signification as the holy precinct.[18] As an historical precursor of *jihād*, the *ḥerem* of the Hebrew Scriptures seems to have been a divinely-ordained war almost without rules, except that the spoils of war were given to the deity, Yahweh the warrior god, as victor. There was no concept of limitation in the conflict or restriction to the violence, which is the main distinction between *ḥerem* in the Hebrew Scriptures and the *jihād* of Islām. Nor in the concept of *ḥerem* was there any serious attempt to call the people who worshipped other gods or idols to repentance. The lesson we learn is that the divine legislator required a separation of the people of Israel from marriages with the indigenous population that would lead to idolatry. Where, as under Solomon, such marriages take place, they stand condemned because they bring idolatry in their wake and, because Yahweh's resulting anger leads ultimately to the partition of the kingdom under his son (1 Kings 11:7–14).[19] It is difficult to perceive of 'consecration through destruction' as anything other than, in modern terms, xenophobia and ethnic purity being justified as the attempt to bring about God's holiness on earth. In this respect, the *jihād* of Islām in the classical era is more peaceful and integrationist (though not necessarily any more accommodating) in its purposes.

Part One

Text and Meaning

1
Jihād in the Qur'ān and the *Sunnah*

> Allāh is the Light of the heavens and the earth. The similitude of His light is as a niche wherein is a lamp. The lamp is in a glass. The glass is as it were a shining star. [This lamp is] kindled from a blessed tree, an olive neither of the East nor of the West, whose oil would almost glow forth [of itself] though no fire touched it. Light upon light. Allāh guideth unto His light whom He will. And Allāh speaketh to mankind in allegories, for Allāh is Knower of all things.
>
> Q.24:35

Islām means 'submission' in the sense of submission to the living God. Abdulaziz Abdulhussein Sachedina suggests that a verse of central importance in the Qur'ān, which is read today as 'faith (*dīn*), in the eyes of God, is in truth Islām' (Q.3:19) originally may have signified something closer to 'to behave duly before God (*dīn*) is to surrender (*islām*) to Him'.[1] The word '*islām*' refers to the act of surrender to God rather than to the name of a specific religion. Two other verses in the same chapter of the Qur'ān (Q.3:83–4) are instructive:

> Wh[y] do they desire another *dīn* [way of conduct] than God's, [when] to Him has surrendered (*aslama*) whoso is in the heavens and the earth, willingly or unwillingly, and to Him they shall be returned?
>
> Say, 'We believe in God, and that which has been sent down to us, and sent down on Abraham and Ishmael, Isaac and Jacob, and the Tribes, and in that which was given to Moses and Jesus, and the Prophets, of their Lord; we make no distinction between any of them, and to Him we surrender (*muslimūn*).'

The Qur'ān

Muslims believe that the Qur'ān is the literal Word of God revealed directly to the Prophet. This distinguishes it from the Bible, which Christians have believed historically was divinely inspired but for the most part was written by human agents (an exception was Ezekiel, who was handed a scroll of 'lamentations, mourning and woe' but told to eat it prior to speaking to the people of Israel; 'it was in my mouth as honey for sweetness': Ezekiel 3:3). The Qur'ān is unlike the Bible in that it is not a chronological history of God's people to be read from cover to cover. It is better understood as a source of guidance for Muslims with historical references as proofs of God's actions on behalf of humanity. Muslims therefore consider that it is possible to open the Qur'ān at any passage and understand it. In addition, the Qur'ān itself states that it serves as a correction to, and fulfilment of, the Torah and New Testament. According to the doctrine of *taḥrif*, the Jews and Christians received genuine books from God, but their adherents corrupted the text to such an extent that the books in their possession no longer reflect the divine will when they differ from the Qur'ān. Nevertheless, the Qur'ān declares itself to be part of an Abrahamic tradition of monotheism that defines Jews, Christians and Muslims as 'People of the Book'. The religion revealed to the Prophet Muḥammad was in essence the same as that which had been revealed to former prophets such as Noah (Nūḥ), Moses (Mūsā), Jesus ('Īsā), and especially Abraham (Ibrāhīm). Ultimately, God in his wisdom knew what was best for each community, and some of the laws that were revealed varied accordingly (cf. Q.5:48: '...to every one of you We have appointed a right Way and an open road. If God had willed, He would have made you one community...').[2]

In order to understand the meaning of *jihād*, and its status as a key aspect of Islām, we have first to appreciate some of the elements of the faith and the historical context in which it arose. This means that we must address issues involved in studying the Qur'ān, and in particular the vexed question on which scholars disagree – the extent to which a particular verse of the Word may or may not have been superseded by the revelation of a later verse (the doctrine of 'abrogation'). This introduces a very important issue, which is that different scholars read texts in different ways. There may be a consensus (*ijmā'*) of scholars on some matters, but by no means all. This has a crucial bearing on how we interpret *jihād* today. Disagreement about some aspects of the doctrine is long-standing, and it is mistaken to believe that consensus ever means unanimity or uniformity. As Morton Smith has argued with regard to the Bible, every statement in favour of a particular position suggests the existence of counter-propositions.[3]

The gradual transmission of the text to the Prophet

The Holy Qur'ān is the first source of Islamic jurisprudence because for Muslims it is the central repository of faith as the Word of God. This may be defined as the entirety of the revelations of God to the Prophet Muḥammad in Arabic. It was God's revelation to the Arabs in their own language, much as the Torah was sent to the Jews in their own language. Because the Qur'ān is believed to be the literal

Word of God, throughout history it has been memorized in Arabic, regardless of the native language of the student. The jurist Shāfi'ī (d. 204/820) pronounced that 'of all tongues, that of the Arabs is the richest and most extensive in vocabulary. Do we know any man except a prophet who apprehended all of it?' He added that God had 'given evidence... that His Book is [in] Arabic'. According to this writer, the 'extensiveness of the... [Arabic] tongue' was one of the reasons for the communication of the Qur'ān in Arabic.[4] The divine instructions[5] recorded by the Prophet were designed to put the Qur'ān together in the final format intended for God's Final Testament to the world (Q.75:17).

The Divine Revelation occurred gradually over time. There is thus a distinction to be made (though Muslims have not traditionally regarded it in this way) between the process of revelation via the medium of speech[6] and the written book (*muṣḥaf*) which has been transmitted to the Muslim community by continuous testimony. This is not to claim that the Qur'ān was 'created': this argument, espoused by the sectarian Mu'tazilites, was famously opposed by Aḥmad ibn Ḥanbal (d. 241/855), founder of the Ḥanbalī school of law, who argued that it was 'uncreated from cover to cover'. At the time Ibn Ḥanbal was imprisoned for his views, though they have subsequently become an integral part of the main tradition of Islām. Allāh was and is pre-existent to everything known and unknown. His Word is an integral part of His being. It could no more be created than Allāh himself could be created.[7]

The definitive edition of the Qur'ān has 114 chapters (*sūrat*) and 6235 verses (*āyat*) of unequal length. It seems to have been collected together after the death of the Prophet (*Rasūl*) in 11/632: although the accounts of its compilation differ, they all attest to the fact that the Prophet died before the collection was undertaken.[8] In one account, the first successor of the Prophet, Abū Bakr, arranged that a fair copy of the text of the Qur'ān should be made in the form of a book (*muṣḥaf*). The order of the verses was to remain as prescribed by the Prophet.[9] The final, authoritative, version in seven copies (which fixed even the spelling) was completed under the direction of the third caliph, 'Uthmān, within 20 years of the Prophet's death.[10]

In an important verse, the Qur'ān explains the principle of graduality, whereby the Word of God was revealed to the Prophet in a particular dialect of Arabic[11] over a period of 23 years rather than all at once: this was done 'so that your hearts may be strengthened, and We rehearse it to you gradually, and well-arranged' (Q.23:32). It was, in other words, revealed piecemeal to an illiterate, or largely illiterate, society so as to avoid hardship to believers:[12] 'and there are among them illiterates, who know not the Book, but [see therein their own] desires, and they do nothing but conjecture' (Q.2:78). The purpose of the Qur'ānic revelation spreading over such a vast time period was to enable Muḥammad to request direct guidance from God as the community's needs changed over time. This also helps to explain the differences in the Meccan and Medinan verses.

As to whether the Prophet was or was not illiterate, there is controversy within the tradition, between Sunnīs and Shī'a and within each community.[13] Though the matter remains highly contentious, the Prophet clearly dictated to scribes.[14] There is also evidence in the tradition (*ḥadīth*) recorded by al-Bukhārī that the Prophet amended documents and was prepared to write down documents with his own hand in his final illness (although in the event he did not do so).[15] The sense of verses Q.7:157, 158 is contested as to whether they specifically refer to illiteracy on the part of the Prophet (should the translation be 'unlettered' or 'unscriptured'?).[16]

The Qur'ān was revealed to the Prophet in two distinct periods of his mission, the first part (85 chapters), which emanated from Mecca (Makka), being mainly concerned with matters of belief (the Oneness of God, the necessity of the prophethood of Muḥammad, the hereafter, debate with unbelievers and the invitation to accept Islām), while the second part (29 chapters) comprised legal rules and regulated various aspects of life in the new environment of Medina (Madīna).[17]

A holistic interpretation of the text or the replacement of some texts by others ('abrogation')?

It is necessary to reconstruct the two periods, Meccan and Medinan, in order to understand the incidence of 'abrogation' (*naskh*) in the Qur'ān.[18] Literally, '*naskh*' means 'obliteration' or 'annulment'. It is the suspension or replacement of one ruling by another, provided that the latter is of subsequent origin, and that the two rulings are enacted separately from one another (that is, in separate texts).[19] We have the evidence of a tradition (*ḥadīth*) recorded by al-Bukhārī that, during the compilation of the Qur'ān, texts were included even when there was reason to suppose that an earlier text had been abrogated by a later one: 'Uthmān bin 'Affān stated that he would not 'shift anything… from its place'.[20] No abrogation can take place if the text has precluded the possibility of abrogation. The *ḥadīth* which proclaims that '*jihād* shall remain valid till the day of resurrection'[21] precludes the possibility of another *ḥadīth* abrogating the permanent validity of *jihād*.[22] Two verses of the Qur'ān (Q.2:106; Q.16:101) make it clear, at least in the minority opinion of the jurist al-Shāfi'ī,[23] that abrogation is a wholly internal phenomenon: no *ḥadīth*, in other words, can abrogate a verse in the Qur'ān.[24]

The case is more complex, however, when verses of the Qur'ān are abrogated by later verses, since the verse order of the complete text is not regarded as chronological. In Mohammad Hashim Kamālī's words, 'the broad sweep of [abrogation] (*naskh*) was… taken so far as to invalidate a major portion of the Qur'ān'. This is precisely the case with regard to one of the two *ayat* of the sword (*āyat al-sayf*) which reads, in the relevant part: 'and fight the polytheists all together as they fight you all together, and know that God is with those who keep their duty [to Him]' (Q.9:36; cf. Q.9:5). *Sūrah* 9 was almost the last of the chapters of Qur'ān to be revealed in most accounts, 27th out of the 28 Medina *sūrat*, or 113th out of the total of 114 chapters.[25] Jurists were not in agreement as to how many earlier verses were abrogated by the verses of the sword, but Muṣṭafā Abū Zayd considered that they abrogated no less than 140 *āyahs* in the Qur'ān; others suggested 113, 114 or 124.[26] Kamālī comments that the jurists of the second century of Islām 'considered war as the norm, rather than the exception' in relation to non-Muslims.[27] There was, he contends, a degree of 'exaggeration in the use and application of *naskh*'.[28] The reason behind this attitude was the need, which was then prevalent, to be in a state of constant readiness for battle in order to protect Islām. 'Under such political circumstances, it is not difficult to understand how abrogation was utilised as a means by which to strengthen the morale of the Muslims in facing their enemies.' Historically, *naskh* was not perhaps the tool of political expediency that Kamālī suggests, but the issue remains open as to whether this tool for understanding apparently conflicting passages in the holy scripture of Islām remains valid.

Revelation by substitution? The four stages in the development of the Qur'ānic concept of *jihād*

The traditional reading of the Qur'ān outlines four 'stages' which arose from the historical development in which the Prophet found himself. Here is one such traditional reading, that of Shamseddin al-Sarakhsī (*c.* 400/1010–482/1090), one of the greatest jurists of the classical age, whose 30-volume *Mabsūṭ* is considered to rank among the world's leading legal works (see Chapter 2):[29]

> [1] At the beginning, the Prophet was enjoined to propagate the message of Islām peacefully and to avoid direct confrontation with the unbelievers. This is disclosed in the Qur'ānic texts… [Q.15:94 and Q.15:85, respectively 'be firm in what is commanded and stay away from the idol worshippers' and 'deal with them [that is, the unbelievers] in a just and fair manner']
>
> [2] And then Allāh enjoined the Prophet to confront the unbelievers by means of argumentation, which is clearly expressed in the Qur'ān, 'call to the path of your Lord with wisdom and fair exhortation, and argue with them in good manner' (Q.16:125). However, void argumentation is strictly prohibited, and it is only permissible if it is done in good spirit, meaningfully and effectively to disclose the truth. (Q.29:46)
>
> [3] And then Allāh granted permission to the Prophet and his followers to fight their enemies with the Divine command, 'Permission to fight their enemies is granted upon those who were unjustly wronged…' (Q.22:39). They were then enjoined to wage war against those who initiated aggression against Muslims (Q.2:193). And then the Muslims were enjoined to wage war against the unbelievers, with one condition that it must be waged after the end of the

> Forbidden Months (*al-ashhur al-ḥurum*)[30] with the divine command 'fight you the unbelievers immediately after the end of the Forbidden Months'. (Q.9:5)
>
> [4] The final stage came with the Divine command of Allāh enjoining the Prophet and his followers to wage war against the unbelievers unconditionally. It is expressed in the Qur'ānic text, 'Fight you all in the path of Allāh, and be aware that Allāh is all-knowing' (Q.2:244). This command will remain as such and its unconditional nature implies that its realization is imperative upon Muslims. Unless war is made imperative, attempts to bring about the superiority of Islām and the inferiority of unbelief will not be a success. [cf. Q.58:39]

Ayatullah Murtazá Muṭahharī summarizes the traditional reading thus:[31]

> One series consist[s] of those verses that tell us unconditionally to fight, so if we had ears and heard only these and not the others, it would be possible for us to think that Islām is a religion of war. The second series consist[s] of verses that give the order to fight but with certain conditions: conditions such as the opposing side being in a state of war with us, or a mass of Muslims or non-Muslims having been placed under the heels of a group from amongst themselves which has trampled on their freedom and rights. The third series of verses make[s] it perfectly clear to us that the call of Islām is not sounded with any force of arms. And in the fourth group Islām decisively announces its love of peace.

These traditional stages have been recently described by Reuven Firestone,[32] who distinguishes between stage one, or 'non-confrontation' (Q.15:94–5), and stage two, or 'defensive fighting' (Q.22:39–40a). The two subsequent stages are stage three, or 'initiating attack within the ancient strictures' (Q.2:217) and the final stage, stage four, 'the unconditional command to fight all unbelievers' (Q.2:216: 'fighting is commanded upon you even though it is disagreeable to you. But it is possible that you dislike something which is good for you and that you love something which is bad for you. God knows, but you know not').

The objection to the substitution of the so-called 'earlier' verses by the 'later' ones is twofold. Firstly, it prevents a holistic reading of the Holy Qur'ān. If we take a comparison from the Hebrew Scriptures, there are thought to be three authors of the Book of the Prophet Isaiah: but we do not read the third author in preference to the first two; instead, we read the book as a single document.[33] A modern holistic approach to reading the Qur'ān, quite different from that of the jurists of classical times, is provided by Dr Amīr 'Alī, who suggests twelve senses of *jihād* which are to be found in the Qur'ān and *ḥadīth*: recognizing the Creator and loving him most (Q.9:23, 24); resisting pressure of parents, peers and society (Q.25:52); staying on the straight path steadfastly (Q.22:78); striving for righteous deeds (Q.29:69); having courage and steadfastness to convey the message of Islām (Q.41:33); defending Islām and the community (Q.22:39–40); helping allied people who may not be Muslim; removing treacherous people from power (Q.8:58); defence through pre-emptive strikes (Q.2:216); gaining freedom to inform, educate and convey the message of Islām in an open and free environment (Q.2:217); freeing people from tyranny (Q.4:75); and, after victory, removing tyranny, treachery, bigotry, and ignorance and replacing them with justice and equity (Q.4:58; Q.5:8; Q.7:181; Q.16:90). For Amīr 'Alī, Islām is not a religion of religious coercion (Q.2:256), while '*jihād* in Islām is striving in the way of Allāh by pen, tongue, hand, media and, if inevitable, with arms. However, *jihād* in Islām does not include striving for individual or national power, dominance, glory, wealth, prestige or pride.'[34] Needless to say, modern radical Islamist writers such as Sayyid Quṭb reject this interpretation, deny that every verse of the Qur'ān is the final principle of Islām and assert instead the principle of abrogation ('the various stages through which this movement develops').[35]

A second objection to the revelation by substitution theory is that it cannot be verified by objective evidence.[36] In general terms there is a divine prerogative to 'erase what He wills and endorse... what He wills. With him is the master copy of all the revelations' (Q.13:39). But though there are references to textual change (Q.2:106; Q.16:101), or the replacement of revelation by revelation, the Qur'ān itself does not expound a theory of *naskh*. If the 'generally recognized meaning of *naskh* in relation to the Qur'ān is... the nullification of the original ruling, while the original wording is recorded', the fact remains that both texts are extant in the surviving [written copy of the Qur'ān] (*musḥaf*)': hence the potential confusion.[37]

On the analogy of external *naskh*, that Islām, the latest of the Divine Revelations, sets aside certain of the social and ritual laws of the earlier religious systems, there was support for internal *naskh*. There was also a practical necessity for it: two conflicting statements might be seen as incapable of reconciliation and hence simultaneous implementation.[38] Beyond this, there is less agreement. There is no statement in the Qur'ān or among the *ḥadīth* that a particular verse had been substituted by another; nor did classical Muslim scholars possess any clear indication that one verse was later or earlier in revelation than another. They merely asserted that this was so, often without giving clear reasons, so that we now cannot know how it is 'possible to distinguish the verse which is the sole valid source for obligatory action from the verse whose ruling has been abandoned...'[39]

Different senses of *jihād* in the Qur'ān

Reuven Firestone proposes 'a new reading' of the Qur'ān which rejects the alleged 'evolutionary theory' of 'just war', and which instead proposes a different four-fold grouping.[40] The first group of verses is of those which express a non-militant means of propagating or defending the faith. Of particular importance here are the references to other 'peoples of the Book', such as Q.29:46 and Q.42:15. In these passages, it is not the role of the Holy Prophet or the Muslim community to inflict punishment or escalate the conflict with opponents such as Jews and Christians: 'Our God and your God is one, and it is Him to whom we surrender' (Q.29:46); 'Allāh is our Lord and your Lord. We have our works and you have yours. There is no argument between us and you. God will bring us together, for the journey is to Him' (Q.42:15). This principle is brought out the verse Q.109:6: 'to you your religion and to me mine', which is capable of interpretation in the sense of an acceptance of religious pluralism,[41] although its original purpose was probably to take 'cognizance of the unbridgeable gap between Islām and the religion of the Meccans'.[42]

A second group comprises those verses which express restrictions on actual fighting. An example is Q.2:190 ('fight in the path of God those who fight you, but do not transgress limits; for God does not love transgressors'). Such transgressions might include the killing of women and children and other non-combatants, as well as some pre-Islamic customs that were expressed in new Islamic terms.[43] Two other divisions express conflict between God's command and the reaction of the followers of the Holy Prophet; and verses which strongly advocate war for God's religion. The greatest number of verses fall within the third group, the conflict between God's command and the response of the people. Thus Q.2:216 ('fighting is commanded upon you even though it is disagreeable to you') is an obvious example, but there are several other passages which suggest a tension within the early Muslim community over the issues related to raiding and warfare (Q.3:156; Q.3:167–8; Q.4:72–4; Q.4:75; Q.4:77; Q.4:95; Q.9:38–9; and Q.9:42).

Finally, there is the group of verses which strongly advocate war for God's religion. Thus Q.2:191 is one of the passages frequently cited by radical Islamists seeking to justify attacking non-Muslims to establish a caliphate ('kill them wherever you find them and turn them out from where they have turned you out, for sedition (*fitnah*) is worse than killing…').

There are 35 occurrences of the word *jihād* or its equivalent in the Qur'ān.[44] The verses have been analysed in the order suggested by Cherāgh 'Alī in the nineteenth century, without necessarily following his interpretation.[45] The conclusion is highly pertinent. There are just four verses which use derivations from *jihād* and are clearly 'warlike' in intention or which, given the context, are open principally to a 'warlike' interpretation.[46] In contrast, there are eleven verses which are pacific in intent or seem to be open principally to a pacific interpretation.[47] Twenty of the verses are capable of different interpretations: they are open to a pacific reading, but they could be read as having a 'warlike' intent.[48] In addition, there are a number of verses (Q.2:190–3; Q.8:59–70; Q.9:5;

Q.9:12; Q.9:30; Q.9:38–9; Q.61:4) regularly cited by radical Islamists to justify attacking non-Muslims in order to establish a Caliphate.[49]

Differing views among modern commentators on the Qur'ān

The central issue is the status of the 'verses of the sword' (Q.9:5; 9:36)[50] and whether they abrogate other statements in the Qur'ān. Here it is important to note the views of relatively recent 'theological' commentators on the Qur'ān, chosen randomly according to availability. Muḥammad Asad (1317/1900–1412/1992),[51] commenting on Q.2:190, asserts that only a war of self-defence in the widest sense of the word can be considered a war 'in God's cause', that is, 'in the cause of the ethical principles ordained by God'. The defensive character is confirmed by the expression 'those who wage war against you' and is further clarified by Q.22:39 ('permission [to fight] is given to those against whom war is being wrongfully waged'). Asad notes that 'according to all available Traditions, [Q.22:39] constitutes the earliest (and therefore fundamental) Qur'ānic reference to the question of *jihād*'.[52] He adds that 'this early, fundamental principle of self-defence' is maintained throughout the Qur'ān, as is evident from Q.60:8, as well as from the concluding sentence of Q.4:91, 'both of which belong to a later period'. In view of the preceding ordinance, the injunction to 'slay them wherever you may come upon them' is valid only within the context of hostilities already in progress in a war of self-defence or a war of liberation, 'for oppression (*fitnah*) is even worse than killing'. For Asad, the translation of *fitnah* as oppression is 'justified by the application of this term to any affliction which may cause man to go astray and to lose his faith in spiritual values'.[53] Q.2:194 is interpreted to mean that 'although the believers are enjoined to fight back when they are attacked… they must, when fighting, abstain from all atrocities, including the killing of non-combatants'.[54]

Muḥammad Asad, commenting on Q.9:5 ('the verse of the sword'), contends that, read in conjunction with the two preceding verses, as well as with Q.2:190–4, the verse relates to warfare already in progress with people who have become guilty of a breach of treaty obligations and of aggression. Asad asserts that 'every verse of the Qur'ān must be read and interpreted against the background of the Qur'ān as a whole'. There can be no question of a meaning of '"conversion or death", as some unfriendly critics of Islām choose to assume'.[55]

For a second modern 'theological' commentator of the Qur'ān, Shaykh Muḥammad al-Ghazālī (1335/1917–1416/1996), 'Islām is a tolerant and accommodating religion', which when it comes to matters of religious conviction and belief 'specifically forbids coercion and compulsion'.[56] By the imperative of their religion, Muslims are taught not to impose their beliefs on others by force.[57] In his commentary on Q.2:216, he states that 'peace is to be welcomed when rights are protected and beliefs are respected; but if peace means abject surrender and subjugation, it cannot be easily defended on moral or realistic grounds'. Sedition is a greater threat than killing (Q.2:217). Do not commit aggression (Q.2:190) is 'an eternal principle' for Shaykh Muḥammad al-Ghazālī, and 'everything else the Qur'ān has to say on this subject agrees with it'. There can be no waging war against those who do not commit aggression.[58] Even defensive war is legitimate only if it is undertaken for the cause of God and not for personal glory or to gain a special advantage. Al-Ghazālī contends, with regard to *sūrah* Q.9 that it has been misinterpreted and 'maliciously misconstrued' to suggest that there was a declaration of war on non-Muslims without exception. Instead, the historical war in question was to be prosecuted 'specifically against those groups who had aided the enemies of Islām or violated the rights of Muslims'.

Jihād in our time, the same author continues,

> encompasses a whole range of activities including inventiveness, development, and construction on land, in the sea and in outer space. It implies research in all fields to gain wider and deeper understanding of the world around us and all the phenomena associated with it.[59]

It is necessary also to 'revive faith in God, to weed out corruption and evil, and to ensure the health and well-being of society'.[60] Muslims are called to reciprocate the respect and tolerance shown them by other faiths; but the author recalls the hostility of Judaism and Christianity in the past and the opprobrium that Muslims who are in the process of rediscovering and reasserting their Islamic identity receive in the contemporary era.[61] Islām affirms the value of human life, yet human life is too lowly valued in many Muslim countries.[62] The Prophet was no despot coming into the world to change people's minds by force; instead, 'no use of force can ever be justified to compel people to accept a particular religion or creed'. Muslims are thus in general non-belligerent, but they will 'stand up to aggression and defend their beliefs to the last'.[63]

So much for modern 'theological' commentators. A second type of commentary on the Qur'ān is provided by the authors of 'contents guides'. Again, two examples are taken, chosen at random because of availability. The first is that of Faruq Sherif, who considers that 'all religions, philosophies, laws and ideas are the product of a particular time and place, and cannot be properly understood and judged except in the light of the circumstances in which they came into being'. To maintain in legislating today, he argues, that 'commandments laid down 14 centuries ago are invariable and binding for all time is to defy the primordial law of evolution and to ignore the spirit of the Qur'ān which attributes the quality of permanence only to spiritual values'.[64] Nevertheless, Faruq Sherif takes a robust view of *jihād* as fighting. 'The great expansion of Islām in the short time after its inception was largely due to the militant spirit of the new faith', he argues. 'A great many verses of the Qur'ān enjoin on Muslims to take up arms against polytheists, unbelievers and hypocrites.' He accepts that Q.2:186 implies that fighting is only justified when the enemy has attacked, but contends that

> this is by no means the general rule. Nor is there any substance to the argument which is sometimes advanced to the effect that *jihād* should be understood primarily in the sense of moral endeavour and self-discipline in the cause of service to Islām, and only secondarily in that of holy war.

Instead, in his view, 'the emphasis is distinctly on warring against non-believers with the object of propagating Islām, this being, by the express injunction of the Qur'ān, one of the primary duties of Muslims'.[65] In a number of verses the command to fight is supported by the promise of rewards. Other verses show God's displeasure with those who shirk their duty of fighting. Moreover, with the exception of a few verses which are revealed with reference to particular events such as the battles of Badr and Uḥud, 'all the texts concerning *qitāl* and *jihād* have a general import. The obligation to engage in holy warfare is meant to persist, in the words of the Qur'ān cited above, until God's religion reigns supreme.' Thus the Muslim world is at all times in a position of 'potential hostility' towards the non-Muslim world.[66]

A second 'contents guide', that provided by Fathi Osman, takes a quite different approach. This discusses *jihād* in the context of 'universal relations'.[67] Peace is the general rule for Muslims, but peace cannot be secured unless it is based on justice. For Fathi Osman, 'fighting in the way of God, *jihād*, and for His cause which secures justice and peace for all, is restricted to defending human rights, whether related to the human life or the homeland, or to… opinion, belief and expression'.[68] Islām does not allow Muslims to fight to impose their faith by force, which the misleading translation of *jihād* as 'holy war' might lead some mistakenly to think. Citing Q.2:194, it is argued that, even if it was sometimes practised in history,[69]

> fighting for the sake of fighting or mere expansion of land or imposition of beliefs is forbidden by the principles of Islām, and whenever fighting becomes legitimate for self-defence, it is restricted to those who are fighting on the other side. The use of weapons that lead to mass destruction and indiscriminate killing, and thus hurt non-combatants, cannot be allowed according to Islamic moral and legal principles.[70]

However, fighting against oppression is no less legitimate than fighting against aggression (Q.2:193; Q.8:39).[71] Indeed, in the Qur'ān, the acceptance of oppression is condemned: the state of injustice has to be changed, the oppressors confronted. 'This is the *jihād* as instructed by the Qur'ān, not as it might be claimed or practised by… expansionist invaders.'[72]

No perpetual war between the world of Islām and the world of unbelief (*Kufr*) in the Qur'ān

A final approach to the Qur'ān, that adopted by Qamaruddin Khan, is to discuss its political concepts by objective analysis, what he calls 'the spirit of free and honest inquiry'. The credentials of this scholar are impeccable, since he has studied *inter alia* al-Māwardī's theory of the state[73] and the political thought of Ibn Taymīyah.[74] In Islamic history, Khan contends, 'the Muslim state has often been equated with the Islamic faith' and it is asserted that the one exists for the other. 'This attitude has given the impression that Islām is a political device rather than a moral and a spiritual force.'[75] The reality is quite different, Khan asserts. The state was a 'circumstantial event'; it did not follow a set pattern, divine or human, but grew out of history:[76]

> The main concern of the Muslims was to propagate the new faith. To realize this aim they had to unite themselves into an organization which gradually developed into a state. The functions of this state were more or less the same as of other states in the world. The reference in the Qur'ān to the problems of war and peace are therefore incidental, and do not constitute the essentials of statecraft and do not provide any basis for a political theory... the Islamic state is neither the creation of a divine injunction nor is it equivalent to Islām.

Thus, for Qamaruddin Khan, the Prophet did indeed establish a political regime, but this was 'incidental' to his historical situation, and 'not the essential aim of his Prophetic mission'. The Qur'ān thus provides a set of Islamic values 'and not the structure of the state'. There is no such thing as a permanent Islamic constitution, since Islamic political theory is a 'changing and developing concept, adapting itself to the exigencies of time and place'. The term 'Islamic state' was not used before the twentieth century; nor is there any warrant in the Qur'ān for the imamate theory. There is a prevailing misconception in the minds of many Muslims based on Q.16:89, Khan argues, that the Qur'ān contains an exposition of all things. This particular verse was intended to explain that the Qur'ān contains information about every aspect of moral and social guidance; but it is not an inventory of human knowledge. Instead, Q.6:149 ('to God belongs the arguments conclusive; for had He willed, He would have guided you all') suggests that God in his wisdom 'omitted to provide in the Qur'ān a rigid constitution that would have become unworkable after some time and brought positive discredit to Islām'. For Qamaruddin Khan, Islām must progress in the world 'as an independent spiritual and moral force, conquering not lands, rivers and mountains, but the hearts and souls of men'. Moreover, Islamic values can be developed within different political and social conditions and under different political systems.[77]

As far as *jihād* is concerned, 'there is absolutely no implication of any political theory' in the Qur'ān.[78] Elsewhere, Khan emphasizes the idea of a temporal *jihād*, determined by historical circumstance:[79]

> ...there is no Qur'ānic sanction for the theological division of the world into the *Dār al-Islām* and the *Dār al-Ḥarb*. According to the Qur'ān, the world is divided between believers and non-believers. It repeatedly says that the believers together constitute one people and the disbelievers together constitute another people, as in 'the believers are brethren of one another' (Q.49:10) and 'those who disbelieve are friends of one another' (Q.8:72).
>
> But the Qur'ān nowhere demands that the Muslims should remain permanently at war with the non-believers. The verses (for instance, Q.4:89 and Q.9:5) which seem to give the impression of perpetual war between the world of Islām and the world of *Kufr*, are decidedly topical and circumstantial in their import, and cannot be taken as permanent injunctions of God... [The Qur'ān] enjoins the incessant struggle until the whole world has been submitted to the message of Muḥammad. But the struggle is to be done by *da'wah* [persuasion and preaching]. Resort to force is allowed only as a defensive or self-protective measure...

It is clear from the preceding discussion that modern or relatively interpretations of the Qur'ānic passages concerning *jihād* may differ substantially from the viewpoints of the classical interpreters. If we take the example of Islamic attitudes towards other religions, then it is possible to take two different approaches: the first is to deal with the 'laws themselves and with the various ways in which they were explained, interpreted and related to the Qur'ān and *ḥadīth*, the two textual sources of the *sharī'ah*'; the second approach is to study the classic material essentially as a background and then to proceed to modern Muslim views on religious liberty and the attitude to the 'other'.[80] Neither approach has been excluded in the previous discussion, where we have started with the classical viewpoint and then proceeded to more modern interpretations. The time has now come to proceed to the traditions, the second source of the *sharī'ah*.

The *Sunnah*

The second primary source for the Muslim believer and the historian is the *Sunnah*:[81] 'nor does he say (aught) of (his own) Desire. / It is no less than inspiration sent down to him' (Q.53:3–4). '...So take what the Messenger assigns to you, and deny yourselves that which he withholds from you...' (Q.54:59). One of the most authoritative of the compilers of *ḥadīth*, al-Bukhārī, records on the witness of Hudhaifa that 'the Prophet told us that the virtue of honesty descended in the roots of men's hearts [from Allāh] and then they learned it from the Qur'ān and then they learned it from the *Sunnah* [the Prophet's traditions]'.[82] The highest category of *ḥadīth* were the *ḥadīth qudsī*, or sacred *ḥadīth*, in which God is the speaker. A collection of 40 *ḥadīth qudsī* contains no reference to *jihād*,[83] though it does denounce false claims to martyrdom.[84] However, an authoritative collection of 90 *ḥadīth qudsī* by William A. Graham does contain a reference, and an important one, to *jihād* (Saying 46), where martyrdom becomes the rapid road into the presence of God:[85]

> [The Prophet] said: 'God answers him who goes forth [as a warrior] in his cause: "Only faith in Me and complete trust in the veracity of My Apostles causes him to go forth. I shall send him back with what he has gained in the way of rewards or booty, or [else] I shall cause him to enter paradise."' [Muḥammad said:] 'If it were not that I would cause my community hardship, I would not remain behind the troops, and I would wish that I might be killed in God's cause. Then I would live, then be killed; then live, then be killed.'

It has been suggested that the 'moral aspects of *jihād*' in the *ḥadīth* may be categorized as the obligation to fight in the cause of Allāh; the reward for fighting; the reward for martyrdom; divine aid against the enemy; criticism for the 'hypocrites' (those who found excuses for inaction); and, finally, exemptions from fighting.[86]

Jihād in the early collections of the *ḥadīth*

For the early collectors of the authoritative statements of the Prophet, Paradise lay 'under the shadows of the swords',[87] which implied a military purpose for *jihād* (keeping a horse for *jihād* was seen as a deed which in itself would earn a reward).[88] They were equally clear that there was a 'gate of *jihād*' which was one of the entries to Paradise,[89] reflecting the divine reward for he who was martyred[90] and the guaranteed place in Paradise for those who carried out *jihād* in God's cause.[91] The Prophet said of the martyrs of Uḥud: 'are we not their brothers? We entered Islām as they entered Islām and we did *jihād* as they did *jihād*.'[92] The commitment to *jihād*, however, had to be voluntary[93] and the motivation had to be pure: no booty[94] and 'nothing but *jihād* in [God's] cause and belief in his Word…'[95] There was no reward for he who desired some worldly advantage[96] or fighting for vain show and seeking to acquire much[97] (though there might be a reward of legitimate booty).[98] Instead, a *jihād* in which a man placed himself and his property in danger for the sake of God, and returned without them, was a superior deed.[99] There could be no lagging behind on the way to *jihād* (let alone bleating like goats),[100] since the wounds gained in the fighting would be seen on the Day of Judgement. Believers should be prepared to fight and be killed as if it was happening three times in all.[101] The reward in Heaven, indeed, would be 'better than the world and all that is in it'.[102] All earthly sins, except debt, would be forgiven.[103] When asked what act gave a man in Paradise the highest possible distinction, the Prophet replied '*Jihād* in the way of Allāh! *Jihād* in the way of Allāh!'[104] His companions considered that partaking in *jihād* with the Prophet himself was a particular cause for divine reward.[105] Those who accompanied the Prophet 'were large in numbers but there was no proper record of them… Few were the persons who wanted to absent themselves…'[106]

However, the narrators cited by the compilers of the Prophet's tradition were not in agreement on the worth of *jihād* as against other duties incumbent on a believer. 'Abdullah bin Mas'ūd, Al-Walīd bin 'Ayzar and 'Abdullah Ibn Mas'ūd (as reported by al-Bukhārī),[107] and Abū Harayra, 'Abdullah bin Mas'ūd and Abū 'Amr Shaībanī (as reported by Muslim)[108] claimed that saying prayers at their stated times and being dutiful to one's parents came before *jihād* in Allāh's cause. The logic of this order was given by Abū Hurayra (as reported by al-Bukhārī and Muslim): to believe in Allāh and His Apostle, the Prophet, had to precede *jihād* for that act of duty to be carried out truly in Allāh's cause; performing pilgrimage (*Ḥajj*) was the third meritorious deed within this scheme.[109] Anas (again as reported by al-Bukhārī) recalled the lifelong commitment to Islām, the Prophet and *jihād* made by those pledging allegiance to Muḥammad on the day of Khandaq, the battle of the Trench.[110] Ibn 'Abbās narrated that after the conquest of Mecca, the Prophet declared that there was no further need for migration (*hijrah*); but there remained *jihād* and good intentions.[111]

In the event of a *jihād*, the victory was that of God alone:[112]

> whenever Allāh's Messenger (may peace be upon him) came back from the battle or from expeditions or from [the major pilgrimage (*Ḥajj*)] or [the minor pilgrimage ('*umrah*)] and as he reached the top of the hillock or upon the elevated hard ground, he uttered Allāh-u-Akbar thrice, and then said: There is no god but Allāh. He is One, there is no partner with Him, His is the sovereignty and His is the praise and He is Potent over everything. (We are) returning, repenting, worshipping, prostrating before our Lord, and we praise Him. Allāh fulfilled His promise and helped His servant, and routed the confederates alone.

Once the call was given for *jihād*, an immediate response was required,[113] except for two categories of people: the disabled (Q.4:95)[114] and those whose parents were still alive and who therefore owed them a duty of service.[115] There could be a different form of *jihād* for parents, too. 'Abdullah b. Mas'ūd and Abū 'Amr Shaybanī narrated that prayer and kindness to parents preceded earnest endeavour (*jihād*) in the cause of Allāh.[116]

As reported by al-Bukhārī, Nāfī' narrated the judgement of Ibn 'Umar that Islām was founded on five principles: belief in Allāh and his Apostle (the Prophet), the five compulsory prayers, fasting in Ramaḍān, payments for charity (*zakāt*) and pilgrimage (*Ḥajj*).[117] When asked for the meaning of Q.49:9 ('if two groups of believers fight each other, then make peace between them, but if one of them transgresses beyond bounds against the other, then you all fight against the one that transgresses'), Ibn 'Umar was reported to have said:

> we did it during the lifetime of Allāh's Apostle when Islām had only a few followers. A man would be put to trial because of his religion; he would either be killed or tortured. But when the Muslims increased [in number], there were no more afflictions or oppressions [of this kind].[118]

The early usage of 'martyrdom' in the traditions of the Prophet

Any investigation of the Prophet's tradition in the area of 'struggle' has also to taken into account the early usage of the term 'martyr' (*shāhid*), since this is of critical significance for the moral underpinning of *jihād*: he who is recruited for *jihād* has to understand what benefits accrue to him in the eventuality of death or martyrdom (*shahādah*). The Prophet himself undertook 19 military campaigns (*ghazawāt*)[119] and expected to be among ten martyrs whom he could name in advance.[120] 'Being slain is but one way of meeting death, and the martyr is the one who gives himself, expectant of reward from Allāh.'[121] Only a true believer could enter Paradise.[122] By definition a *jihādī* had to be a Muslim since he fought 'in the way of Allāh'; because of this, even a former murderer could become a martyr.[123] Yet salvation was not assured if he fought for reasons of worldly pride, for example the wish to be regarded as a 'brave warrior': Hellfire might await such a liar.[124] Suicide prompted by the seriousness of wounds received in battle would not guarantee a place in Paradise: 'verily [the rewards of] the deeds are decided by the last actions'.[125] (This was distinct from accidentally killing oneself in battle, which did guarantee martyr's status.)[126] Angels protected the body of the martyr with their wings until he was buried.[127] The Prophet prayed over a martyr who had been disfigured after being killed.[128] The garments of the martyr were unwashed, since he was buried in the garments in which he was slain.[129] Even after six months, the body of the martyr could be exhumed and found to be largely free from decomposition.[130] The cut limbs of a martyr were blessed.[131] The *jihādī* who had been martyred would be willing to return to the world because of the great merit of martyrdom that he had witnessed; Anas bin Mālik narrated that a martyr would be willing 'to be killed ten times for the sake of the great honour that has been bestowed upon him'.[132] The intercession of a martyr would be accepted for 70 members of his family.[133]

The Prophet stated that 'the martyrs of my *Ummah* will be small in number'.[134] Nevertheless, martyrdom for the *jihādī* was not a unique status. There are surprisingly numerous of types of martyrdom which were recognized in the Prophet's lifetime. According to the Prophet, as related by Jābir ibn 'Atīq, there are seven types of martyrdom in addition to being killed in Allāh's cause: death from plague,[135] drowning,[136] pleurisy or cholera,[137] 'an internal complaint' or abdominal disease,[138] being burnt to death, having a building fall down on one, and, for a woman, dying in childbirth, were all causes of martyrdom.[139] Even this list was not exhaustive. Abu Mālik heard from the Prophet that being killed by being thrown from a horse or camel, being stung by a poisonous creature, or death in bed 'by any kind of death Allāh wishes' would lead to martyrdom and entering Paradise.[140] Death through the defence of one's property was another recognized cause of martyrdom,[141] as was the defence of property, family, 'or his blood or his religion'.[142] The early tradition of the Prophet seems to have wished to include as many believers as possible within the ranks of the martyrs guaranteed a place in Paradise: the distinctive merits of the *shāhid* were inevitably reduced as a consequence.

The issue of canonicity among the collections of *ḥadīth* literature

The analysis presented above has been drawn principally from the two collections of *ḥadīth* literature by al-Bukhārī and Muslim, both of whom gained the reputation of soundness (*ṣaḥiḥ*). Al-Bukhārī, whose full name was Abū 'Abdullah Muḥammad bin Ismā'īl bin Ibrāhīm bin al-Mughīra al-Ja'fī (194/810–256/870), spent 16 years compiling his collection, and ended up with 2602 *ḥadīth* (9082 including repetitions. The actual number, according to the most recent scholarship, is even higher: see Table1.1). His criteria for acceptance into the collection were amongst the most stringent of all the scholars of *aḥadīth*. He insisted on evidence that any two men named consecutively in the chain of authority (*isnād*)[143] must have met in person.[144] His contemporary Muslim, whose full name was Abūl Ḥusayn Muslim bin al-Ḥajjāj al-Nisapūrī (206/817–261/874), compiled 3033 *ḥadīth*; he drew upon al-Bukhārī's work, but differed in methodology, since his criterion for acceptance was less onerous, in that two individuals in the chain of authority had only to have been contemporaries who could have met. 434 named persons in al-Bukhārī do not appear in Muslim, while 625 names appear in Muslim whom al-Bukhārī does not name. However there is clear overlap between the two authorities: al-Bukhārī names 208 companions and contemporaries of the Prophet, and Muslim 213, of whom 149 are common between them.[145] Soundness (*ṣaḥīḥ*) after these two writers was defined in the order as follows: 1) any report found in al-Bukhārī and Muslim; 2) any report found in al-Bukhārī alone; 3) any report found in Muslim alone; 4) any report which matched their criteria, even if they did not include it; 5) any *ḥadīth* in accordance with the criteria of either of the two men.[146]

Table 1.1 indicates the number of *isnād* and *ḥadīth* in each work of the main *ḥadīth* collections,[147] according to the *Ḥadīth* Encyclopedia of the Thesaurus Islamicus Foundation. One of the purposes of indicating the considerable number of *aḥadīth* is to emphasize that relatively few of them were concerned with the theme of *jihād*.

Table 1.1 Numbers of *ḥadīth* and *isnād* in the main collections

Ḥadīth Collection	Date of death of collector	Number of *ḥadīth*	Number of *isnād*, not including repetitions	Number of *isnād*, including repetitions
Al-Bukhārī	256/870	7658	7224	10315
Muslim bin al-Ḥajjāj	261/875	7748	9965	14188
Abū-Dāwūd	275/888	5276	7045	7865
Al-Tirmidhī	279/892	4415	5645	6530
Al-Nisā'ī	303/915	5776	6174	7131
Ibn Māja	273/886	4485	5242	5991
Mālik bin Anas, *Al-Muwaṭṭa'*	179/795	1861	970	2140
Total		37219	42265	54160

Source: <www.cmeis.cam.ac.uk/ihsan/> For the dates of death of the collector: Graham, *Divine Word and Prophetic Word in Early Islām*, 83. Graham adds Aḥmad bin Ḥanbal (d. 241/855) and ad-Dārimī (d. 255/869) to the list of canonical or nearly canonical authors.

The fourth 'rightly-guided' caliph, 'Alī, affirmed that false traditions of the Prophet were already current in his lifetime, so that Muḥammad had stated: 'whoever attributes falsehoods to me makes his abode in Hell'.[148] The chain of authority (*isnād*) had become a critical issue by the time of al-Bukhārī and Muslim, but the relatively late date of their compilations, some century and a half after the death of the Prophet, leaves the enquirer with the question: what had happened earlier and how do we know that fabricated traditions that had started at an early date were not, unwittingly, incorporated in their collections by these great men? The *isnād* tradition did not exist in the first century after the Prophet: there was a 'barrier' beyond which it is difficult to penetrate. And what happened if there was a proliferation of *isnāds*?[149] The Prophet had feared sectarianism in Islām after his death: there would be 73 sects, he was thought to have said; 72 destined for Hell, with only those determined to maintain the unity of Islām destined for Heaven.[150]

'Weak' and 'false' *ḥadīth* and the implications for the *jihād* traditions

Orientalists[151] such as Ignaz Goldziher in the nineteenth century denounced 'fabrications' in the traditions, which he considered were a weapon of debate by the various groups competing for control of the Islamic movement. Even Muslim legal scholars such as Professor Kamālī accept that the development of serious and persistent differences in the community by the year 40/660, marked by the emergence of the Khārijīs and the Shī'a led to 'distorted interpretation of the source materials, or... outright fabrication'.[152]

Moderate followers of Goldziher in the later twentieth century, such as G. H. A. Juynboll, argued that 'fabrication or forgery' may have begun 'almost immediately after the Prophet's death, if not on a small scale even already during his lifetime'. Juynboll wrote:

> Too many Companions, especially Anas, Abū Hurayra, Ibn 'Abbas and Jabir b. 'Abd Allāh to name but a few of the most important alleged *ḥadīth* transmitters among them, were 'credited' with such colossal numbers of obviously forged traditions that it is no longer feasible to conceive of a foolproof method to sift authentic from falsely ascribed material.[153]

In particular, Anas (d. 93/711), who allegedly lived to the age of 103, was a convenient source for forgers because of his longevity.[154]

Abū Harayra stands at the head of the list of *ḥadīth* transmitters, with 5374 'channels through which *aḥadīth* were transmitted';[155] he is said to have instructed 800 students. 'Abdullah bin 'Umar had 2630 narrations, Anas ibn Mālik 2286, 'Ā'ishah Umm al-Mu'minīn 2210 and 'Abdullāh ibn 'Abbās 1660, Jābir ibn 'Abdullāh 1540 and Abū Sa'īd al-Khudrī 1170.[156] M. Z. Ṣiddīqī talks of a 'crisis of authenticity' with fabrications spread by heretics, sectarians, storytellers and even devout traditionalists, 'the most dangerous type of *ḥadīth* forgers'. Aḥmad ibn Muḥammad al-Bāhilī (d. 275/888) was generally venerated for his piety, but admitted that he had forged traditions in order to make the hearts of the people tender and soft. Abū-Dāwūd found 400 of his traditions to have been forged.[157] In a field where 'accepting the traditions mean[t] knowing the men',[158] Ibn Ḥajar (156/773–237/852) placed scholars in twelve categories: Companions of the Prophet (*Ṣaḥāba*); the most truthful and accurate scholars; trustworthy or accurate scholars; those who were truthful; those who were truthful but sometimes committed mistakes; those who were acceptable (there being little evidence of their reliability); those who were considered weak (that is, where some scholars had spoken against them); those who were unknown; those who had committed mistakes; those who did not meet the legal requirement of righteousness or were stupid; those charged with forgery; those who were both liars and forgers.[159] The *aḥadīth* themselves could be rejected owing to a defect in the narrator, to a discontinuity of the chain of authority (*isnād*), or because of weakness for various incidental reasons; and finally because they were considered false *ḥadīth*.[160]

Given the potential pitfalls in discerning the true tradition from the false, it took the genius of Muḥammad Idrīs al-Shāfi'ī (150/767–204/820) to establish the principles by which the various legal doctrines could be synthesized into a coherent system. In *Al-Risāla,* which laid down the basis for such a synthesis, al-Shāfi'ī established the overriding authority, next only to the Qur'ān, of the

Sunnah of the Prophet Muḥammad as transmitted in the traditions.[161] Al-Shāfi'ī emphasized the uniqueness of the Prophet and the superiority of his comments to those of the Companions.[162] In addition, he asserted the overriding character of the *Sunnah* as a law deduced from the *ḥadīth*. If Muslims had judged on the basis of the Qur'ān alone, he argued, without taking into account the *Sunnah* of the Prophet, they would have 'cut off the hand of every thief and they would have flogged every fornicator'.[163] Kamālī argues that

> in the pre-Shāfi'ī period, *ḥadīth* was applied to the statements of the Companions and their Successors... It thus appears that *ḥadīth* began to be used exclusively for the acts and sayings of the Prophet only after the distinction between the *Sunnah* and the *ḥadīth* was set aside.[164]

General rules can thus be elucidated for the overall acceptability of a *ḥadīth*, which is determined by the weakest element in its proof. Kamālī comments:[165]

> Thus the presence of a single weak narrator in the chain of *isnād* would result in weakening the *ḥadīth* altogether. If one of the narrators is suspected of lying whereas all the rest are classified as trustworthy (*thiqah*), and the *ḥadīth* is not known through other channels, then it will be graded a weak. In scrutinising the reliability of *ḥadīth*, the '*ulamā*' of *ḥadīth* are guided by the rule that every *ḥadīth* must be traced back to the Prophet through a continuous chain of narrators whose piety and reputation are beyond reproach. A *ḥadīth* which does not fulfil these requirements is not accepted. A weak or *ḍa'if ḥadīth* does not constitute a *shar'ī* proof (*ḥujjah*) and is generally rejected.

Two traditions relating to *jihād* fall potentially within the scope of this ruling on the acceptability of *ḥadīth*. The first concerns the alleged tradition of the Prophet, cited by President Musharraf of Pakistan in his speech against terrorism on 12 January 2002, that *Jihād-e-Ashgar* (the smaller *jihād*) is over but *Jihād-e-Akbar* (the greater *jihād*) has begun. This tradition means that armed *jihād*, that is, the smaller *jihād*, is over. As Rudolph Peters notes, this interpretation was 'hardly touched upon' in pre-modern legal writings on *jihād*.[166] Ibn Taymīyah, who had his own warlike axe to grind against the Mongols, refuted this tradition categorically:

> There is a *ḥadīth* related by a group of people which states that the Prophet... said after the battle of Tabūk: 'We have returned from *Jihād Asghar* to *Jihād Akbar*.' This *ḥadīth* has no source, nobody whomsoever in the field of Islamic Knowledge has narrated it. *Jihād* against the disbelievers is the most noble of actions, and moreover it is the most important action for the sake of mankind. (Chapter 4)

Such a reading, if accepted in the version of Ibn Taymīyah, would tend to undermine the more pacific interpretation of *jihād*. However, most classical and modern commentators, including Ibn Taymīyah, accept the legitimacy of the Ṣūfī path of 'Purification of the Self' (*taṣawwuf*) which this tradition enshrined.[167]

Operating on the other side, in restricting the claims made by warlike interpreters of *jihād* in modern times, is the dubious case resting on the Qur'ānic commentary (*Tafsīr*) of Ibn Kathīr (d. 774/1373)[168] and the *Sunan* (religious rulings based on the customs of the Prophet) of *Imām* al-Tirmidhī (209/824–279/892),[169] one of the seven canonical books of *aḥadīth*,[170] that the martyr for the cause of Allāh would be rewarded in Paradise with 72 'black-eyed virgins'.[171] The story is suspect because of a short chain of authority; it is an isolated *ḥadīth* (just one out of a huge number of traditions), and one moreover which is not to be found in al-Bukhārī or in Muslim or in the Qur'ān. Because it is contained in one of the six or seven canonical books, it does not follow that the tradition is correct.[172] The main reason for opposing the validity of the *ḥadīth* is because of its content. Although the intercession of the martyr for 70 members of his family is to be found in the collection of Abū-Dāwūd, in other respects this story contradicts the accepted teaching of the Prophet. According to this teaching, the martyr should not be seeking higher reward while fighting in the way of Allāh, but be pure in motivation 'nothing but *jihād* in [God's] cause and belief in his Word…' The definitive answer to the false *ḥadīth* of the 72 black-eyed virgins as reward for the martyr is to be found in the *ḥadīth qudsī*, where those 'killed in the cause of God' are not reckoned as dead, but are fed a heavenly sustenance with their Lord (Q.3:169).[173] Proximity to God, not worldly physical pleasure, has to be the main reward for the martyr. It therefore follows that God will not wish to have in his close proximity, or feed a heavenly sustenance, to those who have disobeyed his commands.

The story of the 72 black-eyed virgins thus should be dismissed as a false *ḥadīth*,[174] a forgery inspired probably by the urge of professional storytellers and preachers 'for popularity through arousing an emotional response in their audience'.[175] There are nevertheless *imāms* today[176] who seek to convince gullible Muslim males (what of Muslim females?) that the work of a suicide bomber will be rewarded with 72 'black-eyed virgins' in Paradise so that he needs to take practical measures to protect his genitals with additional towelling.[177] Such an interpretation is doubly exploitative: it is the exploitation of the humiliation and frustration, especially sexual frustration, of potential recruits who, because of their unemployment, cannot afford to get married and thus enter into, in the teaching of Islām, a legitimate sexual relationship.[178] It is also the exploitation by privileged elites of those less fortunate than themselves.[179] Given the levels of unemployment among the Palestinian people, having a university degree provides no guarantee of employment and economic betterment. The repetition of the false *ḥadīth* of the 72 black-eyed virgins may receive some local or regional support for the purposes of recruitment of suicide bombers or to assuage the anguish of the bereaved, but can do nothing but disservice to a true understanding of Islām as one of the world's great religions. 'Verily [the rewards of] the deeds are decided by the last actions.'[180]

The importance of historical context in understanding the Prophetic traditions

In his analysis of the more than 400 'political' *aḥādīth*,[181] many of which he considers to have been forged by different religio-political sects and schools to suit their own aims and purposes,[182] Professor Qamarrudin Khan nevertheless argues that 'a good number of the *jihād aḥādīth*', particularly those reported by al-Bukhārī, are genuine. He continues:[183]

> in all these *aḥādīth*, the idea of fighting for the cause of God is most apparent. And the emphasis is on personal sacrifice and seeking the pleasure of God, rather than on fighting itself... So in the eyes of the Prophet it is only fighting in the way of God that is *jihād*. And the way of God means the defence of religion, and it does not mean war for territorial conquest, and it also does not mean defence of secular power. And in any case *jihād* does not mean aggressive war. The Prophet did not support the theory that 'might is right', clearly reflected in the theory of *jihād* almost unanimously advocated by Muslim jurists and '*ulamā*' [scholars]. He only upheld the principle that 'right is right'. And other things may also be right, but he fought only for the rightness of religion, and only in self-defence and when there was no way out. But he did not develop or preach a generalized theory of war...
>
> To think that the Prophet who had been raised among mankind to bring peace and happiness to them, and sent to them as a mercy (Q.21:107) would establish a *Sunnah* of War, is simply preposterous and a contradiction in terms. Even the thought of this idea [is] an affront to his high office and calling.

In Islamic history, Qamarrudin Khan contended, 'the Muslim state has often been equated with the Islamic faith' and it is asserted that the one exists for the other. 'This attitude has given the impression that Islām is a political device rather than a moral and a spiritual force.'[184] The reality is quite different, Khan asserted. The state was a 'circumstantial event'; it did not follow a set pattern, divine or human, but grew out of history.[185] He emphasized the idea of a temporal *jihād*, determined by historical circumstance.[186]

Khan argued that just as there was no principle of state to be discerned in the Qur'ān, so also there was no such principle in the *Sunnah*. 'The Prophet was appointed only as Prophet; he received no assignment from God also to build a state.' The Prophet, he claimed, 'left no political *Sunnah* for the Muslim *Ummah*', for the functions of prophethood and kingship are 'entirely different'. Islām has prescribed no principle of state or form of government. In Khan's view, Muslims have never developed the principles of historical criticism or applied them to the life of the Prophet.

Two pieces of historical evidence point to an early understanding of the relationship between the faiths which is quite different from those proponents of the Islamic state who seek hegemony for Islām. The first is the amnesty offered to Muslims by the Christian ruler (Negus) of Abyssinia, a place of refuge commended by the Prophet himself, which offers a dramatic early limitation on the concept of *jihād* and the concept that Islām was by its very existence in a state of war against other states. 'We have come here, O King, to your land seeking your protection and we do hope that we shall not be dealt with unjustly.' The ruler refused to give them up or betray them to the Qurayshīs. The refugees testified to the fact that 'they worshipped there according to their rites, and celebrated daily services, and nobody maltreated them or abused them by unpleasant words'.[187]

The second piece of evidence is the Constitution of Medina or the Medina Agreement (*Ṣaḥīfāt al-Madīnah*), the so-called 'first written constitution in the world' as Muḥammad Hamidullah called it.[188] The Charter granted non-Muslim citizenship rights in what was called the *ummah* (not to be confused with the later Islamic understanding of that term), and specified their obligations on an equal footing with the Muslim fellow-citizens, so long as they agreed to co-exist peacefully with them.[189] Clause 25 is striking in this respect. 'The Jews of Banū 'Awf are a community (*ummah*) along with the believers. To the Jews their religion (*dīn*) and to the Muslims their religion (*dīn*)...'[190] The Indonesian writer Munawir Syadzali, who has written a study on Islām and the administration of the state (1990), argues that since the Constitution of Medina did not mention Islām as the religion of the state, the Prophet did not actually call for the establishment of a theocratic state in which Islām would serve as its sole basis.[191]

The Prophet: a spiritual but not a political leader?

A later Muslim (ibn Isḥāq, d. 150/767), when reviewing the life of the Prophet commented:[192]

> prophecy is a troublesome burden – only strong, resolute messengers can bear it by God's help and grace, because of the opposition which they meet from men in conveying God's message. The Apostle carried out God's orders in spite of the opposition and ill-treatment which he met.

The task of reassessing the Prophet's life is in any case difficult, but it is made much more so by revivalist tendencies in Islām 'which look upon him more as a political leader than as a Prophet'.[193] If we make comparison with studies of 'the historical Jesus', it is clear that an enormous proliferation of writing[194] may produce relatively little clarification. Jesus was a pious Jew guilty of nothing that would carry the death sentence on religious grounds; he was not an anti-Roman agitator or a pretender to the throne of the royal Messiah,[195] though he did believe in the imminence of the kingdom of God. Though undeserving of crucifixion, 'in the unsettled political and religious circumstances of inter-testamental Palestine someone could easily lose his life without actually committing any culpable act against the Jewish law or the Roman state'.[196]

Subsequently there also came about a Muslim version of aspects of the life of Jesus, termed by Tarif Khālidī a 'Muslim Gospel'.[197] It is clear, for example, that there was early Muslim access either to a Gospel translated into Arabic or to a lectionary.[198] The Qur'ān affirms (Q.4:150) that a true belief must include belief in all prophets. In the case of Jesus, he is said to be a 'word' from God and a 'spirit' from Him. The Qur'ān emphasizes his ministry as one of 'cleansing' (Q.3:55), which works in two directions: Jesus is to be cleansed from the perverted beliefs of his followers (the crucifixion and the doctrine of the trinity are denied, while Christians are destined to sectarianism and mutual antagonism until the Day of Judgement: Q.5:14), just as he himself sought to cleanse Judaism of its perceived deficiencies. It is Jesus' ascension, not his crucifixion, which confirms the miracle of his pure birth (Q.43:61), a sign of God's omnipotence. The legacy of Jesus is gentleness, compassion and humility (Q.19:33). By the time of al-Ghazālī (d. 505/1111), whose *The Revival of Religious Sciences* (*Iḥyā' 'Ulūm al-Dīn*) contains the largest number of sayings ascribed to Jesus in any Arabic Islamic text, 'Jesus was enshrined in Ṣūfī sensibility as the prophet of the heart *par excellence*'. Since the full understanding of the mysteries of the heart and its innermost nature was beyond the reach of human intellect, metaphors and parables (*amthal*) were needed to express these mysteries. Hence the prominence of the sayings of Jesus in al-Ghazālī's writings.[199]

Mawlānā Amīn Aḥsan Iṣlāḥī's comment that 'the Prophet Jesus (peace and blessings be upon him) exhorted his disciples to wage *jihād*'[200] might surprise Christians until it is understood that the Mawlānā was thinking in terms of 'striving' as in Q.53:39 ('man can have nothing but what he strives for'). The Qur'ān records Jesus asking 'who shall be my helpers for Allāh?' (Q.3:52). There is no precise parallel for this remark recorded in the synoptic gospels, but Matthew, Mark and Luke all record Jesus' requirement of his followers to deny themselves and take up their cross to follow him – a different image, but unquestionably an image of the disciples 'striving' to follow his path to God.[201]

There is no Christian parallel, of course, to the so-called 'Muslim Gospel' of Jesus, but there is, or at least there needs to be, a Christian understanding of the Prophet. The Prophet was a political and military as well as a spiritual leader. As has been seen above, the Prophet himself undertook 19 military campaigns[202] and expected to be among ten martyrs whom he could name in advance. He authorized over 70 military encounters, ranging in intensity from pitched battles in defence of Medina, to sieges, raids and skirmishes against enemy targets.[203] This militaristic aspect of his life and work has been compared unfavourably by some commentators with Jesus' essentially non-militaristic approach to contemporary affairs. Moreover, the Prophet's actions, particularly his approval of Sa'd ibn Mu'ādh's act of vengeance for the treachery of the Banū Qurayẓah which resulted in between 400 and 900 executions,[204] has been seen as 'a source of embarrassment to Muslims, particularly in the modern period'.[205] Perhaps it was this incident which prompted the extraordinarily ill-judged comment on 6 October 2002 from the Rev. Jerry Fallwell, a Baptist minister based in Lynchburg, Virginia, that the Prophet was a 'terrorist' (a statement which he subsequently appeared to retract but which understandably outraged Muslim opinion).[206]

Two propositions need to be accepted at the outset. The first is that the self-understanding of Jesus and the Prophet about their ministries was quite different. We do not perhaps know as much about the spiritual background to the Prophet's ministry in the way that we are informed about the Judaic background to the ministry of Jesus. The process of revelation was also a more gradual one in the case of the Prophet, extending over 23 years, whereas for Jesus revelation seems to have been immediate following upon John's baptism of him in the River Jordan. The second proposition is that the historical context of the two ministries was quite different. In the case of Jesus, any rising by the Jews against Roman rule would have been ruthlessly suppressed. The idea of a Messiah, 'he which should have redeemed Israel' (Luke 24:21), or liberated it in a political sense, nevertheless remained strong. There had been military risings against Seleucid occupation before Jesus' time, which had resulted in a period of independence under Hasmonean rule; but Judas Maccabeus had called in the Romans as allies and erstwhile allies became political and military masters. After Jesus' death, the First and Second Jewish Wars of 66–70 and 132–135 CE ended in failure. A Christian separatist political movement would have been destined to failure; but the early Christian community was distinguished in any case by its non-violence. Because they argued against war, the early Christians were accused of collaboration by those organizing armed resistance against Roman occupation.[207]

Contrast the historical context of the Prophet's lifetime. No foreign army of occupation was in place, although potentially the Arabian peninsula was the scene of rivalry between Rome and Sāsānian Persia. Prior to the arrival of Islām, the tribe enjoyed a degree of autonomy if not sovereignty, acknowledging no political authority above or beyond itself. Alliances between tribes were formed and broken as a norm.[208] Was it true, as the first theoretician of *jihād*, 'Abdullāh Ibn al-Mubārak, claimed nearly 150 years after the Prophet's death, that Muḥammad's sole mission was *jihād*? The words attributed to the Prophet were that he 'was sent with the sword before the Hour [of Resurrection]', that his subsistence was laid down for him 'under the shadow of [his] spear', while humility and debasement were imposed on those who opposed him.[209] It was the Prophet himself, according to Muslim, who initiated the requirement of commanders to call upon opponents to embrace Islām, pay the protection tax (*jizya*) or else face battle. In general terms, his commanders were required to 'fight with the name of God and in the path of God. Combat [only] those who disbelieve in God. Fight yet do not cheat, do not break trust, do not mutilate [and] do not kill minors.'[210]

For the jurist Shāfi'ī (d. 204/820), writing nearly a century after the death of the Prophet, certain verses of the Qur'ān (Q.9:112, Q.9:36, Q.9:5 and Q.9:29) demonstrated that God had 'imposed [the duty] of *jihād* as laid down in His Book and uttered by His Prophet's tongue'. God had made it known that 'going into battle was obligatory on some, not all'; provided some went forth, so that a sufficient number fulfilled the collective duty, the others who remained did not fall into error.[211] Furthermore, while on campaign, God had 'distinguished between fear and secure prayers in order to protect the followers of His religion from a sudden attack'. If he feared a sudden attack by the enemy, the *imām* should 'quickly perform the prayer… upon a warning from the guarding party'.[212]

Certainly pre-Islamic and early Islamic Arabia was a militaristic society, though the teaching of the Prophet sought to restrain such militarism. Inscribed on the hilt of his sword were the words: 'Forgive him who wrongs you; join him who cuts you off; do good to him who does evil to you, and speak the truth although it be against yourself.'[213] As sedentary tribes became larger and more powerful and engaged in lucrative commerce, the Qurayshīs becoming more prosperous than the others but also becoming divided into two groupings.[214] The Prophet was an unlettered Arab Prophet (Q.62:2). In the words of the fourth 'rightly guided' caliph, 'Alī,

> Allāh sent him with a sufficing plea, a convincing discourse and a rectifying announcement. Through him Allāh disclosed the ways that had been forsaken, and destroyed the innovations that had been introduced. Through him He explained the detailed commands.[215]
>
> Allāh sent him with undeniable proofs, a clear success and open paths. So he conveyed the message of declaring the truth with it. He led the people on the [correct] highway, established signs of guidance and minarets of light, and made Islām's ropes strong and its knots firm.[216]

The jurist al-Qarāfī (626/1228–684/1285) contended that while the Prophet functioned also as head of state (*imām*) and judge (*qāḍī*), the majority of his time he spent informing 'the people, on behalf of God, of the rulings' he had found in divine revelation.[217] He was thus simultaneously the divine Messenger (*Rasūl*), a religious leader dealing with juridical questions (*muftī*) and a propagator of the faith (*muballigh*).[218]

The revelation was given in Arabic to Arab society; the Arabs were to spread the faith 'unto other people as soon as they c[a]me into contact with them'; yet this was a society where military exploits were glorified as the essential virtue.[219] In the words of Weiss and Green,[220]

> the *ummah*, as the Prophet left it, was potentially a religious community open to any who would declare their acceptance of Islām. In outward form, however, it was still a confederation of clans and tribes, membership in which was based on affiliation with a member clan or tribe. In this outward aspect, the *ummah* was roughly coterminous with the Arab nation, although not absolutely so owing to the existence of some Christian tribes. Its inner essence required, however, that this purely Arab character eventually be abandoned so as to make possible the full inclusion of persons of different races, thus fulfilling the universal mission which was the *ummah*'s original justification.

Did the Prophet envisage or plan for the expansion of Islām beyond Arabia?[221] As with Jesus' ministry, this is closely related to the question as to whether he conceived his ministry as limited to his own people or to be universal. Without a common activity – expansion in the name of the faith – it was unlikely that the tenuous system of alliances he had created would be held together indefinitely; the risk was internecine warfare. There is also evidence that the Prophet was particularly interested in the situation on the northern borders of Arabia, where he made treaties with Jewish and Christian communities. Though he may have issued a letter to the principal leaders of the day, calling upon them to embrace Islām, he also established the first written constitution, and a pluralist one at that, in Medina.[222]

Perhaps the Prophet's ministry is best summed up in the call for unity among believers, when God addresses them with the words 'let there be one community (*ummah*) of you, calling to good, and commanding right (*ma'rūf*) and forbidding wrong (*munkar*); those are the prosperers' (Q.3:104). This injunction is found in seven further Qur'ānic verses (Q.3:110; Q.3:114; Q.7:157; Q.9:71; Q.9:112; Q.22:41; Q.31:17). As Michael Cook argues, 'the phrase "commanding right and forbidding wrong" is firmly rooted in Qur'ānic diction'.[223] The themes which appear in conjunction with commanding right are essentially the duties of the faithful: performing prayer; paying alms; believing in God; obeying God and His Prophet; keeping His bounds; reciting his signs; calling to good; vying with each other in good works; and enduring what befalls one.[224] Commentators disagreed on how restrictively the duty was to be applied (whether to all believers or only a sub-group such as the scholars) but none sought to restrict the scope of the application of the verse.

Michael Cook concludes that Qur'ānic exegesis 'put most of its weight behind the interpretation of Q.5:79 as a reference to the mutual forbidding of wrongs committed within the community'.[225] It may be good to risk one's life in seeking to command good or forbid wrong, but there was no consensus on this point, except to deny that the apparent negation of the duty in Q.5:105 ('O believers look after your own souls. He who is astray cannot hurt you, if you are rightly guided') is applicable to present times.[226] It is a basic value of humanity that when one encounters someone engaged in wrongdoing (for example, rape), one should do something to stop him. Pre-Islamic Arabia, as evidenced by *jāhilīyyah* poetry, knew the terms 'right' and 'wrong' and even paired them; but it did not possess the notions of 'commanding' or 'forbidding' them. Nor did any culture outside Arabia have any direct influence on this Islamic concept, an innovation in the Qur'ān itself and thus of the Prophet's ministry.[227] There is, needless to say, no Qur'ānic basis for an extension of the idea of 'forbidding wrong' to the duty of *jihād*, a development for which Ibn Taymīyah was responsible and which has been taken further by those who claim to be his followers (see Chapter 4).

However, it was in a letter of al-Walīd II dated 125/743, which, while corrupt in places, is generally assumed to be authentic, that the relationship between Allāh, the Prophet and the worldly succession is most clearly elucidated. The letter amounts to a salvation history divided into two eras; one of the prophets, the other of the caliphs.[228] God chose for Himself and mankind a religion which He chose to call Islām. The messenger Muḥammad did not preach anything new, but confirmed the message of previous prophets:

> through him God made guidance clear and dispelled blindness, and through him He saved [people] from going astray and perishing. He [that is, God] elucidated the religion through him [that is, the Prophet] and He made him a mercy to mankind. Through him He sealed His revelation. He gathered unto him everything [with] which He had honoured the prophets before him, and He made him follow their tracks…

After God sealed His revelation with Muḥammad, the era of prophethood came to an end, to be replaced by that of the caliphs, who were viewed by al-Walīd II as the legatees of the prophets. There could be no community (*ummah*) without an *imām*, for it is the leader who constitutes the community and ensures that God's ordinances are implemented:

> Then God deputed his caliphs over the path of His prophethood… for the implementation of His decree, the implementation of His normative practice (*sunnah*) and restrictive statutes and for the observance of His ordinances and His rights, supporting Islām, consolidating that by which it is rendered firm, strengthening the strands of His rope,[229] keeping [people] away from forbidden things, providing for equity among His servants and putting His lands to right, [doing all these things] through them…
>
> So the Caliphs of God followed one another, in charge of that which God had caused them to inherit from His prophets and over which He had deputed them.

> Nobody can dispute their right without God casting him down, and nobody can separate from their polity (*jamā'a*) without God destroying him...

The shortcoming of this text, and of most juristic analysis, is that it serves to minimize the spiritual example of the Prophet. Yet it is this aspect of the Prophet, whose spiritual path is to be followed, and who was sent as a 'mercy [or grace] to the worlds', which is emphasized in the Qur'ān (Q.7:158; Q.21:107). Without Muḥammad's own mysticism, the subsequent mystical trend within Islām in which the believer is 'annihilated in the Prophet' before he can hope to reach God, could not have developed (see Chapter 3). Al-Ghazālī's *The Revival of Religious Sciences* correctly makes the Prophet the central figure in the book.[230] Here is one of the descriptions of the Prophet as a spiritual figure from chapter 20 ('the book of the conduct of life as exemplified by the Prophetic character'), which corrects the image of the proto-terrorist that appears in some recent publications in the English language:[231]

> Muḥammad was the most forbearing, honest, just, and chaste of men... He was the most generous of men. Neither a *dīnār* nor a *dirham* was left him in the evening. If something remained, and there was not anyone to whom he could give this excess – night having fallen unexpectedly – he did retire to his lodging until he was able to give this excess to who was in need of it.
>
> Muḥammad did not take of those things which Allāh gave him, except his yearly provisions. He gave the remaining excess of his small quantity of dates and barley to charity...
>
> Muḥammad was the most bashful of men and did not stare into anyone's face. He answered the invitation of the slave and the freeborn... He became angry for Allāh and not for his own sake. He exacted the truth even though it brought harm to him and his companions.
>
> Muḥammad, while fighting certain polytheists, was offered the help of other polytheists. However, he replied, 'I do not seek assistance in conquest from a polytheist', even though he was with few men and in need of anyone who could increase his numbers...
>
> Because of hunger he at times tightened a stone around his stomach. He often ate what was at hand, did not reject what was available, [nor] did he refrain from lawful food... He did not eat reclining nor from a footed tray. He used his sole as a napkin. Until the time of his death, he did not dislike to eat wheat bread three days in succession as a sign that one [should] choose neither poverty nor avarice.
>
> He attended feasts, visited the sick, attended funerals, and walked alone without a guard amongst his enemies.

> He was the humblest of men, the most silent without being insolent, and the most eloquent without being lengthy. He had the most joyful countenance, none of the affairs of the world awing him…
>
> He visited the sick in the farthest section in the city… A moment did not pass without his doing an action for Allāh or [doing] that which was indispensable for the soundness of his soul. He went to the garden of his companions. He did not despise a poor man for his poverty and misfortune, nor he did not fear a king because of his power; rather, he urged them equally to Allāh.
>
> Allāh combined in him virtuous conduct and perfect rule of people, though he was untaught, unable to read or write, grew up poor amongst the shepherds in the land of ignorance and desert, and was an orphan without father and mother. Allāh taught him all the fine qualities of character, the praiseworthy paths, the reports of the first and last affairs, and those matters through which there is [obtained] salvation and reward in the future life and happiness and reward in the world. Allāh taught him to cleave to that which is obligatory and to forsake the useless.
>
> May Allāh direct us to obey Muḥammad in his commands and to imitate him in his actions. Amen, O Lord of the worlds.

The only weakness of al-Ghazālī's description lies in the fact that it was written over four centuries after the Prophet's death. But there was also praise from contemporaries of the Prophet. Perhaps the most eloquent contemporary encomium was delivered in a sermon of 'Alī, the fourth caliph:[232]

> …one should follow His [that is, God's] Prophet, tread in his footsteps and enter through his entrance… Certainly, Allāh made Muḥammad – the peace and blessing of Allāh be upon him and his descendants – a sign for the Day of Judgement, a conveyor of tidings for Paradise and [one who warned] of retribution. He left his world hungry but entered upon the next world safe. He did not lay one stone upon another [to make a house] until he departed and responded to the call of Allāh. How great is Allāh's blessing in that He blessed us with the Prophet as a predecessor whom we follow and a leader behind whom we tread.

The last word, however, should rest with the Prophet himself. Two years or so before his death, in Ramaḍān of year 8/December 629, he set off on his final campaign against Mecca, to rescue Allāh's sanctuary from the grip of polytheism. The fighting was fierce; the killing on a significant scale. Abū Sufyān, leader of the Qurayshīs, eventually appealed to him in the words: 'O Prophet of Allāh, the majority of the Quraysh is annihilated. There is no more Quraysh after this day', whereupon the Prophet called off the fighting. The 'helpers' of the Prophet (the *anṣār*, those believers at Medina who had helped him after his exile from

Mecca) then began to murmur that he was 'moved by love of his relatives and compassion on his clan'. Muḥammad addressed them with the words: 'I am the slave of Allāh, and His prophet. I have migrated to Allāh, and to you. My life is your life; my death is your death.' Immediately on hearing this, the followers began to weep, such was the power of his words on them.

Muḥammad then proceeded on a circuit of the Ka'ba, until he came to an idol at the side of it. He began to stab at the idol, saying: 'truth has come and falsehood has vanished; it is the property of falsehood to vanish'. His commitment to end polytheism was one of the determining principles of his life. Finally, on the mount of aṣ-Ṣafā, he raised his hands and praised Allāh. The Qurayshīs recognized that he had 'succeeded', that is to say, he had finally overcome their resistance. The Prophet then said: 'my answer is that given by my brother Joseph (son of Jacob): "no blame be on you this day. Allāh will forgive you; for He is the most merciful of the merciful"' (Q.12:92).

This account of Aḥmad ibn Yaḥya al-Balādhurī (d. 274/829)[233] is a compelling one for several reasons. It confirms the analysis of scholars that the most powerful factor in Muḥammad's career was 'his unshakeable belief from beginning to end that he had been called by Allāh',[234] a conviction which did not admit of doubt and which exercised an incalculable influence on others called to join him in his *jihād fī sabīl Allāh*, '*jihād* in the way of Allāh' against polytheism. Similarly, the certainty with which he came forward as the executor of Allāh's will gave his words and ordinances 'an authority which proved finally compelling'.[235] According to Tirmidhī, he was preferred to the other prophets because of his ability to speak concisely; by the fear with which Allāh struck his enemies; by the fact that the taking of spoils was made legal for him; by the fact that the earth was made for him into a mosque and purifying substance; that he was sent to all people; and finally because the prophets were 'sealed' with him, that is to say, that he was the last Prophet and there would be no prophet after him.[236] In his person, the chronology of the appearance of the earlier prophets was overturned: the former prophets, whose laws were superseded by Islām, were henceforth to be considered followers of the last Prophet in spite of the fact that they had been sent by God long before him.[237] Finally, Allāh permitted the defence and ultimate victory of his cause over the polytheistic Meccans and their gods al-Lāt, al-'Uzzā and Manāt. The culmination of Muḥammad's career was the Farewell Pilgrimage to Mecca, the first 'reformed' pilgrimage (10/March 632) in the last year of his life. His sense of exultation must have echoed the divine pronouncement: 'today I have perfected your religion and completed my favours for you and chosen Islām as a religion for you' (Q.5:3). Islām's immunity from abrogation has become an essential component of its self-proclaimed superiority over all other religions.[238]

The puritanical jurist of the early fourteenth century, Ibn Taymīyah (see Chapter 4), provided as part of his polemic against Christians his own gloss on the Prophet's career and the extent to which *jihād* was an integral part of his mission. It was an interpretation which, in the light of the evidence surveyed above, seems to be at variance with the practice of the Prophet; but in terms of the later self-definition of the Muslim community it contains elements of truth, particularly the idea of the unswerving commitment of the community to what it *perceived* to be Muḥammad's example. The Prophet, he stated, had summoned Christians to Islām[239]

> and waged *jihād* against them, and commanded others to summon them and wage *jihād* against them. This is not an innovation which his community invented after Him, as Christians did after Christ. Muslims do not allow a single person after Muḥammad to change a thing of His Law – to permit what he forbade, to forbid what he permitted, to necessitate what he eliminated, to eliminate what he necessitated. Rather, what is permissible (*al-ḥalāl*) among them is what God and His Messenger permitted, and what is forbidden (*al-ḥarām*) is what God and His Messenger forbade. Religion is what God and His Messenger legislated, as opposed to the Christians, who introduced innovations after Christ…

2
'*Jihād* of the Sword': Carrying the Message Abroad (*Futūḥāt*)

> Allāh gave the Prophet Muḥammad four swords [for fighting the unbelievers]: the first against the polytheists, which Muḥammad himself fought with; the second against apostates, which Caliph Abū Bakr fought with; the third against the People of the Book, which Caliph 'Umar fought with; and the fourth against dissenters, which Caliph 'Alī fought with.
>
> Al-Shaybānī

Al-Shaybānī's theory of the 'four swords',[1] quoted above, suggests that *jihād* was a doctrine for survival and affirming the power of the four 'rightly-guided' caliphs (*al-Khulafā'-ar-Rāshidūn*, 10/632–40/661). It has been said that 'tribal states must conquer to survive'.[2] Raiding and warfare were essential for the economic survival of tribesmen.[3] Tribal politics involved *ghazw*, the practice of collecting booty by conducting raids on traders' caravans, on rival tribes or on peaceful and poorly-defended communities.[4]

Yet there were also tribal alliances for particular purposes and perhaps, even before the rise of Islām, there was some sense of an Arab community without yet an overarching political authority.[5] In the longer term, Islām proved 'particularly successful in uniting tribal societies and in motivating militant struggle in the interests of the *ummah* as a whole'.[6] This suggests for some a more pacific aspect of 'carrying the message abroad' (*futūḥāt*), in which the propagation of the Word was an important factor and military conquest was coincidental.

Patricia Crone, in her recent analysis of early Islamic political thought, makes no concessions to such a viewpoint: 'with or without conversion, the conquerors' understanding of Islām was particularist', she writes.

> Religion was being used to validate the dominion of a single people, or to expand their ranks, not to unite mankind in a single truth above ethnic and political divisions… Whatever Muḥammad may have preached, *jihād* as the bulk of the Arab tribesmen understood it was Arab imperialism at God's command. Their universalism was political.

If this disposes of the stereotyped misconception of Islām as a 'religion of the sword', in Crone's view, it 'lands us with the opposite problem of explaining how the jurists could see holy war as a missionary enterprise at all. *Jihād* was still in the nature of divinely enjoined imperialism'.[7]

Hearts and minds are not won by conquest and invasion alone,[8] yet hearts and minds clearly were won over to Islām over a period of time. Muslims may have become the majority of the population by *c.* 209/825 in Iran, and *c.* 286/900 in Egypt, Syria and 'Irāq.[9] For the emphasis upon voluntary conversion to Islām, we have the evidence even of a hostile Christian writer on 'Umar's negotiated capture of Jerusalem in 14/638. The caliph had not wanted to pray in the Christian holy places because

> the Christians would have lost those places, which would have been made oratories [for] my people. I did not want this to happen, but preferred that the Saracens pray in the place where I prayed. I do not want the Saracens to gather and pray to the detriment of the Christians…[10]

'Umar subsequently granted the Christians of Jerusalem a special dispensation known as the Covenant of 'Umar, which reaffirmed the Qur'ānic precept that 'there shall be no compulsion for these people in the matter of religion'.[11] A negative piece of evidence is also of importance: with the exception of the early *riddah* wars under Abū Bakr, there was little or no sustained rebellion against Islām and in favour of polytheism. Where rebellions occurred, they tended to be of a Muslim sectarian kind, or else associated with opposition to the ruling dynasty.

Historians recognize the Muslim conquests of the Sāsānian/Persian empire and large swathes of the Byzantine empire as 'a fundamental watershed of world history'. Yet this was may not have been 'clear to those who lived through the conquests'.[12] Moreover, at first, the divisions within the young Muslim community following the death of the Prophet seemed so great that expansion abroad and thus the survival of the new Muslim regime seemed an unlikely outcome. These divisions were twofold. First and foremost, there were tribes which returned to their former state before the advent of the Prophet. This led to wars of 'return' (*riddah*) or apostasy. There were no clear instructions from the Prophet for the campaign against apostasy led by Abū Bakr,[13] but he insisted on the payment of taxes to Medina as the distinguishing mark of membership of the *ummah*.[14]

He had the support of the people of the Ḥijāz, the Qurayshī and the Thaqīf, who proved to be better united than their opponents.[15]

The first *jihāds* against unrighteousness: the Khārijī revolts as exemplars of the rejectionist community

> He who obeys me has obeyed God and he who disobeys me has disobeyed God; he who obeys the commander has obeyed me and he who disobeys the commander has disobeyed me. The *imām* is only a shield behind whom fighting is engaged in and by whom protection is sought; so if he commands piety and acts justly he will have a reward for that, but if he holds another view he will be held guilty. (*Ḥadīth* based on Muslim 20/4542)

There are many *aḥādīth* which reflect a profound pessimism about the quality of leadership that was to await the Muslim community after the death of the Prophet and his Companions. 'Condemnation by God and His Prophet of unjust, tyrannical and corrupt rule', comments Abdullah Schleifer, 'runs like a thread through the canonical literature and great punishment awaits on the Day of Judgement the leader who misuses his command.'[16] The responsibility of the Muslim was to try to turn the tyrannical ruler away from his evil conduct, but such criticism had to stop short of violence or rebellion: 'the most excellent *jihād* is when one speaks a true word in the presence of a tyrannical ruler', is a frequently quoted *ḥadīth*. The tyrant should be deterred, if possible, by hand, word or in the believer's heart. It was mandatory, however, to pursue *jihād* alongside the *imām*, irrespective of his personal qualities or lack of them: 'the injustice of the tyrant or the justice of the just matter little', Ibn Ḥanbal declared.[17] Ultimately, in Sunnī juridical consciousness the threat to the unity of the community was a more important danger than using *jihād* as an instrument for the purification of Muslim rule – the believer was required to fight whoever separated from the community or whoever rebelled against the *imām*. *Jihād* was to be directed by the *imām*, or in modern terms the state, or not at all.

The schismatics, known as the 'seceders' or Khārijīs (Khawārij) (though they rejected this term since they saw themselves as reformers),[18] took a diametrically opposed view. Instead of accepting the rule of the caliph, just or unjust, they took to heart the Qur'ānic injunction to command right and forbid wrong.[19] Many of them were Qur'ānic fundamentalists (they used an expurgated Qur'ān without *sūrah* 12);[20] they were also exclusivists, who believed that they were the only true Muslims. The oral profession of faith (reciting the *shahādah*) was not enough to assure a person's status as a 'true' Muslim worthy of salvation; instead, it had to be coupled to a life of righteousness and good deeds. This stress on works led the Khārijīs to forbid all luxuries such as music, games and ornaments. Intermarriage and relations with other Muslims were also discouraged.

They espoused the doctrine of *takfīr*, classifying believers as unbelievers (*kufr*), if the view of the erstwhile believer did not accord with their own predilections – even though such a course of action appeared to contradict a specific verse of the Qur'ān (Q.4:94) as well as a number of the *aḥādīth*.[21] Any Muslim who committed major sins became thereby a non-Muslim. They believed that *jihād* should be waged against those who did not accept their view of Islām. Indeed, for them, *jihād* was regarded as the fifth – not even the sixth – pillar of the faith (*jihād* could take precedence over prayers, alms-giving and the pilgrimage).[22] Furthermore, in contrast to the orthodox Sunnī position, *jihād* was described as an individual duty which could not be avoided rather than a collective responsibility, discharged by some but not others.[23] They described the action of a Muslim in 'going out' from a corrupt community as a *hijrah*, referring to the escape of the Prophet and his companions from Mecca which marked the start of the Islamic era. The *hijrah*, in other words, was for the Khārijīs not simply an historical event, but 'a model for proper Muslim behaviour'.[24] Modern militant Islamists, such as Sayyid Quṭb and even Osama bin Laden, who have advocated this procedure, are thus following the tradition of the Khārijīs.

They follow the tradition in another way, too, for each group of Khārijīs 'was at once a terrorist band and a fanatical religious sect. They were held together by the conviction that they were the only true Muslims, and that their rebellions had profound religious justification.'[25] Courage in war was taken as the paramount condition for the election of the Khārijī *imām*, in contrast to its relegation as the sixth of seven requirements for a ruler in later Sunnī theory. Indeed, there could be no obedience due to an *imām* who deviated from the law of God as they interpreted it; instead, he could justly be deposed.[26] The Khārijīs sometimes tolerated the existence of more than one *imām*, electing two – one to lead in war, the other to lead in prayer.[27] A Muslim who failed to recognize the *imām* of the Khārijī or to accept their doctrine when invited to do so was generally classified as an 'apostate' (*murtadd*), the punishment for which was to be put to the sword, women and children as well as men, non-combatants as well as combatants. The most extreme Khārijī groups argued that whoever committed apostasy in this way could not be allowed to be repent and must be killed, along with his wife or wives and children. The Khārijī excess of ethical intent in attempting to 'Islamize' both the state and society produced a doctrine capable of justifying atrocities clearly condemned by the *aḥādīth* as well as in the political theory of the Sunnī jurists. The Khārijī doctrine of religious murder (*isti'rāḍ*), applied against non-Khārijī Muslims from the earliest uprisings, appears never to have been practised against non-Muslims.[28] Indeed, the group, or perhaps more accurately, collection of between 14 and 21 independent sub-sects (depending on one's definition),[29] seem to have been relatively indifferent to the cause of *jihād* against unbelievers and surprisingly tolerant of other religions.[30]

The Khārijī doctrine of *jihād* as a permanent armed struggle to preserve the community from the corruption of misrule became a *jihād* against the very body of the Muslim community. Indeed, the struggle was essentially conducted against the political and religious leadership and against the emerging concept of consensus of the community (*ijmā'*). Instead, the Khārijīs' maxim was that 'judgement belongs to none except God'.[31] In the words of 'Alī, the fourth caliph, these were words 'of truth which [were] given a false meaning'.[32] Calling for an oath of allegiance on 'the book of God and the *sunnah* of His Prophet',[33] 'Alī was forced to declare war (*jihād*) on the Khārijīs in 37/658; in revenge, 'Alī was murdered at the mosque in Kufa in 40/661.[34] 'Alī is recorded by al-Bukhārī as stating:[35]

> I heard the Prophet saying, 'in the last days [of the world] there will appear young people with foolish thoughts and ideas. They will give good talks, but they will go out of Islām as an arrow goes out of its game, their faith will not exceed their throats. So, wherever you find them, kill them, for there will be a reward for their killers on the Day of Resurrection.'

Chase F. Robinson has recently suggested a significant remodelling of our vision of Khārijism on the basis of a detailed examination of the phenomenon in Jazira.[36] Firstly, though the sects (such as the Ibāḍī, Ṣufrī, Azraqī) and sub-sects (such as the Bayhasiyya and Murji'at al-Khawārij) of the Khārijīs are evident, for Robinson the movement was seen by the state as a reasonably united one of 'revolutionary tribesmen' in that it possessed both a programme and a tradition. Secondly, Robinson suggests that the leaders of the Khārijī rebellions were often disgruntled military commanders who had been dropped, for one reason or another, from the army rolls (the *dīwān*). 'The state thus produced its own opposition', in Robinson's interpretation, since the commanders embraced Khārijī ideas only after their dismissal and as a consequence of their disaffection. 'Khārijism… held out the prospect of both continued employment and high status in warfare legitimized by the piety and holiness of those prosecuting it.' It was a reassertion of 'primeval, conquest-era Muslim identity…'[37]

Robinson depicts Khārijism as a fusion of asceticism and revolution, citing as example a sermon from the rebellion of Ṣāliḥ bin Musarriḥ, whose revolt commenced in 76/May 695. The sermon evinces separatism as the necessary course of action for the Khārijī ascetic:

> I charge you with fear of God, abstinence in this world, desire for the next, frequent remembrance of death, separation from the sinners, and love for the believers. Indeed, modesty in this world makes [God's] servant desirous of what is God's and empties his body for obedience to God; frequent remembrance of death makes one fearful of God, so that one entreats him and submits to him. Departure from the sinners is a duty on all believers; God said in His Book:

'do not pray over any one of them who dies, ever, nor stand at his grave; they denied God and his messenger, and they have died sinners'. (Q.9:84)

'Alī's alleged killing of 'Uthmān was regarded as one his 'greatest acts of obedience to God'; where he had gone wrong was in accepting arbitration with people he should have simply fought. This had suggested that he doubted his own entitlement to the caliphate.[38] Rebellion and tyrannicide were thus not only lawful, but obligatory. Dissociation from 'Alī and his party and successors meant not seclusion but active rebellion:

So prepare yourself – may God have mercy upon you – to fight (*jihād*) against these enemies aligned [against Islām] and the oppressive leaders of error, and to go out from the transient to the eternal world, and to join our believing, resolute brothers who have sold this world for the next, and who have expended their wealth, seeking to please God in the hereafter.

God has prepared the higher reward for those who fight: family and possessions are transient, and are to be sacrificed for everlasting life in the hereafter.[39] The sermon is the *locus classicus* for the historical origins of the modern exhortation to 'global *jihād*' or the committing of a terrorist outrage. It was a polemic for an activist Khārijism. There were quietist Khārijīs who were content to remain in *Dār al-Kufr*, but such people clearly got in the way: Ṣāliḥ bin Musarriḥ was content to have anyone who did not accept his call, and the requirement of exile (*hijrah*), assassinated.[40]

Khārijism forced the religious establishment to define an orthodox position on the divisive issues. It required the jurists to concentrate the utilization of *jihād* as an armed struggle in the hands of the caliph, and ultimately in the hands of the state. In this way the movement had a major impact on Islām. The zeal and militancy of the Khārijīs has also been a model for many later Islamic, and radical Islamist, movements. As Abdullah Schleifer expresses it, 'the Khārijī understanding of *jihād* as a revolutionary model for Islamizing the state and society has continued to haunt Sunnī Islām to the present day'.[41]

The succession issue and the choice of the right *imām*

The second type of division (*fitnah*) was over the succession, which led to civil war. It has been correctly observed that 'choosing the right *imām* (or more precisely proving that the *imām* chosen was the right one) was a matter of vital importance for salvation; disputes over his identity thus precipitated the formation of sects and [the] declaration of belief in the legitimacy of one's own [*imām*] came to form part of the creed'.[42] The fact that it was around the caliphate that Muslim sects crystallized is inexplicable except by recognizing that, at least until

234/848, the caliph was both a religious and a political leader.[43] The Khārijī Abū Ḥamza delivered a sermon in the course of his rebellion in the Ḥijāz towards the end of the Umayyad period, in *c.* 129–30/746–47, in which the so-called deputies of God, the caliphs, 'came across as anything but rightly guided'. 'Uthmān was said to have fallen short of his two predecessors[44] and the second part of 'Alī's caliphate failed to 'achieve any goal of what was right'. The Umayyads were denounced as 'parties of waywardness' whose might was 'self-magnification. They arrest[ed] on suspicion, [made] decrees capriciously, kill[ed] in anger and judge[d] by passing over crimes without punishment.' However, the Khārijīs were unique in that they 'rejected not only the Umayyads themselves, but also the caliphal office which they represented'.[45]

In spite of later Shī'a assumptions, there does not seem to have been much contemporary support for the idea of hereditary succession to the Prophet; in any case he left no direct male descendants.[46] The role of leader was an uncomfortable one: three of the four 'rightly-guided' caliphs – 'Umar, 'Uthmān and 'Alī[47] – were assassinated.[48] Who, among the Prophet's followers, best fitted the requirements of excellence for the leader of the new movement? The Prophet's death had been unexpected; it is virtually certain that he had not made arrangements before his death for the subsequent organization and leadership of the community.[49] There were no explicit or unambiguous prophetic directives, for the Prophet had said 'not I but God appoints a successor over you…'[50] In a sermon at Ghadīr Khumm on his way home after performing his last pilgrimage, he had stated: 'I say unto you that whoever whose Master (*mawlā*) I am, 'Alī is his Master.' Though the words were not contested, the sense of *mawlā* was. While the Prophet had said that 'Alī was his 'trustee and heir' who would discharge his debt and fulfil his oath, the meaning of this remark, too, was contested.[51] The *ḥadīth* that 'the best reader/reciter of the Qur'ān will lead you' was one of the proof-texts adduced to point to the greater qualification for the office of the caliph/*imām*. Piety and moral excellence were seen as essential.

Abū Bakr was the first caliph, whom some have seen as the Prophet's friend and implicitly designated heir apparent.[52] His acceptance of Islām as a middle-aged man was said by some to be more significant than that of 'Alī, the fourth caliph and a direct relative of the Prophet, whose submission to Islām was as a young and inexperienced boy.[53] Others stated the reverse, that 'Alī's prior acceptance of Islām shamed the older Abū Bakr into following suit.[54] On the other hand, the Prophet himself had asked Abū Bakr to lead the community prayers,[55] though the Shī'a claimed that 'Alī was a better reader of the Qur'ān than Abū Bakr.[56] Similarly, the Prophet had asked Abū Bakr to collect alms, to lead the pilgrimage (*Ḥajj*) and several military expeditions in which he had protected the Prophet with his own life: these responsibilities testified to his knowledge of prayer, alms-giving, pilgrimage and *jihād*: 'these are the support of religion'.[57] While

'Alī was said to be the 'most excellent among [the followers] in legal decision-making', Abū Bakr was 'the most merciful'.[58]

It was Abū Bakr who took control of the situation at the death of the Prophet and, when others were panicking and arguing that his death could not be announced to the young Muslim community for fear of an adverse reaction, he calmly affirmed that the Prophet was dead, 'for death spares no one'.[59] Leadership over the Arabs as a whole could be provided only from the Qurayshīs, Abū Bakr argued, since the Arab tribes would not submit to anyone else. The Saqīfa assembly, which endorsed Abū Bakr as caliph, was nevertheless some way short of being a legitimate consultation (*shūrā*), since most of the prominent *muhājirūn* were absent. Rather, it should be seen as a coup masterminded by Abū Bakr and 'Umar, his successor as caliph, to prevent either the successful candidacy of 'Alī or the secession of the 'Emigrants' (*Muhājirūn*) and the 'Helpers' (*Anṣār*) under their own leaders.[60]

For the first six months, Abū Bakr was only a part-time leader of the new community and his powers as *Khalīfat Rasūl Allāh*, Successor of the Prophet of God, were in reality quite limited.[61] He succeeded in stabilizing the new regime, approached the problem of apostasy by launching military campaigns (*riddah*) against the groups who had reverted to their former faiths; and by sheer determination was able to hold firm and pass on the succession to his nominee, 'Umar bin al-Khaṭṭāb on his death in 13/634.[62] Three remarks of Abū Bakr concerning *jihād* have been recorded. The first was an encomium of the *jihādī*: '...every step of the warrior of God merits him seven hundred pious deeds, raises him seven hundred grades and effaces for him seven hundred sins'.[63] According to one transmission of the tradition, he instructed his commanders: 'do not embezzle, do not cheat, do not break trust, do not mutilate, do not kill a minor child or an old man of advanced age or a woman, do not hew down a date palm or burn it, do not cut down a fruit tree, do not slaughter a goat or cow or camel except for food...' According to a different transmission of the tradition, he enjoined upon his commanders 'the fear of God. Do not disobey,' he stated, 'do not cheat, do not show cowardice, do not destroy churches, do not inundate palm trees, do not burn cultivation, do not bleed animals, do not cut down fruit trees, do not kill old men or boys or children or women...'[64] It was under Abū Bakr that Syria was conquered by the Muslim forces.[65]

'Umar (13/634–23/644) reversed Abū Bakr's policy and allowed the former apostates to be recruited into the army. An Arab empire became, for the first time, conceivable.[66] Though later theorists such as al-Māwardī dismissed the possibility of defeat as non-existent,[67] the risks of expansion must have been considerable. The conquests first of Syria and then Egypt were spontaneous and haphazard. In the case of the conquest of Egypt, the victorious commander 'Amr bin al-'Āṣ, simply set off on campaign on his own initiative with some 3500 tribesmen, which led to 'Umar sending him reinforcements. The conquests were considered to belong to the Arab tribesmen who had won the victory; there was no thought of them being made in the name of a king, a *khalīfah* or even the new faith.[68] 'Amr remained governor until he was deprived of this post by 'Uthmān and 'Alī. He supported Mu'āwiya's rebellion in return for a promise of reinstatement. In the year after the battle of Ṣiffīn (37/657), he returned to Egypt as governor and remained there until his death five years later.[69] The disaffection of powerful regional rulers was a serious threat to the regime, and there is little doubt that support for 'Alī's cause began to disintegrate after the loss of Egypt.

The most important early Arab conquests occurred during the caliphate of 'Umar. Since Arabia had been largely pacified, it was only by directing raiding and warfare outside Arabia that Medina's hold over the Arab tribes could be preserved.[70] The armies seem to have fought as amalgamations of tribal units, each with their own banner.[71] Infantry predominated, because it was easier to levy, more effective and less expensive: a cavalryman received three times the infantryman's share of the booty (the same rate for his service, but two further shares for the mount).[72] 'Umar instructed his commanders to fear God and to 'march with the assistance of God and victory':[73]

> Persevere in right conduct and endurance. Combat, in the path of God, those who disbelieve in God; yet do not transgress, because God does not love those who transgress.
>
> Do not show cowardice in an encounter. Do not mutilate when you have the power to do so. Do not commit excess when you triumph. Do not kill an old man or a woman or a minor, but try to avoid them at the time of the encounter of the two armies, and at the time of the heat of victory, and at the time of expected attacks. Do not cheat over booty. Purify *jihād* from worldly gain. Rejoice in the bargain of the contract that ye have made [with God] and that is the great success.

In spite of 'Umar's instructions, men probably fought for a combination of motives – for their religion, for the prospect of booty and because their fellow tribesmen were doing it. There was no recording of their names on a register (what was to become later on the *dīwān*), or payment of salaries (*'aṭā'*), until the later years of 'Umar's caliphate. By then, rates of pay were determined by 'precedence' (*sābiqah*), that is the date of conversion to Islām and the resulting number of years of service in the army.[74] By 20/641 all the lands of the fertile crescent, 'Irāq, Syria, Palestine and Egypt[75] had been conquered. Two key battles, both probably in 15/636, shattered the Byzantine and Sāsānian empires, respectively at Yarmūk in Syria and al-Qādisiya in 'Irāq.[76] 'Umar was arguably the founder and organizer of the expanded Muslim Arab state. According to one tradition, he adopted *in toto* the Persian revenue laws when that empire was absorbed into the Muslim state.[77]

Apart from the set-piece battles, resistance to the Arab invaders seems to have been muted. Given the importance of Egypt, Syria and 'Irāq to early Christianity this was surprising. The Nestorian church in 'Irāq remained a dynamic community after the conquest, but the Greek-speaking Christian elite in Syria and Egypt fled. Though the Copts in Egypt, the Monophysites in Syria and the Nestorians in 'Irāq had long had troubled relations with their overlords, disaffection was probably important only in cases where Christian Arab border tribes and military auxiliaries joined the conquerors, or where fortified cities capitulated.[78] Zoroastrians were not granted the same rights and status as Jews and Christians and the Zoroastrian faith quickly collapsed in the wake of the Arab conquests,[79] though it was displaced not simply through the process of conversion but also through the later settlement of Iran by Muslim Arabs, a development which continued into the 'Abbāsid period.[80]

Crucially, the early Arab invaders made no attempt to impose their faith on their new subjects and discouraged conversions by non-Arabs. 'Umar allowed the Christian Arab tribes to retain their own faith and they did not have to pay the poll tax (*jizya*) to which non-Muslims were subject; they did, however, have to pay alms (*ṣadaqah*) at twice the rate of their Muslim fellow tribesmen.[81] Conversions led to the demand for tax privileges which cut down revenues and also resulted in conflicts over status. All this suggests that the early conquests were not intended to advance Islām by the sword, except and in so far as the beneficiaries of the change of regime, the new ruling elite, were Arab Muslims.[82] With the exception of Syria, where many of the indigenous population were Arabs who accepted Islām, the Arabs maintained a social distance from their newly conquered populations and little assimilation took place.[83] Only in the Arabian peninsula itself, as a result of a ruling of 'Umar in 20/641 based on the statement of the Prophet that 'two religions cannot coexist in the peninsula', was Islām proclaimed the sole religion, with Jews and Christians to be removed from all but the southern and eastern fringes of Arabia.[84]

Carrying the message abroad (*futūḥāt*) and proto-*jihād*

To what extent were the early conquests motivated by a nascent ideology of *jihād*? The Prophet had set the precedent with the Prayer of Fear (*ṣalāt al-khawf*: Q.4:101–3), in which one row of believers was to keep watch with weapons in hand while a second row performed the prostration (*sujūd*). When battle was about to be joined, the Prophet would pray 'O God, Thou art my protection, my Giver of victory, my Giver of help! O God, by Thee I attack and by Thee I fight.' At the battle of Uḥud, he prayed: 'O God, to Thee belongs all praise; to Thee all [pleas] are addressed; Thou art the Helper.'[85] Assuming there was time for preparation, the Qur'ānic passages from the chapter on the spoils of war (*Surāt al-Anfāl*: Q.8), which emphasized the spiritual and material benefits to the *jihādī*, were almost certainly recited before the battle.[86] The month of fasting of Ramaḍān was also perceived as a month of *jihād*, a month in which Allāh grants military victories to His believers. The Prophet gained two of his greatest victories, Badr in 2/624 and the reconquest of Mecca in 630, during the month of Ramaḍān. We cannot know to what extent this model was replicated in other skirmishes during the early Arab conquests. There were also exhortations from commanders, as was the case with Sa'd ibn Abī Waqqāṣ' exhortation before battle of al-Qādisiya against the Persians: 'if you renounce this world and aspire for the hereafter, God will give you both this world and the hereafter'.[87] It was commendable, but not compulsory, to have one of God's 99 names as the war cry, though we know that tribes also used the names of commanders as their rallying call in battle.[88] Tribal war cries probably dated back to the wars of the period of pre-Islamic 'ignorance' and 'barbarism' (*jāhilīyyah*).[89]

All this may amount to '*jihād* for the sake of (or in the path of) Allāh' (*jihād fī sabīl Allāh*), or as the Prophet called it, fighting 'in the name of God, in the way of God, and in conformity with the Messenger of God'.[90] But we also know what it did not amount to. It did not constitute a fully-fledged ideology of *jihād* as was later established by the jurists. For this we have to wait about 150 years (the first surviving treatise on *jihād* is that of 'Abdullāh ibn al-Mubārak, who died in 181/797). His treatise compiled 206 of the Prophet's traditions dealing with the subject.[91] He had taken part in many *jihād* campaigns and emphasized his loyalty to successive *imāms*, without whom 'roads would not be secure for us and the weak among us would have been prey for the strong'. Stressing the superiority of *jihād* over the devotional practices of ascetics, he stated that the most virtuous deed was to guard believers in far-off places.[92] Appropriately enough, he was a partisan of the dissemination of knowledge through appointments to the '*ulamā'* and the study of the *ḥadīth*.[93] The practice preceded the theory. The jurists provided a *post facto* rationalization of the Arab conquests, 'a legal justification for the rapid expansion of the Islamic empire that occurred in the decades following the Prophet's death'.[94] In this respect, the jurists' ideology of *jihād* is no different from the development of Sunnī political theory in general. In the words of Professor H. A. R. Gibb, this political theory was 'the rationalization of the history of the community… all the imposing fabric of interpretation of the sources is merely the *post eventum* justification of the precedents which have been ratified by the consensus of the community (*ijmā'*)'.[95]

If *jihād* was the motivating ideology at the outset of the conquest, it can at most be said to have been a proto-*jihād*.[96] Perhaps we come closest to this sense of a proto-*jihād* in the text of a sermon allegedly delivered by 'Alī, the future fourth caliph, to 'Umar to discourage him from campaigning in person in Iran:[97]

> In this matter, victory or defeat is not dependent on the smallness or greatness of forces. It is Allāh's religion which He has raised above all faiths, and His

> army which He has mobilized and extended, [until] it has reached the point where it stands now, and has [reached] its present position. We hold a promise from Allāh, and He will fulfil His promise and support His army.
>
> The position of the head of government is that of the thread for beads, as it connects them and keeps them together. If the thread is broken, they will disperse and be lost, and will never come together again. The Arabs today, even though small in number are big because of Islām and strong because of unity. You should remain [at home] like the axis for them, and rotate the mill [of government] with [the help of] the Arabs, and be their root. Avoid battle, because if you leave this place the Arabs will attack you from all sides and directions [until] the unguarded places left behind by you will become more important than those before you…
>
> If the Persians see you tomorrow, they will say 'He is the root [that is, chief] of Arabia. If we do away with him, we will be in peace.' In this way it will heighten their eagerness against you and their keenness to aim at you… As regards your idea about their [large] number, in the past we did not fight on the strength of large numbers, but we fought on the basis of Allāh's support and assistance.

'Umar broke with the practice of the Prophet and the wish of the commanders in his refusal to continue distributing the lands of 'Irāq and Syria among the Companions. In spite of their protestations, 'Umar argued that if he continued to distribute the lands, he would have no resources from which to maintain an army to protect the new borders and newly-conquered towns.[98] The Companions finally agreed with him and remarked that 'yours is the correct opinion'. Though he later found a justification for his action from Q.59:6–10, he had departed from those other Qur'ānic injunctions which commanded the distribution of booty. This was an early example of the principle of *istiḥsān*, the theory of 'just preference' to justify the departure from an established rule in the interest of equity and public welfare. 'Umar preferred the general benefit of the Muslim community to that of individuals who traditionally had drawn advantage from the division of spoils of war and recently-conquered lands.[99] Henceforth, any land conquered by the Muslim army was considered *kharāj* land, that is, subject to the land tax.[100] The term *istiḥsān* itself, it seems, was first used by the jurist Abū Ḥanīfa (d. 150/767), the founder of the Ḥanafī school of Islamic law, and the first jurist to write a monograph on international law (*siyar*), though it has not survived.[101] 'Umar, according to his later critics, not only violated an indisputable Islamic principle when he refrained from dividing the conquered lands, but in addition he imposed a new fiscal system of which there was no mention in the Qur'ān and the Tradition. Thus his action should be considered an innovation (*bid'ah*) in the divine law. 'Alī, the Shī'a critics of 'Umar contended, would have reversed the provision, but widespread land confiscation would have led to revolt and thus he took no action. Such acquiescence does not, however, prove that 'Alī endorsed 'Umar's policy.[102] During the ten years of 'Umar's reign, the nature of the caliphate or the Muslim state, had been transformed. The army had a stake in the imperial policies of the Qurayshīs and the caliphate, which had been precarious under Abū Bakr, was firmly established.[103] 'Umar pronounced that the caliphate belonged to all of the Qurayshīs and could not be monopolized by any particular family. Within a fortnight of this pronouncement, he was struck by an assassin.[104]

His successor, 'Uthmān, was chosen by an electoral council (*shūrā*), but it was scarcely a 'democratic' election: there were just six electors. He was chosen as the only strong counter-candidate to 'Alī.[105] The policy of 'Uthmān (23/644–35/656) formed a break with the past. He was committed to a return to clan government, the dominance of the Qurayshīs and the Umayyad clan over the Muslim community. His attempt to reconvert communal (*ṣawāfī*) land into crown property marked a significant step towards turning the caliphate into a traditional monarchy.[106] 'Uthmān's regime was one of nepotism, and this was thought to undermine the principle of consultation or *shūrā*;[107] but though there were many complaints at the time, by the standards of the abuses of his successors, his wrongdoings appear relatively trivial (there were, for example, no murders authorized by him).[108] Nevertheless, such were the abuses of his regime that by the year 34/654–5 there were calls by former Companions of the Prophet for a *jihād* against the caliph,[109] later moderated to a call for his abdication and the appointment of an alternative caliph.[110] His assassination set a bad precedent for the future. As he told his assassins, 'if you kill me, you put the sword to your own neck, and then Allāh will not lift it from you until the Day of Resurrection. And if you kill me, you will never be united in prayer, and you will never divide the booty amongst you, and Allāh will never remove discord from amongst you.' It has recently been argued that this set the precedent for acts of extremist violence that have tended to cause division within the Muslim community,[111] although the same could be said about the earlier murder of 'Umar.

In contrast, the policy of his successor 'Alī (35/656–40/661), who was not elected by an electoral council (*shūrā*),[112] was to emphasize the equality of all believers and to stress that the spiritual leader (*imām*) should be more than a tyrannical tax-gather and guardian of vested interests. Instead, the practice (*sīrah*) of 'Alī was that any surplus revenue (*faḍl*) could be removed from the provinces only with consent. Allegiance to 'Alī and his memory defined a particular anti-centralist fiscal position in 'Irāq.[113] Unwisely, 'Alī opened up the treasury and disbursed the money to the common people (unwisely, because he was facing insurrection from those who accused him of moral responsibility for the murder of 'Uthmān) and he insisted on deposing all of 'Uthmān's provincial governors.[114] These populist measures won him some popular support, but lost him the support of the provincial governors, who were crucial to the survival of his regime. Furthermore, his refusal to make financial concessions to the nobility and tribal chiefs left them vulnerable to bribery from Mu'āwiya.[115] 'Alī's son, al-Ḥasan, was counselled to coax his companions and[116]

> appoint the men of distinguished houses and nobility to offices, for you buy their hearts with that. Follow the practice of the *imāms* of justice of conjoining hearts [that is, paying bribes to influential men] and restoring concord among the people... You know that the people turned away from your father 'Alī and went over to Mu'āwiya only because he equalized among them in regard to the [proceeds of taxation (*fay*')] and gave to all the same stipend. This weighed heavily upon them.

These structural miscalculations, rather than 'Alī's weakness in conceding arbitration after the battle of Ṣiffīn or his massacre of the Khārijīs at al-Nahrawān, mistakes though they were, seem to have destabilized his regime.

Sayed 'Alī Reza or Razī (359/969–404/1013) records 'Alī's sermon on *jihād*, which must count as one of the most eloquent (even if ultimately unsuccessful) appeals for support:[117]

> Now then, surely *jihād* is one of the doors of Paradise, which Allāh has opened for His chief friends. It is the dress of piety and the protective armour of Allāh and his trustworthy shield. Whoever abandons it Allāh covers him with the dress of disgrace and the clothes of distress. He is kicked with contempt and scorn, and his heart is veiled with screens [of neglect]. Truth is taken away from him because of missing *jihād*. He has to suffer ignominy and justice is denied to him.
>
> Beware! I called you [insistently] to fight these people night and day, secretly and openly exhorted you to attack them before they attacked you, because by Allāh, no people have been attacked in the hearts of their houses but they suffered disgrace; but you put it off to others and forsook it until destruction befell you and your cities were occupied...
>
> How strange! How strange! By Allāh my heart sinks to see the unity of these people on their wrong [path] and your dispersion from your right [path]. Woe and grief before you. You have become the target at which arrows are shot. You are being killed and you do not kill. You are being attacked but you do not attack. Allāh is being disobeyed and you remain agreeable to it. When I ask you to move against them in summer you say it is hot weather. Spare us [until] the heat subsides from us. When I order you to march in winter you say it is severely cold; give us time [until] the cold clears from us. These are just excuses for evading heat or cold because if you run away from heat and cold, you would be, by Allāh, running away [in a greater degree] from [the sword; that is, war]...

In another sermon, 'Alī denounced those who found pretexts for inaction at the time that a *jihād* had been called:[118]

> O people, your bodies are together but your desires are divergent… The excuses are amiss like that of [a] debtor unwilling to pay. The ignoble cannot ward off oppression. Right cannot be achieved without effort. Which is the house besides this one to protect? And with which *Imām* would you go… fighting after me?
>
> By Allāh deceived is one whom you have deceived while, by Allāh he who is successful with you receives only useless arrows. You are like broken arrows thrown over the enemy…

On another occasion, 'Alī observed that the frontiers of the land of Islām were being eroded, but there remained no enthusiasm for a campaign in Syria (Syria was the base of Mu'āwiya's rebellion).[119] 'Does not faith join you together, or [a] sense of shame rouse you?', he remarked in another sermon. 'No blood can be avenged through you and no purpose can be achieved with you.' His followers were likened to camels with stomach ache.[120] The Companions of the Prophet had fought with vigour: 'if we had behaved like you, no pillar of [our] religion could have been raised, nor [could] the tree of faith… have borne leaves'.[121] It was impracticable for the *imām* to embark of every campaign of *jihād* 'like [a] featherless arrow moving in the quiver', for the *imām* was 'the axis of the mill': 'it rotates on me while I remain in my position. As soon as I leave it the centre of its rotation would be disturbed and its lower stone would also be disturbed…'[122]

Valour was a question of gritting one's teeth so that swords skipped off the skull and closing one's eyes because it strengthened the spirit and gave peace to the heart. Above all, the banner of the regiment had to be guarded. There must be no retreat:[123]

> By Allāh, even if you run away from the sword of today you [will] not remain safe from the sword of the next world. You are the foremost among the Arabs and great figures. Certainly in running away there is the wrath of Allāh, unceasing disgrace and lasting shame. And certainly a runner-away does not lengthen his life, nor does anything come to intervene between him and his day [of death]. Who is there to go towards Allāh like the thirsty going to the water? Paradise lies under the edge of spears. Today the reputations [of warriors' valour] will be tested.

Martyrdom was an inevitable consequence of the call to *jihād*. When the Companions had been called to *jihād*, they had responded and trusted in their leader and followed him. By implication those called to fighting (*al-jihād*) in the sermon should do likewise: 'he who desires to proceed towards Allāh should come forward'.[124] As the sermons demonstrate, 'Alī was not lacking in moral courage, and after a low point following the arbitration after the battle of Ṣiffīn, his fortunes had seemed to be verging towards recovery at the time of his assassination.[125] What is undeniable is his moral stature in comparison with Mu'āwiya, whose lack of commitment to Islām and unscrupulousness were proverbial (he was called 'the most infidel and abominable of men').[126] The coup may have brought an end to the *fitnah*, the inter-Muslim war, but it was followed by an era 'biting kingship' (*mulk aḍūd*).[127] The Umayyad dynasty had to have 'Alī and his followers cursed from the pulpits in order to create its own sense of legitimacy.[128]

The Muslim armies and their conduct

The coup of Mu'āwiya bin Abī Sufyān following 'Alī's assassination in 40/661 was a victory of the Quraysh and their Syrian followers over the 'Irāqis and, within 'Irāq itself, it was a victory for the tribal leaders (*ashrāf*) over a divided Muslim elite.[129] Wars of expansion on all fronts were launched to help divert the attention of the tribesmen to foreign soil.[130] The Syrian army was the backbone of the new Umayyad regime. Syrian loyalties seem to have been more important than tribal or dynastic ones.[131] Eventually the Syrian army became disgruntled at the prospect of campaigning in all parts of the empire simultaneously and mounted a coup against al-Wālid II in 126/744. For all practical purposes, this amounted to the end of the Marwānid regime: 'the very basis of its rule was destroyed when it lost the support of the Syrian army'.[132] The disposal of provincial surpluses remained a live issue until the end of Umayyad rule. At the same time the concept and practice of payment for military service ('*atā*') was gradually developed so that the last Umayyads and the first 'Abbāsid rulers had a professional army at their disposal. It became possible 'to speak of the Muslim army, rather than the Muslim community in arms'.[133]

Yet not all the parts of that army were controlled from the centre of the empire. The West was cut off militarily and politically from the East. The first expedition across the Pyrenees took place in 99/717. Narbonne was captured and converted into a base of future operations in 101/719–20. It was not until 114/732 that Amīr 'Abd al-Raḥmān al-Ghāfiqī was defeated by Charles Martel in a battle at an uncertain location between Poitiers and Tours, in what proved to be a decisive encounter. Narbonne was evacuated in 142/759 and the Muslim threat to Francia receded. With the downfall of the Umayyads in the east, the Muslim territory of Spain (al-Andalus, that is, based on Cordoba)[134] became independent of the rest under its own Umayyad dynasty, its distinctive juristic traditions and separate army. This lasted for more than three centuries (138–422/755 or 756–1031), and a strong case can be made for the survival of the imāmate after 422/1031. What had changed, however, was the pretension of al-Andalus (Cordoba) to provide the focus for a united state of Islām in the Iberian peninsula. The tendency towards fissiparity, evident earlier in the east, had by this date overcome the western outpost of Islām. When the Berber Almoravid confederation was invited in to sort out the divisions between the Muslims in Spain, the invitation came from *qāḍīs* and *faqīhs* who adduced the godlessness of their rulers' ways of life and manner of rule. The Almoravids specifically recognized the *imām* in the east by placing the title *abd allāh* on their coins, and then, from 535/1140 or 1141, by adding the expression *al-'abbāsī*.[135]

Thus, before the period in which any of the treatises on *jihād* were written, a development of fundamental importance had occurred with regard to the organization of the army which explains the geographical extent of the Muslim conquest. When Hārūn invaded the Byzantine empire in 165/782 he took with him no supply train but a vast amount of cash. Muslim armies were expected to buy their supplies from traders and peasants at markets. The army thus acted as a vital infusion to a region's money supply.[136] It is true that the troops were only obliged to act in this orderly manner in Muslim territory. On the other hand, the knowledge in frontier regions that a prompt surrender would prevent pillage and would positively boost the local economy were powerful inducements for a transfer of loyalties.

Warfare was increasingly the occupation of a professional army, and there was accordingly a restriction in the requirement of a military *jihād* on the part of individual Muslims. In a *ḥadīth* recorded by the radical Ibn al-Jawzī (d. 597/1200), Q.2:216 was said to be 'in force and… the requirement of *jihād* is necessary for everyone', but it was a collective (*farḍ al-kifāyah*) and not an individual obligation.[137] For Ibn al-Jawzī, Q.9:122 ('the believers should not all go forth') did not abrogate Q.2:216, but merely qualified it: the distinction was between the requirement on every male of fighting age to fight (*farḍ 'ayn*) and the reality that not everyone was obliged to respond to the call unless needed (*farḍ al-kifāyah*). The jurist al-Shāfi'ī (150/767–204/820), the founder of the Shāfi'ī school of Islamic law, seems to have introduced this term '*farḍ al-kifāya*', of which there is no evidence before him. It was described by him as a collective obligation, which if 'performed by a sufficient number of Muslims, the remaining Muslims who did not perform it would not be sinful'.[138] For al-Shāfi'ī, therefore, the performance of *jihād* required a sufficient number of agents rather than devolving upon every individual. Only in an emergency did the duty become obligatory (*wājib*) on all Muslims individually.[139]

One of the earliest books of *ḥadīth* on taxation under Islām, that of Yaḥya (d. 203/818), asserted the principle that the ownership of land was vested ultimately in Allāh and the Prophet, from whom the Muslims received it; therefore the state, as representative of the whole Muslim community, was the owner of the land.[140] Most later scholars, including al-Ghazālī, claimed that the *imām* could impose any tax within the bounds of the general interest of the Muslim community. *Kharāj*, the land tax established by 'Umar, was an instance of this general rule.[141]

Nonetheless, it is clear from the disputes between al-Awzā'ī (d. 157/773) and Abū Yūsuf (d. 182/798) that different arguments could be deduced from events such as the Prophet's failure to expropriate the properties of Muslims after the fall of Mecca. Abū Yūsuf, grand *qāḍī* of Baghad from 166/782, dedicated his principal treatise, *Kitāb al-kharāj*, to Hārūn al-Rashīd (r. 169/786–193/809), the fifth of the 'Abbāsid kings. It was no accident that the theory of the imāmate should be included in a work primarily devoted to taxation, since there is a close connection in Islamic theory of government between taxation and just rule.[142] Abū Yūsuf called Hārūn al-Rashīd 'commander of the faithful' and extolled the legitimacy of 'Abbāsid rule; but in his theory there was no necessary connection between the ruler's personal qualifications and the exercise of authority. If the ruler was tyrannical, the burden of sin was his alone; the moral responsibility of the individual was patience, though he could seek to reprove and correct such evil conduct. In distinction to Shī'a activism and propensity for rebellion (revealed by the revolt of 145/762), quietism became almost a criterion of Sunnī orthodoxy for Abū Yūsuf.[143] Hārūn al-Rashīd asked Abū Yūsuf whether an invitation should be extended to infidels to embrace Islām before waging war upon them. In response, the jurist recounted the instructions which the Prophet used to give to the Companions before battle and also enlightened him on the practice of Abū Bakr and 'Umar.[144] The early 'Abbāsid rulers (132/749–*c*. 218/833) were 'enthusiastic participants' in the study of *ḥadīth*, which served to bind 'the community of scholars with the Prophet and his Companions, the scholars with each other, and the caliphs not just with the Prophet, or their own predecessors, but also with the scholars'.[145]

Mālik ibn Anas (d. 179/795) was the most distinguished jurist of Medina in his day[146] and founder of the Mālikī school of Islamic jurisprudence. In his collection of *ḥadīth* entitled *Al-Muwaṭṭa'*, he tended to relate first the relevant *ḥadīth* from the Prophet, then from one of the Companions, and lastly the practice and opinions of the lawyers of Medina.[147] 'The ruler is God's shadow on earth and his spear', states an utterance of the Prophet reported on the authority of Mālik but not recorded in his collection.[148] Mālik records 51 *aḥādīth* in chapter 21 of his *Al-Muwaṭṭa'*, the book of *jihād*. This in turn is divided into 21 subsections, covering themes such as stimulation of desire for *jihād*, booty from war, awarding bonuses from the tax of one-fifth, martyrs in the way of Allāh, things in which martyrdom lies, and how to wash the body of the martyr.[149] Mālik confirms the reward that will await the martyr: 'when the Day of Rising comes, blood will gush forth from his wound. It will be the colour of blood, but its scent will be that of musk'.[150] Henceforth, after Mālik's *Al-Muwaṭṭa'*, 'practically no Islamic *corpus juris* was devoid of chapters on international law, entitled variously *siyar*, *dimā'* (or *siyar ad-dimā'*, conduct with regard to bloodshed), [the military campaigns of the Prophet (*maghāzī*)] and *jihād*'.[151]

The standard actions that were to be performed before a battle against non-Muslims hardly vary between the sources, be they Sunnī or Shī'a.[152]

> Fight in the name of God, in the way of God, and in conformity with the religion of the Messenger of God. Do not begin to wage war until you have invited the enemy to bear witness that there is no deity other than God, and that Muḥammad is the Messenger of God, and to accept the message you have brought from God.
>
> If they accept your message, then they are your brothers in faith. Thereafter, call upon them to transfer themselves from their abode to that of the Emigrants. If they do so [they will have the same rights and responsibilities as the Emigrants]. Otherwise inform them that they are like the country Arabs, and that the ordinances of God will be applicable to them to the same extent as they are to the Muslims, but that they shall not be entitled to a share in the [revenue derived from conquest] (*fay'*) or [spoils of war] (*ghanīmah*).[153]
>
> If they refuse to accept Islām as their religion, then call upon them to render the poll tax (*jizya*) readily and submissively. Should they accept this condition, accept it from them and refrain from harming them. But if they refuse [to pay *jizya*], then ask God for His help against them and then wage war with them. Do not kill children, elderly men, or women if they do not offer any resistance. Do not mutilate them, or act unfaithfully [in relation to the spoil], or act treacherously towards them.

Differences on *jihād* between the classical jurists

What does differ between the jurists is the extent to which classical scholars of the Mālikī school, unlike the others, tended to espouse moderate opinions on *jihād*. For the Syrian jurist *Imām* Abu Sufyān al-Thawrī, the Medinan jurist Ibn Shibrimah, and the other Mālikī scholars including the founder of the school itself, *Imām* Mālik bin Anas (d. 179/795), *jihād* is not the principle (*al-aṣl*) that determines the nature of relations between Muslims and non-Muslims. On the contrary, they espoused non-aggressive principles, namely reconciliation, peace, mutual cooperation to achieve common interests based on justice, fairness and truth, the freedom of religious expression and dissemination. Al-Thawrī was even more categorical when he said that

> fighting the idol-worshippers is not an obligation unless the initiative comes from them. If that is the case, they must be fought in fulfilment of Allāh's command 'if they [the unbelievers] fight you, kill them' and His saying 'and fight all the idol-worshippers as they fight you all'.[154]

For the moderate school of classical jurisprudence, unbelief (*kufr*) did not denote an act of aggression (*'udwān*) against others.[155] Belief was a matter of faith and in one of the Medinian texts the Qur'ān declares that 'there is no compulsion in religion' (Q.2:285). This was interpreted as having a wider meaning than a mere recognition of one's liberty to choose one's own religion. Non-Muslims living in *Dār al-Islām* must be left free to exist and practise their religion without interference from others, including the state. This school did not distinguish non-Muslims as the enemies of Islām. Exponents of this school came predominantly from the Ḥijāzī scholars of second-century Islām (that is, the school of Mecca and Medina), which was basically a continuation of the juristic tradition of the renowned jurists of the late-first-century Medina, namely Sa'īd bin al-Musayyab (d. 94/712), and his disciple and close associate 'Aṭā' bin Abī Rabāḥ (d. 114/732). Their views on peace and war in Islām were adopted and reinterpreted by the later important jurists including Ibn Jurayḥ (d. 150/767), 'Amr bin Dinār (d. 172/788), the founder of the Malīkī school of jurisprudence, Malīk bin Anas, and others. For these scholars, unbelievers should not be subjected to war because of their unbelief, for this would be tantamount to aggression (*'udwān*) against freedom of religion, the universal principle which was to be strictly upheld by Islām. For some scholars of the moderate school, the war of extermination explicitly expressed in the 'verse of the sword' was only applicable to Arab unbelievers during the times of the Prophet. The rule was inapplicable against the 'people of the book' (Jews and Christians) and even against the Magi (*majūsī*) and non-Arab unbelievers.[156] However, they did not object to declaring *jihād* against unbelievers who had been legally identified as enemies of Islām. The war was not only justified but legitimate if the unbelievers themselves had first committed aggression and hostility against Muslims. The argument was based on the Qur'ānic text which urged Muslins not to commit aggression (Q.2:190). Elsewhere the Qur'ān exhorts Muslims to fight aggressors among unbelievers, who have been identified as enemies until 'there is no sedition (*fitnah*) and the religion is only for Allāh' (Q.2:193; Q.8:39).

For the Ḥijāzī scholars, the undertaking of *jihād* was a religious duty obligatory upon the Muslims, but it was only legitimate when applied against those unbelievers who had been identified politically as the enemies of Islam because of their aggression or hostility. They also recognized that when war was declared, it would continue until enemies refrained from aggression and there was no further sedition and persecution of believers (*fitnah*). Thus the rationale for war was political: to safeguard Muslim rights to determine their political existence and practise their religion (an early form of self-determination?); and to resist external aggression which threatened to undermine the territorial sovereignty of *Dar al-Islām*.

It is clear that for the Ḥijāzī school, whose viewpoint was also shared by the renowned Syrian jurist and traditionalist of the second century of Islām, Sufyān al-

Thawrī (d. 161/778), unbelief (*kufr*) was not the underlying reason for a military *jihād* against unbelievers. Nor should they be regarded as enemies without any genuine justification. The basis of this argument lay in the interpretation of *fitnah* in the Qur'ānic texts. Unlike their Syrian and Egyptian counterparts, the Ḥijāzī scholars interpreted the phrase 'there is no *fitnah*' in the verses to have nothing to do with the complete elimination of belief.[157] To them, 'free from *fitnah*' denoted a condition of affairs in which Muslims were safeguarded from persecution, and enjoyed total freedom to exist and practise their religion without intimidation.

To support this argument, the Ḥijāzi scholars relied on Ibn 'Umar's rebuttal of the criticism levelled against him by opponents for his refusal to support Ibn al-Zubayr's revolution to topple Mu'āwiya's regime. In a heated discussion with Ibn 'Umar, the supporters of Ibn al-Zubayr reasoned that the legitimacy of their revolution was justified on the ground that it was waged to 'free Muslims from sedition (*fitnah*)' to which Ibn 'Umar cynically responded: 'in the past we have fought [against the enemies] until there is no sedition (*fitnah*) and the religion is only for Allāh. But today you have sought to fight against each other until there is an escalation of *fitnah* and the religion is for other than Allāh!'[158]

For the majority of classical Muslim scholars, particularly of the second century of Islām, the notion of unbelief in the Qur'ān was always perceived as tantamount to injustice (*ẓulm*), aggression (*'udwān*) and sedition (*fitnah*). This view led them to the general assumption that all unbelievers must be the enemies (*al-'adūw*) of Muslims, without further investigation as to whether they were or not the actual perpetrators of injustice, aggression, and sedition. Two eminent jurists of the Ḥanafī school of jurisprudence, al-Shaybānī (132/749 or 750–189/805) and al-Sarakhsī were the leaders of this hard-line school of *jihād*. Al-Shaybānī quoted the Qur'ānic 'verses of the sword' (*āyat al-sayf*) (Q.9:123; 9:39; 2:190; 39:79), which call upon Muslims to wage all-out war against unbelievers unconditionally, particularly those who were geographically nearer to the frontline of *Dār al-Islām*. The underlying assumption of his views was that *jihād* was to be conducted perpetually until there was a complete elimination of religious *fitnah*, that is polytheism and unbelief.[159] Al-Ṭabarī (224–310/839–923), in his interpretation of Q.2:193, concluded:

> this is the Divine instruction revealed upon the Prophet… in order to wage war against the unbelievers [who waged war against the Muslims] until there is no *fitnah*, i.e. until there is no polytheism (*shirk*) and the worship is only for Allāh, and until there are no deities or equal rivals (*andād*) set beside Allāh as objects of worship, obedience, trust and love.[160]

In contrast, for the radicals among the hard-line scholars, the possibility of truce of a peace with unbelievers was totally inconceivable. This was founded on the assumption that the revelation of the 'verses of the sword' had brought about a total abrogation or annulment of all non-aggressive texts in the Qur'ān, including the text that strongly encourages Muslims to be inclined to peace (Q.8:61). Among the leading exponents of this view was the Baṣra-based scholar and Qur'ānic exegete Qatāda, known as Abū al-Khaṭṭāb (60/679–117/735), who contended that Q.8:61 had been abrogated by the verse of the sword (Q.9:5). Since peace or a diplomatic solution was in principle no longer applicable, *jihād* was the underlying principle buttressing Muslim external relations. *Jihād* against non-Muslims was as a consequence both a religious and a political imperative. Muslims were under a permanent obligation to wage an unconditional and all-out war against non-Muslims until they embraced Islām or paid the poll tax payable by non-Muslims (*jizya*), in accordance with Q.9:29, as a token of submission and loyalty to a Muslim government.[161] Differences within the hard-line school can be perceived in their view of peace treaty or truce (*muwāda'ah* or *muṣālaḥah*). Al-Shāfi'ī (d. 204/820), the founder of the Shāfi'ī law school, held that a truce should not normally exceed four months, or one year at most. This was based on the Qur'ānic verse Q.9:12. Al-Qushayrī argued that a truce must not exceed one year, especially when Muslims are certain of the superiority of their forces. Al-Shāfi'ī was reported by al-Qurṭubī to have argued that it must never exceed ten years because otherwise this would undermine the underlying principle of *jihād* against unbelievers. Attempts at renewal of an expired treaty were not recommended except as necessity (*ḍarūrah*) to protect the general interests of the Muslim community.

However, al-Shāfi'ī's hardline position was not representative of other schools of thought. Abū Ḥanīfa is quoted by Ibn Qudāma to the effect that a ten-year peace treaty can be extended 'as a contract' with no time restriction; units of ten years may be taken as signifying that longer periods are permissible; since a treaty is a contract it can be negotiated without time limit. The interest of Muslims might be served as well by peace as war.[162] Ibn Qudāma and Ibn Rushd attributed to Mālik, Abū Ḥanīfa and Ibn Ḥanbal (d. 241/855) such views, which placed a primacy on the interests of the Muslim state. Ibn Rushd's view, which is quoted below, is definitive on the split between the jurists.[163]

The developed ideology of *jihād*

A good example of the developed ideology of *jihād*, albeit at a relatively late date, is provided by the ninth chapter of the *Da'ā'im al-Islām* of al-Qāḍī al-Nu'mān. This was the official lawcode of the Fāṭimid state of Egypt issued by its ruler, al-Mu'izz li-Dīn Allāh, around the year 349/960. This lawcode is still recognized by all courts in the Indo-Pakistan subcontinent in personal and family matters as the definitive source of Ismā'īlī law.[164] The 'Alīd emphasis of the chapter is clear, with many sayings of the fourth rightly guided caliph included, a whole treatise written for his followers on how to exercise judgement with regard to subordinate officers (the so-called '*Ahd* of 'Alī, a mirror for princes document), and the claim that it was 'Alī, not Abū Bakr, who took to Mecca the revelation of *Sūrah Barā'a* (chapter 9 of the Qur'ān), the chapter which included the 'verses of the sword'.[165] The other sections are much as to be expected, concerning the obligation to wage *jihād*; the inducements for waging *jihād*; the actions to be performed before battle; how to wage war; war with idolaters; the rules governing captives in war; the security and protection of aliens; on peace, covenants and *jizya*; an account of enemy property captured in war; the distribution of the booty; fighting with rebels; the rules regarding booty captured from the rebellious party; the rules concerning relations between two disputing factions and those Muslims against whom fighting is permissible. War must be waged against those who deny Muḥammad's Prophethood or repudiate his messengership. Allāh has strengthened Islām and helped His Messenger by making *jihād* in the cause of God obligatory.[166] The statement attributed to the Prophet was quoted that 'the root of Islām is prayer; its branch the alms tax; and the apex of the tree (or hump of the camel: *ṣanām*) is *jihād* in the way of God'.[167] The authority of 'Alī was brought into play for the Prophet's statements that 'faith has four foundations: patience, certitude, justice and *jihād*'; that '*jihād* in the way of God is a gate among the gates of Heaven'; and that to achieve martyrdom in the way of God was one good act above all others.[168] *Jihād* was indeed 'one of the gates of Paradise. He who abandons it earns the contempt of God, and He will make him the target of calamity and dishonour'.[169] Enemy property captured in war was to be divided into fifths, 'and the fifth is for us, the People of the Prophet's House, and it is for the benefit of the orphans among us and the destitute and the wayfarers'.[170]

The views of Shamsūddīn al-Sarakhsī (*c.* 400/1010–482/1090), one of the greatest jurists of the classical age, on the various stages of *jihād* have been quoted above (Chapter 1).[171] He took the four 'rightly-guided' caliphs[172] as the precedent or legal justification for all subsequent action. Abū Bakr's wars against the apostates (the *riddah* wars) justified subsequent action of this kind. As for rebellion (*fitnah*), this resulted in political disintegration and chaos and had to be opposed by force. Under the third caliph, 'Uthmān, the *fitnah* was already serious; his assassination further worsened the situation. The undertaking of *jihād* was a responsibility of the *imām* on behalf of the whole community. The *imām*'s powers covered the conduct of both hostile and peaceful relations with non-Muslims. Drawing upon the *Sunnah*, Ibn Rushd had insisted on the prior invitation to Islām as the essential prelude to warfare ('go to the enemy and call them to Islām and tell them what they should do. I swear to God', the Prophet had said, 'if one person becomes a Muslim as a result of your effort, this is better for you than everything under the Sun').[173] In contrast, al-Sarakhsī drew upon a specific historical incident, the siege of Banū al-Muṣṭaliq under the command of the Prophet, to justify pre-emptive hostilities against the enemy without a declaration of war or prior invitation to Islām. This latitude for the *imām* concerning the onset of hostilities is repeated by al-Sarakhsī when it comes to making peace. Whereas al-Shāfi'ī had allowed the *imām*, if necessary, to disregard the ten-year limit on treaties, al-Sarakhsī allowed him to make the treaty on terms which were initially favourable to the enemy, but he was also prepared to allow him to renege on the treaty unilaterally when the Muslims had gained supremacy.[174]

According to al-Sarakhsī, the objective of a legitimate war against non-Muslims was to honour the religion and defeat the polytheists. It was a duty not to be neglected by the *imām*. Subject to there being a strong Muslim army, the legitimacy of the offensive war depended essentially on the expectation of victory or chances of success. An offensive war was legitimate in the sense that it sought to achieve a just objective. Neutrality was a possibility only if the Muslim forces were weak, and no victory was foreseeable; or if the Muslim forces were to intervene on one side or the other in a war between polytheists. In the case of a defensive war, neutrality was no longer a possibility. All Muslims (including Muslims resident in the hostile country) should join in the conflict.[175] According to this theory, Muslims in Britain and the United States in 2003 should have helped Saddam Hussein to defend 'Irāq.

Al-Sarakhsī accepted that an unjust *imām* might be in power. 'If the *ṣultān* rules justly, the subjects should give thanks to God and the *ṣultān* will be rewarded by God; but if the *ṣultān* rules unjustly, the subjects should show patience and the *ṣultān* should bear the responsibility against God.'[176] If the just *imām* was defeated, al-Sarakhsī's primary preoccupation was the unity of the Muslim community, not the leadership itself. While support for the *imām* was essential to prevent dangers of *fitnah*, in the end it was authority itself which needed support, not the *imām*.[177] (Ibn Taymīyah would later state that 'sixty years of an unjust *imām* were preferable to one day with no authority'.)[178] Al-Sarakhsī's idealization of the past thus served the needs of the community for a continuity of power structure, irrespective of the type of regime, or whether there had been a *coup d'état* against the previous ruler. In al-Sarakhsī's theory, coexistence with non-Muslim states occurred only when the forces of the Muslims were weak. In terms of foreign relations, he failed to distinguish between ends and means. Since the ends of the Muslim state were legitimate, the argument ran, the means adopted must also have been legitimate. In this sense, the unilateral breaking of a treaty at a moment convenient to the Muslim state could be justified, even though the implication was that the motivation in making the treaty in the first place had been insincere. In reality, most Muslim rulers did keep to their treaty obligations, in spite of al-Sarakhsī's argument that they were not obliged to do so.

Scholars such as Ibn Qudāma and al-Shaybānī also argued that only the *imām* could declare a *jihād*.[179] What happened if he failed to do so, when such a declaration was necessary? What happened if he declared as a *jihād* a war which it was illegitimate to describe in such terms, because of some worldly or other interest of the ruler? Al-Nasā'ī (d. 303/915) reported the *ḥadīth* of the Prophet that 'the greatest *jihād* is a just word to a tyrant ruler (*imām jā'ir*)'.[180] Al-Qurṭubī (d. 671/1273) argued that it was politically imperative for scholars (the '*ulamā*'), that is, the people who loose and bind, to ensure that only a righteous and knowledgeable person was nominated to the highest office. What if the leader was found to be immoral, unjust and oppressive? Radical scholars argued that revolution or fighting (*qitāl*) was only recommended if a coup could be mounted successfully against the unjust political regime, without causing unnecessary destruction to the people themselves. One of the most famous scholars of this school was the Ash'arite al-Juwaynī (d. 478/1085), who explained that morality and injustice, like insanity, were defects in the quality of leadership. For him, any ruler found guilty of these 'moral defects' must be removed from power. If the ruler acted in a manifestly unjust fashion, or did not respond to verbal admonition, then it was for 'the people of binding and loosing' (that is the '*ulamā'*) to prevent him, even if it resulted in doing battle with him. In a second work by the same author, this time on the imāmate, however, there was no mention of the issue of the unjust ruler.[181] The Persian jurist and philosopher al-Shahrastānī (479/1086–548/1153) had similarly radical views to those of al-Juwaynī. Al-Zamakhsharī (d. 538/1144), a Persian-born theologian of the Mu'tazilite school, criticized political quietism, describing alliances with the corrupt ruling class as counter-productive to the cause of justice and truth, a subordination to the forces of evil and tantamount to forming an alliance with those who spread tyranny (Q.10:113). What is significant about this school is that, although it represents the radical strand of the classical scholars on the theory of government and administration in Islām, its leading theoreticians were not inclined towards the use of the sword as the only practical means for political reform. The military option was the option of last resort only when diplomatic options for the peaceful transfer of power had been found to be unworkable; when there was a conviction that the revolutionary option would be a success; and finally, when there was a conviction that undertaking a coup would not result in the escalation of bloody civil strife in the Muslim community.

Perhaps the most militant exponent of this school was the Andalusian scholar Ibn Ḥazm (Abu Muḥammad 'Ali ibn Aḥmad ibn Sa'īd ibn Ḥazm, d. 456/1064). He criticized the unjust leadership in the Umayyad imāmate of Andalusia, and called for immediate change through whatever means necessary, either through political reform or armed struggle. He claimed that this view was a common one shared by leading Companions of the Prophet and by leading founders of the schools of jurisprudence such as Abū Ḥanīfa, Mālik bin Anas, al-Shāfi'ī and others. He argued that the traditions of the Prophet used by conformist or pacifist scholars to justify a quietist position were no longer applicable but had been abrogated by other traditions which called for a revolt (*al-khurūj*) against unjust leadership. The principle of 'absolute obedience to the ruler be he just or unjust, righteous or corrupt' was no longer relevant, because it was clearly contradictory to the Qur'ānic texts that enjoined what was right and prohibited what was evil (Q.3:104; Q.9:71; Q.39:41). Ibn Ḥazm considered the revolutionary struggle to oppose a corrupt and unjust leadership as the party of *jihād*. He denounced quietism (*al-sukūt*) as tantamount to cooperation in a sinful act (*al-ithm*) and aggression (*al-'udwān*). Such an attitude, according to Ibn Ḥazm, was absolutely forbidden in the Qur'ān. However, his 'message, though appropriate, was scarcely heard by posterity'.[182]

The dangers of a coup against an unworthy ruler leading to the partition of Islamic lands had become evident to later jurists. Abū Ḥanifa, though he did not deny that the duty might in principle make rebellion mandatory, sought to override such an alarming implication 'by invoking the likely costs of such action'.[183] The Ḥanafī Mu'tazilite al-Ḥākim al-Jishumī (d. 494/1101), alone among the classical scholars, linked forbidding wrong with rebellion against unjust rule and did so 'in a tone of marked enthusiasm'.[184] It was presumably this theorist that Ibn Taymīyah was thinking of when he stated that the Mu'tazilites regarded 'war on the leaders as one of their religious principles'.[185]

The *Ordinances of Government* (*al-Aḥkām al-Sulṭaniyya*) of the Shāf'ite Abū al-Ḥaṣan al-Māwardī (361/972–449/1058),[186] which were written to the command and for the use of the *Imām* al-Qādir Billah (r. 381/991–422/1031), were designed to argue that a duly elected *imām* cannot be displaced in favour of a worthier candidate, for there had been many historical examples of unworthy rulers but few depositions.[187] Al-Māwardī rejected the Ash'arī view, expounded by al-Baghdādi (sometimes known as Ibn Ṭāhir, *c.* 369/980–429/1037), that two *imāms* could coexist, albeit in widely separated lands (he stressed that their territories should be separated by sea – a condition which applied to the Umayyads of Spain).[188] Al-Māwardī's opposition to this concept reflected the refusal of the 'Abbāsids and their supporters to admit the claims of dangerous rivals, the Fāṭimids of Egypt and the Umayyads of al-Andalus.[189] However, he did consider the circumstances that might lead to the forfeiture of the imāmate, including evil conduct or heresy, infirmity of mind or body and, most significant of all, curtailment or loss of liberty. Within the last category, the case of an *imām* placed under restraint, 'control over him having been seized by one of his auxiliaries, who arrogates to himself the executive authority', described the situation of the 'Abbāsid rulers for the previous century or so during which time the Būyid or Buwayhid *amīrs* had usurped their power (they entered Baghdād unopposed in 334/945).[190] In a similar vein, but at a later date, Abū Ḥāmid al-Ghazālī (450/1058–505/1111) acknowledged that the Seljūq Turks, not the *imām*, held actual authority (*shawkah*). It was they who could be relied on by the caliph to wage *jihād* against the infidels.[191]

Al-Māwardī distinguished between an emirate (*imāra*) freely conferred, with defined territorial jurisdiction, and one seized by conquest or usurpation. An *amīr* appointed by the *imām* was not divested of office on the death of the theoretical overlord. If his government included a frontier area, he was entitled to undertake *jihād* (defined by al-Māwardī as one of the ten distinguishing powers of the *imām*) and to divide the booty among the combatants. Only an *amīr* appointed by the *imām*'s representative, the *wāzīr*, was not so entitled: he had to obtain prior authorization from the *imām* before undertaking *jihād*. Al-Māwardī defined seven conditions required of an *amīr* who had seized power; but by the very nature of the weakening of the emirate which had permitted the seizure to happen in the first place, there was no mechanism by which such conditions could be enforced.[192] Professor Gibb calls the arrangement a 'sort of concordat, the caliph recognizing the governor's sole control of policy and civil administration, in return for recognition of his own dignity and right of administration of religious affairs'; but, at least in the compact theory, a concordat implies an arrangement between two independent authorities, yet the imāmate had lost its independence.[193]

It is clear that, with the passing of time, the differences between the jurists on the issue of *jihād* and its relations to questions of political power widened rather than diminished. The Mālikī jurist al-Qarāfi (626/1228–684/1285), who worked in Ayyūbid–Mamlūk Egypt, produced his chief work around the year 660/1262.[194] In this, he asserted that disagreement was not confined to the jurists, but went to the heart of government: the overwhelming majority of the head of state's pronouncements constituted *fatāwā* and were open to challenge since the divine protection (*'iṣmah*) enjoyed by the Prophet did not extend to the *imām* or *sulṭān*:[195]

> Among their discretionary actions are their *fatāwā* concerning the rulings on such things as religious observances and the like... or the obligation to wage *jihād*, etc. None of their pronouncements regarding these matters constitute binding decisions. On the contrary, anyone who does not believe these statements to be correct may issue a *fatwā* in opposition to that of this judge or caliph. Likewise, if they command us to perform an act which they believe to be good, or they forbid us to perform one which they believe to be evil, it remains the right of anyone who disagrees with them not to follow them... other than [in circumstances where it is feared that] opposing the *Imām* will constitute an act of sedition...

The acceptance of juristic disagreement: Ibn Rushd

These differences between jurists were further highlighted by the Cordoban jurist Averroës (Abū al-Walīd Muḥammad Ibn Muḥammad Ibn Rushd, 520/1126–595/1198), another member of the Mālikī school, who wrote his principal legal handbook, *The Beginning for him who interprets the sources independently...* (*Bidāyat al-Mujtahid wa-Nihāyat al-Muqtaṣid*) around 564/1169 when he became a judge (*qāḍī*) in Seville. For Ibn Rushd, scholars of the different schools, basing their interpretation on Q.2:216, were agreed that *jihād* was a collective and not a personal obligation.[196] The obligation applied to adult free men who had the means at their disposal to go to war and who were healthy enough to do so. For young men, except in an emergency when there was no one else to carry out the duty, the obligation was conditional on prior permission having been granted by parents. Scholars, he contended, were in agreement that 'all polytheists should be fought' (Q.8:39), with the exception of the Ethiopians and the Turks, an exception which was based according to Mālik on a tradition of the Prophet. Non-combatant women and children were not to be slain. Ibn Rushd asserted that most scholars were of the opinion that the *imām* could pardon captives, enslave them, kill them, or release them either on ransom or as non-Muslim subjects of the Islamic state (*dhimmī*). There was controversy on the matter, he noted, because the Qur'ānic verses seem to contradict one another on the subject, the practice of the Prophet and the first caliphs was inconsistent, and the fact that the interpretation of the Qur'ān was at variance with the Prophet's deeds (Q.8:67; Q.47:4). There was agreement, however, that it was only permissible to slay the enemy if a safe-conduct (*amān*) had not been granted, though there was debate as to whether slaves and women could grant such a safe-conduct. However, the *amān* did not afford protection against enslavement. Whereas al-Shāfi'ī (d. 204/820) argued that hermits, the blind, the chronically ill, the insane, the old, peasants and serfs might be slain, Mālik (d. 179/795) sought to exempt these categories. Ibn Rushd explained the contradiction in that some of the traditions were at variance with the Qur'ānic injunction (Q.9:5) and that the 'verse of the sword' was itself at variance with Q.2:190. The source of the divergence, in his view, was the motive for killing the enemy:[197]

> Those who think that this is because they are unbeliev[ers] do not make any exceptions for any polytheist. Others, who are of the opinion that this motive consists in their capacity for fighting, in view of the prohibition to slay female unbelievers, do make an exception for those who are unable to fight or who are not as a rule inclined to fight, such as peasants and serfs.

In no circumstances should enemies be tortured or their bodies mutilated; there was disagreement on whether death through burning was acceptable because there was an authoritative tradition according to which the Prophet had declared 'do not burn him'. There was further disagreement on the destruction of buildings and the felling of trees, which Abū Bakr had prohibited.

Two issues, the nature of the truce and the aims of warfare, received particular attention from Ibn Rushd. The controversy about the conclusion of a truce arose from the contradiction between the Qur'ānic verses Q.9:5 and Q.9:29 on the one hand and Q.8:61 (the peace verse) on the other. There were conflicting views as to whether a truce should last for three years, four years or even ten years. Ibn Rushd concluded:

> those who considered that the verse of fighting abrogates the verse of peace [or truce] did not approve of peace except out of necessity. Those who considered that the verse of peace places limits on that verse [of fighting] [approved] of peace if the *imām* was in favour of it.[198]

On the question of the aims of warfare, the controversy between jurists arose from the fact that a general rule based on Q.2:139 and Q.8:39 conflicted with a particular rule given in Q.9:29. The general command to fight the polytheists was found in the *Sūrah Barā'a*, which was revealed in the year of the conquest of Mecca (8/630), while the tradition of the Prophet dated back to before the conquest of the holy city. Other scholars argued that general rules should always be interpreted by particular rules, and therefore that the poll tax (*jizya*) should be payable by any polytheist and not just from non-Arab 'People of the Book' (Jews and Christians; Zoroastrians were to be treated in a similar manner to them).

According to Ibn Rushd, the jurists were divided on the issue of whether it was permitted for the *Imām* to promise a reward to the troops before battle. The disagreement arose from the conflict between the purposes of war and the apparent meaning of the Prophet's tradition, that the troops should actively pursue the enemy. If the *Imām* offered a reward before battle there was 'apprehension that the warriors will spill their blood for a cause other than seeking Allāh's favour',[199] that their sacrifice might be vitiated by an apparently worldly motive. Were a convert's children, wife and wealth safe from an invading army in the *Dar al-Ḥārb* if he himself migrated? The jurists were divided, some saying that what he left behind had the protection of Islām, while others argued that there was no sanctity whatsoever for his property. Mālik argued for sanctity on the grounds that the Prophet had stated that when individuals pronounced the profession of faith in Allāh and his Messenger (the *shahādah*) 'their blood and wealth stand protected from me'.[200] Mālik, al-Shāf'ī and Abū Ḥanīfa were also divided on the question of how the land conquered by the Muslim army should be divided up, the reason for their disagreement arising from an apparent conflict between Q.8:41 and Q.59:10. 'Umar did not divide up the lands of 'Irāq and Egypt that were conquered in his time by force of arms. Ibn Rushd concluded that a reconciliation of the two verses resulted in the opinion that 'land acquired as part of the spoils should be kept intact, undivided, but division should apply to whatever is besides land'.[201]

It is evident from the preceding discussion that Ibn Rushd's text is a work which juxtaposes the controversies of the different legal schools (*ikhtilāf*). Had there been no preceding development of legal schools the work could not have taken this form. Many private schools had eventually been amalgamated into associations of schools along geographical lines; eventually these collapsed in the third/ninth century into numerous personal schools which eventually dwindled to four.[202] Ibn Rushd effectively stresses the role of independent reasoning (*ijtihād*) in resolving the divergences between the schools, particularly since the divergences often arose from apparent contradictions in the sacred texts of Islām.[203] Significantly, Ibn Rushd drew upon Aristotle's exposition of equity as a rectification of legal justice to explain the adjustment of a defective general law of *jihād*. The command, in the form of a general law, utterly to destroy the enemy had proved injurious to the interests of the Muslim community in view of the impossibility of fulfilling it. Therefore God had ruled that sometimes peace was preferable to war. It was thus the intention of the lawgiver to counteract the absolute obligation by commending peace and leaving the ultimate decision to those in authority.[204]

Others also questioned some of the basic premises underlying the doctrine of *jihād*, in particular the presumption in favour of its justice as 'just war'. Ibn al-Farakh al-Fārābī (259/870–339/950, known as al-Pharabius in Europe), was one of the companions of the ruler at the Ḥamdanid Amīr Sayf al-Dawla's court in Ḥalab (Allepo).[205] He was also one of the first to urge a rationale for waging war on grounds of justice, without exclusive reference to the duty of *jihād* (though this duty was not denied).[206] Unjust wars were judged by al-Fārābī to be wars motivated by the ruler's personal advantage such as lust for power, honour or glory; wars of conquest waged by the ruler for the subordination of peoples other than those he ruled before the declaration of war; wars of retribution, the object of which could be achieved by means other than force; and wars leading to the killing of the innocent for no other reason than the ruler's propensity or pleasure for killing. Only the *imām*, who was conceived of as both a philosopher and lawgiver,[207] had the legitimate authority to proclaim a just war. Even if declared by legitimate authority, just wars were restricted to certain types of action: wars in defence of the city/state against foreign attack; wars to assert valid claims against a foreign people who failed to honour these rights; wars against a foreign people who refused to accept a public order considered by the state declaring the war to be suitable for them; and finally wars against a foreign people whose most suitable place in the world was that of slavery.[208] These conditions are far from those which would be considered grounds for a 'just war' in modern times, but they did serve to rule out any future wars of conquest such as the Arab conquest of Egypt and north Africa, Syria and 'Irāq in the first Muslim century. *Jihād* had not been laid to rest, since later writers in the 'just war' tradition such as Ibn Khaldūn (732/1382–808/1395) conceded that *jihād* was in itself just war; though he acknowledged that most rulers who embarked on war did so for non-religious reasons such as lust for power and personal ambition.[209]

Other religionists as second-class citizens: dhimmitude and the payment of *jizya*

Yohanan Friedmann argues that

> as long as the only idolaters encountered by the Muslims were inhabitants of the Arabian peninsula, the Muslims fought against them without compromise. This was caused not only by the Qur'ānic attitude to idolatry, but also by the ardent desire of early Islām to achieve religious uniformity in the peninsula.

However, once Islām had become the sole religion in most of the peninsula, and the newly-converted Muslim Arabs triumphantly emerged from their historical habitat, 'the religious considerations that demanded unflinching struggle against idolatry and other non-Muslim religions were replaced by the requirements of running a state and building an empire'.[210]

The discussion of the early Muslim conquests and the subsequent theory of *jihād* would be incomplete without a consideration of the payment of *jizya* by non-Muslims, their second-class status as *dhimmīs* and the modern controversy over the significance of *dhimma* or 'dhimmitude'.[211] There are two related issues concerning 'dhimmitude'. The first is the amount of tax payable by non-Muslims and whether they were exploited fiscally because of their refusal to convert. The second concerns the nature of their inferior status within the Muslim polity. Objectionable today under modern conceptions of the equality of human rights, was the status regarded as quite so degrading in the medieval and early modern periods?

Firstly, with regard to the payment of the *jizya*, Ibn Rushd reports a disagreement between the jurists over the annual amount of *jizya* that was due by a *dhimmī*. Mālik contended that the amount due was that imposed by the second caliph, 'Umar, which was four *dīnārs* (48 *dirhams*) for those whose transactions were in gold and 40 *dirhams* for those whose transactions were in silver, along with a requirement to host Muslims for three days. Al-Shāfi'ī stated that the minimum was fixed at 1 *dīnār*, but that the maximum was not fixed and depended upon negotiation. Another group of jurists, including al-Thawrī, contended that nothing at all was fixed and that all was left to the independent reasoning (*ijtihād*) of the *imām*. Abū Ḥanīfa and his disciples, including Abū Yūsuf and Shaybānī, argued that the rate of the tax varied between categories of non-Muslim taxpayer, and was payable at rates of 4, 2 and 1 *dīnār*, according to the presumed capacity of the taxpayer to pay.[212]

The precise levy, which did not 'become due except after the passage of one year', varied according to period and also location within the Muslim empire, but for the early period it has been argued that the levy was modest, 1 *dīnār* being equivalent to about a fortnight's pay for a day labourer.[213] (For another comparison, Shaybānī took it for granted that a fit and healthy female slave could be bought for 1000 *dirhams*.)[214] However, the real burden of the tax would have depended on a number of key factors such as the success or otherwise of the harvest, the extent of monetization in the area, or the particular inflationary circumstances of a siege.[215] If the harvest was poor, or there was a lack of currency, then clearly (as with any other tax in the medieval and early modern period), the levy could become oppressive, especially if the administration was corrupt or refused tax remissions in cases of need.[216]

Significantly, when Aurangzīb reimposed the *jizya* in Mughal India in 1089/1679 (it had been abolished by Akbar in 971/1564),[217] he chose rates of levy which were exactly the same as those recorded by Abū Ḥanīfa and his disciples, including Abū Yūsuf. Without any upward revision of the value of the coinage,[218] the tax on non-Muslims would have become progressively lighter over time, which may account for Montesquieu's comment that 'instead of being subjected to an endless series of fines which entered the rich imagination of greedy rulers', non-Muslims preferred 'to submit to the payment of a minimal tax which can be fulfilled and paid with ease'.[219] (Montesquieu's was a theory of Islamic conquest which explained the transfer of allegiance on the grounds of fiscal oppression by previous rulers. While insufficient in itself, this argument may nevertheless contain an element of truth, particularly in relation to Muslim acquisitions in the Balkans.)

Comparison between the burden of the *jizya* on non-Muslims and other taxes levied on Muslims is problematic.[220] One calculation, for the early period of 'Abbāsid rule, suggests wide regional variations within the empire, but an overall tax rate per Muslim inhabitant of at least 17 *dirhams*, with the probability that the rate was equivalent to at least 20 *dirhams* and may have been as high as 30 *dirhams* when levies paid in kind were included.[221] With regard to the non-Muslim population, we do not know how many taxpayers were included in each category of the levy (4, 2 and 1 *dīnār* or 48, 24 and 12 *dirhams* respectively), though we may assume that most taxpayers would have fallen into the lowest of the three rates. It seems reasonable to conclude that it was the number of non-Muslim taxpayers which made the *jizya* a significant source of income for the ruler, not the oppressive nature of the levy.[222] Jewish merchants paid 10 per cent of their turnover in *jizya*.[223] Since imprisonment for non-payment was the only penalty against non-Muslims that was allowed,[224] it may be argued that the *jizya* was less oppressive than the arbitrary Christian levies on Jews in later periods. However, the exemptions may not have been honoured and there may have been other abuses in collection. Like the Christian levies on Jews, compositions might be payable which would have become heavier if the rate remained fixed but a number of the non-Muslims converted over time.[225] Moreover, other taxes, such as *kharāj*, were payable when non-Muslims cultivated the land.

What is clear is that these taxes were levied on non-Muslims in return for protection and because they were not required to participate in *jihād* (the equivalent to non-combatants, for example, women, the old, the young, the sick and priests and monks, who were exempt from the *jizya*). Abū Yūsuf reported the following incident concerning the defence of non-Muslims in Syria (a similar story was told by al-Balādhurī).[226] News of an impending attack prompted Abū 'Ubayda to instruct officials to repay the *jizya* and *kharāj* to non-Muslims: 'we hereby return to you the money you have paid us, because of the news of the enemy troops amassed to attack us; but, if God grants us victory against the enemy, we will keep to the promise and covenant between us.' Abū Yūsuf reports that, on receiving back their returned tax payments, the *dhimmīs* allegedly told the Muslims: 'May God bring you back to us and grant you victory over them!'[227]

Sūrat at-Tawba ('on repentance') enjoins fighting against non-Muslims until they agree to 'pay the exemption tax (*jizya*) with a willing hand, after having been humbled [in war]' (Q.9:29). A number of degrading social and cultural requirements were imposed on the *dhimmīs*, the most notable of which was wearing round the neck the receipt for payment (*barā'a*) of the *jizya*.[228] The tax had to be paid in person, not through an intermediary, in a standing posture while the tax collector sat. (One tradition stated that the tax collector had the right to seize the individual by the throat and demand payment with the words 'Pay your tax, *dhimmī*!') It was standard practice for the *dhimmī* to receive a blow and be pushed aside after making payment, 'so that he will think that he has escaped the sword through this [insult]'.[229]

Clearly such stipulations, in Majīd Khaddurī's phrase 'hardly left a respectable position for the *dhimmīs*'. What they did have, however, was self-rule under their own religious head who was, in turn, responsible to the Muslim authorities. This was the so-called *millet* system, 'the result of the extension of the idea of extraterritoriality to religious groups'.[230] While the Ottomans used the uniform term '*dhimmī*' to refer to all non-Muslims, they recognized some differentiation within this broad category which was formally expressed by the creation of the *millets*. Meḥmed II, the Conqueror, selected a respected scholar and anti-Catholic George Scholarios (who later took the name Gennadios) in 858/1454, to be the head (*millet basi*) of the Orthodox millet (*Rum millet*). Subsequently, other communities such as the Armenian, Jewish, Serbian and Bulgarian *millets* were added.[231] The most important organizational form in the Ottoman Empire, therefore, was the *millet* system. In order to facilitate the control of the Ottomans' vast non-Muslim population, it was necessary to use some elements of the pre-existing infrastructure to reduce costs and facilitate relations. In most cases, the only institutions to survive were religious: religious institutions were exceptionally well suited for indirect rule because they possessed a centralized system which theoretically reached down to the local level.[232]

Thus, while the *dhimmī*'s rights were respected within his own community, he suffered clear disabilities within society at large and was reduced to the status of a second-class citizen. He was under a legal disability with regard to giving testimony, and under the criminal law, marriage law and inheritance law.[233] If 'dhimmitude' was so unattractive, why was it that, once other options (such as Portugal and southern Italy) had been closed off, the majority of the 150,000–300,000 Jews expelled from Castile in 897/1492 went to the Islamic lands?[234] And if there was no prospect for economic advancement, why was it that the *jizya* became payable by Jews in cloth in the Ottoman lands? Finally, does not the Ottoman record in the early modern period stand up to scrutiny when we bear in mind that those 300,000 or so Muslims who chose the option of conversion and assimilation after 897/1492 (the Moriscos) were eventually expelled from the kingdom of Valencia in 1017/1609?

Selective memory rather than historical reality: Crusades and Saladin's 'counter-crusade'

The expulsion of the Christians from the Holy Land in 689/1291 was far from marking the end of the Crusading movement. Crusades continued for three more centuries over a vast area stretching from Morocco to Russia and played an important role in the politics and society of late medieval Europe. The last Crusade is usually taken to be the failed attack by the Portuguese king Sebastian on the kingdom of Morocco which met disaster at Alcazar in 985/1578.[235]

Since the era of the Crusades is perceived by some historians as the first great phase of warfare between the Islamic world and the West, it merits particular consideration. Resistance to the Frankish incursion into Palestine (the so-called First Crusade) was sufficiently weak that by 492/1099 the Crusaders had captured Jerusalem. Little help was forthcoming from Syria, Egypt or 'Irāq; on the contrary, two years earlier the Fāṭimids of Egypt had offered a treaty to the Franks which in effect would have partitioned Syria. The Franks noted that the Ismā'īlī Fāṭimids were more friendly towards them than the orthodox Sunnīs. For a time, an alliance between the Franks and Damascus held firm; but this was broken by the Crusaders before the arrival of the Second Crusade (542/1148). The first stirrings of a '*jihād* of the sword', a counter-crusade, were not evident before the campaign of 'Imād al-Din Zangī (476/1084–541/1146),[236] who captured the north Syrian fortress of Edessa (al-Ruha') in 539/1144.

Zangī's son Nūr al-Dīn, who ruled for almost 30 years (541/1146–569/1174), was defeated in 558/1163 in a campaign near Krak des Chevaliers, which was ascribed by his religious critics to the presence of music and liquor in his camp. After this setback, a more puritanical drive towards a '*jihād* of the sword' characterized Nūr al-Dīn's policy. For him, Egypt had to be wrested from the Fāṭimids because this would mean 'an immediate and substantial accretion of military and financial resources for the war in Syria'.[237] Three invasions of the Syrian army under the Kurdish general Shīrkūh, acting in the service of Nūr al-Dīn, resulted on the third occasion in his becoming *wazīr* to the Fāṭimid caliph. This was short-lived, since Shīrkūh died two months later and was succeeded by his nephew, Ṣalāḥ al-Dīn al-Ayyūbī (Saladin), who on the death of the last Fāṭimid caliph in 567/1171 had the *khuṭbah* recited in the name of the 'Abbāsid caliph. This return to Sunnī orthodoxy arose from Saladin's own convictions, though it may have been prompted by Nūr al-Dīn's orders (Saladin later stated that 'we have come to unite the word of Islām and to restore things to order by removing differences').[238] However, the rift between the two men was immediately evident. Saladin's concerns at first were to build up a dependable army in Egypt: the Fāṭimid army was disbanded or massacred in 564/1169; new fiefs (*iqṭā's*) were granted out to his Turco-Kurdish forces.[239] Once control of the army had been secured, Egypt had to be defended against pro-Fāṭimid attacks from within and Crusader attacks from outside.[240]

After the death of Nūr al-Dīn, Saladin spent some dozen years building up his power against his Muslim adversaries. As late as 577/1181, the reformed Egyptian cavalry comprised only 8640 men, of whom 111 were *amīrs*.[241] Throughout his reign, relations with the free-born *amīrs* remained problematic, and there were serious crises with them in 587/1191 and 588/1192. The territorial expansion of the state promised them material rewards (Syrian fiefs were assigned to his ablest, oldest and most ambitious supporters);[242] but these rewards could be quickly dissolved with military defeat. Above all, the Ayyūbī state risked leaving no successor state except to Saladin's relatives (an Islamic statement recognizing the hereditary principle argued that 'kings nurture the growth of their kingdoms for their children');[243] but these relatives wanted to share a collective, patrimonial, sovereignty with the head of the family. The political difficulties after Saladin's death, and the three civil wars affecting the Ayyūbī state, were consequences that were inherent in a state built upon family confederation: while Egypt remained a unified realm, his Syrian lands 'broke up after his death into a mosaic of small principalities ruled by his sons, nephews and cousins'.[244] The quasi-empire built up by Saladin lasted less than 70 years after his death.[245]

Saladin's doctrine of *jihād* made the Syrian *amīrs* 'the very kernel of the state'.[246] Damascus had to be the centre of Saladin's state, because Egypt was too distant; but Syria lacked the military and financial resources of the Egyptian kingdom.[247] There could be no question of residence in a comfortable palace, since this might compromise the permanent commitment to *jihād*, whose abandonment was a 'sin for which no excuse can be brought to God'.[248] Peter Partner suggests that Saladin's appeal to the Almohad ruler Ya'qūb al-Mansūr for help in the *jihād* in Palestine shows that 'after a very long period in which the central area of the Islamic world had in effect left [*jihād*] to be the concern of the frontier *ghāzi* fighters on the periphery' this idea had been restored to mainstream

Islām.[249] Yet, aside from the fact that the gesture received no response, Saladin's *jihād* propaganda is viewed by Lyons and Jackson as an attempt, 'conscious or unconscious, to canalize energy and direct it outwards', an attempt which failed. The *jihād* propaganda and the continuous self-justification in his letters to Baghdād amounted to 'coloured rhetoric in which everything is shown in extremes and internal contradictions are glossed over or ignored'. Saladin's *jihād* doctrine did not provide 'an immediate, practical and coherent policy', for which there was no substitute.[250] Saladin seems to have come close to taking his own *jihād* propaganda at face value: 'Imād al-Dīn al-Iṣfahānī claimed that Saladin could not stop himself from reading the volume on *jihād* he had written for him.[251]

Though Saladin professed his loyalty to the *imām*, and his determination to 'complete the conquests of the commander of the Faithful', his assumption of one of the caliph's titles as 'the victor' (*al-Nāṣir*) was treated with fury at Baghdād, while his objectives were regarded with trepidation.[252] For these ambitions were capable of almost infinite extension; territories had to be conceded to him in the interests of Islām, while any sharing of power was rejected as a 'weakening of unity'.[253] Towards the end of his life, Saladin admitted the distrust of other Muslim rulers which had delayed the commencement of his *jihād* by some eight and a half years until the capture of Aleppo and had been a preoccupation for twelve years of warfare: the Muslim rulers sat 'at the top of their towers', and would refuse to come down to join the struggle until the Muslim cause was lost.[254]

Saladin's fears were correctly founded, for after his death in 589/1193 *jihād* propaganda and support for the 'counter-crusade' evaporated almost overnight.[255] Three issues seem to have loomed large over Saladin's successors in the Ayyūbī state. The first was a consequence of the succession problems and repeated civil wars resulting from the system of collective, patrimonial, sovereignty. It was not until 647/1249, just a few months before the demise of the regime, that the title '*al sulṭān*' was adopted on the Egyptian coinage.[256] This was partly because the Egyptian rulers had not asked for the authorization of the *imām* at Baghdād. Yet the oversight is explicable because, in the collective sovereignty of the Ayyūbī state, 'many members of the dynasty simultaneously had the right to claim the title' since 'they all shared to some extent the right to rule in their own names'.[257] This meant that only the emergence of a charismatic figure such as a second Saladin, who could subsume the competing interests within the dynasty under a greater cause, was likely to lead to decisive action.

A second issue was that even relative success in the '*jihād* of the sword' carried very heavy costs. The Egyptian treasury was said to have been emptied; more than the income from the land had been mortgaged and the wealth of Muslims had been dissipated. Salaries existed in name rather than reality.[258] In a much repeated expression, Saladin was said to have 'spent the wealth of Egypt to gain Syria, the revenues of Syria to gain Mesopotamia [and] those of Mesopotamia to conquer Palestine'.[259] Six years of almost unbroken combat prior to the 588/1192 truce[260] had brought Saladin's empire almost to its knees. And all this was after the massive success of Ḥaṭṭīn in 582/1187, in which the Frankish land army had been destroyed, and the resulting capture of Jerusalem. The costs of military defeat would have been as great. The investment simply carried too great a risk.

A third reason for Saladin's successors following a non-ideological policy was that, if successful, *jihād* risked provoking a military reaction. Saladin was unable to expel the Crusaders, who remained in the Levant for another century. Worse, he found himself in a weakened position to fight off the Third Crusade which his earlier victories had provoked. Military and political setbacks such as the loss of Acre in 587/1191 tested the loyalty of the *amīrs*, on whose support the Ayyūbī state depended.[261] Peace treaties with the Franks might be controversial, and denounced in 'pietist circles deeply imbued with the duty and sanctity of *jihād*',[262] particularly those of 638/1240 and 650/1252 which were directed against another Muslim state. But the continuation of truces was uncontroversial, since this had been a practice of Saladin himself. Moreover, his successors might well fear the formation of a coalition of Muslim rulers against them if they were successful: Saladin himself had been threatened with a league of 'all the kings of the east' had he pursued his campaigns against Mārdīn and Mosul.[263] The key point was that territorial expansion tended to be at the expense of neighbouring Muslim dynasties. This had been the lesson of the rise of the Ayyūbī state; it was also the story of its demise and displacement by the Mamlūks in Egypt in 648/1250.[264]

The history of internecine quarrels and the willingness of the Ayyūbī state to negotiate with the Mongols meant that the seriousness of Hülagü's slow advance from Karakorum in 651/1254 was underrated. There had been time to prepare and negotiate a coalition against the invaders, but the opportunity was wasted because of the tradition of rivalry between the rulers of the Muslim territories. There was no new Saladin to repeat his call for unity; indeed, there was no expectation that the Mongols intended more than another of their short-lived raids into the Islamic lands. Too late, Imām al-Musta'ṣim at Baghdād appealed for support from al-Nāṣir Yūsuf in Syria by sending him the robes and diploma of investiture as *al-sulṭān*. In return he sought tangible support against what had emerged as the most serious threat to the caliphate in its five centuries of existence.[265] It was to no avail. Baghdād surrendered to Hülagü, but this did not prevent the massacre of its population, the razing of its monuments and murder of the last caliph and his family.

These traumatic events produced differing responses among the jurists. In early Mamlūk Egypt, the Shāfi'ī jurist Badr al-Dīn Ibn Jamā'a (638/1241–733/1333) capitulated to the status quo and declared military power pure and simple as the essence of rulership: the *imām* must engage in *jihād* at least once a year, while the *sulṭān* must defend the area delegated to him and undertake *jihād* personally.[266] A quite different, quasi-constitutionalist, thesis was propounded by the Mālikī jurist al-Qarāfi (626/1228–684/1285), whose views we have already encountered.

A third, and more enduring, juristic treatment, was propounded by Ibn Taymīyah (661/1268–728/1328), who concentrated on the problem of the defence of the Islamic lands and the obligation of collective defence. We will return to his views in Chapter 4, but first we must turn in Chapter 3 to alternative, more pacific, spiritual conceptions of *jihād* which may be distinguished from the '*jihād* of the sword' which had emerged in the classical period.

3
Jihād al-Nafs: The Spiritual Struggle

> The Prophet, peace be upon him, said: 'Shall I tell you something that is the best of all deeds, constitutes the best act of piety in the eyes of your Lord, elevates your rank in the hereafter, and carries more virtue than the spending of gold and silver in the service of Allāh, or taking part in *jihād* and slaying or being slain in the path of Allāh?' They said: 'Yes!' He said: 'Remembrance of Allāh (*dhikr*).'
> *Jihād* – A Misunderstood Concept from Islām. A Judicial Ruling [*fatwā*] issued by Shaykh Hisham Kabbani, Chairman, Islamic Supreme Council of America and Shaykh Seraj Hendricks, Muftī, Cape Town, South Africa[1]

'What have the Arabs ever done for us?', the British columnist Robert Kilroy-Silk asked in January 2004 in a misjudged article which resulted in a national furore.[2] One answer may have been missed in the plethora of responses: a spiritual path. Many considerable specialists of the Muslim and Arab world in the classical period have had difficulty in accepting that it was capable of 'real' spirituality. Instead, it may be suggested that out of the nucleus of pious people around the Prophet there emerged a threefold relationship between Islām, *īmān* and *ihṣān*. Islām is the complete surrender of the faithful to God's will. *Īmān*, faith, constitutes the interior aspect of Islām. As for *ihṣān*, to do well or serve God constantly, to strive hard in God's cause (itself a form of *jihād*), the Qur'ān itself asserts that mercy is 'with those who practise' it (Q.29:69).[3] Drawing upon the traditions recorded by al-Bukhārī and Muslim, it may be contended that being a good Muslim is to practise *ihṣān*, which means worshipping God as if you see Him, in full awareness that even if you cannot see God, He oversees you all the time.[4] The early Ṣūfīs were careful to record the chain of narrators in the best traditions of the science of *ḥadīth*. By so doing, they attempted to prove that the early sacred traditions of Islām 'demonstrate both the importance and the transmission of the Prophetic spiritual example'. What came to be known as

Ṣūfīsm had 'deep roots in early Muslim spirituality and the prophetic revelatory event itself'.[5] It was an 'endogenous [movement], a spontaneous development from within Islam's own rich fund of spirituality… based… on the legacy of the sacred scripture [and] the divinely revealed Law…'[6]

The nature of early Ṣūfīsm

The term '*ṣūfī*' was first used to describe Muslim ascetics clothed in coarse garments of wool (*ṣūf*). From this arises the word '*taṣawwuf*' meaning mysticism.[7] Of Ibrāhīm ibn Adham it was said that, as a king's son, he was out hunting one day. A voice from the unseen called: 'O Ibrāhīm! Is it for this that you were created? Is it to this that you were commanded!' On hearing the voice again, Ibrāhīm dismounted, met one of his father's shepherds, took the man's woollen garment, put it on, and gave him in exchange his horse and all he had with him. Then he went into the desert.[8] The Ṣūfīs, it has been said,

> represent a domain of piety to which neither religious law nor religious politics are central… The Ṣūfī persuasion can take any form from a scrupulously observant asceticism to a wild antinomian mysticism, from an abject political quietism to a ferocious political activisim…[9]

Primarily it is a path or way (*ṭarīqah*) along which mystics walk, a path which emerges from the *sharī'ah*. A tripartite way to God is explained in a tradition attributed to the Prophet: 'the *sharī'ah* are my words, the *ṭarīqah* are my actions, and the reality (*ḥaqīqah*) is my interior state…'[10] New orders and fraternities were called 'the Muḥammadan path' (*ṭarīqah Muḥammadiyyah*).[11] To proceed on the Path, one begins with repentance and renunciation and the rest is a constant struggle against the flesh, the baser instincts or the lower self (*nafs*). This is the 'greater *jihād*', for 'the worst enemy you have is [the *nafs*] between your sides'.[12] For al-Ghazālī,

> religion consists of two parts, the leaving undone what is forbidden and the performance of duties. Of these the setting aside of what is forbidden is weightier, for the duties or acts of obedience… are within the power of every one, but only the upright are able to set aside the appetites. For that reason Muḥammad… said: 'the true… *Hijrah* [emigration] is the flight from evil, and the real… *Jihād* is the warfare against one's passions.'

The statement is from al-Ghazālī's *The Beginning of Guidance* (*Bidāyat al-Hidāyah*).[13] Abū Ḥāmid al-Ghazālī (450/1058–505/1111) has sometimes been acclaimed as 'the greatest Muslim after Muḥammad' because he was the leader in Islām's encounter with Greek philosophy, from which Islamic theology emerged enriched, and because he brought orthodoxy and mysticism into closer contact; as a result of this closer contact, 'the theologians became more ready to accept the mystics as respectable, while the mystics were more careful to remain within the bounds of orthodoxy'.[14] In this respect, al-Ghazālī confirmed the work of earlier writers such as Abū Ṭālib al-Makkī (d. 386/998)[15] and Abū'l-Qāsim 'Abd al-Karīm bin Hawāzin al-Qushayrī (376/986–465/1072) 'who had already done much to make moderate Ṣūfīsm respectable for orthodox Sunnites'.[16]

For Abū Bakr al-Kalābādhī (d. 385/995), the science of the Ṣūfīs were 'the sciences of the spiritual states'; every station had its own science, and every state its own 'allusion' (or mystical hints, *ishārāt*, 'the science *par excellence* of the Ṣūfīs').[17] Contemplations enjoyed by the heart and revelations accorded to the conscience cannot be expressed literally. Instead, they are 'learnt through actual experience of the mystical, and are only known to those who have experienced these mystical states and lived in these stations'.[18] Al-Kalābādhī insisted that the Ṣūfīs were orthodox in every respect, including their commitment to '*jihād* of the sword' and to *Ḥajj*. They held that the caliphate was true, and resided in the house of Quraysh. They were in agreement on the precedence of the four 'rightly-guided' caliphs, Abū Bakr, 'Umar, 'Uthmān and 'Alī. They held that it was not right to 'take the sword against governors, even though they commit wrong'. (Abū Ḥafṣ al-Ḥaddād [d. *c*. 260/874 or 270/883] stated that 'rebellion is the messenger of unbelief, as fever is the messenger of death'.)[19] It was the duty of all 'so far as they are able, to do good, and to refrain from doing evil, with kindness, mercy, considerateness, compassion, goodness and gentleness of speech'.[20] It all seems as though Abū Bakr al-Kalābādhī was trying a little too hard to convince. The execution of Ḥusayn ibn Manṣūr al-Ḥallāj in 309/922 ('the martyr *par excellence* of Islām')[21] an event which must have occurred in his childhood, and the threat of outlawing Ṣūfīsm,[22] were too recent for a distinctive Ṣūfī view of *jihād* to emerge. It would seem also that al-Kalābādhī was reflecting the realities of the early period of Islām, in which Ṣūfīs had taken part in, and preached in favour of, '*jihād* of the sword'. We can cite the examples of Ibrāhīm ibn Adham (d. *c*. 160/777), Ibn Sa'īd al-Tawrī (d. 161/778) and Abdullāh ibn Mubārak (d. 180/797). But it is important to stress the early date of these cases.[23] Of these figures, Ibrāhīm ibn Adham ('the key of mystical sciences') was particularly influential among the Ṣūfīs and his conversion story was well known.[24] He it was who described the Path as the closing of the door of ease and the opening of the door of hardship and enumerated five other doors to be opened and closed on the way, a viewpoint cited later by al-Qushayrī.[25] Ibn Sa'īd al-Tawrī figures in al-Kalābādhī's list of 'famous men among the Ṣūfīs' (as does ibn Adham).[26] Abdullāh ibn Mubārak was not listed by al-Kalābādhī, but is known to have been a leading Khurāsān ascetic.[27]

Al-Qushayrī's *Treatise* and the genre of the Ṣūfī textbook

Al-Qushayrī's *Treatise* (*Risālah*) was written 'to the Ṣūfī community in the domains of Islām' in 437/1046 in an attempt to 'adapt Ṣūfīsm to Ash'arī metaphysics'.[28] It became 'the most widely disseminated handbook of Ṣūfīsm in the Islamic world'.[29] Al-Qushayrī took it as axiomatic that the beliefs of the Ṣūfī *shaykhs* were 'in agreement with Sunnī teaching on questions of the fundamentals of faith'.[30] Part one of his work, 'On the *Shaykhs* of This Way: How Their Lives and Teachings Show Their Regard for the Divine Law', enumerated 83 Ṣūfī saints who had 'guarded and helped Islām with proofs of religion'. Part two is an explanation of 28 expressions in use among the Ṣūfīs 'with a clarification of what is obscure in them'. Let us single out three for closer examination. Number 13 is entitled 'erasure of self (*maḥw*) and affirmation of true being (*ithbāt*)'. The affirmation of true being, al-Qushayrī contends, 'is the establishment of the principles of the life of service'. This comprises the expulsion of blameworthy qualities and introduces praiseworthy actions and states that replace them.[31] Number 25 is witness (*shahīd*), which derives from the term 'testimony' (*shahādah*). The author notes that Ṣūfī discussion frequently mentions the witness of knowledge, the witness of ecstasy, the witness of mystical state. In such terminology, 'the witness' is used to describe 'the thing which inhabits the heart of a human being… anything whose remembrance takes possession of a person's heart is his witness'.[32] As for number 26, the ego or soul (*nafs*), Ṣūfīs only mean by this 'those qualities of the servant that are diseased, and whatever there is in his character and actions that is blameable'.[33]

Part three of al-Qushayrī's *Risālah* describes 40 stations and states, the penultimate of which is Ṣūfīsm and the last of which is model behaviour (*adab*), the conduct and discipline of the Ṣūfī in relation to his *shaykh* and associate Ṣūfīs. The first of these states is repentance (*tawbah*), 'the first station for spiritual travellers and the first stage of development in seekers'.[34] The second is 'striving' (*mujāhadah*), and it is here that Ṣūfī spirituality makes its distinctive contribution to a non-belligerent understanding of *jihād*.[35] In the Qur'an (Q.29:69), God states that, 'those who strive (*jāhadū*) for us, we will certainly guide in Our ways; God is with the doers of good'.[36] The authority of Abū Sa'īd al Khudrī was cited. He asked the Prophet, 'which is the best *jihād*?' The Prophet replied, 'To speak the word of justice in the presence of a tyrant authority.' Tears came to Abū Sa'īd's eyes when he recounted the story.[37]

Two further quotations are particularly pertinent for al-Qushayrī's purpose. The first of these is a saying from Abū 'Alī al-Daqqāq to the effect that 'whoever adorns his appearance with *mujāhadah* [striving], God will harness his inner self with *mushāhadah* [the vision of God]', which rested on Q.29:69. And he added: 'know that whoever does not strive [exercise *mujāhadah*] from the beginning will never find the slightest trace of this Way.' A second statement was that of Abū

'Uthmān al-Maghribī, that 'he who thinks that he could know the secrets of the [Ṣūfī] path (*ṭarīqah*) without *mujāhadah*' is mistaken. From these two quotations, it is clear that the spiritual gain was considered immense, but the secrets of the path were not easily learned and required immense commitment. From other quotations given by al-Qushayrī, we can gather that the path could be very long indeed. Abū Yazīd al-Bisṭāmī, the tenth of the *shaykhs* of the way, was quoted to the effect that he endured 22 years as the 'blacksmith' of his ego and 30 years in the struggle for knowledge and the capacity for prayer![38]

There were three perceived Ṣūfī characteristics: fasting (eat only when you are starving), watchfulness (sleep only when sleep overtakes you) and silence (do not speak unless it is necessary). Six difficult things had to be accomplished, or six mountains scaled, before righteousness/sainthood could be achieved. Ibrāhim ibn Adham was cited as the authority for the perception of this struggle as the closing and opening of doors. The doors to be closed were those of ease, honour, comfort, sleep, prosperity and hope or imagining the future. The corresponding doors to be opened were, respectively, those of difficulty or hardship, shame or humiliation, struggle or effort, wakefulness, poverty and readiness for death.[39] Citing the maxim 'anyone whose ego has been honoured has had his religion debased', al-Qushayrī provides the rationale for the struggle (*mujāhadah*) as to[40]

> wean the ego from what is familiar to it and to induce it to oppose its desires [passions] at all times. The ego [animal soul] has two traits that prevent it from good: total preoccupation with cravings [attraction to pleasure] and refusal of obedience [avoidance of pain/harm]. When the ego is defiant in the pursuit of desire, it must be curbed with the reins of awe of God…

Al-Naṣrabādī was cited as the authority for the statement 'your ego is your prison. When you have escaped from it, you will find yourself in eternal ease.' Abū Ḥafṣ talked of the secret of the lamp:

> the self is entirely darkness. Its lamp is its secret. The light of its lamp is inner direction from God. The result of success is prayer. Whoever is not accompanied in his secret self by such direction from his Lord is in total darkness.[41]

Al-Qushayrī proceeds to discuss further stations and states such as consciousness of God. Here he cites once again Abū Sa'īd al Khudrī, this time his account of a man approaching the Prophet, who advised him to be wary or conscious of God (Q.3:102). The Prophet added, 'take upon yourself war for God's sake, for it is the monasticism of a Muslim. Take upon yourself the remembrance of God, for it is a light for you.'[42] Fasting was considered 'one of the characteristics of the Ṣūfīs', 'the first pillar of the spiritual struggle'.[43] Trust in God was another of the stations and states. Abū 'Alī al-Daqqāq was once more cited, this time to the effect that

> Trust is the attribute of believers, surrender the attribute of the friends of God and self-abandonment the attribute of those who know unity. Thus trust is the attribute of the majority, surrender is that of the elite, and self-abandonment is that of the elite of the elite… Trust is the attribute of the prophets, surrender the attribute of the Prophet Abraham and self-abandonment the attribute of our Prophet Muḥammad.

In the station or state of contentment, al-Qushayrī quoted the very important *ḥadīth*: 'wish for others that which you wish for yourself, and you will be a believer. Treat your neighbours well and you will be a Muslim.'[44] On the station or state of will power, he makes it clear that God's treatment of those who aspire to Him is mostly concentrated on the preparation for struggle. The disciple or aspirant (*murīd*) is perceived as a labourer; while the shaykh (*murād*) is 'soothed and gently treated'.[45] One of the key prerequisites was steadfastness. Here the Prophet was quoted speaking in a dream to one of the followers that *sūrah* 11 (*Hūd*) of the Qur'ān had turned his hair white. Which part of it affected you in this way?, he was asked. The answer was verse Q.11:112: 'continue steadfast as you have been ordered'.[46] Another pre-requisite was truthfulness. Quoting Q.4:69 ('those whom God has blessed: the prophets and the truthful'), al-Qushayrī considered this 'the supporting pillar of Ṣūfīsm. In truthfulness this Way finds its perfection and balance. It is a degree next to prophethood'.[47] A further prerequisite was spiritual chivalry (*futuuwah*).[48] Al-Naṣrabādī was cited as the authority for the view that the companions of the Cave (Q.18:13) were called spiritual warriors because they placed their faith in their Lord without intermediary.[49] In Ṣūfīsm spiritual chivalry is an ethical ideal which places the spiritual welfare of others before that of self. It is altruism: 'spiritual chivalry is to deal fairly with others while not demanding fairness for yourself'. Spiritual chivalry was to follow the practice of the Prophet.[50] There is also a recognition of the possibility of sainthood (*wilāya*), a saint (*walī*) being a friend or protégé of God. Following upon the Qur'ānic verse (Q:10:62, 'no fear is upon them, nor do they grieve'), the Ṣūfīs argued that 'one of the traits of the saint is that he has no fear'.[51]

Towards the end of his fourth part, on other Ṣūfī characteristics, al-Qushayrī inserts an exhortation to new adherents:[52]

> The principles of the Ṣūfīs are the soundest of principles, and their *shaykhs* are the greatest, their scholars the most learned of men. If the student who has faith in them is a spiritual traveller capable of progress towards their goals, he will share with them in the inner discoveries that distinguish them. He will have no need of childish dependence on anyone outside this community. If

> the student is properly a follower without autonomy of state who wishes to advance through the realms of imitation until imitation becomes real, then let him imitate his forefathers. Let him proceed upon the path of the Ṣūfīs, for they will serve him better than anyone else…

The international network of schools or colleges (*madrasahs*) which helped to unify the formation of Sunnī society emanated from the model of Karrāmiyya[53] spiritual schools established in al-Qushayrī's Khurāsān.[54] The Ṣūfī orders, the teaching lines that 'would carry the culture and attitudes of the great *shaykhs* through the whole of the Islamic world', originated there as well. The most effective of their textbooks was al-Qushayrī's *Risālah*.[55]

The Pre-eminent al-Ghazālī

Though less significant for the development of Ṣūfīsm, al-Ghazālī is more famous than al-Qushayrī in the Western world. His purpose in writing was quite different. Unlike his younger brother,[56] al-Ghazālī was an ethical theologian, or theorist of ethical mysticism, and not truly a Ṣūfī.[57] He recognized that what was most special about the Ṣūfīs 'cannot be learned but only attained by direct experience, ecstasy and inward transformation'.[58] Al-Ghazālī frequently compared the spiritual exercise of 'recollection', designed to render God's presence throughout one's being (*dhikr*, a spiritual concentration attained through the rhythmical repetitive invocation of God's names), to *jihād*. He provided an extensive commentary on the Prophet's saying that 'whoever dies waging the greater *jihād* will share the rank of *shahīd* with the martyrs of the lesser *jihād*'. Both, according to al-Ghazālī, had sealed their belief, severing all ties except to Allāh by dying at the moment of sacred combat, and it was this blessed sealing state that assured them Paradise.[59] For al-Ghazālī, *nafs* had two principal meanings. Firstly, it meant

> the powers of anger and sexual appetite in a human being… and this is the usage mostly found among the people of *taṣawwuf* [that is, the Ṣūfīs], who take *nafs* as the comprehensive word for all the evil attributes of a person. That is why they say: one must certainly do battle with the ego (*mujāhadat al-nafs*) and break it.

For al-Ghazālī, the second meaning of *nafs* was that of 'the soul, the human being in reality, his self and his person'. However, it was differently described according to its various states. If calm under command and removed from disturbance caused by the onslaught of passion, it was called 'the satisfied [or the tranquil] soul' (*al-nafs al-muṭma'innah*). When it failed to achieve calm, yet set itself against the love of passions, it was called 'the self-accusing [or the reproachful, admonishing] soul' (*al-nafs al-lawwāmah*), because it rebuked its owner for his neglect of the worship of his master. If it gave up all protest and surrendered itself in total obedience to the call of passions and the devil, it was named 'the soul that enjoins evil' (*al-nafs al-ammārah*), which might be taken to refer to the ego in the first meaning.[60]

The *Revival of the Religious Sciences* (*Iḥyā' 'Ulūm al-Dīn,* from which these preceding remarks are derived), is a comprehensive work of 40 chapters – 40 was the number of patience and trial, 'the number of days of seclusion that the adept undergoes at the beginning of the Path'.[61] Unlike other Ṣūfīs, al-Ghazālī was prepared to provide a full-scale account of the duty to forbid wrong. There were, he contended, five levels (*marātib*) of performance of the duty: informing; polite counselling; harsh language; physical action against objects; and the threat or the use of violence against the person. There must, however, be sufficient power to perform the duty. As long as one was not compelled to participate in wrongdoing, for example, by rendering assistance to an unjust ruler, then there should be no need to resort to emigration (*hijrah*) to avoid an ineffective exercise of the duty that might cause one harm. Alternatively, there could be an effective exercise of the duty, but one which caused the individual harm, such as speaking out in the presence of an unjust ruler. Here al-Ghazālī drew an analogy with war 'in the way of the faith' (*jihād*). A lone Muslim might hurl himself at the enemy and be killed; but since the morale of the enemy might be harmed, this might be said to be advantageous to the Muslim community as a whole. It might be said to be similarly advantageous to the community if someone were killed while trying to right a wrong, discredit the wrongdoer or encourage the faithful; but to be justified, such action had to be successful.

Al-Ghazālī went further than most of his contemporaries by arguing that collecting armed helpers (*a'wān*) was legitimate, and did not require the permission of the ruler, in cases where the duty of righting a wrong could not be accomplished by the individual acting on his own. The formation of armed bands was thus permissible. Just as in *jihād*, any who were killed trying to forbid wrong would be considered as martyrs. However, the use of force or violence might lead to disorder (*fitnah*) and to consequences worse than the original wrongdoing. Moreover, the anecdotes recounted by al-Ghazālī suggest that while the risk of martyrdom had been accepted and the penalty suffered by earlier righteous individuals who had condemned tyrants who had not feared God, scholars of his own time had either failed to speak out or had been ineffectual, because of their love of worldly advantage. Critical in al-Ghazālī's approach was that the individual, as was the case with the military *jihādī*, had to carry out the deed simply for God's sake, without personal or worldly motives of any kind. Most of al-Ghazālī's contemporaries and successors modified his account to require the permission of the ruler for the formation of armed bands.[62] Only Jayṭālī (d. 750/1349) followed al-Ghazālī's arguments yet exceeded him in his enthusiastic endorsement of righteous rebellion against the evil ruler.[63]

The location of Ṣūfī spiritual activity: the 'dervish lodge' (*khānqāh*) and the fortified monastery (*ribāṭ*)

Ṣūfī spiritual activity did not, in itself, require a physical location; but for continuity of the activity, and for the development of a tradition under the guidance of a *shaykh*, a physical location, including the tombs of prominent *shaykhs*, was essential. Through the lodge a new hierarchy of authority was established, dependent upon that of the *shaykh*. It was in such buildings that some of the pivotal works of Ṣūfī philosophy and literature were written.[64] In Anatolia, whereas the *madrasah* was located near the citadel, and was thus associated with established power, the dervish lodge was placed in an accessible popular location, which demonstrated that it was outside the control of existing political and religious institutions as well as facilitating alliances with local groups of residences. Music and dance were encouraged to gain additional popular support,[65] while the tomb chamber was a conspicuous feature with large windows which permitted the public to view the site from outside.[66]

Three early Ṣūfī centres studied by Ethel Wolper, Sivas, Tokat and Amasya, were the main centres of the Türkmen revolt of Bābā Rasūl in 600/1204. In a number of accounts, a dervish lodge in or near Amasya served as a meeting place for Bābā Rasūl and his followers. Bābā Ilyās, a prominent Ṣūfī who was considered to have been so influenced by Christianity to have become a Christian convert, was said to have been one of the instigators of the revolt.[67] Thus from relatively early times, even in urban areas, Ṣūfī centres became suspect to the political and religious authorities as potential hotbeds of revolt and heterodoxy. Ethel Wolper talks in terms of 'hybridization' rather than 'syncretism' with regard to the Ṣūfī engagement with Christianity,[68] though it is doubtful whether purists such as the great Ḥanbalī scholars Ibn al-Jawzī (d. 597/1200) and Ibn Taymīyah would have regarded such a faith as Islām at all (the latter notwithstanding his own Ṣūfī connections).[69] For his part, the great Ṣūfī Jalāl al-Dīn Rūmī (d. 672/1273), founding father of the Mawlawīs, stated that 72 sects 'hear their mysteries from us'. 'We are like a flute', he remarked, 'that, in solo mode, is in accord with two hundred religions.' At his funeral, Jews and Christians were present, carrying the Torah, the Psalms, and the Gospels. The Christians claimed that the example of Jalāl al-Dīn Rūmī had helped them to comprehend 'the true nature of Jesus, of Moses, and of all the prophets. In him we have found the same guidance as that of the perfect prophets about whom we have read in our books.'[70] The relations with Christians of Ḥājjī Bektāsh, founder of the Bektāshīya order in the Ottoman lands, were even closer than those of Rūmī, and many of his early followers seem to have been Christians. The political importance of the Bektāshī arose from its connection with the Janissaries; the receptivity of the Janissaries in turn may be explained by their Christian origins.[71] It was this tendency to attract a spiritual following from other religions that marked the Ṣūfī experience in India.

The purpose of the second type of Ṣūfī building initially seems completely different: this is the *ribāṭ* or fortified monastery, found well away from urbanized areas on the frontier of the lands of Islām. Imām Abū Ḥafṣ al-Bakrī stated that

> the people in the *ribāṭ* are the *murābiṭūn* who agree on the same goal and corresponding conditions… the *ribāṭ* is established so that its inhabitant may have the qualities which Allāh stated in *Sūrat al-Anfāl* (Q.8:60): 'make ready against them whatever force and war mounts[72] you are able to muster, so that you might deter thereby the enemies of God, who are your enemies as well, and others besides them of whom you may be unaware…'

The *ribāṭ* might start off as no more than a watch-tower and small fort, but it became a work of piety for individuals to enhance the building and strengthen it at their own expense.

Allegedly the first *ribāṭ* in north Africa was established by Harthama ibn A'yan in 179/795. The military significance of the building was such that local *amīrs* took it upon themselves to establish a series of *ribāṭs* in their lands. Ibn Khaldūn (733/1332–809/1406) reported that the Aghlabid *amīr* Abū Ibrāhīm Aḥmad built a total of 10,000 in his lands in North Africa! Even if this ruler built to such an extent, the *ribāṭs* were unlikely to have been on the scale of that of Soussa, built by *Amīr* Ziyādat Allāh of Qayrawān in 205/821.[73] The minaret here served as a fortified lookout tower. Circular towers defended the building, while the central courtyard was surrounded by vaulted galleries of arcades. Numerous chambers opened from the galleries, providing cells for the residents, the *murābiṭūn*, both as living quarters and as studies. A prayer room (*masjid*) occupied the first floor of the southern half of the building: this comprised eleven aisles covered with barrel vaults, with a small dome raised above the general roof level of the *ribāṭ*. This remarkable edifice, with its innovative barrel vaulting, demonstrates the double character – military and religious – of the life of the *murābiṭūn*.[74] The ideas of the Ṣūfī and *mujāhid*, 'Abdullāh ibn Yāsīn (*c.* 405/1015 or 410/1020–451/1059) can be seen in his address to his followers:

> company of *murābiṭūn*, today you number about a thousand, and a thousand will not be overcome by less. You are the nobles of your tribes and the leaders of your clans. Allāh has put you right and guided you to His Straight Path. You must command the correct and forbid the bad, and strive for Allāh as He should be striven for.[75]

The transmutation of one branch of Ṣūfīsm: Shāh Walī Allāh and the caliphate

Al-Ghazālī had an influence on most subsequent Muslim thinkers, and he was the only one to whom the Islamic revivalist Quṭb al-Dīn Aḥmad ibn 'Abd al-

Raḥīm, popularly known as Shāh Walī Allāh (1114/1703–1176/1762), paid tribute and who was specifically acknowledged in the introduction to one of his writings.[76] Without purification of the heart, it was not possible to overcome the moral degeneration which permeated the individual and collective life of the Muslim community and he advocated *taṣawwuf*, which, for him, meant a direct approach to the heart. However, Shāh Walī Allāh departs from the characteristic Ṣūfī position by laying great stress on the state as an agency for the moral reform and ideological guidance of the people through its role in enjoining good and forbidding evil. He followed al-Farābī in considering the state to be a social necessity and al-Māwardī in his treatment of the caliphate. He sought to integrate and reconcile existing traditions within Islamic thought rather than to delineate a new direction as such, and in this task was prepared to use writers from seven centuries earlier and the example of the first two 'rightly-guided' caliphs.[77] The establishment of a caliphate he regarded as a collective religious obligation on the Muslim community.[78]

Shāh Walī Allāh's *magnum opus*, *The Conclusive Argument from God* (*Ḥujjat Allāh al-Bāligha*), is considered to be 'among the most profound works of Islamic scholarship'.[79] In this, he defined the caliphate or Islamic state in a comprehensive formulation, which included a strong emphasis on '*jihād* of the sword' as one of its most important duties:[80]

> It is the general authority to undertake the establishment of religion through the revival of religious sciences, the establishment of the pillars of Islām, the organization of *jihād* and its related functions of maintenance of armies, financing the soldiers, and allocation of their rightful portions from the spoils of war, administration of justice, enforcement of [the limits ordained by Allāh, including the punishment for crimes (*ḥudūd*)], elimination of injustice, and enjoining good and forbidding evil, to be exercised on behalf of the Prophet...

Shāh Walī Allāh formulated a new concept of an extraordinary caliphate (*khilāfah khāṣṣah*), which comprised both temporal and spiritual authority. Muslims were obliged to obey whatever command was issued by the caliph in the interests of Islām and the Muslim community. Only if the ruler committed 'evident infidelity', and openly rejected, condemned or placed in disrepute any of the 'essential postulates' of the true faith was it permissible, indeed obligatory, to struggle for his deposition. Such a struggle would then be considered a true *jihād*. However, such a struggle should be preceded by an individual, preferably in private, seeking to persuade the ruler to command right and abstain from committing evil: this individual remonstration, provided it was not accompanied by violence, would be regarded as the highest form of *jihād*.[81]

For Shāh Walī Allāh 'the most complete of all prescribed codes of law and the most perfect of all revealed religions is the one wherein *jihād* is enjoined'.[82]

Indeed, he effectively argued that no religion was complete if it did not stipulate and prescribe *jihād*.[83] Shāh Walī Allāh sought to achieve the supremacy of Islām over other religions and the primacy of the Muslim community over non-Muslims. In his view, such an outcome was

> inconceivable without contemplating among the Muslims a *khalīfah*, who can [place in open disrepute] those who might transgress the ideological frontiers, and commit acts which have been prohibited by their religion or omit their obligations under it…[84]

Shāh Walī Allāh quoted a reported tradition of the Prophet: 'what a marvel of God's will it is that there are people who enter Paradise in chains'.[85] The tradition could be interpreted in a number of ways, but he took it to mean that God's mercy to mankind requires the fullest opportunity for all to follow the straight path ('perfect mercy toward mankind requires that God guide them to virtuous conduct, deter the oppressors among them from their oppressive acts… without cutting off the sick part from the body, no human being can attain health'). Thus if a 'little amount of strong action necessarily leads to greater good, it ought to be taken inevitably'.[86] Whereas the traditional Ṣūfī path was an individual one, and the only obstacle on the way was that of sin and the obstacles set by the individual's baser nature, Shāh Walī Allāh argued that compassion for each individual requires that they should not be left alone in their sinful condition. In his terms, the only inducement to enter a true Islām and enjoy its blessings was to remove active opposition to the faith. In his remarks on the bitter medicine administered to a sick man, Shāh Walī Allāh came perilously close to forcing consciences and contravening the Qur'ānic precept that there should be no compulsion in religion:[87]

> …it is no mercy to them to stop at intellectually establishing the truth of Religion to them. Rather, true mercy towards them is to compel them so that Faith finds way to their minds despite themselves. It is like a bitter medicine administered to a sick man. Moreover, there can be no compulsion without eliminating those who are a source of great harm or aggression, or liquidating their force, and capturing their riches, so as to render them incapable of posing any challenge to Religion. Thus their followers and progeny are able to enter the fold of faith with free and conscious submission.

Through '*jihād* of the sword', Shāh Walī Allāh contends, Islām was 'brought into full prominence and pursuing its path is made like an inevitable course for humanity'.[88] The Prophet pursued *jihād* against 'those who opposed them [the early Muslims] until the command of God was fulfilled despite their unwillingness'.[89] *Jihād* was legislated for to promote the word of God and make sedition cease (Q.8:39).[90] God 'empowered certain of His sincere worshippers participating in the *jihād* to perform deeds which the mind would not imagine possible for that number of physical bodies'.[91] The use of force in *jihād* was construed as forming part of the scheme of divine mercy:[92]

> He orders one of God's prophets to make war against them [= the states where 'they do not believe in God and conduct themselves in sin'] so that the motivation to wage the *jihād* is inspired into the hearts of his people and they become 'a people brought out for mankind' and the divine mercy comes to include them. Another case is that a group becomes aware through the comprehensive outlook of the goodness of saving the oppressed ones from the predatory ones and undertaking the punishment of the disobedient ones and forbidding evil, so that this becomes a cause for the peace and contentment of the people and thus God rewards them for their action.

Shāh Walī Allāh's reading of Islamic history was thus predestinarian and triumphalist:[93]

> *Jihād* made it possible for the early followers of Islām from the *Muhājirūn* and the *Anṣār* to be instrumental in the entry of the Quraysh and the people around them into the fold of Islām. Subsequently, God destined that Mesopotamia and Syria be conquered at their hands. Later on it was through the Muslims of these areas that God made the empires of the Persians and Romans to be subdued. And again, it was through the Muslims of these newly conquered realms that God actualized the conquests of India, Turkey and Sudan. In this way, the benefits of *jihād* multiply incessantly, and it becomes, in that respect, similar to creating an endowment, building inns and other kinds of recurring charities… [In viewing the expansion of the faith as akin to an economic investment, Shāh Walī Allāh seems to be following an argument propounded by Ibn Taymīyah.][94]
>
> …*Jihād* is an exercise replete with tremendous benefits for the Muslim community, and it is the instrument of *jihād* alone which can bring about their victory… The supremacy of his Religion over all other religions cannot be realized without *jihād* and the necessary preparation for it, including the procurement of its instruments. Therefore, if the Prophet's followers abandon *jihād* and pursue the tails of cows [that is, become farmers] they will soon be overcome by disgrace, and the people of other religions will overpower them.

In was not just in theory but also in practice that Shāh Walī Allāh supported '*jihād* of the sword'. His two heroes were Maḥmūd of Ghazni and Aurangzīb, precisely the two rulers most closely associated with the violent assertion of

Muslim power in India (see Chapter 5).[95] What India needed, in his view, were pious *ġāzīs* who would pursue *jihād* in order to root out polytheism at its core.[96] Nādir Shāh of Persia had invaded India in 1151/1739 and reached Delhi; in Shāh Walī Allāh's judgement, he 'destroyed the Muslims and left the Marāthās and Jāts secure and prosperous. This resulted in the infidels regaining their strength and in the reduction of the Muslim leaders of Delhi to mere puppets.'[97] Aḥmad Shāh Durrāni, *amīr* of Afghanistan and founder of the Sadozai dynasty of the Abdālī tribe, had been involved in this plunder and devastation, and had himself wrought havoc in Delhi by his own invasion in 1170/1757. This did not stop Shāh Walī Allāh appealing to him to invade:[98]

> We beseech you in the name of the Prophet to fight a *jihād* against the infidels of this region. This would entitle you to great rewards before God the Most High and your name would be included in the list of those who fought for *jihād* for His sake. As far as worldly gains are concerned, incalculable booty would fall into the hands of the Islamic *ġāzis* and the Muslims would be liberated from their bonds…

Shāh Walī Allāh obtained his wish, and Durrāni defeated the Marāthā army at the third battle of Pānīpat in 1174/1761, but this was only one of the threats faced by the Muslims. The victory at Pānīpat was the high point of Aḥmad Shāh Durrāni's – and Afghan – power. Afterward, even prior to his death, the empire began to unravel. By the end of 1174/1761, the Sikhs had gained power and taken control of much of the Punjab. The following year, Durrāni crossed the passes from Afghanistan for the sixth time to subdue the Sikhs. He assaulted Lahore and, after taking their holy city of Amritsar, massacred thousands of Sikh inhabitants, destroying their temples and desecrating their holy places with cows' blood. Within two years the Sikhs rebelled again. Durrāni tried several more times to subjugate the Sikhs permanently, but failed. By the time of his death in 1186/1772, he had lost all but nominal control of the Punjab to the Sikhs, who remained in charge of the area until defeat by the British in 1265/1849.

Thus far, we have seen that Shāh Walī Allāh overturned the traditional, peaceful, Ṣūfī path in favour of a '*jihād* of the sword'. He was the Ibn Taymīyah of the Indian subcontinent,[99] but his was an internal *jihād* against polytheists not against external (Muslim) invaders. The depredations of Nādir Shāh of Persia were denounced because he was a Shī'a, while the depredations of Sunnī invaders like Durrāni were passed over in silence. What was needed was a Sunnī *jihād* against Shī'īsm, particularly the influence of Safdar Jang, the Shī'a *wazīr* of the Emperor Aḥmad Shāh between 1160/1748 and 1166/1753.[100] Like Ibn Taymīyah in the fourteenth century, Shāh Walī Allāh was critical of Ṣūfī innovations and those who sought to accommodate unorthodox traditions.[101] If the argument had rested there, then there would be little point in regarding him even as a transitional

Ṣūfī theorist, because force had predominated in the argument.[102] However, his writings are more complex than this and reflect the Ṣūfī preoccupation with the heart: 'one's heart', he contended, 'can only be inspired to *jihād* if one is able to develop an attitude identical with that of the angels'. The *jihādī* has to have purity of heart, and be farthest away from a base animal nature. In the Prophetic tradition, he is akin to one who fasts with the utmost devotion; indeed, it is better than fasting and praying for a whole month.[103] '*Jihād* of the sword' conforms, Shāh Walī Allāh asserts, to the Divine scheme and inspiration. So much so that 'the act of killing is not attributed to its human agents. This is like attributing the act of killing of a traitor to the ruler rather than the executioner, as the Qur'ān says: "...so you slew them not, but God slew them"' (Q.8:17).[104]

For Shāh Walī Allāh, when a martyr appears on the Day of Judgement, 'his act shall be manifest on him, and he shall be granted bounties in some form similar to his act'. Explaining the Qur'ānic verse Q.3:169 ('think not of those who are killed in God's way as dead. Nay they are being provided sustenance from their Lord'), the Prophet stated that their spirits reside inside the bodies of green birds; the beauty of the green bird symbolizes the martyr's soul. Martyrs in the cause of God have two qualities. Their souls are 'fully gratified and filled with spirituality'. They are, secondly, 'overwhelmed by the Divine mercy which encompasses the entirety of the cosmic system, including the holy enclosure and the angels present in the Divine proximity'. The martyr's soul is in some mysterious way attached to the Divine throne and 'love, bounty and happiness are constantly showered upon him'.[105]

In Shāh Walī Allāh, Ṣūfī preoccupations were radicalized and politicized in the cause of '*jihād* of the sword'. The dual authorities of Ibn Taymīyah (see Chapter 4) and Walī Allāh's close contemporary Muḥammad ibn 'Abd al-Wahhāb (1115/1703–1206/1791: see Chapter 6) exercised a baleful influence, respectively on his thought and legacy. For some, Shāh Walī Allāh's thought is 'still a major obstacle [to] the modernization of Indian Muslims'; for others, his 'acute grasp of the collective psychology of nations and his penetrating analysis of the political behaviour of mighty states, evidences the profundity of his political genius'.[106] To the extent that he viewed life as a moral struggle in which harmonious individuals achieve a balance between their angelic and animalistic dispositions in the hope of attaining a place in the holy enclosure (*ḥāẓīrat al-quds*), the rendezvous for the spirits of great human beings after emancipation from their corporeal bodies, Shāh Walī Allāh remained a Ṣūfī.[107] To the extent that the holy enclosure includes a place for martyrs from '*jihād* of the sword', he has been transmuted from a Ṣūfī into something else. It was the defeat of the Marāthā kingdom by Aḥmad Shāh Durrāni of Afghanistan at the third battle of Pānīpat in 1174/1761 that gave Shāh Walī Allāh his posthumous reputation as the saviour of Islām in India. Subsequently, his influence, and the influence of his son Abd al-Aziz on the *jihād* movement led by Sayyid Aḥmad Shahīd (d. 1246/1831) was considerable.

Though imbued with Ṣūfī ideas and practices, the writings of Shahīd leave no doubt that he thought in terms of Ṣūfī heresy: he condemned the 'innovations (*bid'āt*) of the "Ṣūfīstic polytheists"', 'heretics in Ṣūfīstic garb' and 'polytheists in Ṣūfīstic garb'.[108]

Most of the 'revivalist' *jihāds* of the nineteenth century were led by individuals who emanated from Ṣūfī orders (see Chapter 7). There were three principal reasons for this. One reason was that the Ṣūfī *shaykhs* obtained oaths of allegiance (*bay'ah*) from their followers that were personal in character (as against the institutional oath, for example, in Christian monastic orders).[109] This meant that if a Ṣūfī *shaykh* decided to declare a *jihād* then, subject to the plausibility of the case, he was able to rely on a body of loyal support for the enterprise. The Ṣūfī *shaykh*, writes Mark Sedgwick, 'often play[ed] a special role in Islamic societies because he [was] a major figure whose position [was] independent of almost all other interests'. His position depended almost exclusively on his own prestige, or at least the prestige of the order he led, which was also independent of other structures. *Shaykhs* were thus frequently called on by those who were not their followers to act as arbitrators, or in times of crisis, to provide a wider leadership of the community.[110]

A second reason was the existence of different tendencies within the Ṣūfī movement. These have to be thought of as in some respects comparable to the tensions within the Christian monastic orders, for example, between 'reformed' Cistercians as against 'unreformed' (that is, seen as 'worldly') Benedictines in the High Middle Ages. The founders of new Ṣūfī orders were almost always the leaders of breakaway movements, which rejected the approach of the founding order. Ṣūfīsm, both in its learned and popular varieties, has commonly been presented and has often presented itself as anti-modernist and anti-reformist, but it is striking that it has been precisely in modern and modernizing settings that Ṣūfīsm has made some of its greatest gains.

A third reason is perhaps the most compelling: this is that the Ṣūfī movement was in origin, as well as in its later development, a transnational phenomenon. It thus had the ability to make comparisons between Islamic practice in one country as against another and could determine whether corruption and false practices had gradually emerged. In that sense, transnational Islām is not a new feature of the faith communities of Islām at all, although some of the implications of transnationalism are only now being considered.[111] As Azyumardi Azra, Martin van Bruinessen and Julia Howell have argued,[112]

> a number of entirely different arguments concerning the relationship of Ṣūfīsm and modernity have been made in connection with the worldwide wave of Ṣūfī-led *jihād* movements against colonial powers and/or indigenous elites of the late nineteenth and early twentieth century. Evans-Pritchard's well-known explanation (1949) of how the Sanusi order provided an integrating

structure to the fissiparous Bedouin tribes of Cyrenaica and thus played a role in Libyan nation building easily lends itself to adaptation in other segmentary societies. Ṣūfī orders appear in these cases to adopt a new political role, as predecessors and progenitors of modern nationalist movements. Their militancy in these cases contrasts sharply with the peace loving, tolerant and inclusivistic attitudes commonly attributed to Ṣūfīsm. This gave rise to the concept of 'neo-Ṣūfīsm', launched by scholars (most prominently Fazlur Rahman) who felt that a number of important changes in the nature of Ṣūfīsm had taken place in the late eighteenth and early nineteenth centuries. 'Neo-Ṣūfīsm' was claimed to distinguish itself by increased militancy, stronger orientation towards the *sharī'ah* and rejection of *bid'ah*, and a shift from efforts to achieve unity with God to imitation of the Prophet. The debate on neo-Ṣūfīsm raises questions relevant to an understanding of the resurgence of Ṣūfīsm in modern urban environments and its relation to Islamic reformism.

Part Two

Contextual Theorists and State Systems

4
Ibn Taymīyah and the Defensive *Jihād*: A Response to the Crusades and the Mongol Invasions

> Anger drops from the pages of his books, formulated so beautifully, in such general terms, that when a modern Muslim reads it, or even when I read it myself, it is impossible not to think of present-day Muslim society. The effect of his work is electrifying. His books are banned in several countries around the Islamic world, although they can always be found under the table. From their own point of view, Muslim governments which forbid this fourteenth-century propaganda are right. Because it is inflammatory material… I am sure Ibn Taymīyah didn't think of collateral damage in the modern meaning of the word. He was confronted with a military situation in which both armies comprised Muslim military professionals. He had to develop a theory that justified fighting against other Muslims…
>
> Interview given by Professor Johannes J. G. Jansen, 8 December 2001.[1]

The credentials of the Shaykh al-Islām

No other Muslim writer, medieval or contemporary, has exercised as much influence on the modern radical Islamist movement as Ibn Taymīyah (661/1268–728/1328).[2] There are several reasons for this. The first is that from the remarkably young age of 19, he became a professor of Islamic studies and was already well versed in Qur'ānic studies, *ḥadīth*, *fiqh*, theology, Arabic grammar and scholastic theology. He was an extraordinarily prolific scholar who was given the title Shaykh al-Islām by his supporters:[3] his pupil, Ibn Qayyim al-Jawziyya, compiled a list of the great man's writings which contains 350 works. A modern compilation of his legal rulings alone comprises 37 printed volumes.[4] As a professor of Ḥanbalī law,[5] the most conservative of the four major Sunnī legal schools (the others were the Ḥanafī, Mālikī and Shāfi'ī), he started issuing legal opinions (*fatwās* or *fatāwā*) on religious matters without necessarily following any of the traditional legal schools – itself the hallmark of Ḥanbalī scholarship. He was a member of a politically disadvantaged legal school and did not aspire to serve in government. Thus, with Ibn Taymīyah, the emphasis of his writing shifted from the concerns of his predecessors with the imāmate and the sultanate, to the effective operation of the *sharī'ah*. Though few heeded his call, his reassertion of the idea of the *ummah* living by the *sharī'ah* is basically sound, 'and alone promises stability and permanence amid the [transitory nature] of the political organization in the form of a caliphate…'[6]

The second reason is that this independence of thought (*ijtihād*), though praised by some (one supporter called him 'the leader of *imāms*, the blessing of the community, the signpost of the people of knowledge, the inheritor of Prophets, the last of those capable of independent legal reasoning, the most unique of the scholars of the Religion') earned him many enemies.[7] Ibn Taymīyah was a ferocious opponent of innovation (*bid'ah*): 'the more an innovator tries to be original, the further he distances himself from God', he remarked.[8] 'Anything new is *bid'ah* and every *bid'ah* is an error.'[9] Yet there is no doubt that he himself was an innovator, and was so regarded by others: 'he is a misguided and misguiding innovator [*mubtadi' ḍall muḍill*] and an ignorant who brought evil [*jāhilun ghālun*] whom Allāh treated with His justice. May He protect us from the likes of his path, doctrine, and actions!'[10] He was imprisoned on several occasions and indeed died in imprisonment. An example of innovation which has been repudiated is Ibn Taymīyah's view that fighting with 'Alī against Mu'āwiya was neither a duty nor a *Sunnah*. This product of Ibn Taymīyah's independent reasoning was found invalid because of the existence of a clear Qur'ānic text to 'fight the group that is a transgressor', along with the Prophet's *ḥadīth* warning 'Ammār bin Yāsir, a companion of Muḥammad and 'Alī, about the faction of transgressors who would kill him. Contrary to Ibn Taymīyah's interpretation, 'the faction of transgressors' was that of Mu'āwiya, while fighting on 'Alī's side was a duty and *Sunnah*.[11] Against his accusers, Ibn Taymīyah contended that on intellectual questions of universal concern such as exegesis, *ḥadīth*, Islamic jurisprudence (*fiqh*) and the like, 'the correctness of one view and the incorrectness of the other' could not be established by the ruling of a judge.[12]

The third reason for Ibn Taymīyah's significance arises from the context in which he lived and worked. It was an age of profound spiritual and political upheaval. In 656/1258, the 'Abbāsid Empire was defeated by the invading Mongol armies, leading to the capture of the great city of Baghdād. For most Muslims, the defeat of the ruling dynasty was an unmitigated disaster. Baghdād was a renowned city of Islamic learning that had suffered the fate of being looted and pillaged. The city's decline was political, economic, demographic and social. The regular trade with Syria and Egypt was cut off; trade with India was disrupted. For more than 500 years since 132/750, Baghdād had been the capital of a weak and notional Islamic empire under the ‘Abbāsid dynasty. The last caliph and all his family were executed by the Mongols in 656/1258 and the dismemberment of the empire followed rapidly upon this event. Baghdād was reduced to a provincial capital, ruled from Tabrīz.[13] It is true that two years later, at the battle of Ayn Jālūt in Palestine, the Mongol advance was stopped by the Mamlūk army commanded by Qutz (Qutuz), king of Egypt (who was killed) and Baybars (his successor), which resulted in Syria falling under the control of the Mamlūks rather than the Mongols. The Mongols continued to press the Islamic borderlands, and in the year 667/1268 Ibn Taymīyah's father fled for the greater security of his family from Harran in Mesopotamia to Damascus. Yet the security of Damascus was only relative: it had been besieged twelve times between the death of Saladin in 589/1193 and the entry of the Mongols into an undefended city in 658/1260.[14]

The problem was not merely with the Mongol invaders but also with the Christian communities, some of which were prepared to support the Mongols. The three invasions of the region around Damascus by Maḥmud Ghāzān in the years 699/1299–703/1303 – Ghāzān (Qazān) was the Khān in Iran (ruled 694/1295–704/1304), a Muslim convert with Shī‘a leanings[15] he is considered to have been one of the greatest of the Ilkhān dynasty – found support among the Christians as well as the Druze and ‘Alawī Shī‘a population. Ibn Taymīyah took part in a *jihād* against the people of Kasrawān (now in Lebanon) in 699/1300, at which time he may have issued a *fatwā* (now lost) authorizing fighting against Christians and ‘Maltese Christians’ (presumably Maltese Knights) allied to the Mongols. The same year he went to Cairo to exhort the Mamlūk *ṣultān* to undertake a *jihād*.[16]

In 696/1297, he had been employed by the Mamlūk government of Egypt to preach a *jihād* against the last ‘Crusader state’ on the mainland, the kingdom of Cilician Armenia.[17] According to Ibn Taymīyah, ‘Ṣūfīsm, Shī‘a imāmism, tomb veneration and saint intercession’ paralleled the errors of Christianity. In his view, *jihād* against the ‘People of the Book’ was superior to *jihād* against idolaters, since they were seen as active agents of unbelief.[18] For Ibn Taymīyah, Christians had gone to excess[19]

> in laxity, so that they have failed to command the good and prohibit what is forbidden. They have failed to do *jihād* in the way of God and to judge justly between people. Instead of establishing firm limits, their worshippers have become solitary monks. Conversely, the rulers of the Christians display pride and harshness and pass judgement in opposition to what was handed down by God.

Ibn Taymīyah's concept of the true believer

Ibn Taymīyah's proclaimed that ‘to fight the Mongols who came to Syria’ was ‘a duty prescribed’ by the Qur'ān and the example of the Prophet.[20] His courage was evident when he went with a delegation of '*ulamā*' to talk to Qazān, the Khān of the Mongol Tatars, urging him to halt his attack on the Muslims. None of the '*ulamā*' had dared to raise the matter except Ibn Taymīyah. He declared: 'you claim that you are Muslim and you have with you [those who call the faithful to prayers (*mu'adhdhins*)], judges, *Imāms* and *shaykhs* but you invaded us and reached our country for what?' While your father and your grandfather, Hülagü,[21] were non-believers, he argued, 'they did not attack the land of Islām; rather, they promised not to attack and they kept their promise. But you promised [us one thing] and broke your promise.'[22]

It is clear from Ibn Taymīyah's words that an important distinction was to be made between a true believer in Islām and a partial convert, a lapsed believer or an apostate. Ibn Taymīyah believed that the ideal Muslim community had been the original community in Medina, surrounding the Prophet. Ever since then, the quality and morality of Muslims had declined. Muslim leaders, in particular, bore much of the burden for not encouraging the proper faith and attitudes among the people and thus for the political divisions which had facilitated the Mongol advance. His strongest condemnations were reserved for the Mongols: according to Ibn Taymīyah, the mere act of conversion was insufficient to make a person a 'true' Muslim. The Mongols, for example, still relied on the *Yāsā* code of law derived from their polytheistic tradition instead of the *sharī'ah*. At his acquisition of supreme power in 602/1206, Chinghis-Khān already had prepared his Great *Yāsā*, which continued to be developed during his lifetime.[23] The word *Yāsā* means 'order, decree'. The Great *Yāsā* was a compilation of his laws, rules, and words of wisdom. The work was written in the Uīghur script that Chingis himself had introduced as the written language of the Mongols. It was written on scrolls that were bound in volumes, and kept in secret archives to which only the supreme ruler and his closest associates had access. According to Juwainī,

> Chinghis-Khān did not belong to any religion and did not follow any creed, he avoided fanaticism and did not prefer one faith to the other or put the ones above the others. On the contrary, he used to hold in esteem beloved and respected sages and hermits of every tribe, considering this a procedure to please God.

According to Makrīzī, 'he ordered that all religions were to be respected and that no preference was to be shown to any of them. All this he commanded in order that it might be agreeable to Heaven.' Furthermore, 'He forbade them [his commanders] to show preference for any sect, to pronounce words with emphasis, to use honorary titles; when speaking to the Khān or anyone else simply his name was to be used.'[24] Such religious syncretism on the part of an 'infidel, polytheistic' ruler would scarcely have served to inspire respect or obedience from a conservative Muslim such as Ibn Taymīyah.

In 702/1303 Ibn Taymīyah issued a *fatwā* against the Muslims of Mārdīn (who had surrendered to the Mongols in 658/1260),[25] arguing that they were neither Muslims nor unbelievers as evidenced by their apathy about the law imposed by the Mongols and their refusal to undertake *jihād* against the Mongol occupation.[26] He stated:[27]

> everyone who is with them in the state over which they rule has to be regarded as the belonging to the most evil class of men. He is either an atheist (*zindīq*) and hypocrite who does not believe in the essence of the religion of Islām – this means that he [only] outwardly pretends to be a Muslim – or he belongs to that worst class of all people who are the people of [heretical innovations (*bid'ah*)]… They place Muḥammad [in a position] equal to [the position of] Chinghis-Khān; and if [they do] not [do] this they – in spite of their pretension to be Muslims – not only glorify Chinghis-Khān but they also fight the Muslims. The worst of these infidels even give him their total and complete obedience; they bring him their properties and give their decisions in his name… Above all this they fight the Muslims and treat them with the greatest enmity. They ask the Muslims to obey them, to give them their properties, and to enter [into the obedience of the rules] which were imposed on them by this infidel polytheistic King…

This ruling[28] created a precedent whereby so-called apostates and their like may be considered worthy targets of violent revolution, even if they provide legitimate (and apparently Muslim) political leadership. Ibn Taymīyah provided a rationale for this viewpoint in his treatise on *Public Policy in Islamic Jurisprudence*:[29]

> It has been established from the Book, from the *Sunnah*, and from the general unanimity of the [Muslim] nation that he who forsakes the Law of Islām should be fought, though he may have once pronounced the two formulas of Faith [in Islām]. There may be a difference of opinion regarding rebellious groups which neglect a voluntary, but established, piece of worship… but there is no uncertainty regarding the duties and prohibitions, which are both explicit and general. He who neglects them should be fought until he agrees to abide [by these duties and prohibitions]: to perform the five assigned prayers per day, to pay the *zakāt* [alms], to fast during the month of Ramaḍān, and to undertake pilgrimage to the Ka'ba [at Mecca]. Furthermore they should avoid all forbidden acts, like marriage with sisters, the eating of impure foods (such as pork, cattle that has died or was unlawfully slaughtered, etc.) and the attack on the lives and wealth of Muslims. Any such trespasser of the Law should be fought, provided that he had a knowledge of the mission of the Prophet, Peace be Upon Him. This knowledge makes him responsible for obeying the

orders, the prohibitions and the [authorizations]. If he disobeys these, then he should be fought.

Reviving the duty of '*jihād* of the sword'

Ibn Taymīyah expressed his arguments most clearly in the chapter on 'The Religious and Moral Doctrine of *Jihād*' in his book *Governance According to Allāh's Law in Reforming the Ruler and his Flock* (*al-Siyāsa al-shar'iyya fī iṣlāḥ al-Rā'ī wa'l-Ra'iyya*). For him, the command to participate in *jihād* and the mention of its merits occur innumerable times in the Qur'ān and the *Sunnah*. Therefore 'it is the best voluntary [religious] act that man can perform'. All scholars agreed that it was better than the greater pilgrimage (*Ḥajj*) and the lesser pilgrimage (*'umrah*), better than voluntary *ṣalāt* and voluntary fasting, as the Qur'ān and the *Sunnah* indicated. The Prophet had stated that 'the head of the affair is Islām, its central pillar is the *salāt* and the summit [literally, the tip of its hump, an allusion to the camel] is the *jihād*'.[30] Ibn Taymīyah cited Muslim, who had reported another *ḥadīth* of the Prophet to the effect that remaining at the frontiers with the intention of defending Islamic territory against its enemies (*ribāṭ*) was better than one month spent in fasting and vigils.[31] Both al-Bukhārī and Muslim had reported the *ḥadīth* that fasting without interruption and spending the night in continuous prayer were the only acts equal to military *jihād*.[32]

Ibn Taymīyah extolled the benefits of *jihād* in terms that are still quoted today by militant Islamists:[33]

> ...the benefit of *jihād* is general, extending not only to the person who participates in it but also to others, both in a religious and a temporal sense. [Secondly,] *jihād* implies all kinds of worship, both in its inner and outer forms. More than any other act, it implies love and devotion for Allāh, Who is exalted, trust in Him, the surrender of one's life and property to Him, patience, asceticism, remembrance of Allāh and all kinds of other acts [of worship]. And the individual or community that participates in it, finds itself between two blissful outcomes: either victory and triumph or martyrdom and Paradise. [Thirdly,] all creatures must live and die.
>
> Now, it is in *jihād* that one can live and die in ultimate happiness, both in this world and in the Hereafter. Abandoning it means losing entirely or partially both kinds of happiness. There are people who want to perform religious and temporal deeds full of hardship in spite of their lack of benefit, whereas actually *jihād* is religiously and temporally more beneficial than any other deed full of hardship. Other people [participate in it] out of a desire to make things easy for themselves when death meets them, for the death of a martyr is easier than any other form of death. In fact, it is the best of all manners of dying.

Ibn Taymīyah sought categorically to refute the tradition of the 'greater' or peaceful *jihād*:

> there is a *Ḥadīth* related by a group of people which states that the Prophet... said after the battle of Tabūk: 'we have returned from *Jihād Asghar* to *Jihād Akbar*.' This *ḥadīth* has no source, nobody whomsoever in the field of Islamic Knowledge has narrated it. *Jihād* against the disbelievers is the most noble of actions, and moreover it is the most important action for the sake of mankind.[34]

Ibn Taymīyah extended the concept to incorporate a central maxim of the Qur'ān:[35]

> Allāh said by way of description of Prophet Muḥammad...: 'He orders them with that which is good and forbids them that which is bad. And he makes allowed for them that which is clean and good, and forbids them that which is unclean and detestable' (Q.7:157)...
>
> Since *jihād* is part of the perfection of enjoining right and prohibiting wrong, it, too, is a collective obligation. As with any collective obligation, this means that if those sufficient for the task do not come forward, everyone capable of it to any extent is in sin to the extent of his capability in that area. This is because its obligation when it is needed is upon every Muslim to the extent of his/her ability, as the Prophet... said in the *ḥadīth* found in Muslim:
>
> 'Whoever of you sees wrong being committed, let him change it with his hand (i.e. by force). If he is unable to do that, then with his tongue, and if he is unable to do that, then with his heart.'[36]
>
> This being the case, it is clear that enjoining right and prohibiting wrong is one of the greatest good works that we have been ordered to do.

His maxim was that the benefit secured by performing the duty must outweigh any undesirable consequences,[37] a consideration, which as Michael Cook states, 'rules out attempts to implement it through rebellion'. Forbidding wrong, for Ibn Taymīyah, is part of what God's revelation is all about, and is closely linked to the duty of *jihād*. The purpose of all state power is to carry out the duty.[38] It seems fairly clear that Ibn Taymīyah's analysis does not provide, in modern times, bin Laden and his followers with a clear justification for their action: if the risk is that the disadvantages outweigh the benefit, then we are warned not to proceed with enjoining good and forbidding wrong, and in any case the emphasis, Michael Cook tells us, is on civility (*rifq*).[39] Nevertheless, for some of his modern critics, the 'problem today is with Ibn Taymīyah himself... despite his great knowledge and although he was well read, [Ibn Taymīyah] was emotional and inclined to sentimentality. His *fatāwā* were sometimes issued as an unbalanced response…'[40]

Forbidding wrong: the Guiding Book and the Helping Sword

For Ibn Taymīyah, the priority was not to wage war in the *Dār al-Ḥarb*. It was to turn inwards, and purge the Sunnī world of infidels and heretics. His *jihād* was to be 'a force which at the same time would renew individual spirituality and create a united society dedicated to God which could then triumph over the world'.[41] His main concern was to ensure that all fit male Muslims of age should be prepared to fight to defend their territory against any internal or external military challenge:

> So, whoever avoids the fighting which Allāh has ordered so as not to be exposed to temptation, has already fallen to temptation, because of the doubt and sickness which have come into his heart, and his neglecting of the *jihād* which Allāh has ordered him to undertake.
>
> Contemplate this very carefully, for it is a very dangerous question. People, in this regard, are in two categories:
>
> 1) One group enjoins and forbids and fights in order, as they imagine, to remove chaos and temptations. There actions are a greater chaos or temptation than that which they seek to remove. This is the example of those who rush to fight in the conflicts which arise among the Muslims, such as the Khārijīs.
>
> 2) Another group leaves enjoining right and forbidding wrong and fighting in the path of Allāh by which religion may become solely for Allāh, and His word may be uppermost, in order that they may avoid being tempted. They have already fallen into and succumbed to temptation.

The chapter on 'The Religious and Moral Doctrine of *Jihād*' is important both for what is restated and the passages which are omitted. There is no discussion, for example, of radical abrogation in the Qur'ān; no discussion either of the 'verses of the sword'. The Ḥanbalī school of law generally did not rely on the concept of 'abrogation', but preferred to place the verses of the Qur'ān in their context. Of the 19 citations from the Qur'ān, only five directly concern *jihād* though all come from the Medina *sūrahs*. The first of the passages cited ('that ye believe in Allāh and His Messenger, and that ye strive [your utmost] in the Cause of Allāh, with your property and your persons: That will be best for you, if ye but knew!': Q.61:11) stresses two kinds of worship, bodily, which also includes mental effort; and pecuniary or monetary. The passage can be interpreted in either a pacific or a warlike sense. Similarly, 'fighting for the Faith' (Q.8:72) may be taken to mean 'application of oneself and one's substance or wealth' in the cause. Another passage cited is that 'Allāh hath granted a grade higher to those who strive and fight with their goods and persons than to those who sit [at home]' (Q.4:95). The terms 'strive and toil' and 'strive and fight' are all translations of the same verb and are open to pacific as well as warlike interpretations. A further passage concerns the willingness of believers to 'suffer exile and strive with might and main, in Allāh's cause, with their goods and their persons'. Without the interpretation provided by the *ḥadīth* of the Prophet,[42] Q.9:19 could be said to be no more than acting strenuously or 'taking pains' for the faith. Exile, 'striving' and commitment with goods and persons (Q.9:20) are open to a pacific as well as a more aggressive interpretation. In the next passage cited (Q.5:54), 'fighting in the way of Allāh' can be interpreted in a pacific sense as 'striving' or 'contending' as well as in a more warlike sense.

In none of the *jihād* passages cited is the sense unequivocally warlike. Instead, Ibn Taymīyah short-circuits the argument on the nature of *jihād* with the following formulation:

> Since lawful warfare is essentially *jihād* and since its aim is that the religion is Allāh's entirely (Q.2:189; Q.8:39) and Allāh's word is uppermost (Q.9:40), therefore, according to all Muslims, those who stand in the way of this aim must be fought.

Within Ibn Taymīyah's political thought, the fundamental duties of government are trust and justice.[43] 'Religion and the state', he declared, 'are indissolubly linked. Without the power of coercion (*shawkah*) of the state, religion suffers. Without the discipline of revealed law, the state becomes a tyrannical structure.'[44] The first two duties of the ruler were to preside over prayer and to direct the *jihād*.[45] 'Right religion must have in it the Guiding Book and the Helping Sword.'[46] For Ibn Taymīyah there were two types of *jihād*, voluntary and involuntary. The first type of *jihād* 'is voluntary fighting in order to propagate the religion, to make it triumph and to intimidate the enemy, such as was the case with the expedition to Tabūk[47] and the like'. But such occasions had passed. Islām was on the defensive. There thus could be no question of an offensive *jihād*.

> We may only fight those who fight us when we want to make God's religion victorious, Ibn Taymīyah wrote. 'God, Who is exalted, has said in this respect: "and fight in the way of God those who fight you, but do not commit aggression: God loves not aggressors"' (Q.2:190)...[48]

The Muslim lands were occupied by the half-pagan Mongols. Living within the lands of *Dār al-Islām* were some who had 'not yet embraced Islām' or who lived in ignorance.[49] The involuntary *jihād* 'consists in defence of the religion, of things that are inviolable, and of lives. Therefore it is fighting out of necessity.' When the young Muslim community was attacked by the enemy in the year of the

Trench (5/626), Allāh did not permit anybody to abandon *jihād*, although He did allow them to take no further part in the *jihād* after the siege was lifted in order to pursue the enemy. If war is offensive on the part of Muslims, it is a collective duty (*farḍ al-kifāyah*), which means that if it is fulfilled by a sufficient number of Muslims, the obligation lapses for all others and the merit goes to those who have fulfilled it. In contrast, if the war is defensive (that is, the enemy attacks first), then repelling them becomes a duty for all those under attack and it is the duty of others to help them.

It has been argued that Ibn Taymīyah and other Ḥanbalī jurists were not as opposed to Ṣūfīsm as was once believed; indeed, some Ḥanbalī '*ulamā*' were well-known Ṣūfīs, and Ibn Taymīyah was himself an initiate of the Qādiriyyah *ṭarīqah*.[50] He was prepared to accept a form of Ṣūfīsm based on Islamic legalism and tradition, but he vigorously repudiated Ṣūfī pantheism and innovations such as the worship of saints and pilgrimages to their shrines. Ibn Taymīyah considered the consensus of the faithful (*ijmā'*) to be impossible to achieve, since they could not all be brought together to pronounce on a unanimous *fatwā*;[51] he opposed the Shī'a, whom he considered 'more dangerous than the Jews and the Christians and… more to be feared since they acted treacherously within the community';[52] his position with regard to anthropomorphism (*al-tashbīh*), that is, ascribing human attributes to God, is unclear and the accusations against him were rejected by his followers.

Above all, Ibn Taymīyah has been criticized in modern times as a radical reactionary, seeking to regain the alleged purity and uniformity of the early followers of the Prophet, and for the degree of intolerance that he has inspired among contemporary radical Islamists toward expressions of the faith that are different from their own. While requiring strict standards to differentiate Muslims from the Mongol Tatars, Ibn Taymīyah was reluctant to use the term *takfīr*, the charge of unbelief levelled against other Muslims who do not conform, as is done frequently by some contemporary militant Islamists. He nevertheless considered it appropriate to determine a Muslim's 'Muslimness'. Ibn Taymīyah is viewed as an exponent of 'extremism' who was sent to prison by four judges representing the four schools of law for alleged anthropomorphism or a literalist interpretation of the Qur'ān.[53] Thus, for some recent critics, Ibn Taymīyah is not worthy of the title 'Shaykh al-Islām', as his followers have called him,[54] while for militant anti-establishment Islamists his several arrests[55] and death in prison only serve to confirm his radical credentials.

For all that he is viewed as a forerunner of violent Islamism, Ibn Taymīyah's conception of *jihād* was essentially that of a 'just war' waged by Muslims whenever their security was threatened by infidels. Such a just war was very different from a 'holy war' seeking religious conversions. In Ibn Taymīyah's view, for an unbeliever to be killed if he did not become a Muslim would constitute 'the greatest compulsion in religion' and would contravene the Qur'ānic injunction

(Q.2:257). *Jihād* was, for him, a just and defensive war launched and waged by Muslims whenever their security was threatened in the *Dār al-Islām* by infidels or heretics. Lawful warfare was the essence of *jihād*, the aim of which was to secure peace, justice and equity.[56] As Majīd Khaddūrī comments,

> no longer construed as war against the *Dār al-Ḥarb* on the grounds of disbelief, the doctrine of the *jihād* as a religious duty became binding on believers only in the defence of Islām. It entered into a period of tranquillity and assumed a dormant position to be revived by the *Imām* when he believed Islām was in danger…[57]

The principle to enjoin what is right and to prohibit what is evil was incumbent on scholars and administrators but also required the active participation of the whole community. *Jihād* in this sense required immediate participation according to one's ability, position and authority. Ibn Taymīyah justified the argument on the basis of the tradition, narrated on the authority of Abū Sa'īd al-Khudrī (d. 74/693), in which the Prophet is reported to have said:[58]

> Anyone of you confronting the evildoing should do his utmost to change it with his hand. If he is not able, he should change it with his tongue. If he is still not able, he should then denounce it with the heart. And the last one is the indication of the weakest state of his faith.

Ibn Taymīyah thus should be seen as a revivalist of the doctrine of *jihād* and perhaps its last great theoretician in the Middle Ages.[59] His *fatwā* regarding the Mongols (see Appendix) established a precedent: in spite of 'their claim to be Muslims, their failure to implement *sharī'ah* rendered the Mongols apostates and hence the lawful object of *jihād*. Muslim citizens thus had the right, indeed duty, to revolt against them, to wage *jihād*.'[60] While Ibn Taymīyah's writings were known to Muḥammad bin 'Abd Al-Wahhāb, the later writer distanced himself from some of Ibn Taymīyah's more extreme views on violence and killing. The 'conscious adoption' of Ibn Taymīyah's writings into the Wahhābī world view occurred later, in the nineteenth century, when the Wahhābīs had 'a theological and legal need for the strict division of the world into Muslims and unbelievers and the overthrow of rulers who were labelled as unbelievers…'[61]

The contemporary violent Islamists' distortion of Ibn Taymīyah's thought

Ibn Taymīyah's standpoint was not always a moderate one at the time, let alone when interpreted in the light of modern inter-faith relations. In his short treatise *On the Status of Monks*, he argued that those in the religious orders who were found outside their monasteries might be killed; they might also be killed if they had dealings with people outside their monastic community rather than living a completely isolated life. This tract was reprinted in Beirut in 1997 by Nasreddin Lebatelier (the Belgian Muslim convert Jean Michot) under the title *Le Statut des Moines*, with an introduction quoting from the Groupe Islamique Armé's (GIA's) communiqué number 43, which stated that it was justifiable under Islamic principles to take the lives of the seven Trappist monks killed in Algeria in 1996. This led to Jean Michot's removal from a Professorship at Louvain and, upon his appointment at Oxford, to a formal request for a disavowal from the Anglican Secretary for Inter-faith Relations. Michot issued a statement which made clear that he had 'never developed any kind of apology for murder' in his writings or statements. He 'completely endorsed the condemnation of the GIA by the consensus of the Muslim community' and had always considered that 'these killings were a particularly tragic event in Islamo-Christian relations'.[62] But the damage had been done and the GIA's reliance on Ibn Taymīyah's text to justify its atrocity reinforced the impression that this jurist's intolerance extended to Christians as well as to those he considered Muslim heretics. Though he did not adhere to the extreme Almohad viewpoint that the Prophet's concessions to religious minorities had lasted five centuries and had lapsed by the twelfth century, there is no doubt that Ibn Taymīyah was prepared to argue that the *dhimmīs* should benefit from Muslim protection only to the extent that it was in the interest of the community that this should happen. However, if Muslim protection came to an end the fate of the minorities would be exile, as 'Umar had exiled Jews and Christians from the Arabian peninsula.[63]

For Osama bin Laden, Ibn Taymīyah, along with Shaykh Muḥammad Ibn 'Abd al-Wahhāb, is one of the great authorities to be cited to justify the kind of indiscriminate resort to violence which he terms *jihād*. In particular, Ibn Taymīyah was cited twice in sermons and communiqués in 2003. In his sermon published on 16 February 2003, bin Laden said of Ibn Taymīyah:[64]

> The most important religious duty – after belief itself – is to ward off and fight the enemy aggressor. Shaykh al-Islām [Ibn Taymīyah], may Allāh have mercy upon him, said: 'to drive off the enemy aggressor who destroys both religion and the world – there is no religious duty more important than this, apart from belief itself'. This is an unconditional rule.

He returned to the same subject in his speech posted in English on 18 July 2003.[65] No true Islamic state was currently in existence, he contended. To attain it, five conditions were needed:

> a group, hearing, obedience, a *hijrah* [that is, detachment from the world of heresy to establish and strengthen a community of believers outside it, in

> the path of the Prophet Muḥammad] and a *jihād*. Those who wish to elevate Islām without *hijrah* and without *jihād* sacrifices for the sake of Allāh have not understood the path of Muḥammad…

Once more the name of Ibn Taymīyah was brought in to support the cause:

> If *jihād* becomes a commandment incumbent personally upon every Muslim, it [*jihād*] rises to the top of the priorities, and there is no doubt of this, as Shaykh al-Islām [Ibn Taymīyah] said: 'nothing is a greater obligation than repelling the aggressive enemy who corrupts the religion and this world – except faith itself'.

That the prominent jurist is regularly cited by bin Laden to support his cause is therefore not in doubt. As early as August 1996 he had praised him for 'arousing the *ummah* of Islām against its enemies'.[66] The question is whether or not the citations are justified. Does bin Laden in reality not take Ibn Taymīyah out of context and distort his thought? Ibn Taymīyah's preoccupation, it has been seen, was with Muslim decline in the period of the Mongol invasions. It is true that he encouraged resistance to the foreign invader but that this was a genuinely defensive response cannot be doubted. An organization such as al-Qaeda, which has justified world-wide acts of terrorism, and in particular the events of 11 September 2001, can claim with only an extraordinary feat of intellectual dishonesty that it is waging a defensive *jihād*. It does argue this; but the simple chronology of cause and response denies the validity of this argument. Only an excessively long period of American involvement in 'Irāq would give bin Laden a justification for the argument of a defensive *jihād*. The defence of Saddam Hussain did not qualify in this respect, since Saddam's was not an Islamic state (although he had called for a *jihād* and the call received some support abroad).[67] The opportunity to establish an Islamic state had existed with the Ṭālibān in Afghanistan, but this opportunity was lost, according to bin Laden, because of the failure of Muslim countries to support the *jihādī* cause.

Thus, within bin Laden's world view the rulers of the Arab states have betrayed Allāh, the Prophet and the 'nation'. The second barrier to his proposed *jihād* are 'the *'ulamā'* and preachers who love truth and loathe falsehood, but refrain from participating in *jihād*; they have devised interpretations and have turned the young people against taking part in *jihād*'. Here bin Laden seeks to present himself as a latter-day Ibn Taymīyah, at odds with the orthodox Muslim clerics of his time. His critique is one of bitter polemic, just as Ibn Taymīyah's had been in his own time. Yet there are key differences. The first difference concerns the means to the end. For bin Laden, violence is a legitimate means to achieve the end. The 'righteous' minority clearly have, in practice, the right to coerce the majority to accept their leadership. For Ibn Taymīyah, however, in the words of Qamaruddin Khan, the historian of his political thought:[68]

> the state… is… neither a divine commission nor a power-state based on sheer military might; it is a cooperation between all members of the community to realize certain common ideals – the recognition of *tawḥīd*, one God, the Creator, the Provident, the Law-giver, and of the Prophet, the intermediary between God and man, and the submission to a common law, the *sharī'ah*.

So consensus, not coercion, must prevail according to Ibn Taymīyah. And there is another, equally important, distinction to be made in the means to achieve the end. For al-Qaeda supporters, indiscriminate violence is justifiable and essential on the basis of the Qur'ānic text Q.9:52:[69]

> this torture will not, in any way, be carried out by means of preaching (*da'wah*), because preaching is activity of exposure, aimed at clarifying the truth in a way that makes it more easily acceptable. Preaching has nothing to do with torture; *jihād* is the way of torturing [the infidels] at our hands.
>
> By means of *jihād*, Allāh tortures them with killing; by means of *jihād*, Allāh tortures them with injury; by means of *jihād*, Allāh tortures them with loss of property; by means of *jihād*, Allāh tortures them with loss of ruling. Allāh tortures them by means of *jihād* – that is, with heated war that draws its fire from the military front…

Such an encomium to violence would have been inconceivable to Ibn Taymīyah. For in the chapter subsequent to his discussion of *jihād*, Ibn Taymīyah considered the penalties on murderers and the rights of relatives to retaliation. The Qur'ānic injunction 'let him not exceed the limit in slaying' (Q.17:33) was cited in his discussion of premeditated killing and the retaliation which follows. The murderer may have started the aggression, but he has stirred up a desire for retaliation in others who behave and are prepared to act as the pre-Islamic pagans used to behave, that is, without restraint.[70] All Muslims are equal and subject to this requirement of the law of retaliation. But Ibn Taymīyah also cites Q.5:32 and the *ḥadīth* that 'on the last day, Allāh will judge among the people by the blood' which they have shed in this world.[71] The perpetrator of violence against other innocent Muslims must take this injunction seriously. Elsewhere, though he remained firm about the need for compliance with the command on the believer to pursue *jihād* when necessary, Ibn Taymīyah cited Q.5:8 and Q.2:190 as justification for 'fairness and lack of animosity in the *jihād*'.[72] Armed struggle against the leader came within the general rule enunciated by Ibn Taymīyah: 'whenever there is conflict or competition between benefits and disadvantages, between the good and the bad, it is necessary to prefer what is best on balance'.[73]

For bin Laden, in his statement in August 2003, the realm of modern unbelief (*jāhilīyyah*) extends not merely to the West, but also to those Muslim states which ally with the West or who otherwise fail to answer the call to *jihād*:

> The region's rulers deceive us and support infidels and then claim they still cling to Islām. What increases this deceit is the establishment of bodies to lead the people astray. People may wonder how it is that bodies engaging in [studying] Islamic law and jurisprudence play this role, whether wittingly or unwittingly…
>
> For example, when the regime decided to bring the American Crusader forces into the land of the two holy places [i.e. Sa'ūdī Arabia], and the youth raged, these bodies [the unfaithful clerics]… issued *fatāwā* and praised the behaviour of the ruler…

Bin Laden contends that 'Islām ceases to exist when the ruler is an infidel' and in such circumstances 'there must be an act that will elevate a [believing] *imām*'.

The bin Laden thesis can find no real justification from within the political theory of Ibn Taymīyah. Rather, Ibn Taymīyah had argued that there is no basis in the Qur'ān or the *Sunnah* for the traditional theory of the Caliphate (*khalīfah*) or the divine theory of the imāmate (*imāmah*). He was particularly critical of the Shī'a theory of the divine right of *imāms*, whose alleged 'grace and benevolence' he called 'mere deception'.[74] The ideal and perfect union of personal qualities in the righteous leader had been found only in the first era of Islām. This particular providential dispensation would not be re-enacted.[75] Instead, Islām should be viewed as a social order where the law of Allāh must reign supreme. *Imāms* might be good or wicked, but in no circumstance was armed revolt or deposition permitted. Instead, in all deeds that conformed to the principle of obedience to God, the *imām* was to be obeyed, though his subordinates might be disobeyed in limited cases of notorious scandal or incapacity.[76]

For the historian of his thought, Qamaruddin Khan, the weakness of Ibn Taymīyah's theory was precisely this insistence on obedience when it is obvious that persistent and universal tyranny cannot be endured indefinitely. This is the principal reason, he claims, 'why democratic institutions could not develop in the Muslim community despite the thoroughly republican spirit of Islām'.[77] There may well be other reasons for the difficult coexistence of Islām and democracy; but that is beside the point. The rejection of rebellion in virtually all circumstances is clear. Thus for bin Laden to cite Ibn Taymīyah to support one particular view (his endorsement of *jihād*) while simultaneously inciting rebellion against the rulers of states with majority Muslim populations (a viewpoint Ibn Taymīyah specifically rejected) is as historically misleading as it is politically mischievous.

It is also clear that Ibn Taymīyah would have had no understanding of the extremist language used by bin Lāden's supporters in defending *jihād* as a trial by suffering for infidels which will bring them to the path of righteousness.[78] It may thus be concluded that modern violent Islamists may cite Ibn Taymīyah's name as endorsement for their far-fetched theories, but either they are ignorant of his true views or they deliberately deceive the public in the Islamic world by calling upon his name. In reality, while he certainly criticized those who shirked their obligations to wage a defensive *jihād*, Ibn Taymīyah made distinctions and set limits. His comment on the distinction between cowardice and courage might be applied to the contemporary circumstance of a global threat of terrorism which knows no moral or geographical limits:[79]

> the commendable way to fight is with knowledge and understanding, not with the rash impetuosity of one who takes no thought and does not distinguish the laudable from the blameworthy. Therefore the strong and valiant is he who controls himself when provoked to anger, and so does the right thing, whereas he who is carried away under provocation is neither courageous nor valiant.

5
Jihād as State System: the Ottoman State, Ṣafavid Persia and the Mughal Empire

A *ġhāzī* is one who is God's carpet-sweeper
Who cleanses the earth of the filth of polytheism
Do not imagine that one who is martyred in the path of God is dead
No, that blessed martyr is alive.

Thus wrote the poet and moralist Aḥmadī (734/1334?–814/1412) in the 1390s in his *History of the Ottoman Kings*, which he described as a 'book of holy wars'. For Aḥmadī, all Ottoman rulers were *ġhāzīs*; from being mere raiders they had become holy warriors.[1] The same was true, in origin, of the Ṣafavids and Mughals.

In making a comparison between three Muslim states of the early modern period, there is no case for arguing for complete chronological congruence. Clearly, the Ṣafavid state was of the shortest duration, from 906/1501 to about 1132/1720; the Mughals lasted considerably longer, from 932/1526 to 1273/1857, while the Ottoman state lasted far longer than the others, from *c.* 699/1300 to 1341/1923. In conventional historiography, a fairly quick *coup de grâce* was delivered to the Ṣafavids, while there was a period of Mughal 'twilight' of some 150 years, and an interminable period of Ottoman 'decline', from 973/1566 according to some commentators or 1094/1683 according to others. Gábor Ágoston's recent depiction of Ottoman history as constituting a period of 'Islamic gunpowder empire' to 973/1566, a period of the empire on the defensive between 973/1566 and 1110/1699 and a period of retreat and reform between 1110/1699 and 1241/1826 is much to be preferred.[2] Each empire tended to develop its own distinctive culture – an amalgamation of their common Turco-Mongolian political heritage, their Islamic identity and the regional political cultures they inherited; but there was a great deal of cultural exchange across the new political and religious boundaries. Military technology spread from one culture to another: Akbar's artillery force in

Mughal India was greatly strengthened by military experts sent by the Ottomans.[3] Even the political conflicts and sectarian differences between Ṣafavid Persia and the Ottomans did not lead to a firm cultural barrier between the two states.[4] The resolution of some of the rival claims to supremacy was quite late, however. It was not until about 1137/1725 that it became accepted that two *imāms* coexisted, the Ottoman *sulṭān* and the Mughal emperor, 'whose separate existence was made possible by the ocean which divided their separate dominions'.[5] There was never any real prospect of an alliance of the Sunnī Ottomans and Mughals against Shī'a Persia. Instead of uniting against heresy or the enemies of Islām, Sunnī rulers had *jihād* declared in order to fight each other,[6] for example in the Ottoman onslaught on Mamlūk Egypt. Too late, the consequence of this failure to unite became evident in the nineteenth century, when the Ottoman Empire was left as the only major independent Muslim state.[7]

Jihād as a factor in the rise of the Ottoman state

From the time of Murād I (763/1362–769/1389), the title of caliph (*khalīfah*) was used by the Ottomans as one of their general titles, 'the title having by then lost its original meaning'.[8] 'In its Ottoman version, *ġhāza*, *jihād* (*cihād* in Turkish) became the official *raison d'être* of the Ottoman Empire.'[9] Peter Sugar's depiction of the conceptual framework underlying the expansion of the Ottoman state is deceptively straightforward. In fact, in recent times, historians have debated the nature of the ideology of the *ġhāzīs*, or holy warriors for Islām, in the formation of the Ottoman state. Given that the Ottoman warriors were often allied with Christians and incorporated 'infidels' into their ranks, either they were not *ġhāzīs* or the term *ġhāzī* did not always mean what we think. In predatory raids, launched jointly by Muslims and infidels, both shared in the booty.

The most recent historian of the phenomenon, Cemal Kafadar, supports the latter view. *Ġhāzā* really meant raiding, not divinely-commanded war (*jihād*); as a result, it was not constrained by the legal norms of *jihād* and could serve to emphasize expansion as well as the acquisition of booty and glory.[10] Yet the two ideas of duty and expansion could be linked, as in the two independent reports of Meḥmed II's speech to the council which decided to attempt the conquest of Constantinople (in the end, this was achieved in 856/1453 after a siege of 54 days):[11]

> The *ġhāzā* is our basic duty, as it was in the case of our fathers. Constantinople, situated as it is in the middle of our dominions, protects the enemies of our state and incites them against us. The conquest of this city is, therefore, essential to the future and the safety of the Ottoman state.

After his triumph, he proclaimed that he was the only Muslim ruler who could 'fit out the people waging the holy wars of *ġhazā* and *jihād*'.[12] He fought the *ġhāzā* without a break, 'to a degree that even a contemporary historian found excessive'.[13] Meḥmed II had a strong interest in military science and established at Constantinople an Imperial Cannon Foundry, Armoury, Gunpowder Works and Arsenal, which made it 'probably the largest military–industrial complex in early modern Europe, rivalled only by Venice'.[14] It was mainly with the help of cannon founders, artillerymen and miners from Germany, Hungary and the Balkans that Western technology was successfully appropriated by the Ottomans. Between 40 and 50 Germans were employed in the state cannon foundry almost a century later, in 950/1544.[15] The Imperial ambassador commented that 'no nation in the world has shown greater readiness than the Turks to avail themselves of the useful inventions of foreigners, as is proved by the employment of cannon and mortars, and many other things invented by Christians', though he added an important caveat that there remained a religious impediment to a complete transfer of technology from the West.[16]

In his collection of sultanic laws (*Kanunname*) the Conqueror observed the need to delegate. He defined the grand vizier as

> above all the head of the viziers and commanders. He is the greatest of all. He is the absolute deputy in all matters. The head of the treasury (*defterdar*) is the deputy for my Treasury, and he [the Grand Vizier] is the supervisor. In all meetings and in all ceremonies the grand vizier takes his place before others.[17]

Meḥmed also established the rule of succession which was to last until 1026/1617. On his accession in 854/1451, he had an infant brother murdered to prevent any possibility of a rival to the throne; he had fought a pretender at the siege of Constantinople. As a result of these incidents earlier in his reign he therefore issued a law 'for the order of the world' (by which he meant peace in his dominions) that on his accession to the throne, a new *sulṭān* should execute his brothers. This 'law of fratricide' to prevent the fragmentation of the state[18] was implemented with vigour and even enthusiasm by his successors: Selīm I 'the Grim' (*Yavuz*) deposed his father in 918/1512 and extended the law of fratricide to include the murder of his nephews, while on his accession in 1003/1595, Meḥmed III killed off 19 brothers, and for good measure, 20 sisters as well. The law did not end succession disputes, however. Indeed, it made them more likely, since rebellion was the only alternative to execution. Meḥmed II's successor, Bāyezīd II (886/1481–918/1512) faced a disputed succession until the death of his exiled younger brother Jem in 900/1495. Yet the Ottoman state recovered from three great wars of succession (885/1481–886/1482, 917/1511–919/1513 and 965/1558–968/1561), and the law ensured the triumph if not of the ablest or most suitable, at least of a ruthless *sulṭān*.

In 906/1501, Ismāʿīl Ṣafavid had routed the army of the Ak-Koyunlu dynasty, entered Tabrīz and proclaimed himself Shāh Ismāʿīl I,[19] the first ruler of the Ṣafavid dynasty of Persia (although at first based on Āżarbāyjān only: the complete conquest of the kingdom took him a period of ten years). One of his first actions on his accession was to proclaim the Shīʿa form of Islām as the religion of the new state, thus clearly differentiating Persia from the Ottoman state which might otherwise have tried to incorporate it within its dominions.[20] The names of the twelve Shīʿa *Imāms* were mentioned in the *khuṭbah*.[21] Bāyezīd II implored Ismāʿīl to return to 'orthodox Islām' and cease the massacre of Sunnī Muslims, but did not otherwise intervene in the internal policies of the Ṣafavid state. Selīm was more aggressive, and before he deposed his father he had led raids on Ṣafavid territory from Trebizond, where he was governor. Bāyezīd's inaction was the main justification for the deposition by his son.[22] It was therefore clear what Selīm's policy would be as *sulṭān*. In eastern Anatolia there were many Turcomans who were actual or potential supporters of Shīʿīsm (for the first decade of his reign, Ismāʿīl relied on Turcoman *amīrs* for his support). Selīm saw these as heretics who were potential 'fifth columnists' for the new Ṣafavid state. He ordered their execution before he set off on campaign in 920/1514 against Ismāʿīl, having obtained written *fatāwā* (*fetvās*) from the '*ulamā*' that it was his duty to have the Shāh killed as a heretic and infidel.[23] Thus, although the wars between the Ottoman state and Ṣafavid Persia could be termed dynastic and territorial conflicts (the new Persian state, for example, sought alliances with the Ottomans' enemies such as Venice, and later the Habsburgs), the ideological divide between Sunnīsm and Shīʿīsm was such that each campaign was regarded by the Ottomans as a new *ġhāzā*.[24] At his accession, Shāh Ismāʿīl had declared himself the vicar of God and claimed a share of divinity: he was the spiritual master (*murshid-i kāmil*), possessing the power of interpretation or independent reasoning ('*ijtihād*).[25] His troops were dubbed 'red heads' (*Qizilbāsh*) because of their distinctive red headgear with twelve gores or folds commemorating the twelve Shīʿa *Imāms*. The *Qizilbāsh* were simultaneously spiritual and military–political supporters of the dynasty, though eventually the Shāh had to curb the powers of the *Qizilbāsh* tribal *amīrs* in the interest of protecting the unity of the state.[26]

At Chāldirān in 920/1514, Selīm's forces routed the Persian army. The superiority of the Ottoman forces was a result of their possession of hand guns and artillery, as Selīm's dispatch to his son made clear. The Ottomans possessed 500 cannons, the Ṣafavids none.[27] Why Ismāʿīl's forces lacked these weapons has been considered 'one of the puzzling features of the period'.[28] Richard Knolles' account, first published in 1603, emphasized this factor, especially 'the terror and violence of the Turks' artillery', and makes clear the reason for Ottoman supremacy. Whereas his account of the two rulers was essentially to Selīm's detriment, when it came to their relative resource base, there was no comparison between the two rulers. Ismāʻīl had seized power in Persia relatively recently. He needed to secure the good will of the population, and thus heavy taxation was out of the question: 'his coffers being empty, and wanting money, the sinews of war, he was not able to raise so great an army as he might out of those populous kingdoms and countries'. In contrast, Selīm's forces received their daily wages or monthly pay in ready money from his paymasters. He always had 'a great mass of coin' stored in 'seven towers at Constantinople', and his annual revenue exceeded expenditure by one-quarter.[29]

Yet superior resources did not guarantee success. Although he proceeded to Tabrīz, where a fortnight after the victory the *khuṭbah* was read in his name in the mosques, his attempt to winter in Persia was a failure. The Janissaries refused to winter so far from home and Selīm had to retreat to Constantinople with substantial losses.[30] Tabrīz could be captured quite easily by the Ottomans (it was taken by them in 920/1514, 940/1534, 954/1548, 992/1585 and 1044/1635) but the rest of Āżarbāyjān, the original centre of the Ṣafavid state, was always a more difficult proposition and was used as a rallying-point by the opponents of Turkish conquest. The disaster of 920/1514 taught the Ṣafavids to avoid open battle with the Ottomans. They relied instead on a scorched earth policy. Once the Caucasus mountains were crossed, there were no physical barriers to deter or detain the Ottomans in the summer, but the terrain and climate forced them to withdraw each winter. This suggests a strategic imperative for the Ottomans. It was almost impossible to fight simultaneously in east and west because each campaigning season had to begin at Constantinople, and the contingents had to winter in their fiefs, replacing their men and equipment. The overriding need therefore was for the *sulṭān* to alternate between the eastern and western theatres of war, keeping his enemies off balance. For this reason, the invasion of Persia could not be pursued to a definitive conquest, and for all the support for Sunnī orthodoxy from the Ottoman dynasty Shīʻīsm could not be defeated.

Selīm's other great conquest, in 922/1516–17, was of Syria and Egypt. Here the issue was not one of heterodoxy, for the Mamlūk dynasty was Sunnī. Instead, the weakening of the Mamlūk state, which was under assault from the Portuguese in the Red Sea, and its reliance on Ottoman aid, had led to a justifiable fear that it would be unable to defend the Holy Places from Portuguese attack. More dubious was Selīm's claim, as justification for his invasion, that the Mamlūks had proved incapable of protecting the pilgrim route in the Ḥijāz from Arab robbers.[31] In 922/1516, a delegation from Mecca and Medina was refused permission by al-Ghawrī, the last Mamlūk ruler, to proceed to Constantinople. After marching down the Euphrates valley, Selīm's army routed the Mamlūk army near Aleppo. A second victory six months later on the outskirts of Cairo led to the fall of the Egyptian capital. Selīm had the last ʻAbbāsid caliph, al-Mutawakkil III, sent to Constantinople. There, in a ceremony held at the mosque of Aya Sofya, the caliph allegedly transferred to Selīm and his heirs all rights to the caliphate,[32] which served to link the caliphate with the Ottoman dynasty until its extinction. Süleyman later claimed at his accession that God had brought him to the throne of the sultanate and the position of the Great or Exalted Caliphate.[33] No Ottoman *sulṭān* ever performed pilgrimage (*Ḥajj/hac*) to Mecca during the more than six centuries of the dynasty (684/1286–1341/1922).[34] For Selīm and his successors after 922/1517 it must now have seemed unnecessary, since the caliphate had come to Constantinople instead.

In Egypt, the old Mamlūk order was allowed to subsist, with its laws, and administration; but the independent fiefs (*timars*) were abolished, since the Ottomans required grain and other provisions from Egypt which the *timar* system would have consumed. In Syria, new fiefs were established along the lines of Ottoman practice elsewhere. Selīm profited from the hostility of the local population to the excesses of the last years of Mamlūk rule. His two new acquisitions, Egypt and Syria, yielded about 100 million aspers in revenue out of a total Ottoman revenue of about 530 million. Most of south-western Arabia was conquered, too, except the Yemen (which was left until 975/1568). Selīm thus appropriated the title 'servant and protector of the holy places' following his acquisition of the Ḥijāz, including Mecca and Medina. Since a naval base was established at Suez, and the Portuguese were immediately challenged in the Indian Ocean, the importance of the Asian spice trade in Ottoman strategic thinking is evident.[35] Finally, as a result of the conquest, Ottoman overlordship was gradually extended into the Maghrib, starting with Algiers, through an alliance with the corsair Hayreddin (Khayr ad-Dīn) Barbarossa, who captured Algiers in 923/1518 and declared his allegiance to Selīm the following year:[36] henceforth he was termed a *ġhāzī* in Ottoman sources.[37] Selīm and his successors claimed that it was through the will of Allāh that they had acquired the titles of 'Inheritor of the Great Caliphate, Possessor of the Exalted Imamate [and] Protector of the Sanctuary of the Two Respected Holy Places' which gave them superiority over all other Muslim rulers. Selīm had been 'succoured by God', and was the divinely-appointed Shadow of God, even the Messiah of the Last Age in one text. He was 'Master of the Conjunction', or World Conqueror. He hewed a garden from a disorderly world; but it was left to his son and heir Süleyman to enjoy its fruits.[38]

The greatest Sunnī ruler? Süleyman the Law-Giver (*al-Qānūnī*)

It was under Süleyman the Law-Giver (*al-Qānūnī*, 926/1520–974/1566) that the Ottoman Empire 'reached its regional frontiers and was able fully to assert political legitimacy within its own sphere'. It was in this period that it found its 'characteristic ideological and cultural expression'.[39] His was the longest reign of any Ottoman *sulṭān* and he ruled for considerably longer than his main

European rivals. In modern terms, his state ranked as a superpower, by virtue of its geopolitical situation, its enormous territory and population (except in comparison with India and China), its economic resources and the administrative structure capable of mobilizing those resources.[40]

In 931/1525 the forces of Francis I of France were shattered by the army of the Emperor Charles V at Pavia. The French king was captured and taken to Madrid. He appealed to his exact contemporary, the Ottoman *sulṭān* Süleyman,[41] for support and received a reply that it was not befitting for rulers 'to cower and be imprisoned'. Süleyman proclaimed himself '*sulṭān* of *sulṭān*s, the leader of the lords, the crown of the sovereigns of the earth, the shadow of God in the two worlds [that is, the caliph of Islām], the *sulṭān* and *padişah* [that is, chief among rulers, *şahs* or *shāhs*] of the Mediterranean, Black Sea' and various lands. His 'glorious ancestors' had never refrained from 'expelling the enemy and conquering lands'. He followed in their footsteps, 'conquering nations and mighty fortresses with my horse saddled and my sword girthed night and day'.[42]

The initial response was not particularly favourable, but the basis of an Ottoman–Valois alliance had been formed. Süleyman had already perceived that the political and religious divisions[43] of Reformation Europe had provided a unique opportunity for the Ottomans to expand their position in mainland Europe. Under Louis II Jagiellon, the Hungarian army had been disbanded, and the nobility divided into pro- and anti-Habsburg factions. The opportunity for the Ottomans to establish a client Hungarian state was too good to be missed. On his third campaign, Süleyman marched into Hungary (932/1526) and crushed the forces of Louis II Jagiellon at Mohácz, where once again the Ottoman artillery proved their superiority, this time over the Hungarian cavalry (see jacket illustration).[44] The Ottomans employed between 240 and 300 cannons at Mohácz, whereas the Hungarians had only 85, of which 53 were used in the battle.[45] Advances in Ottoman gunpowder manufacturing, small arms production and gun casting demonstrate the Ottomans' early success in adopting Western military technology and introducing indigenous innovations. The superiority in Ottoman firepower in the sixteenth century forced their enemies to modernize their armies and defences. The Ottomans were not the slow and imperfect recipients of a supposedly superior Western military technology and tactics, as most historians of the Eurocentric school maintain; rather they were important participants of the dynamics of organized violence in the Eurasian theatre of war. Their armies were usually larger than those of their opponents: at Mohácz, Süleyman had a force of at least 60,000 men, whereas the Hungarian army was only 26,000 men.[46] Moreover, for most of the sixteenth and seventeenth centuries their system of defence proved cheaper and more cost-effective than that of their rivals such as the Austrian Habsburgs in Hungary.[47]

After occupying Buda, Süleyman retired from the devastated country and factionalism once again became rife in Hungary. John Zápolyai was elected king by a majority of the Hungarian diet and he was crowned at Stuhlweissenburg in 933/1526. A minority of the nobles, under the leadership of Istvan Bátory and the dowager Queen Mary, summoned a counter-diet which elected King Ferdinand I of Habsburg, the younger brother of the Emperor Charles V. In 933/1527, Ferdinand I defeated Zápolyai at Tokay and had himself crowned at Stuhlweissenburg. Although Zápolyai had been overthrown, he refused to give up his claim to the Hungarian throne, and he appealed to Süleyman for recognition in return for payment of tribute. In 934/1528 Süleyman accepted Zápolyai as vassal and a second Ottoman invasion began in his support in 935/1529. This resulted in the failed siege of Vienna (936/1529): Süleyman withdrew because of the approach of winter, and the fact that his heavy artillery was in Hungary; but Ottoman raiding parties had swept through Austria and Bohemia with impunity.

Ferdinand I was left in control of so-called Royal Hungary, a narrow band of territory to the west and north of Lake Balaton representing no more than 30 per cent of the late medieval Hungarian kingdom (indeed a declining proportion since the Ottoman advance was to continue, reaching its fullest extent only in 1074/1664). Ferdinand tried to extend his possessions by besieging Buda in 937/1530, but this only convinced Süleyman of the need for a further campaign, which began in 938/1532. Ferdinand signed a truce with Süleyman in 939/1533, by which he recognized the *sulṭān* as his 'father and suzerain', agreed to pay an annual tribute, and abandoned any claim to rule beyond so-called Royal Hungary. By 948/1541, Süleyman was again encamped at Buda, and this time direct Ottoman control was implanted on that province (*beylerbeylik*). Hungary was divided into three parts, a division which lasted until 1110/1699. The western portion, Royal Hungary, was largely unaffected by the campaign and remained under the rule of Ferdinand of Habsburg. The largest, central portion was transformed into the Ottoman-controlled province of Buda. The somewhat smaller eastern principality of Transylvania was ruled by John Sigismund Zápolyai as a puppet of the Ottomans.

The pretence of an independent Transylvanian principality was therefore at an end, but it did not stop a desultory war being fought over it between the Habsburgs and Ottomans in the years 959/1552–969/1562 and 971/1564–975/1568. The Ottomans had the upper hand throughout the period, as the treaties following the wars reveal. In 953/1547, Ferdinand agreed to pay annual tribute to the *sulṭān* of 30,000 Hungarian ducats for his possession of Royal Hungary. These terms were repeated in 969/1562, and also in the treaty of 975/1568 between Maximilian II and Selīm II, but this last treaty also contained territorial adjustments which favoured the Ottomans.

If the conditions for the truce imposed by Süleyman on his Christian adversary, Ferdinand I, followed traditional juristic concerns that such a treaty should be of short duration and include a financial arrangement favourable to the Muslim power, his alliance with Francis I broke completely new ground for an offensive alliance system. It is true that during the period of the counter-crusades, the Muslim forces had sometimes allied with Crusader powers against other Crusaders (see Chapter 2). In this case, such alliances could be construed as defensive in nature, the defence of Islamic lands being paramount. It would be difficult to interpret Süleyman's alliance with Francis I, from the Ottoman point of view, as other than offensive in nature. The aim was to create a diversion so that Charles V would be unable to support his brother's cause in Hungary. Ottoman naval superiority in the eastern Mediterranean was to be reinforced and extended, if possible, to the western Mediterranean. In particular, Charles V's capture of Tunis in 941/1535 was not to be repeated at Algiers in 947/1541. Furthermore, in return for France obtaining a permanent treaty from the Ottomans in 941/1535, conferring trading advantages (itself based on the precedent of a treaty with Venice signed in the first year of Süleyman's reign),[48] Francis I was required to extend the fullest naval cooperation to Barbarossa, which resulted in Toulon becoming a Muslim-controlled port within the kingdom of France for eight months in 950/1543–951/1544.[49] Ottoman dynastic interest had by this date clearly prevailed over the juristic tradition that arrangements with Christian powers were of a temporary nature only. In effect, Süleyman had determined that 'mine enemy's enemy is my friend', whether in the case of Valois France (clearly preferable to the Habsburg dynasty, since further away from Ottoman territory) or the Protestant nobility of eastern Europe (the bastion against Austrian Habsburg militant Counter-Reformation Catholicism). Under Ottoman rule, Calvinism was propagated freely in Hungary and Transylvania, which became a Calvinist and Unitarian stronghold.[50]

Ibrahīm Pasha, Süleyman's grand vizer extraordinary between 929/1523 and 942/1536, referred to the *sulṭān* as the 'universal ruler and refuge of the world' and 'universal ruler of the inhabited world'.[51] Selīm had almost become a world conqueror (*sahib-kiran*). Süleyman would exceed even the achievements of his father; he would become the personification of the Ottoman dynasty.[52] In addition, millenarian expectations were rife that a great 'renewer of religion' (*müceddid*) would appear by the year 960/1552–1553. Süleyman was thus regarded by some as the 'World Emperor and Messiah of the Last Age'. As the divinely designated ruler of the world, he had a messianic mission and enjoyed special support from unseen saints. His ultimate victory and the establishment of the universal rule of Islām would be ensured by this army of invisible saints who fought at his side.[53] The world was filled with injustice, but as the messianic ruler in the tenth century of the Muslim era, Süleyman's commitment to perfect, impartial justice, would restore order and justice.[54]

Gradually, after the execution of Ibrahīm Pasha in 942/1536, there was a greater emphasis on the modification, compilation and codification of imperial ordinances and their reconciliation with the dictates of Islamic law (*sharī'ah* or *şeri'at*) thus earning Süleyman the epithet of *al-Qānūnī*. The collections of legal texts (*Kanunnames*) were a vital resource for the greatly strengthened provincial administration of the Ottoman state.[55] Once the Islamic Millennium had arrived, it was no longer an appropriate theme for emphasis. The last decade of the reign, from the Peace of Amasya with Persia in 962/1555, which inaugurated a period of over 30 years of peace with the Ṣafavids, implicitly recognized that the dream of world conquest was illusory.[56] (After two further wars, the Peace of Zuhāb of 1049/1639 confirmed Süleyman's acquisition of Baghdād, Baṣra and ʻIrāq and thus access to the Persian gulf, but left Tabrīz, eastern Georgia and Āzarbāyjān as Ṣafavid territory. This treaty brought about a long period of peace with Persia which lasted until 1188/1726, after the fall of the Ṣafavid dynasty.) In the earlier part of the reign, a theory of the universal sovereignty possessed by the *sulṭān* as caliph was close to being established: according to this idea, Süleyman was the '*sulṭān* of the people of Islām', and the Ṣafavids and any others who disobeyed his commands were no more than 'rebels'.[57] The Peace of Amasya was more realistic, tacitly acknowledging the parity of Ṣafavid dynastic legitimacy and negotiated geographical boundaries as a legal definition of statehood.

Problems of factionalism and the disintegration of military power under the later Ottomans

When the era of conquest came to an end, Ottoman revenues were unable to increase further in a period of rapid inflation. This led to the rapid debasement of the currency and consequential reduction in the real value of the revenues. The finances of the state were poorly administered. There was a deficit in all but three years for which accounts survive in the period between 972/1565–973/1566 and 1111/1700–1112/1701, and some years such as 1005/1597–1006/1598 were real years of crisis.[58] Each new accession saw a period of largesse which imposed a crippling burden of debt on the later years of the reign. In the seventeenth century, before the rise of the Köprülü dynasty in 1066/1656, the only period of relative financial stability had been the years of Murad IV's majority and its aftermath (1041/1632–1051/1642). The Köprülü grand viziers reduced the deficit while at the same time constructing and arming the fleet and financing the war with Crete.

Undoubtedly one of the fundamental causes of the growing financial problems of the Ottoman state was the doubling of the number of state pensioners and paid troops between 970/1563 and 1017/1609, above all the increase in the number of Janissaries after 981/1574, when they were allowed to enrol their sons into what was becoming an hereditary militia.[59] Without their frequent and excessive claims, the Janissaries could not have survived, for it is estimated that while the cost of living in the Ottoman Empire rose tenfold in the years 750/1350–1008/1600, official Janissary pay had risen only four times. Increasingly, they supplemented their income by engaging in artisan and small-scale trading activities. New recruits were not needed and in any case could not be paid: consequently the *devşirme* levy of slaves in the Balkan region was abandoned after 1046/1637. The system of *timars* providing a cavalry force also fell into disarray, since the fiefs were too small to enable the knights (some 201,000 of them in 1063/1653)[60] to finance participation in campaigns. The military ineffectiveness of the *sipahis* was fully revealed in 1004/1596 when some 30,000 abandoned the battlefield at Mezö-Keresztes before the final Ottoman victory and were subsequently dismissed from their holdings in Anatolia. Many fiefs were confiscated by the treasury and farmed out to produce as much revenue as possible, while others were illegally converted into private property by their holders. As a result of these social changes there gradually emerged a powerful group of provincial notables (*a'yāns*), who often served as revenue farmers and drew economic benefits from the demise of the earlier form of Ottoman administration.

Such social and economic changes might not in themselves have proved disruptive, but they exacerbated social tensions which came to a head in a series of sustained rebellions, notably the bandit (*jelali*) movement in Anatolia (1004/1596–1018/1610),[61] the revolt of Abaza Meḥmed in Erzerum (1031/1622–1037/1628) and in subsequent revolts in 1056/1647, 1064/1654–1065/1655 and 1067/1657–1068/1658. The difficulties experienced by the Ottomans in their long wars after 985/1578 were the chief reason for these rebellions. When the frontiers were expanding, the army was kept content by the prospect of booty or the opportunity to settle the new territories as fiefs. Stable frontiers brought discontent in the army. The unemployed soldiers took to brigandage as a means of livelihood; the prospects for Ottoman victory against Ṣafavid Persia at the end of the sixteenth century were diminished by the need to divert military resources to the suppression of discontent in Anatolia. The Ottoman Sultans were thus hoist by their own petard. Without 'long wars' they could not hope to keep the army content; but there could be no guarantee of launching a successful war, and failure made the problems of government worse.

Matters came to a head in the succession crisis of the first third of the seventeenth century. Aḥmad I succeeded in 1012/1603 at the age of 13, and because of his young age the law of fratricide was not applied; nor was it applied systematically thereafter. On Aḥmad's death in 1026/1617, none of his sons was of age; he was succeeded by his brother Muṣṭafā I, who was deposed in 1027/1618 because he was reclusive to the point of madness. This brought to power 'Oṣmān II, the eldest son of Aḥmad, who was then about 14 years of age. Under the influence of grand vizier Dilawar Pasha, he showed – in Sir Thomas Roe's words – 'a brave and well-grounded design… of great consequence for the renewing of that decayed empire'. The aim was, under pretence of defending the borders against Persia, to raise a new army of 60,000 men in the provinces of Asia Minor and Syria, with Kurds providing half the number and the backbone of the new force. This new army would be powerful enough to allow the *sulṭān* to dispense with the Janissaries, who were now proving unreliable. But before the plan could be brought to fruition, the Janissaries rebelled because of lack of pay, and the *sulṭān*'s loss of prestige following the failure of the war with Poland and the siege of Chocim in particular. Roe contended that 'Oṣmān II would not have suffered the rebellion had he not 'lost that awe and reverence which always attendeth upon Majesty'. 'Oṣmān II was assassinated in 1031/1622. This led to Muṣṭafā I's restoration for a period of 15 months until his second deposition in 1032/1623.

Sir Thomas Roe identified two important issues which this brief crisis exemplified. The first was that the *sulṭān* wished to 'settle a new government' for good reason. This was that he stood 'at the devotion of his own troops for peace, or war, life or death, and [was] in effect nothing but the steward or treasurer of the Janissaries'. For their part, the Janissaries and other paid troops were paid extra at every change of reign, in order to secure their loyalty; they had 'tasted the sweet [taste] of prosperous mutinies' and had been rendered insubordinate. 'Oṣmān II had stated that he was 'subject to his own slaves, upon whom he spent great treasures, and yet they would neither fight in war, nor obey in peace, without exacting new bounties and privileges'.[62]

The second issue exemplified by the crisis under 'Oṣmān II was the descent into factionalism. Every three or four months, 'by the change of the vizier, the provinces were destroyed', Sir Thomas Roe reported, because 'they placed and displaced the [provincial] governors according to their own factions'. Contrary orders were sent out throughout the Ottoman lands so that 'no man knew who was king [*sulṭān*] or vizier, nor whom they should obey'.[63] Allowing for some exaggeration in the report, it is nevertheless clear that periods of strong rule, for example under Süleyman (three of whose grand viziers were very able men and held power for nearly half his reign) were also periods of stability for the grand viziers. In contrast, periods of weak rule were also made worse by instability in this key office. One of the principal features of Ottoman instability in the years 986/1579–1026/1617 was the very rapid turnover of grand viziers, who on average survived less than two years in office. This was a consequence of increased court and harem intrigue after Süleyman's death. When Köprülü Meḥmed Pasha was appointed in 1066/1656, he was the eleventh grand vizier in a reign that had lasted only eight years. It is from his period that firm rules were established about the role of the vizier.[64]

All of which raises the question of the effectiveness of the Ottoman army, which had been the great strength of the state in the period of Selīm and Süleyman. It is true that 300,000 men, infantry and cavalry could be paid at the outset of the Polish campaign in 1620.[65] Yet size was not everything, because this campaign failed. Central resources provided for the upkeep of up to 190,000 troops during the war leading to the loss of Hungary in 1110/1699, at a cost of nearly 60 per cent of total expenditure. But these formed only part of the costs, since there were in addition soldiers paid by the provincial treasuries as well as the *timar*-holder cavalry and other costs that were assigned on tax-farm revenues.[66] From being an instrument of Ottoman power, the army became something of a liability and certainly a factor in inertia in the state. The Janissaries were already the principal obstacle to reform, as the assassination of 'Oṣmān II had shown; yet they were not abolished until 1241/1826, and then in bloody circumstances.[67]

While there is evidence of Ottoman literature of the seventeenth century which, in the manner of the Spanish *arbitristas*, bemoans the 'decline' of institutions in the state such as the *timar* system and the lack of leadership from the top of the structure,[68] the striking feature of the Ottoman polity is its resilience and recuperative powers.[69] It took the Austrian Habsburgs 16 years to recapture Hungary after the failure of the second Ottoman siege of Vienna in 1094/1683. Even at the Peace of Carlowitz in 1110/1699, the Ottoman state lost only part of Süleyman's territorial gains (Hungary and Transylvania as well as Podolia). The Ottomans had lost territories before and recovered them. The state still represented the full embodiment of coordinated Muslim power. It was far too early to talk of Ottoman decline or the 'sick man of Europe'. After 1130/1718 an unprecedented period of 50 years of peace followed, interrupted only by a brief campaign in which Serbia and western Wallachia were recaptured (1148/1735–1152/1739). No event before the middle of the eighteenth century was likely to shake the Ottomans' 'reliance on the power of their sword, the justice of their rule or the righteousness of their faith'.[70] The success of Ottoman armies against Russia in 1122/1711 and against Austria in 1149/1737–1152/1739 is often forgotten.[71] It was not poor military performance, but the failure to exploit the opportunities of peace in order to implement far-reaching structural reforms which was to prove highly damaging in the long term. This became evident in the Ottoman–Russian war of 1182/1768–1188/1774, followed by the Russian annexation of the Crimea in 1197/1783.

Mughal exceptionalism

Tīmūrid India far outstripped the Ottoman and Ṣafavid state in terms of its resources. Under Akbar (963/1556–1014/1605), the empire tripled in size. By 1008/1600 it had a population of about 110 million and a land mass of about 2.5 million km^2. In the course of the seventeenth century, it is likely that the population increased somewhat (perhaps to 150 million), as did the land mass (to about 3.2 million km^2).[72] In contrast, the Ottoman state at the death of Süleymān I in 974/1566 had a much bigger land mass of some 9 million km^2 but a population of perhaps only 25 million.[73]

It was not merely sovereignty over this huge population which gave the Mughal dynasty under Akbar its power. For nearly five decades the emperor and his advisers drew upon the profits of military conquest and the existing administrative traditions to establish an efficient system of finance, especially from the land revenue system. In the judgement of John F. Richards, 'when working properly, the system acted to spur commercial activity, to enhance production of foodstuffs and industrial cash crops, and to increase the state's revenue base'. The land revenue system of the Mughals in the seventeenth century surpassed the revenue structures of the contemporary European states in its scale of operation and its organizational cohesion. Above all, unlike the European states, from early on in Akbar's reign until the last decade of the seventeenth century, except for highly unusual years, income far exceeded expenditure.[74]

The tradition of Muslim invasions of Hindustan had begun as early as 14/636 and continued on the basis of almost one raid per decade until 601/1205.[75] Maḥmūd of Ghaznī (388/998–421/1030) undertook 17 forays (407/1017–417/1027).[76] The idiom of these raids was that of *jihād* fought by 'men who considered themselves *ġhāzīs* or fighters for the faith against heterodoxy and polytheism'. Slaves, treasure, ransom and tribute, were the tangible benefits of such *jihād*.[77] There was no concern with establishing a government or ruling infrastructure. In the words of Richard Eaton,

> the predatory nature of these raids was… structurally integral to the Ghaznavid political economy: their army was a permanent, professional one built around an elite corps of mounted archers who, as slaves, were purchased, equipped, and paid with cash derived from regular infusions of war booty taken alike from Hindu cities in India and Muslim cities in Iran.[78]

The Ghaznavid *sulṭān*, Eaton comments, 'never undertook the responsibility of actually governing any part of the subcontinent whose temples he wantonly plundered'.[79]

Subsequently, a more tolerant era in the sixteenth century placed a gloss on the raids. 'Fanatical bigots', it was said, had represented India as 'a country of unbelievers at war with Islām' and had 'incited his unsuspecting nature to the wreck of honour and the shedding of blood and plunder of the virtuous'.[80] Regrettably, there is little reason to suppose that Maḥmūd of Ghaznī was deceived in this way. However, what became the norm was less *jihād* than 'world subduing' (*jahangiri*, a Persian term),[81] or later still, in Abū al-Fazl's circumlocution, 'the extension of the tranquillity of mankind'.[82] Mughal rulers 'treated temples lying within their sovereign domain as state property; accordingly, they undertook to protect both the physical structures and their Brahmin functionaries'. Richard Eaton contends that, as a result, 'the Mughals became deeply implicated in institutionalized Indian religions'.[83]

The Mughal Emperor drew his dynastic legitimacy from the Mongol succession. At Samarkand (in modern Uzbekistan) in 771/1370, Tīmūr placed on his head 'the crown of world conquest'.[84] Tīmūr had claimed to be the 'promoter and renovator of the religion of Muḥammad' and read the *khuṭbah* in his own name in the mosque.[85] The founder of the Mughal empire was Ẓahīr-ud-Dīn Muḥammad Bābur, known simply as Bābur ('lion'). He called himself 'ruler of the empire' (*Pādshāh*) from 912/1507, and was also known as 'carrier of the world-illuminating light' from the time of his capture of Hindustan (932/1526–935/1529) with only 13,000 troops, which seemed to provide proof of divine aid to his cause.[86] The notion that Bābur's officer Mīr Bāqī destroyed a temple dedicated to Rama's birthplace at Ayodhya and then obtained approval from the emperor for the building of a mosque on the site – the Bāburi Masjid destroyed by Hindu nationalists in 1413/1992 – is almost certainly fictional.[87] Though its authenticity has been denied, Bābur's testament of 933/1526 or 935/1528 to his son Humāyūn was categorical on the need to respect India's diversity of religious traditions and to render justice to each community according to its customs. Islām, he contended, could progress by noble deeds rather than terror, but needed to avoid the dispute between Sunnī and Shī'a which was 'the weakness of Islām'.[88]

Bābur remained committed to the Tīmūrid tradition of creating appanages for his various sons, which greatly weakened the position of his successor, Humāyūn (937/1530–947/1540; second period of rule 962/1555–963/1556),[89] which no amount of emphasis on the theory that the king was the shadow of God on earth could counteract.[90] Humāyūn was eventually forced into exile in Iran by Shēr Shāh Sūr, who became the Afghan ruler of north India after successive victories in 945/1539 and 946/1540. It was only a short-lived period of rule (Shēr Shāh was killed in an accidental gunpowder explosion in 951/1545), though his sons remained as rulers until 962/1555. As ruler, Shēr Shāh sought to preserve intact 'the main instrument of his power: a well-recruited, well-paid, trained and disciplined army of horse and foot'.[91]

Though there were important similarities in the ideological premises on which the Ottoman, Ṣafavid and Mughal monarchies were founded, there were also significant differences in governing practice which can most clearly be perceived in the era of Akbar the Great (b. 948/1542; r. 963/1556–1014/1605). Firstly, with regard to the army. The Ottomans recruited non-Muslim military recruits by the slave *devşirme* levy; the slaves were then converted to Islām. In contrast, in India under the Mughals, non-Muslims were not forcibly converted, but were given full admission into the Mughal officer corps as non-Muslims. Akbar reduced the importance of his role as Muslim overlord, and became instead 'the greatest of the Rājpūt masters',[92] the commander of a Hindu warrior force. As Richard Eaton observes:

> what bonded together Mughal officers of diverse cultures was not a common religion... but the ideology of 'salt', the ritual eating of which served to bind people of unequal socio-political rank to mutual obligations: the higher-ranked person swore to protect the lower, in return for which the latter swore loyalty

> to the higher. Such bonds of loyalty among Mughal officers not only ran across religious or ethnic communities, but persisted over several generations.[93]

'As we have taken the salt of Jahāngīr', one group of officers told their besiegers, 'we consider martyrdom to be our blessings [*sic*] for both worlds. You will see what [feats] we perform before you till our death.'[94]

A second, highly significant, difference between the Mughal dynasty and its Ottoman and Ṣafavid counterparts (which were vigorously committed to the defence of their respective orthodoxies), was the tendency towards syncretism in the ruling dynasty, particularly from the time of Akbar's assumption of the role of supreme interpretative guide (*mujtahid*) in 986/1579. The political theorists and Islamic scholars surrounding Akbar were deeply influenced by Shī'a Islām. In particular, they subscribed to the Shī'a notion that God had created a Divine Light that was passed down in an individual (the *imām*) from generation to generation. Akbar was particularly interested in the Chishtiyyah order of Ṣūfīsm and after 969/1562 regularly visited the tomb of the founder of the order, Khwāja Mu'īniddīn Chishtī. In 977/1570, following the birth of his son, he walked the 228-mile distance from Agra to Ajmer to worship at the tomb and give thanks for the birth of his son Salīm. He eventually asserted royal control over the administration of two of the most important shrines belonging to the Chishtiyyah order.[95]

The central theorist of Akbar's reign was Abū al-Fazl ibn Mubārak (958/1551–1010/1602), who joined Akbar's court in 981/1574. He believed that the Imāmate existed in the world in the form of just rulers. The *imām*, the just ruler, had a secret knowledge of God, was free from sin, and was primarily responsible for the spiritual guidance of humanity. Influenced by Platonic ideas, Abū al-Fazl viewed Akbar as the embodiment of the perfect philosopher-king. In Akbar's theory of government, as influenced by Abū al-Fazl, the ruler's duty was to ensure justice (*'adālah*) for all the people in his care no matter what their religion. All religions were to be equally tolerated in the administration of the state, a principle known as *sulahkul*, or 'universal tolerance'; hence the repeal of the *jizya* and the pilgrimage taxes levied on Hindus. For Abū al-Fazl, the prime reason for levying the tax in ancient times was 'the poverty of the rulers and their assistants'. Since the ruler now had 'thousands of treasures in the store-chambers of the world-wide [*sic*] administration', why should a just and discriminating mind apply itself to collecting this tax?[96] Rather than propagating a '*jihād* of the sword', Abū al-Fazl argued that territorial expansion was for the 'repose of mortals' since it extended the 'benefits of peace'. In practice, Akbar had to overcome the rebellion of those who sought to propagate a '*jihād* of the sword' in favour of a rival candidate to the throne.[97] World-wide rule (*jahanbani*) was equivalent to a form of guardianship, which rejected forcible conversion. Akbar himself recognized that he was 'the master of so vast a kingdom' yet acknowledged that the 'diversity of sects and creeds' created problems for 'the conquest of empire'.[98]

The policy of 'peace with all' was not, however, an assertion of Akbar's divinity or anything of the kind. 'There is no God but God, and Akbar is God's *khilāfah'* was the terminology used at court, though its use was restricted. The declaration (*maḥzar*) of 987/1579 was needed because of the 'wide expanse of the divine compassion', that is, the extent of Akbar's lands, which incorporated a 'confusion of religions and creeds', unlike the Sunnī majority of the Ottoman lands or the Shī'a majority of the Ṣafavid state. It was therefore up to Akbar to 'untie the knot', that is, resolve any divergence of opinion between the '*ulamā'* in the interest of his subjects and his administration and in order to create a sense of confidence in royal justice.[99]

Sir Thomas Roe commented that Akbar had thought he might prove as good a Prophet as Muḥammad himself' and issued a new law 'mingled of all'.[100] The expansion of the imperial presence to include both the spiritual and the temporal world was relatively uncontroversial, and parallels developments in Ottoman Turkey and Ṣafavid Iran. The new syncretism, Akbar's 'religion of God' (*dīn-i ilāhī*), was much more controversial. This was essentially a court phenomenon, and a small inner circle at court of less than 20 adherents at that,[101] though it was enough to alarm the Muslim religious establishment. Akbar's half-brother, Mīrzā Muḥammad Ḥakīm, the governor of Kabul, had already led a revolt in 971/1564; on this occasion, he issued a *fatwā* enjoining 'true' Muslims to revolt against Akbar. The revolt was contained by forces under the command of Prince Salīm, the future Emperor Jahāngīr. Ḥakīm was forced to flee to Uzbekī territory and never mounted another challenge to the throne;[102] however suspect Ḥakīm's motivation, the rebellion suggests that any attempt to propagate the new syncretism beyond the court would have led to fierce resistance.[103]

A third difference between the Ottoman and Ṣafavid states and Mughal India was that, in 971/1564 Akbar abolished the *jizya* tax, which was the sign of *dhimmī* status. (In contrast, it was not until 1299/1882 that Shāh Nāṣir al-Dīn of Iran abolished the *jizya* on the Zoroastrian community.)[104] Subsequently, long after Akbar's 'religion of God' had been forgotten as a failed experiment, the *jizya* was reintroduced by Aurangzīb (1068/1658–1118/1707) in 1089/1679. This reflected both a stricter adherence to Muslim principles and increased fiscal pressure. The *jizya* was easier to collect than most taxes, so there was always pressure to levy it rather than remit it though rival policies were in competition for the future of the Mughal empire.[105] Notwithstanding this pressure, it is a remarkable fact that the *jizya* was collected on average less than once every three years: in the 197 years of Mughal rule from the foundation of the empire in 932/1526 until the invasion of Nādir Shāh of Persia in 1151/1739, it was collected for only 57 years. Satish Chandra comments that

> the forces which made for mutual toleration and understanding between the Hindus and the Muslims, and for the creation of a composite culture in which

> both Hindus and Muslims cooperated, had been silently at work for the past several centuries, and had gathered too much momentum to be lightly deflected by temporary political difficulties.[106]

At the end of the thirteenth year of his reign (1082/1671), Aurangzīb was faced with a financial deficit: the financial problems worsened in the long period of warfare after 1101/1690. He remained in the Deccan for the last 25 years of his reign (1092/1682–1118/1707), because his nobles would not have obeyed his orders had he absented himself.[107] The nobles preferred profitable deals with the Marāthās to carrying out the Emperor's policy. Finally, in 1110/1699, Aurangzīb broke completely with Akbar's tradition of compromise and declared the war on the Marāthās to be a *jihād*.[108] The growth of cliques and factions reflected a lack of confidence in Aurangzīb's policies. There was opposition to the reimposition of the *jizya* and to the new religious orthodoxy.[109] In the words of Athar Ali, while Aurangzīb's attempt to give a new religious basis to the Empire may indicate that he felt a change was called for, 'the complete failure of this policy showed that religious revivalism could be no substitute for a thoroughgoing overhaul of the Mughal administrative system and political outlook'.[110]

The *manṣabdārī* system was the nucleus of the Mughal nobility. The *manṣabdārs* received their pay in cash or in the form of land, called *jāgīrs*. The *jāgīrs* by their nature were transferable except in the case of the former patrimonial lands of territorial chiefs who had entered Mughal service (*watan jāgīrs*).[111] The system of *jāgīr* transfer was necessary for the unity and cohesion of the empire. While the *jāgīrs* were transferable, the *zamindarī* was permanent and hereditary. The *jāgīrdārs* were responsible for collecting the revenues, even from the *zamindars*. The French traveller François Bernier, writing in 1080/1670 following upon his travels in 1066/1656–1078/1668, attributed the downfall of the Mughal empire to the system of the transfer of *jāgīrs*: 'why should we spend more money and time to render the land fruitful when ultimately it will neither benefit us nor our children?', the *jāgīrdārs* asked themselves.[112] The system could work satisfactorily only if there was enough land available for distribution or if the number of *manṣabdārs* holding *jāgīrs* was kept under control. If *jāgīrs* were granted on a reckless scale, the stage would soon be reached when there would be not enough lands. Scarcity of *jāgīrs* led to an inflation of the estimated income of the land, excessive values which could not be realized in practice.[113] The growing financial pressure took the form of a crisis of the *jāgīrdārī* system and affected every branch of state activity.[114] The Deccan wars took a heavy toll and Aurangzīb granted *manṣabs* to the Deccani nobles generously in order to win their loyalty and in the process exhausted all the available land. The Emperor then cancelled the existing assignments in order to make fresh allotments. This situation gave rise to intense factionalism, which continued into the following reigns and undermined the position of the Emperor. Under Farrukh Siyar

(r. 1124/1713–1131/1719), transfers of *jāgīrs* were on paper only, and could not be realized in practice.[115] Satish Chandra concludes that not 'even the wholesale abandonment of Aurangzīb's policies could... save the Mughal empire from disintegration'.[116] Mughal despotism could be replaced only by the despotism of another dynasty, or by a series of states which held together in a federation governed by a balance of power. From being a force for integration in the empire, the Mughal nobility had become a force for disintegration. Shāh Walī Allāh had wanted the 'annihilation' of the Marāthās and had considered this an easy task;[117] in reality, it was the failure to accommodate the Marāthās and to adjust their claims within the framework of the empire, which carried with it the breakdown of the attempt to create a composite ruling class in India. This undermined the stability of the Mughal Empire well before the East India Company projected itself as an alternative governing structure.[118]

Reviving the state and faith under the later Ottomans

Though invariably referred to as the 'sick man of Europe' from the mid-nineteenth century, and by historians reading this concept backwards to the state after 1094/1683, what is really striking about the Ottoman state is its capacity for survival which implies a capacity for adaptation which is often denied. Daniel Goffman argues that[119]

> the secret to Ottoman longevity and the empire's ability to rule over a vast and mixed collection of territories was not its legendary military, its loyal bureaucracy, its series of competent rulers, or a particular system of land tenure. Rather, it was simply its flexibility in dealing with this diverse society... It fashioned a society defined by diversity (although certainly not equality) of population and flexibility in governance.

The Russian annexation of the Crimea in 1197/1783 was met with a vow from Abdülhamid I that it would be retaken by a *ġhāza*.[120] Bonaparte's capture of Egypt in 1212/1797 was met with a declaration of *jihād* against the 'infidel savages' who had occupied that land. For Selīm III, the French invasion of Egypt endangered Mecca and Medina, and would result in territorial fragmentation and the extirpation of Muslims from the face of the earth (this in spite of Bonaparte's claim that he was the friend of 'all true Muslims': 'have we not destroyed the Pope, who preached war against the Muslims?', he asked rhetorically). Other religions and sects were not safe under the French, Selīm claimed, since their doctrine of liberty was hostile to religion.[121]

If words were to be matched by action, then long-delayed reforms had to be implemented. Selīm sought to transform the *timar* fiefs into leases held by tax farmers (*a'yāns*) without a requirement of military service; he wanted to establish a New Order Janissary corps modelled on the French style of dress. The first proposal was deeply unpopular with the peasantry, since the *a'yāns* were allowed wide latitude in the methods used to collect their revenues. The second met with outright rebellion by the Janissaries and resulted in Selīm's deposition (1222/1807).[122] The following year, the *a'yāns* deposed Muṣṭafā V, his successor, and imposed a document entitled the Pact of Alliance, under the terms of which the government recognized the hereditary nature of their rights to the lands they controlled. Their capacity to make and unmake sultans was short-lived, however. After the end of the Napoleonic Wars in 1230/1815, Maḥmud II succeeded in abolishing the *a'yāns*. After 1246/1831 he abolished the 'remaining but completely dysfunctional' *timars*.[123] In 1241/1826, after strengthening the artillery command, the Janissary corps was abolished, seen as it was as the main impediment to the revival (*ihyā'*) of the state;[124] and those who rebelled were ruthlessly executed. 'The Sultan must show that he can sheath the sword when justice is satisfied', commented the British ambassador.[125]

In making these reforms, there can be little doubt that the Ottomans were reflecting the widespread hostility to the abuses of the Janissaries which had been revealed in the spontaneous first Serbian revolt of 1218/1804. Powerful conservative Muslim elements would not allow Selīm to make concessions to the Christian rebels. Nor could Selīm guarantee that the Serbs would not face revenge from the Janissaries if they laid down their arms. The succession of revolts eventually led to autonomy in 1230/1815 and *de facto* independence in 1245/1830. The Greek revolution of 1236/1821 was, in contrast, an uprising planned to take place in three different locations, including Constantinople. As later defined in 1242/1827, the Greek revolution was a war 'against the enemies of Our Lord', defensive in character, of justice against injustice, of 'reason against the senselessness and ferocity of tyranny', a true war of the Christian religion against the Qur'ān.[126] Maḥmud II came to believe that he was the intended victim of an Orthodox Christian conspiracy, backed by the Russians, and sought a *jihād* declaration from the Chief Mufti (*şeyhülislām*) against the Greek Christians. To his credit, although the act of resistance led to his eventual replacement and execution, the Chief Muftī consulted the Patriarch (who opposed the rebellion) and refused the *jihād* declaration. The Patriarch, three bishops and two eminent priests were executed.[127] Greek independence was recognized in the London Protocol of 1245/1830 and by the Ottomans in the Treaty of Constantinople of 1248/1832.

After the War of Greek Independence, the Ottoman state developed a consciousness of its decline and attempted to change its administrative, educational, and military structures as a matter of urgency. Maḥmud II had destroyed some of the key institutions of the Ottoman State. Something had to be put in their place; above all there had to be a defining set of concepts and the means to implement them in order to replace what had been lost. The Reorganization or Restructuring

(*Tanzīmāt*) Edict of 1254/1839 enunciated the principle of 'equality' without distinction of religion and tried to forestall further Christian rebellions by making it clear that individual property rights would be assured in the future.[128] The government sought to create a new bureaucracy and a new army which depended upon a reformed central fiscal system; there would need to be a new system of schools to train the necessary personnel, and a reformed system of law. The main aim was 'Ottomanism', in the sense of political, social and economic integration which would result in a new shared political identity for citizens.[129]

The Reform Charter (*Islāḥat Farmani*) of 1272/1856, drafted in large measure by Stratford Canning, the British Ambassador, sought to achieve in practice the 'equality' between Muslims and Christians that had been promised in 1254/1839. Nevertheless, it did not lead to equal shares for all communities in the burden of national defence. The edict promised the abolition of the discriminatory poll tax (*cizye* or *haraç*) paid by Ottoman Christians and Jews. The tax was indeed abolished, but in practice it was replaced by an exemption tax, which was first called 'military assistance' (*iane-i 'askerī*), and later 'military payment-in-lieu' (*bedel-i 'askerī*). (This should not be confused with the cash payment *bedel-i naqdī*, the sum of money which could be paid by Muslims in lieu of military service.) The latter was far higher and really only affordable for members of the elite. The net result was that non-Muslims continued not to serve in the army; the 1287/1871 regulations clearly took this situation for granted.[130]

The *millet* system was to be reformed to meet the changed needs of community. The Greek Orthodox *millet* was given a new constitution in 1278/1862, the Armenian Orthodox *millet* the following year and the Jewish *millet* in 1281/1864–65. These changes brought others in their wake, including a sharpening of the distinctions between Muslims and non-Muslims (now it was the Muslims who appropriated the term '*millet*' as a religious–national identity), and a strengthening of the idea of 'nationality'. The Ottoman government was perceived increasingly as a 'Turkish', that is, Muslim majority, government. *Sulṭān* Abdülhamid ('Abd ül-Ḥamīd) II argued that the Ottoman state had rested on four principles: the ruler/dynasty was Ottoman; the administration was Turkish; the faith was that of Islām; and the capital was Istanbul. The foundations of the state would be weakened if any of these principles was undermined.[131] In contrast, the minorities were mostly Christians who considered themselves deprived of freedom and the right of self-determination, and in particular the Orthodox *millet* fragmented into nascent nation states.[132] The Bulgarian insurrection of 1291/1875, the war against Serbia and Montenegro in 1292/1876 and the later insurrections in Herzegovina (1299/1882) and Crete (1314/1897) were movements in which nationalist aspirations were predominant.[133]

The fourth Russo-Turkish war of the nineteenth century, which was launched in 1294/1877 by Alexander II 'for Orthodoxy and Slavdom',[134] lasted ten and a half months. The Russian purpose was to 'neutralize, if not liquidate, the

Ottoman state and the caliphate as a political and cultural–religious force'. It was intended that Bulgaria would become entirely dependent on Russia and its outlet through Salonica to the Mediterranean.[135] With the full backing of the Chief Muftī (*şeyhülislām*), a defensive *jihād* was proclaimed by Abdülhamid II. Parliament proposed that the pacifist *sulṭān* should adopt the title *ġhāzī*, though he had no wish to use it.[136]

The war was a disaster for the Ottoman state. Over 300,000 Muslims were massacred and a million people were uprooted in the Balkans and Caucasus. The Balkan provinces of Serbia, Romania, Montenegro and Bulgaria were lost to the Ottomans (though Bulgaria gained autonomy rather than independence). The Berlin Conference of 1295/1878 produced a settlement which destroyed the illusions that the Ottoman state was an eternal, unchanging, great power. Some two-fifths of the territory of the state and one-fifth of the population were lost. The British, who since 1253/1838 had appeared to act as guarantors of the Ottoman state, changed policy: Cyprus was acquired by the British in 1295/1878 at the expense of the Ottomans, and Egypt in 1299/1882. The Ottoman priority was now self-development and self-reliance, essential requirements if the Muslim state (*devlet-i Islām*) itself was to survive the designs of the partitioning powers.[137] Abdülhamid II's absolutism was reinforced by strong criticism that his interference in the military command had led to the defeat in 1295/1878: as a result, he closed the second Parliament and suspended the constitution of 1293/1876 indefinitely.[138]

Abdülhamid II had two principal policies with which to retrieve what seemed a nearly fatal situation. The first was internal, a reorganization of the army under German guidance. He was forced reluctantly to declare *jihād* against Greece in 1314/1897 because of its absorption of Crete. The war lasted a mere 30 days and was an overwhelming victory for the Ottoman forces, which had been reformed under the influence of Colmar Von der Goltz, though serious problems of military organization remained.[139]

In 1307/1890, Abdülhamid II authorized the levying of light cavalry regiments (*hamidiyyah*) among the Kurds to act as a militia maintaining order in the provinces, following ethnic disturbances between Armenians and Turks at Erzerum.[140] The Armenian *millet* had been sufficiently friendly to the Ottoman government in the nineteenth century that it had been called the 'faithful nation' (*millet-i sadika*). As late as 1312/1895, 2633 Armenians were still in government service, a relatively high proportion of the 1.2–1.4 million Armenians.[141] However, the first voice for autonomy had been raised by the Patriarch of Armenia during the war of 1294/1877. The Armenian rising in the autumn of 1312/1894 served little purpose unless it was to make the Armenian case in Europe by achieving martyrdom following the harsh Turkish reprisals implemented by the *hamidiyyah*.[142] 'The aim of the Armenian revolutionaries is to stir disturbances, to get the Ottomans to react to violence, and thus get the foreign powers to intervene', claimed Currie, the British Ambassador in Istanbul, in 1311/March 1894.[143] By the winter of 1313/1895–96, it was reported that over 30,000 Armenians had perished in the bloodshed of the past two years.[144] The true figure for the number of Armenians killed between 1894 and 1896 is likely to be between 80,000 and 100,000, though whether these killings can be attributed to a preconceived 'plan' is unclear. Sulṭān Abdülhamid was aware of, and consented to, the massacres and in certain instances ordered severe repression – though he may have been kept unaware of the true extent of the pogrom.[145]

Abdülhamid II's second policy was to assert his role as caliph, and to threaten to launch an international *jihād* against imperialism if Muslim interests were seriously damaged by the great powers. Instead of giving priority to an empire which was identified with the Turks, the title of caliph (*Amīr al-Muslimīn*) ought to be emphasized at all times, he considered, since this placed the stress on Muslim unity. Relations with Muslim countries must be strengthened, Abdülhamid wrote:[146]

> As long as the unity of Islām continues, England, France, Russia and Holland are in my hands, because with a word [I] the caliph could unleash the *cihād* among their Muslim subjects and this would be a tragedy for the Christians… One day [Muslims] will rise and shake off the infidel's yoke. Eighty-five million Muslims under [British] rule, 30 million in the colonies of the Dutch, 10 million in Russia… altogether 250 million Muslims are beseeching God for delivery from foreign rule. They have pinned their hopes on the caliph, the deputy of the Prophet Muḥammad. We cannot [therefore] remain submissive in dealing with the great powers.

Friendship between Britain and the caliph helped in defusing tensions during the Indian Mutiny of 1273/1857, or so it was claimed in Turkey.[147] A miscalculated *jihād* could backfire; it was the properly manipulated threat of *jihād* alone which might produce suitable results for Abdülhamid II.[148]

Abdülhamid II was convinced that the European powers, which had seized much of his territory and had engineered the 'liberation' of other parts of his empire, had embarked on a new 'crusade'. In using this term, he was echoing the terminology of writers who made the comparison of contemporary colonialism to the earlier Crusading era. His language was taken up in the pan-Islamic press. The first Muslim history of the Crusading movement, published in 1316/1899, drew attention to the fact that 'our most glorious *sulṭān*, Abdülhamid II, has rightly remarked that Europe is now carrying out a crusade against us in the form of a political campaign'.[149]

The deposition of Abdülhamid II in 1327/1909, following the restoration of constitutional monarchy nine months earlier, brought to power the Young Turks, with a commitment to prevent the formation of political groupings bearing the name of nationalities or races. Rigid adherence to the policy of nationalism (*Kavmiyet*) risked destroying the Ottoman state and provoking Arab separatism. The rapid development of Arab separatist movements after 1328/1911 and the Arab revolt of 1334/1916 destroyed the cooperation underlying the idea of the Muslim *millet* and the underlying *raison d'être* of the Ottoman system.[150]

From World War I *jihād* to genocide: the Young Turks and the Armenian genocide

Ottoman alienation from Britain and France led to increasing dependence on Germany. Enver Pasha, the Minister of War and predominant political figure in the Young Turks government (Committee of Union and Progress [CUP] or *Ittihad ve Terakki Jemiyeti*), and a few like-minded leaders dragged a reluctant cabinet into World War I on the side of the Central Powers.[151] In 1332/November 1914 the call to *jihād* was issued in five separate *fetvās* calling especially on the 'Muslims of Turkish stock in Kazan, Central Asia, Crimea, India, Afghanistan and Africa to rise against their Russian and European masters'. But the call 'elicited… very little Muslim response… for the reason that they had no compelling interest in fighting for one European power against the other'. Furthermore, as Karpat comments, 'the call did not emanate from a free caliph dedicated to the faith but from a small clique who controlled the state and acted in concert with their German ally'. The call made little impact on the war, but served to discredit the caliphate since the declaration smacked of opportunism.[152] In India, news of the *jihād* declaration created little stir, as the British had predicted.[153] 'Abbās Ḥilmī II, Khedive of Egypt since 1309/1892, was in Constantinople at the time of the *jihād* declaration and backed the proclamation: every Egyptian should rebel against British rule, he commanded. None did so. Instead, Britain established a protectorate over Egypt in 1333/December 1914, deposed 'Abbās Ḥilmī II, and proclaimed his uncle Ḥusayn Kāmil '*sulṭān* of Egypt'.[154] In one respect, the *jihād* declaration was of immediate importance: it legitimized the formation of irregular (*chete*) units, which would ultimately be used in bringing about the Armenian genocide.

Atrocities against Armenians commenced within three days of the *jihād* declaration, following a false rumour that they had rebelled and joined the Russian cause. Mehmet Talat Pasha, Minister of the Interior (1331/1913–1335/1917) and Grand Vizier (Prime Minister, 1335/1917–1336/1918) later revealed the motivation of the government in an interview with Henry Morgenthau, the American ambassador:[155]

> We base our objections to the Armenians on three distinct grounds. In the first place, they have enriched themselves at the expense of the Turks. In the second place, they are determined to domineer over us and to establish a separate state.

> In the third place, they have openly encouraged our enemies. They have assisted the Russians in the Caucasus[156] and our failure there is largely explained by their actions. We have therefore come to the irrevocable decision that we shall make them powerless before this war is ended…
>
> It is no use for you to argue… we have already disposed of three-quarters of the Armenians; there are none at all left in Bitlis, Van, and Erzeroum. The hatred between the Turks and the Armenians is now so intense that we have got to finish with them. If we don't, they will plan their revenge…
>
> We care nothing about the commercial loss… We have figured all that out and we know that it will not exceed five million pounds. We don't worry about that. I have asked you to come here so as to let you know that our Armenian policy is absolutely fixed and that nothing can change it. We will not have the Armenians anywhere in Anatolia. They can live in the desert but nowhere else.

The governments of France, Great Britain and Russia issued a declaration, in 1333/May 1915, denouncing the atrocities as 'crimes against humanity and civilization' for which all the members of the Turkish government would be held responsible, together with its agents implicated in the massacres.[157]

Armenian males between the ages of 20 and 45 were drafted into the regular army, while younger and older age groups were put to work in labour battalions. Then, in the aftermath of the disastrous outcome of Enver Pasha's winter offensive at Sarikamis, the Armenian soldiers in the regular army were disarmed out of fear that they would collaborate with the Russians. The order for this measure was sent out in 1333/ February 1915. Finally, the unarmed recruits were among the first groups to be massacred. These massacres seem to have started even before the decision was taken to deport the Armenians to the Syrian desert. Many of those Armenians who had been recruited into the regular army units were transferred to the labour battalions as well.

What started happening in 1333/April 1915 was of an entirely different nature. The massacres were aimed primarily at the Armenian male population. In the labour battalions there were tens of thousands of Armenian men, who were already assembled and guarded by armed soldiers. Vehip Pasha, the commander of the Caucasus front, instigated court martial proceedings against those responsible for killing 2000 Armenian labourers. But once the fury was unleashed, rational arguments, even if based on the interests of the army, fell on deaf ears. The German ambassador affirmed in 1333/April 1915 that 'the government is indeed pursuing its goal of exterminating the Armenian race in the Ottoman Empire'. Morgenthau cabled the US State Department, informing them that 'a campaign of race extermination is in progress under a pretext of reprisal against rebellion'. Donald Bloxham argues that the 'provisional law' promulgated on 27

May, permitting the military authorities to order deportations in the interests of 'security' and 'military necessity', removed any further barrier to genocide:[158]

> the very nature of the deportations is sufficient evidence of genocidal intent... [The Armenians] were sent, defenceless and without provision or the means of subsistence, to desert regions where natural attrition could take its deadly toll... the desire of the radicals for massacre was also fulfilled as irregulars and Kurdish and other Muslim tribesmen, alongside some units of the army, descended on the deportees at strategic points. Barely 20 per cent of the deportees from this phase of the deportation programme would reach their desert destinations. The twin track of measures – deportation and accompanying massive killing – was repeated throughout the expulsions from eastern Anatolia, though not in the western provinces, where the deportees passed relatively unmolested to their desert fates.

In 1334/December 1915 a circular telegram clarified that the purpose of the deportations was annihilation of the Armenians. Instructions were issued advising against slowing the deportations and urging the dispatch of the deportees to the desert.[159] At the end of the year, in a decision without precedent in the history of '*jihād* of the sword', Armenians desiring to convert to Islām were to be notified that their conversion could only take place after they reached their final destination. In view of the earlier instructions clarifying the purpose of the deportations as annihilation, the new instructions implied that Armenians were no longer to be allowed to escape destruction for any reason, including even conversion to Islām. Undoubtedly religious fanaticism was an impelling motive for the Turkish and Kurdish rabble who slew Armenians in what they may have believed misguidedly was service to Allāh; but the men who really conceived the crime had no such motive. Practically all of them were atheists, with no more respect for Islām than for Christianity, and with them the one motive was cold-blooded, calculating state policy. No one knows how many Armenians were killed in the massacres and forced to march to the Syrian desert. If the estimate of 1.3 million is correct, then 'as many Armenians were slain as were soldiers serving the French Republic'.[160] The figure would have amounted to approximately half the Armenian population.

Regrettably, only one Turkish government, that of Damad Ferit Pasha, has ever recognized the Armenian genocide for what it was.[161] That government held war crimes trials and condemned to death the main leaders responsible. The court concluded that the leaders of the Young Turks government were guilty of murder: 'this fact has been proven and verified'. It maintained that the scheme of genocide was carried out with as much secrecy as possible; that a public façade was maintained of 'relocating' the Armenians; that they carried out the killing by a secret network; that the decision to eradicate the Armenians was not a hasty decision, but 'the result of extensive and profound deliberations'. Ismail Enver Pasha, Ahmed Cemal Pasha, Mehmed Talat Bey, and others were convicted by the Turkish court and condemned to death for 'the extermination and destruction of the Armenians'.[162]

Following the War of Liberation of 1337/1919–1341/1922, *ġhāzī*[163] Muṣṭafā Kemal's army established a Grand National Assembly. There, in 1341/November 1922, Kemal announced that temporal power would henceforth be vested in the sovereignty of the Turkish people:[164]

> Sovereignty and sultanate are taken by strength, by power, and by force. It was by force that the sons of Osman seized the sovereignty and sultanate of the Turkish nation; they have maintained this usurpation for six centuries. Now the Turkish nation has rebelled, has put a stop to these usurpers, and has effectively taken sovereignty and sultanate into its own hands...

Considering his life in danger, the deposed *sulṭān*, Mehmed VI, took refuge with the British government and requested his transfer 'as soon as possible from Constantinople to another place'. He was deposed as caliph for colluding with Turkey's enemies, and Abdülmecid II was appointed in his place. Finally, in 1344/March 1924, in an act which purported to 'enrich the Islamic religion', the caliphate was abolished by the National Assembly, Abdülmecid II was formally deposed, and all members of the former ruling dynasty were expelled from the Turkish Republic.[165] The last Ottomans were put on board the Orient Express and packed off to Europe. Thus ended over four and a half centuries of Ottoman history from 856/1453 in which state-controlled *cihād* was present at the outset and subsequently had never been very far from the centre of political affairs.

6
Muḥammad Ibn ‘Abd al-Wahhāb and Wahhābism

> The state of Shaykh Muḥammad Ibn ‘Abd al-Wahhāb [Sa‘ūdī Arabia] arose only by *jihād*. The state of the Ṭālibān in Afghanistan arose only by *jihād*. The Islamic state in Chechnya arose only by *jihād*. It is true that these attempts were not perfect and did not fill the full role required, but incremental progress is a known universal principle. Yesterday, we did not dream of a state; today we established states and they fall. Tomorrow, Allāh willing, a state will arise and will not fall…
>
> Abū ‘Abdallah Al-Sa’dī, al-Qaeda’s *Voice of Jihād* Magazine, Issue No. 9: Memri Special Dispatch 650, 27 January 2004

Few figures in the history of Islām have attracted such controversy as Muḥammad Ibn ‘Abd al-Wahhāb (*c.* 1115/1703–1206/1791).[1] For some American authors, particularly those writing in the aftermath of the events of 11 September 2001, the legacy of Muḥammad Ibn ‘Abd al-Wahhāb is entirely negative. The majority of the suicide bombers involved in the attack on the Twin Towers and the Pentagon were of Sa‘ūdī origin. The Sa‘ūdī state is inextricably linked with Wahhābism. Therefore the evil of 11 September 2001 is attributed to the Wahhābī tradition and even to the views of Muḥammad Ibn ‘Abd al-Wahhāb himself (though there is no necessary congruence between the ideas of the founder of a movement and his successors). For Stephen Schwarz, himself a Ṣūfī, anything of Ṣūfī origin is automatically acceptable (even though historically Ṣūfīs, too, have led ‘*jihāds* of the sword’: see Chapter 7). He talks of ‘Wahhābī obscurantism and its totalitarian state’, ‘fundamentalist fanaticism’ as well as describing it as ‘Islamofascism’.[2] Muslims from other traditions denounce Wahhābīs because they call themselves ‘the asserters of the divine unity’, thus laying exclusive claim to the principle of monotheism (*tawḥīd*) which is the foundation of Islām itself. This implies a dismissal of all other Muslims as tainted by polytheism (*shirk*). Thus Hamid

Algar, Khomeini's official biographer, argues that Wahhābism is 'intellectually marginal', with 'no genetic connection' with movements that subsequently arose in the Muslim world. In his judgement, it should be viewed as 'an exception, an aberration or at best an anomaly'.[3]

In the most recent discussion of Muḥammad Ibn 'Abd al-Wahhāb's views, and the first full analysis of his writings which have not received scholarly analysis to date, Natana DeLong Bas takes a more measured view. In her judgement, Ibn 'Abd al-Wahhāb 'taught a more balanced perspective involving the need for the Muslim to express both correct belief (orthodoxy) and correct practice (orthopraxy)'. He defined *tawḥīd* 'as a broad concept encompassing the requirement of recognizing God alone as the Creator and Sustainer of the universe and recognizing God's uniqueness'. *Shirk* comprised 'any word or deed that would violate either monotheism or God's uniqueness'. It is true that he thought that the practices of Shī'a and Ṣūfīs constituted *shirk* 'and thus could not be considered true Islamic practices', but he 'did not exclude such people as unbelievers (*kuffār*) who were outside Islām, although he did consider them in error and in need of correction'. According to Natana DeLong Bas, Ibn 'Abd al-Wahhāb emphasized educational means (dialogue, discussion and debate) rather than 'conversions of the sword' as the means of spreading the faith. *Jihād*, in his view, was defensive in nature and did not glorify martyrdom. He did not consider that it should be allowed to descend into a tool for state consolidation (a criticism of the Ottoman use of *jihād*). Its main aim was to win adherents to the faith,[4] not to be a tool for aggression. Thus, his teachings stand in marked contrast to contemporary radical Islamists, most notably Osama bin Laden. If bin Laden is considered a Wahhābī, then 'at the dawn of the twenty-first century, it is clear that there is more than one type of Wahhābī Islām'.[5]

Ibn 'Abd al-Wahhāb's teaching and the practice of *jihād* in his lifetime

Natana DeLong Bas does not deny Ibn 'Abd al-Wahhāb's 'puritan' tendencies, therefore, or the fact that he considered his version of Islām as the only one that was 'true'; what is at issue is whether he espoused violence to achieve his objectives in his lifetime. (If his followers chose to espouse violence after his death, this is another matter. It might be considered that they had misinterpreted the teachings of the father figure of the tradition.)[6] Since Ibn 'Abd al-Wahhāb's teaching was concerned with eradicating polytheism (*shirk*), it might have been expected that he would lay heavy emphasis in his writings on forbidding wrong. Surprisingly, according to Michael Cook, he did not. The two most prominent occasions when he referred to this duty were in a letter to his followers at Sudayr and in a discussion of the duties of scholars. To his followers he said that it was important to perform the duty with tact. If the offender was a ruler (*amīr*), it was important not to criticize him in public. Minimizing the demands of the duty did not, in Ibn 'Abd al-Wahhāb's view, damage the integrity of the mission. In the second case, in earlier times scholars had carried out their duty of commanding right and forbidding wrong, pitting themselves against heresy. The struggle against polytheism was of a different, and more fundamental kind.[7]

Ibn 'Abd al-Wahhāb made it clear that it was the responsibility of every individual to engage in direct, personal, study of the Qur'ān and the *ḥadīth*. He cautioned against using unclear Qur'ānic passages to justify conflict with other Muslims, as the Khārijīs[8] and Mu'tazilites had done. True authority over the community, in his view, was based on a shared faith in God and a brotherhood of all believers. He eschewed the cult of the personality: education was to be progressive, with violence a means of last resort.[9] Ibn 'Abd al-Wahhāb was heavily influenced by Aḥmad ibn Ḥanbal (d. 241/855), founder of the Ḥanbalī school of law, as reinterpreted by Ibn Taymīyah (661/1268–728/1328). From Ibn Taymīyah he gained the view that it was polytheism (*shirk*) to introduce the name of a prophet, saint or angel into a prayer (indeed, it was *shirk* to seek intercession from any but Allāh); but Ibn 'Abd al-Wahhāb seems to have been unaware of the dialogue between Ibn Taymīyah and a prominent Ṣūfī of his time on this issue:[10]

> *Ibn 'Ata'Allāh*: Surely, my dear colleague, you know that *istighāthah* or calling for help is the same as *tawassul* or seeking a means and asking for intercession (*shafā'ah*); and that the Messenger, on him be peace, is the one whose help is sought since he is our means and he the one whose intercession we seek.
>
> *Ibn Taymīyah*: In this matter, I follow what the Prophet's *Sunnah* has laid down in the *Sharī'ah*. For it has been transmitted in a sound *ḥadīth*: 'I have been granted the power of intercession' [al-Bukhārī and Muslim, *ḥadīth* of Jābir: 'I have been given five things which no prophet was given before me…'] I have also collected the sayings on the Qur'ānic verse: 'It may be that thy Lord will raise thee (O Prophet) to a praised estate' (Q.17:79) to the effect that the 'praised estate' is intercession… As for seeking the help of someone other than Allāh, it smacks of idolatry.
>
> *Ibn 'Ata'Allāh*: With regard to your understanding of *istighāthah* as… seeking the aid of someone other than Allāh which is idolatry, I ask you: is there any Muslim possessed of real faith and believing in Allāh and His Prophet who thinks there is someone other than Allāh who has autonomous power over events and who is able to carry out what He has willed with regard to them? Is there any true believer who believes that there is someone who can reward him for his good deeds and punish him for his bad ones other than Allāh? Besides this, we must consider that there are expressions which should not be taken just in their literal sense. This is not because of fear of associating a partner with Allāh and in order to block the means to idolatry. For whoever

seeks help from the Prophet only seeks his power of intercession with Allāh as when you yourself say: 'this food satisfies my appetite'. Does the food itself satisfy your appetite? Or is it the case that it is Allāh who satisfies your appetite through the food?

As for your statement that Allāh has forbidden Muslims to call upon anyone other than Himself in seeking help, have you actually seen any Muslim calling on someone other than Allāh? The verse you cite from the Qur'ān was revealed concerning the idolaters and those who used to call on their false gods and ignore Allāh. Whereas, the only way Muslims seek the help of the Prophet is in the sense of *tawassul* or seeking a means, by virtue of the privilege he has received from Allāh… or seeking intercession, by virtue of the power of intercession which Allāh has bestowed on him.

As for your pronouncement that *istighāthah* or seeking help is forbidden in the *Sharīʻah* because it can lead to idolatry, if this is the case, then we ought also to prohibit grapes because they are means to making wine, and to castrate unmarried men because not to do so leaves in the world a means to commit fornication and adultery…[11]

Apart from intercessory prayer, Ibn ʻAbd al-Wahhāb's main doctrinal differences with other Muslims were to assert that all objects of worship other than Allāh were false, and that those who worshipped such were deserving of death; the bulk of mankind were not monotheists, since they sought to win God's favour by visiting the tombs of saints; it was *shirk* to make vows to any other being; it involved unbelief (*kufr*) to profess knowledge not based on the Qur'ān, the *Sunnah* or the necessary inferences of reason; it involved unbelief and heresy (*ilḥād*) to deny the Divine initiative (*qadar*: 'due measure and proportion': Q.54:49) in all acts; finally that it was unbelief to interpret the Qur'ān in the light of hermeneutics (*ta'wīl*). Additionally, Ibn ʻAbd al-Wahhāb's system is said to have departed from that of Ibn Ḥanbal in making attendance at public prayers (*ṣalāt*) obligatory; in forbidding the smoking of tobacco, the shaving of the beard and the use of abusive language; in making alms (*zakāt*) payable on secret profits; and in stressing that the mere utterance of the Islamic creed was insufficient to make a man a true believer.[12]

Three points are worthy of comment here. The first is that the utterance of the creed had always previously been taken as evidence of conversion in *jihād*, except, that is, by Ibn Taymīyah. The second is with regard to Ibn ʻAbd Al-Wahhāb's rejection of interpretation and heremeneutics. Hamid Algar argues[13]

…to imagine that the meanings and applications of the Qur'ān and *Sunnah* are accessible, in any substantial and usable fashion, by disregarding the virtual entirety of post-revelatory Islamic tradition, is unrealistic. It is equally illusory to suppose that either individual or society is a blank space on which the

> Qur'ān and *Sunnah* can be authentically imprinted without admixture from either historical or contemporary circumstance.

This is precisely the clash of views, in contemporary Christianity, between the established churches and the independent (or so-called 'free') evangelical churches, with their primacy on the Word and their rejection of tradition and interpretation.

The third point concerns the visiting of tombs of saints and intercession using the name of a prophet, saint or angel. Here there was a danger that Ibn 'Abd al-Wahhāb's viewpoint not only ignored practices which were enshrined in tradition, consensus and *ḥadīth* but also confused means and ends: it is not the case that what is sought from God through the intercession or by means of a person, living or dead, is actually sought from that person, to the exclusion of the divine will, mercy and generosity.[14] Here, perhaps, a comparison between the Catholic tradition in Christianity and the viewpoint of the Protestant reformers, who were virulently opposed to intercession by the saints, is instructive. Notwithstanding the very great divisions with Christianity over the last 500 years or so, a diversity or plurality of traditions is now recognized as the consequence of different types of spirituality and different theological emphases – though Protestant iconoclasm brought about permanent and damaging change to many churches and religious monuments, much as Wahhābī influence has done.[15]

It became increasingly clear to Ibn 'Abd al-Wahhāb that, in spite of his clear preference for missionary work (*da'wah*) as the means of gaining adherents, 'mere persuasion unaided by political power might prove effective in the case of an individual, but it was difficult to bring about any radical change in a people's outlook without the backing of a political force'.[16] He therefore looked to an alliance with Muḥammad Ibn Sa'ūd (d. 1179/1765), the chief of Dir'iyya, one of the larger Najdī[17] oases. This agreement was struck in 1157/1744: 'you (Ibn Sa'ūd] will perform *jihād* against the unbelievers. In return you will be *imām*, leader of the Muslim community and I [Ibn 'Abd al-Wahhāb] will be leader in religious matters.'[18] 'The alliance was based, as it still is,' wrote Ameen Rihani in 1346/1928, 'upon the sword of Ibn Sa'ūd and the faith of Ibn 'Abd al-Wahhāb.' In the same year (1157/1744) as the treaty,[19] or perhaps somewhat later,[20] the alliance declared *jihād* against polytheism, that is against all who did not share its understanding of *tawḥīd*; the movement was also directed politically against the control of the shaykh of Riyāḍ, Dahhām bin Dawwās.

The *jihād* was to last 30 years[21] until 1187/1773, when Riyāḍ was captured. The essence of Wahhābism, Michael Cook writes,[22]

> was to pit against polytheism a political dominance created by military force. In principle this… could be seen as an instance of forbidding wrong… [but] it was simpler and more effective to identify the militant monotheism of the Wahhābīs

as holy war against the infidel. It was by bringing the frontier between Islām and polytheism back into the centre of the supposedly Muslim world that Wahhābism contrived to be a doctrine of state-formation and conquest.

There seems to have been a contradiction between theory and practice during the *jihād* of the first Wahhābī state. Ibn ʻAbd al-Wahhāb's depiction of *jihād* was intended to set it apart from pre-Islamic practices, especially raiding.[23] Intent was to be the critical motivating factor in undertaking *jihād*: piety and devotion to God ensured that the ultimate purpose of *jihād* was not to eliminate the enemy by the sword, but to persuade him to submit to Islām. Those captured had the choice of submitting to the Muslim authority and paying the *jizya* or death.[24] Following Ibn Ḥanbal, Ibn ʻAbd al-Wahhāb considered it preferable to keep women and children captive so that they became Muslims themselves, rather than to ransom them to the enemy. While the *amīr* was the political and military leader of the *jihād* expedition, the *imām* was responsible for issuing the call to *jihād*, ensuring the spiritual guidance to Muslims during the campaign, and also the preservation of life and property. (Thus, for example, the beheading of enemies or the amputation of hands and feet were prohibited.)[25]

According to a letter from Ibn ʻAbd al-Wahhāb to the people of Qaṣīm, in effect an early Wahhābī creed, he asserted:[26]

> …I am a *Walī* of the Prophet's companions: I mention their good qualities, seek [Allāh's] forgiveness for them, refrain from mentioning their shortcomings, stay idle regarding what happened between them and believe in their virtues…
>
> I assert that *jihād* will always be valid under the *Imām*'s leadership, whether [he is] righteous or sinner; praying behind [sinner] *imāms* is also permissible.
>
> As for *jihād*, it will always be performed and valid from the time that Allāh sent Muḥammad… until the last of this *ummah* fights the [false Messiah (*Dajjāl*)].
>
> *Jihād* cannot be stopped by the injustice of the unjust or even the fairness of those who are just.
>
> I believe that hearing and obeying Muslim rulers is [mandatory (*wājib*)], whether they are righteous or sinners, as long as they do not enjoin Allāh's disobedience.
>
> And he who becomes the Caliph and the people take him as such and agree to his leadership, or if he overpowers them by the word to capture the *Khilāfah* [until he captures it], then obedience to him becomes a necessity and rising against him becomes *ḥarām*.
>
> I believe that people of *bidʻah* should be boycotted and shunned until they repent.

> I judge people of *bid'ah* according to their outward conduct and refer knowledge of their inward [state of faith] to Allāh…

Widespread killing prevented the ultimate purpose of *jihād* – conversion, according to Ibn 'Abd al-Wahhāb,[27] unlike most previous commentators and jurists – from being accomplished. Nor was there licence to take whatever property was seized or to engage in the deliberate destruction of property, the killing of animals or the razing of crops. Minerals or treasure found buried in the earth – the current Sa'ūdī regime of petrodollars, beware! – were, according to Ibn 'Abd al-Wahhāb, to become the collective property of Muslims.[28] He argued that the spoils of war were also collective property and, affirming the preservation of human life as the guiding principle, prohibited any 'cult of martyrdom'. Ibn 'Abd al-Wahhāb denied any requirement for a period of migration or exile in the wilderness (*hijrah*) as a precondition for adherence to the movement: what was needed was an end to disbelief and the cessation of fighting against the forces of monotheism.[29]

Though influenced by Ibn Taymīyah, Ibn 'Abd al-Wahhāb differed from him in two important respects. The first was in the extent to which non-combatants might be drawn into the violence of *jihād*: unlike his predecessor, he stressed that monks should be called to Islām rather than killed. He did not call for the annihilation of Jews or Christians, but wished them to have a *dhimmī* relationship with the Muslim state.[30] Secondly, unlike Ibn Taymīyah, he did not regard anyone who did not adhere to his teachings to be an unbeliever (*kāfir*) who had to be fought. Instead, basing his view on Q.9:66, he argued that only an apostate was truly a *kāfir*. Apostasy could only reasonably be said to have taken place if there had been prior instruction in the Qur'ān and *ḥadīth*, followed by a rejection of the faith on the part of believer. Even so, the prophetic example made fighting against the apostate permissible but not an immediate or absolute requirement.[31] However, the entire Muslim population, with the exception of Ibn 'Abd al-Wahhāb's own followers, were guilty of 'associationism' and thus potentially fell under the term of 'unbelief'.[32]

The emphasis of Ibn 'Abd al-Wahhāb's theory was that *jihād* was not an offensive activity, but a method of last resort to defend the Muslim community from aggression and to allow for proselytism to take place.[33] There seems to have been a difference between the practice of *jihād* before 1201/1787 and afterwards. Three previous British attempts to seize Kuwait had met with failure because of stout resistance. In 1202/1788, the British joined forces with the Wahhābīs in the occupation of Kuwait and received it as their reward for joining the alliance and supplying them with weapons and money:[34]

> It was a well known fact that this Wahhābī campaign was instigated by the British, for [the] Al Sa'ūd were British agents. They exploited the Wahhābī

> [school (*madhhab*)], which was Islamic and whose founder was a *mujtāhid*, in political activities with the aim of fighting the Islamic State and clashing with the other [law schools (*madhāhib*)], in order to incite sectarian wars with the Ottoman state. The followers of this *madhhab* were unaware of this, but the Saʻūdī *Amīr* and the Saʻūdīs were fully aware. This is because the relationship was not between the British and… Ibn ʻAbd al-Wahhāb, but between the British and ʻAbd al-ʻAzīz… and then with his son Saʻūd…

Even if this significant difference in the practice of *jihād* after 1201/1787 is minimized, however, there remain problems with this theory when applied to the formative period of the Wahhābī state. Jerzy Zdanowski argues that the first Wahhābī state was established and expanded as a tool for looting, which became 'both the basis and prerequisite for its existence'. Ecological conditions, together with a minimal potential for productive development and the relatively small volume of expendable produce, especially in the case of nomads, made looting the surest and most effective means for the acquisition of assets. Another source of income was the ransom that was imposed on all settlements and tribes which were subdued by force. Some communities, in realizing that they had no chance to preserve their independence, proposed to pay the ransom out of their own accord. In doing so, they hoped that ransom payments would at least be spread over future years; other cases involved paying contributions in order to buy exemptions from military service. Ransom, whether paid in money or kind, did not differ from the tribute paid by weaker tribes to the stronger ones in pre-Islamic Arabia. When ransom was imposed after conquering an enemy settlement and requisitioning the inhabitants' weapons, armour and horses, ransom did not differ from ordinary looting. The conquering of settlements and adjoining palm groves was often connected to the appropriation of homesteads and land, thereby enabling the Wahhābīs to make trading profits by selling dates and other agricultural produce. After conquering Riyāḍ in 1187/1773, numerous homesteads and palm groves of the inhabitants who escaped from the Wahhābīs passed into the hands of ʻAbd al-ʻAzīz ibn Saʻūd. An especially precious item was the estate of the conquered *amīr*, Dahhām bin Dawwās.[35] There is evidence that Ibn ʻAbd al-Wahhāb had been personally involved in the destruction of a celebrated tomb at Jubaila before 1156/1744:[36]

> One day the Shaykh told the Prince [ʻUthmān bin Muḥammad bin Muammar]: 'let us demolish the dome at the grave of Zaid bin al-Khaṭṭāb… It is erected on deviation. Allāh would not [i.e. does not?] accept it. And the Prophet… had forbidden building domes or mosques on the graves. Moreover, this dome has enthralled the people and replaced their creed with polytheism. So it must be demolished.' The Prince acceded to his suggestion. Then the Shaykh remarked that he was afraid that the people of al-Jubaila would revolt against this action.

> Al-Jubaila was a village close to the grave. 'Uthmān then mobilized an army of six hundred soldiers and marched towards the grave in order to destroy the dome. The army was accompanied by the Shaykh…
>
> The Shaykh thus strove in his preaching and *jihād* for fifty years from [1157/1744] until he died in 1206[/1791]. He resorted to all the methods in his mission – *jihād*, preaching, resistance, debates and arguments, elucidation of the Qur'ān and *Sunnah* and guidance towards the legal ways shown by the Prophet… until people adhered to obedience, entered the Religion of Allāh, demolished the domes and mosques built by them on the graves and agreed to run their affairs in accordance with Islamic Law, discarding all rules and laws which had been applied by their fathers and forefathers…

It would be incorrect to assume that Ibn 'Abd al-Wahhāb was other than a controversial figure in his own lifetime. Muhammad ibn Sulaymān al-Madanī ash-Shāfi'ī (d. 1194/1780), concluded that

> this man is leading the ignoramuses of the present age to a heretical path. He is extinguishing Allāh's light. But Allāh… will not let His light be extinguished in spite of the opposition of polytheists, and He will enlighten everywhere with the light of the '*ulamā*' of *Ahl as-Sunnah*.

In his *Book of Monotheism*, Ibn 'Abd al-Wahhāb wrote:[37]

> Pronouncement alone [that 'there is no God but Allāh; Muḥammad is his messenger'] does not ensure immunity of life and property, nor does the understanding of the meaning of the evidence, nor the pronouncing and acknowledgement of it, nor appealing in prayers (*namazes*) to the one and only Allāh, who has no companions. The property and life of a man are immune only when everything mentioned above is complemented by a complete rejection of all objects of worship except Allāh. Any doubt or hesitation deprives a man of immunity of his property and his life.

On this issue, and the related one of calling Muslims heretics, Ibn Sulaymān al-Madanī argued that 'if a person calls a Muslim an "unbeliever", one of the two becomes an unbeliever. If the accused is a Muslim, the one who accuses [him] becomes an unbeliever.' Against the presumption that a believer was a true Muslim, Ibn 'Abd al-Wahhāb stated: 'we do not care about the words. We look for the intentions and meanings.' He thus contradicted or opposed two *aḥādīth*, one of which declared 'we judge according to the appearance we see. Allāh… knows the secret', while in the other the Prophet refuted the assertion that a dead individual was not a true Muslim and instead asked the question: 'did you dissect his heart?' Ibn Sulaymān al-Madanī repudiated another of Ibn 'Abd al-Wahhāb's arguments by concluding that 'it is certainly permissible to have recourse to the mediation of pious men while it is permissible to make so of good deeds'. 'It should not be forgotten that the wolf will devour the lamb out of the flock', he concluded, with Hell as the punishment for those who reject the Prophet's teaching after right guidance (Q.4:115).[38]

Muslims are divided on how to regard Ibn 'Abd al-Wahhāb and his achievement. 'Shall we deny him the title of reformer?', asked Ameen Rihani in 1346/1928:[39]

> He wrought of a certainty a great reform in Najd; but he did not in a higher sense even point the way to a Reformation in al-Islām. He harks back with a vengeance to the days of the Prophet; destroys the superstitions, that is true, under which succumbed the vital truth of the oneness of God, but rakes up in the process all the old inhibitions which make Wahhābīsm insufferable. Shall we then call him a teacher? He was more than that; for, in addition to teaching the people of Najd a religion which they had forgotten, he infused into them a spirit which, locked as they are in the heart of Arabia, gave them the power to expand and to express their superiority with the austerity, the confidence, and the arrogance of the followers of the Prophet. And he could do this only by sticking to the Qur'ān, cleaving often to the surface meaning of its word...
>
> But how shall we know the real polytheist (*mushrikūn*) from those who have but half-way strayed from orthodoxy? For non-orthodoxy in supplication, for instance, is according to Ibn Taymīyah, of three degrees... in the first degree only, according to Ibn Taymīyah and Ibn 'Abd al-Wahhāb, are the blood and the property of a man forfeit; while in the other two degrees, the guilt might be denoted as a misdemeanour or what is called, in Roman Catholic theology, a venial sin. Now, how are the [Wahhābī agents of enforcement (*Ikhwān*)], in battle with those whom they consider *mushrikūn*, to distinguish the one from the other? This question did not seem to occur to either Ibn Taymīyah or Ibn 'Abd al-Wahhāb...

Wahhābī *jihād* after Ibn 'Abd al-Wahhāb's death: three Sa'ūdī regimes

The first Sa'ūdī regime was of relatively short duration after the death of Ibn 'Abd al-Wahhāb in 1206/1791. It was brought to crushing defeat by the forces of Muhammad 'Alī (r. 1220/1805–1264/1848), ruler of Egypt, who was encouraged by the Ottomans in 1226/1811 to take direct action against the Sa'ūdīs. Before then, the regime had declared its true colours of anti-Shī'īsm in its sack of Karbalā' in 1216/1802 (an attack that was recalled in the aftermath of the Karbalā' bombings of March 2004)[40] and puritanical iconoclasm in its first occupation of Mecca in 1217/1803. The Muftī of Mecca, Aḥmad Zaynī Daḥlān al-Makkī ash-Shāfi'ī wrote:[41]

> In 1217/1802 they [the Wahhābīs] marched with big armies to the area of at-Tayf. In Dhu-l-Qa'dah of the same year, they laid siege to the area [where] the Muslims were, subdued them, and killed the people: men, women, and children. They also looted the Muslims' belongings and possessions. Only a few people escaped their barbarism.
>
> They [the Wahhābīs] plundered what was in the room of the Prophet…, took all the money that was there, and did some disgraceful acts.
>
> In 1220/1805 they [the Wahhābīs] laid siege to Mecca and then surrounded it from all directions to tighten this siege. They blocked the routes to the city and prevented supplies from reaching there. It was a great hardship on the people of Mecca. Food became exorbitantly expensive and then unavailable. They resorted to eating dogs…

Mecca capitulated a second time in 1220/February 1806 and the Wahhābī regime set about the destruction of the ornamental embellishments of the tombs of all the great Muslim leaders, including that of the Prophet himself.[42] Moral 'reform', or the public censure of error, followed in the wake of the Wahhābī takeover. To ensure that the community of the faithful would 'enjoin what is right and forbid what is wrong', enforcers of public morality known as *muṭawwi'īn* (literally, 'those who volunteer or obey') were integral to the Wahhābī movement from its inception. *Muṭawwi'īn* served as missionaries, as enforcers of public morals, and as 'public ministers of the religion' who preached in the Friday mosque. Pursuing their duties in Jedda in 1220/1806, the *muṭawwi'īn* were observed to be 'constables for the punctuality of prayers… [who,] with an enormous staff in their hand, were ordered to shout, to scold and to drag people by the shoulders to force them to take part in public prayers, five times a day'. In addition to enforcing male attendance at public prayer, the *muṭawwi'īn* were also responsible for supervising the closing of shops at prayer time, for looking out for infractions of public morality such as playing music, smoking, drinking alcohol, having hair that was too long [men] or uncovered [women], and dressing immodestly.[43]

Michael Cook's contrast between the first and second (1238/1823–1305/1887) Sa'ūdī states, the second being much more concerned with forbidding wrong in Wahhābī society because the opportunities for an offensive *jihād* were significantly reduced, may thus be somewhat overdrawn. The first Sa'ūdī state had already shown some of this preoccupation with what Michael Cook calls 'turning righteousness inwards'.[44] In reality, considering that it lasted over 60 years, the second Sa'ūdī state deserves fuller attention than it has received. Stephen Schwarz notes that the second state was 'unstable', but adds little to our understanding of its structure.[45] Madawi al-Rasheed talks of a 'fragile Sa'ūdī revival' in this period.[46]

The third Saʻūdī state was created after 1319/1902, when ʻAbd al-ʻAziz ibn Saʻūd (r. 1319/1902–1373/1952) captured Riyāḍ. By 1327/1910 ʻAbd al-ʻAziz ibn Saʻūd was using the traditional method of sending out *muṭawwiʻīn* to the desert tribes 'to kindle in them a zeal for *jihād*'.[47] In 1330/1912, the Muslim Brotherhood (*Ikhwān*) was formed.[48] Agricultural communities called *hujra* were settled by Beduin who came to believe that in settling on the land they were fulfilling the prerequisite for leading Muslim lives; they were making a *hijrah*, 'the journey from the land of unbelief to the land of belief'. It is still unclear whether the *Ikhwān* settlements were initiated by ʻAbd al-ʻAziz ibn Saʻūd or whether he co-opted the movement once it had begun, but the settlements became military cantonments in the service of his consolidation of power. Ameen Rihani reported about 70 of them, each with a population from 2000 to 10,000, which had sprung up in ten years. He also noted that flogging was common in Riyāḍ for those who smoked, for non-attendance at prayer and other offences against the Wahhābī code.[49] He described the *muṭawwiʻīn* as 'fired with the militancy' of the unitarian faith: 'every one… is a Peter the Hermit… these recent recruits to Wahhābīsm, the emigrants of Allāh, are the material of which the *Ikhwān* are made'.[50] It was the duty of every Wahhābī to wage *jihād* against the *mushrikūn* ('polytheists', in this context all non-Wahhābī Muslims).[51] As newly converted Wahhābī Muslims, the *Ikhwān* were fanatical in imposing their zeal for correct behaviour on others: for Rihani, they were 'the roving, ravening Bedu of yesterday, the militant Wahhābīs of today… the white terror of Arabia'.[52] They enforced rigid separation of the sexes in their villages, for example, and strict attention to prayers, and used violence in attempting to impose Wahhābī restrictions on others. Their fanaticism forged them into a formidable fighting force, driven by a strict discipline in the distribution of booty;[53] with *Ikhwān* assistance, ʻAbd al-ʻAziz ibn Saʻūd extended the borders of his kingdom into the Eastern Province and the Ḥijāz. Ultimately, the fanaticism of the *Ikhwān* undermined their usefulness; the failure of the *Ikhwān* rebellion (1346/1928–1348/1930) led to their eclipse.[54]

Had Britain defended the Hashemites in the Two Holy Places, Stephen Schwarz argues that Wahhābīsm might have 'remained an obscure, deviant cult, and the Peninsula would very likely have developed modern political institutions'.[55] In 1343/October 1924, Abd al-ʻAzīz ibn Saʻūd's forces occupied Mecca, and in December the following year they took Medina and Jedda. The possession of the Ḥijāz offered the Saʻūdī state a lucrative source of income from the pilgrim traffic; but this financial consideration did not stop the Wahhābī-influenced destruction of tombs at Mecca and Medina.

The best source for this is Eldon Rutter's account, since he visited the area in 1344/1925 shortly after the Saʻūdī takeover.[56] He noted that 'in their hearts all the town-dwellers and most of the Ḥijāzī Beduin hated the Wahhābīs' because of what had happened.[57] One of the reasons why the Meccans did not worship in the Ḥaram more often, he claimed 'was their hatred of the Wahhābīs, whom they accused of altering the form of the service'.[58] At the Prophet's birthplace, Mawlid al-Nabī in Mecca, the Wahhābīs, 'true to their principles, demolished the dome and minaret of the building and removed draperies and other ornaments from it…'. When the place was mentioned in a gathering of Meccans, 'faces grew grave, and here and there among the company a bitter curse would be uttered against the Najdīs'.[59] At Fāṭima's birthplace (Mūlid Sitna Fāṭima), both the Prophet's praying place and the birthplace of Fāṭima itself had been covered by small domes before the occupation but these were demolished and lay in ruins. Important stones such as the one which it was claimed had spoken to the Prophet were 'indistinguishable from the other stones composing the wall, as the whole had been whitened by the obliterating hands of the Wahhābīs'.[60] A number of birthplaces of Companions of the Prophets had had small mosques built over them, but 'nearly all had been partially destroyed'. The cemetery of El Maala had formerly had many tombs 'crowned by small but handsome domes, but these, without exception' had been demolished, 'together with most of the tombstones'. Eldon Rutter concluded that 'no dome which has the faintest connection with any dead person may continue to exist under the stern Wahhābīte order'.[61] He also visited the Baqī' cemetery of the Prophet's Companions near Medina:[62]

> It was like the broken remains of a town which had been demolished by an earthquake… All was a wilderness of ruined building material and tombstones… Demolished and gone were the great white domes which formerly marked the graves of Muḥammad's family, of the third *Khalīfah*, 'Uthmān, of Imām Mālik, and of others. Lesser monuments had suffered a like fate…

Eldon Rutter provides compelling evidence of the intolerance of the Wahhābī *jihād* against perceived manifestations of polytheism in Islamic traditions other than their own. He called them 'intolerant Puritans'[63] and noted also their intolerance towards others in practice. 'The only point in which the Najdīs do not follow the Prophet', he contended, 'is in their hatred of nearly all modern Muslims save their own community. On account of this one matter it may truly be said that the Wahhābīs do constitute a new sect.' Their dictum with regard to opponents within Islām was, he contended: 'if they be strong, shun them; or if they be weak annihilate them'. Rutter, who could scarcely conceal his loathing for the ideology, accused the Wahhābīs in general, and the *Ikhwān* in particular, of ignorance.[64] Yet while there may have been many ignorant Wahhābīs at the time, Rutter's own discussion shows that in matters of dispute with the remainder of the Islamic world, 'in every instance the verdict of the conference [of the '*ulamā*'] agreed with Wahhābī practice'.[65] The only difference was in the hatred of practices which the Wahhābīs claimed were contrary to Islām. The key point was that they refused to accept the diversity of the Islamic tradition. Hence their refusal to mix with other Muslims 'in prayer or in social intercourse'.[66]

The iconoclasm and puritanical zeal of the new regime lost the Sa'ūdīs friends in the Muslim world. The Indian Khilafatists divided into pro- and anti-Sa'ūdī camps as the news percolated through to the subcontinent.[67] Promoting Wahhābīsm was an asset to 'Abd al-'Aziz ibn Sa'ūd in forging cohesion among the tribal peoples and districts of the peninsula. By reviving the notion of a community of believers, united by their submission to God, Wahhābīsm helped to forge a sense of common identity that superseded pre-existing or parochial loyalties. By abolishing the tribute paid by inferior tribes to militarily superior tribes, Abd al-'Azīz ibn Sa'ūd undercut the traditional hierarchy of power and made devotion to Islām and to himself as the 'rightly guided' Islamic ruler the cement that would hold his kingdom together. The unity of the Muslim *ummah* under al-Sa'ūd leadership was the basis for the legitimacy of the Sa'ūdī state, although this presupposed acceptance of the Wahhābī doctrinal interpretation. Acceptance was to be enforced by a new institution in Mecca in 1345/1926, the Committee for Commanding Right and Forbidding Wrong, which was designed initially to check the aggressive behaviour of the *Ikhwān* towards the local population and foreign pilgrims. A similar committee was set up in Jedda in the same year, and the pattern was followed elsewhere in the Sa'ūdī state. Within four or five years, these committees were taking a strong line, for example in the enforcement of prayer discipline, backed up by groups of Najdī soldiers.[68]

Madawi al-Rasheed argues that Sa'ūdī state was 'imposed' on a people without an 'historical memory of unity or [a] national heritage that would justify their inclusion in a single entity'.[69] The population was divided by tribal, regional and sectarian (that is, Sunnī–Shī'a) differences. (Rihani stated in 1346/1928 that there were about 30,000 Shī'a in al-Hasa alone.)[70] Essentially this population was conquered by an indigenous Najdī leadership allied with Wahhābī religious proselytizers and sanctioned by a colonial power (Britain). Tribal and regional histories and cultural traditions that did not conform to the image of the inevitable rise of the Wahhābī movement and of al-Sa'ūd ascendancy were suppressed. 'Abd al-'Aziz ibn Sa'ūd's virulent hatred of Shī'īsm, however, was revealed by his discreet British alter ego, Philby:[71]

> A suggestion had been put forward at Bahrain that Ibn Sa'ūd was interested in the question of reopening the 'Irāq shrines to Shī'a pilgrimage on account of his Shī'a subjects in the Hasa. His answer to my very tentative question on the subject was decisive: 'I would raise no objection', he replied, 'if you demolished the lot of them, and I would demolish them myself if I had the chance'… The straight-spoken iconoclast of [1336/]1918 was yet to experience the limitations imposed on him by his growing strength and by his sense of responsibility towards the great world of Islām, of which the Wahhābī sect might perhaps be regarded as the kernel or, at most, as a leaven destined to leaven the whole lump…

'Petrodollar puritanism' and the issue of tolerance of diversity in Islām

Wahhābism did not gain its present significant position with Islām because of its inherent strength as a tradition within the faith. It gained its position because of the wealth of the Sa'ūdī monarchy. In 1340/1922, Philby noted that Kuwait and Baḥrayn were the only important commercial outlets of the Wahhābī territories, neither of which was under Wahhābi control. To remedy this unsatisfactory state of affairs was, he thought, 'Abd al-'Aziz ibn Sa'ūd's 'main preoccupation'.[72] If petroleum deposits were confirmed in the state, Philby recognized that this would make the regime incalculably rich. Standard Oil (now Chevron) struck oil in Bahrein in 1350/1932 and hired Philby to negotiate an arrangement with the Sa'ūdī regime.[73] By 1352/mid-1933 a deal for a 60-year oil concession was reached. The consortium, enlarged by the addition of the Texas Oil Company (now Texaco) in 1354/1936 changed its name to Aramco (Arabian American Oil Company) in 1363/1944. Following the substantial increase in oil prices in 1392/1973, the Sa'ūdī government acquired a 25 per cent interest in the company, rising to 100 per cent in 1400/1980.[74] Without doubt, Sa'ūdī oil revenues have paid for the spread and dissemination of Wahhābism. Loretta Napoleoni goes so far as to term the process 'the financing of Sa'ūdī Arabia's religious imperialism'.[75] With a quarter of the world's proven oil reserves (261.8 thousand million barrels in 2002), Sa'ūdī Arabia is likely to remain the world's largest oil producer for the foreseeable future.[76] In principle, therefore, the dissemination of Wahhābism by means of Sa'ūdī resources, or what we would call the process of 'petrodollar Puritanism', seems set to continue, subject to the survival of the regime itself.[77]

The perception of Wahhābism among others, within Islām and outside, is that of intolerance. Some Muslims, indeed, are among the sternest critics of the sect.[78] In Chapter 1, it was noted that the Prophet had feared sectarianism in Islām after his death: there would be 73 sects, he is thought to have said, 72 destined for Hell, with only those determined to maintain the unity of Islām destined for Heaven. Intolerance of others is not a Wahhābī monopoly; many other groups share this characteristic, though perhaps not to the same degree. On 4 March 2004, in the aftermath of the bomb attacks on Shī'a worshippers on the day of Ashura at 'Karbalā in 'Irāq and Quetta in Pakistan, Yoginder Sikand noted the use made of this *ḥadīth* by those seeking to emphasize their group's claims to represent the 'authentic' Islamic tradition against others. A prominent Barelwi scholar argued to Sikand that 'if we try [to] promote unity between the sects that would be going against the saying of the Prophet himself. And that would be a very grave crime indeed!' On another occasion, Sikand was told by a teacher associated with a different group: 'Islām says that our sole purpose must pronounce the truth (*ḥaqīqah*), no matter what the cost.' 'And the truth', he added, 'is what I have written in these books about the other groups that call themselves Muslims. They have actually wilfully or otherwise distorted Islām and are far from the path of the Prophet.' He continued: 'we have to speak out against them, no matter what the consequences. The truth must be clearly distinguished from error.'

Yoginder Sikand correctly concludes:

> every Muslim group claims to be the one saved sect, and implicitly or directly argues that the other groups are, by definition, aberrant, not really Muslim, and hence destined to doom in hell. This firm conviction of having a monopoly over religious truth inculcates an unshakable self-righteousness that dismisses all other truth claims, whether of non-Muslim religious communities or of other Muslim groups.

While noting that there is 'no Islamic counterpart of the Christian ecumenical movement that in recent years has made bold moves to promote understanding and cooperation among different Christian groups',[79] Sikand argued the urgency of 'the need for Muslim ecumenism'. 'The much bandied-about slogan of Islamic brotherhood based on the notion of the pan-Islamic *ummah* falls flat in the face of continued Muslim sectarian rivalry', he concluded.[80]

The problem is more serious than this. The inherent diversity within the Islamic tradition has been denied by Wahhābism, although there may be signs of a change of attitudes among younger scholars.[81] A prominent Wahhābī scholar of the old school comments that there is only one true Islām, the rest being false paths:[82]

> This religion has one path, one direction and is based on one methodology – that which the Prophet of Islām… followed along with his comrades. This religion which Allāh has chosen for mankind is not subdivided into different sects nor does it divert into different paths. However, a number of people have gone astray and corrupted the religion, forming many different groups that bear no relation to Islām… (cf. Q.6:153: 'and [know] that this is the way leading straight unto Me: follow it, then, and follow not other ways, lest they cause you to deviate [literally, 'become scattered'] from His way').

While refutation has always been part of religious education in Islām, it is only relatively recently that it has been recognized that *madrasah* education has, in some areas such as Pakistan, become a source of hate-filled propaganda against other sects and a potent mechanism widening the sectarian divide.[83] By far the greatest increase in the numbers of *madrasahs* in Pakistan (from 1779 out of a total of 2801 in 1988 to 7000 out of a total of 9880 in 2002) has occurred within the Deobandi tradition, which though arising quite separately, has been heavily influenced by Wahhābism in recent times. Of the Deobandi students interviewed, 46 per cent favoured the Ṭālibān as their model. Prominent among the views taught, and among the ideas received by students, are militant views of *jihād* as well as intolerance of others.[84] In the words of another report, 'sectarian tensions are… bound to increase so long as the *jihādī madrasah* is allowed to preach religious intolerance'. The report adds that, 'for the students of these schools, *jihād* against members of other sects is as much a religious duty as *jihād* against non-Muslims'.[85]

Given the imperative of bridging doctrinal and interpretative differences among Muslims so as to encourage mutual accommodation and intra-Muslim dialogue, it is necessary to find a role model from within Muslim tradition itself. Once more, the pre-eminent al-Ghazālī (see Chapter 3) comes to our rescue, since in *The Decisive Criterion for Distinguishing Islām from Masked Infidelity*[86] he provided a compelling defence of the centrality of intellectual freedom, dialogue, and reasoned discourse to the construction of religious knowledge. In this work, al-Ghazālī contended that 'not everyone who embraces senseless hallucinations must be branded an unbeliever, even if his doctrines are clearly absurd', a view that exemplified his belief that no-one can monopolize the truth.[87] Who, he asked, could lay claim to 'this monopoly over the truth… Why should one of these parties enjoy a monopoly over the truth to the exclusion of the other?'[88] Al-Ghazālī argued that those who claimed such a monopoly were merely conflating their own 'interpretation with revelation'. They failed to recognize that their doctrines were grounded in interpretative presuppositions that were historically-determined.

Al-Ghazālī maintained that the only way to decide between the legitimacy of different readings of the Qur'ān, and to reduce interpretative conflict, was through the adoption of an appropriate methodology. He questioned whether consensus (*ijmā'*) could be used as the yardstick for judging what is acceptable or not, given the difficulties of defining consensus consensually. He argued that the task for theologians was to

> establish among themselves a mutually agreed-upon criterion for determining the validity of logical proofs that enjoys the recognition of them all. For if they do not agree on the scale by which a thing is to be measured, they will not be able to terminate disputes over its weight.[89]

Tradition could not merely be imitation: al-Ghazālī pronounced himself thankful not to have been 'afflicted by that blindness that condemns people to being led around by others (*taqlīd*)'.[90]

Finally, he questioned the authority of religious scholars and jurists to pass judgements about who was, and who was not, *kufr*:[91]

> Those who rush to condemn people who go against… any [particular]… school as unbelievers are reckless ignoramuses. For, how [can] the jurist, purely on the basis of his mastery of Islamic law (*fiqh*), assume this enormous task? In

what branch of the law does he encounter [the necessary] skills and sciences? So when you see the jurist who knows nothing but law plunging into matters of branding people unbelievers or condemning them as misguided, turn away from him and occupy neither your heart nor your tongue with him. For, challenging others with one's knowledge is a deeply ingrained human instinct over which the ignorant are able to exercise no control.

As al-Ghazālī put it, 'you must impose restraint on your tongue in regard to the people who turn towards the *qibla* [that is, the direction of the Ka'ba at Mecca]'.[92] Even Ibn Taymīyah, who spent a great deal of effort combating al-Ghazālī's views, came close to this idea when, in his commentary on the 112th *sūrah* of the Qur'ān, he stated that Mu'tazilites, Khārijīs, Murji'tes as well as moderate Shī'a were not to be regarded as infidels. They were in error in their interpretation, but they did not threaten the principle of the law. He was not prepared to be so lenient to the Jahmīya, because they rejected all the names and attributes of God, or to the Ismā'īlīs because they denied the value of ritual law.[93] For Ibn Taymīyah, divergence (*khilāf*) within the community was inevitable, but was minimal among the traditionalists and became greater only as one moved further away from orthodoxy. The main point is that the Muslim community was, and remains, in agreement on the primacy of the *sunnah* and of the *ḥadīth*. The consensus of scholars on this point is reaffirmed at the very moment they are in disagreement on other matters: to settle the question, they all appeal to these sources. Islamic theology is about faithfulness to origins and defending formulations against doubters and detractors. Like jurisprudence, theology is the study or foundations of religion, based on the *sunnah* and the *ḥadīth*, as against *kalām*, which is viewed as a theology of rationalist inspiration.[94]

Wahhābism is here to stay and cannot be wished away by those traditions within Islām which disagree with its interpretation. There has to be an accommodation; and, however difficult it may be to implement or to accept the accommodation, its form has to include the acceptance of diversity within the mainstream traditions, a diversity which, as we have seen, even Ibn Taymīyah accepted. Wahhābīs are entitled to their 'puritan' views, but they are not entitled to impose their views on others, or to destroy sites which are the memorials or places held in spiritual importance by other faiths or by other traditions within Islām. The extreme Deobandi (and perhaps Wahhābī-inspired) destruction of the giant Buddhist statues at Bāmiān by the Ṭālibān in Afghanistan in March 2001 was a religious disaster for the Hazara people and for Buddhism[95] but also for the Islamic tradition of tolerance;[96] it was in addition a cultural heritage disaster for the world at large, though there are hopes that they may be rebuilt.[97] There can be no place in the future mainstream of Islām for such intolerance or for a *jihādī* world view which seeks to impose its views by force and greatly enlarges the scope of *jihād* propounded in the writings of Ibn 'Abd al-Wahhāb himself.[98]

7
Jihād in the Period of the Colonial Powers: Islamic Revivalism and Politicization of the Masses?

> Allāh prescribed the *jihād* upon us in order to remove us... from the harm which arises from leaning towards this impure world and clinging to its things which actually [amount to] nothing and [from which] the only benefit one can obtain is regret. Thus, my beloved ones, support Islām by your souls in order to gain his acceptance... (Muḥammad Aḥmad known as the Mahdī).[1]

'God will send to this *ummah* [that is, the Muslim community] at the head of each century those who will renew its faith for it.' The idea of revival (*tajdīd*) expressed in this *ḥadīth* was, in John O. Voll's expression, 'a longstanding and continuing dimension of Islamic history'. This campaign for revival (*iḥyā'*) or reform (*iṣlaḥ*) was 'an effort of socio-moral construction or re-construction' of the values of both religious and also socio–political life based on the Qur'ān and the *Sunnah*.[2] Karpat notes that in the nineteenth century there were some 24 or 26 revivalist movements which[3]

> started as local or regional movements seeking a return to the basic foundations of Islām – the Qur'ān and the *Sunnah* – and gradually, or in some cases simultaneously, became militant movements of resistance against Russian, Dutch, French, English, and Italian occupation of Central Asia and the Caucasus, the East Indies, North Africa, and Egypt...
>
> Most of the revivalist movements were led by a new brand of Ṣūfīs, whose militancy contrasted sharply with the peaceful, pious, and socially reclusive attitude of classical Ṣūfīsm. The transformation of the Ṣūfīs into guerilla fighters, army commanders, and even state leaders resulted from their belief that *jihād* was not only an effort at personal spiritual enhancement but also a

struggle against *fitnah*, the lapse and degradation of the faith and, ultimately, against those who caused it, be they foreign occupiers or Muslim rulers. The Ṣūfīs believed that in order to achieve self-transcendence the soul must be eternally free and independent of any oppression, limitation, and encroachment – all likely burdens under foreign rule and emulation of Europe. Because any individual Muslim is free to call the *jihād*, the Ṣūfīs did exactly that in order to achieve the 're-Islamization' of society...

Clearly, we cannot encompass all the revivalist movements of the period up to and including the nineteenth century but only some of the more important ones.[4] Before we do so, however, we need to place these revivalist concerns against a longer perspective.

Jihād movements in pre-colonial Africa

As far as is known, the earliest *jihād* in Africa south of the Maghrib was that of Askia Muḥammad I, the ruler of Songhay (897/1492–934/1528). His predecessor, Sunnī 'Alī Ber, had captured Timbuktu and other territories along the River Niger. In spite of his name, he was not, however, a strict Muslim. Askia, in contrast, performed the pilgrimage (*Ḥajj*) to Mecca in 902/1496–97. There, he received a green turban cap (*qalansuwa*), a white turban and a sword and was appointed the *sharīf* of Mecca's deputy *Khalīfah* over Takrur (Western Sudan). (He formally handed over these symbols of authority to his son in 943/1537.)[5] He also received the blessing (*barakah*) of the pilgrim, which gave him the spiritual power, on his return, to declare *jihād* against the Mossi in 903/1498: they were considered both a political and religious threat, even though Sunnī Alī Ber had defeated them 15 years earlier. Al-Sa'dī chronicles that 'there had been no other *jihād* in this region except this expedition' in 903/1498. 'The *jihād* was conducted according to Islamic law', Nehemia Levtzion comments. 'Askia had first sent an ultimatum to the Mossi king, calling him to accept Islām. After consulting his ancestors' in accordance with tribal practice, 'the Mossi king rejected the ultimatum. Askia Muḥammad invaded Mossi country, destroyed towns, and took prisoners (who became Muslims).' Mossi had not been subjugated, however, for three later expeditions took place between 955/1549 and 985/1578.[6] Finally, in 999/1591, the Moroccans invaded with about 4000 troops, mostly musketeers,[7] and destroyed 'the already crumbling political structure of the Songhay empire and... what religious equilibrium there was. Islām was then to become identified, at least in the early years of Moroccan rule, with a tyrannical alien ruling group.'[8]

A second example of *jihād* in Africa took place in Ethiopia after 937/1531. This movement is better known than many, since it had its contemporary Yemeni historian, Shīhāb ad-Dīn Aḥmad bin 'Abd al-Qādir (known as 'Arab Faqīh), whose *History of the Conquest of Abyssinia*, was written in *c*. 947/1541, that is, while the *jihād* was still in progress.[9] The origins of the movement are to be found in the arrival of a Portuguese ambassador, Dom Rodgrio de Lima, at the court of Emperor Lebna Dengal (Dawit II) of Abyssinia in 926/1520 and the attempt of Portugal to establish an alliance with the ruling Christian dynasty. Such an alliance would have potentially outflanked the Ottomans, who had only recently extended their rule to Syria and Egypt. The Ottoman governor of Zabīd in the Yemen undertook to supply a Muslim *jihād* against the Christian alliance with the necessary firearms and support, including troops from the Ottoman army.[10]

The leader who was found for the *jihād* was Imām Aḥmad Grāñ ('the left-handed': Aḥmad Ibn Ibrāhīm al-Ghāzī, c. 911/1506–949/1543), an Adal of Somali origin – not an Ethiopian – who had secured power in Adal and converted it into an Ottoman satellite state. He carried out a series of successful raids and forays into Abyssinia on an ever-increasing scale, until in 933/1527, when only 21 years old, he won a really substantial victory at Eddir over the Emperor's brother-in-law, Degalhan. Guns had not yet reached Abyssinia (two were first brought in by Arabs in 936/1530), and consequently the relative fighting strength of the Muslims was much greater than that of their Abyssinian adversaries, a disproportion which was further increased by Grāñ's real skill as a general and by the indomitable fighting spirit of the *jihādīs* of which 'Arab Faqīh provides eloquent testimony.[11] Emperor Lebna Dengel gained a preliminary victory at Samarna, but in 935/March 1529, the Abyssinians suffered a crushing defeat at Shembera-Kourey, when thousands of their best men were slain, and an enormous amount of booty fell into the hands of Grāñ.

The effects of this battle were decisive; for over a decade the Muslim army pillaged and ravaged the kingdom from end to end. By 941/1535 *imām* Grāñ had conquered the southern and central areas of the state and had even invaded the northern highlands, leaving a trail of devastation behind him. Emperor Lebna Dengal's first son, Prince Fiqtor, was killed in battle against the Muslims at Dewaro in Showa in 942/1536, and three years later a further disaster occurred: the royal Amba of Geshen, in which all the royal princes were held except Lebna Dengal's immediate family, and the vast accumulated treasures of generations of kings, was captured by treachery; the entire population was massacred, and the incalculable wealth stored therein was carried off. At least 50 of the principal churches and monasteries were sacked in the course of the *jihād*. At Mekana Salassie, the church was decorated with sheets of gold and silver, on which there were incrustations of pearls and there were gold statues. Grāñ permitted his troops to set to work with a thousand axes, the chronicler tells us, from mid-afternoon to night. Each man took as much gold as he wanted and was rich forever. The church of Atronsa Maryam was pillaged from midday until the following morning. The *jihādīs* tore out rich brocaded velvets and silks, gold and silver in heaps, gold cups, dishes and censers, a *tābūt* (ark of the covenant) of gold on four feet, weighing more than 1000 ounces, an illuminated Bible bound in sheets of gold, and countless other riches, until they were tired of carrying their loot and loading it up. Much still remained, so they set fire to the church and the store-houses and burned everything.

In 944/1538, Grāñ proposed a marriage alliance with the ruling dynasty to help consolidate his power, but this was refused by the Emperor Lebna Dengal, on religious grounds:

> I will not give [my daughter] to you, for you are an infidel: it is better to fall into the Lord's hands than into yours, for his power is as great as his pity. It is he who makes the weak strong and the strong weak.

The Emperor had placed his trust in the Portuguese alliance to restore his fortunes, but died in 947/1540 before assistance arrived. Instead, it was his son, Emperor Galawdewos (Atnaf Sagad) who benefited from this alliance with the arrival of 400 Portuguese musketeers. The combined force succeeded in defeating and killing Grāñ at Fogera in 949/February 1543. Galawdewos was able to regain his kingdom, though the conversion of most of his subjects to Islām and their reversion to Christianity may have made the effectiveness of his rule problematic at first. In the longer term, the failure of the *jihād* led 'to a great efflorescence of Abyssinian and Christian influence… and to a tradition of religious antagonism between the Christian peoples of the highlands and the Muslims of the lowlands and coast'.[12]

Less is known about some of the later *jihāds*, notably that of Nāṣir al-Dīn, a *marabout* (a member of a brotherhood or teacher),[13] *c.* 1070/1660. It is thought that his followers were mostly Berbers from present-day southern Mauritania, who sought converts to Islām and also to take control of the slave trade. He was killed in battle in 1084/1674. Though he did not participate in Nāṣir al-Dīn's *jihād*, this example is sometimes said to have influenced Mālik Dawda Sy, who launched a *jihād* of his own in Senegambia in 1101/1690. Following the success of this campaign, he founded the dynastic state of Bundu[14] located on the trade route between the Niger and the Gambia, a state which he ruled until his death in 1110/1699. His influence, in turn, is often thought (incorrectly) to have contributed to the *jihād* in Futa Jalon.[15]

The five Fulānī *jihāds* of West Africa

There were five Fulānī-dominated *jihāds* of the eighteenth and nineteenth centuries. These may be regarded as 'religiously-inspired eruptions of reformist zeal, as secular conquests won in the name of Islām, or as Fulānī reactions to Hausa domination, or more plausibly as a mixture of these and other motives'.[16] The *jihād* in Futa Jalon after 1137/1725 led to the creation of an imāmate there, with its capital at Timbo. Subsequently, another imāmate was established at Futa

Toro in 1189/1775–76. There followed three further *jihāds* in the nineteenth century, those of 'Uthmān dan Fodio in Hausaland (1218/1804–1225/1811), of Sheku Hamada in Masina (1230/1815–1236/1821), and of al-Ḥājj 'Umar in the Bambara states of Nyoro and Segu (1268/1852–1280/1864). David Robinson calls the first four 'revolutionary *jihāds*' because of the 'qualitative and permanent changes accomplished by [the] indigenous communities…':[17]

> a number of Fulbe clergy and laity created a self-conscious community, developed their autonomy from the local political establishment, and took up arms when that establishment began to threaten their existence. They succeeded in setting up most of the structures of an Islamic state at the national level and Islamic culture at the local level. In the process they created important new institutions… which blended their Fulbe and Islamic identities and reinforced a sense that they were chosen for holy action in the holy land of the savannah. Subsequently these Fulbe often expanded into other areas, with varying degrees of success, but their strength resided in the core areas where they had reversed the political and social equation…

Karamoko Alfa (Alfa Ibrāhīm Sambegu) returned from a pilgrimage to Mecca inspired with religious zeal. Travelling across the desert, suffering terrible thirst and in danger of losing his life to brigands, he promised Allāh that if He permitted him to return to his home in safety, he would undertake to convert all the infidels in the Futa Jalon to Islām. The first Fulānī *jihād* of 1137/1725 was the fulfilment of his promise. The instrument he chose was his cousin, Ibrāhīm Suri, who had already proved himself a capable general. United by their faith, the Fulānī forces overcame resistance to them and established a theocratic state from 1139/1727, which Karamoko Alfa ruled as *imām* until his death in 1163/1750. The Guinea theocratic state survived until the nineteenth century.

Another imāmate was established at Futa Toro in 1189/1775–76, following seven years of *jihād*. In the Senegalese Futa, a struggle for power arose between Torodbe Muslims and their pagan Fulānī rulers, creating unrest, which was compounded by the raiding of aggressive Moors from the north side of the Senegal River. Under Sulaymān Bal, the Torodbe began a *jihād* which unseated the Fulānī Denyanke dynasty and installed the Torodbe as the new aristocracy with 'Abd al-Qādir as leader. The imāmate continued until the French occupation in the later nineteenth century.[18]

The most significant of the Fulānī *jihāds* was that under the leadership of Shehu 'Uthmān dan Fodio (Shaykh 'Uthmān ibn Fūdiī, 1168/1754–1232/1817), which was launched against the Hausa state of Gobir in 1218/1804.[19] The Shehu had begun to preach as early as 1187/1774,[20] so it was 30 years before the acculturation of ostensibly Muslim rulers with the idolatrous rites of animism forced him to declare *jihād*.[21] These included sacrifices and libations to objects of worship; the failure to observe the Islamic food provisions and prohibited degrees of marriage; the survival of inheritance through the female line in defiance of Islamic law; bawdy songs and an addiction to dance and traditional music; and praise to the rulers drawn from non-Fulānī (or Habe) dynasties which was idolatrous and vainglorious.[22]

The Shehu later recalled that, when he was aged 40 years and 5 months (1208/1794), he met the Prophet in a vision, was turbaned by him, addressed by him as '*imām* of the saints' and commanded to 'do what is approved of' and 'forbidden to do what is disapproved of'; he was also girded with the Sword of Truth (or Sword of God, *sayf al-ḥaqq*).[23] He always denied that he was the Awaited One (*Mahdī*), 'but I am the one who comes to give tidings of the *Mahdī*'.[24] The decision to apply a '*jihād* of the sword' was made more urgent by the decision of the chief of Gobir, Nafata, to proclaim in *c.* 1216/1802 that no one could be a Muslim unless his father had been one; and that without permission no man could wear a turban nor any woman a veil. His successor, Yunfa, continued this anti-Muslim policy so that, in 1218/February 1804, Shehu 'Uthmān proclaimed the 'essential duty' of withdrawal (*hijrah*) from the lands of the heathen (he had already written a tract on this subject some two years earlier).[25] The Hausa rulers who claimed to be Muslims in reality were polytheists and heathen. That the call to *jihād* was based on the Shehu's understanding of the consensus of the community is evident from a selection of arguments drawn from the 27-point manifesto of the movement:[26]

1) That the commanding of righteousness is obligatory by consensus (*ijmā'*);
2) And that the prohibition of evil is obligatory by consensus;
3) And that flight (*al-Hijrah*) from the land of the heathen is obligatory by consensus;
4) And that the befriending of the Faithful is obligatory by consensus;
5) And that the appointment of Commander of the Faithful is obligatory by consensus;
6) And that obedience to him and to all his deputies is obligatory by consensus;
7) And that the waging of... *al-Jihād* is obligatory by consensus;
8) And that the appointment of *amīrs* in the states is obligatory by consensus;
9) And that the appointment of judges is obligatory by consensus;
10) And that their enforcement of the divine laws... is obligatory by consensus;
11) And that by consensus the status of a town is the status of its ruler; if he be Muslim, the town belongs to Islām, but if he be heathen the town is a town of heathendom from which flight is obligatory;

> 12) And that to make war upon the heathen king who will not say 'there is no God but Allāh' is obligatory by consensus, and that to take the government from him is obligatory by consensus;
> 13) And that to make war upon the heathen king who does not say 'there is no God but Allāh' on account of the custom of his town…, and who makes no profession of Islām, is [also] obligatory by consensus, and that to take the government from him is obligatory by consensus;
> 14) And that to make war upon the king who… has abandoned the religion of Islām for the religion of heathendom is obligatory by consensus, and that to take the government from him is obligatory by consensus;
> 15) And that to make war against the king who is an apostate – who has not abandoned the religion of Islām as far as the profession of it is concerned, but who mingles the observances of Islām with the observances of heathendom, like the kings of Hausaland for the most part – is [also] obligatory by consensus, and that to take the government from him is obligatory by consensus;
> 16) And that to make war upon backsliding Muslims… who do not own allegiance to any of the emirs of the faithful is obligatory by consensus, if they be summoned to give allegiance and they refuse, until they enter into allegiance…

The *jihādīs* claimed to be fighting 'in the way of God' and to possess superior motivation than their enemies,[27] one element of this being the propagation of what we consider to have been the false *ḥadīth* about the 72 black-eyed virgins as the reward for a martyr in Paradise.[28] Equally important as the ability to recruit support was the ability to win battles: Muḥammad Bello, the Shehu's son and chief commander, had an unrivalled mastery of cavalry tactics,[29] which gave the Muslim army the edge over its opponents. In 1219/June 1804, 'the prince of Gobir [with Tuareg allies] came out against us and met us in a place called [Tabkin] Kwotto, and God routed them'.[30] Not all the campaigns went as well as this one, however, but gradually the Shehu's authority was no longer confined to the areas his armies had conquered but was accepted by Muslim communities throughout Hausaland. In Hiskett's judgement, 'the main elements of an Islamic state, owing allegiance to an imāmate centred at Gwandu, were already in being several years before the founding of Sokoto, which later became the capital of the Fulānī empire' in 1223/1809–1226/1812.[31]

In the last years before his death, after 1223/1809, Shehu 'Uthmān dan Fodio began to reflect and write on the ideology of the *jihād*. He asserted its moral purpose as combating 'every cause of corruption' and forbidding 'every disapproved thing'. He denied categorically that the campaign was fought for temporal reasons ('I swear by God, I did not accept temporal office in any way'). The Muslim rebels had to be treated as apostates, that is, renegade Muslims who must be slain and buried without washing or prayer in unhallowed graves. The

Shehu seems to have been 'a zealous but wholly orthodox Mālikī theologian' of limited originality,[32] a Ṣūfī in the Qādiriyyah order, who was concerned to defend Ṣūfī practices such as the acceptance of miracles attributed to holy men (*walīs*). Qādirī ideology 'became both the motive force and the rationale that induced' a militant attitude in reformers such as the Shehu.[33] Because of the great distance involved in taking directions from Constantinople, the Shehu backed the idea of an independent caliphate. He emphasized the need for the appointment of good rulers (men of 'outstanding learning, keen insight and extensive study') who would be influenced by the learned. The *imām* exercised essentially a moral authority over the *amīrs*, with whom political and military power remained, though some of the emirates were more closely supervised by Sokoto than others. There were eventually some 15 emirates owing allegiance to Sokoto.[34] When Muḥammad Bello, the Shehu's son and eventual successor died in 1254/1837, he left an empire 'defended by its fortress cities, united, and at the highest peak of power it was ever to attain'.[35]

Even in the lifetime of Shehu 'Uthmān dan Fodio three future trends of great importance were discernible. The community created by his *jihād* began to see itself as a distinct sub-division of the Qādiriyyah order. It had a distinct esoteric litany (*wird*), revealed by God to the Shehu in 1204/1789–90, which 'became the community's sacred patrimony'.[36] In this, the Shehu recalled that, when he was 36 years of age

> God removed the veil from my sight, and the dullness from my hearing and my smell, and the thickness from my taste, and the cramp from my two hands, and the restraint form my two feet, and the heaviness from my body. And I was able to see the near like the far, and hear the far like the near, and smell the scent of him who worshipped God, sweeter than any sweetness; and the stink of the sinner, more foul than any stench… Then I found written upon my fifth rib, on the right side, by the Pen of Power, 'Praise be to God, Lord of the Created Worlds', ten times; and 'O God, bless our Lord Muḥammad, and the family of Muḥammad, and give them peace' ten times; and 'I beg forgiveness from the Glorious God' ten times; and I marvelled greatly at that.

Finally, after his death in 1232/1817, a shrine was built to act as a religious focus for the community, and miracles began to occur, demonstrating in death that the Shehu was indeed a *walī* or holy man and retrospectively justifying the *jihād*.[37] Historians nevertheless stress the ambiguity of his *jihād*. For Mervyn Hiskett, 'the Islamic *sharī'ah* is an ideal. When men try to realize an ideal, it is always possible to cry failure.'[38] The most effective resistance against the Fulānī came from Bornu under the leadership of Shaykh Muḥammad al-Amīn al-Kanimī, who was himself a Muslim reformer (and unlike the Shehu had performed the

Ḥajj to Mecca), but he could see no justification for the Fulānī *jihād* besides political ambition.[39] For his part, M. G. Smith comments that[40]

> no one who has studied the Shehu's writings or life can doubt his primary religious commitment. His *jihād* was successful through a skilful combination of religious and political factors; yet it is precisely this combination which lends it an ambiguous character… This pattern is a general characteristic of Islām, enshrined in the doctrine of *ijma*', by which consensus legitimates necessary changes… The ambiguous character of Shehu dan Fodio's *jihād* derives from the ambiguous character of *jihād* itself.

The *jihād* of al-Ḥajj 'Umar in the Bambara states of Nyoro and Segu of Western Sudan (1268/1852–1280/1864), the fifth in the series of Fulānī *jihāds*, was of a quite different character from its predecessors: it was in essence an imperial war. As a theologian, 'Umar had only slender claims to originality;[41] moreover, the relationship between theory and practice in his *jihād* was much looser than in the earlier campaigns. David Robinson writes:[42]

> The '*jihād* against paganism' was an imperial war, an extension of the Fulbe *Dār al-lslām* into new areas. It was a [campaign][43] not to liberate a Jerusalem or protect persecuted minorities, but to destroy the offensive temples of 'infidelity'. It was an outlet for frustration at societies that could not fulfil the spiritual and material goals of their founders and an opportunity for the truly faithful to start afresh, with a new community, land, slaves, and position. The *talibés*, the 'disciples' and soldiers of the new movement, joined the Tijāniyya, the new order which 'Umar propagated. They fought against notorious warriors and watched many of their own die. They reigned over strange lands and people whom they did not understand and could barely control. Their success and their predicament intensified their consciousness as a chosen people…

For about two decades al-Ḥajj 'Umar received some 1500 to 1800 small arms every year through Bakel and Medine, and this enabled him to maintain a weapons differential over most of his foes. But only the loyalty of a surviving core of *talibés* and other supporters explains the endurance of his garrison state until the French conquest. The *jihād* was constructed around 'Umar's leadership, Fulbe consciousness, and strong religious conviction and managed to survive twelve years of offensive and three decades of defensive warfare. More than any other African *jihād* leader, 'Umar had a broad and long political apprenticeship extending over some 30 years and thousands of miles.

For the Senegalese, 'Umar and his *talibés* were heroes in the cause of Islām against the infidels. Conversely, the Malians regard their ancestors as defenders against invaders who used Islām as a cloak for their imperialism and personal greed. There were accusations that the movement had lost its focus on eliminating idolatry and was instead becoming an occasion for settling scores, waging civil war, and grabbing booty. This argument of *fitnah*, 'trouble' or 'sedition', was invoked by Amadu III and the Kunta. The Kunta, in the person of al-Bekkay, carried this position to its logical extreme by declaring 'Umar an impostor and evil-doer in 1269/1863 in a counter-*jihād* launched against what they called the 'false' *jihād*.

It is true that 'Umar did not mobilize the indigenous inhabitants; nor did he extend commands to local supporters. Rather, he recruited thousands of outsiders, like himself, to conquer and colonize. They concentrated on the destruction of the most visible aspects of 'pagan' religion, not on the administration or education of non-Muslim subjects. They did not stop to consolidate gains, train successors, or reflect on their experience. Thus, in David Robinson's judgement,[44]

> the imperial *jihād*, however necessary in the minds of a generation determined to extend the *Dār al-Islām*, was decidedly less successful in the spread of the faith than its revolutionary predecessor [that of Shehu 'Uthmān dan Fodio]... The defenders clung to their traditional allegiance. Only where colonization and the absorption of women and children were massive did Islām advance...
>
> In so far as an 'Umarian model of state formation existed, it was based on colonization from west to east: an immigrant group settled on the land, administered the state, waged war, brought in new slaves, and exploited the productive capacities of the indigenous inhabitants. While the new ruling class might express themselves in the language of Islamic law, they did not operate in ways qualitatively different from the warrior elites which preceded them. They had the additional stigma of being perceived as foreign... the 'Umarian conquest probably delayed the expansion of Islām because it temporarily associated the Muslim faith with an imperial thrust and intensified loyalty to indigenous institutions.

Robinson considers that al-Ḥajj 'Umar's most lasting contribution to Muslims in Senegal and West Africa was his call to *hijrah* during a recruitment crisis of 1275/1858–59.[45] By attaching the Islamic conception of emigration to the 'pollution' brought on by French expansion, the Shaykh articulated a response to European intrusion that fell between the futility of fighting and the humiliation of surrender. It was used time and time again during the period of the Western nations' 'Scramble for Africa'. *Hijrah* was refusal, non-cooperation, not resistance as such. It assumed an independent if beleaguered Muslim authority to which 'true' Muslims could migrate. The pressure to accomplish *hijrah* was acutely felt by Muslim rulers: Albury emigrated from Senegal in 1307/1890, Amadu from Bandiagara in 1310/1893, and, following the victory of the British at Burmi, so too did Caliph Attahiru from Sokoto in 1320/1903. For many Fulānī, the last of these events is spoken of as a *jihād*, *ṣulṭān* Attahiru as a martyr (*shāhid*), and the exodus as a *hijrah*.[46]

The prototype of the anti-colonial *jihād*: the *jihād* of 'Abd al-Qādir in Algeria

The French invaded Algeria in 1245/June 1830 but met stiff resistance from the outset led by 'Abd al-Qādir (1222/1808–1300/1883) as *amīr* and coordinated by the Ṣūfī Qādiriyyah order. From his capital in Tlemcen, 'Abd al-Qādir set about building a territorial Muslim state based on the communities of the interior but drawing its strength from the tribes and religious brotherhoods. In 1249/1834, his authority was recognized by the French in Western Algeria; but two years later his forces were defeated by the French under the command of Thomas Robert Bugeaud de la Piconnerie. In 1253/June 1837, however, Bugeaud entered into a treaty with 'Abd al-Qādir (the treaty of Tafna), for which he was criticized in France, since it recognized two-thirds of Algeria as remaining under the *amīr*'s control. The *jihād* was resumed two years later in what was in effect a territorial dispute between the colonial and anti-colonial states in Algeria. As a result of the French adopting a ruthless scorched earth policy, 'Abd al-Qādir was obliged in 1259/May 1843 to seek refuge with the Moroccan *ṣulṭān*.

The exile to another state, one which he did not control, altered the nature of 'Abd al-Qādir's *jihād*. Prior to this, the main focus of the *jihād* was on the primary duty of exile to *Dār al-Islām*, the requirement that Muslims should not collaborate with the colonial regime but oppose it in all respects. The *'ulamā'* in the Algerian colonial state appear not to have considered emigration obligatory for Muslims; but 'Abd al-Qādir obtained a *fatwā* from an Egyptian scholar to the effect that it was, while he himself wrote a treatise in his year of exile to Morocco affirming that the obligation to emigrate from *Dār al-Kufr* to *Dār al-Islām* 'will remain in force until the sun rises from the West'.[47] However, he gained no satisfactory answer as to whether Muslim collaborators with the French could be considered 'apostates', so lacked any really decisive coercive principle against the defection of tribes in Algeria to the service of the colonial master.

Once 'Abd al-Qādir was installed in Morocco, the French launched a war against *ṣulṭān* 'Abd al-Rahmān, to force him to renounce support for the *jihād* and to hand over the *amīr*. After a campaign lasting just over a month, the French secured his compliance by the treaty of Tangiers of 1260/September 1844. 'Abd al-Qādir could not conceal his bitterness and sought (to no avail) a *fatwā* against the 'legally abominable deeds' of the Moroccan *ṣulṭān*, which had 'caused us great damage'.[48] Notwithstanding this ultimately fatal setback to the *jihād*, 'Abd al-Qādir won a significant victory at Sidi Brahim near Oran in 1261/September 1845, which required the return of Marshal Bugeaud to command the French forces. In the event, the Moroccan defection and the ruthless French offensive proved decisive: 'Abd al-Qādir surrendered to General Lamorcière and the duc d'Aumale in 1264/December 1847. Thus ended the French conquest of Algeria. 'Abd al-Qādir was treated with respect by the French and released in 1269/1852 by Louis-Napoléon, the president of the Second Republic, with a pension of 150,000 francs. His victory against the odds at Sidi Brahim remains commemorated by a monument in Oran.

The *jihād* of the Mahdī in the Sudan

> When God wanted to make the people of the thirteenth century[49] blissful and to link it with the [first] century [of the *hijrī* calendar] which was honoured by the existence of the Prophet, he caused the Mahdī to be manifest in spirit and in body from the world of concealment. Through him, he revived Islām after it had become merely a trace, nay, a name. God singled out the Sudan for the manifestation of the Mahdī so as to strengthen its people who are, spiritually the weakest people of all the countries…

Thus wrote Ismā'īl bin 'Abd al-Qādir al-Kurdufanī, the Mahdī's biographer, in 1305/1888.[50] It can hardly have been a coincidence that it was on a significant date in the Muslim calendar – the year 1300/1882 – that a new call to *jihād* was issued, this time by Muḥammad Aḥmad bin 'Abd Allāh, known as al-Mahdī (1260/1844–1302/1885), the second son of a ship's carpenter. He declared himself to be of the Prophet's family and issued his manifesto in the following terms:[51]

> The eminent lord [the Prophet Muḥammad], on whom be blessing and peace, several times informed me that I am the *Mahdī*, the expected one, and [appointed] me [as] successor to himself, on whom be blessing and peace, to sit on the throne, and [as successors] to their excellencies the four [rightly-guided caliphs (*Khilāfah'*)] and Princes [of the Faith]… And he gave me the sword of victory of His Excellency [the Prophet Muḥammad] on whom be blessing and peace; and it was made known to me that none of either race, human or *jinn*, can conquer him who has it… He ordered me [to take my exile (*Hijrah*)] to Jebel Kadeer close by Masat, and he commanded me to write thence to all entrusted with public offices. I wrote thus to the Emirs and Sheikhs of religion, and the wicked denied [my mission], but the righteous believed… this is what the eminent Lord [the Prophet Muḥammad] on whom be blessing and peace, said to me, 'He who doubts that thou art the *Mahdī* has blasphemed God and His Prophet'… If you have understood this, we order all the chosen ones to [make their *Hijrah*] unto us for the *jihād*… in the cause of God, to the nearest town, because God Most High has said, 'slay the infidels who are nearest to you'… Fear God and join the righteous, and help one another in righteousness, and in the fear of God and in the *jihād*… in the cause of God, and stand firm

within the boundaries of God, for he who transgresses those boundaries will injure himself. Know that all things are in the hand of God. Leave all to Him and rely on him. He who makes God his support has been guided into the straight way. Peace [be with you].

Strictly speaking, the idea of the Awaited Divinely-Guided One (*Mahdī*) is uncanonical, since there is no justification for it in either the Qur'ān or in the collections of *aḥādīth* of al-Bukhārī or Muslim;[52] nor can the Mahdī's *jihād* be regarded as truly Islamic because, in order to support the dogma that loyalty to him was essential to true belief, he was prepared to modify Islām's five pillars and also the declaration of faith (the *shahādah*). In the case of the *shahādah*, the Mahdī added the declaration 'and Muḥammad Aḥmad is the Madhī of God and the representative of His Prophet'. In terms of the five pillars of faith, *jihād* replaced the *Ḥajj* or pilgrimage to Mecca as a duty incumbent on the faithful. Almsgiving (*zakāt*) was transformed into a tax paid to the Mahdīya, the Mahdī's state.[53] Orthodox Muslims condemned the Mahdist movement and sought to refute the Mahdī's claims. The Mahdists were rebels against the legitimate authority of the Ottoman caliph and fighting against them was allowed: 'in order to protect your religion and safeguard your wealth, you must fight these rebellious charlatans and slay them wherever ye find them (cf. Q.9:5).'[54]

Yet the technical issues of legitimacy and canonicity were of no interest to 'the masses (*al-kāffa*) of the people of Islām' who, as Ibn Khaldūn (d. 808/1406) had affirmed more than four centuries earlier, 'commonly accepted… that there must needs appear in the End of Time a man of the family of Muḥammad who will aid the Faith and make justice triumph; [and] that the Muslims will follow him and that he will reign over the Muslim kingdoms and be called al-Mahdī'.[55] Moreover, such expectations had been heightened by the propaganda and preaching in the Sokoto *jihād*. Though Shehu dan Fodio had disclaimed that he was the 'awaited deliverer', the preaching he authorized by his second son Muḥammad Bello quite clearly heightened popular anticipation of the advent:[56]

The Shehu sent me to all his followers in the east among the people of Zanfara, Katsina, Kano and Daura… I conveyed to them his good tidings about the approaching appearance of the Mahdī, that the Shehu's followers are his vanguard, and that this *jihād* will not end, by God's permission, until it gets to the Mahdī. They listened and welcomed the good news.

In addition, there was considerable migration from the Sokoto empire to the Sudan and Nile valley, probably as a result of the military disturbances but also on the part of people seeking the 'expected Mahdī'.[57] Muḥammad Bello's grandson affirmed that allegiance had been sworn to the Mahdī even before his 'manifestation was perceived'. Shehu dan Fodio 'recommended us to emigrate to you, to assist you and to help you when you were made manifest'.[58] There was thus likely to be an enthusiastic response among a hardcore of refugees who had chosen the Sudan as their place of refuge in anticipation of the advent. One such migrant was ʻAbdullāhi bin Muḥammad, who became the Mahdī's chief lieutenant and successor. He came originally from the Niger–Chad region and had been seeking to transfer his loyalty to an expected Mahdī as early as 1289/1873. Significantly, Muḥammad Aḥmad did not proclaim that he was the Mahdī until ʻAbdullāhi 'twice [fainted] at the sight of [him and] greeted [him] as the expected Mahdī'.[59]

Muḥammad Aḥmad was a prominent Ṣūfī, who was appointed *shaykh* of the Sammānīyah order (*tarīqah*) around 1284/1868. He began to gather supporters, bound by oaths of fealty (*bay'ah*), who were committed to 'purify the world from wantonness and corruption' as well as to fight 'the infidel Turks'. The hill of Gadīr in Dār Nūba became the central location for his secret propaganda before, in 1298/July 1881, he made his first public appearance as Mahdī. Once the *jihād* was proclaimed in 1300/1882, it was accompanied by military success, and thus the campaign became endorsed by victory and support grew rapidly. The *jihād* culminated in 1301/1884, when the Mahdī's forces (the *anṣār*) reached Khartoum. The city fell in 1302/January 1885 and Gordon, the commander of the defensive forces, was killed (possibly against the Mahdī's orders).[60]

We cannot know for certain what objectives the Mahdī would have pursued, for within six months he was dead, probably from typhus. Perhaps the conquest of Sudan was to have been attempted, followed by that of Egypt, Mecca, Syria and Constantinople.[61] ʻAbdullāhi bin Muḥammad took command of the reins of power, and between 1303/1886 and 1306/1889 the *jihād* was conducted along three frontiers – against Abyssinia, Darfur in the west, and along the Egyptian frontier. In August 1889, the long-awaited invasion of Egypt by the *anṣār* was crushed at the battle of Ṭūshkī (Toski).[62] Thus the Mahdist state failed to expand as expected and instead became preoccupied with its own problems of internal disorder. Eventually it went down to bloody defeat to Kitchener's forces at the battle of Kararī near Omdurman (1316/September 1898), a battle in which the young Winston Churchill participated and which he wrote up as *The River War: An Historical Account of The Reconquest of the Soudan,* a work published the following year. The Mahdist state (Mahdīya) had thus survived 13 years without the Mahdī, in spite of the dire predictions of what would follow his death. Even then, the problems were not over for the Anglo–Egyptian condominium. Hardly a year passed during the first generation of the new regime without a Mahdist rising, invariably spearheaded by an individual who claimed to be 'the Prophet Jesus' (*Nabī* ʻIsā), whose role was to kill Dajjāl, the Antichrist – taken to mean the British – and rule according to the law (*sharīʻah*) of Muḥammad.[63] Though none of them commanded overwhelming following, these 'neo-Mahdist' risings continued to harass the 'infidel' colonial government for more than two decades.

Three final comments about the Mahdīya will suffice. The first concerns a phenomenon which we have already encountered, which is that the *jihād* was not only about securing a territorial state: it was about securing conversions, if necessary by force. The Mahdī proclaimed its moral purpose, but maintained that his movement was not a religious order that could be accepted or rejected at will; instead, it was a universal regime, which challenged man to join it or else to be destroyed. The movement was therefore profoundly sectarian: those who denied his Madhīship were unbelievers (*kuffār*).[64] Defeat by the Mahdist forces meant certain death unless conversion was immediate: among the forced conversions or conversions under duress in the aftermath of the massacre of Hicks' forces (1301/November 1883) were those of Rudolf Von Salatin, an Austrian officer and governor of Darfur province and Lupton, a British officer who had been the governor of Baḥr al-Ghazal province.[65] The Mahdī therefore subscribed to the classical formulation that for polytheistic prisoners 'nothing is accepted from them except Islām or the sword'.[66]

A second issue concerns the Mahdī's philistinism, which was on a scale greater even than that of the Ṭālibān in recent times.[67] He alienated the four law schools by ordering the burning of all books on law (*fiqh*) in addition to the *sunnah* and books of Qur'ānic interpretation (*tafsīr*). Apart from the Qur'ān and the Madhī's own proclamations only two works (his own collection of prayers, the *Rātib*, and an incomplete selection of *aḥādīth* of his own) were allowed to remain in circulation. There were a number of regulations which prohibited adornment, music, extravagance at weddings, and tobacco and wine. There were also regulations against the worship of saints and sorcery.

Thirdly, and finally, there was the Mahdī's attitude towards his religious tradition of origin: Ṣūfīsm. Once he had established a broad base of support, the Mahdī outlawed all the Ṣūfī orders, no doubt seeing them as a potential rival power base.[68] There was thus a contradiction between the early and later development, between Muḥammad Aḥmad the Ṣūfī and Muḥammad Aḥmad the Mahdī. The justification for the change was divine revelation. The Prophet had told him three times at the moment of his appointment: 'who does not believe in his Madhīship does not believe in Allāh and his Prophet'.[69] This was tantamount to conferring absolute power on Muḥammad Aḥmad, an attribution of exclusive authority which he sought to implement in his lifetime, regardless of the opposition it aroused. It was diametrically opposed to the principle of consensus[70] on which Shehu dan Fodio had tried to build his *jihād*. Autocracy in the application of the *sharī'ah* typified the Mahdī's Islamization programme and foreshadows the sort of Islamic state envisaged by contemporary extreme Islamists.

The *jihād* of Imām Shāmil in Russia

The sustained *jihād* against the colonial policies of nineteenth-century Russia was also closely linked with the phenomenon of Ṣūfīsm or neo-Ṣūfīsm.[71] The Chechen term for sanctified violence is *gazavāt*, and the Russian invasion of

Chechnya and other Muslim-held lands launched a full century of *gazavāts*, in which the resistance struggle was led by Ṣūfī religious leaders, shaykhs and *imāms*, whose warrior troops were called *murīds* (that is, Ṣūfī disciples). The first significant *murīd* leader was Shaykh Manṣūr (1144/1732–1208/1794), who was chosen by the elders in 1199/1785 to be the first *imām* of the North Caucasians. Karpat notes that

> the oppressed peasantry and the tribes responded enthusiastically to... Mansur's call to *gazavāt*..., to fight against the surviving elements of paganism and animism, against social inequalities, and against the Russians and, especially, their local followers, who were regarded as the source of evil.[72]

After some striking military success, Manṣūr was captured by the Russians in 1205/June 1791 and executed four years later.

General Aleksey Ermolov, supreme commander in the Caucasus region in the years 1232/1817–1242/1827, developed a plan for the 'total subjugation of the Caucasians', which was to be implemented first in Chechnya; it included the construction of fortresses, destruction of rebel villages by 'fire and sword' campaigns and the elimination of their inhabitants. After Ermolov was recalled in 1242/1827, the policy of 'elimination' continued to be implemented. As a result of this onslaught, the various tribes of Circassians, Avars, Lezgis and so on began to act together as a Daghestani–Caucasian entity under Ġhāzī Muḥammad ibn Ismā'īl al-Gimrāwī ('Gazimulla'), who was recognized as *imām* in 1244/1829. Once more the *gazavāt* was declared, this time to convert pagan tribes such as the Galgan, Kists and Ingush. Gazimulla was killed in battle against the Russians in 1247/1832. After a short period of rule as *imām* by Hamzad Beg, who was assassinated two years later, leadership of the movement fell to the most famous of the *murīd* warriors, Shāmil (1211/1797–1287/1871).

Shāmil was no stranger to war with Europeans. While performing the *Ḥajj* in 1243/1828, he had met *amīr* 'Abd al-Qādir, the leader of Algerian resistance against the French, who shared with him his views on guerrilla warfare. For 25 years of continuous fighting, from 1249/1834 until his capitulation in 1275/1859, he led the struggle of the mountain people, building on the ideological, psychological and organizational foundations laid by the Ṣūfī Naqshbandī Order in the Eastern Caucasus several decades before.[73] The British traveller John Baddeley, writing in 1325/1908, talked of Imām Shāmil as the indirect 'protector of the British empire and India' because of his role in having kept tied down significant numbers of Russian troops – some 350,000 men – over such a long period.[74] Shāmil succeeded in establishing an independent theocratic state, with a centralized structure, in his domain which covered a large part of Daghestan and the whole Chechen region. His imāmate was grounded in the basic principles of the *tarīqah* of late Ṣūfīsm: on the hierarchical structure and cohesiveness of the brotherhoods; on the cult of the leader and the unconditional obedience of the *murīds*.[75] Public life in Shāmil's domain was governed by an extreme form of puritanism, essentially alien to the people of the Caucasus; yet it proved to be a necessity for the guerrilla army, for it was a system which could concentrate all forces in defence and bring about the discipline of the people that was required for war. One of the reasons Murīdism appealed so broadly to people was that it preached the need not only for deepened faith, the *sharī'ah* and *gazavāt*, but for social justice as well.[76]

Shāmil learnt from a significant military setback at Akhoulgu in 1254/1839 to avoid sustained encounters with the Russian forces. But gradually, the Russian policy of enforced population deportation, the deliberate deforestation of the region (which removed the cover for the guerrilla fighters) and military exhaustion on the part of the Murīdists wore down the resistance. When the Crimean War broke out in 1269/1853, Shāmil was unable to assist the Ottoman cause; the Murīdists were almost totally isolated from the Ottomans and left to fight on their own resources. The end of the Crimean War in 1272/1856 allowed the Russians to concentrate their forces and, three years later, Shāmil made his last stand at Gunib. Shāmil was captured and eventually, in 1286/1870, permitted to leave for Constantinople and then to proceed on pilgrimage to the Holy Places. He died at Medina in 1287/February 1871. His compatriots were less fortunate: Karpat estimates that a million Caucasians were forced to migrate to the Ottoman lands in 1278/1862–1281/1865,[77] a brutal example of ethnic cleansing *avant le nom.*

The Qādirī order, with its origins in twelfth-century Baghdād, first appeared in the Caucasus in 1277/1861 headed by a Daghestani shepherd named Kunta Ḥājī Kishiev. Based in Chechnya, Kunta Ḥājī taught a mystical practice that, unlike the Naqshbandīs, allowed vocal *dhikr*, ecstatic music and dancing. At first, he counselled peace with the Russians.[78] His popularity surged but soon his following, swelled by many *murīd* fighters from Shāmil's former army, so alarmed the Russians that he was arrested and exiled in 1280/1864. In 1280/January 1864 at Shali in Chechnya, Russian troops fired on over 4,000 Qādirī *murīds* ('the dagger fight of Shali' has remained in the memory of the Chechen people ever since), killing scores and igniting a fresh wave of violence. Kunta Ḥājī died in enforced exile in the province of Novgorod in 1284/May 1867.

The brotherhood, whose remaining leaders all claimed spiritual descent from Kunta Ḥājī, became implacable Russian foes and struck deep roots in the Chechen countryside. Together with the rejuvenated Naqshbandīs, the Qādirīs rose up against the Romanovs in 1281/1865, 1293/1877, 1296/1879 and the last decade of the nineteenth century and plagued Tsarist rule in the Caucasus through the Bolshevik Revolution. The revolutionary years were especially bloody in Daghestan and Chechnya. The Qādirīs, and a Naqshbandī movement led by Shaykh Uzun Ḥājī, battled for eight years against the White and the Red armies to create a 'North Caucasian Emirate'. The pious, uncompromising Uzun Ḥājī

– whose tomb remains an important pilgrimage site for Chechen Muslims – saw little difference between the Tsarist Russians and the atheist Communists. His uprising in Daghestan was finally suppressed in 1338/February 1920, when he was killed, although the struggle was continued by Avars and Chechens under Sheikh Najmuddin Gotsinski for another five years and even longer on a more intermittent basis.[79]

In 1346/1928 a new civil war in the Caucasus was precipitated by the Bolsheviks, which lasted until 1354/1936. A further Ṣūfī-led revolt[80] began in 1359/1941 and lasted until 1366/1947. After the mass deportation of the Chechens to Siberia ordered by Stalin in 1363/1944, the Kunta-Ḥājī brotherhood regained strength. Far from destroying the Ṣūfī brotherhoods, the deportations actually promoted their expansion. For the deported mountain people the Ṣūfī orders became a symbol of their nationhood in the lands to which they were exiled. Moreover, these orders proved efficient organizers, thus ensuring the community's survival.

The *jihād* of Amīr Ya'qūb Beg in Chinese Central Asia

Mountain ranges, rivers and deserts divide East Turkestan, now known as the Xinjiang Uīghur Autonomous Region, into three distinct regions – northern (*shimaliy*), eastern (*ghärbiy*), and southern (*janubiy*) – and shape the ecological areas within these regions. The political subdivisions of modern Xinjiang are more complex and fragmented than these basic regions: nominally autonomous districts and counties are associated with the Qazaq, Qirghiz, Hui, Mongol, Tajik and Xibo nationalities.[81] By 1170/1757, China under the Manchu Qing Emperor Qian Long had invaded Zungharia and East Turkestan and established indirect political and military control. Many exiles and refugees from the Qing conquest settled in the Khoqand Khanate of the Ferghana region west across the Pamirs from Kashghar. These exiles supported frequent rebellions against the Qing Empire during the nineteenth century, when East Turkestan became the object of power struggles between Khoqand, China, Russia, and Britain.

The *jihād* in Xinjiang, which began in 1280/June 1864, shows some of the characteristics of Islām in India and the opposition it could arouse. As had the Mughals in India, the Chinese had sought to bring about cross-religious alliances under an 'obligation of salt' (*tūz*), the ritual eating of which served to bind people of unequal socio-political rank to mutual obligations. Chinese rule thus could be said to have led to a degeneration of the Muslim spirit.[82] Chinese rule had been linked to a grouping of Naqshbandī Ṣūfīs known as Isḥāqīs (followers of Muḥammad Isḥāq Walī [d. 1007/1599], or *Qara Taghliqs*, 'Black Mountaineers').[83] For the Āfāqīs, another group of Naqshbandī Ṣūfīs (they were followers of Kwāja Āfāq [d. 1104/1693 or 1105/1694], and were known as *Aq Taghliqs*, or 'White Mountaineers'), zealous Muslims in exile in Khoqand, such cross-cultural political alliances smacked of religious syncretism, which was to be opposed at all costs by *jihād* or *ghazawāt*.[84]

The Muslims of Xinjiang had suffered from alien domination for over a century[85] and rebellion was likely to take an anti-Qing dynasty and anti-Chinese form. An earlier invasion in 1241/1826 led by Jahāngīr, an anti-Qing Muslim leader (*khwāja*) based in Khoqand, had required troop reinforcements of 36,000 men from China.[86] In spite of the importance of this precedent, the underlying grievance of Xinjiang 40 years later was less religious than fiscal. As a result of the Taiping rebellion in the Shanxi and Gansu provinces of China, the Xing could no longer send subsidies to Xinjiang to defray its military costs: the result was an increase in the tax burden on the local people.[87]

The precipitant in the revolt was the rumour of a massacre of Tungan (Hui or Chinese-speaking) Muslims; from this moment on, it was likely that the unifying factor in the resulting revolt would become Islām itself. Most of the Muslim population of Xinjiang, regardless of their ethnic and social background, participated in the rebellions, which were organized initially by Tungan minority in the cities throughout Kashgharia and Ili.[88]

In 1282/1865 ʻĀlim Quli, the Khān of Khoqand, sent Yaʻqūb Beg (who was probably an Uzbek in origin) at the head of a small army to install a new leader in Yarkand as a puppet governor for Khoqand.[89] In the next two years Yaʻqūb Beg managed to wrest control of the cities of Kashgharia from the Qing, Tungan and Turki Muslim forces. Divisions within the short-lived regime of the Kuchean *khwājas* made it relatively easy for Yaʻqūb Beg to impose his will.[90] Under his rule east Turkestan was held fairly securely, but when he attempted to extend his rule north into the Ili and Jungarian regions, and east to Turfan and Urumchi,[91] he was far less successful, and the resulting wars devastated these regions. There is no doubt that Yaʻqūb Beg slaughtered more Muslims than he did infidels, but in each case the violence was excessive.[92] A few of the Muslims were heretics or quasi-heretics, such as two *shaykhs* from Badakhshan, one of whom claimed to be the 'Mahdī of the Last Day', who were placed in a pit and stoned to death;[93] most were not, but simply opponents of Yaʻqūb Beg's puritanical theocratic state and extensive territorial ambitions. A. N. Kuropatkin, the leader of a Russian mission to the area, commented that the *jihād* had been launched on a declining revenue base, which inevitably resulted in increased taxation:[94]

> He has acted as though he would turn the country into one vast monastery, in which the new monks must, whilst cultivating the soil with the sweat of their brow, give as much as possible – nay, the greater part of their earnings – into the hands of the Government, to devote to warlike impulse.

Yaʻqūb Beg's self-proclaimed title of *amīr* (one of several designations he used, including *ataliq ġhāzī* or 'fatherly warrior for God') was confirmed by the

Ottoman caliph in 1290/June 1873: Abdulazīz also referred to him as governor (*ḥākim*) of Kashghar and expressed the pious hope that he should not enter into unnecessary conflict with neighbouring countries.[95]

Closer relations with the Ottomans brought some much-needed additional armaments, but Ya'qūb Beg's need for armaments and military training for his forces of some 40,000 men was so pressing that relations with other powers were needed too. Treaties were entered into with Russia in 1289/June 1872 and with Britain in 1291/April 1874.[96] When Afghanistan failed to provide rifles in sufficient quantities, the British stepped in and appear also to have provided workmen for an armaments factory of sorts in Kashgharia. Officers from the Ottoman army were not employed as extensively as might have been expected, reflecting the limits on Ya'qūb Beg's capacity to implement military reforms. How much time and effort had been spent by the Ottoman sultans to discard the Janissaries and build a new army?, he mused. He was able to make a start with reforming the infantry; but the cavalry was left unreformed, dominated as it was by the Khoqandians, the mainstay of the regime, from whom he had most to fear.[97]

When the end came to the independent *jihād* state in Chinese Central Asia, it came as a damp squib. Ya'qūb Beg died suddenly and unexpectedly in 1294/May 1877; by 1294/early January 1878 the Chinese conquest was complete. This contrasts markedly with the Chinese experience in conquering the Muslim rebellion of Shanxi, Xining and Suzhou which took Zuo Zongtang some seven years from 1284/1867 to 1290/1873.[98] The collapse of the *jihād* state also confounded speculation abroad, which had assumed that Ya'qūb Beg's state was secure and that a Chinese military assault was unlikely to meet with success. Foreign observers had underestimated Zuo Zongtang's belief that Xinjiang was essential to the security of Mongolia, which in turn was essential to the security of Peking; this 'domino theory' left no choice but to campaign for the recovery of Xinjiang.[99] For his part, Ya'qūb Beg had sought a negotiated settlement with the Qing court which would have left him in overall control of his state, while recognizing Chinese suzerainty. This may account for the otherwise extraordinary last order Ya'qūb Beg issued to his troops that they were not to fire upon the advancing Chinese forces.[100]

The Chinese success owed more to the disarray of the enemy than to the strength of its army or the triumph of its military strategy. The origins of rebellion lay in fiscal pressure resulting from the inability of the Qing to transfer subsidies on the scale required for its military establishment; but the fiscal pressure had grown, and not diminished, in a period of 13 years of continuous warfare. Moreover, for all the willingness of many Muslims to be mobilized by the *ghazawāt* and to submit to the rules of *sharī'ah*, the abuses of power by the Khoqandians and the fact that they were, in essence, a foreign elite (one moreover which even Ya'qūb Beg had been unable to control), posed the question as to whether Xinjiang had not been better off under Chinese rather than Khoqandian rule.[101] The succession dispute following the death of Ya'qūb Beg evolved rapidly into civil war and a tripartite partition of east Turkestan. Divided and fatally inactive in the face of the Chinese advance, Xinjiang was incorporated into the system of Qing provincial administration, which was followed by the extensive immigration of Han Chinese. At the time of writing, the Uyghurs are still numerically ascendant over the Han, but it is likely that the demographic balance will shift before long in favour of the Han in what has now become a permanent part of Chinese Central Asia.

Jihād in the era of the Indian Mutiny

The reassertion of the idea of *jihād* in the context of an anti-colonial struggle in India was begun by the *ṣulṭān* of Mysore, Haidar 'Alī, and continued by his son, Feth 'Alī, known as Tīpū (1163/1750–1213/1799):[102] they claimed to be of Qurayshī descent and sought the status of *pādshāh* of India. Tīpū entered into alliances with Afghanistan and France but was defeated by the British and killed in battle at Seringapatam in 1213/May 1799. He had proposed that the Ottoman caliphate should become the real political centre of Islām, and be mobilized to oppose European encroachment.[103] Tīpū's defeat brought an end to an independent Mysore sultanate. Karpat notes that the British concluded from this incident that the caliphate 'could be used to tame the Muslims under its rule and to establish an Islamic front directed against its enemies, Russia in particular'.[104] Thereafter, the British tried to use the Ottoman caliph to tame the Sepoy rebels in 1273/1857 and to persuade the *amīr* of Afghanistan in 1294/1878 to cease his opposition to the British and to help mobilize the Muslims of Central Asia against the Russians.[105]

After the elimination of Mysore, the propagation of *jihād* in India is associated with the life, writings and campaigns of Sayyid Aḥmad Barelvi (1201/1786–1246/1831), known as *shahīd* after his death, who was a disciple of Shāh Walī Allāh (see Chapter 3). For Sayyid, the British were, quite simply, treacherous (*daghābāz*)[106] and to be opposed by the faithful who should proceed into exile and establish a *Dār al-Islām* within the subcontinent, which would oppose colonialism and religious syncreticism alike.

Sayyid Aḥmad's *hijrah* began in 1241/January 1826. Following the battles of Akora and Hazru, where the *mujāhidīn* acted 'like a leaderless band', it was decided that 'the successful establishment of *jihād* and the dispelling of disbelief and disorder could not be achieved without the election of an *imām*'. Sayyid Aḥmad's conception of the *jihād* state was a combination of the secular authority of the *ṣulṭān* with the religious authority of the *imām*, the latter having a general supervisory role over the former.[107] Sayyid Aḥmad's power in the tribal area of what would become the North-West Frontier Province fluctuated wildly, and his forces were finally defeated at Balakote by the Sikh army of Kunwar Sher

Singh in 1246/May 1831. Sayyid Aḥmad's body was identified and burnt by the Sikhs.[108]

The doctrine of *hijrat* continued to predominate in *jihādī* thinking in the years before the great Sepoy rising or 'mutiny' of 1273/1857. The establishment of an independent imāmate beyond the borders of British-controlled India enabled the proclamation of *jihād* against it as a *Dār al-Ḥārb*, a 'land of war' against which the struggle could be carried on as between two states, backed up by the existence of district centres inside British India.[109] The *jihādīs* had been fighting the British continuously since 1268/1852 and had already conceived of the strategy of seeking to 'tamper with the allegiance of the army'.[110] Though the rising of 1273/1857 presented them with opportunities for continuing their campaign and gaining further support, the rising was not a pre-planned movement and lacked any unity of purpose or coordinated plan of action.[111] An invasion from the north-west would have enhanced the chances of success of the Sepoy rising,[112] but the movement lacked the material resources to go on the offensive.[113]

The nearest to coordination during the rebellion came from the activities of Azimullah Khan, the Muslim secretary of Nana Sahib (Dhundu Pant), leader of the rebellion at Kanpur, where the British garrison and colony was massacred. Azimullah printed pamphlets that called for a *jihād* against the infidel and gathered together disaffected Indian officers, whether Hindu or Muslim, and presented seditious ideas to them. Azimullah expressed the extent of his intrigues and seditious plans to the Turkish general Umar Pasha in 1272/1856, in an attempt to gather Ottoman support. It was Azimullah who formed an infrastructure of Indian agents to distribute seditious anti-British propaganda and not Russian agents, as the British believed. But it was not enough; and events moved too rapidly for the Sepoy rebellion to produce distinctive but unified political ideas of resistance which could bridge differences between the various religious communities. It is true that Bahādur Shāh, the last Moghul ruler, was encouraged to assert his claim to sovereignty in 1273/May 1857; but the rebel council established at Delhi ignored the Mughal emperor, and he was deposed and prosecuted by the British the following year. It is true that an independent sovereign could have declared a legitimate *jihād*, which might have gone some way towards bringing to pass the Muslim prophecy that foreign rule would last a hundred years (the rising was timed to coincide with the centenary of the victory at Pīnipat in the Hijrī calendar, 1174/1273 ME); but none of this would have resolved the longer-term problem of sustaining an alliance with the Hindus and Sikhs in order to oust the British.[114]

The majority of the Indian '*ulamā*' contested the idea of a *jihād* against British rule in the decade or so following the failure of Sepoy rebellion in 1273/1857:[115]

> the supreme tribunals of Islām have unanimously and solemnly declared that India under its present tolerant and equal government is certainly not *Dār al-Ḥārb* ('the country of the enemy'), upon whose rulers war should be waged by the faithful; and consequently, no Indian Wahhābī who has not utterly broken with the orthodox portion of his Church [*sic*] can be disloyal on merely religious grounds…
>
> [Quoting Hunter:][116] the Mussulmans here are protected by Christians, and there is no *jihād* in a country where protection is afforded, as the absence of protection and liberty between Mussulmans and infidels is essential in a religious war, and that condition does not exist here. Besides, it is necessary that there should be a probability of victory to Mussulmans and glory to Islām. If there be no such possibility, the *jihād* is unlawful…

Notwithstanding this distancing of the Indian '*ulamā'* from the *jihādī* position, the tribal alliances of the north-west frontier secured the position of the rebels even against British campaigns on an increasing scale in 1274/1858 and 1279/1863, the first of which saw the deployment of Lee Enfield rifles which gave the British greatly increased firepower over their opponents.[117]

In the second campaign in 1279/1863, the tribes in that most volatile of regions initially showed concerted support. It seemed that the British had

> underrated the hold which the Fanatical Colony had acquired over the Frontier tribes.[118] Those who had joined them for the sake of Faith were burning with hopes of plunder or of martyrdom, while the less bigoted clans were worked upon by the fear of their territory being invaded by the British.[119]

It was only when tribal dissensions, coupled with financial inducements, began to work their effect that this apparent unity collapsed. The British succeeded in restoring order only by co-opting sub-tribes such as the Bonairs, Amzais and Khudikhels who, 'having once committed themselves openly against the fanatics' were certain not to realign with them.[120] Nevertheless, further British expeditions were required in 1284/1868, 1305/1888 and 1308/1891. The period of Abdullāh's *imārat* spanned four decades from the battle of Ambeyla in 1279/1863 until his death in 1318/1901.[121]

The development of modernist arguments in opposition to Muslim revivalism

The classic *jihād* declarations against foreign invasion were those of Egypt against the British in 1299/July 1882[122] and of Libya against the Italians in 1330/January 1912.[123] However, such resistance was rarely successful. Muslim intellectuals such as Jamāl al-Dīn al-Afghānī (1254/1839–1314/1897) and the young Muḥammad 'Abduh (1264/1848–1322/1905) were, in contrast, impressed by the unorthodox but relatively successful Mahdist movement in resisting the semi-colonial Anglo-Egyptian condominium.[124] Both Muḥammad 'Abduh and Muḥammad Rashīd Riḍā (1281/1865–1354/1935) conceived of *jihād* as essentially defensive in nature, or in Riḍā's words, 'defence against enemies that fight the Muslims because of their religion': it could be used as an instrument against colonialism.[125] In al-Afghānī's view, the Ottoman caliph should become more aware of his moral authority and use it 'in an intelligent manner' on issues such as the Egyptian question 'which is actually an Ottoman or Islamic question'.[126] In reality, as has been seen in Chapter 5, Abdülhamid II certainly sought to initiate a pan-Islamic policy where it was both possible and prudent to do so.

Yet politics is the art of the possible. Other Muslim intellectuals, notably the Indian educational reformer Sayyid Aḥmad Khān (later Sir Sayyid, 1232/1817–1315/1898) deduced, in the aftermath of what came to be known as the Indian Mutiny, that resistance to overwhelming military power was impossible and might in certain circumstances be undesirable.[127] He wrote in 1287/1871:[128]

> First, what is *jihād*? It is war in defence of the faith *fī sabīl Allāh*. But it has conditions, and except under these it is unlawful. It must be against those who are not only [unbelievers (*kuffār*)] but also 'obstruct the exercise of the faith' (Q.47:1)… there must be *positive* oppression or obstruction to the [Muslims] in the exercise of their faith; not merely want of countenance, negative withholding of support, or absence of profession of the faith; and further, this obstruction and oppression which justifies *jihād* must be, not in civil, but in religious matters; it must impair the foundation of some of the 'pillars of Islām'… positive oppression (*ẓulm*) and obstruction of the exercise of the faith (*ṣadd*) can alone justify *jihād*.

Cherāgh 'Alī adopted a similar approach in his treatise on *jihād* discussed in Chapter 1, and the view was shared by many contemporaries. As Rudolph Peters argues, these modernists 'introduced a separation between the religious and political spheres, an obvious innovation with regard to a religion which claims to dominate all domains of human activity'.[129] It is thus ironic that the revivalist movements up to and including the nineteenth century not only failed to achieve their purpose in many respects but that in their very failure they stimulated a modernist discourse – even if this was a discourse which opponents such as al-Afghānī characterized as divisive and likely to lead, within their definition, to the abandonment of a truly Islamic position. This debate remains ongoing in contemporary Islām.

Part Three

Ideological Interpretations

8
Sunnī Political *Jihādists* of the Twentieth Century: Mawdūdī, Ḥasan al-Bannā', Quṭb

The world has never been introduced to an ideological state... From the very beginning up to now, Islām is the only system of thought that [has] provided a pure ideological system without a trace of nationalism.

Mawdūdī, 1940[1]

...It requires a realistic and practical code and system of life; and any theories that are presented to it will be judged on the benefits that are proved to result from their application in the realm of practical life. This has always been the criterion in passing judgement on social ideologies, and it will be the way in which the non-Muslim world will judge the ideology of Islām.

Sayyid Quṭb[2]

Islām must be sterilized of other ideologies.

Abū Bakr Baṣīr, 1996[3]

Two of the three Islamist writers considered in this chapter, Mawdūdī and Quṭb, have exercised a profound influence over contemporary Islamic thought because of their numerous writings but especially their long and detailed commentaries on the Qur'ān. Their statements on *jihād* in their commentaries cannot be taken as entirely objective analysis of the Qur'ānic text; rather, they are detailed justifications of their Islamist political thought.[4] Of the three writers, Mawdūdī needs to be considered first since his writings were known to Ḥasan al-Bannā' and Quṭb. It should be noted that whereas the last two writers were based in Egypt and reflected the viewpoint of a *jihādist* struggle within a Muslim majority country this was not the case with Mawdūdī. On the contrary, his views emerged and were developed in the context of the then largest Muslim minority in the world, the

Muslim community of the pre-partition Indian subcontinent. His conversion to the cause of Pakistan was very much that of a latecomer. As late as 1361/1943, he was campaigning against Pakistan and is reported to have stated that 'to demand Pakistan [as a separate state] is a sin against Allāh and the Prophet'.[5]

Mawdūdī

For all his inconsistency and changes of political viewpoint, Sayyid Abu'l-A'la Mawdūdī (1321/1903–1399/1979) has exercised more influence over modern Islamic, and especially radical Islamist, thinking than perhaps any other figure.[6] His supporters made exaggerated claims for him, that he was the 'Ibn Taymīyah of his era', 'the founder of a school of thought' on a par with one of the four main Sunnī schools of law.[7] For his critics, such as the editors of the website called Islamist Watch, Mawdūdī 'summarizes the entire Islamist plot and some of its justifications in the Qur'ān'.

Islamic *jihād* as world revolution

Therefore Mawdūdī's writings, and especially the summary address that he delivered on 'War in the cause of Allāh' (*Jihād fī sabīl Allāh*) on Iqbāl Day (13 April 1939)[8] serve as 'an excellent (and nearly comprehensive) summary of Islamist ideology'.[9] This statement made it clear that the purpose of *jihād* was none other than world revolution, since Islām knows no national boundaries and accepts no other system than its own:[10]

> …the objective of the Islamic *Jihād* is to eliminate the rule of an un-Islamic system, and establish in its place an Islamic system of state rule. Islām does not intend to confine this rule to a single state or to a handful of countries. The aim of Islām is to bring about a universal revolution. Although in the initial stages, it is incumbent upon members of the Party of Islām to carry out a revolution in the state system of the countries to which they belong, their ultimate objective is none other than a world revolution.

How could such an astonishing claim have been made? In 1939, Islām could hardly have seemed on the march. It was the secular ideologies, Nazism, fascism and Marxist–Leninism, which seemed to be making progress at the expense of the world's religions. In order to understand the reasoning behind this claim, we need to return to an earlier period in Mawdūdī's life and the origins of his thought and writings on *jihād*.

Mawdūdī's biographer, Nasr, places the 'turn to an Islamic ideological perspective' in the years between 1932 and 1937. Prior to 1930, events had 'pointed him in the direction of Islamic revivalism' but he had not yet begun to look to Islām itself and the revival of its values as the 'key to reversing the decline of Muslim power in India'.[11] Mawdūdī trained with two Deobandi '*ulamā* at the Fatihpuri mosque's seminary in Delhi and received his certificates to teach religious sciences (*ijazahs*) in 1344/1926. Yet in his lifetime he never publicized his Deobani training or his ties to the '*ulamā*. The existence of his *ijazahs* remained unknown until after his death.[12]

Instead, politics and the independence movement were his main preoccupations. The earliest organized expression of Muslim communalism, the Khilāfat movement, to which Mawdūdī had belonged, collapsed in 1342/1924 and with it the hopes and aspirations of many of the Muslims of India. Mawdūdī continued to believe in the desirability of the caliphate, but recognized now that it would have to be built anew for religious reasons. What had been lost when the Turkish government abolished the caliphate could not easily be revived. Mawdūdī, in Nasr's expression, 'saw the demise of the caliphate as a consequence of the machinations of Westernized Turkish nationalists on the one hand, and as the betrayal of Islām by Arab nationalists… on the other'. His deep-seated suspicion of nationalism and Western influence dates from this moment.[13]

The immediate background to Mawdūdī's writing his *Islām's Law of War* (*Islām ka qānūn-i jang*), which appeared in serialized form between February and May 1345/1927 and was subsequently republished in book form as *Jihād in Islām* (*Al-Jihād fī'l-Islām*) in 1348/1930 was what he conceived of as a threefold threat. The first was from the Congress movement in India. Gandhi had masterminded the Khilāfat agitation and though the movement had failed, Congress retained the leadership of political agitation against British colonial rule. A Khilāfat agitator such as Mawlānā Abū'l Kalām Azād argued that, within the ambit of Congress, Muslims should not merely participate in the struggle for independence but, through revitalizing their religious heritage, should act as its leaders. But, for Mawdūdī, apart from the fact that Azād was too pacific in his interpretation of *jihād*,[14] the central question was how much political power Muslims would enjoy within such a movement. When Gandhi taunted the Muslims in 1347/1929 that 'we will win freedom with you or without you, or in spite of you', this served to confirm his suspicion that the answer was 'very little'.[15]

The second threat came from hardline Hindu support for the Hindu Mahasabha, the Arya Samaj and the *Shuddi* reconversion movement. This campaign brought to a fore latent tensions between the majority Hindu and minority Muslim communities which had only partly remained dormant during the Khilāfat campaign. In a sustained attack on some of the central beliefs of Islām, the leader of the *Shuddi* campaign, Lala Munshi Ram (Swami Shraddhanand) had alienated Muslim opinion. Shraddhanand had claimed in 1341/July 1923 that 'Muslim aggression' was the main reason for the 'enmity' between Hindus and Muslims, displayed especially in the communal disturbances at Malabar and Multan. 'Many of the Muslim religious leaders have said in their speeches that the snake and the mongoose can be friends, but there can be no unity between Hindu *kaffirs* and Muslims.' Thus, in his view, the Hindus must organize and emerge as 'the strongest'.[16] Three pamphlets attacking the Prophet were published in India in 1341/1923, 1342/1924 and 1345/1927.[17]

In 1345/December 1926 Shraddhanand was assassinated. Hindus were outraged by this event, criticized Islām as a militant faith and even Gandhi, who had criticized the *Shuddi* movement, seemed to waver in his long-standing hostility to the idea that Islām was a 'religion of the sword':[18] 'the sword is yet too much in evidence among Musalmans', Gandhi declared in 1345/December 1926. 'It must be sheathed if Islām is to be what it means – peace.'[19] Orthodox Islām needed to be defended against its critics, as the Khilāfat agitator Muhammad 'Alī remarked in a sermon at the Jami' mosque attended by Mawdūdī in 1345/1926.[20] However naively, Mawdūdī seems originally to have entered the political fray with the aim of halting the rise of Hindu power and converting the whole of India to Islām – to end forever the uncertainty of the Muslim place in the polyglot culture of India.[21]

The third threat motivating Mawdūdī was the challenge to orthodoxy within Islām itself. In 1924 Aḥmadī missionaries to Afghanistan had been arrested, brought to trial on charges of apostasy and executed. The British criticized the executions and the religious laws which sanctioned them. The Aḥmadī were opposed to a warlike *jihād* on the grounds that it was incompatible with the spirit of Islām as a religion of peace and that it sought to propagate the Islamic faith through violence.[22] For Mawdūdī, it became imperative to issue an orthodox defence of *jihād*, lest the Aḥmadī view (which was already attractive to the British) began to gain ground with Muslims who opposed the Arya Samaj and yet criticized Shraddhanand's murder.[23]

Jan Slomp terms Mawdūdī's study of *jihād* 'probably the most comprehensive book on this subject ever written by a Muslim', providing as it did an interpretation that was 'extremely influential'.[24] Mawdūdī stresses that the period before Islām, the time 'of not knowing the truth', the period of *jāhilīyyah*, was extremely violent. In contrast, *jihād* and Islām stood for respect for life (Q.5:32). God used *jihād* to protect people from each other in an Islamic version of Hobbes' 'dissolute condition of masterless man'; freedom of religion was also protected by *jihād* (Q.22:40). 'There is no place for compulsion in religion' (Q.2:256), and thus unbelievers (*kuffār*) may not be forced to change their faith. In the Indian context, Hindus, for example, are entitled to remain Hindus. The original goal of Islamic war, Mawdūdī contends, was not use to force to make people accept Islām, but to liberate them from injustice (*fasād*) and violence (*fitnah*). A stable and appropriate political structure was needed for Muslims to carry out the 'original and principal service' for which God had placed them on earth, namely 'enjoining what is of honour and forbidding what is reprobate' (Q.3:104).[25] Mawdūdī makes a sufficiently sweeping claim for justice and righteousness as the chief objectives of government that, in Slomp's phrase, 'many Muslims could interpret this recommendation as a direct invitation to an Islamic revolution':[26]

> Therefore, in order to eradicate evil and prevent wrong, Islām has prescribed that by systematic endeavour, *jihād* – and if the necessity should befall, by war and bloodshed – all such governments should be wiped out. In their place a just and equitable system of government should be erected which is founded upon the fear of God and based upon the canons He ordained.

The seizure of power by the party of God (*Ḥizbu'llah*)

'The objective of Islamic *jihād*', Mawdūdī claimed, 'is to put an end to the dominance of the un-Islamic systems of government and replace them with Islamic rule.'[27] Non-Muslims must live as subject peoples within an Islamic state, but 'they cannot be allowed to impose their spurious laws on God's earth and thus create evil and strife'.[28] The seizure, or at least the attainment, of political power was the logical objective of faith. The quest for a virtuous order transformed the community of the faithful (the *ummah*) into what Mawlānā Azād had called, following the Qur'ān, the *Ḥizbu'llah*, the party (or followers, partisans) of God (Q.58:22),[29] a term also applied by Mawdūdī:

> those who affirm their faith in this ideology become members of the party of Islām and enjoy equal status and equal rights, without distinctions of class, race, ethnicity or nationality. In this manner, an International Revolutionary Party is born, to which the Qur'ān gives the title of *ḥizb-Allāh*.

He contended that Islām as a God-given system is perfect: if Muslims commit crimes, Islām itself is not at fault. Even so, he tried to exonerate Muslim armies from mass killings and the enslavement of populations in the distant past by offering sweeping apologias which do not necessarily stand up to historical scrutiny. He denounced war in modern civilization, and argued that Islamic laws and rules concerning warfare were superior to their Western equivalents. Comparison of modern international laws governing war with much earlier Islamic laws revealed that as far as humane principles were concerned, nothing had been improved by the West, especially since there was insufficient distinction between legitimate and illegitimate goals of war and the means of warfare were not defined in Western laws. He was clear that *jihād*, as a medium of conducting positive warfare, was an essential duty of the Islamic state.[30]

In his statement on 'War in the cause of Allāh' (13 April 1939), Mawdūdī refined the definitions of *jihād* ('to exert one's utmost efforts in promoting a cause')[31] as well as both 'religion' and 'nation':

> Islām is not the name of a mere 'Religion', nor is Muslim the title of a 'Nation'. The truth is that Islām is a revolutionary ideology which seeks to alter the social order of the entire world and rebuild it in conformity with its own tenets and ideals. 'Muslims' is the title of that 'International Revolutionary Party' organized by Islām to carry out its revolutionary programme. *Jihād* refers to that revolutionary struggle and utmost exertion which the Islamic Nation/Party brings into play in order to achieve this objective.

For Mawdūdī,

> Islām is not merely a religious creed or a name for a collection of a few acts of worship. It is a comprehensive system which seeks to annihilate all evil and tyrannical systems in the world, and enforce its own programme of reform, which it deems best for the well-being of mankind.

Unlike the historical practice of Islām, for example in medieval Spain, Mawdūdī was no pluralist,[32] rejecting the idea that two or more systems of belief and government could coexist harmoniously within the same state. On the contrary, a single system or polity was not only greatly to be preferred but was indispensable to the Islamic cause:[33]

> Apart from reforming the world, it becomes impossible for the Party itself to act upon its own ideals under an alien state system. No party which believes in the validity of its own ideology can live according to its precepts under the rule of a system different from its own. A man who believes in Communism could not order his life according to the principles of capitalism whilst living in Britain or America, for the capitalistic state system would bear down on him and it would be impossible for him to escape the power of the ruling authority. Likewise, it is impossible for a Muslim to succeed in his aim of observing the Islamic pattern of life under the authority of a non-Islamic system of government. All rules which he considers wrong, all taxes which he deems unlawful, all matters which he believes to be evil, the civilization and way of life which he regards a wicked, the education system which he views as fatal... all these will be so relentlessly imposed on him, his home and his family, that it will be impossible to avoid them.

The notions of an offensive or a defensive *jihād* were also deemed irrelevant by Mawdūdī, given his definition of it as a revolutionary struggle which had nothing to do with conflicts between states.[34] He accepted also that the sort of Islamic state which he defined did not exist at the time of writing (1357/1939). In essence, he blamed bad Muslims ('advocates of an all-too-easy salvation') for this. It was they who had 'diverted the Muslims away from their real mission'.[35] There is no doubt that such advocates of an all-too-easy salvation were, in Mawdūdī's mind, those committed to the principle of *jihād al-nafs*, that is, by and large, those from the Ṣūfī tradition.[36]

With the prospects of partition of the Indian subcontinent strengthening after 1358/1940, Mawdūdī came to believe that the new homeland for the Muslims had to be rescued from the control of Jinnah and the Muslim League. What the Muslims needed was a cadre of dedicated, morally upright, and religiously exemplary men who would both represent the ideals of the Islamic order and be prepared to achieve it. A new party was necessary for the success of a movement for the renewal of Islām (*tajdīd*).[37] He called Jinnah's Muslim League a 'party of pagans' (*jamā'at-i jāhilīyyah*):

> no trace of Islām can be found in the ideas and politics of Muslim League… [Jinnah] reveals no knowledge of the views of the Qur'ān, nor does he care to research them… yet whatever he does is seen as the way of the Qur'ān… All his knowledge comes from Western laws and sources…[38]

As Nasr comments, 'the term *jamā'at-i jāhilīyyah* was no doubt coined to make the contrast between the Muslim League' and Mawdūdī's proposed *Jamā'at Islāmīyyah* 'more apparent'. In 1360/ August 1941, 75 individuals joined Mawdūdī's new political party. He told them that 'Islām is none other than *jamā'at,* and *jamā'at* is none other than *imārat* [emirate]', but his own power was more curtailed than he would have wished for. He was nevertheless prepared to resign as *amīr* of the *Jamā'at Islāmīyyah* in 1391/1972, some seven years before his death.

In his Friday sermons in 1356/1938, Mawdūdī had emphasized that man could not be obedient to two systems simultaneously. Instead there had to be one principal obedience (*dīn*): true believers (*al-Mu'minūn*: cf. Q.33) were identified by their efforts to obliterate the false *dīn* and establish in its place the true *dīn*, irrespective of their success or failure. Sufferers of trouble in the path of Allah were thus who ultimately established the true obedience with their sacrifices, but Mawdūdī recognized that *jihād fī sabīl Allāh* was the function 'of only those persons who have the will to carry out this onerous task, and such persons are always few in number'.[39]

The lordship of God? The search for the Islamic state

In the section on *jihād* from *Let us be Muslims*, Mawdūdī makes it clear that 'the real objective of Islām is to remove the lordship of man over man and to establish the kingdom of God on Earth'. 'To stake one's life and everything else to achieve this purpose is called *jihād*'; other religious acts (prayer [*ṣalāt*], fasting, *Ḥajj* and alms-giving [*zakāt*]) are all 'meant as a preparation for this task'. In order to bring people to 'the path of well-being and righteousness after rescuing them from the path of destruction, there is no other remedy except to set right the mutilated shape of government'. No scheme of reform for the people could be implemented without acquiring control of the government machinery. For Mawdūdī, the reform needed in the very basis of the government was that 'there should not be lordship of man over man but that of God over man'. Those who run the government must not become supreme sovereigns but, must instead recognize God as their Sovereign:[40]

> And then taking over the leadership and superintendence of God's servants, conduct the affairs of the government in accordance with God's laws and with belief in their responsibility and accountability in the Hereafter as also in God being the Knower of the unseen. The name of this striving is *jihād*.

Unlike many other Islamist theoreticians, Mawdūdī took care to define what he meant by an Islamic state in his *Islamic Law and the Constitution*, the first draft of which was written while he was imprisoned in 1953–55.[41] Nasr comments that though Mawdūdī's Islamic state grappled with the notion of democracy, it 'remained at odds with it'; the Islamic state was to be judged by its adherence to the faith (*dīn*) and not by its mode of government. Mawdūdī gave the state broad coercive powers and a monopoly over such key Islamic doctrines as *jihād*. In Nasr's view, 'the contradiction inherent in applying democratic mechanisms to an ideological state structure' was 'self-evident'.[42] For Mawdūdī, Islām and the Islamic state are synonymous. Religion was in effect reduced to something which was socially useful in 'inculcating the habits and discipline that assist in the project of striving for an Islamic state'; instead of making politics a part of religion, his Islamism made Islām a political religion.[43]

Among recent commentators, A. G. Noorani denies the validity of the central concept of Mawdūdī, the search for the Islamic state:

> in truth, there is no such thing as an Islamic State. Indeed there simply cannot be one: Islām shapes the personality of man and of society through him. It does not provide for the institutions of the government for these vary with time, whereas the fundamentals of faith are valid for all time.[44]

Undoubtedly Noorani is correct. Qamaruddin Khan observed 30 years ago that the term 'Islamic state' was 'never used in the theory or practice of Muslim political science before the twentieth century' (see Chapter 1). There is, he observed, 'no such thing as a permanent Islamic constitution'. Islamic political theory was a 'changing and developing concept, adapting itself to the exigencies of time and place'.[45] The constitution of an Islamic state is nothing other than the scheme of Abū al-Ḥaṣan al-Māwardī (361/972–449/1058) 'devised ten centuries ago' in his *Ordinances of Government* (*al-Aḥkām al-Ṣulṭaniyyah*).

For Mawdūdī, Islām is less about personal faith than organized faith leading to political power. The 'reforms', that is, changes which Islām wants to bring about, he asserted, 'cannot be carried out merely by sermons. Political power is essential for their achievement'. The seizure of political power by the Islamist party was thus 'positively desirable', indeed 'as such obligatory'.[46] He was even prepared to admit that the state that he envisaged bore resemblance to the fascist and Communist state but (or at least, so he claimed) without the specific aspect of dictatorship.[47] The Islamic state he envisaged was 'universal and all-embracing'. No field of affairs could be considered personal and private. The state instead sought to 'mould every aspect of life and activity in consonance with its moral norms and programme of social reform'.[48]

Mawdūdī rejected the term 'theocracy' for the state he envisaged, terming it instead a 'theo-democracy', a democratic caliphate that would not be ruled by any particular religious class but by the whole community of Muslims.[49] Since the definition of the state was ideological, nationhood too was to be derived from ideological (that is, religious) convictions. At best non-Muslims might have a system of separate communal electorates from which they would vote for a candidate from their own community.[50]

Among the most important of modern human rights is the right to freedom of belief. Freedom of belief implies the freedom to change one's religion. Not so for Mawdūdī. He writes:

> as regards Muslims, none of them will be allowed to change creed. In case any Muslim is inclined to do so, it will be he who will be taken to task for such a conduct, and not the non-Muslim individual or organization whose influence might have brought about this change of mind.[51]

Ishtiaq Ahmed accuses Mawdūdī of dishonesty on the question of freedom of belief. In his pamphlet *Human Rights in Islām*, which was first published in the UK in 1976, Mawdūdī directed his arguments towards a Western audience and made no mention of the doctrine of apostasy. Instead, he concentrated on the Qur'ānic injunction that there should be no coercion in matters of faith (Q.2:256).[52] Ishtiaq Ahmed notes that in his statement before the Court of Inquiry set up to investigate the anti-Aḥmadī riots in Pakistan in 1953, Mawdūdī (along with the rest of the '*ulamā*') declared that apostasy was punishable by death in Islām.[53] For Ishtiaq Ahmed, Mawdūdī emerges as 'an ideologue of state might and an opponent of human freedom and equality', who sought to establish 'an all-embracing doctrinal state based on an ideology which is believed to originate in divine revelation' – the issue here is whether a 'political system' is to be discerned in the Qur'ān, which commentators such as Ishtiaq Ahmed and Qamaruddin

Khan deny – but one which, for all practical purposes, was 'dependent on the interpretation of pious experts'. Rather than a 'theo-democracy', it could 'more accurately be described a modern-day theocracy'.[54]

Islām refuses minority status

Mawdūdī utilized his commentary on the Qur'ān to buttress his arguments about the centrality of *jihād* for Islām and the struggle to create an Islamic state. His comment on Q.2:218 noted that

> *jihād* denotes doing one's utmost to achieve something. It is not the equivalent of war, for which the Arabic word is *qitāl*. *Jihād* has a wider connotation and embraces every kind of striving in God's cause… '*Jihād* in the way of God' is that strife in which man engages exclusively to win God's good pleasure, to establish the supremacy of His religion and to make His word prevail.[55]

In his gloss on Q.4:97 and Q.4:100, and probably drawing on al-Māwardī,[56] Mawdūdī introduced the un-Qur'ānic concepts of *Dār al-Islām* and *Dār al-Ḥarb*; in certain circumstances, a believer who continues to live in a land where an un-Islamic order prevails 'commits an act of continuous sin'. This would include all Muslims living in Europe and the United States unless they struggle to put an end to the 'hegemony of the un-Islamic system and to have it replaced by the Islamic system of life' and unless they lived there 'with utmost disinclination and unhappiness'.[57] For Mawdūdī, Islām cannot long accept minority status.

Mawdūdī notes that Q.9:38 'formed the basis of a legal ruling issued by the jurists regarding *jihād*'. Without a summons, it was merely a collective duty of all Muslims (*farḍ al-kifāyah*). Once Muslims are called upon 'by their leader' (the term is ambiguous), *jihād* becomes obligatory on every Muslim. Those who fail to perform this duty without legitimate excuse should be regarded as ceasing to be Muslims ('such a claim [to be a Muslim] will not be entertained').[58] On Q.9:72, Mawdūdī commented that 'this declaration… marked the end of the period of leniency showed to the hypocrites'.[59] Exemption from *jihād* depends not only on physical disability, sickness or indigence, but loyalty to God and His Messenger. God will not pardon those who thank the heavens for their sickness since it has provided them with a timely excuse to stay away from the war-front (Q.9:91).[60] When a true believer is summoned to *jihād*, he simply cannot enjoy the 'cosy comfort of his home' but must move about the earth and exert himself 'so as to make the true religion prevail' (Q.9:112).[61]

In his commentary on Q.22:78, while Mawdūdī maintains that 'there are forces of resistance which obstruct people from serving God and pursuing His good pleasure', nevertheless he also accepts the *ḥadīth* of the greater *jihād*, 'a man's striving against his own self'. 'One finds the world full of those who have rebelled against God and who incite others to rebellion. To strive against these forces and to devote all the power of one's mind, heart and body in this connection in what is required of man', a requirement embodied in this particular verse.[62]

Mawdūdī's attitude towards *jihād* led him at various times to be imprisoned and courted by successive regimes in Pakistan. In 1367/1948, although officially observing a ceasefire with India, Pakistan had resumed support for the insurgents in Kāshmīr by dispatching armed paramilitary units. It allowed this conflict to be called a *jihād* at least for the purposes of recruitment. Mawdūdī denounced the government's position as hypocrisy. Either the ceasefire had to be observed or the government had openly to declare war by calling it a *jihād*. His stand, and that of his party, the *Jamā'at al-Islāmīyyah*, were made extremely difficult. Mawdūdī was imprisoned between 1367/1948 and 1369/1950, while the party was charged with sedition.[63] The government did not change its standpoint. The prime minister, Liaquat 'Ali Khan, in a speech in 1370/August 1951, noted that there was talk of *jihād* in Pakistan but not of war: '...*jihād* really means to strive for justice and truth whereas war means to fight others for territorial ambitions...', he contended.[64] On another occasion, in 1384/1965, Ayub Khan, a determined enemy of the *Jamā'at al-Islāmīyyah*, publicly appealed to Mawdūdī to support the war against India by declaring a *jihād*.[65]

For Mawdūdī, *jihād* was 'used particularly for a war that is waged solely in the name of Allāh against those who practise oppression as enemies of Islām'. As late as 1379/1960, he referred to *jihād* as 'the supreme sacrifice of life' which 'devolves on all Muslims'. When an Islamic state was attacked by a non-Muslim power 'everybody must come forward for the *jihād*'. If the country was not strong enough to defend itself, neighbouring Muslim states must assist;

> if even they fail, then the Muslims of the whole world must fight the common enemy. In all such cases *jihād* is as much a primary duty of the Muslims concerned as are the daily prayers or fasting. One who shirks it is a sinner. His very claim to be... a Muslim is doubtful. He is a hypocrite whose '*ibādah* and prayers are a sham, a hollow show of devotion.[66]

For all Mawdūdī's ideological standpoint on *jihād* as revolution in Islām, he was prepared to concede a peaceful political role for the Islamist movement in Pakistan. The *Jamā'at* contested elections and won some, though not many, seats. It had to engage with the messy compromises that politics entails, including support for the anti-Ayub Khan candidacy of Fāṭima Jinnah in the presidential elections of 1965: the endorsement of Fāṭima Jinnah ran counter to Mawdūdī's views on the social role of women.[67]

Linked to Mawdūdī's endorsement of the electoral process was his rejection of armed revolt as the means of bringing about the Islamic state: what was needed instead was an 'intellectual and moral revolution'. This was made clear in an interview given in 1388/February 1969 to *The Muslim* newspaper in London. His views were expressed sufficiently lucidly and categorically to repay close attention:[68]

> a lasting and perennial revolution cannot be brought about in any society unless the people among whom such a revolution is being achieved are generally prepared, intellectually and morally, to imbibe it and live up to it. In a nation where this preparation has not been done, efforts towards armed revolution can serve no purpose. The idea is not just to have a change, but a change for which the society has been prepared. There is no short-cut to it… Mere *coup d'état* cannot serve that purpose. But that is only one aspect of the matter; there are many others from which such an effort may be positively harmful. I refer to a few points in this respect:
>
> (a) The forces that are opposed to the Islamic movement possess control over armed forces, police and administration… A clash in such a position can only lead to the destruction of the movement, and not to the destruction of our foes.
> (b) Even if control over the organs of the state is achieved through an armed revolution, it would not be possible to run the state and carry on its affairs in accord with the Islamic way, for the simple reason that the society and its different sections have not been properly prepared for [the] moral transformation that Islām wants. And if the un-Islamic ways persist and continue to pollute the society in its multifarious aspects, while the Islamic Movement holds the reins of power, this may disillusion the people from the Movement and even from Islām as such.
> (c) Armed revolution as a means to power would be open to others as well… the danger is that the Muslim countries will remain ensnared in a vicious circle of revolutions and counter-revolutions, as they are caught up [with] today… the Islamic Movement would also become a party to this unwholesome game and would have to shoulder her share of people's wrath and hatred.
> (d) If you want to bring about an armed revolution, it is indispensable that you will have to organize your movement on the pattern of secret societies. Secret movements have a temperament of their own. They admit of no dissent or disagreement. The voice of criticism is silenced in them… Those who lead and run such movements become, through the internal logic of this method of work, cruel, intolerant and despotic… by the time such persons succeed in bringing about revolutions, they themselves have turned into tyrants, sometimes even greater tyrants than the ones they have been trying to remove.
> (e) Similarly another demand of the inner logic of this technique of work is to permit its workers to resort to deceit, lies, forgeries, frauds, bloodshed

and many other things which are forbidden in Islām – they are not only allowed to do so, but, if success is really contemplated, they are trained to do all that. This, in fact, makes them [believe] in the dictum that '[the] ends justify [the] means'... Is it really reasonable to expect that such persons, when they come to power, will honestly and scrupulously follow the principles of Islām and run the entire society according to the Islamic code of behaviour?

(f) It is also in the nature of revolutions brought about by the bullet, that they can be maintained only through the bullet. This produces a climate wherein peaceful switch-over towards an Islamic Order becomes virtually impossible. One despotism is replaced by another despotism. Hands change, but the system persists, [w]hile the objective of the Islamic movement is to change the system as such, and not merely to change the hands.

I would invite all those who are interested in establishing an Islamic order to reflect seriously and ponder over these aspects of the problem. I think they cannot avoid the conclusion that the Islamic Revolution can be brought about in its own way, and not by falling prey to the 'short-cuts' which can, in fact, only cut short the poise and tranquillity of society.[69]

Ḥasan al-Bannā'

Unlike Mawdūdī or Sayyid Quṭb, who rejected gradualism and compromise, the discourse of Ḥasan al-Bannā' (1323/1906–1368/1949) showed less enthusiasm for the forceful overthrow of un-Islamic regimes. In 1346/1928 he founded the first fully-fledged Islamic fundamentalist movement in the Arab world, the Muslim Brotherhood (*al-Ikhwān al-Muslimūn*). His treatise *On Jihād*[70] has to be viewed against the broader objectives of the Muslim Brothers.

The Muslim Brotherhood (*Ikhwān*)

In about 1358/1940, al-Bannā' defined an eightfold activity for the Muslim Brotherhood: it was a Salafī movement (*al-salaf al-ṣālīḥ*) seeking to take Islām back to its good ancestors and pure sources; it adopted an orthodox Sunnī method, committed as it was to implement the purified *Sunnah* in everything, especially in belief and worship; it was grounded in a Ṣūfī reality,[71] acknowledging that the foundation of goodness is the purity of the soul, the cleansing of the heart, constant work, abstinence, love for the sake of Allāh, and the objective of commanding right; it was also a political organization; an athletic group; a scientific, educational, and cultural association; a business corporation and a social [service] idea.[72]

This protean definition was rather more than all things to all men:[73] the multifaceted activities of the Brotherhood help explain the genuine mass support it elicited and the extraordinary growth of the movement to 2000 branches by the time of al-Bannā''s death in 1368/1949. The Brotherhood sought to transform the assertion of the oneness of God (*tawḥīd*; 'there is nothing that could be compared with him': Q.112:4) into a doctrine covering all spiritual and social activities and leading to fundamental change in the individual, society and state.[74] In his writings, al-Bannā' certainly emphasized the 'burning, blazing, intense faith fully awakened in the souls of the Muslim Brotherhood';[75] following the example of the Companions of the Prophet, they were to be monks by night and knights by day;[76] but it was above all to be a non-sectarian movement.[77] Political conflicts ought not to be turned into religious wars and should be resolved by dialogue.[78] For al-Bannā', the principle of individual involvement – to enjoin good and forbid evil – was the origin of pluralism, leading to the formation of political parties and social organizations, or the democratization of social and political processes. Because the ultimate source of the legitimacy of consultation (*shūrā*: Q.3:159; on his death bed, the second caliph 'Umar had appointed such a *shūrā* to choose his successor)[79] was the people, its representation could not be restricted to one party that usually represented only a fraction of the population. A continuous ratification by the community was required, because governance is a contract between the ruled and the ruler.[80] This requirement serves to rule out accusations of authoritarian dictatorship against al-Bannā''s theory,[81] quite apart from his own rejection of the Nazi, fascist and Communist systems as 'based on pure militarism'.[82] During World War II, the British nevertheless suspected him of pro-Italian sympathies.[83]

Jihād: the road to salvation from Western colonialism

Al-Bannā' advanced the doctrine of God's sovereignty or divine governance (*ḥākimīyyah*) and his lordship (*rubūbiyyah*) as the organizing principles for government and a potent symbol of political Islām.[84] These terms are used by militants, who contend that this legitimation for rule over a Muslim society requires a strict application of the *shari'āh*.[85] For al-Bannā', both Islamic government and the authority of Islamic law over society and the people was based on Qur'ānic guidance (respectively Q.4:105; Q.24:51; Q.5:48; Q.5:44–5; Q.5:47; Q.5:50 for Islamic government, and Q.5:49 and Q.4:65 for the authority of Islamic law).[86] Non-adherence to this divine plan of governance is perceived by radical fundamentalists as ingratitude to God (*kufr*) and asserting a form of partnership with God in His lordship (*shirk*).[87] There is nostalgia in al-Bannā' for a return of the caliphate, abolished by Kemal Atatürk in 1342/1924: 'where is the Caliph these days?', he asked.[88]

Clearly there was no realistic prospect of an immediate return of this overarching authority. Instead, divine governance could be implemented by an application of the Qur'ānic injunction to enjoin good and forbid evil.[89] Islamic sovereignty was superior to that of other systems, and it was the duty of Muslims to guide other nations into Islām: 'we… are neither communists, nor democrats nor anything similar to what they claim', he wrote. 'We are, by God's grace, Muslims, which is our road to salvation from Western colonialism.'[90] Al-Bannā' did not, however, reject every Western doctrine or idea or advocate a closed system. Muslims needed to develop to keep up with the pace of change in the world. Nor could he rule out that a secular government, rather than an Islamic one, might be chosen by the people. Ahmad S. Moussalli argues that, by insisting on the legitimacy of consultation (*shūrā*), al-Bannā' made a real advance 'at a time when one of the major practicalities of real politics is the authoritarian nature of politics exercised in the Muslim world'.[91]

In an important supplement to his treatise on *jihād*, al-Bannā' denied the authenticity of the *ḥadīth* recording the Prophet's extolling of the 'greater *jihād*', the *jihād* of the heart or the spirit. With regard to the commanding of good and the forbidding of evil, he also minimized the tradition that 'one of the loftiest forms of *jihād* is to utter a word of truth in the presence of a tyrannical ruler'. On the contrary, supreme martyrdom only arose when one killed or is killed 'in the way of God'.[92] The tract on *jihād* reveals the militant side of Ḥasan al-Bannā'. Death comes to us all, inevitably; therefore, to strive for an honourable death is to win perfect happiness. The 'honour of martyrdom' was thus a gift bestowed on us by God.

Thus, all Muslims must undertake *jihād*: *jihād* is an obligation from Allāh on every Muslim and cannot be ignored nor evaded. Islām was 'concerned with the question of *jihād* and the drafting and the mobilization of the entire *ummah* into one body to defend the right cause with all its strength'. The 'purity of language, the clarity of exposition, the lucidity of ideas and the force of spirituality' of the Qur'ānic texts and the *ḥadīth* did not, in al-Bannā''s view, require further elucidation. 'Forgiveness' and 'mercy' were associated with slaying and death in Allāh's way in Q.2:216 (cf. Q.2:190–3). In Q.4:71–8, Allāh urged Muslims to remain alert and to acquire experience in warfare, in armies and troops, or as individuals, as circumstances might dictate. Warfare was associated with prayer and fasting, and established as one of the pillars of Islam. Chapter 8 (*Sūrat al-Anfāl*) was in its entirety

> an exhortation to *jihād* and a command to remain steadfast once engaged upon it… It is for this reason that the first Muslims… adopted it as a war chant which they would chant whenever their apprehensions mounted and the battle grew grim.

Chapter 47 (*Sūrat al-Qitāl*, commonly known as *Sūrat Muḥammad*) mentions two key factors that form the foundation of the military spirit: obedience and discipline. Chapter 48 (*Sūrat al-Fatḥ*) is also dedicated in its entirety to one of the military campaigns of the Messenger of Allāh; this 'was a special occasion of *jihād*, which took place under the shadow of a tree where an oath of allegiance unto steadfastness and death was taken, and this bore the fruit of tranquillity and victory'. He cited the authority of 31 *ḥadīth* without discrimination, and lent credence to the myth of the 72 black-eyed women as a reward in Paradise for the martyred:[93]

> On the authority of al-Miqdām ibn Ma'dīyakrib, who said: 'The Messenger of Allāh… said: "The martyr possesses six distinctions with respect to Allāh: he is forgiven, amongst the first to be forgiven; he is shown his place in Paradise; he is not punished in the Grave; he is secure from the supreme terror of the day of judgement; the crown of dignity is placed on his head, a single ruby of which is more precious than the entire world and all it contains; he is wedded to seventy-two of the women of heaven; and he may intercede for seventy of his relatives."' (Transmitted by al-Tirmidhī and Ibn Māja)

Al-Bannā' asserted 'the consensus of [legal] scholars on the question of *jihād*' and cited six authorities, from various legal schools, to demonstrate this. What are we alongside such a history?, he asked. When then do Muslims wage war?[94]

> People have been for some time stigmatizing Islām because of the religious ordinance of *jihād* and the [divine] permission to wage war until the [message of] the precious Qur'ānic verse is fulfilled (Q.41:53). And now here they are acknowledging that it is the surest way to peace![95] God ordained *jihād* for the Muslims not as a tool of oppression or a means of satisfying personal ambitions, but rather as a defence for the mission [of spreading Islām], a guarantee of peace, and a means of implementing the Supreme Message, the burden of which the Muslims bear, the Message guiding mankind to truth and justice. For Islām, even as it ordains *jihād*, extols peace (Q.8:61). The Muslim would go forth to fight, one concern within his soul – to strive to his utmost until 'God's word is the most exalted' (Q.9:40).

For al-Bannā', the goal of the Islamic *jihād* was 'the most noble of goals' and the means employed were 'the most excellent' since they were defensive (Q.2:190). There was thus a concept of mercy within *jihād*. Women, children and old men were not to be killed, nor were the wounded, monks and hermits and 'the peaceful who offer no resistance'. *Jihād* became one of the ten fundamental pillars of the oath of allegiance (*bay'ah*) delineated by al-Bannā' to infuse loyalty to the members of the *Ikhwān* to its struggle and mission. It was considered by him to be 'a permanent duty that is binding upon Muslims till Doomsday'.

Al-Bannā' thus bequeathed two alternative strategies to the Muslim Brotherhood. The first, as has been seen, was a moderate fundamentalism which was prepared to engage with the democratic process and accept the legitimacy of consultation (*shūrā*) with the people. But he had also a fully-fledged, conservative, theory of military response. In May 1948, the founding committee of the Muslim Brotherhood held a meeting under al-Bannā''s chairmanship which adopted the key decision to demand on the Egyptian, and other Arab governments, 'to declare *jihād* against the Zionists [who had occupied Palestine] and to adopt all measures which would guarantee the deliverance and liberation of Palestine'.[96]

In late 1942 or early 1361/1943, al-Bannā' created a section of the Muslim Brotherhood called the 'secret apparatus' (*al-jihāz al-sirrī*) or 'special organization' (*al-niẓām al-khaṣṣ*). Al-Bannā' justified the participation of the *Ikhwān*'s Special Apparatus in military attacks on the British Army in the Suez Canal as self-defence against external aggression. 'Freedom of existence', he once argued, 'is the natural right of every human being, and nobody is entitled to infringe on this right. Anyone who transgresses against it is nothing less than an unjust dictator or despotic tyrant.'[97] The problem with such a structure was that if the command structure was cut by the imprisonment or house arrest of al-Bannā' himself then an unauthorized act of violence could take place. When the Egyptian government dissolved the Brotherhood in 1368/December 1948, and placed al-Bannā' under house arrest, this is precisely what occurred. A member of the Muslim Brotherhood warned him of the possibility of an assassination attempt. He replied:[98]

> What should I do? They have imprisoned my brothers and left me alone. I asked them to imprison me but they refused. I said 'if the *Ikhwān* is a criminal gang, I am their leader, so you have a right to kill me'. They cut my phone service; they confiscated my licensed gun; and they imprisoned my brother 'Abd al-Bāsiṭ, who used to help me travelling. They took my car and prevented me from going abroad. I asked to visit a friend of mine in Banha, but they refused. I even asked to visit my brothers in prison, but they refused.

The assassination of Prime Minister Mahmud Nuqrashi Pasha on 28 December was repudiated by al-Bannā'. Yet he was blamed for it by the government, and in 1368/February 1949 he was assassinated by government agents in an act of revenge.

Sayyid Quṭb

The assassination of Ḥasan al-Bannā', and the 'happy and joyous American reception' of it in particular,[99] was one of three critical events in the life of Sayyid Quṭb (1323/1906–1385/1966). The second was his visit to the United States in *c.* 1948–50, where he was shocked by its materialism, racism and apparent sexual permissiveness. As Moussalli notes, a hardening of the fundamentalist line is apparent in his writings from this date.[100] A third influence, embittering his thought and making it absolute and irreconcilable, was the imprisonment he endured under Nasser's regime, which lasted more than eleven years (1954–64, 1965–66), during which he suffered from torture.

Some have argued that Quṭb, justifying his response on the Qur'ānic verse (Q.2:194), contended that the suppression of the Muslim Brotherhood (*Ikhwān*) removed all choice but to use force in self-defence, repelling aggression with fighting and war (*'illat al-qatl wa al-qitāl*). Dr Adil Salahi, the editor and translator of Quṭb's greatest work, his commentary *In the Shade of the Qur'ān* (*Fī Ẓilāl al-Qur'ān*), denies categorically that Quṭb advocated violence. It is true that several new groups sought revenge 'for the injustice perpetrated by the Nasser regime on the Brotherhood in 1954'. Their leaders had arranged for arms to be smuggled into Egypt; Quṭb convinced them that arms shipments must be diverted, and that there should be no thought of their being used in the future. Dr Salahi contends that 'the wave of persecution that started in 1965 had no justification other than the existence of an organization that aimed to advocate the message of Islām peacefully and educate people in how to follow the Prophet's guidance'. He and six others were sentenced to death, but there was no justification for the sentence.[101] Quṭb and two others were executed in August 1966, but the other four death sentences were commuted to life imprisonment. He could have escaped his execution by a false declaration sufficient to satisfy the authorities.[102] This he would not do. Nor could he condone it in others. The changing of sides and backing of a dictatorship determined to suppress the Islamic message was, in Quṭb's view, the ultimate betrayal.[103]

Jihād as the perpetual revolutionary struggle against the forces of unbelief, injustice and falsehood (*jāhilīyyah*)

The most important book to precede the hardening of Quṭb's views after his visit to the United States was *Social Justice in Islām* (first edition in Arabic, 1949).[104] The concern with social and economic justice found in all his writings is expressed in two quotations: 'those who consider themselves Muslim, but do not struggle against different kind of oppression, or defend the rights of the oppressed, or cry out in the face of [a] dictator are either wrong, or hypocritical, or ignorant of the precepts of Islām';[105] 'when millions of a nation cannot find a mouthful of pure water to drink, it is undeniably luxury that some few people should be able to drink Vichy and Evian, imported from overseas'.[106] At this date, Quṭb was still prepared to talk of 'permanent and fundamental rights of humanity', with 'no difference… between one religion and other'. The same principle, he argues, is extended to cover human relationships in general. 'When Islām commands war against infidel peoples, the command refers only to defensive war which is aimed at stopping aggression' (Q.22:40; Q.2:186). This is war 'solely to defend the Muslim world against physical aggression, so that its members may not be seduced from their faith', a war to remove 'all material obstructions from the path of the gospel [*sic*: faith], that it may reach out to all men'.[107]

Two years later, in *Islām and Universal Peace* (1951), the attitude had hardened somewhat. The sole objective of the Islamic mission was that of liberating man, individually and collectively, from the unjust rule of certain 'earthly lords' (*al-arbāb al-arḍiyyah*), be these individuals, regimes, systems or institutions. It meant to bring about a universal peace (*al-salām al-'ālamī*) in the real sense, not only at international level among nations but also within the internal boundary of the nations suffering injustice. In this regard, Islām could not restrain itself from taking action against injustice, wherever it was found. It would have been unwise and illogical for Islām to make a 'superficial peace' (*al-silm al-kādhib*) with a non-Muslim country, while simultaneously allowing this country to impose a system that was founded on a 'lordship other than that of God' upon its people, which deprived them of their rights to judicial and social justice and was inherently oppressive. Regardless of the religion or form of governance that predominated, it was a religious duty for Muslims to deliver such people from injustice so that they might live in peace. *Jihād* was the means of achieving this objective.[108]

Quṭb's interpretation of *jihād* was that of a perpetual revolutionary struggle against the forces of unbelief, injustice and falsehood, or in short, *jāhilīyyah.*[109] The term '*jāhilīyyah*' in the Qur'ān was translated by the nineteenth-century scholar Goldziher as 'barbarism' rather than 'ignorance',[110] a term which is close to Quṭb's meaning for this fundamental concept, which he applied also to those parts of the Muslim world not considered truly Islamic in polity. There could be no coexistence between Islām and *jāhilīyyah*, or between *Dār al-Islām* and *Dār al-Ḥarb.*[111] It has been argued that Quṭb's was a critique of post-Enlightenment secular rationalism itself.[112] Yet the critique is potentially reducible to simple terms: any land that is not ruled by Islamic law is the abode of war (*Dār al-Ḥarb*) regardless of the people's religion.[113]

The supremacy of the 'final texts' over the 'transitional texts'

Quṭb's 'Islamic system' was characterized by seven features: the unicity or oneness of God (*tawḥīd*);[114] the eternal and everlasting nature of God's divinity (*al-'ulūhīyyah*); the constancy (*al-thabāt*) of the Islamic vision, which acted as a bulwark against Westernization and the appropriation of European ideas, customs and fashions; the comprehensiveness (*al-shumūlīyyah*) of Islām, a unity which was manifested in 'thought and behaviour, vision and initiative, doctrine and system, source and reception, life and death, striving and movement, life and means of livelihood, this world and the next';[115] the balance or equilibrium of Islām (*al-tawāzun*), between that which is revealed and that which is accepted by faith since man lacks the capacity to comprehend it; the positiveness (*al-'ījābīyyah*) of the Islamic faith, the motivating force in the life of individuals to create an *'ummah*; finally, the realism or pragmatism of Islām (*al-wāqi'īyyah*), which was grounded in the reality of life yet which sought to establish the highest and most perfect system for mankind. Quṭb's system aimed at reviving Islām; he denied the importance of its long history, though his viewpoint was consistently teleological:[116] Muslim leaders who had failed had done so because they either did not understand, or they disregarded, the spirit of Islām: Mu'āwiya, for example, turned the *ummah* into a nation governed by a dynasty.[117]

Quṭb's views represent a synthesis between classical theology and the modern fundamentalist interpretation of Islām. The two principal influences on his view of *jīhad* were Ibn Qayyim al-Jawziyya (691/1292–751/1350), a reputable Ḥanbalīte scholar of the eight century of Islām, and Mawdūdī, whose works were published in Arabic after 1370/1951.[118] Through his reading of Ibn Qayyim's work *Provisions for Final Destiny* (*Zād al-Ma'ād*), Quṭb discovered what he called the dynamic nature of Islām as a system of life and a revolutionary movement. He quoted at length Ibn Qayyim's analysis of the evolution of the concept of *jihād* from its embryonic stage in Mecca to its birth, and the successful establishment of the political community of Islām, in Medina.[119] From 'the great Muslim scholar' Mawdūdī,[120] Quṭb was reinforced in his view that the conflict between Islām and *jāhilīyyah* was not only inevitable but perpetual. There was no possibility of peaceful coexistence between the two. Secondly, Islām was a revolutionary ideology. Islām demanded (and continues to demand) the elimination of all governments that are contradictory to its ideology. The universal nature of Islām means that it seeks to govern the whole world, not just a portion of it, according to the will of God and for the salvation of the whole of mankind. Thirdly, Muslims must seek to turn their countries into Islamic states; all non-Islamic governments must be vanquished, and the world must be ruled according to the requirements of Islām (viewed as a more comprehensive conception than just the application of the *sharī'ah*). Finally, the permanent duty of *jihād* must not be confused with forceful conversion. In Mawdūdī's doctrine, Muslims are obliged to do their utmost to dislodge unbelievers from positions of power and force them to live in submission to the Islamic system.[121] For Quṭb, however, as long as all people have a free and unfettered right to receive the message of Islām, a non-Muslim authority need not be removed. The phrasing of his argument is nevertheless open to misinterpretation:[122]

> Islām is a general declaration of the liberation of man on earth from subjugation to other creatures, including his own desires, through the acknowledgement of God's Lordship over the universe and all creation… this declaration signifies a total revolution against assigning sovereignty to human beings, whatever forms, systems and situations such sovereignty may take. There are two essential prerequisites for this great goal. The first is to put an end to all oppression and persecution which targets the followers of this religion who declare their own liberation from human sovereignty and submit themselves to God alone… The second prerequisite is the destruction of every force that is established on the basis of submission to human beings in any shape or form. This will

> guarantee the achievement of the first goal and put into effect the declaration that all Godhead and Lordship on earth belong to God alone...

Quṭb's view of *jihād* evolved as a result of his reading of Mawdūdī and the Qur'ānic texts, and in the course of a sustained polemic against Western Orientalists and Islamic modernists such as Muḥammad 'Abduh, Muḥammad Rashīd Riḍā' and Muḥammad 'Izzat Darwazah.[123] He could not accept the interpretation of the modernists who sought to diminish the 'final texts' compared to the transitional texts. He claimed that they relegated the 'final texts' to a trivial and insignificant position which was subject to interpretation in the light of the 'transitional texts'. According to the modernists, Islām was to conduct itself in effect 'like a snail looking after its own affairs within its own boundary'. Islām, they claimed, need not seek to eliminate the forces of materialism and *jāhilīyyah*. Islām would resort to force only in a war of self-defence. In contrast, for Quṭb when the Qur'ānic text *al-Anfāl* (Q.8:61) was revealed.[124]

> God instructed his Messenger to remain at peace with those groups who refrained from fighting him and the Muslims, whether they entered into a formal treaty with the Muslims or not. The Prophet continued to accept a peaceful relationship with unbelievers and people of earlier revelations until *Surāh* 9 was revealed, when he would only accept one of two alternatives: either they embraced Islām or paid *jizya* which indicated a state of peace. Otherwise, the only alternative was war, whenever this was feasible for the Muslims to undertake, so that all people submit to God alone.

Quṭb stressed the 'serious realism' of the Islamic approach, which confronts human situations with 'appropriate means'. Islām seeks to 'provide an environment where people enjoy full freedom of belief' and abolishes oppressive political systems which seek to deprive people of this freedom. There could be no compromise over the essential principle, which was to enable people to worship God in his oneness or unicity. Finally, there is a legal framework governing relations between the Muslim community and other societies; anyone who places impediments in the way of the message of Islām 'must be resisted and fought by Islām'. Quṭb emphasized that Islām is a religion of peace, 'but this must be based on saving all mankind from submission to anyone other than God'. The ultimate aim is 'to destroy all forces that stand in its way of liberating mankind from any shackle that prevents the free choice of adopting Islām'.[125]

Quṭb refutes the argument that the verses are 'valid only in the case of Arabia at the time of revelation'. What happened to Muslims at the time of the fall of Baghdād in 656/1258?, he asks. They were massacred. What happened to Muslims at the time of the Partition of the Indian subcontinent? They were massacred. What has happened to the Muslim minorities in Communist Russia,

China and Yugoslavia? They have been oppressed and experienced atrocities. This exemplifies God's statement in Q.9:10 ('they indeed are the aggressors'). He concludes that it is the ability of Muslims to put the rulings into effect that counts, not the particular circumstances that led to their revelation.[126] 'Nowadays [presumably the 1950s and early 1960s] Muslims do not engage in *jihād*', he contends, 'because there is practically no Muslim community in the real sense of the term.'[127]

Indeed, the enemies of Islām were seeking to quash the Islamic revivalist movements the world over. The clearest example was the case of Atatürk and his movement in Turkey which was 'uncompromising in its enmity to everything Islamic'. 'Atheists and religious enemies of Islām' cooperate in supporting such regimes, which 'try to achieve for them the task left unfulfilled by the Crusades old and new'. A false 'Islamic' appearance also 'exercises a sedative influence on the Islamic revivalist movements'. The unmasking of regimes that are in fact committed to *jāhilīyyah* was thus a basic objective of the Islamist movement. A 'very sly, shrewd and cunning Orientalist' – that is the historian of religions Wilfrid Cantwell Smith – had even described the Atatürk movement, which was of 'atheistic orientation', as 'the greatest and wisest movement of Islamic revival'.[128]

For Quṭb, 'joining a *jihād* campaign for God's cause represents freedom from the shackles of this earthly life and physical pleasures'. Whenever a community abandons *jihād* and refuses to fight for God's cause, 'it is bound to suffer humiliation. Its eventual loss is much greater' than fighting in the true spirit in the first place.[129] The concept of a defensive *jihād* is repudiated: Q.9:123 advocates fighting those who live next to the land of Islām, 'without reference to any aggression they might have perpetrated. Indeed their basic aggression is the one they perpetrate against God… it is this type of aggression that must be fought through *jihād* by all Muslims.'[130] Contrary to what might be expected, Quṭb claims that the idea that those who do not participate in a *jihād* campaign are the 'ones who devote time to studying and understanding' the faith is false. Instead, 'every *jihād* campaign is a means to acquire a better understanding' of Islām.[131]

As to the contention that the Islamic faith managed to spread only because it used the sword, this is rejected absolutely. *Jihād*, on the contrary, 'seeks to guarantee freedom of belief' (Q.2:256). People who argue otherwise forget that it is meant to serve God's cause:[132]

> it aims to establish God's authority and to remove tyranny. It liberates mankind from submission to any authority other than that of God (Q.8:39). *Jihād* does not aim to achieve the hegemony of one philosophy or system or nation over another. It wants the system laid down by God to replace the systems established by his creatures. It does not wish to establish a kingdom for any one of God's servants, but to establish God's own kingdom….

The reasoning is specious. Quṭb certainly argues that God must be in command of all life. But it is a fact of the modern world that not everyone accepts that God exists. They are in error and part of *jāhilīyyah*. Indeed everything which does not accept the supremacy of the Islamic state and its legal system is automatically in error. Far from seeking to avoid hegemony, Quṭb's system is a hegemonic one. The only way in which his reasoning can be justified is by accepting his assumption that God exists, that God's sovereignty (*ḥākimīyyah*) alone, and not any human system, is required for society; and that Islām is the only correct faith. Any alternative vision, value system, concept, feeling, tradition or custom is wrong and part of *jāhilīyyah*.[133] In so doing, he went further than Mawdūdī, and rejected his concept of 'partial *jāhilīyyah*'.[134] Non-Muslims were allowed to enjoy the freedom to practise their religion in a limited sense as a set of beliefs and rituals only, but were deprived of their rights to self-determination, or to translate their 'misconceived' religion into an autonomous world view, social order and society.[135] Clearly Quṭb's system as developed while in prison was neither pluralist nor tolerant.

The need for an Islamic vanguard and its influence on bin Laden

In 1384/1965 Quṭb published his most notorious book, *Milestones* (*Ma'ālim fī al-ṭarīq*), which led to his re-arrest with the accusation of conspiracy against the Egyptian President, Abdul Nasser. He was tried and rapidly sentenced to death based upon excerpts from this publication. The book was critically reviewed by al-Azhar's Committee of *Fatāwā* led by its Grand Shaykh, al-Shaykh Ḥasan Ma'mūn. It contended (erroneously) that instead of being signposts towards the reinstatement of Islām as a dominant order, 'the teachings of Sayyid Quṭb's *Milestones* served as signposts towards violence against every Muslim government on earth and towards the creation of a world of anarchy'.[136]

As Quṭb recognized, most of the argument on *jihād* in *Milestones* was repeated from *In the Shade of the Qur'ān*. Ibn Qayyim was again cited, and there was the same fourfold analysis of the method of Islām. He pronounced that Islamic absolutism and human freedom are compatible:[137]

> this religion is really a universal declaration of the freedom of man from servitude to other men and from servitude to his own desires, which is also a form of human servitude; it is a declaration that sovereignty belongs to God alone and that He is the Lord of all the worlds. It means a challenge to all kinds and forms of systems which are based on the concept of the sovereignty of man; in other words, where man has usurped the Divine attribute.

'This struggle is not a temporary phase but an eternal state – an eternal state, as truth and falsehood cannot co-exist on this earth.' Any restraint on *jihād* was a question of strategy for the movement, not a matter of belief or principle.[138] Sayyid

Quṭb's *jihād* is gradual or transitional (*marḥaliyyah*) from one stage to another. It begins with an ideological struggle to undermine the basic tenets and principles of the *jāhilīyyah* world view. The process of *jihād* then gradually progresses and is converted to its final stage (*nihā'iyyah*) in which an armed struggle is finally applied to overcome the forces of *jāhilīyyah* and the establishment of a 'true' Islamic system. Sayyid Quṭb was not in favour of the institution of armed struggle as a launching pad for *jihād*, since he considered this against the nature of the Islamic movement. In the introduction to *Milestones*, he tried to convince the Muslim activists that the use of force is not an appropriate method to remove *jāhilīyyah* from power, while other transitional (*marḥaliyyah*) methods are not fully exhausted.[139]

If armed struggle is not immediately practicable, what kind of *jihād* should the Islamic vanguard strive for? In response to the question, Sayyid Quṭb proposed the following scheme.[140] Firstly, the creation of an Islamic organization that seeks to unite all the Muslim activists and exponents of Islamic movement under a single leadership. Having established itself as the vanguard of Islām (*al-ṭali'at al-Islāmiyyah*), the organization should set itself up not only as a nucleus but also the catalyst for the creation of an Islamic society that is supposed to replace the existing *jāhilī* society. Secondly, the main duty of the Islamic vanguard is to undertake ideological struggle to oppose *jāhilīyyah* through effective conduct of propagation of the faith (*da'wah*) by means of persuasion, argumentation and education. Propagation of the faith must begin with the call upon people to acknowledge the sovereignty of God (*ḥākimīyyat Allāh*), for He is the ultimate source of authority as epitomized in the first Islamic creed that 'there is no God save Allāh and Muḥammad is His Messenger'. No one is exempted from this call. Thirdly, there must be *hijrah*:[141] a mental, ideological and cultural separation in which the Islamic vanguard and those who are converted to their religio-political idealism are required to keep themselves aloof from *jāhilī* society and to concentrate on consciousness-raising, nurture, growth and strengthening group solidarity.[142] Finally, the establishment of an authentic Islamic society is possible only when the Islamic vanguard and the new 'converts' decide to consolidate their existence and transform their movement into a religio-political unit, completely separable from its *jāhilī* remnant. Unless there is an Islamic society, any attempt to undertake armed struggle to eliminate the forces of *jāhilīyyah* is impossible. The Manichean struggle depicted by Quṭb between good and evil (within his definition, that is) was not a temporary phase, 'a temporary injunction related to changing conditions and transient circumstances'; instead it was a perpetual and permanent war. '*Jihād* for freedom', he declared, 'cannot cease until the Satanic forces are put to an end and the religion is purified for God *in toto*.'[143]

What is particularly striking, and perhaps serves as the true testimony to Sayyid Quṭb's pervasive influence on militant Islamism, is that as recently as 18 July

2003 Osama bin Laden argued for essentially the same process, though in this case violence was the intended end product:[144]

> *Jihād* is the way to attain truth and abolish falsehood. Therefore, the youth… who love the religion and sacrifice [themselves] for Allāh must pay no attention to these civil servants [bin Laden's term for clerics who disagree with his viewpoint] and to those who refrain [from waging *jihād*]…
>
> The land is occupied in the full sense of the word. Yet despite this, people are busy with all sorts of [other] rituals. We must focus on making the starting-point *jihād* for the sake of Allāh, guarding against those who refrain [from exile] (*hijrah*) and *jihād* for Allāh. All these are obligatory in the present situation in order to establish the truth and abolish falsehood.

John L. Esposito argues that 'it is almost impossible to exaggerate the direct and indirect impact and influence' of Mawdūdī, Ḥasan al-Bannā' and Quṭb. He adds that[145]

> [t]heir writings have been published and distributed throughout the Muslim world. Their ideas have been disseminated in short pamphlets and audiocassettes. The leadership of most major Islamic movements, mainstream and extremist, non-violent and violent alike, has been influenced by their ideas on Islām, Islamic revolution, *jihād* and modern Western society. Their recasting of Islām as a comprehensive ideology to address the conditions of modern Muslims produced a reinterpretation of Islamic belief that has been so widely used, it has been integrated unconsciously into the religious discourse of Muslims throughout the world who would normally disassociate themselves from Islamic movements…
>
> Though part of a centuries-old revivalist tradition, all three men were modern in their responses. They were neo-fundamentalist in the sense that they returned to the sources or fundamentals of Islām. But they reinterpreted the Islamic sources in response to the challenges of the modern world. This is apparent in their teachings, organization, strategy, tactics, and use of modern science and technology…
>
> Like Ḥasan al-Bannā' and Mawlānā Mawdūdī, Quṭb regarded the West as the historic enemy of Islām and Muslims as demonstrated by the Crusades, European colonialism and the Cold War. The Western threat was political, economic, and religio-cultural. Equally insidious were the elites of the Muslim world who rule and govern according to foreign Western secular principles and values that threaten the faith, identity and values of their own Islamic societies. Going beyond al-Bannā' and Mawdūdī, Quṭb denounced governments and Western secular-oriented elites as atheists against whom all true believers must wage holy war.

9
The Shīʿa Depiction of *Jihād* and Martyrdom (*Shahādah*)

> God gave preference to Muḥammad before all His creatures. He graced him with prophethood and chose him for His message… We are his family (*ahl*), those who possess his authority (*awliyā'*), those who have been made his trustees (*awṣiyā'*) and his inheritors (*wurathā'*); we are those who have more right to his position among the people than anyone else… We know that we have greater claim to that right, which was our entitlement, than those who have seized it…

The letter of *Imām* al-Ḥusayn to the people of Baṣra reveals the early Shīʿa thinking on legitimate succession.[1] The Shīʿa were, and remain, 'followers' or 'supporters', those who supported 'Alī's claim to both religious and political authority. They believe that the Prophet appointed his son-in-law 'Alī as his successor at Ghadīr Khumm not long before he died (see Chapter 2).[2] They also believe that, on leaving for the Tabūk expedition, the Prophet appointed 'Alī to remain at Medina as his deputy, stating that 'you are to me as Aaron was to Moses'.[3] In the view of the Shīʿa, 'Alī was regarded as the executor (*waṣī*) of Muḥammad and had been designated as such by him.[4] 'Alī sought a return to the Prophet's practice of treating the 'Emigrants' (*Muhājirūn*) and the 'Helpers' (*Anṣār*) in Medina on a par; it was to them alone that the right of consultation (*shūrā*) on the succession to the caliphate belonged.[5]

For Imām Muḥammad al-Bāqir (d. *c.* 120/737 or 122/739), *walāyah*, that is love for, and allegiance to, the *imām*, was a pillar of Islām to which the other six pillars were subordinate ('through it and through the one to whom allegiance should be paid, the knowledge of the other pillars is reached').[6] The *imāms* were conceived by him to be the protectors or guardians of the believers, to whom obedience was owed as an obligatory duty.[7] The *imām*, in his view, enjoyed infallibility (*ʿiṣmah*) and could, after his death, exercise intercessional powers (cf. Q.7:17: 'on the day when we shall call all men with their *imām*…'). The imāmate had an hereditary character, vested in the progeny of the *imāms*, and manifested in the children of ʿAlī's second son, al-Ḥusayn (cf. Q.33:6; Q.43:28). The phrase 'those who hold authority' (*ūlī al-amr*) in Q.4:59 was interpreted to mean the *imāms* 'from the family of Muḥammad'.[8]

Foundational narratives of martyrdom

The founder of the Ḥanbalite school of law, Aḥmad ibn Ḥanbal, was prepared to endorse ʿAlī's claim to be the fourth caliph, but not to go so far as some of his supporters, who preferred ʿAlī to earlier caliphs such as Abū Bakr and ʿUmar. Such a person, he feared, 'might be a Rafīḍī'. Within this definition, a Rafīḍī was someone who abused and cursed Abū Bakr and ʿUmar and believed in the imāmate of *imām* ʿAlī as something ordained by God.[9] It was claimed that the Shīʿa were 'those who followed ʿAlī when the people differed concerning the Messenger of Allāh and who followed Jaʿfar ibn Muḥammad when people differed concerning ʿAlī'. According to a report cited by Ṭabarī, after the battle of Ṣiffīn when ʿAlī returned to Kūfa and the Khārijīs (Khawārij) broke away from him,[10] the Shīʿa remained steadfast on ʿAlī's side and declared that they were bound to him by an additional oath of allegiance: to befriend his friends and to regard his enemies as their own enemies ('on the basis of friendship of his friends and enmity of his enemies') This stress on the second oath of allegiance indicates the Shīʿa character of the remaining group of ʿAlid supporters.[11] They were the Shīʿat ʿAlī, for whom ʿAlī was the 'legatee among legatees [of the prophets]' and 'heir to the knowledge of the prophets'.[12] In contrast, ʿAlī's critics wanted his removal and the appointment of a new caliph by a council (*shūrā*) of the most eminent Early Companions as had been stipulated by ʿUmar. ʿAlī also lacked support among the Qurayshīs, who under Abū Bakr's constitution had been recognized as the ruling class which alone was entitled to decide on the caliphate.[13] Nevertheless, for *Imām* Khomeini, writing in 1391/October 1971, 'the greatest disaster that [ever] befell Islām', greater in his view even than the 'tragedy of Karbalā'', was the usurpation of ʿAli's rule by Muʿāwiya. For Muʿāwiya 'caused the system of rule to lose its Islamic character entirely and to be replaced by a monarchical regime'. Khomeini's declaration centred on the incompatibility of monarchy with Islām.[14]

The group of ʿAlid supporters was fatally weakened by two events following the assassination of ʿAlī in 40/661. The first was the abdication of al-Ḥasan to Muʿāwiya, ostensibly in order to prevent *fitnah* in the body of Muslims (even his original call to *jihād* had been lukewarm: it was a loathsome (*kurh*) duty, he declared, quoting Q.2:216)[15] but probably because of significant defections among his supporters ('he intends to seek a truce with Muʿāwiya', they said to themselves, 'and to surrender the reign to him; he is weak and confounded').[16] Mu'āwiya, the secessionist governor of Syria, did not deny al-Ḥasan's exalted position in relation to the Prophet and his superior place in the world of Islām; what he did instead was to assert that the criteria for leadership of the community were personal power, ability in political affairs, and military capacity, in other words political rather than religious claims. The abdication of al-Ḥasan, which was considered binding in Mu'āwiya's lifetime, was a tacit recognition of this state of affairs. The Shī'a community was quiescent during the nine years between al-Ḥasan's abdication in 41/660 and his death in 49/669.[17] Revolt was precipitated by Mu'āwiya's action in appointing his son Yazīd to succeed him, in spite of his personal failings; this occurred in 60/March 680 because Mu'āwiya had failed to secure the consent of the Shī'a leaders to this arrangement. The second disaster to damage the cause of the Shī'a community was the defeat and death of Ḥusayn bin 'Alī and a total of 72 supporters at the hands of a numerically superior force at the battle of Karbalā' in 61/October 1680. The bodies of Ḥusayn, three of his brothers[18] and his other supporters were trampled by horses and were left decapitated. The event is recalled in the ceremony of al-'Āshūra', the tenth day of the month of Muḥarram, the most solemn day in the Shī'a calendar.[19] Ḥusayn preferred martyrdom to renunciation of his claim to the imāmate and became not merely a *ṣhahīd* but 'the chief [or lord] among martyrs', *Sayyid al-Shuhadā*: 'the sacred cause that leads to *shahādah* or the giving of one's life has become a law in Islām. It is called *jihād*.'[20] Within four years, Ḥusayn's grave at Karbalā' had become the site for 'wild and unprecedented expressions of grief, weeping and wailing for the suffering and tragic death of the grandson of the Prophet'.[21] To the extent that these events gave the Shī'a a martyrology, enabling the passion of 'Alī's sons to be remembered in martyrdom plays (*ta'ziyah*) and permitting the community to indicate its willingness for martyrdom by displays of self-inflicted wounds, they served a long-term purpose: the Shī'a perception of Islām is inseparable from the concept of martyrdom (*shahādah*).

The Imāmiyyah world view

In the short term, however, the leadership of the movement was decimated. Yet even this diminished band of supporters was subdivided according to different tendencies. The majority group, probably from early times and certainly in the modern period, have been the 'twelvers' (Ithnā 'Asharīya), the state religion of Iran since 906/1501, who followed the cult of Twelve *imāms* (see Chapter 5). On the death of the (quietist) fourth Shī'a *imām*, 'Alī Zayn al-'Ābidīn in 95/713, the Zaydīs followed Zayd rather than his half-brother, Muḥammad al-Baqīr, as *imām*.[22] For this relatively small group, which has remained mostly in the Yemen, any descendant of Fāṭima may become *imām*, and it is possible to have periods without an *imām* since appointment to this position is a consequence of exemplary behaviour rather than descent. The second group in terms of its size in modern times has been the 'seveners' (Sab'īyyah, Ismā'īliyyah or Ismā'īlīs), so called both because of their belief that prophets come in cycles of seven and their choice of Ismā'īl as the seventh and final *imām* (as distinct from the Twelvers who chose Mūsā and his successors following the death of the sixth *imām*).[23]

Three political concepts came to predominate among the Shī'a following upon this unfortunate family history and further developments following upon the seclusion or 'first occultation' (*ghayba*) of the twelfth *imām*, 'Alī ibn Muḥammad Simmarī in 260/847. The first was their distinctive concept of the imāmate, which was developed by the sixth *imām*, Ja'far aṣ-Ṣādiq (*imām* 114/732–148/765),[24] who propounded the view that the imāmate and caliphate should be divided into two separate institutions until such time as God made the true *imām* victorious. The second was the doctrine of dissimulation (*taqīyah*) and the third was the idea of the permanent 'occultation' of the twelfth or last *imām*.

Taken together, these concepts might seem like a rationalization of defeat by an unsuccessful minority that had been crushed by the 'Abbāsids. Elaborated as doctrines of the Shī'a faith out of the political needs of the time, these concepts nevertheless came to provide the community with a distinctive identity that helped preserve it from competing ideologies, whether of an extreme separatist tendency or of the Sunnī majority. There were two key elements, *naṣṣ* and *'ilm*, with regard to the imāmate. Belief in 'explicit designation' (*naṣṣ*) defined the imāmate as 'a prerogative bestowed by God upon a chosen person, from the family of the Prophet, who before his death and with the guidance of God, transfers the imāmate to another by explicit designation'.[25] In reality, there were always two *imāms*, the speaking *imām* (*nāṭiq*) and his son and successor, who during the lifetime of his father was silent (*ṣāmit*). Only on his father's death was the son and successor entrusted with the scriptures and the secrets of religion, becoming the 'proof' (*al-Ḥujjah*) for mankind.[26] *'Ilm* was the

> special sum of knowledge of religion, which can only be passed on before his death to the following *imām*. In this way the *imām* of the time becomes the exclusively authoritative source of knowledge in religious maters, and thus without his guidance no one can keep to the right path. This special knowledge includes both the external and the esoteric meanings of the Qur'ān.[27]

The esoteric meanings of religion (*Wilāyat Allāh*) were handed over by the Prophet to 'Alī, and thus became the inheritance of the *imāms* who were bound to keep them secret. The *imām* was the legatee of the Prophet and was infallible (*ma'ṣūm*, hence the doctrine of *'iṣmah* or inerrancy) in all his acts and words. To ignore or disobey the *imām* was equal to ignoring or disobeying the Prophet.[28] This infallibility made him not only the guardian (*hāfiḍ*) but also the interpreter of the law.[29] Ja'far aṣ-Ṣadiq raised the doctrine of dissimulation (*taqīyah*) to that of a condition of faith: 'fear for your religion and protect it with the *taqīyah*, for there is no faith (*imān*) in [the person in] whom there is no *taqīyah*.' One who does not keep *taqīyah*, he asserted, 'has no religion'.[30]

The idea of the permanent 'occultation' of the twelfth or last *imām* became a feature of the Imāmiyya's world view only from the death of the fourth deputy (*safīr* or *wakīl*) in 329/941.[31] The *imām* will remain hidden until God gives him permission to manifest himself; but he remains of profit to mankind[32] in that he continues to give guidance, hears prayers and intercedes on behalf of the faithful. All government in the absence of the *imām* was in a sense usurped although it might acquire a functional legality. This prior political circumstance limited the capacity of the Shī'a community to respond to the issue of *jihād*, since *jihād* was not to be undertaken in the absence of the *imām* or the person he had appointed. If neither were present it was a fault to fight the enemy or to undertake *jihād* under an unjust leader; anyone who did so committed sin.[33] Similarly, rebellion against an unjust government was not permitted in the absence of the *imām*. Until the return of the Awaited One (al-Mahdī) there was no end to dissimulation (*taqīyah*) and no duty of rebellion against an unjust ruler.[34]

The Ismā'īlī caliphate and its rivals, 297/909–567/1171

For the Twelver Imāmīs, as has been seen, because of the 'occultation' of their twelfth *imām*, the *imām* could not represent an active source of law. Not so for the Ismā'īlīs, for whom the 262-year rule of the Fāṭimid dynasty, based at Cairo in Egypt after 362/973, represented an end to the period of concealment (*dawr al-ṣatr*) and the beginning of the promulgation of a distinctively Ismā'īlī lawcode (*madhhab*).[35] It was noted above in Chapter 2 that a good example of the developed ideology of *jihād*, albeit at a relatively late date, is provided in the ninth chapter of the *Da'ā'im al-Islām* of al-Qāḍī al-Nu'mān. This was the official lawcode of the Fāṭimid state of Egypt issued by its ruler, al-Mu'izz li-Dīn Allāh, around the year 349/960.[36] But it was a unique work: before al-Nu'mān there was no independent Ismā'īlī law; what came after him was repetition or restatement.[37] There was an attempt in this work to reconcile certain of the differences in doctrine between the Ismā'īlīs and the Mālikī Sunnī school of jurisprudence, while al-Nu'mān nevertheless sought to 'elaborate the *ẓāhirī* doctrinal basis of the Fāṭimids' legitimacy as ruling *imāms*':[38] it was essential to recognize the rightful *imām* of the time, since the *imām* was the third principal source of Ismā'īlī law after the Qur'ān and the *Sunnah* of the Prophet. (This did not prevent the formation of a Fāṭimid counter-caliphate for a short period.)[39]

The Ismā'īlīs tried to protect their unity by ensuring that initiates pledged to observe the secrecy of the 'inner meaning' (*bāṭin*) of their doctrine or 'wisdom' (*ḥikma*); they were sworn to secrecy prior to initiation by taking an oath (*'ahd*) which gave them access to the party of God (*Ḥizbu'llāh*) and to the party of saints (*awliyā'*). Though elements of doubt remain, because the form of oath has not been preserved in any authentic Ismā'īlī sources, it seems that the oath remained unchanged for centuries.[40] For all that the oath of loyalty and the theory of al-Nu'mān added two more 'pillars' to the five of Islām, devotion to the *imām* (*walāyah*) and *jihād*, the Fāṭimid dynasty was not conspicuous by its pursuit of war in God's cause (*fī sabīl Allāh*). The 'seveners' were further subdivided into the theological subdivisions of Qarmaṭīs, Nizārīs, Musta'līs and so on, as a result of other disputes. These other groups were considerably more violent than the Fāṭimids.

The depredations of the Qarmaṭīs of Baḥrayn reached their peak in the decade following Abū Ṭāhir's attack on a pilgrim caravan returning from Mecca to 'Irāq in 312/April 924. Almost every year Abū Ṭāhir led a raid into Mesopotamia or attacked the pilgrim caravans. In 316/928, he built an abode (*dār*) which he called the 'abode of emigration' and summoned supporters of the imminent *Mahdī*.[41] It was, it seemed to some, only as if by a miracle that Abū Ṭāhir failed to capture Baghdād in that year. Worse was to follow the following year. The Qarmaṭīs occupied Mecca, wrought a massacre of the inhabitants and pilgrims, burned copies of the Qur'ān, defiled the holy city and robbed the Ka'ba of its Black Stone, 'a sacrilege... unparalleled in the history of Islām'.[42]

For such an act of sacrilege Abū Ṭāhir was justly criticized, to which he replied that he was incapable of bringing those people who had been killed back to life, though he promised the return of the Black Stone. It was not returned in his lifetime (he died in 332/944) and it was not returned for more than 20 years after its theft (the restoration occurred in 339/951).[43] The Qarmaṭīs argued that everything that happened was precisely predestined; they alone, as the true believers, possessed knowledge of this predestination (*al-qadar*) and were infallible (as well as invincible in their actions). Notwithstanding this belief, Abū Ṭāhir and his followers allowed themselves to be deluded into believing in Ramaḍān 319/September–October 931 that a young Iṣfahānī imposter was the true Mahdī and thus that the original 'religion of our father Adam' had appeared. Incredibly, this led to the public cursing of Abraham, Muḥammad, and 'Alī and his descendants, all of whom were said to be 'wily deceivers' (*dajjālūn*), while the true 'wily deceiver', the Iṣfahānī, went on a rampage of senseless atrocities.[44] After the short interlude under the influence of the true 'wily deceiver', the Qarmaṭīs reverted to their doctrine about the imminence of the future Mahdī, though quite how this aberration was explained away remains obscure. A further aberration took place in 358/969, when they broke with the Fāṭimids and entered the service of the 'Abbāsids; but they returned to their former allegiance in 375/985.[45]

A further schism occurred in 487/1094, which has come to be known as the Nizārī–Musta'lī schism: this was initially a dispute over the succession following the death of al-Mustanṣir in that year.[46] But the Nizārī secession was also about creating a Persian Ismāʿīlī identity. Even before the schism, the Ismāʿīlī 'summoner' or religio-political missionary (*dāʿī*) and theologian Nāṣir-i Khusraw (d. after 465/1073) had composed all his works in the Persian language. For him, the true *jihād* was the war that must be waged against the perpetrators of bigotry, through spreading knowledge that dispels the darkness of ignorance and nourishes the seed of peace that is innately embedded in the human soul.[47]

As part of his corpus of some 40 or more works (see Chapter 3), al-Ghāzalī wrote a number of treatises in which he attempted to refute the doctrines of the Ismāʿīlīs. The most detailed of these is the *Faḍaʾiḥ al-Bāṭiniyya wa faḍāʾil al-Mustaẓhiriyya* (*The Infamies of the Bāṭiniyya and the Virtues of the Mustaẓhiriyya*) – more commonly referred to as the *Kitāb al-Mustaẓhirī*.[48] The purpose of the work was systematically to analyse and dismantle the Ismāʿīlī doctrine of *taʿlīm*, the 'authoritative instruction or teaching' of the *imām* of the time, as expounded by Ḥasan-i Ṣabbāḥ (d. 518/1124), the head of the Ismāʿīlī *daʿwah* organization in Persia. Written during the formative phase of Ḥasan-i Ṣabbāḥ's activities, just before the *Imām*-Caliph al-Mustanṣir's death, al-Ghāzalī directed his attack on the activities and ideas connected with the infiltration of the Ismāʿīlīs of Alamūt inside the Saljuq empire. In 483/1090, Ḥasan-i Ṣabbāḥ took over the mountain fortress of Alamūt in northern Iran, which was later to become the headquarters of the Nizārī Ismāʿīlī state and *daʿwah*; though it was only subsequently, in 488/1095, that the doctrine of *taʿlīm* became prominent in the consolidation of Nizārī Ismāʿīlism.

No direct evidence remains of Ḥasan-i Ṣabbāḥ's writings on the doctrine of *taʿlīm*. Instead, we are reliant on al-Shahrastānī's text, which purports to be a paraphrase of what he wrote.[49] Al-Shahrastānī (d. 548/1153) refutes al-Ghāzalī's contention that the recourse to an infallible *imām* (hence the doctrine of *taʿlīm*) arose from the invalidity of reason; rather, it acknowledges the limits of reason: 'through our need we come to know the *Imām*, and through the *Imām* we come to know the extent of our need'.[50] On the doctrine of the imāmate (*al-imāma*), al-Ghāzalī contended that the *imām* of the Nizārī Ismāʿīlīs was equal to the Prophet in 'infallibility and knowledge and in knowledge of the realities of the truth in all matters, except that revelation is not sent down to him'; instead, he received revelation from the Prophet.[51] He questioned why the *imām*'s privileged knowledge should be transmitted only to the initiated; if this privileged knowledge was a secret (*sirr*) divulged by the Prophet only to ʿAlī, what were the reasons for this secrecy?[52] For al-Ghāzalī, 'there must be an *imām* in every age, but only he is qualified for the office; therefore he is the rightful *imām*'.[53] He considered that election (*ikhtiyār*) was the only valid source for the designation of the *imām*, although ultimately it was reducible to God's choice and appointment, 'a grace and gift of God, unattainable by any human contriving'.[54] The *imām* was not infallible, in al-Ghāzalī's scheme, but should settle issues after proper consultation with the *ʿulamāʾ* ('exploiting the talents of the *ʿulamāʾ*').[55] In contrast, the Nizārīs contended that since the Sunnī community lacked access to the 'authoritative instruction or teaching' of the infallible *imām* it was likely to be misguided, since it was dependent on fallible conjecture (*ẓann*) and speculation derived from human reason.[56] Notwithstanding these important differences, it was above all in their doctrine of the Resurrection or the Last Day (*qiyāmah*) that al-Ghāzalī considered that the Nizārīs had transgressed the limits of tolerable dissent.[57] This leads us directly to the phenomenon of the Nizārī Assassins and their doctrines.

The first era of sustained group terrorism? The Nizārī Ismā'īlīs, assassination and the doctrine of the Last Day (*qiyāmah*), 559/1164–654/1256

It was Ḥasan-i Ṣabbāḥ who initiated an open revolt against the Saljuq *amīrs* and laid the foundations of an independent Nizārī Ismā'īlī state based on the fortress of Alamūt. The Saljuq vizier (*wazīr*), Niẓām al-Mulk, who was assassinated on 12 Ramaḍān 485/16 October 1092, is thought to have been the first prominent victim of the Nizārī devotees (*fidā'īs*). Thereafter the bulk of the assassinations were of local Saljuq *amīrs* who attempted to resist the Nizārīs or to raid Nizārī Ismā'īlī settlements.[58] Ḥasan-i Ṣabbāḥ did not invent the use of assassination as a political weapon to rid the community of its religio-political adversaries: as has been seen above (Chapter 2), the Khārijīs had earlier resorted to this policy. There can be little doubt, however, that Ḥasan-i Ṣabbāḥ assigned particular importance to these suicide missions.

> Like the contemporary suicide bomber in Palestine or elsewhere, the *fidā'īs* were glorified in the community for their bravery and devotion. Rolls of honour of their names and assassination missions were evidently compiled and retained at Alamūt and probably other fortresses… From early on, the assassinations were countered by the massacres of Ismā'īlīs, or of all those in a town suspected, or accused of being Ismā'īlī. The massacres, in turn, provoked assassinations of their instigators, which led to further assassinations.[59]

The phenomenon was not unlike the cycle of violence in contemporary Israel.

After the Nizārī–Musta'lī schism of 487/1094, and the murder of Nizār the following year, the Nizārīs experienced a period of concealment (*dawr al-satr*), when the *imām* would not be directly accessible to his followers. (There was recognition of the anonymous succession of Nizār's progeny at Alamūt until 556/1161.)[60] Ḥasan-i Ṣabbāḥ, the chief representative (al-*Ḥujjah* or proof) of the inaccessible *imām* and advocate of the 'new preaching', died in 518/June 1124, and the Nizārīs continued their campaigns and strengthened their position against the Saljuq *amīrs*.

The fourth chief representative, Ḥasan bin Muḥammad, who succeeded in 557/1162, was also the first Nizārī *imām* to manifest himself openly in the Alamūt period. On 17 Ramaḍān 559/8 August 1164, Ḥasan declared to a special assembly of Nizārīs held at Alamūt that the *imām* of the time had relieved his followers of the burdens of the *sharī'ah* and had brought them to the Resurrection or *qiyāmah*. Ḥasan thus proclaimed himself the 'lord of the Resurrection' (*qā'im al-qiyāmah*), the *imām* inaugurating the Resurrection (*imām qā'im*), whose mission would be the *da'wah* of the Resurrection. The breaking of the fast of Ramaḍān on what has always been celebrated since by Nizārīs as a day of rejoicing, and the dispensing with the *sharī'ah*, were inevitably controversial decisions. It was too much for Ḥasan's own son-in-law, who assassinated him in 561/January 1166.[61]

The *da'wah* of the Resurrection lasted until 608/1211, when, in an extraordinary volte-face, Jalāl al-Dīn Ḥasan, the sixth lord of Alamūt, publicly repudiated the doctrine and ordered his followers to observe the *sharī'ah* in its Sunnī form. The Nizārīs seem to have followed his lead without demur, regarding this as a return to the era of 'dissimulation' (*taqīyah*), a caution which had been lifted during the period of the Resurrection.[62] The *modus vivendi* with mainstream Sunnī religious and political leadership might have seemed to remove the need for assassination as their chief political instrument, since the Nizārīs had always directed this chiefly at their Sunnī opponents. The Mongols brought the resistance of Alamūt to an end in 654/1256, and the eighth and last lord was murdered on the way to the court of the Great Khān. The Mamlūk *sulṭān* Baibars finished off the movement in Syria in 671/1272, thus ending for ever the political power of this feared sect. It was not quite an end to acts of political assassination perpetrated by Nizārīs, which continued until *c.* 726/1326 but these were no longer directed by an independent Nizārī state.[63]

The Shī'a as a politico-religious minority in Ottoman-controlled 'Irāq

The establishment of the Ṣafavid dynasty in Iran (906/1501–1135/1722) and the gradual conversion of the majority of the population to Twelver Shī'ism in the period of Ṣafavid rule were significant developments for the Islamic world, comparable in the view of William McNeill to the Protestant Reformation in Europe. As has been observed above (see Chapter 5), the Ottoman–Ṣafavid conflict was more than an era of dynastic rivalry, but a fundamental struggle between Sunnī 'orthodoxy' and what was perceived by the Ottomans as Shī'a heterodoxy. For most of the period, however, what would later be known as 'Irāq fell under Ottoman control, yet it possessed the Ḥawza,[64] the religious seminaries based at the two holiest Shī'a shrines of Karbalā', the shrine of the third *imām*, al-Ḥusayn; and Najaf, the latter traditionally regarded as the burial place of Adam and Noah. From the late eighteenth century onwards, the shrine cites, and later Samarra, emerged as the training ground for Shīʻa clergy world-wide and a place of exile for Iranian and Indian dissidents.[65] Juan Cole remarks that the ʻIrāq border between the Ottoman and Ṣafavid states ʻconstituted a frontline in the two powers' tug of war, and the loyalties of the Twelvers in Baghdād, the shrine cities, and Baṣra were always suspect'. Numerically, the Shīʻa of ʻIrāq were weak, however: as late as *c.* 1214/1800, ʻIrāq probably had a population of only 1 million, a fifth of that of Iran, by no means of all of whom were Shīʻa.[66]

Two trends in Ottoman policy towards the minority creed are discernible. The first, in spite of Sunnī hostility to what was regarded as a heretical sect, was one of modest encouragement. The Shīʻa shrines attracted pilgrims; the income gained was a valuable source of revenue and thus wealth to ʻIrāq. Thus, at the very least, the shrines had to be protected. After the Ottoman governor not only failed to protect Karbalā' from being sacked by the Wahhābīs in 1216/1802 (an attack that was recalled in the aftermath of the Karbalā' bombings of March 2004: see Chapter 6) but actually fled before the Wahhābī advance, he was executed in an act designed to show solidarity with the Shīʻa and to deflect criticism from Iran, which threatened to annex the shrine cities if the Ottomans failed to protect them.[67]

A second trend in Ottoman policy was much more ominous. It has been seen that the *Qizilbāsh* were simultaneously spiritual and military-political supporters of the Ṣafavid dynasty, though eventually the *shāh* had to curb the powers of the *Qizilbāsh* tribal *amīrs* in the interest of protecting the unity of the state (see Chapter 5). The Ottomans feared the *Qizilbāsh* more for their support for Ṣafavids than their esoteric Twelver doctrines, and sought to pursue drastic policies against them such as summary executions, population transfers and even their attempted extermination.[68] Subsequently, after this threat had receded, the Ottomans sought to assert direct control over the shrine cities. In 1240/1824, Dāwūd Pasha besieged Karbalā' for eleven months, but failed to take the city and accepted a lump sum payment as a compromise.[69] The second siege in 1258/January 1843 was brief and bloody: the Ottomans lost 400 men, while some 5000 inhabitants were killed, about 15 per cent of the population, inside the city or outside its walls. A Sunnī governor was appointed, and a Sunnī preacher to deliver the sermons after Friday prayers.[70] Dissimulation (*taqīyah*) once more became the order of the day among the Shīʻa. Millenarian expectations about the coming of the promised *Mahdī* in 1260/1844 led to some support for Sayyid ʻAlī Muḥammad Shīrāzī's claim to be the ʻgate' (*bāb*) to the twelfth *Imām*; but his followers, known as Bābīs, were a source of internal upheaval within Shīʻīsm rather than a signal for unified resistance to Sunnī ascendancy.[71] Bābīsm won considerable support in Iran, however, although Shīrāzī was executed by the government there in 1266/July 1850. Thereupon the remaining Bābīs gave their support either to Ṣubḥ-i Azal, the Bāb's designated successor and an advocate of continued militancy, or to Mīrzā Ḥusayn ʻAlī Nūrī, a former Bābī exiled to ʻIrāq, who declared himself the *Bahā'u'llāh* – the prophet–founder of the new Bahā'ī faith – and advocated obedience to the government and an end to militancy.

The 'Great 'Irāqī Revolution' or *jihād* of 1338/1920 and its implications in Iran

In 1335/March 1917, British troops took Baghdād under the leadership of Major General Stanley Maude. The British claimed that they had come to 'liberate' the 'Irāqī people from Ottoman imperial tyranny, and promised to give the 'Irāqī people independence and the right to choose their own government as soon as the war was over. Suspicious of British intentions, Shī'a leaders in the holy city of Najaf began to agitate against the occupation. This culminated in the formation of *al-Nahdha al-Islāmīyyah* (the Renaissance or League of Islamic Awakening Party), which was one of three parties formed at about the same time.[72] On the same day that Baghdād fell to the British, a limited uprising against the British occupation took place in Najaf. It was swiftly and brutally crushed by the British Army, which surrounded the city and bombed one of its important quarters. Eleven 'Irāqīs were executed in retaliation for the murder of one British officer.[73]

An Arabic copy of President Woodrow Wilson's Fourteen Points was published in 1336/October 1918, and widely circulated in 'Irāq. Point 12 received special attention: 'the Turkish portion of the present Ottoman Empire should be assured a secure sovereignty, but the other nationalities which are now under Turkish rule should be assured an undoubted security of life and an absolutely unmolested opportunity of autonomous development'. However, for the British, the fate of 'Irāq had been sealed by the Sykes–Picot Pact of 1334/May 1916.[74] In 1337/November 1918, the British Viceroy of India sent a telegram to Sir Arnold Wilson, the administrator of 'Irāq, stating: 'let it be known to all that it is in the [Paris] Peace Conference that the fate of the 'Irāqī sectors would be decided'. And, anticipating the requirement of a referendum on the mandate, the Viceroy ordered Wilson to carry out a controlled plebiscite, with only 'Yes' to the mandate as an acceptable answer. The referendum would consist of three questions: 1) do the 'Irāqīs wish to have a united Arab state, extending from north of Mosul to the Persian Gulf, under a British mandate?; 2) do they wish, in this case, to have an Arab leader by name to head this state?; and 3) in this case, who is this leader?

Shaykh Muḥammad Taqī al-Ha'irī al-Shīrāzī, the most prominent religious leader of the Ḥawza in Karbalā', brought such discussion to an abrupt end by issuing a *fatwā* that 'no Muslim can choose or elect anyone to position of power and government other than a Muslim'. Al-Shīrāzī, together with Shaykh al-Shar'īa al-Aṣfahānī of Karbalā', wrote to President Wilson, in 1337/February 1919:

> All peoples rejoiced for the declared purpose of participating in the European wars; namely, the restoration to the oppressed nations their rights, and opening the way for them to enjoy independence according to the terms you have declared. Since you were the initiator of this project, the project of happiness and general peace, it is appropriate that you be the resort for lifting the obstacles from its accomplishment. There is indeed a strong obstacle, preventing most of the 'Irāqī people from expressing their aspirations, in spite of the declared desire of the British government that all 'Irāqīs should express their views. The general opinion amongst them is that since they are a Muslim nation, it should enjoy a judicial freedom and choose a new, independent Arab–Islamic state headed by a Muslim king, who is bounded by a national assembly. As for the talk about [taking up the issue after] the post-Peace Conference period, we would like to inform you that we are responsible for bringing hope to the 'Irāqī people and removing all obstacles in their way to express their views and aspiration to a sufficient degree to allow the international public opinion to see the truth about the purpose of what you have outlined, in complete freedom. To you, thus, will be the eternal honour in history and in its current modern civilization.

In 1337/August 1919 the British, alarmed by the state of political agitation in the country, had the army arrest six leaders of Karbalā''s 'Islamic Society' who were working closely with al-Shīrāzī and his son. The exile of these leaders, the British thought, would deter the rest, especially al-Shīrāzī. Instead, al-Shīrāzī demanded the release and return of the exiles and, when this was refused, he announced that if the exiles were not brought back to 'Irāq, he would leave for Iran and declare *jihād* against the British from there. As the supreme religious leader of the Shī'a in 'Irāq and Iran, and also of the Shī'a minorities in India, there was a strong likelihood that the Iranian people would rally around him for a *jihād* against the British, who were also seeking to control the Iranian ruler, Aḥmad Shāh, the last ruler of the Qājār dynasty.

By the winter of 1338/1919–20, the political opposition in Iran and the Shī'a Ḥawza in Qom had mobilized to seek the abolition of the Anglo-Persian Treaty of 1337/August 1919, as violating the constitution. Pressure from within Iran and the urging from al-Shīrāzī in 'Irāq, who sent one of his colleagues to meet Aḥmad Shāh while he was on a visit to the holy sites in 'Irāq, induced Aḥmad Shāh to force his Prime Minister to resign, which was followed by the repudiation of the proposed treaty in the following spring. However, these events served only to precipitate Rizā (Riḍa) Khān's *coup d'état* of 1339/February 1921, which was brought about by Major-General Edmund Ironside on behalf of British interests.[75] *Imām* Khomeini later poured scorn on his qualifications for office: 'he was totally illiterate, an illiterate soldier, no more!'[76] Gradually over the next five years Rizā Khān gained supreme power in Persia, finally bringing about the end of the Qājār dynasty and getting himself crowned as Reza Shāh, the first ruler of the Pahlavī dynasty.[77]

When the news from the League of Nations meeting at San Remo in 1338/April 1920, which confirmed the British mandate,[78] reached 'Irāq, demonstrations, protests, and petitioning campaigns were organized across the country. Al-Shīrāzī issued a *fatwā* prohibiting 'Irāqīs from cooperating with the British occupation. This paralysed the whole country and the British administration. There was unprecedented Shī'ī–Sunnī cooperation, and a delegation was formed to meet the British representative in Baghdād. Al-Shīrāzī issued a declaration the same month urging people in all parts of 'Irāq to send delegates to Baghdād for the purpose of demonstrating and negotiating with the British authorities. He called for preserving calm and security, and warned strongly against causing any harm to members of other minorities, such as the Christian and Jewish residents Baghdād. The demands of the delegations, he argued, should be no less than total independence and the establishment of an Arab–Islamic state.

Al-Shīrāzī's *jihād* declaration followed three months later (1338/June 1920): 'it is a duty upon all 'Irāqīs to call for their rights. While they do that, they should make sure that security and peace are preserved. But, they can resort to defensive force, if the British refuse to comply with their demands.'[79] Preparations for an armed uprising had reached their peak, and fighting broke out at the beginning of July. It is estimated that more than 10,000 'Irāqīs were killed in the four months of the uprising. The British, for all their superiority in terms of armaments, suffered 2000 casualties, including 450 dead. The tribal forces, armed with rifles only, launched a series of successful guerrilla-type attacks. They started by cutting the rail lines and bridges connecting towns that housed British garrisons. They laid a successful siege to the British Army base at al-Rumaitha, which was only broken by the massive use of air bombardment. 'Wholesale slaughter', argued Colonel Gerald Leachman, was the only way to deal with the tribes. (Leachman himself was assassinated a few weeks later in southern 'Irāq.) Gertrude Bell commented on a demonstration of the destruction of a mock village from the air:

> the two first bombs, dropped from 3000 ft, went straight into the middle of it and set it alight. It was wonderful and horrible. They then dropped bombs all round it, as if to catch the fugitives and finally firebombs which even in the bright sunlight, made flares of bright flame in the desert. They burn through metal, and water won't extinguish them. At the end the armoured cars went out to round up the fugitives with machine guns… I was tremendously impressed. It's an amazingly relentless and terrible thing, war from the air… The RAF has done wonders bombing insurgent villages in extremely difficult country, but it takes them all their time to keep a sufficient number of machines in the air and now if we are called upon to bomb Rawanduz intensively, our resources will be strained to the utmost…

The *jihād* failed in its purpose and the victors' interpretation thus prevailed. Sir Arnold Wilson, Civil Commissioner in ʻIrāq, told the British Cabinet at the end of 1920 that 'there was no real desire in Mesopotamia for an Arab government; the Arabs would appreciate British rule'. He amplified: 'what we are up against is anarchy plus fanaticism. There is little or no nationalism.' The revolution had rapidly disintegrated into anarchy. It had all been caused by 'outsiders' ranging from Muṣṭafā Kemal, the Germans and Pan–Islām to Standard Oil, the Jews and the Bolsheviks.[80] David Fromkin, one of the historians who cites this viewpoint, argues that the British political leadership at the outbreak of World War I singularly failed to understand the religious motivation of the population in the Middle East. Since Britain ruled over half of the world's Muslims, a *jihād* against Britain was a 'recurring nightmare'.[81] The British political leadership, strongly influenced by Kitchener who was supposed to know about these things because of his experience in the Sudan, seemed to have believed, or succeeded in convincing themselves, that Islām had a centralized, authoritarian structure and thus that the faith itself could somehow be bought, manipulated, or captured by buying,[82] manipulating, or capturing its religious leadership.[83] This failed to allow for the diversity within the Islamic world and the divisions between Shīʻa and Sunnī viewpoints. Following the occupation of Mecca by ʻAbd al-ʻAziz ibn Saʻūd's forces in 1343/October 1924, the British harboured unrealistic expectations of the capacity of the Wahhābīs to unify the Sunnī world (see Chapter 6).[84] Yet, occasionally, such divisions could be overcome: al-Shīrāzī's *jihād*, which combined Shīʻa and Sunnī forces, demonstrated just the extent to which British preconceptions could be misguided.[85] No less an Arab specialist than T. E. Lawrence[86] reported in the *Sunday Times*:[87]

> The people of England have been led in Mesopotamia into a trap from which it will be hard to escape with dignity and honour. They have been tricked into it by a steady withholding of information. The Baghdad communiqués are belated, insincere, incomplete. Things have been far worse than we have been told, our administration more bloody and inefficient than the public knows. It is a disgrace to our imperial record, and may soon be too inflamed for any ordinary cure. We are to-day not far from a disaster.

He added that the British regime was worse than that which preceded it:[88]

> Our government is worse than the old Turkish system. They kept fourteen thousand local conscripts embodied, and killed a yearly average of two hundred Arabs in maintaining peace. We keep ninety thousand men, with aeroplanes, armoured cars, gunboats and armoured trains. We have killed about ten thousand Arabs in this rising this summer. We cannot hope to maintain such an average: it is a poor country, sparsely peopled…

Lawrence's misgivings were not, in the event, proven correct. The British remained influential in 'Irāq long after independence was granted. They made and unmade regimes in 'Irāq and Iran until World War II and continued to do so even later. Khomeini later blamed Britain and the United States for the creation of the state of Israel and the ensuing misery inflicted on Muslims by Israel.[89] And when the British influence faded, that of the United States replaced it: the coup against Mossadeq in Iran in 1953 was masterminded by the CIA with a degree of British and American cooperation,[90] as Khomeini later reminded his audiences.[91] It was to be a profound and enduring weakness of the second and last Pahlavī ruler, Muḥammad Reza Shāh, that from the time of his restoration in 1372/August 1953 he was perceived as owing his position to foreign intervention. Curiously, part of the planning for the 19 August coup against Mossadeq was that the leading Shī'a clerical figure, Ataytollah Muḥammad Ḥusayn Borujerdi, should issue a *fatwā* for a *jihād* against Communism.[92] In the event, no such decree was issued, but it illustrates the willingness of the CIA to manipulate Shī'a religious sentiments in what was above all a political coup mounted against Mossadeq's economic policies which were considered to be detrimental to US interests.

Architects of the Iranian Revolution: I. Muṭahharī and *jihād*

Ayatullah Murtazá Muṭahharī was gunned down by an assassin from Furqān (a group which opposed the involvement of the clergy in politics) in 1399/May 1979, less than four months after the departure of the *shāh* into exile. He was thus one of the first martyrs of the Iranian Revolution. In his funeral oration, Khomeini asserted that 'assassinations cannot destroy the Islamic personality of the great men of Islām... Islām grows through sacrifice and the martyrdom of its cherished ones. From the time of its revelation down to the present, Islām has always been accompanied by martyrdom and heroism.' During Khomeini's long exile in Turkey and Najaf from 1383/1964 to 1398/1978, the battle with the laity for control of political Islām in Iran was largely fought by Muṭahharī, Khomeini's pupil for a few years after 1365/1946. It has been argued in Iran that the writings of Muṭahharī constitute 'the intellectual infrastructure of the Islamic Republic'.[93]

For Muṭahharī there were clear limits to Muslims' dealings with persons of other faiths.[94] They were not debarred from good fellowship or philanthropic dealings with others (how could they be, since the Qur'ān describes its Prophet as a blessing to all the worlds?), but it was important to remember that

> a Muslim must regard himself as a member of the Muslim body politic and a part of the whole. To be a member of a particular society automatically imposes certain conditions and limits. The non-Muslims being members of a different society, the relations of the Muslims with them must be such as may not be

> incompatible with their being members of their own society. They should in no way jeopardize their own independence and integrity. Hence the relations of a Muslim with the non-Muslims cannot be similar to those which he has with the fellow Muslims.

Imām ʻAlī was cited by Muṭahharī as stating that 'the betrayal of the community is the worst treachery and the deceiving of the Muslim leaders is the most abominable fraud'. He considered it evident that deceiving the *Imām* amounted to the deception of all Muslims. Hence the importance of the *ḥadīth*, which was considered to be authentic by both the Shīʻa and the Sunnī, in which the Prophet said: 'the best *jihād* is to say what is true before an unjust *imām*'.[95]

Muṭahharī joined with the ʻIrāqī reformist Ayatullah Muḥammad Bāqir al-Ṣadr to write a doctrine of the *Mahdī*, entitled *The Awaited Saviour*. The utopian vision of a complete Islamic revolution is revealed clearly in chapter 9, entitled 'An Ideal Society'.[96] Islām, they write, provides 'the glad tidings of the *Mahdī*'s revolution', the salient features of which would be:

> the final victory of righteousness, virtue, peace, justice, freedom and truth over the forces of egoism, subjugation, tyranny, deceit and fraud; the establishment of a world government (or one government for the whole world); the reclamation and rehabilitation of the whole earth so that no area remains waste; the attainment of full sagacity by mankind, freeing mankind from its adherence to ideology and emancipating it from animal impulses and undue social restrictions; the maximum utilization of the gifts of the earth; the equal distribution of wealth and property among all human beings; the complete eradication of vices such as adultery, fornication, usury, the use of intoxicants, treachery, theft and homicide; the end of abnormal complexes, malice and ill-will; the eradication of war and restoration of peace, friendship, co-operation and benevolence; and the complete coherence of man and nature.

If this utopian world view was the ultimate aim for mankind, there was still the relationship between, past, present and future to consider. As Muṭahharī expressed it in his first lecture on history and human evolution:[97]

> 1. Has man, in his social life and throughout history, achieved evolution and exaltation? 2. Is human society undergoing evolution and will it reach a fully evolved state in future? 3. If it is undergoing evolution, what is that ideal society, or, as Plato would say, that utopia of man, and what are its peculiarities?... If we do not recognize the future and have no plan for it, and if we pay no attention to our responsibility for making history, we too deserve being reproached by future generations. History is made by man, and not man by history. If we have no plan for the future, and do not realize our

> responsibility for the future of history, no one can promise us that this ship will reach its destination automatically. The least that can be said is that it may either go ahead or turn backwards. This matter of ability to advance or reverse the course of events, the idea that there isn't a blind coercive force that drives events ahead, is in Islām, and especially in Shī'īsm, a question, which from a sociological viewpoint (as I have explained in my book, *Man and Destiny*), may be considered one of the most sublime of Islamic teachings.

In his *Jihād: The Holy War of Islām and its Legitimacy in the Qur'ān*,[98] Muṭahharī discusses the conditions for the legitimacy of *jihād*.[99] These would include the belief that the other side intended to attack; or that it had created a barrier against the call of Islām; or, in the case of a people subject to the oppression and tyranny of a group from amongst themselves, Islām required conflict with such tyrants so as to deliver the oppressed people from the claws of tyranny (Q.4:75). Against the propaganda of Christian pacifists (by no means the majority of Christians, though this point was not noted by Muṭahharī),[100] he denied that war was always bad: if it was in defence of a right, or against oppression, was it bad? 'Obviously not', Muṭahharī concluded. Wars of aggression, carefully defined,[101] were an obvious evil and to be rejected; but when wars of defence became necessary ('when facing imminent attack and the risk of being destroyed') the failure to act would not maintain the peace, but would be an act of surrender. Surrender is not honourable coexistence; it is coexistence that is dishonourable on both sides: on one side, the dishonour is that of aggression, and on the other side, it is the dishonour of surrender in the face of *ẓulm*, in the face of injustice and oppression.

For Muṭahharī, Islām is a religion that sees it its duty and commitment to form an Islamic state:[102]

> Islām came to reform society and to form a nation and government. Its mandate is the reform of the whole world. Such a religion cannot be indifferent. It cannot be without a law of *jihād*. In the same way, its government cannot be without an army. While the scope of Christianity is extremely limited, that of Islām is extremely wide. While Christianity does not cross the frontiers of advice, Islām is a religion which covers all the activities of human life. It has laws which govern the society, economic laws, and political laws. It came to organize a state, to organize a government. Once this done, how can it remain without an army? How can it be without a law of *jihād*?... Islām says: 'peace if the other side is ready and willing to accept it. If not, and it turns to war: then war.' ... The Qur'ān has fundamentally defined *jihād* not as a war of aggression or of superiority or of authority, but of resistance against aggression.

Muṭahharī also considered the question of whether a defensive war applied only to self-defence. What, for example, of defending others who were suffering oppression? Perhaps the oppressive state did not propose an aggressive external war, but was acting in flagrant abuse of its own citizens, whether Muslims or non-Muslims. Not to act, Muṭahharī argued, was tantamount to helping the oppressor's act of oppression. If the oppressed people were Muslims, 'like today's plight of the Palestinians who have been exiled from their homes, whose wealth has been seized, who have been subjected to all kinds of transgression', then the answer was simple: intervention was not merely permissible, but obligatory. In such circumstances, intervention could not be described as a commencement of hostilities, since the hostilities had been commenced by the initial act of oppression.

Muṭahharī also envisaged two other circumstances in which intervention might be legitimate on behalf of non-Muslims. The first case was where the oppressive ruler had prevented the spread of Islām. Since Islām gave itself the right to spread its message throughout the world, the simple fact of an oppressive regime erecting a barrier against the spread of the faith was justification for intervention, whether or not the people of the state had asked for it in the first place. This was not, in his view, the same as justifying coercion of conscience, which was denied in the Qur'ān ('there is no compulsion in religion, for the truth has been made manifest from the false': Q.2:255). If people become Muslims, well and good: Islām states that whoever wants to believe will believe, and whoever does not want to, will not. The issue was the removal of barriers to voluntary conversion. Later on, he answered the question as to why Islām wants *jihād*: 'it does not want *jihād* for the sake of the imposition of belief', he claimed; instead, 'it wants *jihād* for the removal of barriers'.[103]

The second circumstance affecting non-Muslims envisaged by Muṭahharī was more original: fighting for the sake of humanity. Some might argue, he conceded, that 'fighting for the sake of humanity' might seem to have no meaning: 'I do not have to fight for any rights except my own personal rights, or, at the most, the rights of my nation. What have I to do with the rights of humanity?'[104] For Muṭahharī, certain things existed which were superior to the rights of the individual or nation, above all the defence of what is 'right'. Freedom is reckoned as one of the sacred values of humanity, and is not limited to an individual or a nation.

> I do not think anyone has any doubt that the holiest form of *jihād* and the holiest form of war is that which is fought in defence of humanity and humanity's rights… if the defence shifts from a national to a humanitarian cause, it… becomes a degree more holy.[105]

Interestingly, Muṭahharī seems not to have envisaged a cessation of *jihād* until the superiority of Islām was universally accepted, 'they want to live with us under our protection', and other religious communities had agreed to pay the *jizya*. At this time, citing Q.8:61,

> if they have been humbled, and manifest a mind and heart of peace and compromise, then we are not to be severe anymore. We are not to say: 'Oh no. We do not want peace, we are going to fight.' Now that they have come forward to live in peace and concord, we too must announce the same thing.

Muṭahharī's position, ultimately, was not one that would be likely to command itself to persons of other faiths committed to dialogue with Islām on a basis of equality.

For Muṭahharī, *jihād* was the sacred cause that led to martyrdom (*shahādah*); drawing on Q.5:93 and Q.7:26, *Imām* 'Alī had called it the 'garment of piety' and compared it both to armour and the shield.[106] *Shahādah* had two basic characteristics: firstly, the life of the person was sacrificed for a cause; and secondly, the sacrifice was made consciously.[107] The personifications of such voluntary self-sacrifice were *Imām* 'Alī, 'for whom life would have become meaningless' had he lost the hope of attaining *shahādat*;[108] and *Imām* Ḥusayn, 'the chief [or lord] among martyrs', *Sayyid al-Shuhadā'*.[109] Far from such acts of self-sacrifice being regarded as valueless, they were essential for human progress:[110]

> The *shahīd* can be compared to a candle whose job is to burn out and get extinguished in order to shed light for the benefit of others. The *shuhadā'* are the candles of society. They burn themselves out and illuminate society. If they do not shed their light, no organization can shine… The *shuhadā'* are the illuminators of society. Had they not shed their light on the darkness of despotism and suppression, humanity would have made no progress…
>
> The body of a *shahīd* is neither to be washed nor shroud[ed]. It is to be buried in those very clothes which the *shahīd* was wearing at the time of his death. This exception has deep significance. It shows that the spirit and personality of a *shahīd* are so thoroughly purified that his body, his blood and his garments are also affected by this purification… These rules of Islamic law are a sign of the sanctification of the *shahīd*…
>
> A *shahīd* often sacrifices his life to create fervour, to enlighten society, to revive it and to infuse fresh blood into its body. [The sacrifice of the faithful at Karbalā'] was one such occasion.

The blood of the *shahīd* is akin to a blood transfusion for a society suffering from anaemia. Islām is always in need of *shuhadā'*, for the revival of courage and zeal is essential for the revival of a nation. The *shahīd*, moreover, immortalizes himself and may intercede with God on the Day of Judgement.[111] Weeping for the *shuhadā'* 'has been recommended by Islām' and was an 'indisputable doctrine' of the Shī'a.[112] Muṭahharī's statement on martyrdom was brief and portentous for the faithful of the Iranian Revolution, particularly those who would be sacrificed in the war against 'Irāq: there was no allusion to worldly benefits for the family, let alone to sexual benefits in Paradise for the martyr. Instead, the benefits are entirely spiritual. A martyr is the purest of the spiritually pure, one who exercises complete self-negation and self-sacrifice. The door to Paradise through which the warriors for God (*mujāhidīn*) and the *shuhadā'* will enter, and 'the portion of Paradise set aside for them, is the one which is reserved for God's chosen friends, who will be graced with his special favour'.[113]

Asghar Schirazi notes that the Iranian jurists recognized Muṭahharī as 'the greatest reformer' of the Revolution, but that even he seems to have had some difficulties in his lectures in trying to reconcile traditional Islamic jurisprudence with the needs of the times. The issue centred on the distinction between 'the constant' and 'the changeable' Islamic ordinances. Muṭahharī condemned Habib Bourgiba, the former President of Tunisia, who had wanted to forbid fasting for economic reasons. Muṭahharī also upheld the prohibition on the eating of pork, although he admitted that there were no apparent hygiene reasons for maintaining the ban. He remained vague, however, on what constituted Islamic ordinances that were potentially 'changeable', the real point of contention.[114] Muṭahharī was no religious pluralist: he criticized the UN International Charter of Human Rights because it maintained that all beliefs were worthy of respect. He could not accept this: in Islām, the only belief that is free is belief based on thought; there may well exist other forms of belief which arise from emotion and which stand in the way of thought. This, in his view, could lead to idolatry: Abraham's destruction of the idols of the Israelites was justified since he was the only person at the time who thought rationally without emotion. There is unlimited freedom only in the domain of thought. If a human being was truly free, Muṭahharī contended, then he would inevitably choose Islām. If he did not do so, he could not be genuinely free and 'must be a sick human being'.[115] Freedom thus had to be conditioned by divine revelation and prophecy.

Architects of the Iranian Revolution: II. Khomeini's 'Greater *Jihād*' of the 30 million[116]

> I am now living the last days of my life. Sooner or later I will leave you. But I see before me dark black days ahead for you. If you do not reform and prepare yourselves, and if you do not manage your studies and your lives with order and discipline, then, God forbid, you will be doomed to annihilation.

Thus spoke Imām or Ayatullah Ruḥullah al-Mūsawī al Khumaynī, or Khomeini (1320/1902–1409/1989), unquestionably the principal architect of the Iranian Revolution, in an address delivered to the '*ulamā*' and theological students delivered at Najaf, entitled 'The Greater *Jihād*'.[117] Hamid Algar commented: 'the *jihād* the *imām* spoke of in the celebrated lecture delivered in Najaf, was clearly exemplified by him in his own person. Beyond that he declared *jihād* against the outward enemy and not necessarily a *jihād* that involves recourse to weapons; but a *jihād* definitely on occasion if necessary included that. Thus we find in the person of *imām* one who practised *jihād* in all its comprehensive forms.'[118]

For the late Dr Kalim Siddiqui, founder of Britain's so-called Muslim Parliament,

> *Imām* Khomeini took the bull of history and calmly led it out of the paddock of theology. The *imām*'s last testament... is a masterpiece of restatement of his own position within the Shī'a school of thought, and a masterpiece of his statement in the arena of history which unites all schools of thought in Islām. Any attempt to limit him and his relevance to Iran, or one school of thought in Islām, would be a great injustice to him, to Islām and perhaps also to Iran.[119]

Khomeini asserted that his politico-religious testament was written not solely for 'the noble people of Iran', but 'for all Muslim nations as well as the oppressed people of the world, regardless of nationality and creed'. He began by advising 'the noble but oppressed nation of Iran' not to divert from the straight path towards either 'the atheist East nor the oppressor West', but rather to remain firmly committed, loyal and dedicated to the path granted by Allāh.[120]

Khomeini thought it beyond the power of pen and speech to assess the significance of the 'glorious Islamic Revolution', the achievement of 'millions of estimable people', thousands of the so-called 'immortal' martyrs as well as the 'beloved disabled'. He gave no figure for the casualties of the Iran–'Irāq War of 1400/1980–1409/1988,[121] merely noting that the motive which had prompted all saints to embrace martyrdom made a sanguine death sweeter to them than honey: 'your youngsters have taken a gulp of that in the battle fronts and have felt elated at doing so...' Efraim Karsh calls the war 'a costly exercise in futility' and argues that one of its consequences was 'a profoundly sobering and moderating influence' on Iran's revolutionary zeal.[122] If so, Khomeini was not prepared to acknowledge it. On the contrary, he asserted the view, also stressed by Karsh regarding the earlier years of the war, that the 'Irāqī invasion did not endanger the revolution or drive Khomeini towards moderation; instead, 'the clerics in Tehran capitalized on the 'Irāqī attack to consolidate their regime'.[123] Predictably, for Khomeini, 'the secret of the permanence of [the] Islamic Revolution' was the same secret that had caused its victory over the Pahlavī dynasty: 'its divinely-based ideology and the solidarity of the people throughout the country…' He advocated the same spirit and the same goal for other countries plagued by corrupt governments or 'under the yoke' of the great powers. Other Muslim nations should 'follow the example of the Islamic government in Iran and of Iran's struggling people' and should disregard the hostile propaganda of the enemies of Islām and the Islamic Republic.

Iran's progress was seen as a march towards 'a better knowledge of ourselves and towards self-sufficiency and independence in their full scope'. Khomeini acknowledged Iran's need to learn from the advanced industries of foreign countries; but trainees needed to be sent to learn about the advanced industries in countries which were 'not colonialist and expansionist', and above all avoid contact with the American and Soviet blocs. He hoped that eventually the superpowers would 'acknowledge their mistakes and abandon their predatory policies and adopt a foreign policy based on humanity and respecting others' rights'.

On the tenth anniversary of the Iranian Revolution in 1409/February 1989, about four months before his death, Khomeini reflected on 'the tumult of 15 years of struggle before the revolution and 10 years of back-breaking post-revolution events'.[124] Again, Khomeini referred to 'the martyred custodians who carried the pillars of the greatness and pride of the Islamic revolution upon their crimson and blood-stained shoulders…' There had been many obstacles to a successful revolution:

> The spread of the system of thought that the *Shāh* was the Shadow of God, or that one could not stand up against tanks with flesh and skin; or that we were not religiously bound to wage a *jihād* and struggle; or that asked the question of who was to stand accountable for the blood of those killed; and most defeatist of all, the misleading slogan that the government prior to the appearance of the Lord of Age [i.e. the *mahdī*], peace be upon him, was wrong, as well as thousands of other excuses. These were great and tiring problems which could not be prevented through giving advice, negative encounter or sermons. The only solution was to struggle, to engage in self-sacrifice and offer blood, which God made possible.

Khomeini returned to the theme of the Iran–'Irāq War as the mechanism for consolidating the revolution:

> Every day of the war we had a blessing, which we utilized in all aspects. We exported our revolution to the world through the war;[125] we proved our oppression and the aggressor's tyranny through the war. It was through the war that we… recognized our enemies and friends. It was during the war that we concluded that we must stand on our own feet. It was through the war that we broke the back of both Eastern and Western superpowers.[126] It was through

> the war that we consolidated the roots of our fruitful Islamic Revolution. It was through the war that we nurtured the sense of brotherhood and patriotism in the spirit of all of the people. Through the war we showed the people of the world and in particular the people of the region that one can fight against all the powers and superpowers for several years. Our war helped the materialization of triumph in Afghanistan. Our war will lead to victory in Palestine. Our war caused all the chiefs of corrupt systems to feel belittled before Islām…
>
> How short-sighted are those who think that since we have not reached our final aim at the front, then martyrdom, self-sacrifice and courage are all wasted… The people's interest in learning about Islām throughout Europe, America, Asia and Africa and all over the world in fact stems from our eight-year war.

Perhaps mistakes had been made early on in the war.[127] Khomeini also pointed to the deficiencies in weaponry as a factor in limiting Iran's success: implicitly had Iran not suffered sanctions from the Western powers and the Arab powers apart from Syria, and thus had it had 'all [the necessary] means and equipment at our disposal we would have reached even higher goals in the war'; but the main aim, that of 'repulsing aggression and proving Islām's steadfastness' had been achieved. Khomeini, from 1981 onwards, ordered children to be sent in bands of 10,000 into the line of fire and across minefields. In the autumn of 1982, he issued instructions which included authorization that young people willing to go to the war front did not need the consent of their parents. The 'key to Paradise' was hung around the neck of each recruit. Karbalā' was the war cry on their lips: Karbalā' is the pivot of faith for the Shī'a, 'the climax of a divine plan promising rich rewards for all those who take up arms in the name of the martyred *imām*'.[128] 'Yā Karbalā'! Yā Ḥusayn! Yā Komeini' was the battle cry of the human wave attacks.[129] 'We started to count them. We counted them all day. We gave up when we'd got to 23,000 because we were supposed to leave the area before dark… The offensive was called Karbalā' IV or Karbalā' V…'[130]

The Iranian struggle was part of a much larger and enduring struggle of right against wrong.[131] Not all his colleagues agreed with Khomeini's dogged war leadership. One conservative jurist quipped that 'the goal of the war was to obey the *Imām*'.[132] The hostility of the West had been clear from the outset. The overt and covert agents of the *Shāh* and America had resorted to rumours and slanders. They had even accused those who led the revolutionary movement of having abandoned their prayers and even of being Communists or British agents. America and its servant, Pahlavī, had wanted to 'uproot religion and Islām'. Khomeini is famous for his characterization of the USA as 'the Great Satan'.[133] He proclaimed the United States 'the number one enemy of the deprived and oppressed people of the world' and claimed that 'there is no crime that America will not commit in order to maintain its political, economic, cultural and military domination of those parts of the world where it predominates'. International Zionism, he claimed, coordinated the US propaganda campaign. That was why Iran had endeavoured to sever all its relations with the 'Great Satan'. The 'Irāq war against Iran was a US war by proxy, he contended.[134] He asserted that Iran had witnessed the success of the slogan of 'death to America' in the action of its heroic Muslim youth in their capture of America's 'nest of corruption and espionage', that is the capture of the American embassy and the ensuing hostage crisis:[135]

> We understand the Western powers' hatred towards the Islamic world and [its command of Islamic law, *fiqh* (*fiqahat*)]... The issue for them is not that of defending an individual [i.e. the *Shāh*], the issue for them is to support an anti-Islamic and anti-value current, which has been masterminded by those institutions belonging to Zionism, Britain and the USA which have placed themselves against the Islamic world, through their stupidity and haste...

Khomeini celebrated President Carter's decision to sever diplomatic relations with Iran as liberation 'from the claws of the international looters'. He hoped for the 'quick destruction' of 'such puppets' as Anwar Sadat of Egypt and Saddam Hussein of 'Irāq. Saddam, in his view, 'had done to the oppressed Muslims what the Mongols did' (see Chapter 4). Khomeini encouraged 'Irāqīs to rise up and rid their country of 'this criminal', reminding them that they were descendants of 'those who drove the British out of 'Irāq', an allusion to the (failed) *jihād* of 1338/June 1920.[136]

Khomeini declared that as long he was present in the government, it would not fall into the hands of liberals. Moreover, he would 'cut off the hands of the agents of America and the Soviet Union in all fields'. He remained confident the Iranians were all, 'in principle, as in the past, supporters of the system and of their Islamic revolution'. In 1409/February 1989, Khomeini issued his notorious *fatwā* declaring that the British author, Salman Rushdie, should be executed for having insulted Islām in his novel *The Satanic Verses* published late the previous year. Since then, fearing for his life, the author has been living under constant police protection, and all his public appearances are undertaken amid tight security. The existence of the *fatwā* amounts to a threat of arbitrary deprivation of life and a violation of Article 6 of the International Covenant on Civil and Political Rights, to which Iran is a state party. The Rushdie affair has lingered on with equivocal statements from successor regimes to that of Khomeini,[137] and a considerable degree of implicit support from other parts of the Islamic world, though Muslim views were divided.[138] Khomeini showed no qualms about his decision. The publication of the book, he contended, was 'a calculated move aimed at rooting out religion and religiousness, and above all, Islām and [the religious institutions (*rūḥāniyāt*)]'. It was all part of the same conspiracy that he had already outlined:

> God wanted the blasphemous book *The Satanic Verses* to be published [in our time], so that the world of conceit, arrogance and barbarism would bare its true face in its long-held enmity to Islām; to bring us out of our simplicity and to prevent us from attributing everything to blunder, bad management and lack of experience; to realize fully that this issue is not our mistake, but that it is the world devourer's [that is, the USA's] effort to annihilate Islām, and Muslims; otherwise, the issue of Salman Rushdie would not be so important to them as to place the entire Zionism and arrogance behind it.

For Asaf Hussain, Khomeini was 'without any doubt the most revolutionary Islamic thinker' of the twentieth century.[139] What general principles, then, can be deduced from his political thought?[140] For Khomeini, the politicization of the mosque was critical for the future of the revolution. There could be no separation of religion and politics, least of all at the sermon (*khutbah*) during the Friday *jum'a* prayers:[141]

> It is for an *Imām* of Friday prayer to talk about the religious and worldly interests of the Muslims during his Friday prayer sermons, and inform them of the condition of Muslims in other countries and bring to their notice their interests as well. He should talk to them about their religious and worldly requirements, refer to economic and political issues, inform them of their relations with other countries and elaborate on the interference of colonial powers in their countries.

This had to be so, according to Khomeini, because of Islām's all-embracing world view.[142] An Islamic government was essential, not an optional extra, since the capacity to take part in government formed part of the legal competence (*wilāyat*) of a Muslim jurist.[143] According to the Shī'a view, only the infallible *Imāms* and their appointees were entitled to take the helm of political affairs. In their absence, their representatives, that is, qualified Muslim jurists, were responsible for running the political affairs: but a practical mindset was required, or else the jurist was incapable of office.[144] Khomeini stressed the role of the Shī'a in opposing oppressive government and considered this a potential point of difference with Sunnīs, many of whom thought that rebellion against oppressive government was incompatible with Islām (an incorrect interpretation of the Qur'ānic verse on obedience [Q.4:59], in his view); but he stressed that he sought good relations with Sunnīs,[145] notwithstanding the incompetence of their rulers who were responsible for the strategic gains of Israel at the expense of the Arab nations.[146]

Khomeini was no pluralist and his language has regrettably confirmed a trend in certain parts of the Muslim world towards the acceptance of conspiracy theories and the transformation of anti-Zionism into anti-semitism.[147] The two greatest criticisms of the Pahlavī regime were that it had abased itself before American power, in the granting of capitulatory rights to the USA, and that it had become 'a military base for Israel, which means by extension, for America'. Israel itself, indeed, 'derives from America'.[148] He had expressed a similar idea, with customary virulence, in a speech in 1383/June 1963, in which he had denounced the Pahlavī 'white revolution' against the so-called 'black' reactionary Islamic clerical establishment:[149]

> Israel does not wish the Qur'ān to exist in this country. Israel does not want the '*ulamā*' to exist in this country. Israel does not wish a single learned man to exist in this country… It is still assaulting us, and assaulting you, the nation; it wishes to seize your economy, to destroy your trade and agriculture, to appropriate your wealth. Israel wishes to remove by any means of its agents anything it regards as blocking its path. The Qur'ān is blocking its path…

Given that the pivotal event of contemporary Iranian history was the restoration of Muḥammad Reza Shāh by a CIA-sponsored coup in 1372/August 1953, Khomeini's fevered vision of a world of conspiracies to undo Islām and its heritage, as well as to consume the resources that belonged to Islamic lands, was perhaps understandable – even if his use of language was less pardonable. Staggeringly, in view of his rhetoric, Khomeini must have given prior authorization for the proposed acceptance of Israeli arms shipments in the Iran–Contra Affair, presumably on the grounds that there was no alternative to such supplies if the war against 'Irāq was to be bought to a successful outcome.[150] Khomeini certainly imposed silence on eight potential domestic critics of the affair. He simply instructed them 'as to the interests of the nation' and their question was withdrawn.[151] The failure of the covert operation, and Iran's capture of the Fao Peninsula in 1406/February 1986, led the United States to take much more active measures to support 'Irāq than hitherto in the war.[152] Two months after the peninsula was taken, Khomeini ordered that the war was somehow to be 'won' by 1407/21 March 1987, the Iranian new year.[153] It seems clear that, although according to the Constitution the power of declaring war and peace depended on the leader acting on the advice of the Supreme Defence Council, Khomeini was not subject to limits of this kind.[154]

For all the virulence of his rhetoric and his assertion of the need for Islamic revolution, which reminds one of Mawdūdī or Quṭb, Khomeini was in essence a traditional Muslim moral teacher. Yes, corrupt regimes must be overthrown. The war against 'Irāq was not just directed against Saddam but 'against all unbelief'. But why was this necessary? It was because of the traditional duty of the Muslim scholar: 'when evil innovations appear, it is the duty of the scholar' to condemn them. This was a form of the traditional requirement to command right and forbid wrong, which created in its wake 'a wave of broad opposition' on the part of 'all religiously-inclined and honourable people'.[155] If a political regime attacked the fundamentals of Islām, then this amounted to a category of wrongs of 'such relative weight' that the obligation to right them overrode the danger condition, particularly for the clergy. Michael Cook calls this a new doctrine of Khomeini's which was 'inserted without any attempt to integrate it with the old'.[156] Muṭahharī, too, accepted that the danger condition could be overriding when there was a risk of greater danger to Islām. What was at stake was something on which Islām sets a higher value than it does on life, property or dignity, especially if the Qur'ān itself was deemed to be in danger.[157] Khomeini specifically allowed righteous violence to be unleashed by individual members of the clergy; his pupil Muṭahharī merely observed in addition that cooperation was more effective.[158]

Thus it was a duty of Muslims to 'engage in an armed *jihād* against the ruling group in order to make the policies ruling society and the norms of government conform to the principles and ordinances of Islām'. Because of their rank and position, Muslim scholars must take the lead over other Muslims in 'this sacred *jihād*, this heavy undertaking'.[159] *Jihād* was the religion of the revolutionary struggle:[160]

> all segments of society are ready to struggle for the sake of freedom, independence and the happiness of the nation, and their struggle needs religion. Give the people Islām, then, for Islām is the school of *jihād*, the religion of the struggle; let them amend their characters and beliefs in accordance with Islām and transform themselves into a powerful force, so that they may overthrow the tyrannical regime imperialism has imposed on us and set up an Islamic government.

The Islamic Constitution of Iran, which was ratified in 1400/December 1979, achieved Khomeini's main objectives. It was, however, and to some extent remains, controversial. Asghar Schirazi argues that the debates over the Constitution lost the regime votes in the second of the two referendums of 1400/1979. In the first referendum at the end of March, on whether to establish an Islamic state, there were 20.4 million votes cast; in the second referendum at the beginning of December, only 15.7 million votes were cast.[161] The 175 articles of the Constitution established that the state and the revolution that led to it were Islamic; defined the tasks and the goals of the state in accordance with its Islamic character; bound legislation to the *sharī'ah* (although as Asghar Schirazi demonstrates, this was subject in practice to a great deal of debate and interpretation);[162] ensured that positions of leadership would be reserved for Islamic jurists; placed 'Islamic-defined restrictions on the democratic rights of individuals, of the nation and of ethnic groups'; and finally established institutions whose task it was to ensure the Islamic character of the state.[163] Article 8, citing

Q.9:71, made it the duty of the people and the government to 'enjoin upon one another what is just and forbid what is evil', though the mechanism by which it was to do so has been problematic.[164]

The imāmate was to provide the leadership, and play the fundamental role, in the progress of the Islamic revolution. Article 5 stipulated that an individual jurist who was endowed with all the necessary qualities, or a council of jurists, had the right to rule and exercise leadership in the Islamic Republic as long as the 'Lord of Time', that is, the twelfth *imām* of the Shī'a remained 'in occultation' (an event which, as has been seen, had occurred in 329/941).[165] The leader held supreme command over the armed forces, appointing and dismissing the chief of the general staff and the commander-in-chief of the Guardians of the Islamic Revolution Corps (*Sepay-e Padaran*). The army was to be 'ideologically oriented', responsible not merely for the defence of Iran's borders, but to assume the burden of its ideological mission, *jihād* 'to spread the rule of God's law throughout the world'. This was particularly the task for which the *Padaran* had been set up: 'establishing the Islamic state world-wide belongs to the great goals of the revolution'.[166]

Asghar Schirazi considers that in the debate on the Constitution and in the subsequent interpretation of it, the moderate Islamists were inhibited and failed to resist those he calls the 'hierocratic legalists'. The moderate Islamists were 'credulous and submissive rather than independent and self-confident' in their dealings with Khomeini. He 'understood how to intimidate them with his overbearing style of leadership and they were afraid of being damned by him as infidels and of being excluded from government power'.[167] While Khomeini remained as leader 'no one ever wanted [to], or was able to, contest the supremacy of the leader... Not only did he unite all the branches of the state in his hands, he even exercised the power to shape the constitution'.[168] Only under Khomeini's successors, none of whom had forged the Islamic revolution as he had done, has this concentration of power has been diluted.

More surprising, given Khomeini's stand on exporting world revolution from Iran has been President Khatami's proposal, made in a speech to the United Nations in 1419/September 1998, that the year 2001 should be designated as the 'year of dialogue among civilizations'. This has been reinforced in many of his speeches on the subject subsequently and by the establishment in Iran of an International Centre for Dialogue Among Civilizations (ICDAC) in 1419/February 1999. ICDAC is primarily an organization to promote the concept of a global structure based on mutual understanding and tolerance. To achieve that goal, the Centre sets forth its mission statement as to promote dialogue among civilizations and cultures on an international scale as a means of advancing the interpretation of the UN Charter and of improving human well-being; to promote and expand the culture of dialogue at the national level; to promote the culture of peace in order to foster peaceful coexistence and prevent human rights violations; to help establish and broaden international civil society through cultural interaction among nations; to strengthen spiritual, moral and religious culture; and to conduct research on the significance and possible interpretations of Dialogue Among Civilizations. It is too soon to assert with complete confidence that a permanent sea change has overtaken the Iranian Revolution, not a counter-revolution so much as the hand of peace offered to other cultures and faiths. There remain difficulties with the Iranian government's attitude towards the procurement of weapons of mass destruction. Yet there can be no question that Iran in 1425/2004 presented a different face to the world than it did on Khomeini's death in 1409/1989, and that, for all the great man's achievements, the change was overwhelmingly positive in terms of promoting a culture of peace and dialogue within the international community. There remained at least four shadows on the horizon: these were the international community's opposition to Iran's acquisition of nuclear capability; the renewed recruitment in June 2004 of recruits for 'martyrdom operations' (see Chapter 11), though it was unclear how far the second activity was a response to international pressure over the nuclear issue; the accusations from the United States in July 2004 that Iran had in some way been involved with al-Qaeda prior to the events of 11 September 2001;[169] and the claim, also made in July 2004, that if re-elected in November President George W. Bush would pursue the objective of 'regime change' in Iran.[170]

The resurgence of the Shī'a of 'Irāq. I: the *intifāḍah* of 1411/ March 1991

In launching the war on Iran in 1400/1980, Saddam Hussain had made a strategic miscalculation of considerable magnitude. It is true that the result of the war was inconclusive: Iran failed to topple the Ba'th regime and set in train a wave of religious radicalism in the Middle East. But in 1409/1988 'Irāq had emerged substantially weakened from the war, encumbered by huge debts whereas at the outset of the war it had been in surplus. Saddam's solution was that economic recovery would be paid for by the Gulf states, willingly or otherwise: 'let the Gulf regimes know that if they will not give this money to me, I will know how to get it' was the message passed to Sa'ūdī Arabia by the Jordanian monarch. Kuwaiti indifference amounted to 'stabbing 'Irāq in the back'. The demand was for 'an Arab plan similar to the Marshall Plan to compensate 'Irāq for some of the losses during the war'. If Kuwait continued to conspire with 'world imperialism and Zionism' to cut off the livelihood of the Arab nation', then action would be required to ensure the restitution of 'Irāq's 'rights'. The Kuwait leadership refused to bow to the threats but underestimated Saddam's determination: Kuwait itself became in 1411/August 1990 a further casualty of the Iran–'Irāq War.[171]

The first Coalition against 'Irāq comprised Western and Arab states, a broadly-based coalition of the (relatively) willing against Saddam's unilateral abolition of the international border between ʻIrāq and Kuwait and its annexation as an ʻIrāqī province. It won an overwhelming military victory, but because of a prior understanding with Arab Coalition partners, especially Saʻūdī Arabia,[172] the United States and its allies were not at liberty to pursue Saddam's forces once they had withdrawn in disarray to ʻIrāq. This left the opponents of Saddam's regime within ʻIrāq, who had risen in rebellion in 1411/March 1991 partly as a result of Coalition propaganda,[173] vulnerable to repression.[174]

In particular, the Shīʻa, who are estimated to account for at least 55 per cent of the population of the country,[175] had risen because of multiple grievances against the Sunnī monopoly of political power under the Ba'th regime,[176] and with the aspiration of installing an Islamic government in Baghdād. In the north, the defection of much of the government-recruited Kurdish militia, who vastly outnumbered the *peshmerga*, gave considerable force to the revolt of the Kurdish separatists. Unlike the *peshmerga*, the Shīʻa resistance in the south lacked an organized fighting force, although it maintained cells and had carried out armed operations on occasion. The Shīʻa opposition had long enjoyed sanctuary and support from the Iranian regime, although Teheran did not seem to provide significant material or logistical assistance during the 1411/March 1991 uprising. As Human Rights Watch reported, ʻonce the loyalist troops regrouped and mounted their counteroffensive, only massive foreign assistance or intervention could have saved the ill-equipped and inexperienced rebels'.

The same Human Rights Watch report noted that the violence was

> heaviest in the south, where a smaller portion of the local population had fled than in Kurdish areas, owing partly to the danger of escaping through the south's flat, exposed terrain. Those who remained in the south were at the mercy of advancing government troops, who went through neighborhoods, summarily executing hundreds of young men and rounding up thousands of others.

Saddam Hussein's regime claimed that the uprising was an ʻarmed rebellion' that could ʻunder no circumstances' be regarded as an *intifāḍah*, the Arabic term for uprising most often associated with the Palestinian revolt in the Israeli-occupied territories. Replying to the UN Human Rights Committee, the government characterized the Iraqi uprising as a ʻstate of insurrection and extreme lawlessness, to which it had responded legitimately' and with proportionality. In reality, the repression was a bloodbath. Though the USA warned ʻIrāq against the use of chemical weapons during the unrest, it equivocated and failed to act against its use of helicopter gunships against civilians. A so-called United Nations ʻsafe haven' was established in northern ʻIrāq in 1411/April 1991 for the protection of the Kurds; but it was not until over a year later, in 1413/August 1992 that a no-fly zone was established over southern ʻIrāq.

After the failure of the *intifāḍah* of 1411/March 1991, the Grand Ayatullah Sayyid abū al-Qāsim al-Khoei, a renowned jurist and scholar, and spiritual head of the world-wide Shī'a community, was briefly imprisoned and then forced to appear on television.[177] Saddam Hussein tried to pressurize him into issuing *fatāwā*, or religious verdicts, supportive of Saddam and his government. This he failed to do, but as a result was placed under house arrest until his death the following year. Saddam also exiled, imprisoned, or assassinated many of al-Khoei's most gifted students, representatives and distinguished followers and ordered the destruction of their mosques, shrines and libraries particularly those in Najaf and Karbalā'. More than 200 clerics vanished and remain unaccounted for: the Shī'a consider that the defeat of the *intifāḍah* unleashed more than a decade of persecution by the regime of Saddam Hussein. Rami El-Amine comments:[178]

> The Sunnīs, particularly the tribes in and around Tikrit, were rewarded with jobs and repairs to their roads, schools, and hospitals. The Shī'a, on the other hand, were collectively punished for the uprising: the regime killed tens of thousands of Shī'a and destroyed much of the infrastructure in the South, including the Shī'a holy sites. They were of course deprived of any reconstruction funds and the limited rights they had gained in the past were taken away. But this was not enough for a megalomaniac like Saddam Hussein. To inflict long term, permanent damage on the Shī'a, he constructed a canal that siphoned off the water from the marshes in the South where many Shī'a lived. The combination of catering to its Sunnī tribal base while intensifying its persecution of the Shī'a deepened the Shī'a sense of a common identity and accelerated a turn to Islamism. But, as Cole explains, the appeal of Islamism doesn't come out of nowhere but a specific set of circumstances which Iraq's Shī'a found themselves in after the first Gulf War: 'shut out of the circle of patronage, non-Sunnī Iraqis had to find bases on which to mobilize. They could not form secular parties that might try to appeal across ethnic cleavages on economic issues. The regime's relentless surveillance forced them to turn inward, to family, clan and the mosque. As a result, Shī'a movements were able to organize clandestinely in ghettos and among settled tribes in the late Saddam period to make preparations for an Islamic State.' This explains the ascendancy and leadership of Grand Ayatollah Muḥammed Ṣadiq al-Ṣadr [Moqtadā al-Ṣadr's father], the highest-ranking non-Persian Shī'a Ayatollah, in the [19]90s. Initially resistant to clerical involvement in political matters, 'he came to embody the Shī'as' frustration and to express their demands as he increasingly adopted courageous publicly critical positions'. His assassination by the regime in 1999 led to huge protests and riots throughout the South and in the eastern slums of Baghdād where over two million Shī'a live. Since the fall of Baghdād this ghetto has come to be known as Ṣadr City and is a stronghold of the younger al-Ṣadr. The protests were eventually quelled, but

in killing al-Ṣadr the regime only strengthened his legacy of an activist current among the Shī'a clergy.

The resurgence of the Shī'a of 'Irāq. II: the *jihād* of 1425/2004

The basis for the second Coalition intervention in 'Irāq in 1424/March 2003 was quite different from the previous war twelve years earlier. On the second occasion, the regime had not unified international opinion against it by an invasion across an international boundary. It is true that Saddam Hussein's regime had a long history of deception with regard to its programme to establish weapons of mass destruction (WMD) but it had finally, albeit reluctantly, admitted UN inspectors who, in their report to the UN Security Council dated 27 January 2003, noted that they had found no clear evidence of the continued existence of WMD. A discredited Saddam Hussein regime had also produced a substantial dossier on the subject to the United Nations which argued the case that these weapons had been destroyed in the aftermath of the First Gulf War. Only the permanent members of the Security Council – the USA, Britain, France, China and Russia – were granted access to the full 12,000 page document; the ten non-permanent members of the UN Security Council received edited versions of the 'Irāq weapons dossier, with sensitive information that could be used to develop weapons of mass destruction removed. There were serious shortcomings in the intelligence dossier prepared for the British government about 'Irāq's WMD programme, a dossier which was used by US Secretary of State Colin Powell in his report to the UN Security Council on 6 February 2003, which advocated the case for war. Subsequently, in July 2004, reports to both the British and US governments provided severe criticism of the intelligence shortcomings on which the decision for war was based. Some commentators have gone further than the wording of the reports and argued the case for a 'systems failure at the heart of British government',[179] while criticism of the intelligence system in the United States has been even more severe (see Chapter 11).

Public opinion in the West was in large measure hostile to the Second Gulf War. No doubt several competing issues swayed public opinion: hostility to war in general, and fear of civilian casualties; solidarity with Islamic states on the part of Muslims in Europe; suspicion or hostility to America's ruthless pursuit of self-interest in foreign policy, defined in the statement *Rebuilding America's Defenses: Strategies, Forces and Resources for a New Century*, written in September 2000 by a neo-conservative think-tank, Project for the New American Century (PNAC), which was produced for Dick Cheney, subsequently Vice-President under President George W. Bush, and Donald Rumsfeld, subsequently Defense Secretary in the same administration;[180] the belief that America was determined on a course of war resulting from the build-up of its forces in the Middle East; finally and, perhaps, crucially, the view that the case that 'Irāq posed a credible threat to the West because of its secret possession of WMD was unproven. Many Churchmen were among the leading critics of what was seen as an excessively rapid recourse to war,[181] and appealed to Christian 'Just War' theory to set moral guidelines against which the decision to declare war might be viewed.[182] These reservations about an excessively rapid recourse to war, which seemed to justify the view that George W. Bush simply wished to settle a problem which his father had had to leave unresolved, have been confirmed by subsequent revelations of the secret planning for war in 'Irāq: this began before the events of 11 September 2001 and accelerated thereafter[183] and was largely separate from the issue of 'Irāq's alleged possession of WMD, which became a justification for an intervention by the USA that had been predetermined.[184] Prince Turkī al-Faisal, formerly head of Sa'ūdī intelligence, noted that in his period in office, there had never been any link between 'Irāq and the al-Qaeda terrorist network. In his judgement, intervention in 'Irāq would worsen, rather than reduce, the problem of terrorism.[185] These fears were confirmed by subsequent events.[186]

If the rationale for US intervention against 'Irāq in 2003 was quite different from the earlier conflict, so also were the ways in which the new coalition was formed and the opposition it aroused. There was no specific United Nations mandate for intervention, because this could not be extracted from the UN Security Council. The Coalition leaders, President Bush and Prime Minister Tony Blair, argued that it was unnecessary, in that 18 previous resolutions had been defied by 'Irāq and time had run out. There remain legal objections to the British government's justification for intervention, which rested on 'the combined effect of resolutions 678, 687 and 1141'.[187] The leading states in the European Union, particularly France and Germany, were resolutely opposed to intervention and insistent on a specific prior authorization of the Security Council, which France, as a permanent member, stated that it would in any case veto.

Worse still for the prospects of success, with the exception of Kuwait and certain Gulf states,[188] the Arab and Muslim states refused participation in a second 'coalition of the willing', notwithstanding broad international suspicion of, or opposition to, the regime of Saddam Hussein. This made the forces of the coalition vulnerable to propaganda, whether from Saddam himself,[189] a gun-toting Shaykh at Friday prayers in Baghdād,[190] Osama bin Laden (who accused the 'Crusaders' of seeking to establish a 'greater Israel'),[191] or others (such as the scholars at al-Azhar University in Cairo),[192] that they were 'Crusaders' who must be resisted by a legitimate defensive *jihād*. As late as July 2004 the Syrian information minister argued that resistance was legitimate as long as foreign troops remained in occupation of 'Irāq.[193] In calling for military action against 'Irāq in a speech in Nashville in the autumn of 2002, Vice-President Dick Cheney predicted that 'after liberation the streets in Baṣra and Baghdād are sure to erupt in joy in the same way throngs in Kabul greeted the Americans'. Professor Matthew Levinger correctly argued that this view was[194]

> overly optimistic. Although most Iraqis would probably welcome the overthrow of Saddam Hussein, many would also experience humiliation and resentment over the subsequent occupation of 'Irāq by American soldiers, which would be essential for preserving political stability in the region.
>
> The occupation of 'Irāq by U.S. troops would provide an ideal recruiting platform for al-Qaeda and other extremist organizations. Many people of the region would become convinced that America's goal is global domination, not dignity for Muslim people. If President Bush launches a military campaign without the support of the United Nations, the perceptions will be even worse. Rather than upholding the rule of law, America will appear to be acting arbitrarily in its own interest. Despite our leaders' rhetoric about securing freedom and democracy, many will conclude that the United States cares only about preserving its own power.
>
> If freedom and democracy are so important in 'Irāq, why does the administration show so little interest in promoting these values in Egypt or Saudi Arabia?...
>
> We cannot promote democracy through the barrel of a gun. A regime change in 'Irāq that requires an open-ended military occupation of the country is likely to have disastrous consequences for the stability of the Middle East and for America's national security...

Such prescience, and comments of the same kind from others, had no effect on the military planning in the Pentagon or on the timetable for war. At first, the invasion met with overwhelming success as the forces of Saddam Hussein melted away. The destruction to the infrastructure, especially to the supplies of electricity and water, in 'Irāq was considerable. There was criticism of the level of civilian casualties, and at the breakdown of law and order following the collapse of the regime; but there can be no doubt that there was also popular support for the ejection of Saddam's brutal secret police network. There were some recruits from abroad for the *jihād* in 'Irāq, but relatively few;[195] and the level of resistance to the military occupation seemed at first relatively slight and limited to the so-called 'Sunnī triangle', which comprised Saddam's hard-line supporters, the beneficiaries of his former regime.

If there was any hope that the 'Irāq quagmire would not be exploited by Osama bin Laden, this was dispelled by his broadcast on 18 October 2003:[196]

> I greet you, your effort, and blessed *jihād*. For you have massacred the enemy and brought joy to the hearts of Muslims, particularly the people of Palestine. May God reward you for that. You are to be thanked for your *jihād*. May God strengthen your positions and guide you to achieve your targets.
>
> You should be pleased, for America has fallen into the quagmires of the Tigris and Euphrates. Bush thought that 'Irāq and its oil are easy spoils. Now,

he is in a critical situation, praise be God. Today, America has started to cry out and crumble before the entire world.

Praise be to God, who foiled its plots and made it ask for help from the lowliest people and beg for the mercenary soldiers from the east and the west. No wonder you did these deeds to America and made it suffer in this way, for you are the sons of those great knights, who carried Islām eastward until they reached China.

Let it be known to you that this war is a new crusader campaign against the Muslim world, and it is a war that is crucial to the entire nation. Only God knows the extent of its serious repercussions and negative effects on Islām and Muslims. O young people of Islām everywhere, especially in the neighbouring countries and in Yemen, you should pursue *jihād* and roll your sleeves up. Follow the right path and be careful not to support the men who follow their whims, those who sat idle, or those who relied on oppressors. For those would seek to shake you and inhibit you from pursuing this blessed *jihād*…

[Those who cooperate with the Coalition] claim that they are on the right path, but they are doing a great wrong. God knows that Islām is innocent of their deeds, for Islām is the religion of God. The legislative councils of representatives are the religion of the pre-Islām era. He who obeys the leaders or scholars in permitting what God banned, such as joining the legislative councils, or banning what He permitted, such as *jihād* for His sake, would be associating other gods with Him. God is our sole source of strength.

Bin Laden also added a message to the American people and the US soldiers serving in ʻIrāq:

…You are being enslaved by those who have the most money, the most influential ones, and those who have the strongest news media, particularly the Jews, who are dragging you behind them under the trick of democracy in order to support the Israelis and their schemes and hostility to our religion and at the expense of our blood and land, as well as at the expense of your blood and economy.

Events have proven this. The fact that you were driven to the ʻIrāqī war, in which you have no interest whatsoever, is evidence of this. Bush came with his hard-booted, hard-hearted gang.

This gang is a huge evil on all humanity, its blood, money, environment, and morality. They came to deal strong and consecutive blows to honesty that is the basis of morality, each from the position he holds, until they professionally rendered it dead before the world…

On the one hand, he implements the demands of the Zionist lobby, which helped him enter the White House, to destroy the military power of ʻIrāq, which

> neighbours the Jews in occupied Palestine. In so doing, he shows absolute indifference to the repercussions on your blood and economy.
>
> On the other hand, he conceals his own greed and the ambitions of this lobby in ʿIrāq and its oil. He is still thinking in the mentality of his ancestors who used to kill the red Indians to seize their lands and loot their wealth.
>
> He thought that the matter would be easy spoils and a short trip that would not end in failure. However, God had those lions waiting for him in Baghdād, the home of the Islamic caliphate, the lions of the desert who believe that the taste of death in their mouths is better than honey...
>
> Let the unjust ones know that we maintain our right to reply, at the appropriate time and place, to all the states that are taking part in this unjust war, particularly Britain, Spain, Australia, Poland, Japan, and Italy. The Islamic world's states that are taking part in this war, particularly the Gulf states, mainly Kuwait, the land base for the Crusader forces, will not be excluded from this...

The official response for the Coalition was given by US Secretary of State Colin Powell, who argued that the bin Laden threat would lead more countries to begin military or financial aid to the Coalition in ʿIrāq or, if they were already assisting, to increase existing levels of help.

> It might have scared them in the past, but everybody now knows you are not immune, you can't hide from it, you can't walk away. He is a threat to all of us... it just reminds us that this kind of terrorist is still on the face of the earth, and we have to come together, and we have to do even more with the exchange of law enforcement information, intelligence information [and] the use of military forces. We all have to come together.[197]

This verdict underestimated the skill with which bin Laden had crafted this, and subsequent, statements[198] directed towards the ʿIrāqī opposition to the Coalition. Firstly, he appealed to the pride of the ʿIrāqī people, emphasizing as he did that Baghdād had been home to the caliphate of Sunnī Islām before the Mongol invasion of 656/1258. Similarly, he stressed that the ʿIrāqīs were 'the sons of those great knights, who carried Islām eastward until they reached China': there was no recrimination of the poor showing put up by Saddam's regime against overwhelming American military power; the emphasis instead was on the future struggle, on what courage alone could achieve.

Secondly, bin Laden asserted that President Bush had thought that 'Irāq and its oil' would be 'easy spoils'. The mentality was still that of the nineteenth century and the expansion westward at the expense of the American Indians in order 'to seize their lands and loot their wealth'. Doubtless here bin Laden was seeking to elide two different issues: the Coalition's undertaking to restore oil production and exports, so as to help finance the rebuilding of the infrastructure of ʿIrāq, where clear statements had been made that profits would not be siphoned off to the Coalition itself; and concerns about preferential contracts for reconstruction awarded to Haliburton and other enterprises in which President Bush's associates, above all Vice-President Dick Cheney, had formerly had a direct interest (Cheney had been Haliburton's chief executive between 1995 and 2000 but divested himself of all financial interest in the company after the presidential election).[199]

A third sign of bin Laden's ingenuity in crafting a broadly-based opposition to the Coalition in 'Iraq was in his appeal to the religious leaders of the Shī'a to support the campaign for an Islamic state. It was evident that, as a Sunnī, bin Laden could not make an overt appeal without the risk of rejection. Instead what he could do, which was well-calculated to create the effect that he sought, was to suggest that participation in the proposed legislative council was akin to apostasy – such bodies smacked of *jāhilīyyah*, the pre-Islamic state of barbarism from which the advent of Islām had rescued the faithful. Whether or not the equation of collaboration with *jāhilīyyah* was meant to be taken seriously, it was bound to have an impact on the Shī'a clerical leaders and to encourage them to reaffirm their demands for an Islamic state. This, in spite of the fact that, in principle, the al-Qaeda network was considered to be anti-Shī'a.[200]

Finally, bin Laden homed in on the issue of morality.[201] In his statement to the American people and US troops, he stated:

> they did not care about you, went behind your backs, invaded 'Irāq once again, and lied to you and the whole world. It has been said: a nation's strength is in its morality; if the morality of the people of any nation was to deteriorate they would cease to exist.

Already by October 2003, when bin Laden's address was issued, it was clear that it was most unlikely that WMD would be found in 'Irāq. Hans Blix, formerly the UN Chief Weapons Inspector had expressed his doubts about the existence of WMD in a telephone interview for an Australian radio station the previous month. At the end of January 2004, David Kay, previously head of the 'Irāqī Survey Group, told the Senate Armed Services that it was 'important to acknowledge failure' in the search for WMD. Powell in response made an ambiguous statement that the absence of WMD 'changes the political calculus [*sic*]'; later, he appeared to retract, stating that President Bush 'had made the right decision' in going to war.[202] Worse still for the credibility of the Coalition, in March 2004, Blix hit out much harder, charging the Bush administration with invading 'Irāq as retaliation for the terrorist strikes on the United States, even though there was no evidence linking Saddam Hussein to the attackers:[203]

> in a way, you could say that 'Irāq was perhaps as much punitive as it was pre-emptive. It was a reaction to 9/11 that we have to strike some theoretical,

> hypothetical links between Saddam Hussein and the terrorists. That was wrong. There wasn't anything…
>
> I am not suggesting that Blair and Bush spoke in bad faith, but I am suggesting that it would not have taken much critical thinking on their own part or the part of their close advisers to prevent statements that misled the public…
>
> [Vice-President Cheney in effect stated] that if we did not soon find the weapons of mass destruction that the U.S. was convinced ʻIrāq possessed (though they did not know where), the U.S. would be ready to say that the inspectors were useless and embark on disarmament by other means.

Blix contradicted the view of the Bush administration that the war had made the world a safer place: 'if the aim was to send a signal to terrorists that we are determined to take you on, that has not succeeded', he argued. 'In ʻIrāq, it has bred a lot of terrorism and a lot of hatred [of] the Western world.' Even retired Marine Corps General Anthony Zinni, formerly chief of Central Command (CENTCOM), had previously stated that he did 'not believe that Hussein [had] posed an imminent threat or that he [had] possessed chemical, biological or nuclear weapons'.[204] The ʻIrāqī National Congress (INC), and its head Ahmed Chalabi, had, it seems, earlier provided Rumsfeld and Wolfowitz with the story that they had wanted to hear, much of which could not be corroborated.[205] (Chalabi's alleged payments from US sources were to be cut from the end of June 2004, it was reported.)[206] Finally, in an interview with Tim Russert of NBC on 16 May 2004, Secretary of State Colin Powell made a remarkably frank admission that the intelligence assessment on which he had based his case for war to the UN Security Council was faulty:[207]

> I'm very concerned. When I made that presentation in February 2003, it was based on the best information that the Central Intelligence Agency made available to me. We studied it carefully. We looked at the sourcing in the case of the mobile trucks and trains; there was multiple sourcing for that. Unfortunately, that multiple sourcing over time has turned out to be not accurate, and so I'm deeply disappointed.
>
> But I'm also comfortable that at the time that I made the presentation, it reflected the collective judgment, the sound judgment of the intelligence community; but it turned out that the sourcing was inaccurate and wrong, and, in some cases, deliberately misleading, and for that I am disappointed and I regret it.

In an earlier interview with the media on 19 October 2003, Secretary of State Colin Powell had admitted that

> we still have a dangerous environment in 'Irāq. There are still remnants of the old regime who do not want to see progress. But I am confident that our military leaders there and the wonderful young men and women who are serving their nation so proudly will ultimately get the security situation under control.

The death in a shootout of Saddam Hussein's two sons had denied the former regime of much of its capacity for restoration. The capture of Saddam himself, on 13 December 2003, was by far the greatest triumph of the Coalition. President Bush called it 'crucial to the rise of a free 'Irāq. It marks the end of the road for him and for all who killed and bullied in his name.' But though a 'dark and painful era' was over, it did not mean 'the end of violence in 'Irāq'.[208]

Yet for all the successes of the military intervention, they were increasingly offset by setbacks. The most devastating blow to the Coalition presence in 'Irāq came in May 2004, with the revelation of systematic abuse of prisoners chiefly by American military personnel at the Abū Ghraib prison, the former centre for Saddam Hussein's brutal mistreatment of prisoners. It was the equivalent of Saddam's Bastille, in Simon Jenkins' words, which should have been flattened the day after the occupation/liberation of 'Irāq. Jenkins wrote:[209]

> After 14 months there is no room for excuses. Liberation has been followed by a new bondage, that of individual insecurity, public anarchy and, in much of the country, a looming clerical totalitarianism. The prison guards have been indoctrinated to believe that they are in the front line of the Third World War. They must give no quarter against suspected terrorists, preferably revealing them as 'outsiders', al-Qaeda members or at very least Saddamists. When military or private-contract interrogators arrive and demand that prisoners be 'softened up', what are they supposed to do?…
>
> Iraq was invaded illegally on the excuse that someone in London and Washington thought the country posed an immediate threat. Anarchy was created in 'Irāq on the excuse that Ahmed Chalabi boasted he would be welcomed as a liberator. Torture is committed in jails on the excuse that a link must be found with al-Qaeda. Is blinkered idiocy by *château* generals more 'excusable' than obscene antics by prison guards?

Worse still, from documents secured by the National Security Archive, there was little doubt that the practice of 'hard interrogation' had been officially sanctioned for a long period of time and that Dick Cheney had tried to take corrective action in 1992.[210] There were even suggestions that such techniques had been personally authorized by Donald Rumsfeld, the US Secretary of Defense.[211] However, in view of the considerable political storm over prisoner abuse, the classified papers were subsequently made available in June 2004, including Rumsfeld's Memorandum for the Commander, US Southern Command, dated 16 April 2003.

This listed authorized 'techniques A–X set out at Tab[le] A'. These techniques had to be used 'with all the safeguards described at Tab[le] B' and, moreover, use of the techniques was limited to 'interrogations of unlawful combatants' held at Guantánamo Bay, Cuba.[212]

Commentators argued that the Coalition's pledge to introduce democracy had itself been damaged, since democratic values themselves had been eroded in the debate over legalized torture.[213] Even the pro-Coalition London *Times* argued in a leader that the Abū Ghraib prison should be razed,[214] while the international outrage was particularly fierce, with suggestions that the US had done irreparable damage to its image in the Arab world.[215]

At the same time, the Coalition seemed to be losing its way in 'Irāq. Military casualties were increasing, while the prospects of a permanent political settlement, rather than a short-term fix that would permit the withdrawal of Coalition forces, seemed fairly remote.[216] Colin Powell claimed that he believed prospects were good for a UN Security Council resolution backing the 'Irāqī interim government and creating a broader multinational force. However, his response to further questioning was revealing:[217]

> [After 30 June 2004], the only authority really is that interim government… [I]f they actually asked us to leave, would we? [T]he answer is yes. But we don't expect that to be the case… We want to finish our job, turn full sovereignty over to the 'Irāqī people… [and] come back home as fast as we possibly can… But we're not going to leave while the 'Irāqī people still need us.

The Coalition, Powell contended, would 'make arrangements with the new leadership that it's best for us to stay'. In response to a query as to whether the United States would accept an Islamic theocracy as the permanent government of 'Irāq, Powell responded: 'we will have to accept what the 'Irāqī people decide upon', but expressed confidence that that they would opt for a government that 'understands the role of a majority, but respects the role of minorities'.[218] It was a considerable change of tone from late April 2003, when Donald Rumsfeld, angering Shīʿa clerics thereby, had stated with confidence that an Iranian-style Islamic state would not be allowed in post-Saddam 'Irāq.[219]

The greater realism displayed by Powell a year later, as compared to the earlier bravado of Rumsfeld, reflected the changed situation on the ground, and the threatening signs of Sunnī–Shīʿa cooperation against foreign occupation. There had been one spectacular indication of the opposite tendency. Abu Musab al-Zarqawī, the most wanted man in 'Irāq with a $10 million price on his head, was blamed for a series of deadly attacks in Baghdād and Karbalā' at the beginning of March 2004 that killed at least 143 Shīʿa Muslims in an apparent attempt to trigger a civil war. Dick Cheney, the US Vice-President, stated: 'we don't know specifically about this attack yet, but it has the hallmarks, in my opinion, of an attack orchestrated by al-Zarqawi… What we've seen today in these attacks are desperation moves by al-Qaeda-affiliated groups that recognize the threat that a successful transition in 'Irāq represents.'[220] The Coalition Provisional Authority (CPA) was, however, accused of using tactics of divide and rule in politicizing the ethnic and religious divisions of the country.

Surprisingly, in April 2004, there were unprecedented examples of solidarity and mutual aid between the Muslim majority and minority populations. They exchanged messages of support and the Shī'a joined Sunnīs in donating blood and organizing a relief convoy to Fallujah, which was besieged by American troops. Joint prayers were organized. The slogan 'Sunnīs and Shī'a are united against the American occupation' was painted on walls in Sunnī neighbourhoods of Baghdād. And members of the *Mahdī* army – the militia formed by the fiery Shī'a cleric, Moqtadā al-Ṣadr – went to Fallujah to fight alongside Sunnīs.

Al-Ṣadr launched an uprising against the Coalition forces after the CPA announced that he was wanted for the murder of the moderate cleric 'Abd al-Majīd al-Khoei in Najaf on 10 April 2003.[221] On 16 April 2004, he stated: 'we will not allow the forces of occupation to enter Najaf and the holy sites because they are forbidden places for them. I say that they are here to stay and will occupy us for many years and as such compromise will not work.' He stated that he was waging 'the revolution of the Imām Mahdī', the twelfth Imām of the Shī'a Muslims and their expected saviour. 'It is martyrdom that I am yearning for, so support me and know that this is a war on the Shī'a', he said. Al-Ṣadr also had harsh words for the Shī'as serving on the US-appointed 'Irāqī Governing Council, who had been trying to mediate a peaceful solution to the crisis: 'I address the agents of the West… they say that we are delaying the handover of power and the formation of government, but I tell them that we have delayed [the] selling [out of] 'Irāq and creating a government of agents.'[222]

While al-Ṣadr lacked the spiritual stature of Grand Ayatollah 'Alī al-Ḥusaynī al-Sistanī, and his confrontational tactics exasperated moderate Shī'a, he nevertheless commanded the support of thousands of mainly poor, urban Shī'a who admired his father, the Grand Ayatollah who was killed by Saddam Hussein's agents in 1999. Al-Ṣadr had also capitalized on hostility toward the Coalition following revelations of abuse of 'Irāqī prisoners by US soldiers. Despite the fighting, al-Ṣadr continued to deliver his sermons at Friday prayers in Kufa, a holy city that lies six miles to the north-east of Najaf. Al-Ṣadr described President Bush and British Prime Minister Tony Blair as 'the heads of tyranny' and accused them of ignoring the suffering of 'Irāqīs in Coalition prisons while drawing attention to what he described as the 'fabricated' case of Nicholas Berg, an American civilian who was beheaded by militants in a videotaped revenge killing for the abuse of 'Irāqī prisoners of war.

Shaykh Abdul-Sattar al-Bahadli, the representative of Moqtadā al-Ṣadr and leader of the al-Mahdī Army in Baṣra, threatened on 15 May to unleash a brigade of suicide bombers to attack British forces in the city. Arriving at the mosque carrying a rifle, he was greeted with shouts of 'Yes, yes, yes, to *jihād*' from about a thousand young male followers, and undertook to establish separate male and female 'martyrs' departments'.[223] This followed upon serious fighting that broke out the previous day at Najaf in the cemetery near the revered shrine of Imām 'Alī. Both there and at Karbalā', al-Ṣadr's forces sought to draw the Coalition troops into fighting on sacred ground that posed the risk of serious political consequences: if Imām 'Alī's shrine were to be damaged, support for the Coalition would plummet.[224] After a bloody three-week siege of Najaf, and the intervention of Grand Ayatollah 'Alī al-Ḥusaynī al-Sistanī, al-Ṣadr's forces withdrew from Najaf without surrendering their weapons. One of his aides stated that the 'Mahdī Army is now turning to peaceful struggle… Moqtadā will declare his participation in 'Irāq's political process. He will not participate directly in elections but he will appoint and back someone from his side or elsewhere.'[225]

The murder of Abdul Zahra Othman Mohammad, also known as Izzedin Salim, the Shīʿa head of the 'Irāqī Governing Council, by a suicide bomber on 17 May 2004, marked a new low point for the Coalition's fortunes. The initial suggestion was that it was an atrocity carried out by Abu Musab al-Zarqawi's group (in spite of cover provided by a previously unknown organization called the Arab Resistance Movement al-Rashid Brigades).[226] Support for Moqtadā al-Ṣadr's *jihād* remained patchy, if indeed *jihād* was his true purpose,[227] rather than placing pressure on the 'Irāqī Governing Council so as to gain concessions for his group.[228] What already seemed a strong possibility, however, given the disarray of the 'Irāqī parties, was that a determined push by a coherent and credible Shīʿa group arguing the case for establishing an Islamic state in the proposed elections in January 2005 would be difficult to resist and that concessions to Kurdish separatism would be extremely unpopular with the Shīʿa majority. The frank interview given by Secretary of State Colin Powell, which had acknowledged an Islamic state as a possible outcome, may ironically have served to encourage lobbying for this objective. The central issue was whether the campaign to establish an Islamic state would be primarily peaceful or violent in its nature. Rising levels of violence led to the surprise transfer of power to an 'Irāqī provisional government two days earlier than expected on 28 June 2004, and the departure of Paul Bremer, the ex-US Administrator of 'Irāq, from the country. The paradox remained: without the presence of large numbers of Coalition troops in the country, the interim government would be unable to control the violence and would be likely to collapse; yet the presence of the troops was a powerful argument for those who claimed that the violence was justified as a defensive *jihād* against foreign intervention.

Part Four

Context and Distortion of the Text

10
The Crucible: The Palestine–Israel Dispute and its Consequences

> The fact that must be acknowledged is that the issue of Palestine is the cause that has been firing up the feelings of the Muslim nation from Morocco to Indonesia for the past 50 years. In addition, it is a rallying point for all the Arabs, be they believers or non-believers, good or evil…
>
> Ayman al-Ẓawāhirī, Deputy Leader of al-Qaeda, in his memoirs, *Knights Under the Prophet's Banner*, December 2001, ch. 11[1]

The Palestine–Israel conflict is the crucible of the conflicts affecting the Muslim *ummah*. For more than 20 years since 1403/April 1983 it has acted as the epicentre of global *jihād*. Kashmir, Chechnya and other dangerous regional conflicts do not depend solely on a solution to the Middle Eastern conflict because they have their own specific causes and potential settlements. There can be little doubt that the long, nearly 60-year, struggle to rectify the injustice suffered by the Palestinians in 1367/May 1948 is a unifying factor in what might otherwise seem a set of disparate conflicts, or Arab–Israeli wars, affecting the Middle East.

It was also the earliest of the contemporary conflicts to have emerged historically: the Balfour Declaration of 1336/November 1917 was the root cause of the difficulty, though it is often forgotten that this made clear that 'the establishment in Palestine of a national home for the Jewish people' had to be balanced,[2] so that 'nothing shall be done which may prejudice the civil and religious rights of existing non-Jewish communities in Palestine...'[3]

Osama bin Laden himself drew attention to this conflict in his first declaration in 1417/August 1996. His theme was that of Sa'ūdī betrayal:

> The [Sa'ūdī] regime want to deceive the Muslims people in the same manner when the Palestinian fighters, *mujāhidīn*, were deceived causing the loss of

al-Aqṣā Mosque. In 1355/1936[4] the awakened Muslim nation of Palestine started their great struggle, *jihād*, against the British occupying forces. Britain was impotent to stop the *mujāhidīn* and their *jihād*, but their devil inspired that there is no way to stop the armed struggle in Palestine unless through their agent King 'Abd al-'Azīz, who managed to deceive the *mujāhidīn*. King 'Abd al-'Azīz carried out his duty to his British masters. He sent his two sons to meet the *mujāhidīn* leaders and to inform them that King 'Abd al-'Azīz would guarantee the promises made by the British government in leaving the area and responding positively to the demands of the *mujāhidīn* if the latter stop[ped] their *jihād*. And so King 'Abd al-'Azīz caused the loss of the first Qibla of the Muslim people.[5] The King joined the crusaders against the Muslims and instead of supporting the *mujāhidīn* in the cause of Allāh, to liberate the al-Aqṣā Mosque, he disappointed and humiliated them.

'Betrayal' is a strong word to use, but this is typical of bin Laden's mode of argument. For 'Abd al-'Azīz to have 'betrayed' the Palestinian people at the end of the *jihād*, he would have had to have supported them in the first place. There is no evidence that he did so in the years before World War II. The British had encouraged him to view Palestine as 'a purely British problem', which it was not (although it was a British responsibility); for his part, 'Abd al-'Azīz was indifferent towards the problems of the Palestinians at least until the outbreak of World War II, arguing that 'Palestinians know better their own valleys'. In essence, the problem was irrelevant to his primary concern, the preservation and strengthening of his own realm. This meant countering the threat of his Hashemite rivals, which it seemed that British policy was set on strengthening.[6]

Too late,[7] 'Abd al-'Azīz came to recognize the threat posed by Jewish immigration and wrote on behalf of the Palestine Arabs to Franklin D. Roosevelt in 1357/January 1939, denying the Zionists' 'historic claim' and attacking the Balfour Declaration.[8] By 1364/March 1945, his concerns for the region were full of foreboding, arguing that the creation of a Zionist state would be[9]

> contrary to the Arabs' right to live in their homeland, a right guaranteed to them by natural law established by the principles of humanity which the Allies have proclaimed in the Atlantic Charter [of 14 August 1941][10] and on numerous other occasions…
>
> We state frankly and plainly that to help Zionism in Palestine not only means to endanger Palestine but all neighbouring countries.
>
> The Zionists have given clear evidence of their intentions in Palestine and in all neighbouring countries. They have organized dangerous secret military formations. It would thus be a mistake to say that this was the action of a group of their extremists and that it had met with the disapproval of their assemblies and committees.

> If the Allied Governments wished to see the fires of war break out and bloodshed between Arabs and Jews, their support of the Zionists will surely lead to this result...

President Roosevelt certainly gave 'Abd al-'Azīz reassurances in writing as well as those he gave in person at their meeting on the Great Bitter Lake in Egypt aboard USS *Quincy* in 1364/February 1945;[11] but these proved to be valueless after his death, in the light of President Harry S. Truman's desire to conciliate the American Jewish lobby for domestic electoral reasons.[12] (This lobby has proved to be the most powerful in American politics ever since and acts as a formidable constraint on the independence of American foreign policy towards the Middle East.)[13] The worst that could be said of Sa'ūdī policy was not that it constituted 'betrayal', as bin Laden contends, but that its capacity to influence American opinion was exercised too late and that assurances from a dying President were accepted at face value.

The Palestinian *jihād* of 1355/1936–1358/1939 against British mandate policy

Without a clear foreign protector, the Palestinians were left with a profound sense of grievance, however exaggerated their fears in the view of pro-Zionists.[14] In 1335/June 1936, senior Arab civil servants and judges, both Muslim and Christian, presented an unprecedented memorandum to the British High Commissioner in Palestine:[15]

> The underlying cause of the present discontent is that the Arab population of all classes, creeds and occupations, is animated by a profound sense of injustice done to them. They feel that insufficient regard has been paid in the past to their legitimate grievances, even though those grievances have been enquired into by qualified and impartial official investigators, and to a large extent vindicated by those enquiries. As a result, the Arabs have been driven into a state verging on despair; and the present unrest is no more than an expression of that despair.

In the autumn of the previous year, 'Izz al-Dīn al-Qassām and a small band of followers, possibly in secret cooperation with the Grand Muftī of Jerusalem, but without his specific authorization,[16] had proclaimed a *jihād* ('this is *jihād*', was their cry, 'victory or a martyr's death'). Each member of the organization had to attest to his faith in God and recognize that the revolt was directed against imperialism, aggression, apathy, despotism and oppression. The verses of the Qur'ān on '*jihād* in the Path of God and the homeland' had to be memorized.[17] The attraction of Qassām's ideology was simple and direct:[18]

> The martyr leader called for *jihād* on a religious basis, a *jihād* in the Path of God, for the liberation of the homeland and for ridding the population of oppression. In the notion of *jihād* on a religious basis there are no… problems, no ideological or personal complications, no profundities or alienation; and all that pertains to such a *jihād* is dictated in familiar *āyāt* of the Qur'ān. There was one slogan that encompassed all concepts of the rebellion: 'this is *jihād*, victory or martyrdom', and such a *jihād* is one of the religious duties of the Islamic creed.

'Izz al-Dīn al-Qassām was killed in action after the first Palestinian guerrilla act against British forces in 1354/November 1935.[19] His funeral resulted in mass demonstrations against the British government; a band of followers, known as the Brethren of al-Qassām (*Ikhwān al-Qassām*) was formed.[20] (Nearly six decades later, the recognition of his 'martyrdom' as the first in the Palestinian resistance struggle resulted in 1411/1991 in the newly-formed military wing of the resistance unit Hamas calling itself the 'Izz al-Dīn al-Qassām Brigades.)[21] Al-Qassām's followers became the key commanders throughout the period of the revolt.

A Society for the Protection of al-Aqṣā Mosque had already been formed in 1347/October 1928, and there is no doubt that protection of the mosque formed part of the religious propaganda surrounding the *jihād* and a means of stimulating the fervour of the Arab community.[22] (The British had themselves observed that Muslim, rather than Jewish, control of the Holy Places in Jerusalem was likely to be more acceptable to Christians as well as Muslims: Islām recognized the two earlier monotheistic traditions, while Judaism did not recognize the later two.)[23]

After the outbreak of violence in 1355/April 1936 and the declaration of a state of emergency by the government, the Palestinians established an Arab Higher Committee to oversee the movement, especially a general strike or policy of non-cooperation, the missing element in the past – *jihād waṭani*, or patriotic struggle.[24] The Muftī issued statements against partition and urged other Arab leaders to 'work for rescuing the country from Imperialism and Jewish colonization and partition'.[25] A declaration of martial law followed the assassination of Lewis A. Andrews, the Acting District Commissioner of the Galilee, who was sympathetic to the Zionists, in 1356/September 1937. The rebels' coordinating institutions were dissolved and the Muftī was forced to flee abroad to Lebanon, where the French placed him under house arrest.[26] He became as uncompromising towards the British as he had been towards the Zionists. He was joined in exile by other prominent leaders, notably 'Izzat Darwaza, who established a central rebel headquarters at Damascus known as the Central Committee of the *Jihād*. This Committee, also known as the General Command of the Arab Revolt, proclaimed:[27]

> The fighters (*mujāhidīn*) have sold themselves to Allāh and they have set out in [obedience] to Him, only in order to strive for His goal, for the *jihād* in His way… They try to get ahead of one another [in hurrying] to the battlefield of *jihād* and martyrdom, in order to support what is right, to establish justice and to defend their noble community (*ummah*) and their holy country… We call upon any Muslim [or] Arab to set out for *jihād* in the way of Allāh and to help the fighters in defending the holy land.

There is little doubt, however, that the Qassāmite leadership of the rebellion in Palestine itself operated largely autonomously from the rebel headquarters-in-exile. It also instituted an Islamic court which issued rulings based on the *sharī'ah*.[28]

The rebels took some important towns during the summer and autumn of 1357/1938 – notably Jaffa, parts of Nablus and the Arab sectors of Haifa and Jerusalem. Civil administration in Palestine had more or less broken down by this time, as a result of the rebel gains and the cooperation of civil servants, voluntary or coerced, with the rebellion.[29] Nonetheless, the Arab revolt was largely a Muslim and a rural phenomenon, strongest in locations adjacent to the main areas of Jewish settlement, and also anti-Christian in ideology and practice.[30] British troops numbering some 20,000 men were bogged down in Palestine, succeeding only slowly in suppressing the rebellion as the outbreak of World War II loomed.[31] Apart from the agitation for Home Rule in India, the Arab Revolt in Palestine provided the most formidable threat to British colonial power before World War II. The cost to the Palestinians themselves was extremely heavy, however: 5000 dead, 15,000 wounded, 5600 detained during the revolt; up to a quarter of the casualties were inflicted by Arabs upon Arabs in what had descended in some areas into factional rivalry and civil war.[32]

The British government sought to head off the discontent in Palestine with the publication in 1358/May 1939 of a White Paper, a pro-Arab policy statement which was nevertheless rejected by the Muftī in exile in the Lebanon, and following this, by the Arab states which had previously appeared to accept the British position.[33] The White Paper contended that the British government 'could not have intended' when framing the Balfour Declaration that Palestine should be converted 'into a Jewish state against the will of the Arab population of the country'. Churchill's view in the White Paper of 1340/June 1922 was quoted, that the British government had not contemplated 'the disappearance or the subordination of the Arabic population, language or culture in Palestine'.[34] Palestine was not to be converted into a Jewish national home, but a Jewish national home was to be found in Palestine. A future independent Palestinian state, to be founded within ten years, was to have a shared Arab and Jewish participation in its government. After five further years of limited migration, no further Jewish migration would be allowed unless the Arabs of Palestine were prepared to acquiesce in it. In 1358/July 1939, David Ben-Gurion, the Head of the Jewish Agency, denounced the 'disastrous policy' proposed by the British and announced that the Jews would not be 'intimidated into surrender even if their blood is shed'.[35]

The German foreign minister, Joachim von Ribbentrop, speaking in 1358/October 1939 at Danzig, called the failure to create immediately an 'independent Arab state' an example of British perfidy.[36] This represented something of a volte-face, because prior to 1939 the Nazis had consistently rejected Palestinian calls for financial and material assistances against both the British and the Zionists. Germany had remained largely indifferent to the ideals and aims of Arab nationalism; though there were suspicions that German money and weapons were flowing into Palestine in support of the Muftī and the Arab cause, this cannot be proven from the extant documentation. It was not until 1358/July 1939 that Germany reached agreement with Sa'ūdī Arabia on the supply of arms which would then have been transferred secretly to Palestine. This agreement became inoperative with the outbreak of World War II. It was, in any case, too late to assist the Palestinian rebellion, which subsided after the publication of the White Paper in 1358/May 1939.[37]

Ribbentrop's statement would have had no significance but for the attitude of the Grand Muftī of Jerusalem, al-Ḥājj Amīn al-Ḥusayni, who had been expelled from Palestine by the British in 1356/1937 after expressing his solidarity with Germany, asking the Third Reich to oppose the establishment of a Jewish state, help stop Jewish immigration to Palestine, and provide arms to the Arab population. He moved to Nazi Germany in 1941,[38] where he stayed until the end of the war. In 1360/May 1941, on the way to Germany, he issued a *fatwā* in Italy calling for *jihād*,[39] which coincided with a short-lived but important[40] pro-Nazi rebellion in 'Irāq led by Rāshid 'Alī al-Gaylani:[41]

> I invite all my Muslim brothers throughout the whole world to join in the *jihād* [for Allāh's cause], for the defence of Islām and her lands against her enemy. O Faithful, obey and respond to my call. O Muslims! Proud 'Irāq has placed herself in the vanguard of this Holy Struggle, and has thrown herself against the strongest enemy of Islām certain that God will grant her victory…
>
> The vivid proof of the imperialistic designs of the British is to be found in Muslim Palestine which, although promised by England to Sharīf Hussein has had to submit to the outrageous infiltration of Jews, shameful politics designed to divide Arab–Muslim countries of Asia from those of Africa. In Palestine the English have committed unheard of barbarisms; among others, they have profaned the al-Aqṣā Mosque and have declared the most unyielding war against Islām, both in deed and in word… England, adhering to the policy of Gladstone, pursued her work of destruction to Islām depriving many Islamic States both in the East and in the West of their freedom and independence.

The number of Muslims who today live under the rule of England and invoke liberation from [that] terrible yoke exceeds 220 million.

Therefore I invite you, O Brothers, to join in the *jihād* [for Allāh's cause] to preserve Islām, your independence and your lands from English aggression. I invite you to bring all your weight to bear in helping 'Irāq that she may throw off the shame that torments her. O Heroic 'Irāq, God is with Thee, the Arab Nation and the Muslim World are solidly with Thee in Thy Holy Struggle!

The Grand Muftī's hopes of 'Irāq were misplaced, because the British withstood the siege of Habbaniya and were able to retake control of the country relatively rapidly. In retrospect, his *jihād* declaration appears to have been as ineffective as that issued by Ottoman Turkey in 1332/1914, chiefly because the Arab states preferred to stand by the British, as a known quantity, rather than jump ship as the Muftī had done.

The Muftī gained an audience with Hitler in 1360/November 1941, in which he sought to convince him that the Arab cause was the natural ally of the Third Reich. The Arabs, he claimed, were striving for the independence and unity of Palestine, Syria and 'Irāq and the prospects for the formation of an Arab Legion in the Nazi cause were good. Germany, he argued, held no Arab territories as colonies and understood and recognized the aspirations to independence and freedom of the Arabs, just as she supported the elimination of the Jewish national home. Hitler, however, made no open commitment to Arab independence, though secret undertakings were given;[42] but the Muftī was given the title of 'Protector of Islām' in Nazi-occupied Bosnia. In 1362/February 1943, Hitler ordered the creation of the Nazi SS *Hanzar* (or *Handschar* in German)[43] Division and approximately 100,000 Bosnian Muslims volunteered for a campaign that would lead to serious accusations of war crimes by the end of the war.[44] In 1363/January 1944, the Muftī made a second visit to the *Hanzar* Division and spent three days with it, before it departed by rail from Germany to Bosnia. His speech made clear the basis of a Muslim alliance with Nazi Germany:

This division of Bosnian Muslims established with the help of Greater Germany, is an example to Muslims in all countries. There is no other deliverance for them from imperialistic oppression than hard fighting to preserve their homes and faith. Many common interests exist between the Islamic world and Greater Germany, and those make cooperation a matter of course. The Reich is fighting against the same enemies who robbed the Muslims of their countries and suppressed their faith in Asia, Africa, and Europe…

Friendship and collaboration between two peoples must be built on a firm foundation.[45] The necessary ingredients here are common spiritual and material interests as well as the same ideals. The relationship between the Muslims and the Germans is built on this foundation. Never in its history has Germany attacked a Muslim nation. Germany battles world Jewry, Islām's principal

> enemy. Germany also battles England and its allies, who have persecuted millions of Muslims, as well as Bolshevism, which subjugates forty million Muslims and threatens the Islamic faith in other lands. Any one of these arguments would be enough of a foundation for a friendly relationship between two peoples… My enemy's enemy is my friend…

However, it was not just Britain but the whole Allied cause which the Muftī set in his sights. In a radio broadcast from Berlin in 1363/March 1944 he denounced American policy with regard to the establishment of a Jewish homeland: 'no-one [would have] ever thought', he thundered, that 140 million Americans[46] 'would become tools in Jewish hands…' How could the Americans dare to Judaize Palestine? American intentions were now 'clear', he claimed, and amounted to the establishment of a Jewish empire in the Arab world.[47] The Muftī had gambled on an Axis victory in World War II and lost,[48] which clearly did not help the Palestinian cause, itself already divided.

However, Netaji Subhas Chandra Bose had established an Indian National Army to fight in the cause of Japan yet this did not hinder India's path to independence in 1947; but Bose had conveniently died in 1945.[49] The Muftī's longevity was unhelpful to the Palestinian cause, since he remained an embarrassment, even if he was forced to live in exile abroad. Nevertheless, there are other, more important, factors explaining the demise of the Palestinians as a political force. Chief among these was the simplest, but most powerful, explanation: political and military defeat. Ben-Gurion had described 'the great conflict' in 1355/May 1936: 'we and they want the same thing: we both want Palestine. And that is the fundamental conflict.'[50] As a member of the Israeli foreign ministry complained, Ben-Gurion sought 'to solve most of the problems by military means, in such a way that no political negotiations and no political action would be of any value'.[51] Such inflexibility had the advantage of working to Israel's advantage once the state was established: borders, refugees and Jerusalem could all be resolved in time in Israel's interest. Delay brought Israel benefits, 'as the Muftī helped us in the past'.[52] On another occasion, Ben-Gurion stated sarcastically: 'we are not contractors for the construction of an independent Palestinian state. We believe this is a matter for the Arabs themselves.' Peace with the Arabs was a distant third in the list of priorities below the defence of Israel's interests and the nurturing of its relationship with American Jewry (and with the USA since that was their place of residence).[53]

From disaster (*al-Nakbah*) and dispersal to the first military operations, 1367/1948–1374/1955

The unilateral proclamation of the independence of Israel in 1367/May 1948 brought into the open two irreconcilable theses concerning nationhood. For the

Palestinian Arabs, the claim was that of a population to determine the fate of a country which they had been occupying for 'the best part of 1,500 years' (Lord Curzon's phrase in his memorandum of 1336/October 1917).[54] For the Arab Palestinians, 'this right of immemorial possession' was inalienable. It could not be altered by the circumstances of a 400-year Ottoman occupation (1517–1917), the British mandate or a 'Jewish state' established by brute force in 1948.[55] For the Israelis, the vision of a national revival in their own country had commenced with the First Zionist Conference of 1315/August 1897 and the vision of Theodor Herzl. It had been acknowledged in the Balfour Declaration of 1336/November 1917, and enshrined in the circumstances of the Holocaust, which had led survivors to reach 'the land of Israel' (*Eretz Israel*) 'in face of difficulties, obstacles and perils', as well as enshrined in the sacrifices of Jewish soldiers on the side of the Allies during World War II. Finally, the United Nations General Assembly in 1367/November 1947 had adopted Resolution 181 (II) requiring the establishment of a Jewish state in Palestine. This right was stated as 'unassailable': 'it is the natural right of the Jewish people to lead, as do all other nations, an independent existence in its sovereign state'. [56] Abba Eban, Israel's foreign minister at a later date, argued in *The Case for Israel* (1967) that 'there is no greater falsehood in history than that the Arabs are the sole, legitimate heirs to the lands of an Israel that once was Palestine and before that was Canaan'. Four thousand years of interrupted history thus had to take precedence of almost 1500 years of uninterrupted occupation.[57]

The Disaster (*al-Nakbah*) for Palestinians was of a three-fold character. The first disaster was military defeat. The Muftī of Jerusalem and the Arab League had proclaimed *jihād* following the announcement of the United Nations partition plan. A united state would have been set up had the invasion of Israel succeeded.[58] The Arab states declared, in response to the unilateral proclamation of the independence of Israel:[59]

> The Governments of the Arab States emphasize, on this occasion, what they have already declared before the London Conference and the United Nations, that the only solution of the Palestine problem is the establishment of a unitary Palestinian State, in accordance with democratic principles, whereby its inhabitants will enjoy complete equality before the law, (and whereby) minorities will be assured of all the guarantees recognized in democratic constitutional countries and (whereby) the holy places will be preserved and the rights of access thereto guaranteed…

The injustice of the partition scheme seemed clear to the Arabs: the Jews owned less than 10 per cent of the land, were less than one-third of the population, yet were to be awarded 55 per cent of the land area of Palestine.[60]

According to the United Nations Conciliation Committee for Palestine, the Muslim population in 1364/April 1945 in the areas held by the Israeli Defence Army (IDA, later the Israeli Defence Force, IDF) in 1368/May 1949 had totalled 621,030 (the total non-Jewish population, including Christians and others was 726,800).[61] It is generally agreed that between 600,000 and 760,000 Palestine Arabs became refugees between 1367/December 1947 and 1368/September 1949. Many of the 150,000 or so Arabs who remained at first inside the area controlled by the IDA became refugees between 1948 and 1949; thus, by 1369/1950, the number of refugees had swollen to 930,000.[62] The number of Palestinian refugees who are cared for by UNRWA (the United Nations Relief and Works Agency for Palestine Refugees in the Near East) has since grown to 4.1 million.[63] In terms of material losses, valued in US$ at 1984 prices, these have been estimated at $147,000 million.[64] Given the scale of the potential demand, it is scarcely surprising that Israel has not been prepared to consider compensation claims, though Jewish claims against Germany have been assiduously pursued.[65]

In the early years of the state of Israel, the highest number of refugees that Israel might have been willing to allow to return was 100,000, which the Arabs considered wholly inadequate.[66] Ben-Gurion and other proponents of 'the iron wall' doctrine of Israeli defence, could not contemplate a multi-ethnic society, comprising both Jews and Arabs in the same country. At the funeral of a younger farmer murdered by Arab insurgents in 1375/April 1956, Moshe Dayan, another proponent of the 'iron wall' defence doctrine, in an astonishingly frank admission, eschewed any thoughts of conciliation of the refugees:[67]

> What cause have we to complain about their fierce hatred for us? For eight years now, they sit in their refugee camps in Gaza, and before their eyes we turn into our homestead the land and villages in which they and their forefathers have lived… Let us not be afraid to see the hatred that accompanies and consumes the lives of hundreds of thousands of Arabs who sit all around us and await the moment when their hand will be able to reach our blood. Let us not avert our gaze, for it will weaken our hand. This is the fate of our generation. The only choice we have is to be prepared and armed…

The third, and in some ways most depressing, aspect of *al-Nakbah* for the Palestinians was the almost complete lack of concern of the other Arab states for their fate after 1369/1950. They had lost their homeland, even in the areas to which they had been evicted. Egypt denied the Palestinians access to Egyptian nationality and for the most part ignored the refugee population in the Gaza Strip. From Israel's point of view, as expressed by Ben-Gurion, there were no grounds for any quarrel between Israel and Egypt since 'a vast expanse of desert stretches between the two countries and leaves no room for border disputes'.[68] By definition, there could be no Palestinian presence between the two states according to such a viewpoint. In contrast, the Jordanian government sought to integrate the Palestinians into the life of the nation and offered them citizenship; but then the enlarged Parliament in 1369/April 1950 confirmed the union between Jordan and Arab Palestine proposed by King ʻAbdullāh – in effect, the annexation of such territory formerly belonging to the Palestinians that was not already under the control of Israel. It was scarcely surprising that ʻAbdullāh was murdered in a Palestinian plot at the al-Aqṣā Mosque in 1370/July 1951.[69]

One account suggests that about 400 Israelis were killed and some 900 wounded by Palestinian armed fighters ('those who sacrifice themselves', *fedayeen,* or *fidā'iyūn*, after *fidā'ī*, 'redemption' or self-sacrifice in a cause) between 1368/1949 and 1376/1956. However, both Egyptian and Israeli accounts agree that the raids intensified only after 1375/August 1955: it was the Israeli raid on Gaza that prompted Nasser's change of policy.[70] At the end of that month, Nasser had stated:

> Egypt has decided to dispatch her heroes, the disciples of Pharaoh and the sons of Islām and they will cleanse the land of Palestine... There will be no peace on Israel's border because we demand vengeance, and vengeance is Israel's death.[71]

The *fedayeen* attacks were justified by the Committee of *fatāwā* of no less prestigious a body than the University of al-Azhar in Cairo in a *fatwā* issued in 1375/January 1956.[72] This declared that the conclusion of peace with Israel was not legally permissible because it would represent 'the acceptance of the [right of the] usurper to continue his usurpation… [or] the aggressor to continue his aggression'. It continued:

> Peace should not be made in any way that will help them [the Israelis] to stay as a state in the sacred Muslim land. It is incumbent upon Muslims to help each other irrespective of their language, colour or race so that this land is returned to its people, and the al-Aqṣā Mosque, the place of heavenly revelation and the place of worship of the Prophet, whose surroundings have been blessed by Allāh, is protected and the Muslim shrines are protected from these usurpers. It is imperative on [Muslims] to help the *mujāhids* and all other forces to wage *jihād* in this way and to spend all that is possible for them until the land is purged of these aggressor[s]…
>
> In the rule of Islām, whoever shirks this duty or leaves the Muslims without help or calls [for] what may disunite them or help their enemies against them, leaves the community of Muslims and commits the greatest of sins…

Disunity and defeat of the 'state within the state': the Palestinian movement from the Suez–Sinai War of 1376/1956 to the October War of 1393/1973

After the Suez–Sinai War of 1376/1956, Egyptian forces did not re-enter the Gaza Strip. Israel was thus free from *fedayeen* attacks initiated from Gaza between 1376/March 1957 and 1387/May 1967.[73] The later founders of Fateh (Fatḥ) were disturbed by the ability of the Egyptian authorities to order a complete halt to such attacks following the Israeli evacuation in 1376/March 1957 and concluded that independent Palestinian action should be paramount among their concerns, an attitude which was reinforced by Nasser's relatively half-hearted calls for a Palestinian 'entity' but lack of any plan for the liberation of Palestine. The belief that Arab governments sought deliberately to suppress Palestinian identity was central to the thinking of Fateh.

> There was no alternative for the Palestinians but to go underground and adopt absolute secrecy in their organization, until it could impose itself on th[e Arab] reality and force recognition. This... was the real expression of the aspirations and experience of the vast majority of our [Palestinian] communities.[74]

Nasser was instrumental in the establishment of the Palestine Liberation Organization (PLO), which was set up under the auspices of the Arab League in 1384/May 1964, with a constitution which declared the partition of Palestine in 1947 and the establishment of the state of Israel as entirely illegal,[75] denounced Zionism as expansionist, racist and fascist[76] and committed the PLO to the use of force if required:

> We, the Palestinian Arab people, who believe in its Arabism and in its right to regain its homeland, to realize its freedom and dignity, and who have determined to amass its forces and mobilize its efforts and capabilities in order to continue its struggle and to move forward on the path of *jihād* until complete and final victory has been attained...

But the PLO did not control all the separate Palestinian guerrilla organizations, least of all Fateh (the Palestine National Liberation Movement), which had been in existence since 1377/1958 although its 130-article constitution was formulated only in 1384/1964.[77] Syria regarded the PLO as 'nothing more than a cat's-paw for Nasser' and accordingly was the only Arab country to give Fateh free rein to operate from its borders. The first guerrilla raid took place in 1384/January 1965.[78] Jordan was kept busy arresting Fateh activists, though the number of incidents on the Jordanian border continued to increase.[79] A series of provocations, many of them instigated by Israel according to later testimony of Moshe Dayan, eventually led to the escalation of hostilities that became the Six Day War (1386/June 1967). Israel rapidly occupied the entire Sinai peninsula, the West Bank and the Golan Heights. Jerusalem was proclaimed united: 'we have returned to our holiest places, we have returned in order not to part from them ever again', Dayan declared. Fortunately, the proposal of General Shlomo Goren, the chief rabbi of the IDF, to demolish the al-Aqṣā Mosque with explosives was disregarded.[80]

The Khartoum Arab Summit resolutions issued in 1387/September 1967 emphasized the need for the 'unity of Arab ranks', joint action and coordination. All efforts were to be concentrated on eliminating the 'effects of the aggression on the basis that the occupied lands are Arab lands and that the burden of regaining these lands falls on the Arab States'.[81] The guerrilla movement developed apace, with several hundred attacks on Israel each month by 1389/1969 and a formal sanctuary created for guerrilla forces in Lebanon that year. Yezid Sayigh comments: 'the Cairo agreement was to provide the formal basis for Palestinian–Lebanese relations for at least fifteen years, although it was to be observed more in the breach than the rule'. The Maronite Christians condemned the agreement, but in practice had to accept it in order to avoid civil war. The Palestinians began to emphasise the ethos of martyrdom (*shahādah*) as providing their revolution with particular meaning.[82]

The crushing defeat of the Arab states in the Six Day War freed Fateh from what it saw as their 'oppressive control' of the movement. A relaunch of the guerrilla campaign from within the occupied territories, now called a 'popular liberation war',[83] was agreed on at a meeting of the higher central committee in Damascus only a few days after the war. The fear of an Israeli withdrawal from the occupied lands, after a peace settlement from which the Palestinians would be excluded, was crucial to their thinking. When combat operations were resumed in 1387/August 1967, the guerrilla 'fish' had a 'sea' of some 666,000 Palestinians in the West Bank and another 400,000 in Gaza in which they could swim. Hopes of a new Palestinian revolt on the model of the *jihād* of 1355/1936–1358/1939 were misplaced as a result both of a lack of support and the effectiveness of the Israeli counter-insurgency effort.[84] Yezid Sayigh is clear that 'Palestinian hopes of organizing an armed uprising had been completely shattered by the end of 1967'. Failure had 'far-reaching implications. The centre of gravity in Palestinian nationalism moved into exile… the balance was not to shift significantly until the eruption of the *intifāḍah* in December 1987, twenty years later.'[85]

The battle of Karāma of 1387/March 1968 (*karāma* means 'honour' in Arabic), in which Jordanian and Palestinian forces fought an Israeli invasion, is still depicted in the Jordanian press on each anniversary as a Jordanian victory, one of the symbols of Jordan's modern nationalism and her *first* war of independence. The Palestinians, on the other hand, portray it as their victory. Fateh lost 92 dead (compared to 61 dead in the Jordanian army) and the PLF/PLA (Palestine

Liberation Front/Palestine Liberation Army) lost 24 dead.[86] The battle of Karāma turned the disparate Palestinian guerrilla groups into a mass movement.[87]

Nevertheless, guerrilla attacks mounted from the Gaza Strip were four times or so more numerous than those from Jordan. Some 48 Palestinians were killed and 897 wounded in operations against Israel between 1967 and 1970.[88] Palestinian groups were prepared to target Israeli civilians but in their rhetoric tried to distinguish between the physical destruction of Jews and the elimination of Zionism as a political ideology. They wanted to create a climate of chronic insecurity so as to encourage reverse immigration. Moreover, they claimed that the Israeli military reserve system meant that civilians were no more than 'military personnel in civilian clothes'. 'External operations' such as the high-jacking of passenger aircraft were justified on the grounds, as George Habash of the Popular Front for the Liberation of Palestine (PFLP) put it, that 'the world has not heard, for over half a century, the appeals of justice and international law'. Our people, 'have lived as refugees in tents of misery for twenty years, and so we must fight for our rights'. For his part, Yasser Arafat (Yāsir 'Arafāt) declared in 1388/March 1969 that the PLO (which had been taken over by Fateh and had just elected him as chairman) categorically opposed and rejected such attacks on aircraft, 'for they come at a time when we are making world-wide political gains'.[89] (In reality, Fateh was being challenged by the emergence of a new guerrilla group, Islamic Fateh or *Fatḥ al-Islām*, one of whose members was 'Abdullah 'Azzam, later bin Laden's ideological mentor.) Indeed, by this year there were more than nine Palestinian guerrilla groups.[90] Fateh therefore had to proclaim its own commitment to *jihād*, which, it claimed, washed away the shame of the defeat in the 1967 war 'with the blood of our martyrs… Our pledge to you and to God is to bear arms until victory or martyrdom.'[91] '*Jihād*', one follower said, 'is just another word for the Palestinian revolution (*thawrah*)… It does have a meaning in religion, but we don't use it that way… our whole life is now a struggle (*jihād*).'[92] Some even coined the phrase '*jihād* in the path of the revolution'.[93]

The Palestine National Charter of 1388/July 1968 had proclaimed the struggle as one of self-determination (article 9) which was defensive in nature, in terms of international law (article 18):[94]

> armed struggle is the only way to liberate Palestine. This it is the overall strategy, not merely a tactical phase. The Palestinian Arab people assert their absolute determination and firm resolution to continue their armed struggle and to work for an armed popular revolution for the liberation of their country and their return to it. They also assert their right to normal life in Palestine and to exercise their right to self-determination and sovereignty over it…
>
> The liberation of Palestine, from an international point of view, is a defensive action necessitated by the demands of self-defence. Accordingly the Palestinian

> people, desirous as they are of the friendship of all people, look to freedom-loving, and peace-loving states for support in order to restore their legitimate rights in Palestine, to re-establish peace and security in the country, and to enable its people to exercise national sovereignty and freedom…

But what was 'normal life' in Palestine–Jordan? Fateh appeared to pose a threat to the future of the Jordanian monarchy. At the very least it was a 'state within a state'. There was talk of an impending *coup d'état* in Jordan to be mounted by the Palestinian groups. The *fedayeen* found themselves on the defensive throughout Jordan against an aggressive Jordanian security structure seeking to reassert Hashemite control. At the urging of the Arab heads of state, Hussein and Arafat signed a ceasefire agreement in Cairo in 1390/September 1970. The agreement called for rapid withdrawal of the guerrilla forces from Jordanian cities and towns to positions 'appropriate' for continuing the battle with Israel and for the release of prisoners by both sides. A supreme supervisory committee was to implement the provisions of the agreement. Hussein appointed a new cabinet but army officers continued to head the key defence and interior ministries. In 1390/October 1970, Hussein and Arafat signed a further agreement in Amman, under which the *fedayeen* were to recognize Jordanian sovereignty and the king's authority, to withdraw their armed forces from towns and villages, and to refrain from carrying arms outside their camps. In return the government agreed to grant amnesty to the *fedayeen* for incidents that had occurred during the civil war. While Jordanian nationalists perceive the confrontation between the Jordanian army and the Palestinian organizations as their second war of independence, for the Palestinians it is regarded as the massacre of 'Black September'. George Habash claimed that the 'battle against the reactionary regime in Jordan' was the 'central battle' faced by the Palestinian resistance movement. There was no difference between King Hussein and Moshe Dayan, he claimed: 'as we act in Israel, so should we act in Jordan'. There should be 'guerrilla war in the mountains and clandestine war in the cities'.[95] The worst that was achieved was the murder in Cairo of Tal, the Jordanian Prime Minister, by the Black September Organization (BSO), in 1392/November 1971.[96] Otherwise, the Palestinian movement could scarcely conceal its disarray following the defeat in Jordan.

Habash's analysis was correct, but for reasons which were not apparent at the time. Simha Dinitz, director-general of the Prime Minister's office during the period that Golda Meir was Israeli Prime Minister, confirms that continuing Israeli–Jordanian dialogue 'prevented the rise of the PLO as the central force in the Palestinian arena. As long as the dialogue continued, the PLO was prevented from becoming the main spokesman of the Palestinians or the most important spokesman'.[97] Yet, to the profound relief of Arafat, Golda Meir obdurately rejected Hussein's federal plan for a United Arab Kingdom in 1392/March 1972, which was seen as an attempt by the king 'to put the PLO out of business'. An

Israeli–Jordanian peace settlement with Israeli withdrawal from the West Bank would have meant the demise of the PLO: 'sometimes I think we are lucky to have the Israelis as our enemies', Arafat told his biographer.[98]

As Abū 'Ubayd al-Qurashī noted in 1422/February 2002 (see Chapter 11), military history offers few parallels for strategic surprise as complete as that achieved by Egypt and Syria in launching the Ramaḍān/Yom Kippur War in 1393/October 1973. Prior to the war, Moshe Dayan had told *Time* magazine that 'there is no more Palestine. [It is] finished', while Ariel Sharon had claimed that Israel was a 'military super-power' which could conquer 'in one week' the area from Khartoum to Baghdād.[99] The PLO had not been consulted about the timing of the October war, had not participated in it, and had had its eye off the target. It had been preoccupied in defending its position during the state of emergency in Lebanon in 1393/May 1973, a rehearsal for the civil war that was to erupt two years later; but the PLO's political standing improved dramatically in the aftermath of the Ramaḍān/Yom Kippur War.[100] The EEC Foreign Ministers, for example, in 1393/November 1973, called for 'recognition that, in the establishment of a just and lasting peace, account must be taken of the legitimate rights of the Palestinians'.[101] In 1394/June 1974 at a meeting of the Palestinian National Council (PNC) Arafat appeared to back away from the PLO hard-line concepts of 'armed struggle' and 'total liberation' and began to favour a more diplomatic approach, though there remained ambiguity in the rhetoric.[102] He was rewarded at the Rabat Conference of Arab Heads of State in 1394/October 1974, which declared the PLO the sole representative of the Palestinian people as well as reaffirming the right of the Palestinians to self-determination and to return to their homeland.[103] In 1394/November 1974, UN General Assembly Resolution 3236 recognized the right of the Palestinian people to independence and sovereignty, accepted the PLO as the sole representative of the Palestinian people, and granted it observer status at the United Nations. It reaffirmed 'the inalienable right of the Palestinians to return to their homes and property from which they have been displaced'.[104] Arafat travelled to New York to address the General Assembly in 1394/November 1974. There he said: 'I have come bearing an olive branch and a freedom fighter's gun. Do not let the olive branch fall from my hand.' His analysis was both an attempt at correcting an Israeli-centred view of history and a prediction that the Israeli-dominated Middle East could not long continue:[105]

> an old world order is crumbling before our eyes, as imperialism, colonialism, neo-colonialism and racism, the chief form of which is Zionism, ineluctably perish. We are privileged to be able to witness a great wave of history bearing peoples forward into a new world which they have created. In that world just causes will triumph. Of that we are confident. The question of Palestine belongs to this perspective of emergence and struggle. Palestine is crucial

amongst those just causes fought for unstintingly by masses labouring under imperialism and aggression.

The General Assembly partitioned what it had no right to divide – an indivisible homeland. When we rejected that decision. Our position corresponded to that of the natural mother who refused to permit King Solomon to cut her son in two when the unnatural mother claimed the child for herself and agreed to his dismemberment. Furthermore, even though the partition resolution granted the colonialist settlers 54 per cent of the land of Palestine, their dissatisfaction with the decision prompted them to wage a war of terror against the civilian Arab population. They occupied 81 per cent of the total area of Palestine, uprooting a million Arabs. Thus, they occupied 524 Arab towns and villages, of which they destroyed 385, completely obliterating them in the process. Having done so, they built their own settlements and colonies on the ruins of our farms and our groves. The roots of the Palestine question lie here. Its causes do not stem from any conflict between two religions or two nationalisms. Neither is it a border conflict between neighbouring states. It is the cause of a people deprived of its homeland, dispersed and uprooted, and living mostly in exile and in refugee camps…

It pains our people greatly to witness the propagation of the myth that its homeland was a desert until it was made to bloom by the toil of foreign settlers, that it was a land without a people, and that the colonialist entity caused no harm to any human being. No: such lies must be exposed from this rostrum, for the world must know that Palestine was the cradle of the most ancient cultures and civilizations. Its Arab people were engaged in farming and building, spreading culture throughout the land for thousands of years, setting an example in the practice of freedom of worship, acting as faithful guardians of the holy places of all religions. As a son of Jerusalem, I treasure for myself and my people beautiful memories and vivid images of the religious brotherhood that was the hallmark of Our Holy City before it succumbed to catastrophe. Our people continued to pursue this enlightened policy until the establishment of the State of Israel and their dispersion…

Arafat gained a remarkable victory at the United Nations the following year, when in one of five resolutions connected with the Decade for Action to Combat Racism and Racial Discrimination, UN Resolution 3379 of 1395/November 1975 equated Zionism with racism, though the voting was far from unanimous: 75 voted for the resolution, 35 against and 32 abstained.[106] More critical, however, was the fate of the draft UN Security Council resolution of 1396/January 1976 affirming the right of the Palestinians to establish a state – predictably, this was vetoed by the United States.[107] In 1397/March 1977 the Palestine National Council rejected any American-sponsored ‘capitulationist settlement’ and affirmed ‘the stand of the PLO in its determination to continue the armed struggle and its concomitant forms of political and mass struggle, to achieve our inalienable national rights'.[108] In 1397/November 1977, the UN General Assembly passed by an overwhelming majority a restatement of its concern that 'the Arab territories occupied since 1967 had continued, for more than 10 years, to be under illegal Israeli occupation and that the Palestinians, after three decades, were still deprived of the exercise of their inalienable national rights'. Trends seemed to be going in the direction of the PLO until the world was turned upside down by the unprecedented visit of President Anwar al-Sādāt of Egypt to Jerusalem and his speech to the Israeli Knesset in 1397/November 1977.

Sādāt was extraordinarily frank in his assessment to the Knesset. The Ramaḍān/Yom Kippur War had succeeded, he remarked, in demolishing one of the walls between Israel and its neighbours, the wall that 'warned us of extermination and annihilation if we tried to use our legitimate rights to liberate the occupied territories'. Yet there remained another wall, 'a psychological barrier between us, a barrier of suspicion, a barrier of rejection; a barrier of fear, or deception, a barrier of hallucination without any action, deed or decision'. It was a 'barrier of distorted… interpretation of every event and statement', a psychological barrier that constituted 70 per cent of the problem. The remaining 30 per cent of the problem required solution, however, and here Sādāt's proposals were clear: ending the occupation of the Arab territories captured in 1967; giving practical expression to the fundamental rights of the Palestinian people and their right to self-determination, including their right to establish their own state; confirming the right of all states in the area to live in peace within internationally-recognized boundaries with appropriate international guarantees; a commitment of all states in the region to administer the relations among them in accordance with the objectives and principles of the United Nations Charter, particularly the principles concerning the non-use of force and a solution of differences among them by peaceful means; and, finally, ending the state of belligerence in the region.[109]

Sādāt had underestimated the lack of imagination of the Begin government and the inability of most Israelis to conceive of peace secured in return for negotiating away territory. It was only when Sādāt appealed to President Carter that, most unusually for an Arab leader, he found a receptive audience. Carter berated Begin for a lack of progress in the search for peace.[110] The Camp David Accords ('A Framework for the Conclusion of a Peace Treaty between Israel and Egypt') were signed in 1398/September 1978.[111] Surprisingly, with the support of the Labour Opposition, Begin was able to carry a clear majority of the Knesset in favour of peace with 'the strongest and largest of the Arab states', in the expectation that it would lead 'eventually and inevitably' to peace with all Israel's neighbours.[112] Sādāt and Begin shared the Nobel Peace Prize for 1978 for their efforts (1398/October 1978).[113] Though denounced by the Arab League,[114] by Syria and the hard-line Palestinian groups (they called it a 'military colonialist alliance, the alliance of Sādāt, Carter and Begin' that precluded a Palestinian state),[115] peace was signed at Washington DC in 1399/March 1979.[116] Sādāt presented the treaty to a referendum of the Egyptian people (which he won by 99.9 per cent!).

These negotiations froze out the Palestinians. It can scarcely be a coincidence that Yasser Arafat chose the year 1398/1978 to make his pilgrimage to Mecca, where he delivered an unequivocal speech:[117]

> Palestine and Jerusalem have never been just a Palestinian question, or solely an Arab question, but they are a problem for every Muslim… The liberation of Jerusalem is a [personal responsibility: *farḍ ʻayn*] upon every Muslim, for it cannot remain [a collective responsibility: *farḍ al-kifāyah*] under today's circumstances… I declare here, from the land of the Prophet, from the cradle of Islām, the opening of the gate of *jihād* for the liberation of Palestine and the recovery of Jerusalem…

In 1401/April 1981, Arafat reaffirmed the maximalist claim that there could be 'no solution, no stability, and no security in the Middle East without the attainment of the inalienable rights of the Palestinian people, including its right of return, self-determination, and the establishment of an independent state, with Jerusalem as its capital'.[118]

Then, in 1401/October 1981, Sādāt was assassinated. The culprits were not Palestinians but Egyptian members of a movement called Islamic Jihād (al-Jamāʻat al-Islāmīyyah or al-Jamāʻat al-Jihād) who sought to establish an Islamic state in Egypt.[119] The willingness of other Arab leaders to take the personal risk of making peace with Israel had been thrown into question: it was another 13 years before King Hussein of Jordan signed a treaty with Israel.[120] Avi Shlaim comments that 'although Sādāt was reluctant to admit it publicly, his peace initiative had not produced the results he had hoped for… there was an uneasy feeling, at least in some quarters in Israel, that Sādāt's vision of comprehensive peace in the Middle East had expired with him'.[121]

Nevertheless, Sādāt was assassinated by radical Islamists in Egypt not because he had made peace with Israel – bad enough though this was from their viewpoint – but because he had failed to bring about an Islamic state in Egypt itself. This was a new turn to Islamic radicalism which coincided with other events pointing in the same direction – the Khomeini Revolution in Iran and the occupation of the central mosque in Mecca in 1400/November 1979 by forces loyal to Juhaiman al-ʻUtaibi, the grandson of an *Ikhwān* warrior, whose charges against King Fahd of corruption, deviation, and dependence on the West echoed his grandfather's charges against ʻAbd al-ʻAziz ibn Saʻūd. The spiritual leader of the movement, al-Qaḥṭānī, was declared the true *mahdī*, the one who guides, and the whole movement had a millenarian and reformist agenda, occurring as did during Ramaḍān and in the season which corresponded to the advent of the year 1400 ME in the Islamic calendar.[122]

The assassination of Sādāt: Faraj and the 'neglected duty' of *jihād*

President Anwar al-Sādāt was a 'believing President' who was constantly seen at prayer, and who initially had been more favourable to Islamist groups than Nasser: his ostentatious display of piety contrasted with Nasser's Arab Socialism. However, he identified the *sharī'ah* only as '*a* source' rather than '*the* source' for the law in the constitutional debates of 1390/1971;[123] ten years later, prior to his assassination, he was moving towards the active suppression of the radical Islamist groupings that had proliferated in Egypt. Muḥammad 'Abd al-Salām Faraj (1371/1952–1402/1982) was executed, along with four accomplices,[124] for the assassination of President Sādāt. Other accomplices received long prison sentences: one of them, Karam Zodhy, was released only in September 2003, claiming remorse for the event and declaring that Sādāt had been a 'martyr'.[125] A former Muslim Brotherhood member who was disillusioned by its passivity, after belonging to a series of radical groups, Faraj founded al-Jamā'āt al-Jihād in 1979.[126] Faraj recruited for al-Jihād in private mosques in poor neighbourhoods where he delivered Friday sermons. Al-Jihād succeeded in recruiting members from the presidential guard, civil bureaucracy, military intelligence, the media and academia. The movement was not led by a single charismatic leader but by a collective leadership, which was in charge of overall strategy, as well as a ten-member consultation committee (*majlis al-shūrā*). Everyday operations were run by a three-department supervisory apparatus. Members were organized in small semi-autonomous groups and cells. There were two distinct branches, one in Cairo and the other in Upper Egypt. The Cairo group was composed of five or six cells headed by *amīrs* who met weekly to plan their strategy. Al-Jihād became involved in sectarian conflicts and disturbances in Upper Egypt and Cairo. After the assassination of Sādāt, al-Jihād supporters fought a three-day revolt in Asyut seeking to spark a revolution before being defeated. Offshoots managed to regroup, declaring *jihād* against Mubārāk's regime; al-Jihād has continued to be linked to terrorist incidents and outbreaks of communal violence ever since. It seems to have a narrow base of support mainly in the urban centres of northern Egypt, and many of its leaders live in exile in Western countries. One wing seems to be loyal to Abbud al-Zammur, one of the original founders, now imprisoned in Egypt. Another wing is called Vanguards of the Conquest or the New Jihād Group and appears to be led by the Afghan war veteran and right-hand man of bin Laden, Dr Ayman al-Ẓawāhirī.

Though proficient neither in traditional Arabic nor Islamic studies, Faraj wrote a short book called *The Neglected Obligation* (*Al-Farīḍa al-Ghā'iba*) in order to explain his views.[127] The Shaykh or Grand Imām of al-Azhar, Jād al-Ḥaqq (1335/1917–1416/1996), in effect Egypt's Muftī, produced a *fatwā* of refutation in 1402/January 1982, the arguments of which are worth considering in this context.[128] The first issue between them was the scope of *jihād* itself. For Faraj, 'in spite of its extreme importance and its great significance' for the future of Islām, *jihād* had been neglected by the '*ulamā*'. There was, he claimed, 'no doubt that the idols of this world can only be made to disappear through the power of the sword'.[129] Faraj was a clear advocate of the view that Islām spread by the sword: it was obligatory, he considered, for Muslims 'to raise their swords under the very eyes of the Leaders who hide the Truth and spread falsehoods'. Far from the duty of *jihād* coming to an end, the Prophet said that 'it continues until the Day of Resurrection'.[130] In contrast, the Shaykh of al-Azhar denied what he considered to be the Orientalists' view that Islām was a religion of the sword. He contended that the concept of *jihād* in the Qur'ān is not confined to *qitāl*, that is fighting, and affirmed the validity of the *ḥadīth* of the greater *jihād,* denying that it had been fabricated as Faraj had contended.[131]

Faraj cited a Qur'ānic verse (Q.5:44) which he claimed justified the militant interpretation that Muslims who governed by man-made laws were not true Muslims: '[those who do] not rule by what God sent down, they are the unbelievers'.[132] In refutation, the Muftī contended that, from the context of the verse, it was evident that the phrase 'what God sent down' refers to the Torah, and not to the *sharī'ah*. The verse addressed the Jews of Medina in the days of the Prophet, not Muslims in the contemporary Arab world. Faraj was guilty of perpetuating an error of the Khārijīs.[133]

The third point at issue was whether Egypt under Sādāt could be considered an Islamic state. For Faraj, it was evident that it could not. Though the majority of the population was Muslim, it was governed by laws that were not truly Islamic: thus 'the rulers of this age are in apostasy from Islām. They were raised at the tables of imperialism, be it Crusaderism, or Communism or Zionism. They carry nothing from Islām but their names, even though they pray and fast and claim to be Muslim.' Such rulers were 'the basis of the existence of Imperialism in the Lands of Islām'.[134] Jād al-Ḥaqq ridiculed Faraj's argument:

> the prayer ceremonies are [carried out], mosques are open… everywhere, religious taxes are paid, people make the pilgrimage to Mecca, and the rule of Islām is widespread except in certain matters like the Islamic punishments, usury, and other things that are contained in the laws of the country; but this does not make the country, the people, the rulers and the ruled apostates, since we believe that God's rule is better.[135]

The next issue which divided the protagonists was the legitimacy of tyrannicide. Faraj's argument was that Sādāt, though born a Muslim, became guilty of apostasy from the moment he started to rule. Islamic law punishes apostasy with death. For Faraj, *jihād* has to be counted as one of the pillars of Islām; the ruler who fails in his twin duties of undertaking *jihād* and commanding right and prohibiting wrong 'destroys himself and those who obey him and listen to him'.[136] In refutation, the

Shaykh of al-Azhar cited two *aḥadīth*, both recounted by Muslim that, when the companions asked whether unworthy rulers should be resisted or fought against, the Prophet affirmed not – 'so long as they perform [or establish] the prayers'.[137] Instead of picking which parts of the Qur'ān in which to believe, the Muftī cited Q.2:85 ('do you believe in part of the Book and disbelieve in part?') in order to propagate a contextual Qur'ānic exegesis.[138]

The fifth point at dispute between Faraj and the Imām was the status of the *āyat al-sayf*, the 'verse of the sword' (Q.9:5). Faraj claimed that this verse 'abrogated 104 verses in 48 *sūrahs*' (see Chapter 1).[139] The experienced jurist who denounced his views managed to avoid the difficult subject of abrogation (*naskh*) altogether, in view of the ending of the verse: 'how can anyone legalize, on the basis of the verse of the sword (Q.9:5)', the Shaykh asked, 'the murder of a Muslim who prays, pays religious taxes and recites the Qur'ān?' The verse was addressed to pagan Arabs who had no treaty with the Prophet. Could such a verse be taken out of context and serve to justify the assassination of a contemporary head of state? Clearly not, in the view of Jād al-Ḥaqq.[140]

The sixth and seventh points of difference may conveniently be taken together. Faraj equates the ruling classes of Egypt with the Mongols (*al-Tatār*) of the thirteenth century and makes use of Ibn Taymīyah's collection of *fatāwā* to demonstrate his argument. While not denying the destructive force of the Mongol invasions, the Muftī accuses Faraj of selective quotation from his sources, especially Ibn Taymīyah (see Chapter 4 and Appendix). Ibn Taymīyah had seen the camp of the Mongols, observed that they did not perform their prayers, saw no muezzin in their camp and no *imām* to lead them in prayer. Was the Egyptian army in the Ramaḍān/Yom Kippur War in 1393/October 1973 really no better than that of the Mongol invaders of the thirteenth century? The troops had fasted, and every army camp had had its temporary mosque and *imām* to lead the troops in their prayers.[141]

The essence of the difference between Faraj and the Shaykh of al-Ahzar lies in the status of *The Neglected Obligation* as a text, whether it is in some respects a restatement of religious doctrine or just a political tract. Faraj asserted the need for restoring the caliphate;[142] the requirement for the activist minority (not a 'broad base') to found the Islamic state;[143] he emphasized the paramount need to establish God's law, 'beginning in our own country', rather than concentrating on the liberation of Jerusalem and the Holy Land, however worthy that aim might be;[144] that the primary enemy was already present among the Muslims, in the form of the proponents of unbelief, injustice and falsehood (*jāhilīyyah*),[145] a theme where the influence of Sayyid Quṭb (see Chapter 8) is manifest;[146] finally, he proclaimed that the *al-jihād al-Islāmī*, the fight for Islām, required at the very least 'a drop of sweat from every Muslim'.[147] Implementation of the *sharī'ah* thus became the sole criterion for the legitimacy of a regime that could be considered truly Muslim.

As Piscatori comments, in response, the Egyptian Muftī, Jād al-Ḥaqq, argued that the particular form of government was dependent on historical circumstance: 'the *ummah* chooses its ruler (*Ḥakīm*) by whatever form of *shūrā* (consultation) is prevalent at a given time.[148] *Amīr*, caliph or president, the exact title is a matter of historical contingency, not theological imperative.'[149] The title of caliph was a technical one in its time, equivalent to President of the Republic in present-day circumstances. To fight as a member of one of the *jamā'āt* outside the official army of the state was the equivalent of not observing the traditional pledge of loyalty from the days of Prophet. To encourage people to disregard science and the acquisition of knowledge in the pursuit of *jihād* was a 'call for illiteracy and primitivism in the name of Islām, which will encourage young people to forget about their studies, both at schools and at the universities'. The Shaykh perceived strong similarities between Faraj's mode of reasoning and that of the Khārijīs in early Islām (see Chapter 2): they regarded other non-Khārijī Muslims, or Muslims who committed sin, as apostates who had to be killed for their apostasy. Finally, was it true, as Faraj contended, that *jihād* was really a neglected or unfulfilled duty? The character of *jihād* had changed because defence of the country and of the faith had become a duty of the regular army; but while calling (let alone killing) other Muslims apostates was no part of the duty, 'to conquer oneself and Satan' could definitely be considered part of the Muslim duty of *jihād*. In this sense, *jihād* was not a forgotten or neglected duty at all.[150]

Faraj criticized other groups for their gradualist strategy and involvement in the political system. He insisted that active, immediate, above all, violent *jihād* was the only strategy for achieving an Islamic state. In tactical terms, Faraj argued that the assassination of Egypt's president (called the 'evil prince' and 'the Pharaoh') would be an effective first step in a revolution that would seize power and establish an Islamic state. Political assassination and other violent acts would mobilize the masses. A necessary assumption was that the people were already on the side of al-Jihād and were just waiting to be shown the proper example and leadership. Once Sādāt had been punished for his alleged apostasy, God would do the rest. Since God would grant success and the infidel regime's fall would miraculously cure all social ills, there was no need to prepare the ground and establish one's strength beforehand. In this sense, there was no 'plan for further action once the assassination attempt had succeeded',[151] because none would be necessary. In terms of its single issue strategy for success, the al-Jihād programme was naive, simplistic and certain to fail in the circumstances of a semi-autocratic state with a strong military and security apparatus. Only a complete prior penetration of this apparatus would have given Faraj's programme a real chance of success, but this would have greatly altered its character: it would have had to bargain with the actual world and enter into compromises in order to seize and then secure power. There is no real evidence, for all his influence on subsequent radical

Islamists whose political impact has been much greater, that Faraj was prepared to consider dealing with the real world rather than abstract ideals.

Operation 'Peace for Galilee': Ḥizbu'llah's legitimation of 'martyrdom operations' in Lebanon

In a booklet entitled *After Two Catastrophes*, published in 1388/1968, Tawfīq al-Ṭayyib argued that while the disaster of 1367/1948 was a defeat for liberal Arab thought, the disaster of 1386/1967 was a defeat for Arab Socialist and revolutionary ideas. 'The disaster in our ideas', he argued, 'preceded the disaster in our land… and was the prelude and long-term cause, of the disaster in the land.' Political Islamism alone was capable of defending the Arab lands against the Zionist presence in Palestine.[152] Islamic Jihād asserted its belief in the 'comprehensive struggle' and contended that the conservative Arab regimes (including such diverse regimes as those of Syria, Libya, Sa'ūdī Arabia, Jordan and 'Irāq in this definition) were the reason for 'the backwardness and defeat[ism] in the Arab world'. The Arab states had in reality become a 'security belt' for Israel, 'the real tools' of Zionism and colonialism: the conservative Arab regimes were hostile to *jihād*, because this concept would reveal the 'falsehood of these governments and their slogans and ideas'. *Jihād* would leave such governments 'naked before the masses'.

Instead of the conservative Arab regimes, the natural ally of the Palestinian revolution, in the world view of Islamic Jihād, was the Iranian revolution. Jordan, in particular, was accused of taking a negative view of the Iranian Revolution and trying to secure the support of the Muslim Brotherhood both in Jordan and the Occupied Territories to adopt an anti-Iranian and pro-'Irāq position. It was prepared even to argue that the continuation of the Iran–'Irāq War would benefit the Palestinian cause: Iran was the state 'most committed to the Palestinian issue'; its victory over 'Irāq, were this to happen, would 'create a new situation in the region on the way to the establishment of the Islamic state'; a newly-founded Islamic state would itself be 'an asset in the battle against Israel'.[153] The Palestinian conflict has had an overwhelming effect on all Arabs, but in the course of the 1980s it further Arabized and politicized the Shī'a; above all, the Shī'a 'giant' in Lebanon was roused to forge an indelible identity by resisting and defeating an Israeli occupation that lasted 20 years.

The price was a heavy one. The friends of the conservative Arab regimes had wide connections. In 1398/August 1978 the highly respected Lebanese Shī'a cleric and founder of Lebanon's Amal Movement, *Imām* Mūsā al-Ṣadr, disappeared in Tripoli, Libya. He was due to meet the Libyan leader Mu'ammar al-Gadhafi on the day of his 'disappearance'. The Libyan government repeatedly claimed that its distinguished Lebanese 'guest' and his two followers, Shaykh Muḥammad Ya'qūb and Abbas Badreddine, had left for Italy. But the claim was unconvincing and before long an Italian court rejected it, ruling that the three men never set foot on Italian soil. Whatever really happened to al-Ṣadr, his disappearance was a good fit with Shīʻa foundational history, paralleling, as it did, beliefs pertaining to the disappearance of the Twelfth *Imām*. The memory of this event was captured in the doctrine of *ghaiba* ('concealment' of the hidden *imām*), also referred to as 'occultation', a messianic belief that calls for the eventual return of the *imām* as a saviour figure (*mahdī*) who will lead the Shīʻa to victory over their enemies (Chapter 9). Mūsā al-Ṣadr's disappearance and presumed martyrdom, following a pattern of martyrdoms stretching back to the very origins of Shīʻism in the deaths of ʻAlī and al-Ḥusayn, proved a powerful galvanizing force among Lebanese Shīʻa militiamen, who on the fifth anniversary of the *Imām*'s disappearance in the summer of 1403/1983 and in the midst of Lebanon's civil war, paralysed West Beirut, the Muslim half of the city. It was not until 1424/September 2003 that Muʻammar al-Gadhafi gave an official acknowledgement that *Imām* Mūsā al-Ṣadr 'disappeared in Libya' during a visit in 1398/August 1978.[154]

Mūsā al-Ṣadr had been the driving force in the creation of Lebanon's Amal militia force in 1394/1975 (Amal is Arabic for hope, and was also the acronym of Afwāj al-Muqāwama al-Lubnaniyyah, the 'Lebanese Resistance Detachments'). His 'disappearance' provided it with new life and helped it to coalesce as a political force. By 1398/1978, though many Shīʻa had earlier joined the anti-status-quo forces, they had tired of militia warfare. In southern Lebanon, there was growing anger amongst Shīʻa civilians at the armed Palestinian forces whose treatment of the civilian population left much to be desired and who exposed the people to Israeli attacks through their military actions. In many respects the PLO–Amal conflict can be seen as a playing out in Lebanon of tensions between ʻIrāq and Iran. Amal argued, in the formulation of Nabih Birri, its general-secretary, that the Lebanese people should not suffer for the rape of the Palestinian people. In other words, there should be no imposed resettlement of hundreds of thousands of Palestinians in Lebanon.[155]

The Israeli invasion of Lebanon in 1402/June 1982 was a momentous blunder, a 'war of choice' that could have been averted, but a decision in which the irrational fears of Prime Minister Begin[156] were played upon by the autocratic Ariel Sharon as Minister of Defence, pursuing his ambitions for a greater Israel enjoying political hegemony in the Middle East.[157] Even the US envoy Philip Habib was appalled at the brutality of Sharon's plans: 'you can't go around invading countries just like that, spreading destruction and killing civilians. In the end, your invasion will grow into a war with Syria, and the entire region will be engulfed in flames!'[158] Habib's prediction with regard to Syria did not come to pass, but the analysis was in other respects correct. He had 'given Arafat an undertaking that his people would not be harmed'. Sharon, however, ignored this commitment entirely. Sharon's word, Habib contended, 'was worth nothing'.[159] Operation 'Peace for Galilee' was a name worthy of the double-speak of the regime that had inflicted the unspeakable suffering of the Holocaust on the Jewish people. 'The war in Lebanon', writes Avi Shlaim,

> was intended to secure Israel's hold over Judea and Samaria. This was not the war's declared aim, but it was the ideological conception behind it... War was not imposed on Israel by its Arab enemies. The war path was deliberately chosen by its leaders in pursuit of power and some highly controversial political gains.[160]

Israel went to war against the PLO to prevent it gaining sufficient political momentum so that serious negotiations would be required[161] and to expel the Palestinians from their place of refuge in Lebanon.

The larger invasion had already been anticipated in 1398/March 1978, when following a PLO attack on a bus in northern Israel and Israeli retaliation that caused heavy casualties, Israel invaded Lebanon, occupying most of the area south of the Litani river. In response, the UN Security Council passed Resolution 425 calling for the immediate withdrawal of Israeli forces and creating the UN Interim Force in Lebanon (UNIFIL), charged with maintaining peace. Israeli forces withdrew later in 1978, turning over positions inside Lebanon along the border to a Lebanese ally, the South Lebanon Army (SLA) under the leadership of Major Saad Haddad, thus informally setting up a twelve-mile-wide 'security zone' to protect Israeli territory from cross-border attack. The invasion was prepared in earnest by an Israeli missile attack on the headquarters of the PLO in Beirut in 1401/July 1981.[162] The *casus belli* for Israel's invasion was the unrelated attack on Shlomo Argov, Israel's ambassador to Britain in 1402/June 1982.[163]

Though UNSCR Resolutions 508 and 509, passed unanimously (that is to say, without the usual US veto) called upon Israel to withdraw from Lebanon to the internationally-recognized boundaries,[164] Sharon urged the IDF on to Beirut, where the PLO was to make its last stand. The objective was to destroy the refugee camps in Lebanon and secure the mass deportation of at least 200,000 Palestinians from the country. In 1402/August 1982, US mediation resulted in the evacuation of Syrian troops and PLO fighters from Beirut. The agreement also provided for the deployment of a multinational force comprising US Marines along with French and Italian units. A new President, Bashir Gemayel, was elected with acknowledged Israeli backing. In 1402/September 1982, however, he was assassinated. The day after the assassination, Israeli troops crossed into West Beirut to secure Muslim militia strongholds and failed to intervene as Lebanese Christian militias massacred almost 2750 Palestinian civilians[165] in the Sabra and Shatila refugee camps. As Shimon Peres remarked in the Knesset:[166]

> You don't have to be a political genius or a decorated general, it's enough to be a village policeman to understand ahead of time that these [Phalangist]

militias – in the wake of the murder of their leader – were more liable than ever to sow destruction, even among innocent people. Is this surprising? Was this something unprecedented?

Ariel Sharon was held 'indirectly responsible' for the massacre by the Kahan Commission in Israel:[167]

> it is our view that responsibility is to be imputed to the Minister of Defence for having disregarded the danger of acts of vengeance and bloodshed by the Phalangists against the population of the refugee camps, and having failed to take this danger into account when he decided to have the Phalangists enter the camps. In addition, responsibility is to be imputed to the Minister of Defence for not ordering appropriate measures for preventing or reducing the danger of massacre as a condition for the Phalangists' entry into the camps. These blunders constitute the non-fulfilment of a duty with which the Defence Minister was charged…

In 1403/February 1983, the Israeli cabinet decided, by a majority of 16 to 1 to accept the report of the Kahan Commission. Sharon remained in the cabinet as minister without portfolio but was replaced as Minister of Defence.[168] Ḥizbu'llah leaders later pointed to these massacres 'undertaken while the camps were supposedly under the protection of General Ariel Sharon's soldiers, as a major factor in their decision to fight the Israelis'.[169]

The Israeli invasion of Lebanon was a staggering defeat for the Palestinian 'state-in-exile' organized by the PLO. It lost at one stroke its territorial base, its HQ and the bulk of its military infrastructure. Syria announced the confiscation of all Fateh assets in 1402/October 1982.[170] The void was filled by Islamist groups. The Sabra and Shatila massacres had not only victimized Palestinians but Shīʿa Lebanese refugees as well, who represented close to a quarter of those slain.[171] Donald Neff argues that, without anticipating it, and certainly without wanting it, the policy of Israel in Lebanon 'created two of its own worst enemies' – the Ḥizbu'llah (or Hezbollah) and (later, and only indirectly) Hamas movements.[172]

Ḥizbu'llah means 'the Party of God' (*ḥizb Allāh*), which is derived from the Qur'ānic verse, 'those who form the party of God will be the victors' (Q.5:56). It was founded with the guidance of ʿAlī Akbar Mohtashemi, Iran's ambassador to Syria, in 1402/June 1982 and modelled on Iran's revolutionary guards. Indeed, it was the dispatch of 1500 revolutionary guards (*Pesdaran*) to the Biqāʿ (Bekaa) valley in that year which played a direct role in the formation of Ḥizbu'llah.[173] Whether they brought with them a specifically Iranian form of *jihād* organization is uncertain, but a 'general structure of the *jihād* organization' was published in Tehran at the end of December 1982.[174] Support came from a number of separate

Shī'a groups and associations, whose merger had by no means been inevitable. 'The Israeli invasion helped these groups think more about coalescing'; without the invasion it is doubtful whether 'something called Ḥizbu'llah would have been born'.[175]

Membership of the party was confined to Shī'a Islamists[176] committed in principle to the establishment of an Islamic state in Lebanon, though this objective was contingent on Iran's outright victory in the Iran–'Iraq war.[177] However, Ḥizbu'llah was committed from the outset to the Qur'ānic injunction that there is no compulsion in religion (Q.2:256) and was entirely realistic about the prospects of creating an Islamic state: it was not Ḥizbu'llah, but those who obstructed its creation, whom God would call to account for the failure to achieve the Islamic state.[178] In 1405/February 1985, Ḥizbu'llah issued a statement of its ideology in a so-called 'Open Letter to the Downtrodden in Lebanon and the World'. This emphasized that the organization did not wish to impose Islām on anybody; nor did it want Islām to rule in Lebanon by force as the fundamentally oppressive Christian Maronite regime had attempted. The 'hypocrisy, oppression and blasphemy' of the Gemayel regime were the consequence of its association with Israel.[179] There was thus no choice but to change the 'rotten sectarian system', and the Lebanese had to be given the opportunity 'to choose with full freedom the system of government they want', without any interference of foreign powers or a stipulated political hegemony of the Christians, as had been the case under the National Pact of 1361/1943. In this discourse, Ḥizbu'llah did not hide its commitment to Islām and the establishment of an Islamic regime, but stressed that it carried no aspiration 'to impose it by force'.

'We are moving in the direction of fighting the roots of vice and the first root of vice is America. All the endeavours to drag us into marginal action will be futile when compared with the confrontation against the United States.' The term 'Israel' does not exist in Ḥizbu'llah's literature: instead the phrase 'Zionist entity' is substituted; this entity was depicted as 'the American spearhead' in the Islamic world and accordingly must be destroyed. Even tacit recognition of the Zionist entity was to be rejected. The conservative Arab states were deemed 'defeatist and under the influence of America', while the UN and the Security Council were considered to be against the oppressed peoples: the right of veto used by the USA to defend Israel should be abolished and Israel should be expelled from the UN The Iranian Revolution was commended in all aspects. Khomeini was depicted as 'the rightly guided *imām* who combines all the qualities of the total *imām*', the man who had detonated the Muslim revolution, and was bringing about the glorious Islamic renaissance.[180]

Political violence was legitimized in moral and religious terms as defensive *jihād*, since its paramount aim was self-preservation.[181] Before we analyse the means, however, we need to consider the priority that Ḥizbu'llah placed on the ends, which represented a complete reversal of the Sunnī priority as defined by Faraj in *The Neglected Duty*. For Faraj, only when society was sufficiently Islamicized and Islamic rule was instituted, could the external enemy be confronted. Conversely, for Ḥizbu'llah, Shī'a-led resistance[182] against Israel took primacy over the confrontation of the Gemayel regime. The external enemy must first be confronted; only then could society be freed and a free people choose an Islamic, or any other, form of rule. The difference in emphasis on 'the resistance priority' was one reason for Ḥizbu'llah's conflict with Amal, the latter party being much more concerned with the attainment of political power in its own right.[183] The liberation of all Palestine and Jerusalem in particular was for Ḥizbu'llah an Islamic, rather than a purely Palestinian, duty; the party declared itself in complete affinity with 'its prisoners of war in Palestine'. Only a 'combative community' (*ummah mujāhidah*), once created, would be capable of defeating Israel. Thus Ḥizbu'llah's *jihād* was both Islamic and Lebanese. The party was first and foremost a '*jihādī* movement', a 'party of the resistance'. Resistance to Israel was 'the priority of all priorities'.[184] Since Israel only understood 'the logic of force', Ḥizbu'llah's political activity served its resistance, not vice versa: 'the Resistance is Ḥizbu'llah and Ḥizbu'llah is the Resistance'.[185]

The lesser *jihād* was seen as contingent on the greater *jihād* but the latter was also dependent on the former. This interdependence is illustrated by the example of a man who actively seeks to avoid the lesser *jihād*, that is, military combat, but claims to pursue the greater *jihād*: he has necessarily failed the greater *jihād* test.[186] For Ḥizbu'llah, the *jihād* with the self is only greater than the military *jihād* in the sense that it is its precondition, but not in an abstract or absolute sense. Therefore both a greater and a lesser *jihād* are required of each and every Muslim. Although this does not apply to offensive *jihād* (since only the Twelfth *Imām* is entitled to wage such a war), it does apply to the defensive *jihād*, which is one of the five articles of the Islamic faith and one of the eight ritual practices (*'ibādāt*) of Shī'a Islām.[187] This religious observance is grounded not only in the logic of self-preservation, but is also a function of the Shī'a historical preoccupation with the rejection of injustice and humiliation. It is against this backdrop that Ḥizbu'llah insists that, as a re-enactment of *Imām* al-Ḥusayn's defensive *jihād* against oppression and his rejection of humiliation, its resistance to Israel's occupation of South Lebanon was not a 'sacred right' which could be relinquished, but a 'religious legal obligation' (*wājib sharī'*), which could not.[188] This obligation remained incumbent upon all believers, even if Israel did not fire a single bullet, because its very occupation was an act of aggression and a form of subjugation, which necessitated a defensive *jihād*. Ḥizbu'llah continues to maintain that 'it is the right and duty of all people' whose land is occupied to 'resist occupation'. They may not be 'required to do what the Islamic Resistance does', but they must make the resistance their 'priority' in the political, cultural and educational fields.

Jihād was not merely conceived as a religious duty but as a divinely-guided, and thereby divinely-rewarded, course of action. It was in this context that the Qur'ānic concept underlying the name of Ḥizbu'llah acquired even greater resonance: 'as to those who turn [for friendship] to God, His Messenger and the believers, it is the party of God that must certainly triumph' (Q.5:56). Perceived this way, all the sacrifices made by the Resistance and the martyrs lost were rationalized, not as fruitless but ultimately worthwhile. *Imām* al-Ḥusayn's martyrdom both served as an exemplar of defensive *jihād* and a model of self-sacrifice. Karbalā' was considered the benchmark against which all acts of martyrdom were measured (see Chapter 9).[189] Martyrdom was associated with all forms of defensive *jihād* that involved the impending possibility of death, which could thereby be classified as instances of self-sacrifice. The possession of a martyrological will (*irādah istishhādiyyah*) is the attribute which the party believes sets it apart from the Israeli enemy.[190] Saad-Ghorayeb concludes:

> since man is doomed to die anyway, it is far more morally and rationally logical that he makes his death a purposive one that serves the cause of God and ensures him a place in paradise. All that is required of him is to hasten the prospect of his death by engaging in a military or political *jihād* against the oppressor, which today is represented by Israel.[191]

From considering the ends, we can then place the chosen means in clearer perspective. In the judgement of Martin Kramer, Ḥizbu'llah

> owed its reputation almost solely to its mastery of violence – a violence legitimated in the name of Islām. This legitimation may be fairly described as Ḥizbu'llah's most original contribution to modern Islamic fundamentalism. Ḥizbu'llah's vision of an Islamic state and society was derivative, but its methods for inspiring and rationalizing violence displayed a touch of genius.[192]

Violence came in two main forms, 'martyrdom operations' or suicide bombings (in Western parlance), and hostage-taking. In the first category were included the 1403/April 1983 suicide attack at the US Embassy in West Beirut (63 dead, called 'the first punishment' in Ḥizbu'llah's Open Letter); the bombing of the headquarters of US and French forces in 1404/October 1983 (298 dead, called 'the second punishment' in Ḥizbu'llah's Open Letter); the assassination of the President of the American University of Beirut, Malcolm Kerr, in 1404/January 1984; and the bombing of the US Embassy annex in East Beirut in 1404/September 1984 (nine dead).

Hostage-taking was confined to approximately 100 Western hostages, but these received enormous media coverage. Of these incidents, over 87, correctly or incorrectly, have been ascribed to Ḥizbu'llah, in spite of its denials.[193] The stories of the victims were truly harrowing, though some hostages were killed in captivity and were unable to recount them; others have written their personal accounts, which are a tribute to their courage as well as an indictment of man's inhumanity to man.[194] Terry Anderson, who spent six years as a hostage – the longest period of captivity of any of the 18 US hostages kidnapped in the latter stages of the civil war in Lebanon – announced after his release that he was to sue the Iranian government for more than $100 million damages, for allegedly sponsoring his kidnappers in Lebanon.[195] The acts of hostage-taking have been seen, in hindsight, as counter-productive even by the captors;[196] but what needs to be stressed is just how prevalent the phenomenon was at the time in Beirut. By 1407/1987 the International Committee of the Red Cross estimated that 6000 Lebanese had been kidnapped and/or had disappeared since 1394/1975.[197] Tim Pritchard, who produced the documentary account of the Beirut phenomenon of hostage-taking in the award-winning film *Hostage*, comments:[198]

> The logistics of kidnapping, imprisoning, guarding, feeding and transporting so many Western hostages within such a small country could only have been possible with the help of a network of hundreds of supporters. The real key to the hostage crisis though was finally unlocked for us, not through our interviews with Hezbollah, nor by the hostages, nor by the western politicians who tried to get them out, but by the reaction of the wider Lebanese population. Few ordinary Lebanese, whether Christian, Muslim or Druze would condemn the taking of western hostages during the [19]80s. Those that lived through those years would instead ask you to consider the facts: a bloody civil war was raging between Lebanon's different religious sects and family clans, thousands of Lebanese were taken hostage, the Israelis invaded, Iran's Revolutionary Guards got involved, the Americans, French, Italians and British sent in troops, the Syrians attacked. Hostage taking was not the knee-jerk reaction of an isolated group, but the response of a population desperate to regain some power in a world that was rapidly spinning out of their control.

Whatever the views of Lebanese at the time, from the point of view of their perpetrators, 'martyrdom operations' and hostage taking still required justification from within the tenets of Islām. The killing of innocent civilians is not permitted in Islām; therefore the only justification for killing certain of the Western hostages – such as Philip Padfield, Leigh Douglas and Alec Collett among the British; Peter Kilburn and William Buckley among the Americans – can only have been (the dubious one) that they were not truly civilians but military or intelligence personnel. Buckley, the former CIA Political Officer Station Chief in Lebanon, died after 15 months in captivity from illness and torture.[199] In 1406/April 1986, the bodies of three American University of Beirut employees, American citizen Peter Kilburn and Britons John Douglas and Philip Padfield, were discovered near

Beirut. The Revolutionary Organization of Socialist Muslims claimed to have 'executed' the three men in retaliation for an American air raid on Libya two days earlier. Six days after the discovery of these bodies, a Beirut newspaper received a videotape film showing a man being hanged. The Revolutionary Organization of Socialist Muslims claimed the man was British citizen Alec Collett, who had been kidnapped more than a year earlier.[200] Terry Waite's imprisonment may have occurred because, although as the envoy of the Archbishop of Canterbury he had successfully negotiated the release of several hostages, he was falsely thought to have been involved in the US 'arms for hostages' scandal with regard to Iran.[201] Clearly the plethora of groups operating in Lebanon, then as now,[202] makes it difficult to argue that a single group – let alone a single foreign sponsor, *pace* Terry Anderson[203] – had overall control of the phenomenon of hostage-taking. Rather, chaos seems to have prevailed. Specific groups such as Ḥizbu'llah, while in principle hostile to hostage-taking on moral and religious grounds, were responsible for this chaos, although the party itself was split, it seems.[204] The best case that can be made for the kidnap policy was that, by harming the West, they were contributing in some measure to the eviction of Israel and the downfall of the Gemayel regime in Lebanon.[205]

'Martyrdom operations' were in a different class from hostage-taking. Americans and pro-Americans could see no justification for the attack on them since they were part of an international peacekeeping force;[206] but this was not how certain sections of Lebanese opinion viewed their presence. American sponsorship of the detested 1403/May 1983 Lebanese–Israel 'peace agreement', which was clearly to Israel's benefit, 'made the Marines the subject of Shī'a execration'.[207] The attack on the US Embassy is in a different category, since it has been alleged that this was carried out by remote-controlled explosion inside the building itself, not as a 'martyrdom operation' from the outside.[208] If so, then this does not square with the memory of Lance Corporal Eddie DiFranco, who could not describe the face of the driver of the truck carrying the bomb, but remembered clearly that he was smiling just before the explosion.[209] There may, of course, have been more than one bomb. Someone had to have authorized such attacks. Ḥizbu'llah acknowledged Khomeini's concept of the Guardianship of the Chief Jurisprudent or Jurisconsult (*wilāyat al-faqīh* or *vilāyat-i faqīh*). Ultimately the attacks took place because Khomeini authorized them. The 'martyrs' at the US Marines' compound

> martyred themselves because the *Imām* Khumanyī permitted them to do so. They saw nothing before them but God, and they defeated Israel and America for God. It was the *Imām* of the Nation [Khomeini] who showed them this path and instilled this spirit in them.[210]

Reports by the intelligence branches of the Lebanese army and the Lebanese Forces (Phalanges), which were leaked to the American press, alleged without evidence that Sayyid Muḥammad Ḥusayn Faḍlallāh,[211] the leading figure among the Lebanese Shīʿa after the disappearance of *Imām* Mūsā al-Ṣadr, had granted prior religious dispensation to the attackers on the eve of their mission – a charge he denied immediately and consistently. While he recognized that he had a following within Ḥizbu'llah, he disclaimed any membership, let alone leadership of the movement; nor was he its 'spiritual guide'.[212] Faḍlallāh consistently denied that he had authorized 'martyrdom operations', but consistently implied that he had the authority to do so if he wished. Faḍlallāh eventually gave them the fullest possible endorsement short of an explicit *fatwā*. 'Sometimes you may find some situations where you have to take risks,' he said later, 'when reality requires a shock, delivered with violence, so you can call upon all those things buried within, and expand all the horizons around you – as, for example, in the self-martyrdom operations, which some called suicide operations.' For Faḍlallāh, the attacks had been 'the answer of the weak and oppressed to the powerful aggressors'.[213] 'Death', he contended, 'is a step that leads to reaching the martyr's goals. That is why the believer, when he achieves self-martyrdom, lives through spiritual happiness.'[214]

Faḍlallāh justified the suicide bombings on the grounds of what we would now call 'asymmetrical conflict'. No other means remained to the Muslims to confront the massive power commanded by the United States and Israel. In the absence of any other alternative, unconventional methods became admissible, and perhaps even necessary.

> If an oppressed people does not have the means to confront the United States and Israel with the weapons in which they are superior, then they possess unfamiliar weapons... Oppression makes the oppressed discover new weapons and new strength every day.[215]

As Martin Kramer argues, 'the method itself redressed a gross imbalance in the capabilities of the competing forces'. Faḍlallāh reasoned:[216]

> When a conflict breaks out between oppressed nations and imperialism, or between two hostile governments, the parties to the conflict seek ways to complete the elements of their power and to neutralize the weapons used by the other side. For example, the oppressed nations do not have the technology and destructive weapons America and Europe have. They must thus fight with special means of their own. [We] recognize the right of nations to use every unconventional method to fight these aggressor nations, and do not regard what oppressed Muslims of the world do with primitive and unconventional means

> to confront aggressor powers as terrorism. We view this as religiously lawful warfare against the world's imperialist and domineering powers.

'These initiatives', he insisted, 'must be placed in their context.' If the aim of such a combatant

> is to have a political impact on an enemy whom it is impossible to fight by conventional means, then his sacrifice can be part of a *jihād*, a religious war. Such an undertaking differs little from that of a soldier who fights and knows that in the end he will be killed. The two situations lead to death; except that one fits in with the conventional procedures of war, and the other does not.

Faḍlallāh, denying he had told anyone to 'blow yourself up', did affirm that 'Muslims believe that you struggle by transforming yourself into a living bomb like you struggle with a gun in your hand. There is no difference between dying with a gun in your hand or exploding yourself.' What is the difference between setting out for battle knowing you will die *after* killing ten of the enemy, and setting out to the field to kill ten and knowing you will die *while* killing them?, he reasoned.[217] Faḍlallāh described the process by which the weak demoralized the strong:[218]

> The Israeli soldier who could not be defeated was now killed, with an explosive charge here, and a bullet there. People were suddenly filled with power, and that power could be employed in new ways. It could not be expressed in the classical means of warfare, because the implements were lacking. But it employed small force and a war of nerves, which the enemy could not confront with its tanks and airplanes. It appeared in every place, and in more than one way. Thus our people in the South discovered their power, and could defeat Israel and all the forces of tyranny.

The weakness of Faḍlallāh's analysis was that it paid insufficient attention to the distinction between military and civilian targets. While the appropriateness and morality of attacking military targets by means of 'suicide operations' might be disputed, they undoubtedly served to create a more level theatre of warfare between a fourth-generation military power and a resistance movement that lacked such resources. Nothing could justify the attack on civilian targets except the specious reasoning, used in Palestine, that Israeli civilians are all potentially members of the IDF because of the military reserve system. More recently, Faḍlallāh has directly confronted the issues of 'martyrdom' operations in Palestine in an interview which was placed on his organization's website:[219]

> ...intellectually speaking, I believe that there is no difference between the martyr in question and the soldier who joins the battle to fight, where he knows that he might be killed by the enemy, but he is obliged to do so because of some critical circumstances and because of his belief of the legitimacy of this battle. So, when a battle needs man to turn to be a weapon in order to exert pressure on the enemies and to kill them, then this will be a part of the mechanism of the battle. Every soldier who joins a battle believing in its legitimacy keeps in mind that he might get killed after turning to a bomb. The problem is in the mentality of the Westerners who fail to consider martyrdom a part of the battle's mobility. Thus there is no difference between the mentality of the soldier who uses conventional methods and the mentality of the martyr (by the martyr operation); for both of them attempt to achieve their goals they believe in...
>
> It is true that *jihād* is not a duty for women, but Islām has, permitted women to fight, if the requirements of a defensive war necessitate a conventional military operation or a martyrdom operation to be carried out by women. Thus, we believe those martyr women are making a new and glorious history for Arab women. We also express our denial of any reservations concerning the martyrdom operations, which have been carried out by women. And we say that the mentioned operations are just like any other operation in the line of *jihād*, because God hasn't defined the mechanism of *jihād* due to the fact that all mechanisms are imposed by the requirements of the battle.
>
> ...we know that the *mujāhidīn* are not targeting the civilians but the occupier in occupied Palestine. In addition, we don't consider the settlers who occupy the Zionist settlements civilians, but they are an extension of occupation and they are not less aggressive and barbaric than the Zionist soldier. At the same time that we confirm the legitimacy of these operations, we regard them among the most prominent evidence of *jihād* in Allāh's way, and we consider any criticism, whether intentional or not, against this type of operation represents an offence against the confrontation movement led by the Palestinian people, including all parties, against the Israeli occupation. Because his battle requires readiness for a maximal state of challenge and confrontation, knowing that the enemy is utilizing all its capabilities and techniques to suppress the Palestinians and to undermine their public and warlike movement. Thus, we have to be well prepared to confront this enemy by all means to make it the first to suffer.

Without the suicide bombers/martyrdom operations in Lebanon, 'we wouldn't have been able to win', Faḍlallāh remarked in 2000, 'but we don't need them any more'. He was one of the first high-ranking Islamic scholars publicly to condemn the 11 September 2001 attacks:[220]

> Nothing can justify the murder of thousands of innocent civilians. No religion justifies such a thing. The Islamic resistance in Lebanon has never killed civilians. All those who were killed were Israeli soldiers!

In 1403/August 1983, Israel withdrew from the Shouf (southeast of Beirut), thus removing the buffer between the Druze and the Christian militias and triggering another round of brutal fighting. By September, the Druze had gained control over most of the Shouf, and Israeli forces had pulled out from all but the southern security zone, where they remained until 1421/May 2000. The virtual collapse of the Lebanese Army in 1404/February 1984, following the defection of most of the Shī'a and Druze units to the militias, was a major blow to the government (around 60 per cent of the army's rank and file were drawn from the Shī'a).[221] In 1404/March 1984 the Lebanese government cancelled the 1403/May 1983 Lebanese–Israel 'peace agreement'; the US Marines departed a few weeks later. Between 1405/1985 and 1409/1989, factional conflict worsened as various efforts at national reconciliation failed. Heavy fighting took place in the 'War of the Camps' in 1985 and 1986 as the Shī'a Amal militia sought to drive the Palestinians from their remaining Lebanese strongholds. The combat returned to Beirut in 1987, with Palestinians, leftists, and Druze fighters allied against Amal, eventually resulting in further Syrian intervention. Violent confrontation flared up again in Beirut in 1988 between Amal and Ḥizbu'llah. The Ta'if Agreement of 1989 marked the beginning of the end of the war. In 1409/January 1989, a committee appointed by the Arab League, chaired by Kuwait and including Sa'ūdī Arabia, Algeria, and Morocco, had begun to formulate solutions to the conflict, leading to a meeting of Lebanese parliamentarians in Ta'if, Sa'ūdī Arabia, where they agreed to the national reconciliation accord in 1410/October 1989.[222] In 1411/May 1991, the militias (with the important exception of Ḥizbu'llah) were dissolved, and the Lebanese Armed Forces began to slowly rebuild themselves as Lebanon's only major non-sectarian institution. In all, it is estimated that more than 100,000 were killed, and another 100,000 left handicapped, during Lebanon's 16-year civil war. Up to one-fifth of the pre-war resident population, or about 900,000 people, were displaced from their homes, of whom perhaps a quarter of a million emigrated permanently. The last of the Western hostages taken during the mid-1980s were released in 1412/May 1992.

The first Palestinian *intifāḍah*: Palestinian peace-making trends and the emergence of Hamas

The outbreak of the Palestinian *intifāḍah*, or uprising, in 1408/December 1987 was completely spontaneous. It took Israel by surprise, but rapidly voices were heard that an iron fist should be applied to smash the rising once and for all. As Avi Shlaim comments, 'the biblical image of David and Goliath now seemed to be reversed, with Israel looking like an overbearing Goliath and the Palestinians with the stones as the vulnerable David'.[223] Images were somewhat deceptive, however, because, as American military analysts subsequently pronounced, the particular form of organization adopted by the Palestinians was ideally suited to the conditions which are now described as 'fourth-generation warfare'. In a work later to be cited by a senior al-Qaeda planner (see Chapter 11), Lieutenant Colonel Thomas X. Hammes wrote:[224]

> ...why the [Palestinian] *intifāḍah* came into being is not as important as what happened – and how it illustrates the fourth generation of war. Most writers on this subject agree on the following sequence of events. (1) The *intifāḍah* started on 9 December 1987 when Palestinian youths took to the streets in riots against Israel occupation forces. (2) Within days of its ignition, the uprising had spread throughout the occupied territories. (3) Within a month, three levels of leadership emerged on the Palestinian side: neighbourhood leaders of 'popular committees', the Unified National Command of the Uprising (UNCU), and finally key Palestinian academics, journalists, and political representatives. All three leadership groups existed before the *intifāḍah* broke out. Yet, by bringing together the street protesters and the three leadership groups, the *intifāḍah* created a unique organization ideally suited to exploit the advantages of fourth generation war. The local neighbourhood networks dealt with grassroots issues – food, water, and medical care. They maintained the morale and effectiveness of the uprising during various attempts by Israeli forces to stamp it out. The UNCU, consisting of representatives of the four main Palestinian nationalist factions, but excluding the fundamentalists, provided overall direction and coordination to the neighbourhood committees.

The Israeli general election of 1409/November 1988 was fought in the shadow of the *intifāḍah*. The Unified Command of the *intifāḍah* issued an appeal to Israeli voters the previous month, which tried to present the Palestinian case for the first time:[225]

> ...instead of your leaders stating the clear truth, they block every initiative aimed at solving our bloody conflict in an honourable and just way. Your leaders consciously distort our blessed popular uprising, its goals and its democratic and peaceful means by increasing their oppressive measures, violating the most basic human rights and international principles which are aimed at protecting women, children and the aged. Your leaders are working on the illusion that they can demoralize and exhaust our people and therefore crush the uprising... the only guaranteed way to peace is by granting the Palestinian people their legitimate rights of self-determination, the right of return, and to establish the independent Palestinian state in the West Bank and Gaza. [Peace

> will be accomplished by a] guarantee of mutual security for all the people in the area…
>
> The slogan of the uprising has… been clear: the need for Israel to withdraw from the territories occupied in 1967 so that we can create our Palestinian state on them…
>
> [The Unified Command reaffirms] its historic responsibility and its determination to continue the uprising until complete Israeli withdrawal from Jerusalem, the West Bank and Gaza is accomplished and the Palestinian people are able to enjoy their right to self-determination, the right to return and to establish the independent Palestinian state.

The Palestinians, it should be noted, did not lay claim to the borders of 1367/1948. In this sense, they accepted, to use a later expression of Danny Rabinowitz's, the 'original sin'[226] of the destruction of their state and the 400 or so villages which were demolished in the aftermath of the shattering of the Palestinian community. Given the election result, the only conclusion to be drawn is that the Israeli voters, while remaining divided on the issue of war or peace, were on balance unconvinced by the Palestinians' appeal. Of the new Israeli Prime Minister Yitzak Shamir, the political commentator Avishai Margalit quipped that he was a two-dimensional man: 'one dimension is the length of the Land of Israel, the second, its width'. He was not a man prepared to give an inch. Shamir reiterated that the *intifāḍah* was not about territory but about Israel's very existence: its suppression was a matter of life and death. To judge from his comments shortly before his electoral defeat in 1412/June 1992, that war was 'inescapable' because without it 'the nation has no chance of survival', Shamir remained trapped in a mindset that nothing had changed since the war of 1948.[227]

In contrast to a stonewalling Israeli government under Shamir, the *intifāḍah* had two contradictory but equally dynamic effects on the Palestinian movement. The first effect was on the PLO, which encouraged a move towards the recognition of Israel as the price for peace negotiations. The second effect was a radicalization and Islamization of the Palestinian movement, which was synonymous with the founding of Hamas. The second effect was to a considerable extent a response to, and rejection of, the new PLO move towards peacemaking.[228]

The Palestine National Council (PNC), the legislative arm of the PLO, issued two statements at Algiers in 1409/November 1988. The first was a declaration of independence. This called Palestine 'the land of the three monotheistic faiths' and committed the new state to maintain Palestine's 'age-old spiritual and civilizational heritage of tolerance and religious coexistence'. Palestine was proclaimed 'an Arab state, an integral and indivisible part of the Arab nation'. It was committed to the principles and purposes of the United Nations, the Universal Declaration of Human Rights and the Non-Aligned Movement. It further defined itself as 'a peace-loving state, in adherence to the principles of peaceful coexistence', but noted that what was required was a permanent peace based upon justice and the respect of rights. It declared itself to be against terrorism, seeking the settlement of regional and international disputes by peaceful means, in accordance with the UN Charter and resolutions.

> Without prejudice to its natural right to defend its territorial integrity and independence, it therefore rejects the threat or use of force, violence and terrorism against its territorial integrity or political independence, as it also rejects their use against the territorial integrity of other states.[229]

In the second, or political, statement, the PNC called the *intifāḍah* 'a total popular revolution that embodies the consensus of an entire nation – women and men, old and young, in the camps, in the villages, and the cities – on the rejection of the occupation and on the determination to struggle until the occupation is defeated and terminated. The *intifāḍah*, it was claimed, had recognized the PLO as the sole, legitimate representative of 'our people, all our people, wherever they congregate in our homeland or outside it'.[230] It proposed two simultaneous strategies. The first was the 'escalation and continuity of the *intifāḍah*'. This would be achieved by providing all necessary 'means and capabilities'; by supporting the popular institutions and organizations in the occupied Palestinian territories; and by bolstering and developing popular committees and other specialized popular and trade union bodies, 'including the attack groups and the popular army', with a view to expanding their role and increasing their effectiveness. The 'national unity' that had emerged and developed during the *intifāḍah*, would be consolidated, it was not clear how. There would be an intensification of efforts at the international level for the release of detainees, the return of those expelled, and 'the termination of the organized, official acts of repression and terrorism against our children, our women, our men, and our institutions'. There would be an appeal to the United Nations to place the occupied Palestinian land under international supervision 'for the protection of our people and the termination of the Israeli occupation'. Finally, there were calls to the Palestinian people outside the homeland, the Arab nation, its people, forces, institutions, and governments, and 'all free and honourable people worldwide' to support 'our *intifāḍah* against the Israeli occupation, the repression, and the organized, fascist official terrorism to which the occupation forces and the armed fanatic settlers are subjecting our people…'

Arafat followed up the two declarations with a second speech at the United Nations in 1409/December 1988, 14 years after his first. Because of opposition from the United States, the meeting had to take place in special session in Geneva.[231] He called the resolution to hold the meeting at Geneva, with the concurrence of 154 states, not a victory over the US decision but 'a victory for international unanimity in upholding right and the cause of peace in an unparalleled referendum. It is also evidence that our people's just cause has taken root in the fabric of the human conscience.' After the usual historical disquisition, Arafat asked a pertinent question of US foreign policy:

> How can the U.S. government explain its stand, which acknowledges and recognizes this resolution [UNSCR 181] as it pertains to Israel, while simultaneously rejecting the other half of this resolution as it pertains to the Palestinian state? How can the U.S. government explain its non-commitment to implementing a resolution which it repeatedly sponsored in your esteemed assembly? – Resolution 194, which provides for the Palestinians' right to return to their homeland and property from which they were expelled, or for compensation for those who do not wish to return…
>
> The rebel's rifle has protected us and precluded our liquidation and the destruction of our national identity in the fields of hot confrontation. We are fully confident of our ability to protect the green olive branch in the fields of political confrontation. The fact that the world is rallying around our just cause to achieve a just peace brilliantly indicates that the world realises in no uncertain terms who is the executioner and who is the victim, who is the aggressor and who is the victim of aggression, and who is the struggler for freedom and peace and who is the terrorist…

The United States replied on the same date, in response to Yasser Arafat's statement accepting UNSCR 242 and 338, which recognized Israel's right to exist and renounced terrorism, that 'substantive dialogue' could be entered into with the PLO.[232]

If the PLO's shift in policy appeared clear, there were powerful voices in the Palestinian movement that were hostile to the initiative, chief among which was the newly formed Hamas organization, which had been established the previous year as an outgrowth of the Muslim Brotherhood. (In Arabic, Hamas is an acronym for Harakat al-Muqāwamah al-Islāmīyyah or Islamic Resistance Movement and a word meaning courage and bravery.) The propaganda of Hamas was directed against the Palestinian National Council resolutions and against the idea of a US–PLO dialogue. The organization's leader, Shaykh Yāsīn, called the declaration of a Palestinian state premature and proclaimed the PLO Charter null and void since it had proclaimed the armed struggle as 'a strategy and not a tactic'.[233] The only course of action in the circumstances in which the Palestinians found themselves was *jihād*:[234]

> The day that enemies usurp part of Muslim land, *jihād* becomes the individual duty of every Muslim. In face of the Jews' usurpation of Palestine, it is compulsory that the banner of *jihād* be raised. To do this requires the diffusion

> of Islamic consciousness among the masses, both on the regional, Arab and Islamic levels. It is necessary to instil the spirit of *jihād* in the heart of the nation so that they would confront the enemies and join the ranks of the fighters.

In contrast to the position of the PLO, which seemed to be prepared to negotiate on the borders of 1967, article 11 of the Hamas Charter defined its position as seeking the return of all Palestine:

> the land of Palestine is an [inalienable] Islamic [land or endowment (*waqf*)] consecrated for future Muslim generations until Judgement Day. It, or any part of it, should not be squandered: it, or any part of it, should not be given up. Neither a single Arab country nor all Arab countries, neither any king or president, nor all the kings and presidents, neither any organization nor all of them, be they Palestinian or Arab, possess the right to do that…

The fear was that the peace process would merely serve to legitimize the 'Zionist entity' – Hamas refused to call Israel by its name; in its view, there could be a truce (*hudnah*) with it, but never a settlement.[235]

Two central issues divided the PLO and Hamas. Firstly, Hamas – and its tactical ally, Islamic Jihād[236] – could not accept the PLO claim to be the sole representative of the Palestinian people and that the *intifāḍah* had enshrined this position. For Hamas, the Qur'ān was 'the sole legitimate representative of the Palestinian people'.[237] By this it meant that without a commitment to Islamic values, the struggle with Israel would not succeed. The other division was equally fundamental: Hamas wanted the PLO to agree that the aim was the establishment of an Islamic state in the whole of Palestine, 'from the river to the sea', that a Jewish entity there was inconceivable, and that *jihād* was the only way to attain this goal.[238] Already by May 1989, Hamas' commitment to violence had led to widespread Israeli arrests, including the leader himself, Shaykh Yāsīn.[239]

In view of the Israeli crackdown, Hamas had to follow the PLO model of external and internal bases of power to protect its overall position: it chose Amman as place of exile, whereas the PLO had chosen Tunis. In the case of both organizations, this process led to disagreements on policy and, overall, weakened the cohesion of the movement. With regard to the PLO, the division between the Palestinians under occupation – who understood the implications of the US–Israeli proposals at Madrid and withstood them, finally leading to a rejection – and the 'Tunisians', the PLO leadership in exile, who misunderstood the legal implications and rapidly gave away at Oslo in 1414/September 1993 'what the Madrid team had struggled not to concede throughout the two years of the Madrid talks',[240] was to have profoundly damaging consequences which were fully revealed seven years later.

The formation of the military wing of Hamas calling itself the 'Izz al-Dīn al-Qassām Brigades in 1991 and the beginning of 'martyrdom operations' the following year led to serious tensions with the PLO. There were rival executions of each other's followers alleged to be 'traitors' and an attempt in 1412/June 1992 to mend fences proved abortive. In a reference to the relationship of the African National Congress to its junior partner, Inkata, Arafat contemptuously dismissed Hamas as a 'Zulu tribe'.[241] A strategic cooperation agreement between Hamas and Iran, forged in 1413/November 1992, was denounced by the PLO in subsequent negotiations with Hamas at Tunis and then Khartoum.[242]

Mishal and Sela conclude that the deportation of 415 Islamic activists by Israel to Lebanon in 1413/December 1992 was a 'milestone in Hamas's decision to use car bombs and suicide attacks as a major *modus operandi* against Israel'.[243] Ḥizbu'llah seized the opportunity to initiate the Hamas leaders in both the techniques of suicide attacks and in its supposed religious justification.[244] The USA responded the following year by declaring Hamas a terrorist organization. News of the Oslo Agreement of 1414/September 1993 and the historic handshake between Arafat and Yitzhak Rabin, the Israeli Prime Minister, in the White House ended the first *intifāḍah*, but left Hamas in a difficult position. There was a threat to its position in Jordan as a result of Israeli and American pressure, an increasingly negative international perception of the organization as a murderous terrorist movement that targeted civilians. Hamas also faced growing Palestinian criticism since it lacked any positive alternative to the peace process, which in the early years had overwhelming support among the Palestinians. Most of the international community supported the peace process and considered Hamas the main threat to its success. Two particularly serious dangers to the radical Islamists arose from the implementation of the Oslo Accords. Hamas was inferior in military terms to the Palestinian Authority's police and security apparatus, while its infrastructure was dependent on external financial resources which might be curtailed by the Palestinian Authority directly or indirectly.[245]

On 5 May 1994, Hamas issued a statement on the PLO–Israel Agreement to the effect that the 'enemy plans to use us [that is, the Palestinian people] as his tools in order to prolong his presence and achieve hegemony over our Arab and Islamic lands'. The Palestinian self-rule authority would be 'put under trial from the very first moment of its existence. Their performance will determine the fate of our people.' It must not serve the interests of the occupiers. For their part, Hamas would not aim their guns at their own people and would continue to protect the unity of the people. The dangers for the Palestinian people were considered immense:[246]

> We are passing through a very crucial stage of our history. It requires strengthening of our national efforts for its protection on the way to resistance till liberation. We call upon all segments of our people to cooperate in this regard with a view to protect [the] interests of our people and enrich its honour.

> It is well known to everyone that we have sworn, and will fulfil our [oath], a solemn pledge with our people that we will not spare any enemy who will come in our way of *jihād*. We will remain faithful to our solemn pledge till we meet our Allāh Almighty martyred or victorious.

One suicide bomber, Hisham Ismail Hamed, issued a communiqué in November 1994, which was distributed in Gaza:[247]

> Our meeting with God is better and more precious than this life. I swear by the same God that in another place there is a paradise wider than the firmament above the earth, for life today is no more than a diversion, a distraction, and a search for wealth… An operation on behalf of *jihād*, carried out by a fighter whose heart is filled with belief and love of his country, strikes terror into the hearts of the arrogant and makes them tremble.

Tensions between the Palestinian Authority and Hamas reached a peak in mid-November 1994: the willingness of 'internal' Hamas in Palestine to come to an accommodation with the Palestinian Authority only served to aggravate relations with the Hamas 'external' command based in Amman.[248] Attacks by Hamas and Islamic Jihād against Israelis led to Arafat detaining activists, reprisals and counter-reprisals. The situation further deteriorated in the following two years, and in March 1996, after suicide bombings in Jerusalem and Ashkelon which killed 25 Israelis, the Palestinian authority outlawed the 'Izz al-Dīn al-Qassām Brigades and arrested many radical Islamists. But the damage had been done: Shimon Peres lost the Israeli general election largely as a result of the intervention of the Hamas suicide bombers who had sought to destabilize the peace process. Peres' successor, Binyamin Netanyahu, 'employed all his destructive powers to freeze and undermine the Oslo agreements'.[249]

The release of Shaykh Yāsīn and his return to Gaza in October 1997 eased some of the Hamas divisions, but could not resolve the fundamental question facing his movement: if it was to evolve into a political party, would it lose its *jihādī* character? If it did not, would it be eliminated by the Palestinian security structure, working in the interests of Israel and the United States? A year later, on 23 October 1998, the Wye Memorandum was signed at the White House by Arafat and Netanyahu, the Israeli Prime Minister, with President Clinton as witness. The CIA was to supervise a security plan whereby the Palestinian Authority would arrest alleged terrorists and confiscate their weapons. CIA personnel would settle disagreements over the arrest of suspected terrorists, manage border checkpoints and other security issues. An American–Palestinian committee would meet every two weeks 'to review the steps being taken to eliminate terrorist cells'. Shaykh Yāsīn and several hundred Hamas activists were rounded up as a result of the implementation of the Wye Memorandum. An increasingly authoritarian

Palestinian leadership under Arafat had taken an immense risk for peace. At a special meeting of the Palestine National Council in Gaza on 14 December 1998 the PLO goal of destroying Israel was finally laid to rest. President Clinton, who had witnessed this decision, applauded it: the ideological underpinning of the Arab–Israeli conflict appeared to have been rejected 'fully, finally and forever' with the cancellation of the offensive clauses of the Palestine National Charter. This gesture was reciprocated six days later by five new conditions imposed by Israel calculated to torpedo the peace process and to place the blame on the Palestinians. Prime Minister Netanyahu had reneged on an agreement he himself had signed. There was a renewed phase of land confiscations for the purpose of building new Jewish settlements and a network of roads to link them.[250]

Netanyahu's government fell on 23 December 1998 and Ehud Barak won a significant vote in favour of the peace process in the general elections the following May. Yet within less than 18 months the second *intifāḍah* had broken out following the failure of a new round of talks at Camp David. Barak had pronounced 'five nos' on the eve of the summit, which gravely prejudiced negotiations: no withdrawal to the 1967 boundaries; no dismantling of the settlements; no division of Jerusalem; no Arab army west of the river Jordan; and no return of Palestinian refugees.[251] On 24 July 2000, the day before the talks broke down, the Palestinian negotiators alleged that Israel had not acknowledged the principles of withdrawal, Palestinian sovereignty in East Jerusalem, and the right of return of the Palestinian refugees as governed by UNSCR 242 and General Assembly Resolution 194. Arafat was blamed by President Clinton for the failure, but had no alternative if he wished to maintain domestic legitimacy and sustain his rule. The Israeli opposition leader, Ariel Sharon, was already campaigning on the platform that he would reject Barak's apparent concessions to the Israelis. Sharon eventually won the Israeli elections on 6 February 2001 against a discredited Barak. The concessions were more apparent than real. If we take the Oslo Accords to be a 'hegemonic peace', as in the analysis of Glenn E. Robinson, Israel's political stability was brought about precisely by the nature of that peace. Whilst Palestinian opposition discourse rejected as unjust the terms of the peace, writes Robinson, 'the oppositional discourse in Israel, speaking the language of power, rejected as unnecessary any significant concessions at all to a much weaker – and much hated – party'. Sharon's election victory over Barak was, in Robinson's words, 'particularly lopsided. Hegemonic peace with Palestine brought turmoil to Israel in a way that its peace treaties with Egypt and Jordan never did.'[252]

The second Palestinian *intifāḍah*: the affirmation of *jihādī* ideology

The second, or al-Aqṣā, *intifāḍah* broke out in late September 2000. Unlike the first uprising, this one was pre-planned. Yasser Arafat had broken off negotiations at Camp David in July and a return to the armed struggle was inevitable from this date. The timing of the uprising was coordinated with Ariel Sharon's 'provocative' electioneering visit to the al-Aqṣā Mosque (the relatively moderate rebuke was from George Mitchell, head of the international commission investigating the reasons for the al-Aqṣā *intifāḍah*).[253] This was affirmed subsequently by 'Imād al-Falujī, the Palestinian Authority's Communications Minister:[254]

> Just as the national and Islamic Resistance in South Lebanon taught [Israel] a lesson and made it withdraw humiliated and battered, so shall [Israel] learn a lesson from the Palestinian Resistance in Palestine. The Palestinian Resistance will strike in Tel Aviv, in Ashkelon, in Jerusalem, and in every inch of the land of natural Palestine. Israel will not have a single quiet night. There will be no security in the heart of Israel...
>
> We say to the Zionist enemy and to the entire world: 'we will return to the early days of the PLO, to the sixties and seventies; the Fateh Hawks will return, as will the 'Izz al-Dīn al-Qassām [the military wing of the Hamas] and the Red Eagles [the military arm of the Popular Front for the Liberation of Palestine].' A new stage will continue until the rights are returned to their owners... we will strike whoever blames us for the failure of the negotiations, because President Yasser Arafat's patience was greater than [the Prophet] Job's. Arafat has become the Job of the twentieth century, because of what the U.S. and Israel lay on him...

Marwān Barghouti, head of the *Tanẓīm* (a militia made up of veterans of the first *intifāḍah*),[255] admitted his critical role in igniting the October 2000 *intifāḍah* in both the West Bank and Gaza, as well as among the Israeli Arabs:[256]

> I knew that the end of September was the last period [of time] before the explosion, but when Sharon reached the al-Aqṣā Mosque, this was the most appropriate moment for the outbreak of the *intifāḍah*... The night prior to Sharon's visit, I participated in a panel on a local television station and I seized the opportunity to call on the public to go to the al-Aqṣā Mosque in the morning, for it was not possible that Sharon would reach al-Ḥaram al-Sharīf just so, and walk away peacefully. I finished and went to al-Aqṣā in the morning... We tried to create clashes without success because of the differences of opinion that emerged with others in the al-Aqṣā compound at the time... After Sharon left, I remained for two hours in the presence of other people, we discussed the manner of response and how it was possible to react in all the cities and not just in Jerusalem. We contacted all [the Palestinian] factions.

The difference between the two *intifāḍahs* was that the earlier one had arisen without the complication of there being a Palestinian Authority in place. This brought new challenges of shared authority, internal borders, negotiations and, initially at least, a security partnership between the Palestinian police and the IDF. The first casualty of the second *intifāḍah* was the security partnership: the primary reason for Sharon's election victory was the unwillingness of Israeli public opinion to tolerate negotiations whilst there appeared to be open conflict. Moreover, since the Palestinian case was no longer simply calling for an end to the occupation but requiring control of East Jerusalem and the Palestinian right of return, these were issues which found much less resonance with Israeli voters.[257]

Glenn E. Robinson suggests that one significant development to have emerged from the failure of Camp David and the al-Aqṣā *intifāḍah* was the discrediting of the 'Tunisians', the members of the PLO who had driven the Oslo negotiations and had failed to deliver Palestinian rights through the peace process. The PLO understanding of the peace process was the creation in the long term of an 'independent, viable sovereign Palestinian state'. The Israeli understanding, it came to be recognized, was no more than 'provisional borders', attributes of sovereignty, and a miniscule Palestinian state, divided into two and surrounded by concrete walls and electrified fences. Ritchie Ovendale comments:

> in 1947 the United Nations partition vote had awarded Palestinians 47 per cent of their 'homeland'; the Oslo Accords of 1993, with the offer of the West Bank and Gaza, awarded 22 per cent of that 'homeland'; at Camp David in 2000 the Palestinians were offered 80 per cent of the 22 per cent of the 100 per cent of their original homeland.[258]

In March 2001 Hamas instituted a policy of using 'martyrdom operations' to strike within Israel's heartland. The Israeli response was the assassination of known leaders of the resistance and bombardment of suspected 'terrorist' areas with helicopter gunships supplied by the United States. Arafat's own guard was attacked in Gaza and Ramallah by Israel at the end of March. In mid-May, after further Hamas attacks, Israeli used jet fighters against the Palestine Authority positions in Ramallah and Nablus. A helicopter attack on 31 July on the Hamas media office in Nablus killed two prominent leaders. Israeli tanks were sent into Jenin on 14 August and again on 11 September. On 26 September a ceasefire was agreed, but Palestinian guerrillas attacked Rafah and Israel sent tanks into that city. On 17 October the PFLP, in retaliation for the assassination of their leader, shot dead Israel's Minister for Tourism. Sharon likened Arafat to Osama bin Laden. The killing of the minister, Ze'evi, was likened to 'Israel's own Twin Towers'. On 18 October Israeli armoured personnel carriers invaded Ramallah and Jenin, while Israeli tanks besieged Bethlehem. At the end of March 2002, Sharon launched 'Operation Defensive Shield', which led to the occupation of most of the main towns and refugee camps on the West Bank. Arafat was besieged at Ramallah. In late June, Sharon announced 'Operation Determined

Path', called by Ritchie Ovendale 'the indefinite re-occupation of any territory that Israel thought necessary for its security'.[259] The cycles of violence and counter-violence continued: by the end of May 2003, 32 months of the al-Aqṣā *intifāḍah* had claimed 762 Israeli and 2274 Palestinian lives, far more than in the seven years of the first *intifāḍah*.[260]

The attitude of prominent figures in the '*ulamā'* with regard to the legitimacy of 'martyrdom' operations underwent a significant change of emphasis as a result of the political requirement of supporting the *intifāḍah*. Shaykh Yūsuf al-Qaraḍawī, one of the leaders of the Muslim Brotherhood and a recognized authority in Sunnī Islām countered a *fatwā* issued by the *muftī* of Sa'ūdī Arabia, Shaykh 'Abd al-'Aziz ibn Bāz,[261] which had argued that 'martyrdom operations' might be regarded as suicide, and therefore unlawful, in these words:

> These operations are the supreme form of *jihād* for the sake of Allāh, and a type of terrorism that is allowed by *Sharī'ah*… the term 'suicide operations' is an incorrect and misleading term, because these are heroic operations of martyrdom, and have nothing to do with suicide. The mentality of those who carry them out has nothing to do with the mentality of someone who commits suicide… He who commits suicide kills himself for his own benefit, while he who commits martyrdom sacrifices himself for the sake of his religion and his nation. While someone who commits suicide has lost hope for himself and with the spirit of Allāh, the *mujāhid* is full of hope with regard to Allāh's spirit and mercy. He fights his enemy and the enemy of Allāh with this new weapon, which destiny has put in the hands of the weak, so that they would fight against the evil of the strong and arrogant. The *mujāhid* becomes a 'human bomb' that blows up at a specific place and time, in the midst of the enemies of Allāh and the homeland, leaving them helpless in the face of the brave *shāhid* who … sold his soul to Allāh, and sought [martyrdom (*shahādah*)] for the sake of Allāh.

Yūsuf al-Qaraḍawī also justified such operations when the targets were civilians:

> Israeli society is militaristic in nature. Both men and women serve in the army and can be drafted at any moment. On the other hand, if a child or an elderly [person] is killed in such an operation, he is not killed on purpose, but by mistake, and as a result of military necessity. Necessity justifies the forbidden.

His explanation of conflicting *fatāwā* from different Islamic authorities on the subject was dismissive: if the fighters are labelled 'terrorists' in a *fatwā*, then it was 'not issued by an authoritative religious source…' The author of such a ruling, in his view, probably served the regime in power or was an agent of the police! The Shaykh pronounced that 'if everyone who defends his land, and dies defending his sacred symbols is considered a terrorist, then I wish to be at the forefront of terrorists'. And he added, in a powerful message that was spread world-wide through the medium of the internet:[262]

> …America is not neutral: for those who are resisting Israel are considered terrorists by it: Hamas, al-Jihād, the PFLP, Ḥizbu'llah, are all considered terrorists, and today they have added al-Aqṣā Martyr Brigades, a Fateh group. They are terrorists according to America. As for Sharon and his gang, they have the right to defend themselves. Palestinians are the aggressors, the transgressors, the unjust terrorists, and all Arabs and Muslims must fight them, because you are either with us or with terrorism. And Arabs and Muslims must fight against Hamas, Jihād, al-Aqṣā and Ḥizbu'llah because they are all terrorists…
>
> And I pray to Allāh if that is terrorism, then O Allāh make me live as a terrorist, die as a terrorist, and be raised up with the terrorists. For in the eyes of America, Sharon is innocent, a gentle lamb, while the Palestinians are savage wolves, violent beasts, terrorists, people of violence, blood-thirsty, criminals, terrorists… We support our brothers in the land of Palestine. We aid and praise their struggle and salute their determination. We believe they can achieve a lot. 'Fighting has been prescribed for you, while you hate it' (Q.2:246). This is the nature of fighting.

The Scholars of al-Azhar and the al-Azhar Centre for Islamic Research published their own ruling in support of suicide bombings:[263]

> He who sacrifices himself is he who gives his soul in order to come closer to Allāh and to protect the rights, respect and land of the Muslims… When the Muslims are attacked in their homes and their land is robbed, the *jihād* for Allāh turns into an individual duty. In this case, operations of martyrdom become a primary obligation and Islām's highest form of *jihād*… The participation of Palestinian children and youths in the *intifāḍah* is a type of *jihād*… when *jihād* becomes an individual duty, all Muslims must join in, and children must go [to battle], even without asking permission of their parents. Those who sell their soul to Allāh are the avant-garde of the *sḥahids* in Allāh's eyes, and they express the revival of the nation, its steadfastness in struggle, and the fact of its being alive and not dead.

Shaykh Ḥamīd al-Biṭāwī, head of the Palestinian Islamic Scholars Association, an organization affiliated with Hamas, stated that, according to the *sharī'ah,* 'if infidels conquer even an inch of the Muslims' land, as happened with the occupation of Palestine by the Jews, then *jihād* becomes an individual duty…', and therefore 'martyrdom operations' were acceptable. Dr. 'Abd al-'Azīz al-Rantīsī, another leader of Hamas, agreed with Shaykh al-Biṭāwī and added that

> suicide depends on volition. If the martyr intends to kill himself, because he is tired of life, it is suicide. However, if he wants to sacrifice his soul in order to strike the enemy and to be rewarded by Allāh, he is considered a martyr... We have no doubt that those carrying out these operations are martyrs.[264]

Arafat's position in May 2002 seemed one of pragmatism rather than principle:

> Palestinian and Arab public opinion is convinced that such operations do not serve our goals. Rather, they cause disagreements with the international community and unite large parts of it against us. You know better than me that such operations cause dissent.

Other voices called for the 'martyrdom operations' to be limited to the territories occupied in 1967. The Minister for Planning and Regional Cooperation in the Palestinian Authority stated: 'we support martyrdom operations but only of the kind that emerged in Jenin, [which] sought to confront the Israeli aggression'.[265] Muḥammad Mahdī 'Uthmān 'Akef, the leader of the Muslim Brotherhood in Egypt, stated in early 2004 that

> the bombings in Palestine and 'Irāq are a religious obligation. This is because these two countries are occupied countries, and the occupier must be expelled in every way possible. Thus, the Muslim Brotherhood movement supports martyrdom operations in Palestine and 'Irāq in order to expel the Zionists and the Americans.[266]

The focus of Israeli retaliation in the first months of 2004 was Hamas. On 22 March, its longstanding leader, Shaykh Aḥmad Yāsīn, was killed by Israeli soldiers, on the command of Prime Minister Ariel Sharon. His successor, Dr 'Abd al-'Azīz al-Rantīsī, was assassinated on 17 April. On the news of Yāsīn's death, Palestinian cabinet ministers stood as Arafat recited a prayer for the dead. The Palestinian leader, referring to Yāsīn, then added: 'may you join the martyrs and the prophets. To heaven, you martyr.' The Palestinian Prime Minister stated that Palestinians had lost 'a great leader'. On 6 June, five life sentences – 165 years in total – were passed by an Israeli court on the charismatic leader of the *intifāḍah*, Marwān Barghouti, for attempted murder and membership of a terrorist organization. 'This court is just a partner in the war against the Palestinian people', Barghouthi stated in a five-minute address before sentencing. 'A continuation of the *intifāḍah* is the only way to independence', he added. 'No matter how many [people Israel] arrests and kills, they will not break our people's determination', he affirmed.[267]

If the pace of 'martyrdom operations' and Israeli retaliation declined somewhat in 2004, it was not as a result of any détente but because Palestinian *jihādīs* were finding it difficult to penetrate the security fence constructed by Israel. Israeli security considerations had now given some meaning to the Palestinian accusation that the security fence was an instrument of apartheid, and that the government aim was the total separation of the two communities. In mid-December 2000, only a few months into the *intifāḍah*, the renowned Palestinian scholar Edward Said had commented:[268]

> The most demoralizing aspect of the Zionist–Palestinian conflict is the almost total opposition between mainstream Israeli and Palestinian points of view. We were dispossessed and uprooted in 1948; they think they won independence and that the means were just. We recall that the land we left and the territories we are trying to liberate from military occupation are all part of our national patrimony; they think it is theirs by biblical fiat and diasporic affiliation. Today, by any conceivable standards, we are the victims of the violence; they think they are. There is simply no common ground, no common narrative, no possible area for genuine reconciliation. Our claims are mutually exclusive. Even the notion of a common life shared in the same small piece of land is unthinkable. Each of us thinks of separation, perhaps even of isolating and forgetting the other.
>
> The greater moral pressure to change is on the Israelis, whose military actions and unwise peace strategy derive from a preponderance of power on their side and an unwillingness to see that they are laying up years of resentment and hatred on the part of Muslims and Arabs. Ten years from now there will be demographic parity between Arabs and Jews in historical Palestine: what then? Can tank deployments, roadblocks and house demolitions continue as before?

In July 2004 the International Court of Justice and the United Nations General Assembly both declared the security wall of Israel illegal, decisions which were immediately rejected by the Israeli government. The only encouraging development is that there are intelligent and perceptive Israeli historians and Jewish historians living abroad[269] who are prepared to contest the inevitability of conflict between the Israeli state and the Palestinians, who wish to set right the historical injustice that has been done, reach a common narrative of understanding between the two sides and advance realistic proposals for compromise which would involve an ending of the occupation of the West Bank and Gaza.[270] Under a government headed by Ariel Sharon, whose name will forever be associated, by the verdict of Israel's own Kahan Commission, with his 'indirect responsibility' for the Sabra and Shatila massacres,[271] and a Palestinian Authority headed by Yasser Arafat, mired in corruption and tainted by human rights violations and authoritarianism,[272] there is no prospect of any such understanding emerging between the politicians.

11
Osama bin Laden: Global *Jihād* as 'Fifth-Generation' Warfare

We regard the Afghan *jihād* as the mother of *jihād*. Many *jihād* movements in the *ummah* have sprung from it.

Mujāhid Syed Ṣalāḥuddīn, the Ḥizb Amīr,
or Amīr of the *mujāhidīn*, in Kāshmīr, 1995[1]

If the Arab Afghans are the mercenaries of the United States who have now rebelled against it, why is the United States unable to buy them back now? Would not buying them [off] be more economical and less costly than the security and prevention budget that it is paying to defend itself now?

Ayman al-Ẓawāhirī, Deputy Leader of al-Qaeda, in his memoirs,
Knights Under the Prophet's Banner, December 2001[2]

Three Pakistani scholars – Iffat Malik, Farzana Noshab and Sadaf Abdullah – have argued that 'the portrayal of *jihād* in Western and other media is quite removed from reality'. They contended that *jihād* had been 'distorted with deliberate intent' and claimed that 'it is never, as popularly represented, a religiously-motivated aggressive war against "innocent" non-Muslims, with the aim of spreading Islām by force'. This was the conclusion they drew following the examination of the case studies of Afghanistan (1979–89),[3] Kāshmīr (since 1989), Palestine (since the *intifāḍah* began on 8 December 1987), Chechnya (since 1994), and the conflicts in Bosnia and Kosovo. Though the authors did concede that Muslims needed to curb the activities of extremists whose 'hard-line activities and statements merely provide ammunition to those seeking to portray Islam and *jihād* in a negative light', they contended that such extremists lacked any large following. They concluded that *jihād* was essentially defensive. '*Jihād* in reality, wherever it is found, is a struggle for freedom, against aggression and oppression, and for human rights.'[4]

Their words were written before the events of 11 September 2001 and the hasty passage of the United Nations Security Council Resolution 1373 on 28 September, which appeared to many commentators to blur the distinction between a 'freedom struggle' for self-determination and 'terrorism'. By the terms of the resolution, before granting refugee status, all states should take appropriate measures to ensure that the asylum seekers had not planned, facilitated or participated in terrorist acts. Furthermore, states should ensure that refugee status was not abused by the perpetrators, organizers or facilitators of terrorist acts, and that claims of political motivation were not recognized as grounds for refusing requests for the extradition of alleged terrorists. The resolution also noted the close connection between international terrorism and transnational organized crime, illicit drugs, money-laundering, illegal arms-trafficking, and the illegal movement of nuclear, chemical, biological and other potentially deadly materials.

Certain governments, notably that of Pakistan, have questioned the replacement of the term 'freedom fighter' by 'terrorist', since it appears to deny the right of self-determination or the right of an oppressed group to secure an amelioration of its fate.[5] UNSCR 1373 appears, on the face of it, to lack realism since it fails to allow for the motivation underlying so-called 'terrorism': the existence of a prior 'asymmetrical conflict' in which overwhelming power rests with the agency of the state, itself seen as repressive.

Jihād perceived as 'fourth-generation' warfare, 2002

The 'revolution in military affairs' in contemporary strategic thinking is the use of highly sophisticated computer sensor devices, 'whereby the fog of war can be lifted and commanders enabled to view the combat situation clearly and respond with high-precision (often unmanned) weapons. These technologies are expensive, and available only to the richest states.'[6] They can be highly effective in disabling known enemies but are virtually powerless to prevent surprise attacks on 'soft' targets.

We have already seen (above, Chapter 10) that, in the context of the Israeli invasion of Lebanon in 1402/1982, Sayyid Muḥammad Ḥusayn Faḍlallāh had justified 'martyrdom operations' against military targets in order to rectify the military imbalance between the 'imperialist forces' and those of national resistance. The first policy statement (*bayān*) of Osama bin Laden (Usāma bin Ladīn) in 1417/August 1996 demonstrated clearly that the principles of asymmetry had been further analysed and developed. Bin Laden commented:[7]

> it must be obvious to you that, [owing] to the imbalance of power between our armed forces and the enemy forces, a suitable means of fighting must be adopted, i.e. using fast-moving light forces that work under complete secrecy. In other words, to initiate a guerrilla warfare, where the sons of the nation, and

not the military forces, take part in it. And as you know, it is wise, in the present circumstances, for the armed military forces not to be engaged in conventional fighting with the forces of the crusader enemy… unless a big advantage is likely to be achieved; and great losses induced on the enemy side… that will help to expel the enemy defeated out of the country. The *mujāhidīn*, your brothers and sons, request… that you support them in every possible way by supplying them with the necessary information, materials and arms. Security men are especially asked to cover up for the *mujāhidīn* and to assist them as much as possible against the occupying enemy; and to spread rumours, fear and discouragement among the members of the enemy forces… The regime is fully responsible for what had been incurred by the country and the nation; however the occupying American enemy is the principal and the main cause of the situation. Therefore efforts should be concentrated on destroying, fighting and killing the enemy until, by the Grace of Allāh, it is completely defeated.

In February 2002 Abū 'Ubayd al-Qurashī, a top al-Qaeda planner, when claiming credit for the 11 September 2001 atrocities, described the strategy underlying the bombings as a response to 'fourth-generation' warfare:[8]

Western strategists[9]… [claimed] that the new warfare would be strategically based on psychological influence and on the minds of the enemy's planners – not only on military means as in the past, but also on the use of all the media and information networks… in order to influence public opinion and, through it, the ruling elite.

They claimed that the fourth-generation wars would, tactically, be small-scale, emerging in various regions across the planet against an enemy that, like a ghost, appears and disappears. The focus would be political, social, economic, and military. [It will be] international, national, tribal, and even organizations would participate (even though tactics and technology from previous generations would be used)…

…fourth-generation wars have already occurred and… the superiority of the theoretically weaker party has already been proven; in many instances, nation-states have been defeated by stateless nations…

…the Islamic nation has chalked up the most victories in a short time, in a way it has not known since the rise of the Ottoman Empire. These victories were achieved during the past 20 years, against the best-armed, best-trained, and most-experienced armies in the world (the USSR in Afghanistan, the U.S. in Somalia, Russia in Chechnya, and the Zionist entity in southern Lebanon) and in various arenas (mountains, deserts, hills, cities).

In Afghanistan, the *mujāhidīn* triumphed over the world's second [superpower] at that time… Similarly, a single Somali tribe humiliated America and compelled it to remove its forces from Somalia.

A short time later, the Chechen *mujāhidīn* humiliated and defeated the Russian bear. After that, the Lebanese resistance [Ḥizbu'llah] expelled the Zionist army from southern Lebanon…

The *mujāhidīn* proved their superiority in fourth-generation warfare using only light weaponry. They are part of the people, and hide amongst the multitudes…

…with the September 11 attacks, al-Qaeda entered the annals of successful surprise attacks, which are few in history – for example, the Japanese attack on Pearl Harbour in 1941, the surprise Nazi attack on the USSR in 1941, the Soviet invasion of Czechoslovakia in 1968, and the crossing of the Zionist Bar-Lev Line in 1973…

This has an extremely high economic and psychological price, particularly in a society that has not been affected by war since the American Civil War. If the USS Cole incident could happen to the American army [*sic*: navy], which is assumed to be in perfect preparedness, then preparing an entire society for 'terrorist' attacks appears hard to achieve…

[The USA] is baffled by fourth-generation warfare that suits [the] *jihād* avant-garde – especially at a time when the Islamic peoples have re-espoused *jihād*, after they had nothing left to lose because of the humiliation that is their daily lot…

The time has come for the Islamic movements facing a general crusader offensive to internalize the rules of fourth-generation warfare. They must consolidate appropriate strategic thought, and make appropriate military preparations…

We pray to Allāh… to bring forth for this [Islamic] nation a new generation of preachers and clerics, who can meet the challenges posed by fourth-generation warfare.

Thus, within a few months of the article by our three Pakistani researchers, which denied the existence of a global *jihād*, Abū 'Ubayd al-Qurashī had effectively provided a rationale of the movement – a rationale, moreover, which conformed to the original blueprint given by Osama bin Laden himself in August 1996. The difference between the two statements was that al-Qurashī placed more emphasis on the methodology, while bin Laden, though he had given due weight to the 'imbalance of power' between the forces had emphasized nevertheless *jihād* 'in the cause of Allāh'. None of this denied the element of specificity about each regional conflict, which was the point stressed by Iffat Malik, Farzana Noshab and Sadaf Abdullah. Each conflict was indeed a struggle for justice or a struggle for self-determination, or contained elements of both.

The common factor, however, was that each was an asymmetrical conflict in which the Muslims were the party which suffered from the imbalance of power. Hence the search for new means, in fourth-generation warfare, to rectify the imbalance. The optimum weapon found has been the suicide bomber, the morality of which will be discussed at the end of this chapter in the context of bin Laden himself and his objectives. It is sufficient here to note, with Professor John Gray, that the attacks on New York and Washington DC on 11 September 2001 were indeed 'acts of war – but not of a conventional kind':[10]

> they were examples of asymmetric warfare, in which the weak seek out and exploit the vulnerabilities of the strong. Using civilian airplanes as weapons and its operatives as delivery systems, al-Qaeda demonstrated that despite the 'revolution in military affairs' [RMA] that has given America an unchallengeable military superiority over all other states, the U.S. remains vulnerable to devastating attack.

However, there is an important question as to whether the security policy adopted by the USA, which some commentators have argued is an adaptation of a failed Israeli approach to dealing with asymmetrical warfare,[11] has not actually made a bad situation much worse and made the USA, and the world, less rather than more secure. These commentators also argue that 'since the mid-1960s there were two parallel processes going on, the Americanization of Israel and the Israelization of American foreign policy'.[12] However, before we review the doctrine of pre-emption and the costs of its implementation, especially the human costs we must first consider the extent to which the United States created its own problem by its overt and covert backing of the anti-Soviet *jihād* in Afghanistan.

Sowing the wind: the 'fabrication'[13] of the Afghan *jihād*

Shortly before midnight on 24 December 1979, Soviet troops began landing at Kabul airport to prop up an ailing Marxist regime that had seized power on 27 April 1978. By the beginning of the new year there were nearly 85,000 troops in Afghanistan.[14] The decision to intervene in Afghanistan was remarkable, given the earlier history of Soviet failure to eliminate political Islām in Central Asia prior to 1941. The Soviet government had attacked Islām both because its Marxist-Leninist ideology dictated an atheist society and because it could not tolerate any rival basis for power. Thousands of clergy were eliminated, the Arabic script was banned, and mosques had been physically destroyed. Yet the prospect of a Nazi victory in 1941 had forced Stalin to change course. Islām had been damaged by the long campaign, but not defeated.[15] Since the Soviet leadership tended to believe its own propaganda and was contemptuous of objective history, rewriting it as the need arose, the lessons of the failure of the long campaign against Islām in Central Asia were lost on the proponents of military intervention at the end of 1979.

The Soviet invasion took place in a country where *jihād* had already been proclaimed the previous year. Qāzī Amīn, a compromise figure with a reputation for trustworthiness, was appointed in May 1976 as the first *amīr* of Ḥizb-i Islāmī Afghanistan. He it was who issued the first formal declaration (*fatwā*) of *jihād* against the Afghan government. This declaration was distributed widely within Afghanistan in the last year of Da'wūd's regime (1977–78) and spread the name of Ḥizb-i Islāmī throughout the country, so that when the Marxists succeeded in taking power on 27 April 1978, many *'ulamā'* looked to this party and to Qāzī Amīn for leadership after they had fled to Pakistan.[16] The Soviet politburo took the popular opposition seriously. At its meeting on 17–19 March 1979, it heard a report from Gromyko:[17]

> the situation in Afghanistan has deteriorated sharply, the centre of the disturbance at this time being the town of Herat…There, as we know from previous cables, the 17th division of the Afghan army was stationed, and had restored order, but now we have received news that this division has essentially collapsed. An artillery regiment and one infantry regiment comprising that division have gone over to the side of the insurgents. Bands of saboteurs and terrorists, having infiltrated from the territory of Pakistan, trained and armed not only with the participation of Pakistani forces but also of China, the United States of America, and Iran, are committing atrocities in Herat. The insurgents infiltrating into the territory of Herat Province from Pakistan and Iran have joined forces with a domestic counter-revolution. The latter is especially comprised by religious fanatics. The leaders of the reactionary masses are also linked in large part with the religious figures. The number of insurgents is difficult to determine, but our comrades tell us that they are thousands, literally thousands…
>
> As far as Kābul is concerned, the situation there is basically calm. The borders of Afghanistan with Pakistan and Iran are closed, or more accurately, semi-closed. A large number of Afghans, formerly working in Iran, have been expelled from Iran and, naturally, they are highly dissatisfied, and many of them have also joined up with the insurgents…

Andropov prepared a secret memorandum for Brezhnev, probably in early December 1979, which warned the Party Leader of the risks of losing the pro-Soviet gains of the Afghanistan coup the previous year, since the new regime in power appeared willing to shift ideological camp:[18]

> After the coup and the murder of Tarakī in September of this year, the situation in Afghanistan began to undertake an undesirable turn for us. The situation in the party, the army and the government apparatus has become more acute, as they were essentially destroyed as a result of the mass repressions carried out

> by Amīn. At the same time, alarming information started to arrive about Amīn's secret activities, forewarning of a possible political shift to the West. [These included] contacts with an American agent about issues which are kept secret from us; promises to tribal leaders to shift away from the USSR and to adopt a 'policy of neutrality'; closed meetings in which attacks were made against Soviet policy and the activities of our specialists; the practical removal of our headquarters in Kabul, etc. The diplomatic circles in Kabul are widely talking of Amīn's differences with Moscow and his possible anti-Soviet steps. All this has created, on the one hand, the danger of losing the gains made by the April [1978] revolution… within the country, while on the other hand – the threat to our positions in Afghanistan [has increased]… [There has been] a growth of anti-Soviet sentiments within the population…

It now seems likely that the risks of the Afghan regime changing camp were greatly exaggerated, so that the basis for invasion was faulty. Nevertheless, President Carter's National Security Adviser, Zbigniew Brzezinski, has confirmed that aid to the *mujāhidīn* began six months before the Soviet invasion and probably precipitated this action.[19] Within a few years, the Soviet decision to invade was recognized by the decision-makers as a mistake. However, once the decision had been taken, the Soviet occupation appeared likely to succeed because of the divisions among the Islamists. David B. Edwards argues that

> the fissioning of the resistance climaxed after the Soviet pullout in 1989, when the parties were engaged in an armed struggle for control of Kābul. However, the process that culminated then had its origins in the period between the Marxist revolution of 1978 and the founding of the radical (seven-party) and moderate (three-party) alliances in September 1981. This was the formative period of the Islamic resistance – when the fault lines that later sundered it first revealed themselves… the pre-eminent characteristics of the Islamic resistance in Afghanistan were multiplicity, fragmentation, and impermanence.[20]

Gulbuddin Hekmatyar asserted his pre-eminence in the *jihād*, on the grounds of the early militancy of the Muslim Youth, originally against Ẓahīr Shāh (monarch, 1933–73) and then against President Da'wūd (President, 1973–78). Muslim Youth activists were the first to recognize the threat of Soviet Communism in Afghanistan, and they had called for a *jihād* against the government before anyone else. His primary opponent was Burhānuddīn Rabbānī, who stressed his close relationship with Professor Niāzī, the first Afghan to import the ideas of the Muslim Brotherhood in Egypt into Afghanistan, as well as his own early involvement in the Islamic movement inside Afghanistan. The dispute demonstrates the importance given to history and lineal succession as different parties jockeyed for position in Peshawar. In a politically turbulent and uncertain environment, Hekmatyar and Rabbānī each made claims of precedence, rooting these assertions in their connection to venerated ancestors who were dead and could not contest claims made in their names. For Hekmatyar in particular, the issue of precedence was vitally important because he had little else to offer by way of justification for his leading the *jihād*. These serious personal divisions enshrined a split between what was to become the Tājīk-dominated wing of the party (which became known as Jamā'at-i Islāmī Afghanistan) and the Pashtun- or Pakhtun-majority wing (which was to become the Ḥizb-i Islāmī Afghanistan). The seeds of civil war after the expulsion of the Soviet troops had been sown long in advance.[21]

By the spring of 1980 there were more than 100,000 troops in the country. However, fighting spread throughout Afghanistan and violent uprisings temporarily wrested Kandahar, Herat and Jalalabad from the Soviet puppet government's control. The name of Aḥmad Shāh Mas'ūd appeared as the rebel leader in the Panjshir valley; later, because of his steadfastness, he was to be canonized as the 'lion of Panjshir', which tended to overshadow the achievements of other leaders.[22] By 1984 there were also some 10,000 Soviet advisers in Afghanistan, who were in effect running the PDRA (People's Democratic Republic of Afghanistan) government. Substantial economic aid ($350 million in 1980) and arms imports from the Soviet Union ($683 million in 1980) seemed likely to buttress the regime.[23] The reality, however, was that the Soviets by 1983 could control the cities, protect the northern pipelines and keep the roads open at least in the daytime, but they controlled only about 20 per cent of the country. In 1983, in a drive to broaden their area of control, the Soviets changed to the strategy of air war, bombing villages and depopulating areas that provided support to the *mujāhidīn*. By the end of 1984, there had been eight Soviet offensives in the Panjshir valley though the forces of Aḥmad Shāh Mas'ūd had not been quelled.[24] It was not until American 'Stinger' ground to air missiles were employed among the *mujāhidīn* that the tactical balance shifted in their favour. In the ten months before August 1987, 187 Stinger missiles were used in Afghanistan, of which 75 per cent hit aircraft or helicopter targets.[25] At a meeting with Gorbachev on 27 July, Najibullah admitted that the use of Stinger missiles had affected 'the morale of our pilots'.[26]

There had been American reluctance to allow the deployment of Stinger missiles, both because of security fears (the wish to prevent such missiles falling into the hands of the Soviets or the Iranians) and because it would demonstrate publicly that the massive Soviet aid pouring into Afghanistan was not the only outside source of assistance to the war. Yasser Arafat could scarcely conceal his fury at the willingness of the Arab states to back the Afghan *jihād*, which had received $19 billion over a nine-year period according to his estimate in May 1990, whilst the Palestine Liberation Organization had only received $2.6 billion since 1964.[27] Sa'ūdī Arabia matched every dollar allocated by the CIA on behalf of the United States, to an estimated $20 billion each, or $40 billion in total for the war.[28] The cost to the sponsors of the 'fabricated *jihād*' was at least $5 billion a year, most of it channelled through the notorious Bank of Credit and Commerce International (BCCI), founded in 1972 by Āghā Ḥasan 'Abidī, a Pakistani businessman. It was the world's largest Muslim banking institution before its fraudulent collapse in July 1991 with debts of more than $10 million.[29]

The whole process of 'fabricating' the Afghan *jihād* was coordinated by Pakistan's Inter-Services Intelligence (ISI) agency. The coordinator of the Afghan bureau after 18 October 1983, Brigadier Muḥammad Yousaf, records that for an armed resistance movement to succeed it needed:[30]

> first, a loyal people who would support the effort at great risk to themselves, a local population, the majority of whom would supply shelter, food, recruits and information. The Afghan people in the thousands of rural villages met this requirement. Second, the need for the guerrilla to believe implicitly in his cause, for him to be willing to sacrifice himself completely to achieve victory. The Afghans had Islām. They fought a *jihād*, they fought to protect their homes and families. Third, favourable terrain. With over two-thirds of Afghanistan covered by inhospitable mountains known only to the local people, I had no doubts about this. Fourth, a safe haven – a secure base area to which the guerrilla could withdraw to refit and rest without fear of attack. Pakistan provided the *mujāhidīn* with such a sanctuary. Fifth, and possibly most important of all, a resistance movement needs outside backers, who will not only represent his cause in international councils, but are a bountiful source of funds. The U.S. and Sa'ūdī Arabia certainly fulfilled this role…

The reason for the willingness of the US to pay exorbitant sums, much of it lost through corruption, for this enterprise is almost entirely to be explained as 'pay back time': what the Communists had done to the Americans in Vietnam, the militant Islamists would do to the Soviets in Afghanistan. A total of 47,378 US military personnel had been killed in action in Vietnam.[31] Congressman Charles Wilson, an avid supporter of US assistance to the *jihād*, stated: 'there were 58,000 [US dead – a higher figure presumably because of uncertainties over those missing in action] in Vietnam and we owe[d] the Russians one… I ha[d] a slight obsession with it because of Vietnam. I thought the Soviets ought to get a dose of it.'[32]

Though Pakistan has usually been depicted as the willing accessory to the *jihād*, the commitment to the Islamists in Afghanistan carried substantial risks, as Yousaf acknowledged:[33]

The Kremlin, and indeed the Soviet General Staff, understood the fundamental truth that without Pakistan the *jihād* was doomed. When President Zia, acting on the urging of General Akhtar, offered Pakistan as a secure base area, he condemned the Soviets to a prolonged counter-insurgency campaign that they were ill-prepared to fight. Like all armies, guerrilla forces cannot survive indefinitely without adequate bases to which they can withdraw from time to time to rest and refit. They need the means with which to fight, they need re-supplying, they need to train and they need intelligence. Pakistan provided all these things to the *mujāhidīn*. For the Soviets this was extremely frustrating. By 1983 they had launched a well-coordinated campaign to make the cost to Pakistan of supporting the Afghan resistance progressively higher. Their aim was to undermine President Zia and his policies by a massive subversion and sabotage effort, based on the use of thousands of KHAD agents[34] and informers. Every KHAD bomb in a Pakistan bazaar, every shell that landed inside Pakistan, every Soviet or Afghan aircraft that infringed Pakistan's airspace, and there were hundreds of them; every weapon that was distributed illegally to the border tribes, and every fresh influx of refugees, was aimed at getting Pakistan to back off. The Soviets sought with increasing vigour to foment trouble inside Pakistan. Their agents strove to alienate the Pakistanis from the refugees, whose camps stretched from Chitral in the north all the way to beyond Quetta, almost 2,000 kilometres to the south. The border areas of Pakistan had grown into a vast, sprawling administrative base for the *jihād*. The *mujāhidīn* came there for arms, they came to rest, they came to settle their families into the camps, they came for training and they came for medical attention. At the time we in ISI did not appreciate how fine a line President Zia was treading. As a soldier, I find it hard to believe that the Soviet High Command was not putting powerful pressure on their political leaders to allow them to strike at Pakistan.[35] After all, the Americans had expanded the Vietnam war into Laos and Cambodia, which had been used as secure bases by the Viet Cong. The Soviet Union, however, held back from any serious escalation. I had to ensure that we did not provoke them sufficiently to do so. A war with the Soviets would have been the end of Pakistan and could have unleashed a world war. It was a great responsibility, and one which I had to keep constantly in mind during those years…

Yousaf provides telling evidence of the administrative complexity of the ISI operation:

My headquarters was established in a large camp of some 70–80 acres on the northern outskirts of Rawalpindi, 12 kilometres from Islamabad, where General Akhtar had his office in the main ISI buildings. Inside the high brick walls were offices, a transit warehouse through which 70 per cent of all arms and

> ammunition for the *mujāhidīn* came, at least 300 civilian vehicles with garage facilities, several acres of training area, a psychological warfare unit, barracks, mess halls for 500 men and, later, the Stinger training school, complete with simulator. It was called Ojhri Camp[36]... The Afghan Bureau which I controlled could not cope with all aspects of supporting the war – General Akhtar had set up another department, also under a brigadier, responsible for what I would term the 'software' of war – the provision of clothing and rations (in this case rice, pulses and flour) for the *mujāhidīn*. These were purchased in huge quantities throughout Pakistan, with CIA money, for distribution to the guerrillas. I cooperated closely with this department. Over two years after my appointment yet another branch was created on the express orders of the President. Because of the rampant corruption within the Pakistani-staffed Commission for Afghan Refugees (CAR), which was handling the supply of food and clothing for all refugees in Pakistan, the ISI was required to take over these duties for Afghan villagers remaining in Afghanistan. This policy of trying to alleviate the suffering of these people was an attempt to get the population to remain in areas of *mujāhidīn* operations so that they would continue to provide information and succour. It was another brigadier's appointment, but although it was funded largely by the U.S. Congress these funds were separate from the arms money.

A final comment from the ISI perspective was Yousaf's verdict on the strengths and weaknesses of his raw material, the *jihādīs*:[37]

> In summary, the *mujāhidīn* have all the basic attributes of successful guerrilla fighters. They believe passionately in their cause; they are physically and mentally tough; they know their area of operations intimately; they are extremely courageous, with an inbred affinity for weapons; and they operate from mountain areas which give them both sanctuary and succour. These virtues are tempered with the vices of obstinacy, and an apparently insatiable appetite for feuding amongst themselves. To defeat a superpower they needed four things: to sink their differences for the sake of the *jihād*; an unassailable base area, which President Zia provided in Pakistan; adequate supplies of effective arms to wage the war; and proper training and advice on how to conduct operations. It was my responsibility to provide and coordinate the latter two.

The 'insatiable appetite for feuding' was never resolved. Hekmatyar's was the strongest force during the years of Soviet occupation, largely because his Islamic party (Hezb-i Islāmī Afghanistan, or HIA) was the main recipient of the money received by the seven official *mujāhidīn* groups recognized by Pakistan and US intelligence agencies for the channelling of money and arms. Hekmatyar's men ambushed Aḥmad Shāh Mas'ūd's forces in Takhar Province, sparking off a campaign of vengeance that resulted in public executions. That outright civil war could break out among the *mujāhidīn* at a critical juncture in the war, before the expulsion of the Soviet troops, was indicative of the rapid erosion of the small amount of unity left for the *jihād*.

Yet, in spite of such evidence of disunity, the Soviet political command had taken enough punishment. At a meeting of the Politburo on 13 November 1986, Gorbachev noted that the Soviets had been fighting in Afghanistan for six years and, unless a new approach was adopted, they would continue to fight there for another 20–30 years. The war had to be ended in the next two years. Gromyko agreed, pointed to the lack of domestic support for the Soviet position and the poor achievement of the Afghan army, while the Americans had every interest in seeing the war drag on for as long as possible. From the Soviet perspective, delay did not help the resolution of problems.[38] Other speakers agreed with the analysis:

> we have lost the battle for the Afghan people. The government is supported by a minority of the population. Our army has fought for five years. It is now in a position to maintain the situation on the level that it exists now. But under such conditions the war will continue for a long time…

The pressing urgency to improve the lot of peasants in the areas under government control was noted: 'the regions under the control of the counter-revolution are better supplied with goods of first necessity (these goods are shipped there by contraband from Pakistan)'. At a further meeting of the Politburo on 21–22 May 1987, Gorbachev commended a policy of realism, which implied a significant retreat on earlier Soviet ambitions:[39]

> We are obliged to conduct a realistic policy. And this needs to be remembered: there can be no Afghanistan without Islām. There's nothing to replace it with now. But if the name of the [People's Democratic Party of Afghanistan, or PDPA] Party is kept then the word 'Islamic' needs to be included in it. Afghanistan needs to be returned to a condition which is natural for it. The *mujāhidīn* need to be more aggressively invited to [share] power at the grass roots. No one is stopping this from being done. But Najib should speak as President and Chairman of the State Council. The personal factor has great importance there…

Gorbachev added, 'in Afghanistan, whoever is on the side of the *mujāhidīn* will long remember how we were killing them and those who are with Najib, that we put everyone on the level of their enemies with one stroke. And we will not get a friendly Afghanistan'.

At a meeting with Najibullah on 20 July 1987, Gorbachev heard a report from the pro-Soviet Afghan leader that the rebels were slowly coming over to the side of the government as a result of the policy of 'national reconciliation'.[40] Najibullah argued that 'Washington and its allies in the region are continuing to whip up tension in and around Afghanistan and are escalating combat operations. Our country has become one of the main links of a policy of state terrorism being pursued by the U.S.'[41] A degree of 'autonomy and independence' might be granted to middle-level rebel chieftains of the territory which they controlled on the condition of their recognizing the central government. As regards the opposition outside the country, the main target was the more moderate section which would facilitate further splits and dissension within the Alliance of Seven. Gorbachev then revealed that, after a Soviet withdrawal, India shared its concern that Afghanistan should not fall under US or Pakistani influence, which would be 'absolutely unacceptable to them'. India was not keen on a rapid Soviet withdrawal of troops. Najibullah contended that the US had turned Pakistan into a bridgehead for a fight against India and Afghanistan, using the Sikhs and the Afghan counter-revolution for their own interests. Might there not be joint retaliatory actions by India and Afghanistan against Pakistan, he surmised. The Indian leaders seriously considered from time to time 'launching a preventive attack, as a sort of demonstration, on Pakistan; not to occupy its territory but as a show of force'. Gorbachev replied that Rajiv Gandhi had even told him that the Indians had 'plans to dismember Pakistan', to which Najibullah added that if the Indians did this, the Afghans – 'without being directly involved' – could provoke serious disturbances in the border regions of Pakistan where the Pashtun and Baluchi tribes lived.[42] Najibullah's military analysis seemed so optimistic that one wonders whether it was truly the Afghan *jihād* that he was analysing:[43]

> In the military field we will solve the problems of neutralizing the irreconcilable rebel groups and destroying caravans with weapons, fortified regions, and bases. At the same time the implementation of measures to cover the border with Pakistan and Iran will be continued. Our goal is not to let the counter-revolution consolidate their positions, especially in the border zone, which should become a bulwark of people's power.

At a later meeting between the two leaders on 3 November 1987, Najibullah noted that as a result of an earlier decision, to 'create a zone free of rebel bands in the north of Afghanistan', the need had arisen to conduct a 'cleansing' in this region using USSR KGB Border Troops. There had been pressure from the USA and Pakistan to create a unified *jihād* under a single leader, but 'the counter-revolutionaries still have not managed to overcome serious existing differences'. Gorbachev stressed the need for financial aid from the USSR to be handled by Najibullah personally: if the fund was administered by a bureaucrat then it would 'all trickle into the hands of his relatives, through clan and family ties'. The abuse of authority should result in punishment and imprisonment, though Gorbachev wondered whether it was not a sin according to the Qur'ān to embezzle aid received from an atheist.[44]

At a later meeting in April 1988, Najibullah pronounced that, as the programme of national reconciliation advanced, 'the counter-revolutionary movement will increasingly lose [its] nature of political terrorism and become simply criminal'. His upbeat report to Gorbachev continued:[45]

> almost a third of the counter-revolutionaries maintain illegal contact with us. In the process, not only detachments associated with the moderate groups of the 'Alliance of Seven' but also of the groups of Hekmatyar and Rabbānī are entering into contact with us. This process will obviously intensify with the signing of the Geneva Accords. Only 50,000 active counter-revolutionaries oppose us. And when the enemy tries to present the 'Alliance of Seven' as a united force, this is not so…
>
> …altogether the counter-revolutionaries number 270,000 men. A third of them are talking with us; 50,000 are irreconcilable; and the rest are taking a wait-and-see position. Relying on the results of Geneva, we can attract the passive part of the counter-revolutionaries to our side…
>
> After the withdrawal of Soviet troops the situation in a number of regions will without doubt become difficult. Our comprehensive plan envis[ages] that we will conduct work among the population which has fallen under opposition control together with a concentration of the armed forces. We will send in the armed forces in certain cases. In a number of provinces, besides redeployment, we envis[age] the creation of powerful organizational nuclei, including in those regions which border Pakistan…
>
> Analyzing the situation further, I want to note that the enemy continues to strengthen his forces, bring in caravans with weapons, and create his reserves in various regions. We are preparing to launch strikes on bases and depots and intercept caravans. But we associate the larger scale of operations with the results of the talks in Geneva. We are also considering the possibility which you have been talking about: the enemy could create a government in one of the regions of Afghanistan in order to turn to the Americans with a request for recognition…

Gorbachev replied with guarded optimism that when the ethnic groups recognized that the government was showing a degree of concern for them, they would respond. Minimizing the extent to which Afghan society had become militarized, he pronounced: 'in the final account they too are in favour of peace so that their people can quietly till their land. This is a decisive factor which also does not contradict the Qur'ān.'[46]

The two men met again in mid-June. On this occasion, Najibullah reported that the beginning of the withdrawal of Soviet troops had complicated the military and political situation in Afghanistan. The situation had worsened in a number of border provinces; an increase in the infiltration of caravans from Pakistan with weapons was being observed, and depots and bases were being created within Afghanistan. He was, however, able to state optimistically that 'the disputes between the foreign and domestic forces of the counter-revolution are growing stronger'. Of all the rebel groups, the most active were those of the Islamic Party of Afghanistan headed by Hekmatyar. They were concentrating their main efforts on the Kābul axis, trying to sow panic among the capital's population with shelling and terrorist acts.[47]

The meeting between the two men on 23 August 1990 saw Gorbachev in a different mood, castigating the conduct of US foreign policy with regard to the Muslim world:[48]

> The impression is being created that the Americans are actually concerned with the danger of the spread of Islamic fundamentalism. They think, and they frankly say this, that the establishment today of fundamentalism in Afghanistan, Pakistan, and Iran would mean that tomorrow this phenomenon would encompass the entire Islamic world. And there are already symptoms of this, if you take Algeria for example. But the Americans were, and will remain, Americans. And it would be naïve if one permitted the thought that we see only this side of their policy and do not notice other aspects. It is clear that the U.S. is not opposed to fundamentalism becoming the banner of 40 million Soviet Muslims and creating difficulties for the Soviet Union. They object only to it affecting their own interests…

Najibullah agreed that the US was primarily interested in strengthening the positions of Islamic fundamentalism not only among the peoples of Soviet Central Asia but among all Soviet Muslims: 'equivalent retaliatory actions will be required to disrupt similar plans and here, in our view, the interests of the Soviet Union and Afghanistan closely overlap'. Indecisiveness in combat operations against the government of Afghanistan and internal differences among the various groups of the opposition had led to even Pakistan becoming disappointed in their own creation – the so-called 'transitional government of Afghan *mujāhidīn*'. All this was also increasingly influencing the mood of the 5.5 million Afghan refugees (about 3 million in Pakistan, up to 1.5 million in Iran, and 1 million in other countries), who were beginning more insistently to demand their return home. Najibullah compared Pakistan to a boiling kettle which needed to let off steam, first in Afghanistan and now increasingly in Kāshmīr.[49]

Najibullah predicted that 'in the next two [to] three years we will be able to achieve a decisive breakthrough in the cause of complete normalization of the situation in the country'.[50] In reality, his days were numbered.[51] The two superpowers agreed to cut off aid to their respective clients with effect from the beginning of 1992. Abdul Rashīd Dostum's northern Üzbek militia rebelled against the government in February 1992. The *mujāhidīn* moved into Kābul in April 1992 in a relatively bloodless victory. Almost immediately internecine warfare broke out, though had Aḥmad Shāh Mas'ūd reacted differently and set up a government without the Peshawar *mujāhidīn* groups, the outcome might have been different.[52] From April 1992 until early 1995, with the first Ṭālibān advance on Kābul, Hekmatyar's forces attacked the city on a regular basis. Elsewhere, control of Afghanistan fell to the various warlords such as the northern minorities led by Aḥmad Shāh Mas'ūd and Dostum. The country descended into ethnic violence and political fragmentation.

Reaping the whirlwind: post-Soviet Afghanistan, Kāshmīr, Central Asia, Chechnya, Xinjiang and Algeria

Professor Yvonne Haddad calls Islamism 'a designer ideology for resistance, change and empowerment'. Her definition of this ideology is highly pertinent for the following discussion:[53]

> it has been self-consciously formulated by a variety of authors and politicians over a century to provide answers for problems facing various Muslim societies. The worldviews that are prescribed generally tend to be packaged as a response to what is perceived as an outside threat to the society. The ideas reflect the variety of issues that have brought Islamism into being in different countries conditioned by the problems its architects have attempted to address. Thus its ideas have mutated into several forms and theories supported by different organizations and movements in different Muslim countries…
>
> Islamism is the continuing process of de-colonization, a persistent reaction against what is experienced as intrusive foreign domination and intervention, whether political, military, economic or cultural. Its advocates continue to seek political participation in the defining of the destiny of the *ummah*… Islamism is reactive, not a reactionary movement, one that responds to internal as well as external challenges… At its core Islamism is a quest for empowerment, addressing political concerns such as sovereignty, liberation, removal of oppressive rulers, and the quest for democratization…

This 'quest for empowerment' was an important reason for the importance of the militant Islamist movements in post-Soviet Afghanistan, Kāshmīr, Central Asia, Chechnya, Xinjiang and Algeria, each of which will be investigated in turn.

Afghanistan

'For they have sown the wind, and they shall reap the whirlwind…' (Hosea 8:7). The carefully-constructed anti-Soviet *jihād* had led to the spectacular success of enforced Soviet withdrawal from Afghanistan (15 February 1989) and the abandonment of the Najibullah regime, which in turn eventually collapsed three years later. Yet a whirlwind was reaped. An example was set of what could be done. The mythology of the successful *jihād* had been revived; and though Arab participation was less marked in the campaign than is sometimes claimed, it was nevertheless significant. Trained and semi-trained fighters, imbued with the radical Islamist ideology, were available to be exported to the world's other trouble spots where a deep sense of oppression on the part of Muslims was the core issue.

Did the founders of US foreign policy in Afghanistan during the Carter administration (1977–81) realize that in spawning Islamic militancy with the primary aim of defeating the Soviet Union they were risking sowing the seeds of a phenomenon that was likely to acquire a life of its own, spread throughout the Muslim world and threaten US interests? Dilip Hiro argued in January 1999:[54]

> The main architect of US Policy was Zbigniew Brzezinski, President Carter's National Security Advisor. A virulent anti-Communist of Polish origin, he saw his chance in Moscow's Afghanistan intervention to rival Henry Kissinger as a heavyweight strategic thinker. It was not enough to expel the Soviet tanks, he reasoned. This was a great opportunity to export a composite ideology of nationalism and Islām to the Muslim-majority Central Asian states and Soviet republics with a view to destroying the Soviet order. Brzezinski also fell in easily with the domestic considerations of Gen. Mohammad Zia ul-Haq, the military dictator of Pakistan… The glaring contradiction of the US policy of bolstering Islamic zealots in Afghanistan while opposing them in neighboring Iran seemed to escape both Brzezinski and his successors. In the words of Richard Murphy, the Assistant Secretary of State for the Near East and South Asia during the two Reagan administrations, 'we did spawn a monster in Afghanistan'. The 'monster' of violent Islamic fundamentalism has now grown tentacles that extend from western China to Algeria to the east coast of America, and its reach is not likely to diminish without a great deal of the United States' money, time and patience, along with the full co-operation of foreign governments…

Zbigniew Brzezinski admitted in an interview for *Le Nouvel Observateur* in January 1998 that he placed the defeat of the Soviet Union higher in the scale of values than having created some 'stirred-up Muslims'.[55]

> *Question*: When the Soviets justified their intervention by asserting that they intended to fight against a secret involvement of the United States in Afghanistan, people didn't believe them. However, there was a basis of truth. You don't regret anything today?

> *Brzezinski*: Regret what? That secret operation was an excellent idea. It had the effect of drawing the Russians into the Afghan trap and you want me to regret it? The day that the Soviets officially crossed the border, I wrote to President Carter: 'we now have the opportunity of giving to the USSR its Vietnam war.' Indeed, for almost 10 years, Moscow had to carry on a war unsupportable by the government, a conflict that brought about the demoralization and finally the break-up of the Soviet empire.
>
> *Question*: And neither do you regret having supported… Islamic fundamentalism, having given arms and advice to future terrorists?
>
> *Brzezinski*: What is most important to the history of the world? The Ṭālibān or the collapse of the Soviet empire? Some stirred-up Muslims or the liberation of Central Europe and the end of the cold war?
>
> *Question*: Some stirred-up Muslims? But it has been said and repeated [that] Islamic fundamentalism represents a world menace today.
>
> *Brzezinski*: Nonsense! It is said that the West had a global policy in regard to Islām. That is stupid. There isn't a global Islām. Look at Islām in a rational manner and without demagoguery or emotion. It is [a] leading religion of the world with 1.5 billion followers.[56] But what is there in common among Sa'ūdī Arabian fundamentalism, moderate Morocco, Pakistan militarism, Egyptian pro-Western or Central Asian secularism? Nothing more than what unites the Christian countries…

Zbigniew Brzezinski clearly understood Soviet Communism much better than he did the potential of militant Islamism to opt for a strategy which might be called 'global *jihād*'.

The Ṭālibān military coalition comprised hard-core Ṭālibān members (including long-time veterans and new adherents); contingents from warlords (former *mujāhidīn*) and tribal contingents; seasonal village conscripts; and foreign forces. The veteran core of the Ṭālibān may have been no larger than 2000–3000 personnel; adherents after 1995 may have numbered another 6000–8000. Warlord, tribal, and conscripted indigenous troops probably comprised an additional 20,000–25,000 troops; these troops constituted the soft outer shell of the Ṭālibān coalition, likely to defect or desert under pressure. Finally, foreign volunteers probably numbered 8000–12,000 men and young persons. These included 5000–7000 Pakistanis, 1500–2000 members of the Islamic Movement of Uzbekistan, and no more than 2000–3000 foreign fighters imported by the al-Qaeda network – mostly Arabs.[57]

Among the troops controlled by al-Qaeda only a few hundreds are thought to have been core al-Qaeda members, distinguished by having signed a *bay'at*, or oath of allegiance to the *jihād*, as defined by bin Laden. From late 1989 to late 1991, most of al-Qaeda's best-trained and most experienced fighters, between 1000 and 1500 of them, moved to Sudan, in the belief that it would be more active and they could 'go to work again'. Bin Laden retained an extensive training and operational infrastructure in Afghanistan and Pakistan, but most of the international recruits returned to their home countries after their period of training.[58] Among the Ṭālibān adherents, the most important were the veteran cadre from Kandahar and surrounding provinces: these were the core troops that held the Ṭālibān coalition together. The key factor in their rise to power was momentum, which they gained through their early victories and the promise of a return to peace and order by means of defeating the warlord factions. This mobilized former students of the north Pakistan *madrasahs*, brought in new recruits, and helped sway Pashtun warlords and tribal leaders to join the coalition. The support of commercial trading interests and Pakistan also was essential in helping to build Ṭālibān power. Reinforcing the power of the Ṭālibān were the foreign volunteers and the Afghan Arabs (organized through the al-Qaeda network) who combined dedication, experience, and relative fighting prowess.

The Ṭālibān accomplished what none of their *mujāhidīn* predecessors had done: the relatively rapid unification of most of Afghanistan. This they did through their religious discipline and fervour; the mobilization of a grassroots constituency (that is, village religious leaders and students) that cut across many divisions of tribe and locality in the Pashtun belt; and the campaign for the restoration of law and order. Many core Ṭālibān leaders had earlier served as members of the *mujāhidīn* militias during the time of the anti-Soviet war – especially the more traditionalist Islamic parties, Ḥarakat-i-Inquilāb-i Islāmī and Younis Khalis' faction of Ḥizb-i Islāmī. What gave the Ṭālibān their initial organizational and ideological coherence was the common training the students had received in certain of the more radical Pakistani *madrasahs*.

In examining the reasons for the Ṭālibān rise to power, many commentators have argued that they enjoyed the overwhelming support from the Deobandi school and its *madrasahs* from which they originated. There was indeed some support, but it was by no means universal. Yoginder Sikand interviewed Waris Maẓharī, the editor of the Urdu monthly *Tarjumān Dār ul-'Ulūm*, the official publication of the Delhi-based Deoband *Madrasah* Old Boys' Association. Maẓharī was forthright in his condemnation of the Ṭālibān ideology:[59]

> …this approach was fundamentally wrong. You cannot forcibly impose Islamic laws on an unwilling people. The Ṭālibān forced the Afghans to abide by their version of Islamic law, and many people did so, not out of conviction, but simply out of fear. In this way, Islām was reduced to a cruel joke. Hence, it was but natural that many Afghans resisted the Ṭālibān, and the regime finally collapsed. Had the Ṭālibān sought to first convince the Afghans of the need to be ruled by Islamic law, and, after preparing public opinion and gaining

> mass support, sought to establish political authority, they would probably have succeeded.
>
> I myself opposed the way that the Ṭālibān was going about trying to forcibly impose its will. I must admit, though, that many '*ulamā'* did support the Ṭālibān. As I see it, this is a reflection of the lamentable fact that many contemporary '*ulamā'* see the world from a basically political, rather than purely spiritual or religious, perspective… The feeling of revenge that drives them to settle scores against the West for what it has done to the Muslim world also leads them to believe that anyone who speaks out against the West is somehow a great champion of Islām.
>
> I would be the last to deny the oppression that many Muslim peoples and countries have suffered, and continue to suffer, at the hands of the West, but I do not feel this is the way to counter Western hegemony. In fact, it is positively counter-productive…

The qualities that made the Ṭālibān a successful social-military movement in the Pashtun areas – their grassroots orientation and religious discipline – did not serve them as well when they moved farther north. Nor did it assist them well in government. For the Ṭālibān, the essence of government was the promulgation and enforcement of *sharī'ah*: its key ministry, on the Sa'ūdī Arabian model of the Committee for Commanding Right and Forbidding Wrong, became the Ministry for the Enforcement of Virtue and Suppression of Vice. It was this ministry which sought to enforce the wearing of the *burqa*', the stipulated length of beards, mosque attendance, and so on. Conetta argues that

> the expansion of Ṭālibān territorial control and the need to balance the requirements of war and governance also posed a classical problem of over-extension for the regime. This was exacerbated by the regime's failure to open [up] the ranks of leadership.

Throughout the period of its rule, the Ṭālibān regime remained overly dependent on a small core of Pashtun followers, especially from the vicinity of Kandahar. These it circulated endlessly between two broad tasks or fronts: the military front in the north and the 'internal front', which involved maintaining *sharī'ah* (and Ṭālibān power) everywhere else. In this context, the Ṭālibān's embrace of bin Laden and his resources made sense.

Regrettably, though the regime may have created a greater respect for the law[60] because of its harsh public displays of amputations and executions, this respect for the law it did not apply to itself. In particular, no influence was brought to bear in order to restrain Ṭālibān acts of revenge. In 1996, Mullāh Omar proclaimed himself *Amīr al-Mu'minīn*, or 'Commander of the Faithful', a term used in the period of the caliphs which suggested the claim to absolute political and religious authority and to be the undisputed leader of the *jihād*. (The name of Afghanistan was later changed to the Emirate of Afghanistan.)[61] On this occasion, Mullāh Omar, symbolically held aloft the 'Robe of Muḥammad' (a cherished Pashtun relic) for about 30 minutes.[62] It seems unlikely that reprisals were taken against particular ethnic populations without Mullāh Omar's specific consent. At Mazār-i-Sharīf on 28 May 1997, some 600 Ṭālibān were killed in an orgy of fighting conducted by the Shī'a Hazārā population. Mullāh Omar called for more students from the Pakistan *madrasahs* to reinforce the Ṭālibān forces and even visited Kābul for the first time to meet his commanders and raise morale among the troops.[63]

On 8 August 1998, the Ṭālibān retook Mazār and immediately went on a revenge killing spree. Ṭālibān forces carried out a systematic search for male members of the ethnic Hazārā, Tājīk, and Ūzbek communities in the city. The Hazārās, a Persian-speaking Shī'a ethnic group, were particularly targeted, in part because of their religious identity. During the house-to-house searches, scores and perhaps hundreds of Hazārā men and boys were summarily executed, apparently to ensure that they would be unable to mount any resistance to the Ṭālibān. Also killed were eight Iranian officials at the Iranian consulate in the city and an Iranian journalist. Thousands of men from various ethnic communities were detained first in the overcrowded city prison and then transported to other cities, including Shiberghan, Herat and Kandahar. Most of the prisoners were transported in large container trucks capable of holding 100–150 people. In two known instances, when the trucks reached Shiberghan, some 130 kilometres west of Mazār, nearly all of the men inside had asphyxiated or died of heat stroke inside the closed metal containers. Some 2000 were reportedly summarily executed after capture in Shiberghan and other areas, including areas to which prisoners from Mazār were deported. A number of neighbourhoods targeted for searches in Mazār had been among those where earlier resistance to the Ṭālibān had been most marked. In speeches given at mosques throughout Mazār, the Ṭālibān governor, Mullāh Manon Niāzī, had blamed Hazārās for the 1997 killings. Witnesses stated that Ṭālibān forces conducting house-to-house searches accused Hazārās in general of killing Ṭālibān troops in 1997 and did not distinguish between combatants and non-combatants.[64] There were further Ṭālibān massacres of Hazārās at Robatak (May 2000) and Yakaolang (January 2001).[65]

Aḥmed Rashīd notes that in Islām, *jihād* is the mobilizing mechanism to achieve change. He then expresses the revulsion at the Ṭālibān regime which their acts of revenge created among the non-Pashtun population of Afghanistan:[66]

> The Ṭālibān were… acting in the spirit of the Prophet's *jihād* when they attacked the rapacious warlords around them. Yet *jihād* does not sanction the killing of fellow Muslims on the basis of ethnicity or sect and it is this, the Ṭālibān interpretation of *jihād*, which appals the non-Pashtuns. While the

> Ṭālibān claim they are fighting a *jihād* against corrupt, evil Muslims, the ethnic minorities see them as using Islām as a cover to exterminate non-Pashtuns... They fitted nowhere in the Islamic spectrum of ideas and movements that had emerged in Afghanistan between 1979 and 1994. It could be said that the degeneration and collapse of all three trends (radical Islamism, Ṣūfīsm and traditionalism) into a naked, rapacious power struggle created the ideological vacuum which the Ṭālibān were to fill. The Ṭālibān represented nobody but themselves and they recognized no Islām except their own...

Until the eve of the war launched by the United States, Britain and their coalition partners in 2001, the Ṭālibān were heavily dependent on Pakistani support.[67] They also benefited from Pakistan's porous borders. Pakistan's military and intelligence establishments had links to the Ṭālibān at multiple levels, reinforced by personnel inside Afghanistan, who were working side by side with the Ṭālibān. These various contacts and influences gave Pakistan potential leverage to pressurize the Ṭālibān into surrendering bin Laden – an effort that might have borne fruit given a more patient approach by the Coalition in 2001. On the other hand, the proverbial stubbornness of the Afghans did not facilitate the task of separating out the al-Qaeda network from its Ṭālibān hosts. The cruise missile attacks ordered by President Clinton on north-east Afghanistan (as well as Libya) on 20 August 1998 in retaliation for the attack on the US embassies in Kenya and Tanzania demonstrated the price that the Ṭālibān – and the civilian population of Afghanistan – were likely to pay for giving sanctuary to bin Laden. Yet Mullāh Omar responded to the attack by reiterating that bin Laden was a guest of Afghanistan, while he called 'America itself... the biggest terrorist in the world'.

The simplest explanation for Mullāh Omar's stubborn defiance of bin Laden – that he was his father-in-law – has been categorically denied by one specialist.[68] By late 1998, if not earlier, bin Laden was clearly a military asset but a political liability to the Ṭālibān regime; nonetheless, Prince Turkī of Sa'ūdī Arabia, the then head of Sa'ūdī intelligence, was unable to secure his expulsion.[69] (Turkī, who had had responsibility for Afghanistan, resigned from his post on 31 August 2001, just twelve days before the attacks on New York and Washington DC, prompting the question as to how much the Sa'ūdīs knew in advance about the 9/11 attacks.)[70] In reality, bin Laden was still useful to Mullāh Omar's regime: the assassination of Aḥmad Shāh Mas'ūd on 15 September 2001, which was intended to help the Ṭālibān take control of the far north, Afghanistan's last anti-Ṭālibān stronghold, was carried out by al-Qaeda operatives.[71] The spectacular coup backfired, however, since it brought the remnants of the Northern Alliance closer together behind the rallying call of revenge for Aḥmad Shāh Mas'ūd's assassination.

Divisions within the Ṭālibān movement had been evident before the war of 2001, especially with regard to relations with the West, and they became apparent again during the Ṭālibān's twilight hours. These divisions pitted the Kandahar *shūrā* against the generally more pragmatic Kābul *shūrā*, which had direct responsibility for government administration and, supposedly, military affairs. The broader constituency for this group were the 'second generation' members of the Ṭālibān ruling coalition – local leaders, veteran *mujāhidīn*, and new Ṭālibān adherents who had joined the group during its post-1994 rise to national power.[72] Once war came in the last months of 2001, the combination of US air power and Northern Alliance ground troops (guided and assisted by US special operations forces) broke the Ṭālibān defensive positions outside Mazār-i-Sharīf and Kābul. The Ṭālibān recognized the inevitable and began withdrawing select troops before their lines collapsed completely under pressure. This pattern was repeated throughout Pashtun areas that the Ṭālibān had decided to surrender. But the disposition of the released and retreating Ṭālibān troops was quite different in the north and the south. In the south, those Ṭālibān who did not retreat to Kandahar were able to melt into their surroundings. This process was evident in Jalalabad, for instance, where former *mujāhidīn* leader Younis Khalis negotiated the surrender of the city from the Ṭālibān. In contrast, in the north, those Ṭālibān coalition troops left behind could not easily reintegrate locally. This was especially true for Pakistanis and Arabs. Many were pursued into Mazār-i-Sharīf, Konduz, Khānabād, and Taloqan, surrounded, bombed by US air power, and killed or captured: in excess of 800 Ṭālibān coalition troops were killed in reprisals or after capture.

More than a tactical military retreat, the Ṭālibān had executed a strategic withdrawal and reorientation during the second week of November, relinquishing any pretence to power in three-quarters of the area previously under their control. The Ṭālibān also effected a separation from most of the al-Qaeda core, many of whom took refuge in the Tora Bora fortified base near Jalalabad, about 350 miles from Kandahar. Carl Conetta suggests that the Ṭālibān may have hoped that their surrender of the capital and retreat to their provincial base, together with their separation from al-Qaeda, would satisfy the war objectives of the Coalition and permit a negotiated settlement.[73] Although retreat allowed the Ṭālibān to concentrate its best troops in the defence of a much smaller area, it also allowed the United States better to concentrate its air power and reduced flight distances for its bombers. The Ṭālibān also underestimated the effect of their retreat on their political authority. The growth of Ṭālibān power had been based on a core of disciplined, dedicated followers and rapid forward momentum. Defeat and retreat robbed them of their charisma and authority. They were unable to reconstitute their power in the south, although as at June 2004, remnants of the Ṭālibān continued to operate there and remained a troublesome factor for the new regime. There were even reports that the British were trying to open negotiations with them via Pakistani intermediaries so as to permit a withdrawal of Coalition troops.[74]

Kāshmīr

> In 1989, the people of Kāshmīr took their historic stance, and declared it a *jihād* in the path of Allāh to achieve one of the two honours, either victory or martyrdom. *Jihād* missions commenced against the Indian occupation, and the *mujāhidīn* party emerged as a strike force in the midst of the occupation.[75]

The words are those of Muḥammad Yūsuf Shāh, who in April 1991 took the name Mujāhid Syed Ṣalāḥuddīn and was appointed the Ḥizb Amīr (or Amīr of the *mujāhidīn*), a position that he still held in June 2004. He was also the chairman of Jihād Council, an association of 13 militant Kāshmīrī *mujāhidīn* groups.[76] A long-standing dispute between India and Pakistan since 1947 took on a new form in 1990. Sumantra Bose calls the first period of the Kashmīr struggle after 1990 the '*intifāḍah* phase', which he argues lasted until 1995. This freedom or independence (*āzādī*) movement was a conflict between state power and popular insurrection: in January 1990, Bose notes that 'massive demonstrations calling for Kāshmīr's *āzādī* from India erupted in Srinagar and other towns in the valley'. The Indian response was to unleash their paramilitary forces on the unarmed demonstrators: three days of protests left 300 demonstrators shot dead in Srīnagar.[77] The numbers killed may have been many fewer, perhaps only 100 (the vagueness about the figures is itself eloquent testimony to the lack of proper investigation of the incidents); but they were certainly more considerable than the 'death of sixty-nine people and the injury of more than 300'[78] in the Sharpeville Massacre in March 1960 that is seen, in retrospect, as a watershed in the struggle against South Africa's system of apartheid.

Such incidents served only to fuel the violent resistance movement, but it is striking that the movement was in origin supported overwhelmingly by local Kāshmīrī recruits: of 844 guerrillas killed in fighting during 1991, according to the official figures of the Indian counter-insurgency command in Srīnagar, only two were *not* residents of Indian Jāmmū and Kāshmīr.[79] On 24 January 1990 JKLF (Jāmmū and Kāshmīr Liberation Front) gunmen responded to the Srīnagar massacres by killing four unarmed Indian air force officers on the outskirts of the city. Thereafter, 'the Valley was caught up in an escalating spiral of violence and reprisal'.[80]

One of the Kāshmīrī independence leaders, Sayyed 'Alī Shāh Gīlānī, argued that the *jihād* required the participation of all Muslims and not just Kāshmīrīs since the 'extreme oppression of the Kāshmīrī Muslims' was 'an open challenge to the entire Muslim *ummah*', especially the people of Pakistan. The Hindus were not the enemy: the true enemies were the Indian state and its agents. Kāshmīrī separatists who did not seek union with Pakistan were labelled 'enemies of the *jihād*' and foes of Pakistan.[81] Subsequently, more militant groups such as the Lashkar-e-Taiba, or Army of the Righteous, the military wing of the Da'wat ul-

Irshad (Centre for Preaching and Guidance), came to control the *jihād*, aiming to spread it 'to every peak, every forest and every path'. The Lashkar introduced 'martyrdom operations', called 'Ibn Taymīyah Fida'ī missions' after the medieval jurist (see Chapter 4). *Jihād* was seen unabashedly as *qitāl*, violent conflict waged against unbelievers, 'the foreign policy of the Islamic state'. In the philosophy of this movement, the abandonment of *jihād* had led to the degeneration of the Muslim community. The conflict in Kāshmīr had thus become for some radical groups the campaign to 'uphold the flag of freedom and Islām through *jihād* not only in Kāshmīr but in the whole world'. The *shudra* or lowest worker caste of India should be saved from the clutches of the Brahmins. Hostility to Ṣūfīsm, 'the normative expression of [Kāshmīrī] faith and Islamic commitment', is another distinguishing feature of the more radical militant groups. Yoginder Sikand thus sees their propaganda as 'an external agenda that is seeking to impose itself on Kāshmīr, and one that seems at odds, in several respects, with the internal conditions in Kāshmīr itself'.[82] The enlarged *jihād* programme has been defined in all-embracing terms: to end persecution and tumult (*fitnah*); to enforce the Islamic world order; to force unbelievers to pay the *jizya*; to protect and shield the weak from oppression; to avenge the millions of Muslims being slaughtered in various parts of the world; to punish those who have made and then broken covenants with Muslims; to restore to Muslims possession of territories which they formerly occupied such as Spain, India, Jerusalem, Turkestan (Xinjiang); and to defend and protect all Muslims from any offensives mounted by the unbelievers wherever they may occur.[83]

How many *mujāhidīn* were killed in the fighting? In September 1995, Mujāhid Syed Ṣalāḥuddīn claimed that 6000 *mujāhids* had been 'martyred' but that they had killed some 20,000 Indian soldiers. The Indian forces had vented their anger on the civilian population, killing some 34,000 innocent civilians, he alleged.[84] The figures are doubtful, but as Sumantra Bose argues, while 'state-sponsored violations of civil liberties and fundamental democratic rights of citizens had been normal, indeed institutionalized practice in Indian Jāmmū and Kāshmīr for four decades prior to 1990' what unfolded after 1990 was 'of a different order and magnitude – a massive human rights crisis'.[85]

Another dimension of the human rights crisis was the exodus of some 100,000 Kāshmīrī (Hindu) Pandits in the first two months of 1990, relatively few of whom have since returned. Many, no doubt, had been intimidated, although group panic and a degree of collusion or encouragement from the Indian authorities – which sought to stigmatize the *āzādī* movement as sectarian and 'fundamentalist' – were also factors. The Pandit exodus revealed the flaw in the 'independent Kāshmīr' concept, that it would have grave difficulty in accommodating the multiple political allegiances concerning sovereignty and citizenship that existed in Indian Jāmmū and Kāshmīr. The exodus was also an indictment of the JKLF, which dominated the first three years of the insurrection and was committed to an independent Kāshmīr with equality for different faiths, ethnicities and regions.[86]

There was a rapid and alarming proliferation of armed groups in the Valley of Kāshmīr, partly caused by the popularity of the cause and partly because Pakistan's ISI sought to channel resources to those groups which were pro-Pakistan rather than which favoured an independent Kāshmīr. The JKLF suffered the highest casualties of any group as a result of the repression of the Indian counter-insurgency forces. The numbers of guerrillas killed were impressive: 2213 in the first three years of the insurrection; 1310 in 1993; and 1596 in 1994.[87] The JKLF announced a unilateral ceasefire in mid-1994, although it suffered further casualties at the hands of the Indian security forces for a long time after this announcement.

Interviewed in 1995, Mujāhid Syed Ṣalāḥuddīn claimed that there were more than 23 Kāshmīrī organizations; but that 'all of these are symbolic with the exception of two or three of them':[88]

> With the blessing of Allāh firstly, and with the help of the call amidst the brotherhood, the Islamic movement in Kāshmīr established the *Mujāhidīn* Party, which is the strongest of the organizations, the biggest and the most organized. Its base is inside Kāshmīr and it represents approximately 92% of the fighters in Kāshmīr: it is possible for you to enter the camps and conduct your own surveys to prove this. Although there are other parties, however, these are very small. There are also some of the nationalistic groups such as the Jāmmū and Kāshmīr Liberation Front, who have declared that the rifle has served its role, and that it no longer has a role on the political arena which must now be pursued by all. We in turn have rejected this view... We are certain that the solution in Kāshmīr will only be realized through the barrel of the rifle. If the political arena was able to provide a solution, it would have done this during the 43 years of political battles [that is, prior to 1990] which the Kāshmīrī people lived through without any result. On the contrary, India has taken advantage of this opportunity strengthening its choking siege on the Kāshmīrī people...

From 1991, HM (Ḥizb al-Mujāhidīn) under Mujāhid Syed Ṣalāḥuddīn was the main beneficiary of funds and other assistance channelled by the ISI. It is scarcely surprising that by 1993 it had emerged as the leading guerrilla movement, although its ideology that 'Kāshmīr will become Pakistan' may have been a minority cause.[89] The last mass demonstrations in the *intifāḍah* phase of the movement were in May and October 1994 and May 1995. In the following three years, there was a degree of 'demoralization and atrophy' – Sumantra Bose's expression – which overcame the *āzādī* movement.

A new phase of the insurgency, the *fedayeen* (or *fidā'iyūn*) phase, began in July 1999, shortly after the Kargil incident and has continued since. Between that date and the end of December 2002, at least 55 attacks took place, 'usually executed by two-man teams', targeting the police, paramilitary and military camps and government installations.[90] The number of guerrillas killed rose significantly, as did the number of non-Kāshmīrīs among them, though there is reason to suppose that the percentage of non-Kāshmīrīs was exaggerated by the Indian authorities.[91] Two new groups, JeM (Jaish-e-Moḥammad, or Army of the Prophet)[92] and LeT (Lashkar-e-Taiba, or Army of the Righteous),[93] came into prominence. Both were declared terrorist groups by the United States, responding to Indian pressure, after September 2001. In the course of 2002 and until 2003, after two spectacular incidents, the second being the attack on the Indian Parliament on 13 December 2001, India came close to war with Pakistan: using the United States' doctrine of pre-emption, it argued the case for a pre-emptive strike against a country which it alleged harboured terrorist organizations and permitted cross-border terrorism. Whether such 'coercive diplomacy' achieved its purpose is a matter of debate.[94] The US Ambassador in Pakistan was summoned to the Foreign Ministry in January 2003 to explain her comments with regard to the need to prevent cross-border terrorism: Pakistan made it clear that it had taken 'all measures' to prevent infiltration.[95] What the Indian coercive diplomacy certainly served to do was to distract attention away from the fact that the conflict is a production of 'the incendiary infusion of the ideology and tactics of trans-national Islamist militancy into a brutalized, desperate local environment – that is, of a conjunction of internal and external factors'.[96]

By the end of 2002, the frequency of raids had decreased, but the selection of targets had been widened. Sumantra Bose suggests that targets were chosen, and attacks timed, to increase communal tensions in Indian-controlled Jāmmū and Kāshmīr as well as to maintain pressure on the already poor state of India–Pakistan relations. The surprise outcome of the Islamabad summit in January 2004, which committed the two powers to the search for a peaceful settlement of the dispute, has not found a positive response among the *fedayeen* groups. 'Nothing but paperwork' was how Mujāhid Syed Ṣalāḥuddīn, supreme commander of the Ḥizb al-Mujāhidīn, responded to what others have regarded as a landmark deal by India and Pakistan to resume dialogue to solve all issues at dispute including Kāshmīr. 'We have seen dozens of such announcements and agreements in the past but unfortunately India never honoured a single one.' Ṣalāḥuddīn stated that Kāshmīrīs

> could not trust India, which insists the region is an integral part of its territory and has deployed thousands of troops to suppress the insurgency… It seems India wants to gain time, during which it [will]… employ every possible resource to crush the freedom struggle in the occupied territory.[97]

Central Asia

We will begin the survey of Islamist movements in Central Asia with Tājīkistān, where the break-up of the Soviet Union in 1991 led almost immediately to civil war. Tājīk Islamists operated through the Islamic Renaissance Party (IRP) and, in 1992, established a military arm of the organization called the Islamic Movement of Tājīkistān. In 1995, the IRP combined with the United Tājīk Opposition (UTO), which had been operating out of northern Afghanistan. By 1995, the conflict settled into a protracted stalemate. Aḥmed Rashīd argues that 'the civil war had quickly become a battle between clans rather than an Islamic *jihād*', a struggle in which the IRP was 'never able to overcome the problems of regionalization'.[98] In 1996, the Ṭālibān captured Kābul and ousted the Afghan–Tājīk government of Afghanistan. With Ūzbek pressure from the north and Afghan–Tājīks fighting the spread of the Ṭālibān in the south, both President Rahmonov and the IRP leader, Sayed Abdullāh Nūrī, began negotiations for peace.

Brokered by the United Nations special representative, Ivo Petrov, the peace settlement called for a coalition government that included the IRP. Despite the problems of the February 2000 parliamentary elections, in which Rahmonov's party predominated, Nūrī declared that the peace process was 'irreversible'. As one of his aides, Moheyuddin Kabīr said, '*jihād* cannot be the only criterion as advocated by the IMU [Islamic Movement of Ūzbekistān/Ūzbakistān or al-Ḥaraka Islāmiyyah, literally 'the dominance of Islām over all other ways of life']. What is needed is a political structure that can further the cause of Islām'.[99] Sharīf Himmatzoda, the former IRP military commander, commented: 'governments in the region have to change their attitudes towards Islamic movements to give them a legal, constitutional way to express themselves and play a role in state building. If they don't do so, people will join the extremists.'[100]

The presence of the formerly militant Islamists in the constitutional structure as the opposition party did not, however, stop the endemic corruption in this impoverished country.[101] Nor did it prevent President Rahmonov from identifying Tājīkistān's two priorities as fighting against terrorism and religious extremism. He banned the covert HT (Ḥizb at-Taḥrīr), a movement that promoted a non-violent establishment of a pan-Islamic caliphate. 'We do not have connections to Osama bin Laden or any other terrorist organizations, as we pursue different methods of struggle', claimed the movement's leader, Nurullo Majidov. 'We are fighting for our ideas through peaceful means', he added.[102] 'The HT wants a peaceful *jihād*', another spokesman said,

> which will be spread by explanation and conversion, not by war. But ultimately there will be war because the repression by the Central Asian regimes is so severe, and we have to prepare for that. If the IMU suddenly appears in the Fergana valley,[103] HT activists will not sit idly by and allow the security forces to kill them.[104]

Aḥmad Rashīd contends that

> the IMU is reorganizing in Central Asia, as is Ḥizb at-Taḥrīr. Both organizations have a new slogan now which is basically anti-Americanism. They feel that over the medium- and long-term, they will be able to mobilize greater popular support because the Americans now have bases in three countries in Central Asia (Ūzbekistān, Tājīkistān and Kyrgyzstan). The Americans will be seen to be propping up dictatorial regimes and not pushing them hard enough to carry out economic and social reforms.[105]

In a surprise move, President Rahmonov reached agreement with President Putin of Russia on 4 June 2004, which appears to have given Russia the upper hand in Tājīkistān for the foreseeable future (Russian border guards will be in charge of Tājīkistān's frontier with Afghanistan until at least 2006).[106]

The IMU (Islamic Movement of Ūzbekistān) was not formed until 1998, but its precursors are to be found in the groups with which Tahīr Yuldashev and Juma Hojiev, its founders, were associated. After the defeat of the opposition in the Tājīk civil war in December 1992, the Ūzbeks together with the Tājīk Islamic opposition left for Afghanistan, where they formed their main base of operations. In 1998, Yuldashev and Hojiev (now known by his *nom de guerre* of Namangani) announced at Kābul the creation of the Islamic Movement of Ūzbekistān (IMU) and declared *jihād* on the regime of President Karimov. The IMU declared that its ultimate goal was the removal of the Karimov regime by force and the establishment of Islamic rule in Ūzbekistān. In an interview given to *Voice of America*, Yuldashev outlined the goals of the movement:[107]

> The goals of the IMU... are firstly fighting against oppression within our country, against bribery, against the inequities and also the freeing of our Muslim brothers from prison... Who will avenge those Muslims who have died in the prisons of the regime? Of course we will. We consider it our obligation to avenge them and nobody can take this right away from us. We do not repent our declaration of *jihād* against the Ūzbek government. God willing, we will carry out this *jihād* to its conclusion...
>
> We declared a *jihād* in order to create a religious system, a religious government. We want to create a *sharī'ah* system. We want the model of Islām which has remained from the Prophet, not like the Islām in Afghanistan or Iran or Pakistan or Sa'ūdī Arabia – these models are nothing like the Islamic model... Before we build an Islamic state we primarily want to get out from under oppression. We are therefore now shedding blood, and the creation of an Islamic state will be the next problem...

In August 1999, the IMU organized the kidnapping of four Japanese geologists working for a mining company as well as other hostages during a summit held at Bishkek. President Akayev denounced the activities of '400 IMU gunmen' who, he claimed, were trying to undermine the whole of Central Asia. The IMU responded by extending the *jihād* to Kyrgyzstan because Akayev had arrested 'thousands of Muslim Ūzbeks who had migrated to Kyrgyzstan' but who had been handed over to the Ūzbek regime.[108]

The United States has contended that bin Laden supplied most of the funding for setting up the IMU. There was certainly a later meeting at Kābul, in September 2000, which involved the Ṭālibān, the IMU, the HT, Chechen separatists and bin Laden himself.[109] Its precise outcome, however, is not known. The IMU helped to defend the Ṭālibān regime in the war of 2001. Significant numbers of IMU fighters were involved in the battle in the Shāh-i-Kot valley, and there is little doubt that the organization has suffered heavy losses at the hands of US and Coalition forces. Juma Namangani was fatally injured during fighting for the northern city of Mazār-i-Sharīf, where the Ṭālibān were routed on 9 November 2001. He died several days later as a result of his wounds, according to General Daoud Khān of the Northern Alliance. Tahīr Yuldashev came close to assassination by security forces in Pakistan in March 2004, but was travelling in a bullet-proof car with Nek Muḥammad, the tribal leader who was himself killed in June by a remote-controlled missile triggered by his mobile phone. By June 2004 there was evidence that Central Asian militant Islamists had formed a new organization named Jundullah (the Lashkar [force] of Allāh), headed by Ataur Rehman, who was detained by the Pakistan security forces. Ūzbek militants were thought to be operating out of the quasi-autonomous Pakistan tribal frontier provinces and it was said that some 60–70 of them had been killed in the security forces' campaign.[110]

Chechnya

Contrary to a great deal of prior speculation and pseudo-intelligence gathering, no Chechens were captured in the war against the Ṭālibān and al-Qaeda in Afghanistan in 2001–02.[111] With the collapse of the USSR in 1991, Chechen President Dzhokhar Dudaev and his nationalist leadership took advantage of Russian President Boris Yeltsin's offer to the Federal provinces to 'seize as much autonomy as they could', and to declare outright national independence for Chechnya. There was no Islamist or *jihādī* element at work in the first Chechen War of 1994–96, the ideology of the independence movement being secular. It is true that Dudaev is reported to have sent one of his most loyal supporters, Shāmil Basayev, and 40 of his followers, to Peshawar in Pakistan and to the Khost region of Afghanistan (the al-Khaldūn Camp) for military training in 1994; but only a small number of post-Soviet Chechen fighters made their way to train in the ISI-run military camps in the Khost region, while Osama bin Laden was actually living in Sudan at the time (1994). The evidence of a conspiracy with al-Qaeda, therefore, cannot be substantiated from this event. While several hundred transnational Arab *jihādīs* made their way to Chechnya in 1995 under the command of the Sa'ūdī Arabian-born *Amīr* ibn al-Khaṭṭāb (Samir Saleh Abdullah al-Suwailem) to assist the Chechen resistance against the Russians, Brian Glyn Williams contends that the evidence of Afghans in the first Russo-Chechen conflict is non-existent.[112]

The Chechens displayed to the world the moderate–secular future they envisaged for their land by overwhelmingly voting for Aslan Maskhadov, a secular pragmatist willing to work with the Kremlin, as president (1997). Yet for Shāmil Basayev and ibn al-Khaṭṭāb, quasi-autonomy was not the end of the struggle: as Basayev said, '*jihād* will continue until Muslims liberate their land and re-establish the Caliphate'. They tried to overthrow President Maskhadov's government by taking over the city of Gudermes, Chechnya's second-largest city, in July 1998 but were driven out by forces loyal to the government. Maskhadov attempted to ban Wahhābism and expel ibn al-Khaṭṭāb but backed off after several assassination attempts. Ibn al-Khaṭṭāb's fighters then attacked Daghestan in August 1999, which precipitated the second Russo-Chechen War. By 12 November, Russia had replied by seizing Gudermes.

Chechen President Maskhadov declared from his hideout in the mountains of southern Chechnya:

> we didn't ask for any military help from anyone, including Afghanistan, because there isn't [any] such... necessity. We have enough forces and means to sustain a full partisan war with the Russian army. There are no Chechen bases in Afghanistan, nor in Yemen. We don't need any bases, because during the previous war Russian generals taught our people how to fight.

By the spring of 2000, Russian troops had established nominal control over most of Chechnya and large-scale hostilities ceased. They continued to conduct their notorious 'sweep operations', to seek out rebel fighters and ammunition depots. 'Sweep operations' became synonymous with abuse, involving the arbitrary detention of large numbers of Chechen civilians (along with captured fighters), who were then beaten and tortured in detention.[113]

After the events of 11 September 2001, Russia went to great lengths to link the war in Chechnya to the global campaign against terrorism. On 12 September, Russian President Vladimir Putin declared that America and Russia had a 'common foe' because 'bin Laden's people are connected with the events currently taking place in our Chechnya'. On 24 September, he stated that events in Chechnya 'could not be considered outside the context of counter-terrorism', glossing over the political aspects of the conflict.[114] In the immediate months after 9/11, the Chechen resistance found itself pressed as never before, owing to the fact that the West had given the Kremlin carte blanche to ratchet up its ongoing war against

Chechen separatism under the guise of playing its part in the war against global al-Qaeda terrorism. As Andrei Piontkovsky wrote in the *Washington Post* in March 2004,[115]

> the Russian leadership constantly reiterates that it is not fighting Chechen separatists but international terrorists, and this has finally become a self-fulfilling prophecy. Thanks to the methods with which we have waged this war, we have turned practically the whole population of Chechnya into enemies and created for metaphysical terrorism a huge reservoir of living bombs – desperate people ready to carry out the plans of the terrorists.

Not that the Russian security forces were without successes: they bribed a Chechen messenger into delivering to ibn al-Khaṭṭāb a booby-trapped letter containing a poisonous agent. Chechen sources confirmed his death as having occurred on or about 19 March 2002. Photographs of ibn al-Khaṭṭāb's corpse showed no signs of wounds from combat, thus confirming the probable cause of death.[116] The important point, however, is that his place in the structure was immediately taken by another *jihādī* from Sa'ūdī Arabia, Abū al-Walīd ('Abd al-'Azīz al-Ghāmidī).[117]

On 4 August 2003, US Secretary of State Colin Powell designated Shāmil Basayev a threat to US security, adding that Basayev 'has committed, or poses the risk of committing, acts of terrorism' against the United States. Russia had already succeeded in 1999 in placing Shāmil Basayev's name and photograph on the Interpol website for offences of 'terrorism' and 'terrorism attempt'.[118] In spite of being hampered in mobility by multiple wounds and the amputation of his foot, Basayev nevertheless remains capable of mounting deadly strikes at Russian targets. As Andrew Mcgregor comments,

> Basayev is considered one of the most experienced and dangerous practitioners of 'asymmetrical warfare' in the world today. It is often said that the suicide truck-bombers of 'Irāq have adopted Palestinian or al-Qaeda tactics, but it was Basayev who perfected the procedure in a series of attacks on Russian targets that began in 2000. Today Basayev declares himself at war with Russian 'state terrorism,' alleging genocidal intentions on the part of Moscow.[119]

Calling himself *Amīr* of the Islamic Brigade of Martyrs (*Shahīds*), Basayev comments:

> today we can see on TV how rapid-reaction units and special-purpose police units are getting dispatched [to Chechnya], and how their wives, their sisters and their mothers are wishing them a good trip. We will be conducting operations right in those cities and villages where they came from... They are trying to

> accuse us of killing innocent people… They cannot be 'totally innocent', for the simple fact that they are approving of this slaughter, financing it, electing the rulers that are publicly promising to deal with the Chechens, and conducting genocide on the Chechen land…

That *jihād* is alive and well in Chechnya is also indicated by the statement issued on 26 November 2003 by the Sharī'ah State Defence Council (*Majlis al-Shūrā*) of CRI (Chechnya Republic Ichkeria, the independent republic of Chechnya):[120]

> Even though the issue of *Ḥukma* (the Decision) of *jihād* today seems to be clear, we often happen to encounter Muslims who ask a question: 'is *jihād* in Chechnya mandatory (*farḍ 'ayn*) or voluntary (*farḍ al-kifāyah*)?' or even 'is *jihād* going on in Chechnya?'… The mission… of *jihād* is [the] protection and spreading of Islām and spreading the calling, and protection of Muslims and unbelievers, who are under the jurisdiction of Muslims, from… foreign aggression… After reading this, we hope it is clear that the war operations that Muslims are conducting against foreign aggressors in Chechnya are *jihād*…
>
> We acknowledge that many *Amīrs* and Commanders are not an ideal, just like ordinary Muslims are not, they are often far from the examples of the disciples of the Messenger of Allāh (peace be upon him). But the *sharī'ah* texts are undeviatingly demanding that the rulers of Muslims in *jihād* are obeyed, except when they order to commit a sin…
>
> We cannot turn a blind eye to the faults or mistakes of the command, or especially support them in it; we must work hard to make them improve their personal qualities, to extend their Islamic knowledge and to promote the *sharī'ah* of Almighty Allāh in all areas of our life.

At the same time, the danger posed by the Russian policy of 'Chechenization', that is to say mobilizing counter-*jihād* forces from within the Chechen community, was recognized.[121] The resistance was nevertheless divided because of President Maskhadov's repudiation of Shāmil Basayev's tactics in the Moscow theatre hostage operation of October 2002.[122]

Xinjiang

On 2 October 2003, in a raid by the Pakistan armed forces on the tribal South Wazīristān region near the Afghan border, Ḥasan Mahsum (Abū Muḥammad al-Turkestanī) was killed along with seven others. The Eastern Turkestan Islamic Movement (ETIM), a small Islamist group based in China's western Xinjiang Province, is said to be one of the most militant of the ethnic Uīghur separatist groups pursuing an independent 'Eastern Turkestan', which at its greatest extent would include Turkey, Ḳazaḳstan, Kyrgyzstan, Pakistan, Afghanistan, and

Xinjiang. (Between 1944 and 1950 a 'Free East Turkestan' was set up by the Uīghurs, but this state collapsed with the Communist takeover of China.)[123]

The US designation of ETIM as a terrorist organization in 2001 was welcomed by Beijing but received very negatively by the Uīghurs, who considered that Washington, once virtually their sole hope for leverage against Beijing, had sacrificed their cause in order to gain Beijing's support for the 'war on terrorism'. Graham E. Fuller and S. Frederick Starr argue that

> this may cause some of them to conclude that they have no alternative but to embrace more radical philosophies to promote their national struggle. If the forces of Islamic radicalism thus gain strength across the region, it will obviously affect the Uīghur national struggle in Xinjiang and beyond.[124]

All the Uīghur groups deny that they are Islamic fundamentalists or that they seek to establish an Islamic state in Xinjiang.[125] Professor Dru Gladney argues that 'Uīghers in general are not engaged in radical Islām. There is a growing conservatism in the region but not the kind of Ḥizbu'llah, Ṭālibān type of Wahhābist Islām that the government seems to be very much afraid of.' Why then, he was asked, was Beijing pushing so hard for international support in its bid to list four Uīgher groups and eleven individuals as terrorists? Professor Gladney considers that this was linked to the fact that 22 Uīghers were caught in Afghanistan and remain detained by the US in Guantánamo Bay. 'The U.S. government has... admitted their presence and... the Chinese are pressuring the U.S. to return these people to China. The U.S. government clearly believes these people are terrorists' but the issue remains whether to return them to China, which has been criticized by the international community for its treatment of political prisoners.[126]

Algeria

If Xinjiang is an example of a potentially serious problem of Muslim separatism which has not, as yet developed into a *jihād* (in spite of some prompting from bin Laden), Algeria is a clear example of a *jihād*, pronounced on 11 January 1995, independent of the Afghanistan and bin Laden phenomenon. To explain the emergence of this most violent of '*jihāds* of the sword', we need to return to October 1988, when a mass movement led to rioting in the Algerian cities. In Algiers, some 6000 demonstrators chanted Islamic slogans.[127] The FLN (Front de Libération Nationale or National Liberation Front) regime put down the riots with significant loss of life, estimated at between 200 and 500 persons killed.[128] The group established to give voice to the reformist aspirations, the FIS (Front Islamique du Salut, or Islamic Salvation Front), was founded in March 1989 and recognized as a legal party in August the same year.[129] Professor Graham E. Fuller argued later, in 1996, that the FIS has strong ties with the grass roots of the population, and understands mass grievances better than almost any other party, especially among the urban poor, lower middle class, and marginalized educated class – all of which lack housing, jobs, and social services – the legacy of decades of FLN misrule. Despite a FIS grasp of what is wrong with the nation and a high degree of neighbourhood social activism, like many other Islamist movements in other countries, it purveys a message rather long on abstract principles, short on details, and fond of the slogan that 'Islām is the answer'.[130]

The FIS won stunning victories in the June 1990 municipal elections – the first free elections in the 28 years of post-independence Algeria – and subsequently in the first round of the national election in December 1991, when it won over 47 per cent of the popular vote, suggesting that it would gain an absolute majority in the second round.[131] Though many votes were cast for the FIS in protest against the corruption of the FLN rather than in favour of Islām as the solution, the FIS leader 'Abbāsī Madanī saw the victories as a 'mandate from Allāh'. (The party slogan, indeed, had been that 'to vote against the FIS is to vote against Allāh': some have argued from this that its commitment to a democratic outcome once it gained power was uncertain.)[132] No doubt with Khomeini's revolution in Iran in mind, the Algerian military felt threatened by the sudden upsurge of the Islamists. On 11 January 1992, they forced the resignation of President Chadli, called off the second round of elections scheduled for 13 January, and dissolved the FIS itself on 4 March. 'The entire FIS apparatus was dismantled', comments Gilles Kepel, 'thousands of militants and locally-elected FIS officials were interned in camps in the Sahara, and Algeria's mosques were placed under tight surveillance'.[133]

The West tacitly approved the suppression of democratic rule in Algeria, without any forethought that the military coup against an Islamist party which had won a clear expression of the democratic would precipitate a bloody civil war. That civil war lasted for the rest of the decade. By May 2002, it was estimated that 150,000 people, mostly civilians, had been killed since 1992. 'Abbāsī Madanī took a moderate stance but this was rejected by a number of groups which were formed to confront the new military regime. These eventually coalesced into two militant clusters: the Groupe Islamique Armé (Armed Islamic Group, GIA) was formed in December 1992[134] and the Armé du Salut Islamique (Islamic Salvation Army or AIS, the FIS military wing), is thought to have been formed the following year. Muḥammad Boudiaf, a hero of the FLN war for independence from France who had been exiled in Morocco since 1965, was recalled to Algeria on 16 January 1992 to take up the position of chairman of the High Council of State which the military had established to fill the void left by the vacant presidency. He was assassinated on 29 June 1992.

In the GIA's Open Letter and Call to All Muslims Worldwide (11 January 1995), issued under the signature of the *Amīr*, Abū Abdel Rahmān Ameen, the

'Algerian Muslim people' were said to be 'performing a severe *jihād* to cleanse the land from the remaining effects of crusade colonization and to establish a guided Islamic state…' France had occupied the state for 132 years as a colonial power and was now said to be 'a full partner in genocide by paying mercenaries and rewarding its agents and financing arms deals…' *Jihād* was pronounced a *sharī'ah* obligation in Algeria. To further justify its case ('ensure the truth and void the evil'), the GIA entered into detailed considerations of its strategy:[135]

1. The noble reason for *jihād* is to please the Almighty because it is a continuous and non-stop obligation to the day of judgement.
2. The goal of *jihād* is the establishment of Allāh's Rule on earth by reviving the guided Islamic state on the way of prophethood and the guidance of the *salaf*.
3. To reach this goal, the oppressive, un-Islamic regime must be removed.
4. The nature of *jihād* in Algeria is not a battle between two equal forces, but a *jihād* between a regime (which controls all the state institutions and is supported by world powers, especially France), and unarmed Muslim people [whose] *mujāhidīn* sons have stood up with their faith facing the regime's power machine, and using whatever weapons they extracted from the enemy. That is why it is the right of [the] *mujāhidīn* to use all ways allowed by *sharī'ah* to hit enemy personnel and financial resources.
5. [The] *mujāhidīn* consider anyone aiding the oppressive regime, an enemy of Allāh and His Messenger, and as a result he/she becomes a military target of *jihād* and for [the] *mujāhidīn*.
6. [The] *mujāhidīn* are abiding by *sharī'ah* in all of their operations, ways and means. They are obeying the creed of *ahl al-sunnah wal-jamā'at*.
7. [The] *mujāhidīn* are not launching blind terror as portrayed in [the] un-Islamic western media. It is the western regimes which use all means possible and perpetrate the worst crimes. [The] *mujāhidīn* are performing a clear struggle with identified goal[s] and means. It is none of our plans to kill innocent civilians if it was not proven that [the] person is a collaborator or an aide to the regime. Even so, punishment does not come until after repeated warnings and after positive proof is obtained.
8. We pledge in front of Allāh, then our Muslim people, that [the] *mujāhidīn* are strictly abiding by the rule of Allāh and *sharī'ah*, at all levels and in all fields. Based on this if any violations or mistakes are committed against personnel or property, then [the] *mujāhidīn* are obliged to correct the violation by compensation and apology…

In spite of the rhetoric, the GIA were clearly denouncing their opponents as *kuffār*, as apostates, who might be killed. It is also clear, from the justification that it issued in communiqué 44 on 21 May 1996, with regard to the killing of the French monks (see Chapter 4), that this was an extremely bitter civil war with few limits.[136] In June 2004, Algeria's alleged 'leading' terrorist group at that date, said to have links with al-Qaeda, the Salafist Group for Preaching and Combat (GSPC), claimed responsibility for killing more than a dozen soldiers. GSPC leader Nabīl Sahraoui stated that 'the Salafist Group for Preaching and Combat [has decided]… to declare war on everything that is foreign and atheistic within Algeria's borders, whether against individuals, interests or installations'.[137]

In *Knights under the Banner of the Prophet* (December 2001), Ayman al-Ẓawāhirī provides the moral of the Algerian case from the point of view of the militant Islamists:[138]

> The Algerian experience has provided a harsh lesson in this regard [that there is no solution without *jihād*]. It proved to Muslims that the West is not only an infidel but also a hypocrite and a liar. The principles that it brags about are exclusive to, and the personal property of, its people alone. They are not to be shared by the peoples of Islām, at least nothing more that what a master leaves his slave in terms of food crumbs. The Islamic Salvation Front in Algeria has overlooked the tenets of the creed, the facts of history and politics, the balance of power, and the laws of control. It rushed to the ballot boxes in a bid to reach the presidential palaces and the ministries, only to find at the gates tanks loaded with French ammunition, with their barrels pointing at the chests of those who forgot the rules of confrontation between justice and falsehood. The guns of the Francophile officers brought them down to the land of reality from the skies of illusions. The Islamic Salvation men thought that the gates of rule had been opened for them, but they were surprised to see themselves pushed toward the gates of detention camps and prisons and into the cells of the new world order…

The struggle against 'Crusaderism' and the Sa'ūdī regime: Osama bin Laden and his ideological supporters

> Some people wrongly believe that Osama bin Laden took a *jihādī* approach due to the influence of Shaykh 'Abdullah 'Azzām, the leader of the Arab *mujāhidīn* in Afghanistan. 'Azzām used bin Laden's financial help to provide relief services to the *mujāhidīn* in their war against the Soviets. The impact that 'Azzām had on bin Laden was limited to political and geographical issues related to *jihād* against the Soviets. 'Azzām was not interested in clashing with the Arab governments that supported him. Still, 'Azzām's interaction with bin Laden laid the groundwork for Ẓawāhirī's influence.

The judgement was that of Montasser al-Zayyāt, in his critical biography of Ayman al-Ẓawāhirī, which was written as a riposte to al-Ẓawāhirī's accusations in his book *Knights under the Banner of the Prophet* (December 2001).[139] Shaykh 'Abdullah 'Azzām (1941–89), has been called by his supporters 'the reviver of… *jihād* ideals in the modern world'.[140] He it was who coined the most striking phrase associated with militant *jihād*: the war, he said, would be won by '*jihād* and the rifle alone: no negotiations, no conferences and no dialogues'.[141] In his study entitled *Defence of the Muslim Lands*, 'Azzam contended that[142]

> if the infidels (*kuffār*) infringe upon a hand span of Muslim land, *jihād* becomes the greatest obligation (*Farḍ 'ayn*) for its people and for those near by. If they fail to repel the *Kuffār* due to lack of resources or due to indolence, then the obligatory duty (*Farā'iḍ*) of *jihād* spreads to those behind, and carries on spreading in this process, until the *jihād* is *Farḍ 'ayn* upon the whole earth from the East to the West.

In this sense, the aggressive *jihād* could be proclaimed as an act of Islamic self-defence. The role of radical Islamist groups such as al-Qaeda was therefore to be the vanguard, to radicalize and mobilize those Muslims who had hitherto rejected their message. In 1987, 'Azzam had written: 'every principle needs a vanguard to carry it forward and [to] put up with heavy tasks and enormous sacrifices. This vanguard constitutes the strong foundation (*al qā'idah al-ṣulbah*) for the expected society.'[143] It is 'propaganda by deed' to create an international army to unite the world Islamic community (the *ummah*) against oppression. For, as Azzam proclaimed at the first Conference on *jihād*, held at the al-Farook Mosque in Brooklyn in 1988:[144]

> every Muslim on earth should unsheathe his sword and fight to liberate Palestine. The *jihād* is not limited to Afghanistan. *Jihād* means fighting. You must fight in any place you can… Whenever *jihād* is mentioned in the Holy Book, it means the obligation to fight. It does not mean to fight with the pen or to write books or articles in the press or to fight by holding lectures.

The personification of the leader of such a terrorist network is Osama bin Laden, 'one of the major scholars of… *jihād*, as well as… a main commander of the *mujāhidīn* world-wide'.[145] His al-Qaeda group (literally 'base', such as camp or home, or foundation; but it can also meaning a precept, a rule, a principle, a maxim or 'method', that is, a mode of activism or a strategy) is considered to be the world-wide body with a dispersed chain of command that alone is capable of committing atrocities virtually anywhere from Indonesia to Chechnya and from Brazil to Bangladesh.

Although bin Laden established a vanguard organization at Peshawar in Pakistan, seven years elapsed when he was absent in Sa'ūdī Arabia and the Sudan. It was not until his return to the subcontinent, this time to Afghanistan in the years from 1996 to 2001, that al-Qaeda matured into an effective terrorist organization. A key element in this was the emergence of a reformulated *salafī–jihādī* ideology,[146] notably with the 'legitimization and authorization of an all-out terrorist *jihād* against the West' (the description is that of Yossef Bodansky) by means of a statement in the form of a prayer sermon and *fatwā* signed by bin Laden and others issued on 20 February 1998.[147] Bin Laden lacks the scholarly credentials to issue *fatāwā* on his own, as Mullāh Omar somewhat caustically remarked in July 2001, when he stated that 'any *fatwā* issued by Osama bin Laden' declaring *jihād* against the United States and ordering Muslims to kill Americans, was 'null and void'. He added: 'bin Laden is not entitled to issue *fatāwā* as he did not complete the mandatory 12 years of Qur'ānic studies to qualify for the position of *muftī*'.[148]

According to Montasser al-Zayyāt, it was Ayman al-Ẓawāhirī who 'managed to introduce drastic changes in Osama bin Laden's philosophy after they first met in Afghanistan in the middle of 1986, mainly because of the friendship that developed between them':[149]

> Ẓawāhirī convinced bin Laden of his *jihādī* approach, turning him from a fundamentalist preacher whose main concern was relief, into a *jihādī* fighter, clashing with despots and American troops in the Arab world. Ẓawāhirī gave bin Laden some of his closest confidants to help him. They became the main figures in bin Laden's al-Qaeda…
>
> Not only did Ẓawāhirī influence bin Laden, the latter [had an impact on] the philosophy of Ẓawāhirī and of Islamic Jihād. For example, bin Laden advised Ẓawāhirī to stop armed operations in Egypt and to ally with him against their common enemies: the United States and Israel. His advice to Ẓawāhirī came upon their return to Afghanistan, when bin Laden ensured the safety of Ẓawāhirī and the Islamic Jihād members under the banner of the Ṭālibān…
>
> It was natural for bin Laden to lead the International Islamic Front for Jihād on the Jews and Crusaders. He excelled at stirring the feelings of Arabs and Muslims when speaking on the Palestinian cause and when threatening the American presence in the Gulf. He argued that the Jewish lobby that controls the United States weakens the Muslim position. The main mission of the Front was to rid Arab and Muslim lands of American hegemony. Ẓawāhirī accepted bin Laden's offer to form the front, which was established in February 1998. His acceptance was due to the administrative problems that Islamic Jihād had suffered, and its lack of financial resources. While bin Laden was the leader of the Front, Ẓawāhirī was clearly the main architect, along with other Egyptian Islamic Jihād members such as Abū Ḥafṣ, Sayf al-'Adl, Naṣr Fahmī also known as Muḥammad Ṣalāḥ, Ṭāriq Anwar, Sayyid Aḥmad and Tharwat Ṣalāḥ Shahāta.

Ẓawāhirī's alliance with Osama bin Laden changed his philosophy [that is, Ẓawāhirī's] from one prioritizing combat with the near enemy to one of confronting the far enemy: the United States and Israel. This caused some confusion to Islamic Jihād members. Many were reluctant at first, but eventually agreed to be part of the Front in order to benefit from the many advantages it offered. As Naggar put it, no members could refuse to join the Front, except for the asylum seekers in European countries. Anyone who refused to join the Front would find himself alone with only his own resources and contacts.

'The world front for *Jihād* against Jews and Crusaders' – now known as the World Islamic Front – declared its commitment to 'kill the Americans, civilians and military' in retaliation for any further US attack on 'Irāq or any other demonstration of hostility in the Muslim world. The *fatwā* decreed that the US threat was profound and all-encompassing because US aggression affected Muslim civilians, and not just the military:[150]

> The ruling to kill the Americans and their allies – civilians and military – is an individual duty for every Muslim who can do it in any country in which it is possible to do it, in order to liberate the al-Aqṣā Mosque and the holy mosque from their grip, and in order for their armies to move out of all the lands of Islām, defeated and unable to threaten any Muslim. This is in accordance with the words of Almighty God: 'and fight the pagans all together as they fight you all together', 'and fight them until there is no more tumult or oppression, and there prevail justice and faith in God'.
>
> This is in addition to the words of Almighty God 'And why should ye not fight in the cause of God and of those who, being weak, are ill-treated and oppressed – women and children, whose cry is "Our Lord, rescue us from this town, whose people are oppressors; and raise for us from thee one who will help!"'
>
> We – with God's help – call on every Muslim who believes in God and wishes to be rewarded to comply with God's order to kill the Americans and plunder their money wherever and whenever they find it…

The *fatwā* also cited its hard-line Islamic authorities, returning the argument to the medieval period:

> This was revealed by *Imām* ibn Qudāma in *al-Mughnī*, *Imām* al-Kisā'ī in *al-Badā'i'*, al-Qurṭubī in his interpretation, and the shaykh of al-Islām [Ibn Taymīyah] in his books, where he said 'as for the militant struggle, it is aimed at defending sanctity and religion, and it is a duty as agreed. Nothing is more sacred than belief except repulsing an enemy who is attacking religion and life'.

Many would agree with the moderate Muslim who commented that 'the opinion' given in the statement was 'clearly against the directives of the Qur'ān and those ascribed to the Prophet'.

The International Islamic Front for Jihād on the Jews and Crusaders was quickly into action. The FBI regarded Ayman al-Ẓawāhirī as bin Laden's right-hand man and the mastermind of the bombing of the US embassies in Nairobi (Kenya) and Dar-es-Salaam (Tanzania) in August 1998. Ayman al-Ẓawāhirī and his brother Muḥammad ('the Engineer') headed the list of suspects in the 'returnees from Albania' case before a Cairo military court in April 1999. Ayman al-Ẓawāhirī received a death sentence *in absentia* in that case. He had made it clear that he wanted people to know of his association with the operation by issuing a statement on 4 August 1998: 'we would like to tell the Americans that their message has arrived and that the response is being prepared. The Americans should read it carefully, because we will write it, God willing, in the language they understand.'[151]

One reason why Ayman al-Ẓawāhirī and others fell into line behind bin Laden with relatively little argument is undoubtedly his huge personal fortune. Jason Burke has argued that it may not have been as large as the estimated $250 million cited by American officials in 1996:[152] one source close to his family states that his share of his father's estate was a mere $35 million[153] – mere, that is, in terms of the financial needs of global terrorism. Nevertheless, although there were many expenses for the organization, these were offset by new sources of revenue. Senior members of the Sa'ūdī royal family paid at least £200 million to Osama bin Laden's organization and the Ṭālibān in exchange for an agreement that his forces would not attack targets in Sa'ūdī Arabia. The money enabled al-Qaeda to fund training camps in Afghanistan later attended by the 9/11 hijackers.

The Sa'ūdī princes were deeply worried over attacks by Islamic fundamentalists on American servicemen at a US army training facility in Riyadh in November 1995 and at the Khobar Towers barracks in June 1996, in which 19 US airmen died. They feared, correctly, that bin Laden's men, who had recently relocated to Afghanistan from Sudan, would attempt to destabilize the kingdom because of their opposition to the presence of US troops. They therefore decided to come to an accommodation with the terrorist leader, which was reached at Kandahar, Afghanistan, in July 1998. Those present at the meeting included Prince Turkī al-Faisal al-Sa'ūd, then chief of the Istakhbarat (the Sa'ūdī secret service), Ṭālibān leaders, senior officers from Pakistan's ISI and bin Laden. Turkī knew bin Laden well, not just through family connections but because in the early 1980s he had hand-picked the young Sa'ūdī to organize Arab volunteers fighting the Russians in Afghanistan. It was agreed that bin Laden would not use his forces in Afghanistan to subvert the Sa'ūdī government. In return, the Sa'ūdīs agreed to ensure that requests for the extradition of al-Qaeda members and demands to close Afghan training camps by third countries were not carried out. To reinforce the deal, the

Sa'ūdīs agreed to provide oil and financial assistance both to the Ṭālibān and to Pakistan. The documents detail donations totalling 'several hundred millions' of dollars.[154]

Another way in which bin Laden's assets may have grown significantly was through manipulating the global stock market, which was particularly marked before the events of 11 September 2001. The exceptionally high level of trading activity in the week before 9/11 was one indication of this. The substantial rise in oil prices after 9/11 was another: oil prices rose by more than 13 per cent, owing to a 'surge in activity originating from brokers and traders buying oil contracts at lower prices to resell them forward at higher prices. It is reasonable to believe that those people knew that something exceptional was about to happen.'[155] As Loretta Napoleoni remarks, 'armed groups do not finance themselves solely with illegal money, they also have access to legal sources of revenue… assets and profits acquired by legitimate means and even declared to tax authorities can be used to finance terror'.[156] Clearly, such operations have become more difficult with the tighter financial controls in place since 9/11 to prevent money laundering and in the attempt to seize terrorist assets.

If financial power was sufficient to maintain bin Laden's control over the world-wide *jihād* prior to 9/11, there is reason to suppose that some militant Islamist groups revealed their displeasure subsequently. Montasser al-Zayyāt reflects this viewpoint:[157]

> Islamists across the globe were adversely affected by the September 11 attacks on the United States. Even Islamic movements that did not target the United States are paying the price for this folly… Bin Laden's desire to take revenge heedless of the American and international response, and its effect on the future of the Islamic movements in the world, has given the Americans and other governments the power to destroy the Islamists before our eyes… bin Laden and al-Ẓawāhirī lost the Ṭālibān, a government that had protected Islamists for many years…
>
> I emphasize al-Ẓawāhirī because I am convinced that he and not bin Laden is the main player in these events… he must not have expected this strong response. The most basic rule of battle is gauging the response of the enemy before taking any action. His miscalculation led him to believe that the American [response] would be similar to the one engendered by the bombing of the two American embassies in Nairobi and Dar-es-Salaam, which was restricted to bombing a few places in Afghanistan with missiles. He should have realized, that in response to the shock that… September 11 caused, the injured lion would try his best to restore his honour, regardless of his image before others…

Montasser al-Zayyāt's argument begs the question, however, as to whether bin Laden and al-Ẓawāhirī had not sought deliberately to provoke the excessive response from the United States, a likelihood that is proposed below in the context of 'fifth-generation' warfare.

This likelihood is reinforced by Ayman al-Ẓawāhirī's discussion, in *Knights Under the Prophet's Banner* (December 2001) entitled 'Moving the Battle to the Enemy'. Here he argues that

> the Islamic movement and its *jihād* vanguards, and actually the entire Islamic nation, must involve the major criminals – the United States, Russia, and Israel – in the battle and do not let them run the battle between the *jihād* movement and our governments in safety. They must pay the price, and pay dearly for that matter. The masters in Washington and Tel Aviv are using the regimes to protect their interests and to fight the battle against the Muslims on their behalf. If the shrapnel from the battle reach[es] their homes and bodies, they will trade accusations with their agents about who is responsible for this. In that case, they will face one of two bitter choices: either personally [to] wage the battle against the Muslims, which means that the battle will turn into clear-cut *jihād* against infidels, or [to] reconsider their plans after acknowledging the failure of the brute and violent confrontation against Muslims. Therefore, we must move the battle to the enemy's grounds to burn the hands of those who ignite fire in our countries. The struggle for the establishment of the Muslim state cannot be launched as a regional struggle: it is clear from the above that the Jewish–Crusade alliance, led by the United States, will not allow any Muslim force to reach power in any of the Islamic countries. It will mobilize all its power to hit it and remove it from power. Toward that end, it will open a battlefront against it that includes the entire world. It will impose sanctions on whoever helps it, if it does not declare war against them altogether. Therefore, to adjust to this new reality we must prepare ourselves for a battle that is not confined to a single region, one that includes the apostate domestic enemy and the Jewish–Crusade external enemy. The struggle against the external enemy cannot be postponed…

John Kelsay is among those who argue that, in arguing the case for his global *jihād*, Osama bin Laden has acted as the innovator, distorting the Islamic tradition further than it go can without being broken, particularly in the areas of proportionality and the killing of innocent people: the second contravenes the Qur'ānic command in Q.5:32, which indicates that if anyone kills another unjustly, it is as though he or she killed the entire world.[158] Moreover, instead of being defensive, the global *jihād* operates offensively outside the area which would normally be construed as the theatre of war in which the legitimate defence of Islamic lands against outside aggression would occur (for example, Palestine,

'Irāq, Afghanistan). Natana DeLong Bas demonstrates that Osama bin Laden can find little or no justification for his philosophy of action from the sources he cites, that is to say in the authentic writings of Ibn Taymīyah and Sayyid Quṭb.[159] Within Ṭāriq Ramaḍān's criteria, we would call Osama bin Laden a 'political and literalist *Salafī*': of this group, Ramaḍān writes that 'their discourse is trenchant, politicized, radical and opposed to any idea of involvement or collaboration with Western societies, which is seen as akin to open treason'.[160]

In a long statement in July 2003, bin Laden dealt with the criticisms essentially by making new assertions and placing the blame elsewhere. Firstly, the aim of the Islamic state/Islamic caliphate was restated. This could not be achieved without certain specific conditions being met, two of which – exile (*hijrah*) and *jihād* required sacrifice.[161] Afghanistan represented a lost opportunity after the eviction of the Soviet troops: an Islamic state could easily have been established according to Islamic standards, not those of the nation state or geography. The Ṭālibān regime could have provided a second opportunity, but it had not been supported by other Muslim-majority states, chiefly Sa'ūdī Arabia. Clerical figures were denounced as self-interested servants of autocratic regimes in power, who refused to propagate the cause of *jihād*:

> This immense obligation [i.e. *jihād*]… has no place among the clerics today who do not speak of it. They all, except for those upon whom Allāh has had mercy, are busy handing out praise and words of glory to the despotic *imāms* [i.e. Arab rulers] who disbelieve Allāh and His Prophet. They send telegrams praising those rulers who disbelieve Allāh and His Prophet. Their newspapers and media spread heresy against Allāh and His Prophet. Other telegrams are sent from the rulers to these clerics, praising them for deceiving the nation.
>
> The nation has never been as damaged by a catastrophe like the one that damages them today. In the past, there was imperfection, but it was partial. Today, however, the imperfection touches the entire public because of the communications revolution and because the media enter every home…
>
> …faithful clerics possess characteristics described in the book of Allāh… The most prominent characteristics are faith and *jihād* for the sake of Allah… Those who do *hijrah*, and those who support Allāh and His Prophet and wage *jihād* for the sake of Allāh, are the faithful ones…

The tone was that of a traditional Islamic moral reformer – bin Laden cited his two favourite role models, Ibn Taymīyah and Muḥammad Ibn 'Abd al-Wahhāb in the statement; the alleged taking of interest by certain banks in Sa'ūdī Arabia was also denounced. Yet the purpose was quite different. Osama the believer might have wished to see a moral reform in Sa'ūdī Arabia. Osama the politician sought to discredit the Sa'ūdī regime and deny it legitimacy according to its own Islamic credentials. 'Islām ceases to exist when the ruler is an infidel', he pronounced. 'It is inconceivable that there be faith, and that the religion will continue to rule, if the ruler (*imām*) is an infidel. This must be clear: if the *imām* is an infidel… Islām ceases to exist and there must be an act that will [establish in his place] a [believing] *imām*'.

> The region's rulers deceive us and support infidels and then claim they still cling to Islām… When the regime decided to bring the American Crusader forces into the land of the two holy places [i.e. Sa'ūdī Arabia], and the youth raged, these bodies [the unfaithful clerics]… issued *fatāwā* and praised the behaviour of the ruler, whom they called *Walī Amr* [man of authority, in accordance with the Qur'ānic requirement that believers obey those of authority amongst them], while in truth he was not really *Walī Amr* over the Muslims…

The challenge to the religious legitimacy of the Sa'ūdī regime is potentially dangerous, but bin Laden's somewhat shaky evidence was emphatically removed when, in the course of the Second Gulf War, the US troops stationed in Sa'ūdī Arabia were removed by August 2003.[162]

In any case, what bin Laden has yet to prove is the extent to which he can mobilize public opinion in Sa'ūdī Arabia against the regime. His second-in-command, Ayman al-Ẓawāhirī, provided a clear analysis of the need for the international *jihād* to make a genuine, deep-rooted, appeal to the masses:[163]

> The *jihād* movement must come closer to the masses, defend their honour, fend off injustice, and lead them to the path of guidance and victory. It must step forward in the arena of sacrifice and excel to get its message across in a way that makes the right accessible to all seekers and that makes access to the origin and facts of religion simple and free of the complexities of terminology and the intricacies of composition. The *jihād* movement must dedicate one of its wings to work with the masses, preach, provide services for the Muslim people, and share their concerns through all available avenues for charity and educational work. We must not leave a single area unoccupied. We must win the people's confidence, respect, and affection. The people will not love us unless they feel that we love them, care about them, and are ready to defend them. In short, in waging the battle the *jihād* movement must be in the middle, or ahead, of the nation. It must be extremely careful not to get isolated from its nation or engage the government in the battle of the elite against the authority. We must not blame the nation for not responding or not living up to the task. Instead, we must blame ourselves for failing to deliver the message, show compassion, and sacrifice. The *jihād* movement must be eager to make room for the Muslim nation to participate with it in the *jihād* for the sake of empowerment [*al-tamkīn*]. The Muslim nation will not participate with it unless the slogans of the *mujāhidīn* are understood by the masses of the

> Muslim nation. The one slogan that has been well understood by the nation and to which it has been responding for the past 50 years is the call for the *jihād* against Israel. In addition to this slogan, the nation in this decade is geared against the U.S. presence. It has responded favourably to the call for the *jihād* against the Americans…

A clash of 'rival exceptionalisms', not a clash of civilizations. I: aspects of militant Islamist exceptionalism

The clash between militant Islamists and the USA is not 'a clash of civilizations', as Samuel Huntington proclaimed in 1993 (which implies an objective reality)[164] but it is certainly a clash of 'rival exceptionalisms' (which implies a *portrayal* of the reality, or a false consciousness on the part of militant Islamists and their militant equivalent in the United States, the neo-conservative Right). If we take the militant Islamist position first, we can see from the Hamas Charter of 1988 the claim that Islām undertakes to make other religions safe from a position of supremacy (article 6): 'it strives to raise the banner of Allāh over every inch of Palestine for, under the wing of Islām, followers of all religions can coexist in security and safety where their lives, possessions and rights are concerned'. There is, however, no evidence that the other faiths (principally, in this context, Judaism and Christianity) wish to live under this beneficent Islamic supremacy.[165] What they actually want is religious pluralism and acceptance on the basis of equality and common citizenship.

David Zeidan describes militant Islamists' view of life as a 'perennial battle':[166]

> A main marker of [militant Islamists] is the view of life as a constant battle between God's powers of good ranged against Satanic evil powers. [Militant Islamists] call on believers to fight and suffer patiently in God's cause, stressing the militant attitude expected of believers in light of this emergency situation…
>
> Fundamentalisms advocate separatism to varying degrees. This includes separation from personal evil and heretical teachings and systems, leading many to set up their own independent movements and institutions. For [militant Islamists] separation usually means emulating the Prophet's *hijra* from Mecca to Medina interpreted as a temporary separation from the *jāhilī* environment in order to consolidate the community's strength and eventually return in power to destroy the evil system and establish God's rule…
>
> Another interesting hallmark of fundamentalisms is the thriving of conspiracy theories. There is a tendency to identify perceived enemies and unmask secret conspirators. For [militant] Islamists, the perennial enemies of Islām are the Crusading Christians, the Jews, and secularism in its manifold manifestations.

> In contemporary [militant Islamism], anti-Semitic rhetoric plays a dominant role.
>
> [Militant Islamists] view contemporary society as neo-pagan (*jāhilī*) in its repudiation of moral absolutes, its sexual permissiveness and secular-atheistic worldview. They stress the need to resist the ungodly and evil dictates of such systems…
>
> The concept of battle seems to have come full circle… Reformist energies are being subverted and dissipated by venting all the frustrations of the past few centuries of dependency and humiliation on those identified as scapegoats…

This study has sought to avoid the terms 'fundamentalism' and 'fundamentalist' on the grounds that they are potentially misleading. Sayyid Muḥammad Ḥusayn Faḍlallāh, the leading figure among the Lebanese Shī'a, whose relationship with the militant organization Ḥizbu'llah has been discussed in Chapter 10, objected to the term 'fundamentalism' on the grounds that it has 'overtones of exclusivism'. He preferred the term 'Islamist movement', which indicates 'a willingness to interact and live harmoniously with other trends of opinion, rather than to exclude them'. 'In the Western perspective', Faḍlallāh contends, 'fundamentalism' carries an implication of violence. 'Islamists have never chosen violence', he asserts. 'Rather, violence has been forced upon them.'[167]

We have preferred to use terms such as 'radical Islamists' or 'militant Islamists' to make clear the important distinctions between those who seek a fundamental change in the ordering of state and society (radical Islamists) and those who seek to use violence for the same purpose (militant Islamists). Even so, lack of clarity in the definitions is a potential problem. Reacting both to the immediate presence of US troops and the state of Israel in the Middle East, and to the long-term apparent failure of both secularists and religious modernists to reach a functioning synthesis between Islām and modernization, violent Islamists portray the encounter between Islām and West in terms of a diametrical confrontation. Islām is seen as inherently superior to the Western tradition, including both Capitalism and Marxism. This existential confrontation does not allow for accommodation or pluralism, but posits Islām as the only correct and acceptable moral order, with the radical Islamist vanguard as exclusive representatives of the Divine will. Islām is viewed as comprising a set of clearly defined and unchanging principles. This essentialist conception of Islām is combined with the assertion that Islām is not simply a religion but a comprehensive way of life. Moreover, this ahistorical and essentialist Islām is regarded as the sole independent variable determining the behaviour of the ruled and rulers alike.

Viewed from the militant Islamic perspective, Islamic government is considered superior to democracy, and divine law is superior to the man-made laws of the Western legislatures. In July 1997, Ayatollah Muḥammad Yazdī, chair of the Judicial Branch and one of the five most powerful men in Iran, stated that 'religious leadership and political leadership are not separate from each other. Politics, government, and Islamic rule are for the clergy only and the non-clergy lack the right to interfere in politics'.[168] There is a minority, opposition, Shī'a viewpoint which opposes clerical involvement in politics. In contrast, while Sunnī Islamists advocate the implementation of the *sharī'ah* they do not usually propose the rule of the Sunnī '*ulamā*' or *muftīs*.[169] Instead, in ways that remain unclear, the restoration of the Caliphate is said to be the solution to the problems of government in the new Islamic state.

The Western liberal belief that society and government should not interfere in self-regarding acts (actions of an individual that affect no one but one's self) is thought to be harmful to the idealized Islamic moral community. It is argued that a moral and believing society should repress individual activities such as drinking alcoholic beverages, playing games of chance, engaging in premarital sex, and watching decadent television programmes, among others, which are regarded as harmful to the well-being of society. In other words, responsibility to the community should override notions of individual rights and liberties.

Islamic notions of human rights are seen as emanating from a Divine covenant, or in Mawdūdī's words, 'when we speak of human rights in Islām we mean those rights granted by God'.[170] Such a view disregards the impact of modernity and the historical struggles that brought forth the modern notion of human rights. This flaw is elegantly summarized by Fouad Zakaria, who criticizes the theocentric conception of man behind this position and challenges it on two grounds:

> it is non-historical, or rather it freezes a certain moment of history and holds fast to it till the very end, thus doing away with dynamism, mobility and historical development. Finally, it is non-empirical; it... seeks to imitate a theoretical and spiritual ideal, while completely disregarding the effect of practice on this theoretical ideal.[171]

Although Islamic governments and Islamist writers frequently stress the exceptional character of their religion, which necessitates certain limitations or even outright violations of human rights, their arguments neatly fit into similar pronouncements by other, non-Islamic governments trying to justify repressive rule. The attempts to devise particularly Islamic moral norms inevitably lead not to a distinct ethical system, but merely curtail existing rights.[172]

The power of the internet has given a greatly increased capacity of militant Islamic groups to project their ideology and to perform psychological warfare against the West, while at the same time seeking to recruit from within, and to mobilize in its entirety, the Islamic community. Never before have relatively small groups been able to appear plausibly as large groups via an uncensored medium of mass communication. TalibanOnline and Jehad.net, which were observed by Gary R. Bunt, have now been removed from the internet. Jehad.net was 'taken out' by the owner of a large financial services firm in Minnesota when he observed that it was being used by al-Qaeda to post information of its terrorist activities.[173] Recently, Marie-Hélène Boccara has listed 25 extreme Islamist websites whose IP addresses and internet service providers (ISPs) were valid as at 16 July 2004.[174] The fact that websites come and go poses no problem for the Islamists: the material is kept in archived form and merely moved on to a new site once the old one becomes inoperative. Thomas Hegghammer of the Norwegian Defence Research Establishment in Oslo and another colleague found on the internet a 42-page document detailing how terror attacks ahead of Spain's general election in 2004 could drive Madrid to pull its troops out of 'Irāq and thus hurt the US-led Coalition. The text presented 'a detailed understanding of Spanish politics which indicates that the people who wrote the text must at least have lived in Spain for a while', Hegghammer stated. 'The documents also refer[red] to the necessity of carrying out attacks against Spanish interests during the elections. And this was a document which was written months before, so people would have had time to read it.'[175]

Equally remarkable, and it is believed the first time such a document was placed publicly on the internet, was the 'job application' to register an expression of interest in 'martyrdom operations'. This was issued in Tehran in June 2004, following a conference on the theme organized by the Committee for the Celebration of Martyrs of the Global Islamic Campaign. The 'job application' in Farsi reads:[176]

> Preliminary Registration for Martyrdom Operations
>
> I ––, child of ––, born 13 –– [Islamic calendar, Iranian version], the City of: –– proclaim my preparedness for carrying out martyrdom operations:
>
> – against the occupiers of the holy sites [referring to Najaf, Karbāla, and other places in 'Irāq].
> – against the occupiers of [Jerusalem].
> – for carrying out the death sentence of the infidel Salman Rushdie.
>
> Also, I would like to become an active member of the Army of Martyrs of the International Islamic Movement. Yes –, No –
>
> Contact telephone:
> Applicant's address:
> Applicant's signature:

Again, it attests to the power of the internet that within a week of the posting, according to Mohammad 'Alī Samadi, some 10,000 had registered their names to carry out 'martyrdom operations' on the defined targets. When asked about the killing of civilians, Samadi made it clear that it was no fault of the militant

Islamists that 'the Zionists have brought their wives and children to the occupied territories and have turned them into shields for themselves'. 'Salman Rushdie is the only non-military target for us, because we believe his attack against Islām was much worse that a military assault', he added. A hard-line cleric, Ayatollah Aḥmad Jannati, urged worshippers at Friday prayers in Tehran to attack US and British interests. 'It is the duty of every Muslim to threaten U.S. and British interests [everywhere]', he claimed.[177]

Ayelet Savyon correctly argues that this mass recruiting of potential suicide bombers formed part of a wider internal debate within Iran, on how to respond to Western, and particularly American, pressure with regard to its nuclear programme. The first approach, espoused by Iran's conservatives, and particularly by the Revolutionary Guards and circles close to Iranian Leader 'Alī Khamenei, was militant and aggressive, and openly threatened European and US targets and interests, both in the Middle East and in the West. Senior Revolutionary Guards officials claimed a plan existed to 'eliminate Anglo-Saxon civilization' using missiles and suicide bombers against '29 sensitive targets' in the West, which had already been identified by Iranian intelligence. The second approach, espoused by Iran's reformist circles, was more moderate, emphasizing diplomatic channels, and opposed to threatening the Europeans. However, for all their differences, both camps were agreed on Iran's need for an advanced nuclear programme. 'A country like Iran cannot have prestige by acquiring nuclear weapons... Iran would raise more threats against it, not obtain security, by having nuclear weapons...', stated Iran's former representative to the International Atomic Energy Agency, Dr 'Alī Akbar Salehi.

> But nuclear technology is different. If a country has access to cutting-edge nuclear technology, it can be proud. Take Switzerland, which has about 6 million people. Can one compare this country, with the volume of knowledge and technology it has, with another country that can hardly feed its people but boasts that it has a nuclear bomb?[178]

A clash of 'rival exceptionalisms', not a clash of civilizations. II: aspects of American exceptionalism

'It should be clear to all that Islām... is consistent with democratic rule', stated President George W. Bush on 7 November 2003. The words were striking in the context of 25 years of tensions between the US and various Islamic movements. For half a century, US policy had implicitly accepted the concept of 'Islamic exceptionalism' – one aspect of which had been that Islām and democracy were considered basically to be incompatible.[179] For Muslims, the US legacy in the Middle East has been most starkly defined by its unwavering political and wholly disproportionate financial support for Israel (to a grand total of $90.6 billion in aid to Israel from 1949 including fiscal year 2001);[180] by the US intervention in Iran in 1953; and by its failure even to condemn the Algerian military when they aborted the second round of the democratic elections in December 1991 because it would have brought an Islamist party to power. As a result, Washington has a credibility problem in the Middle East and with Muslims generally. 'How do I respond when I see that in some Islamic countries there is vitriolic hatred of America', President Bush mused at a press conference on 12 October 2001. 'I'll tell you how I respond: I'm amazed. I just can't believe because I know how good we are.'[181]

The nature of American exceptionalism is a much bigger issue of concern than the faults of its policies towards the Middle East and Islamic countries, which invariably are characterized by cultural insensitivity and the naked pursuit of either commercial interests or the search for sources of oil. American interventionism abroad is rarely quite what it seems. As Major-General Smedley Butler (1888–1940) of the US Marine Corps stated in a speech delivered in 1933 (he later turned the theme into a book, *War as a Racket* [1935]):[182]

> War is just a racket. A racket is best described, I believe, as something that is not what it seems to the majority of people. Only a small inside group knows what it is about. It is conducted for the benefit of the very few at the expense of the masses... There isn't a trick in the racketeering bag that the military gang is blind to. It has its 'finger men' to point out enemies, its 'muscle men' to destroy enemies, its 'brain men' to plan war preparations, and a 'Big Boss' Super-Nationalistic-Capitalism...
>
> I helped make Mexico, especially Tampico, safe for American oil interests in 1914. I helped make Haiti and Cuba a decent place for the National City Bank boys to collect revenues in. I helped in the raping of half a dozen Central American republics for the benefits of Wall Street. The record of racketeering is long. I helped purify Nicaragua for the international banking house of Brown Brothers in 1909–1912... I brought light to the Dominican Republic for American sugar interests in 1916. In China I helped to see to it that Standard Oil went its way unmolested.
>
> During those years, I had, as the boys in the back room would say, a swell racket. Looking back on it, I feel that I could have given Al Capone a few hints. The best he could do was to operate his racket in three districts. I operated on three continents.

At the end of his book, Smedley Butler commented further:

> The next war, according to experts, will be fought not with battleships, not by artillery, not with rifles and not with machine guns. It will be fought with deadly chemicals and gases. Secretly each nation is studying and perfecting

> newer and ghastlier means of annihilating its foes wholesale. Yes, ships will continue to be built, for the shipbuilders must make their profits. And guns still will be manufactured and powder and rifles will be made, for the munitions-makers must make their huge profits. And the soldiers, of course, must wear uniforms, for the manufacturer must make their war profits too. But victory or defeat will be determined by the skill and ingenuity of our scientists. If we put them to work making poison gas and more and more fiendish mechanical and explosive instruments of destruction, they will have no time for the constructive job of building greater prosperity for all peoples. By putting them to this useful job, we can all make more money out of peace than we can out of war – even the munitions-makers.

Given the scale of the international armaments industry which operates in the interests of the great powers, his words of prophecy still carry a resonance today. However, the Major-General, called by MacArthur 'one of the really great generals in American history', failed to appreciate the economic significance of rearmament for the United States as it struggled to emerge from the Depression in the 1930s. If not war itself, at least the instruments of war, have been and remain not only good for American capitalism, but an essential component of it. Much – if not most – technological innovation is driven by the aim of the US military establishment to remain far ahead of other nations in the cutting edge of military capability.

At the end of January 2003, Henry C. K. Liu, Chairman of the New York-based Liu Investment Group, brought Major-General Smedley Butler's arguments up to date and gave them a firm economic basis:[183]

> The economic benefits of [World W]ar [II] were substantial: full employment, price stability through control and rationing, insatiable war demand that translated into guaranteed markets for the private sector, priority allocation of resources to support the spectacular miracle of war production. The U.S. economy quickly became addicted to these euphoric vitamins from war, at least until market fundamentalism started to take over Washington, beginning in the 1970s. Since then, high-tech warfare can happily coexist with high unemployment and stagnant demand, as its demand on labour and production is narrowly concentrated…
>
> Arms control scholars in U.S. strategic think-tanks actively promoted the need to let the Soviets have the bomb and an effective delivery system, being confident at the same time that the U.S. could stay technologically and quantitatively ahead enough of the Soviets [so as not to] compromise national security. By 1971, total offensive force loading (nuclear warheads) stood at: U.S. 4,600 and the USSR 2,000. MIRV (multiple independently targetable re-entry vehicles) technology and SLBM (submarine-launched ballistic missiles)

were urged on the Soviets by U.S. arms control experts to maintain 'stability'. Moreover, the U.S. system of relying on private defence contractors was in a better position to reap economic benefits from nuclear armament than the Soviet system of state enterprises. This strategy to bankrupt the USSR with arms spending was essentially the one which Ronald Reagan employed to win the Cold War… The Cold War was not won by U.S. democratic ideals. It was won by a U.S. arms control strategy that was sustainable only by a capitalistic system that depended on the private sector to produce weapons systems for profit.

The U.S. now is seeking a replacement of the economic role of the Cold War. The attacks on Iraq and Yugoslavia were part of that search – it is a convenient way to burn off fast technical obsolescence to make room for the next generation of smart bombs and cruise missiles. The economies of Massachusetts, California, Texas and North Carolina are closely tied to smart-bomb production, at a cost of $2 million a pop. The Pentagon announced that in the pending invasion of Iraq, up to 400 cruise missiles a day would be launched, more than all fired in the 1991 war [which at $2 million each, would incur a cost of $800 million a day just for cruise missiles]. With 4–6 per cent unemployment being now accepted as a scientific necessity for long-term economic growth, and the prospect of total mobilization for a massive troop confrontation with another superpower being unrealistic in the foreseeable future, the economic impact of regional wars needed for empire building is expected to concentrate on high-tech innovation and tele-application, to maximize kill–ratio advantage, and on superior long-distance command and control of battlefield tactics via satellites, and not on the general economy…

One B2 bomber now costs over $3 billion due to upgrades, up one-third from its original cost. The construction cost of the World Trade Center towers was $1.5 billion at its completion in 1972, and its replacement cost today would be about the same as the cost of a B2. The Stealth fighter will cost 30 per cent more to make it more stealthy and less vulnerable. The new price will be around $65 million each. Yet military spending meant more back in 1962 when the defence budget was near 10 per cent of GDP. It is precisely the fact that military spending has not been a significant factor in the economy in the post-Cold War decade that gives new incentives for new violent conflicts. Military spending has always had an international strategic dimension…

Underlying Henry Liu's argument is the fact that the USA is content both with the facts of military intervention and the costs of intervention because they pose no serious threat to its hegemonic power status but, at least on the surface, appear to reinforce it. As Liu commented, 'it is increasingly clear that the real issue on whether a nation faces attack from the world's sole remaining superpower rests not on its possession of WMD, but on whether it possesses a cred[i]ble counter-strike force as a deterrence to pre-emptive attack from a nation which itself has steadfastly refused to adopt a no-first-use doctrine on WMD'. It was known in advance, for example, that 'Irāq did not have a credible counter-strike force because of the scale of bombing raids prior to 2003. It was clear to American officials by October 1999 that the Anglo-American forces were running out of targets. By early 2001 the bombardment of 'Irāq had lasted longer than the US invasion of Vietnam.[184]

Violent intervention is a consequence of American exceptionalism: in his effort to justify US military intervention in Cuba against Fidel Castro, Secretary of State Dean Rusk presented on 17 September 1962 a list to a US Senate Hearing of 200 plus 'precedents' (now called 'low intensity conflicts') from 1789 to 1960. The possible connections between war and civilian violence in the United States is still largely unexplored territory. Has war directly or indirectly encouraged an American predisposition toward aggression and the use of violence or has it been the reverse? This question has never been satisfactorily investigated by American historians or other scholars. One feature of American exceptionalism is the right of every adult American to carry firearms to protect oneself against other Americans carrying firearms. The lack of a violent revolutionary tradition in America is the principal reason why Americans have never been disarmed, while in every mainland European nation the reverse is true. As a result of this, the number of murders in the United States is extraordinarily high. Ira M. Leonard writes:[185]

> More Americans were killed by other Americans during the 20th century than died in the Spanish-American war (11,000 'deaths in service'), World War I (116,000 'deaths in service'), World War II (406,000 'deaths in service'), the Korean police action (55,000 'deaths in service'), and the Vietnam War (109,000 'deaths in service') combined. ('Deaths in Service' statistics are greater than combat deaths and were used here to make the contrast between war and civilian interpersonal violence rates even clearer.)

Americans have been imbued with the idea that they are a 'latter-day chosen people' with a providential exemption from the woes that have plagued all other human societies. During the Cold War the positive vision of America – the 'myth of innocence' or the 'myth of the new world Eden' – was an essential tool of US propaganda against Communism. In an open, free, democratic society, graced with an abundance of natural resources, and without the residue of repressive European institutions, virtually any white person who worked hard had the opportunity to achieve the 'American Dream' of material success and respectability. It was only late on, and then only partially, that African Americans were allowed to join this promised land on a basis of equality. Peoples of Hispanic origin are still awaiting full entry to common citizenship in all its social and economic aspects.

Another indication of American Exceptionalism, apparent already before 11 September 2001, but made much more deeply entrenched by subsequent decisions, is the idea that the United States, in Peter J. Shapiro's phrase, 'can pick and choose the international conventions and laws that serve its purpose and reject those that do not. Call it international law *à la carte*.'[186] Shapiro argues that as international relations become increasingly tethered to such international agreements at the International Covenant on Civil and Political Rights (ICCPR) and any institutional enforcement arrangements, 'nothing less than America's position of international leadership is at stake'. Worse still, for Shapiro, is the argument that

> America does not have to play by the rules that everybody else plays by because nobody can make it play by them – and besides, it has its own set of more important ones... U.S. non-compliance with international accords saps its authority to press other nations to respect the rule of international law.[187]

A practical example of Shapiro's argument is provided by the criticism of UN Secretary-General Kofi Annan on 17 June 2004 of continuing to grant peacekeepers participating in United Nations' operations immunity from potential prosecution by the world's first permanent war crimes tribunal. This, he argued, had the potential to hurt the credibility of both the Security Council and the UN itself. First adopted by the UN Security Council in July 2002 and renewed the following year, the resolution effectively shields members of UN peacekeeping missions drawn from nations that have not ratified the Rome Statute – the treaty that established the Hague-based International Criminal Court (ICC) – from investigation or prosecution. The United States, which originally signed the Rome treaty but did not ratify it, introduced a text in May 2004 seeking to renew the exemption for a third consecutive year. 'For the past two years, I have spoken quite strongly against the exemption, and I think it would be unfortunate for one to press for such an exemption, given the prisoner abuse in 'Irāq', Kofi Annan said. 'I think in this circumstance it would be unwise to press for an exemption, and it would be even more unwise on the part of the Security Council to grant it.' Such a move, in his view, 'would discredit the Council and the United Nations that stands for rule of law and the primacy of rule of law'.[188] The prisoner abuse scandals in Afghanistan, 'Irāq and at Guantánamo Bay were still being investigated by the American and British military at the time that the UN Secretary-General spoke.[189] The conclusion to be drawn from the Afghanistan and 'Irāq interventions has become inescapable: far from strengthening US power abroad, they have in fact weakened it.[190]

More could be said on the theme of a clash of rival exceptionalisms rather than a clash of civilizations. For Ḥizbu'llah, the struggle between Islām and the West should not be characterized as a civilizational 'conflict' (*khilāf*), as Samuel

Huntington has termed it, but as a civilizational 'dispute' (*ikhtilāf*). The distinction between the two terms is a crucial one, since the former term 'denotes civilizational irreconcilability' whereas the latter 'implies the possibility of civilizational co-existence and harmony'. Thus the rapprochement with France, which was viewed as hostile to Lebanon under President Mitterrand, is an example of the capacity of the party's willingness to reconcile with any Western state that changes its regional policy.[191] The idea of overarching civilizations is in any case open to question, if we accept, for example, that there are sub-civilizations with Islām, such as 'Arab Islām', 'African Islām' and 'Ind[ic] Islām'. The prefix denotes the notion of subordination to the ideological unity and supremacy of Islām itself, without which there cannot be a universal Islamic *ummah*.[192]

The revenge of al-Qaeda: global *jihād* perceived as 'fifth-generation' warfare

'This great victory was possible only by the grace of God', Ayman al-Ẓawāhirī stated on a videotape in which he was recorded talking to bin Laden about the events of 9/11.

> This was not just a human achievement – it was a holy act. These nineteen brave men who gave their lives for the cause of God will be well taken care of. God granted them the strength to do what they did. There's no comparison between the power of these nineteen men and the power of America, and there's no comparison between the destruction these nineteen men caused and the destruction America caused.[193]

In the discussion which follows, it must be made clear that there is no moral justification, either within Islām or from the perspective of the common good of humanity (a concept also recognized in Islām), for the attitude taken by the al-Qaeda leadership. It is both morally repugnant and disastrous for the future of mankind. As Rachid Ghannouchi (Rāshid al-Ghannūshī), the Tunisian Islamist leader and philosopher, said of the attack on tourists in Egypt in November 1997: 'the crime is a stab in the back of Egypt by a group that claims to belong to Islām... So blind they were [as] not to observe the Qur'ānic rule of individual responsibility...'[194] However, Pandora's Box has been opened. Amoral war, war without any limits in terms of avoidance of civilian casualties, must be discussed in terms of its practical implications, irrespective of the condemnation of such acts as 9/11 for what they are, immoral and nihilistic.

It is not often that al-Qaeda can be accused of false modesty, but there has been a false modesty in its unwillingness to admit to a more decisive breakthrough in warfare than acknowledged by Abū 'Ubayd al-Qurashī in February 2002. This breakthrough has to be perceived as 'fifth-generation' warfare and is a considerable advance in the dangers posed to the West, and to regimes al-Qaeda wishes to see overthrown, than the 'fourth-generation' warfare it has already acknowledged. What needs to be clearly understood that it is not necessarily technological innovation, but ruthlessness and cost-effectiveness (to the terrorist) that characterizes 'fifth-generation' warfare. As Henry Liu commented in relation to the costs of war in January 2003:[195]

> why should terrorists resort to ICBMs that are costly and difficult to launch when a small bottle of biological agent can do more damage at a tiny fraction of the cost? A recent NATO study shows that the costs of conventional weapons ($2,000), nuclear armaments ($800), and chemical agents ($600) would far outstrip the bargain basement price of biological weapons ($1) to produce 50 per cent casualties per square kilometre [prices at 1969 dollars].

In order to understand the nature of generational change in warfare, we need to return to the American military analysts, and in particular the article by Lieutenant-Colonel Thomas X. Hammes cited by Abū 'Ubayd al-Qurashī. Hammes wrote:[196]

> while [earlier military analysts] outlined the tactical changes between the generations of modern war, it is essential we understand what actually caused these generational shifts in warfare. The most commonly cited reason is the evolution of technology. While technological changes clearly have a major impact, attributing the generational changes in warfare primarily to technology oversimplifies the problem. The true drivers of generational change are political, social, and economic factors. Each of these factors was pivotal in the evolution of the first three generations of war… the single example of World War I will illustrate the point.
>
> While the evolution of rifled artillery, machine-guns, and barbed wire brought about trench warfare on the Western Front, these technological developments alone were not sufficient to bring about the firepower-intensive second generational war that evolved from 1914–1917. The second generation required not just improved weaponry, but the evolution of an entire political, economic, and social structure to support it. Second-generation war grew from the society of the times. It required the international political structure that focused on the balance of power, formed the alliances, and stuck to them through four incredibly expensive, exhausting years of war. Further, it required the output of an industrial society to design, produce, and transport the equipment and huge quantities of ammunition it consumed. Finally, it required the development of a social system that brought catastrophic losses. Technology, while important, was clearly subordinate to political, economic, and social structures in setting the conditions for World War I…

Hammes' verdict that 'technology, while important, was clearly subordinate to political, economic, and social structures' is critical for an understanding that a change to 'fifth-generation' warfare has in fact occurred since the events of 9/11. From the point of view of the perpetrator of this new form of offensive warfare, al-Qaeda, there has been no acknowledgement that this generational breakthrough in the nature of war has occurred, either from false modesty (which is unlikely) or for reasons of continuing secrecy about its planning and the priorities of its targeting policy (a more probable explanation).

In the West, 'fifth-generation' warfare has not been recognized for what it is for two quite different reasons. The first is that such an acknowledgement would pose a profound challenge to the position of the leaders at the heart of the defence and security apparatus of the United States. They, and the military-industrial complex behind them, can scarcely admit to the American public that the vast expenditure in technological upgrading of weapons systems has been largely misspent in view of the nature of the threat posed by 'fifth-generation' warfare. The abject failings of American intelligence gathering,[197] both about 9/11 and prior to the 'Irāq war, have required a process of re-evaluation: but there is no evidence, as yet, that this process will recognize the nature of the challenge for what it is: a military challenge for which the world's only superpower is woefully ill-prepared.

The second reason why the profound challenge posed by 'fifth-generation' warfare has not been admitted in the West is that it requires a complete rethink of strategy by the elected politicians. The 'war on terrorism' ('Operation Infinite Justice') was conceived as a pre-emptive strike against 'failed states' which provided safe haven for the terrorists (Afghanistan) or against 'rogue states' which allegedly posed an immediate and pressing threat to the West because of their alleged possession of weapons of mass destruction ('Irāq). Even assuming that the basis of the strategy was correctly formulated – which is highly contestable in the second case – it did nothing to create safer societies within the West.[198] It has been established by the US Commission investigating the events of 9/11 that there was no 'Irāq–al Qaeda link, contrary to the statements of the American administration.[199] The whole intelligence-gathering apparatus needs to be refocused and unified in the various states of the West that are the targets of al-Qaeda so that the population is protected from further terrorist attacks. The Madrid bombings of 2004 demonstrated just how far short the states in the West were from achieving 'homeland security'. Instead, new technological panaceas are proposed: in the United States it is the fingerprinting and photographing of every entrant to the country; in the United Kingdom it is the proposed national identity card, which is in any case years away from being implemented.

The distinguished British columnist Simon Jenkins, writing in *The Times*, argued in May 2003:[200]

> the bombs in Riyadh show that the threat of September 11 is not over. That much is clear. Equally clear is that the present danger is not from rogue states

> or weapons of mass destruction, but from murderous gangs with dynamite and cars. As Afghanistan was followed by Bali, so 'Irāq is followed by Riyadh. After waiting out the razzmatazz of war, reality terrorism is back in business. These killers cannot be eradicated. Though they pose a threat to human lives they do not threaten Western values. They may stir dictatorial tendencies in paranoid politicians. But to imply that such incidents undermine freedom is to lose all faith in democracy. Whatever the motives, these are criminal acts. They should be met by the art of intelligence and the science of security, not by the crass hand of 'regime change'... Policing can offer protection only of last resort. It can claim no spectacular victories, only spectacular defeats... Such protection offers politicians no glamour and contractors no profit. It wins no elections. I do not care. We have had the razzmatazz of war. Now let us have the reality of protection.

'Fifth-generation' warfare achieved its objectives in a spectacular way in the terrorist outrages of September 2001, March 2004 and May 2004. On the first occasion, the aim was to ensure that the United States would adopt an inappropriate policy, an overreaction, which would alienate Muslim opinion deeply and possibly permanently.[201] The mistaken policies adopted in the Afghanistan War of 2001 and Second Gulf War of 2003 have achieved that objective. The best evidence suggests that in Afghanistan the main problems of rebuilding the country after the enormous damage sustained in the war still remain; warlord Abdul Rashīd Dostum's forces overran the province of Faryab in April 2004, while Abdul Salaam Khān's forces overran the central province of Ghowr in mid-June 2004. Both al-Qaeda and Ṭālibān elements continue to operate in the south, to such an extent that the election registration process was imperilled.[202]

In the second attack on the West, the Madrid bombings of March 2004, the aim was to 'punish' a democratically-elected government which had backed the US-led war in 'Irāq. For the first time in the history of a Western democracy, it would appear that the election result was greatly influenced[203] by an act of 'fifth-generation' warfare, otherwise called a terrorist atrocity: an anti-'Irāq war government was elected to replace the right-wing Spanish government which had advocated and supported the US-led 'war on terrorism'. This sets a dangerous precedent for 'fifth-generation' warfare intervening at will to attempt to disrupt, or influence, the Western democratic process. The columnist Anatole Kaletsky, writing in *The Times*, argued that the Spaniards had given al-Qaeda 'the greatest political boost that it has ever received':[204]

> this is why, callous as it may sound, the greatest disaster in Spain this month was not the Madrid bombing, tragic though that was for the victims and their families. It was the election. The Spanish voters' decision to reward al-Qaeda (or whoever was behind the attack) with the swiftest and most dramatic

change of foreign policy in modern European history, has transformed the arithmetic of terrorism. It used to be seen as a futile, nihilist lashing out against established order. Today it is much more akin to Clausewitz's famous definition of conventional warfare: 'the continuation of politics by other means'…

By electing a new prime minister whose first act in office was to announce a troop withdrawal from 'Irāq, Spain has surrendered instantly to the terrorists' main demands. For the first time since the 1930s, a terrorist mass murder has immediately achieved its main political objective. From now on, Islamic extremists will no longer need to debate whether terrorism can work. The only argument will be over how many people need to be killed to achieve any particular end. If it takes 200 deaths to get Spanish troops out of 'Irāq, will it take 100 or 400 to remove the Italians? And if 200 killings are needed to gain control of the Spanish elections, how many will change the British Government or eject George Bush? Such calculations may well be misguided, since some countries would prove tougher to crack than Spain. Americans would probably rally round the President after another terrorist attack and Britain might also prove quite robust, although many voters would probably blame Tony Blair's pro-American policies. But al-Qaeda does not need to know in advance whether it will take one dead Briton or one thousand to achieve the same political results as one Spanish death. The ratio will only be discovered by trial and error – and we can be sure the terrorists will be sorely tempted to try this out. The inescapable conclusion is that further terrorist spectaculars are far more likely now than they were before March 14 – certainly in Europe, but probably also in Australia, America and even Japan.

Another British columnist, Simon Jenkins, again writing in *The Times*, offered a profound critique of Western policy in the 'war on terror', which deserves to be quoted *in extenso*, though as yet his warnings have not been heeded:[205]

…panic is the bread and butter of terror. Scaremongering politicians and bulging jails are what the terrorist wants. He seeks to summon them at the pull of a trigger or the banging of a bomb. They convert a mortuary statistic into what he craves, the true weapon of mass terror, the newsflash and the screaming headline. Why offer him his wishes on a plate?…

The Madrid bombing is said to be 'Europe's 9/11'. To Spain the loss of life and resulting trauma have been comparable to America's twin towers tragedy. But Europeans have one advantage in measuring their response. They can learn from two years ago. The learning may not be easy. It takes an effort of will to recall how America, with Tony Blair rightly alongside, reacted initially to 9/11. A surge of world support was generated and directed at catching the perpetrators… The sea in which al-Qaeda had swum for a decade was ready to be drained. Two years saw that opportunity dissipated, catastrophically…

The one group on the ground with the means and the motive (money) to grab bin Laden, the Taleban, was toppled before being seriously tested. That bin Laden has not been neutralised two and a half years after 9/11 is the pre-eminent scandal of the new world order. Afghanistan has been restored to warlords and heroin traders without achieving the stated objective of that restoration, the arrest of the man responsible for 9/11. Worse, the Taleban is back on the warpath. We can only hope that the latest campaign to 'find Osama' succeeds. That the capture of this dreadful man will be greeted with dismay by America's enemies is a measure of the West's diplomatic failure since 9/11.

If the Arab coalition dissolved over Afghanistan, the Western one dissolved over the invasion of 'Irāq. It stands to reason that assaulting Saddam diverted attention from the campaign against al-Qaeda. It stands to reason that driving al-Qaeda into the arms of a post-Saddam 'Irāq was madness. The shadowy Wahhābīs now said to be moving freely about Baghdad would have been killed instantly by Saddam's militias. If ever there were a time not to topple Saddam it was with al-Qaeda still on the loose. It stands to reason that the 'Irāq venture was always going to aggravate not relieve the so-called War on Terror. Western governments which drop thousands of bombs on foreign cities can hardly be surprised if some of their citizens seek revenge. It stands to reason that 8,500 dead 'Irāqī civilians [by March 2004] would be a recruiting poster for any passing dissident eager to kill an American. One of the more odious arguments I heard in Baghdād last November was that it would be convenient to have all global terrorism concentrated in that one place. So much for a more stable 'Irāq. And tell it to the Spaniards.

The violence, insecurity and administrative chaos visited on 'Irāq by the Pentagon this past year has offered al-Qaeda a new sea in which to swim. The tiny minority of Arabs who might have supported the Wahhābīst *jihād* in 2001 has swollen to a dispersed army, eager to take violent revenge on the West for its aggression in the Middle East. And of course those most involved in the invasion of 'Irāq are in the front line. That too stands to reason…

Osama bin Laden has been awarded an astonishing portfolio of triumphs by Western diplomacy. He has brought about the downfall of two autocracies and one democratic government. He has torn Europe's political unity asunder. He has devastated the American budget. He has toppled Cabinet ministers and BBC bosses. He has turned public buildings into fortresses…

The third attack, in May 2004, was not launched directly against the West, but against Western workers in Sa'ūdī Arabia; yet it had immediate and profound economic consequences for the West in a way that neither of the previous attacks had done, with the potential for long-term damage to the West where it hurts most – the functioning of the 'world' economy, that is, the capitalist system.

In the early months of 2004, al-Qaeda's recently-appointed operational chief for the Persian Gulf, Abd al-'Azīz al-Muqrin, made public the movement's strategy to undermine the Sa'ūdī royal family and with it the Western economy. A statement purportedly from al-Muqrin asserted that al-Qaeda would operate in small groups to attack Western interests and security forces in the Sa'ūdī kingdom. Al-Muqrin discussed the recruitment and planning of attacks in Sa'ūdī Arabia. He noted that the al-Qaeda network in the kingdom was highly compartmentalized in the wake of reversals sustained by the movement in 2003. Working in cities required small groups comprising no more than four people, he noted. The activists must all be residents of the city in question 'to avoid spies and suspicious eyes'. 'Most groups waging *jihād* have made the mistake of telling everyone everything about our operations', al-Muqrin said. 'Only the group leader should know what is going on but everyone else should only be told about their role. For example, those who will conduct explosions should only be told about the explosion.'[206] Boasting about his group's capacity to launch the Khobar attack in May 2004, al-Muqrin claimed:[207]

> the heroic *mujāhidīn* have managed to infiltrate the area housing the oil companies such as Halliburton and her sisters in the city of Khobar in the east of the Arabian Peninsula. The *mujāhidīn* managed to kill and injure many crusaders of different nationalities – among them Americans, one of whom was dragged through the streets of the city; a high-ranking British national of one of the oil companies, and an Italian who was slaughtered and returned as a gift to his government and leader.

Before the Khobar attacks, global oil prices had risen to their highest level in 13 years to nearly US$42 per barrel. As a result, the Sa'ūdī government promised to increase its daily oil production from 8.5 million to 9 million barrels per day and raise production again in the near future, if necessary, to keep prices down. A purported al-Qaeda statement argued that this Sa'ūdī pledge was one of the reasons that militants decided to attack foreign oil-firm employees at Khobar. The statement claimed responsibility for the Khobar attacks and vowed to 'cleanse the Arabian Peninsula of infidels'. It is uncertain whether the disruption of oil production is the main al-Qaeda objective, since the oilfields and pipelines are guarded by some 30,000 troops. The economic damage caused by such attacks was considerable: it was estimated (correctly) that the 'fear factor' of future attacks could cause oil prices to rise by as much as another US$8 per barrel.

> Osama bin Laden wants to destroy the Sa'ūdī state. Osama bin Laden wants to destroy the Sa'ūdī government. And so you should understand these comments in that context, that those who are most critical of Sa'ūdī Arabia in a very

> hostile way in the United States, as well as in Israel, share the same objective as Osama bin Laden and those who committed these acts…[208]

Sa'ūdī Foreign Affairs Adviser Adel al-Jubeir's comments on CNN were endorsed by experts on al-Qaeda, who pronounced that the overthrow of the regime in Sa'ūdī Arabia had become one of the principal objectives in the organization's war against the United States. Al-Qaeda views the Sa'ūdī royal family as an extension of US interests in the Gulf region. 'For them, the control of Sa'ūdī territory is important for achieving success against the U.S. for two reasons', commented B. Raman, a former Indian cabinet secretary and director of the South Asia Analysis Group. 'Firstly, Sa'ūdī Arabia could act as a rear base for the anti-U.S. *jihād* in 'Irāq just as Pakistan had served as a rear base for the anti-Soviet *jihād* in Afghanistan. Secondly, they can use Sa'ūdī oil as a *jihādī* weapon in their attempts to bring about the collapse of the Western economy.' Raman noted that al-Qaeda had been supported in its campaign against the kingdom by Abū Musab al-Zarqawī, considered to be one the most lethal of the insurgents in 'Irāq. Raman claimed that al-Zarqawī had been exercising command and control in anti-Sa'ūdī operations.[209]

Will the conservative and unpopular regime of Sa'ūdī Arabia fall to al-Qaeda, which since 2003 has declared war on it? There were fears that this was imminent in July 2002, when anti-government demonstrations were widespread but brutally suppressed, and the ruling elite was said to be deeply divided on the pro-American stance of the de facto ruler, Prince 'Abdallāh. The Pentagon sponsored a secret conference to look at options for American policy in the eventuality of the fall of the Sa'ūdī ruling family.[210] More than two-thirds of the population is under 30 years of age; perhaps a third of men in that age group are unemployed, which may fuel support for militancy.[211] Michael Binyon, writing in *The Times* on 1 June 2004, pronounced the house of Sa'ūd 'doomed by its contradictions', that is to say, the fact that it could neither jettison either of the two key, but contradictory, policies: appeasement of puritanical Islām and alliance with America. Binyon argues that compromise is impossible; others are somewhat more optimistic that, against the tradition and the odds, the Sa'ūdī ruling family will be able to reform itself and ride out the crisis. If it does not, the consequences are immense. In Binyon's words, 'were Sa'ūdī Arabia to implode in bloodshed, the violence could spread across the Middle East, the economies of the industrial world would be ruined and Islām would face a spiritual and logistical crisis of incalculable proportions'.[212] Reports in June 2004 that members of the Sa'ūdī ruling family had moved their fortunes out of the kingdom and sought a safe haven for their petrodollars were not encouraging.[213] The following month there were reports that the US would be reconsidering its attitude towards the alliance with Sa'ūdī Arabia, irrespective of whether or not George W. Bush won a second term of office. Intelligence analysts were reported to have expressed the view that Sa'ūdī

Arabia could face a major change in the near future, paralleling the fall of the shāh in Iran and his replacement by an Islamic regime hostile to the United States in 1979.[214]

Is 'fifth-generation' warfare, or extreme Islamist terrorism, capable of being defeated? Is an upsurge in terrorist incidents a sign of the success of the Bush–Blair strategy, as Michael Gove and others have suggested?[215] We would like to believe so, but it is, in our judgement, irresponsible to draw such a conclusion. The ineluctable conclusion to be drawn is that 'fifth-generation' warfare cannot be defeated unless there is some remarkable breakthrough in the costs and effectiveness of defence technology as it affects the individual, or the group activity, transportation systems and buildings. We know, from the ability of President Musharraf of Pakistan to evade assassination twice in December 2003, that cars can be protected by such devices.[216] It is claimed that the war on al-Qaeda has bolstered demand for command, control, communications, computers and intelligence (C4I) technology and systems. A report by Forecast International claims that US and allied military operations in Afghanistan and 'Irāq have led to an accelerated procurement of C4I systems. The consultancy firm predicts that the global C4I market will have generated about $84 billion in sales by 2013.[217]

We should, however, be wary of technological solutions alone. Henry Liu argued cogently in January 2003:[218]

> terrorism can only be fought with the removal of injustice, not by anti-ballistic missiles and smart bombs. It is a straw-man argument to assert the principle of refusal to yield to terrorist demands. It is a suicidal policy to refuse to negotiate with terrorists until terrorism stops, for the political aim of all terrorism is to force the otherwise powerful opponent to address the terrorists' grievances by starting new negotiations under new terms. The solution lies in denying terrorism any stake in destruction and increasing its stake in dialogue. This is done with an inclusive economy and a just world order in which it would be clear that terrorist destruction of any part of the world would simply impoverish all, including those whom terrorists try to help. The U.S. can increase its own security and the security of the world by adopting foreign and trade policies more in tune with its professed values of peace and justice for all.

Tariq Ali's 'letter to a young Muslim' voiced the frustrations and sense of injustice felt by many Muslims, particularly those in the Arab world, at what was perceived to be the unjust treatment they had received at the hands of the West, especially the United States:[219]

> The Arab world is desperate for a change. Over the years, in every discussion with Iraqis, Syrians, Saudis, Egyptians, Jordanians and Palestinians, the same questions are raised, the same problems recur. We are suffocating. Why

> can't we breathe? Everything seems static. Our economy, our politics, our intellectuals and, most of all, our religion. Palestine suffers every day. The West does nothing. Our governments are dead. Our politicians are corrupt. Our people are ignored. Is it surprising that some are responsive to the [militant] Islamists? Who else offers anything these days? The United States? It doesn't even want democracy, not even in little Qatar, and for a simple reason. If we elected our own government they might demand that the United States close down its bases…
>
> If we elected our own government in one or two countries people might elect Islamists. Would the West leave us alone? Did the French government leave the Algerian military alone. No. They insisted that the elections of… 1991 be declared null and void. French intellectuals described the… FIS as 'Islamo-fascists', ignoring the fact that they had won an election. Had they been allowed to become the government, divisions already present within them would have come to the surface. The army could have warned that any attempt to tamper with the rights guaranteed to citizens under the constitution would not be tolerated… The massacres in Algeria are horrendous. Is it only the [militant] Islamists who are responsible?

Tariq Ali's *cri de coeur* came from someone who professed himself an atheist, and who described the events of 11 September 2001 as having 'nothing to do with religion'. It was written before, in a much-trumpeted initiative, President Bush pronounced on 6 November 2003 that democracy would be good for Arab states. Predictably, the press reaction in the Arab world tended to reflect the government line, which was hostile since the conservative Arab regimes in alliance with the United States had no wish to relinquish their autocratic power. Two comments from the press deserve quotation as illustrating the problem as perceived in the Arab world:[220]

> the simplest rule of democracy is that it cannot be imposed from without. As first-year college students learn, and as is clearly evident in the Greek term from which the word is derived, democracy means that the people rules itself by itself and for itself. Thus, it is inconceivable that anyone external, whatever their intentions, can come to teach [people] how to rule themselves.

A second comment came in the form of an imagined conversation between President Bush and a delegate from the conservative Arab states:

> The emissary tries to ask the American president about democracy. 'Don't you understand English?!', Bush says to the emissary, who responds, 'we don't understand democracy and democracy does not understand us'. Bush insists that the Arab rulers do something. The emissary says that the U.S. was

occupying 'Irāq, destroying it, and killing its people for the sake of democracy and that therefore 'we ask your permission to do the same... Give us the order, and we will kill half of our people'. Bush asks, 'what about the other half?' 'We will arrest them and put them in jail', the emissary promises, 'and then we will declare democracy and release all the prisoners'.

All of this might suggest to the objective observer that American policy leaves much to be desired, and that the application of the doctrine of pre-emption has had a profoundly damaging effect. Such concerns led a number of prominent former diplomats and military personnel in the United States to form an advocacy group to agitate for change in the international diplomatic policies, security policies and military strategy adopted by the United States. The formation of Diplomats and Military Commanders for Change, as the group called itself, suggests how dangerously isolated the Bush administration had become by the summer of 2004 not just around the world but even from America's own bipartisan foreign policy and military establishments.[221]

In order to assess the effectiveness of 'fifth-generation' conflict to date, we need to consider such statistics as we possess. General Tommy Franks, of US Central Command stated: 'We don't do body counts.' The impartial observer is required to attempt the comparative 'body count' to judge the effectiveness of the strategy of al-Qaeda. The results are shown in Table 11.1.

Table 11.1 Civilians reported killed by US-led military intervention

Afghanistan ('Operation Enduring Freedom')	*'Irāq* ('Operation 'Irāqī Freedom', as at 5 June 2004).
Minimum: 1,000–1,300 (Carl Conetta, 24 Jan. 2002);[222]	Minimum: 9,284;
Maximum 3,000–3,400[223] (Marc W. Herold, to March 2002):	Maximum: 11,147.[224]
Minimum Total of Civilian Casualties (2 wars): 10,284	
Maximum Total of Civilian Casualties (2 wars): 14,547	

'The critical element remains the very low value put upon Afghan civilian lives by U.S. military planners and the political elite, as clearly revealed by U.S. willingness to bomb heavily populated regions': Professor Marc W. Herold's words provide a disturbing, but realistic, criticism of the reason for the high civilian casualty rate ('collateral damage' in military parlance). He concludes:[225]

> first, the U.S. bombing upon Afghanistan has been a low bombing intensity, high civilian casualty campaign [in both absolute terms and relative to other U.S. air campaigns]. Secondly, this has happened notwithstanding the far greater

> accuracy of the weapons because of U.S. military planners' decisions to employ powerful weapons in populated regions and to bomb what are dubious military targets. Thirdly, the U.S. mainstream corporate media has been derelict in its non-reporting of civilian casualties when ample evidence existed from foreign places that the U.S. air war upon Afghanistan was creating such casualties in large numbers. Fourthly, the decision by U.S. military planners to execute such a bombing campaign reveals and reflects the differential values they place upon Afghan and American lives. Fifth, this report counters the dangerous notion that the United States can henceforth wage a war and only kill enemy combatants. Sixth, the U.S. bombing campaign has targeted numerous civilian facilities and the heavy use of cluster bombs will have a lasting legacy borne by one of the poorest, most desperate peoples of our world. In sum, though not intended to be, the U.S. bombing campaign which began on the evening of October 7th, has been a war upon the people, the homes, the farms and the villages of Afghanistan, as well as upon the Taliban and al-Qaeda.

The BBC reported on 3 January 2002 that the number of civilian casualties killed in Afghanistan already exceeded the numbers killed in the 9/11 terrorist attacks.[226] US Defense Secretary Donald Rumsfeld and other Pentagon spokespersons routinely responded to criticism about civilian casualties by arguing that the United States had taken great pains to limit 'collateral damage', but that some amount of it was inevitable in war. On 29 October 2001, for instance, Rumsfeld told reporters: 'war is ugly. It causes misery and suffering and death, and we see that every day. But let's be clear: no nation in human history has done more to avoid civilian casualties than the United States has in this conflict.' Carl Conetta correctly argues that Rumsfeld's defence

> begged several pivotal issues. A fault line in support for Operation Enduring Freedom centred precisely on the question of whether the response to the 11 September attacks should have taken the form of a broad 'war' rather than a much more limited military operation – a 'police action' of some sort – focusing narrowly on the perpetrators of the terrorist attack and their cohorts… What [critics] had questioned was the necessity of pursuing aims as ambitious and broad as those that came to define Operation Enduring Freedom. And critics had especially questioned the necessity of conducting a large-scale bombing campaign that included civilian areas in its sweep.[227]

Let us now consider the number of civilians killed in the three principal incidents of 'fifth-generation' warfare discussed previously. The results are presented in tabular form in Table 11.2:

Table 11.2 Civilian casualties in three principal al-Qaeda incidents: September 2001, March 2004, May 2004[228]

	No. casualties
9/11 New York plus Washington DC	2,996
Madrid March 2004	190
Sa'ūdī Arabia May 2004	22

By these calculations, 3208 civilians have been killed in three of the most significant al-Qaeda terrorist operations prior to June 2004. The US-led military response to the first incident (if we count both the Afghanistan and 'Irāq wars) has inflicted a minimum of 3.4 times the number of civilian casualties and a maximum of 4.8 times the number. Without entering into the morality of so-called precision-bombing (which kills civilians as 'collateral damage') as against the deliberate terrorist targeting of civilians, it is easy to determine that the al-Qaeda policy, however reprehensible, has had a considerably lower 'body count' of civilians. This is not in any way to justify al-Qaeda's morally reprehensible policy, or to condone the wilful misuse of passages from the Holy Qur'ān by those extremist groups which seek to justify the 9/11 terrorist atrocities.[229]

The disproportionate damage inflicted by the terrorists has been achieved at the price of relatively few *jihādī* deaths, whereas the number of US and allied troops killed in action in Afghanistan (estimated at 972 US and Coalition troops to 21 June 2002)[230] and 'Irāq (estimated at 816 US troops to 4 June 2004)[231] has been considerable. Sergei Lavrov, the permanent representative of Russia to the UN, remarked in September 2003:

> the situation in 'Irāq is deteriorating with every day. The other day, our colleagues from the U.N., who used to be in charge of Afghanistan-related issues, made some simple calculations. It turned out that if we multiply daily American casualties in 'Irāq by the number of days the Soviet troops were in Afghanistan, the figure would be about 13,000. We had that many casualties in Afghanistan.[232]

The economic cost has also been considerable. In November 2003, President Bush won from Congress final approval for an additional $87.5 billion (£52 billion) package for military operations and aid in 'Irāq and Afghanistan.[233] The financial cost of the 'Irāq war to the USA alone as at 5 June 2004 was estimated at $116.4 billion dollars or the equivalent of fully-funding world-wide AIDS programmes for eleven years.[234] The economic cost of bin Laden's and al-Qaeda's global strategy has been trivial in comparison. By all counts of cost-benefit analysis – damage inflicted compared to *jihādīs*/soldiers lost or economic cost of the war – 'fifth-generation' warfare has been proven to be extremely cheap for al-Qaeda and alarmingly expensive for the proponents of 'Operation Enduring Freedom'.

'Fifth-generation' warfare will therefore continue, irrespective of the current strength of al-Qaeda, rumoured in June 2004 to be back to a strength worldwide of 18,000 men enrolled in 60 countries, and with large financial resources at its disposal.[235] Even if such reports were unfounded, or even if Osama bin Laden himself were captured, it would not alter the fact that 'fifth-generation' warfare is here to stay. It is immoral and the aims of its proponents are a chimera, if not actually mad. Irrespective of the particular group practising the strategy, 'fifth-generation' warfare is here to stay because it has been proven to be spectacularly successful and the Western powers have few weapons of first choice to combat it.

Let us consider the strategies that might be employed and their actual or potential disadvantages. The first strategy is state-authorized assassination of the presumed terrorist leader. On 22 March 2004, Shaykh Aḥmad Yāsīn, the founder of Hamas, was assassinated on the orders of the Israeli government.[236] On 17 April, his successor, 'Abd al-'Azīz al-Rantīsī, was assassinated on the same authority.[237] On 31 August 2004, two virtually simultaneous Hamas 'martyrdom operations' took place in retaliation at Beersheba, resulting in 16 deaths and more than 80 people injured.[238] Anatole Kaletsky commented in *The Times* on the issues involved:[239]

> Israel claims the right to engage in extra-judicial assassinations, to kill civilians at random and to blow up or bulldoze Palestinian houses because it is defending the only genuine democracy in the Middle East. In a sense this is true. Ariel Sharon does have a democratic mandate for his state terrorism, since Israeli voters have repeatedly rejected the alternative policy of negotiation. In the same way U.S. politicians of all stripes – including fundamentalist Christians and others in no way beholden to the Jewish lobby – use democracy to justifying backing Israel and refraining from criticism in the most egregious cases, such as [the] killing of Yassin. This killing will surely unleash another cycle of terror, just as Sharon's provocative campaign in the Israeli elections four years ago did.

It is of no utility in the long term to proclaim 'the United States, right or wrong', or 'Israel right or wrong'.[240] The current political leadership of any country is perfectly capable of short-sightedness and an historic failure to perceive its true national interest. The US government (Rewards for Justice Program, Department of State) placed a bounty of $25 million for 'information leading directly to the apprehension or conviction' of bin Laden, with a further $2 million offered through the Airline Pilots Associated and the Air Transport Association, in the aftermath of 9/11.[241] A similar bounty of $25 million has been placed for 'information leading directly to the apprehension or conviction' of Ayman al-Ẓawāhirī, bin Laden's deputy.[242] Interestingly, the US Department of State's Rewards for Justice Program makes no distinction between top terrorist suspects: all 22 on the FBI website as at 5 June 2004 had a reward of $25 million for information leading to their apprehension or conviction. All had Muslim names, though not all were Arabs; just one had been 'located'.[243] Leaving aside the undesirable private enterprise form of 'bounty hunting' that may result,[244] the effect of the 'most wanted terrorists' list is to give the impression that the individuals are wanted dead or alive: in other words, that the state authorizes their assassination. This was demonstrated by the 'targeted bombings' of presumed compounds of Osama bin Laden and Saddam Hussein, respectively, during the Afghanistan and 'Irāq wars, though both evaded death by this mechanism. Saddam was captured on 13 December 2003; but 'fifth-generation' warfare has continued without him. The same would be true of the capture or assassination of bin Laden: Osama may not yet have inflicted all the damage of which he is capable; but the genie of 'fifth-generation' warfare is already out of the bottle – in this sense, it no longer matters whether he is alive or dead. As one Afghan warlord, Gulbuddin Hekmatyar, has expressed it, there are potentially hundreds of bin Ladens available for the 'Irāq conflict alone.[245] Others can, and will, take his place.

The disadvantage of state-authorized assassination, therefore, is that it may prove counter-productive in the long term. It may simply serve to deepen and broaden the basis of support for the policies advocated by the assassinated leader. Radical Islamist terrorists are not capitulationists. There is little or no 'peace party' within their ranks which could be strengthened by the removal of a prominent hard-line leader. Yet the danger run by a democratic government in authorizing such a policy is considerable. If part of the 'war on terror' in the West is necessarily a battle of hearts and minds with the general public – ultimately, the electorate – why should the general public be other than contemptuous of a government which resorts to the same despicable terrorist measures as its opponents in 'fifth-generation' warfare? There is nothing to choose between the moral low ground of 'state-sponsored' terrorism and its al-Qaeda variant. The missile attack triggered by the mobile phone of the presumed terrorist – reported to be the way that a prominent Ṭālibān supporter in Pakistan's tribal region, Nek Muḥammad, was killed on 17 June 2004 – may on occasion hit its target, but how many innocent people does it kill at the same time?[246] There are no shortcuts to victory. What is needed is painstaking intelligence work, the arrest of suspects and the application of the full, lengthy, and costly judicial process that applies in other criminal investigations. Any short-cut or 'quick fix' approach and democratic governments in the West may rapidly find themselves in grave difficulties with the general public – and ultimately, the electorate.

A second approach adopted in 'Operation Enduring Freedom' was to hold suspected terrorists out of the reach of the international legal process at

Guantánamo Bay, Cuba, which in June 2004 held about 660 prisoners from 42 countries. The US Defense Secretary, Donald Rumsfeld, announced in September 2003 that 'our interest [*sic* = the presumed US interest] is not [in] trying them and letting them out. Our interest – during this global war on terror – [is in] keeping them off the streets, and so that's what's taking place.'[247] Erwin Chemerinsky, Professor of Public Interest Law, Legal Ethics and Political Science at the University of Southern California, responded to Rumsfeld's comment by remarking that there was no authority in American or international law to hold such prisoners indefinitely without judicial process. Since there was no sign of the war on terrorism nearing an end, this was tantamount to the US government claiming the right to hold people indefinitely, perhaps even for the rest of their lives, without complying with the requirements of international law. In a television interview later in the month, Chemerinsky commented further:[248]

> when you're dealing with people who are caught on foreign battlefields, then it's a question of international law. And international law is quite specific. For example, with regard to those who are being held in Guantánamo, they're entitled to a hearing by an independent tribunal to determine if they're prisoners of war or enemy combatants and depending on the status, different rights attach... I didn't say a civilian tribunal. I said they're entitled to a tribunal under international law... it can't be that, just because there's a very grave threat, anything the government wants to do on civil liberties becomes acceptable...

Under pressure from the British government, a number (but not all) of its nationals were released from Guantánamo Bay. They were all readmitted into the UK without charges being brought against them.

The threat posed by abuses in interrogation techniques of prisoners and abusive conduct towards prisoners is more than the so-called 'loss of the moral high ground' in dealing with terrorism. It is that the states which allow such things to happen have entered the same depraved world of the terrorists and betrayed the best traditions of democracy and respect for the individual's rights. Similar threats are posed to civil liberties in general by the enactment and enforcement of draconian anti-terrorist legislation which in some countries has confounded the innocent with the guilty, yet has been found to be of little real utility in the fight against terrorism.[249] The Executive Director of Human Rights Watch, Kenneth Roth, emphasized the need for 'proper boundaries' in what the Bush administration called its 'war on terrorism':[250]

> By literalizing its 'war' on terror, the Bush administration has broken down the distinction between what is permissible in times of peace and what can be condoned during a war. In peacetime, governments are bound by strict rules

of law enforcement. Police can use lethal force only if necessary to meet an imminent threat of death or serious bodily injury. Once a suspect is detained, he or she must be charged and tried. These requirements – what one can call 'law enforcement rules' – are codified in international human rights law.

In times of war, law enforcement rules are supplemented by the more permissive rules of armed conflict, or international humanitarian law. Under these 'war rules', an enemy combatant can be shot without warning (unless he is incapacitated, in custody, or trying to surrender), regardless of any imminent threat. If a combatant is captured, he or she can be held in custody until the end of the conflict, without being charged or tried…

[Existing] guidelines… were written to address domestic conflicts rather than global terrorism. Thus, they do not make clear whether al-Qaeda should be considered an organized criminal operation (which would trigger law-enforcement rules) or a rebellion (which would trigger war rules). The case is close enough that the debate of competing metaphors does not yield a conclusive answer. Clarification of the law would be useful…

Even in the case of war, another factor in deciding whether law-enforcement rules should apply is the nature of a given suspect's involvement. War rules treat as combatants only those who are taking an active part in hostilities. Typically, that includes members of an armed force who have not laid down their arms as well as others who are directing an attack, fighting or approaching a battle, or defending a position. Under these rules, even civilians who pick up arms and start fighting can be considered combatants and treated accordingly. But this definition is difficult to apply to terrorism, where roles and activities are clandestine, and a person's relationship to specific violent acts is often unclear…

As of June 2004, the debate over the failure of intelligence in the West with regard to 9/11 and the evidence of WMD in 'Irāq still continued, without a clear sign of how it would be resolved. Without doubt, the failings of Western intelligence, compared to some supposedly less advanced countries (for example, Pakistan with its unified ISI) is a consequence of the abject political failure to resolve potentially fatal divisions of intelligence procurement by separate agencies. This is most serious in the USA, with four major procurers of intelligence (the military, CIA, Homeland Security, FBI: up to 2002, the United States spent well over $30 billion annually on 13 intelligence agencies, with a Director of Central Intelligence (DCI) responsible for foreign intelligence and an FBI responsible for domestic intelligence); but it is also a concern in Britain with three main procurers of intelligence (the military, MI5 and MI6);[251] the estimated costs per annum of intelligence integration in Germany have been put at €200 million.[252] The historic diffusion of US intelligence procurement into separate agencies was highlighted in a report in 1996.[253] As late as June 2002, the databases of the FBI and the CIA could not exchange data on terrorists.[254]

Richard A. Stubbing and Melvin A. Goodman proposed in June 2002 that intelligence should be

> reshaped to combat terrorism. Intelligence on counter-terrorism must supplant military intelligence as America's top priority. Foreign and domestic intelligence efforts should be combined to fight this threat, with the creation of a new post – Director of National Intelligence – to coordinate foreign and domestic agencies in combating terrorism.

The authors argued that while the Pentagon must have ample military intelligence, the military could not be allowed to dominate strategic intelligence and such collection and analysis agencies as the National Security Agency, the National Reconnaissance Office, and the National Imagery and Mapping Agency. They specifically proposed the establishment of a new director of national intelligence charged with the analysis and publication of all intelligence estimates. This director would be responsible for coordinating all foreign and domestic intelligence and would report findings on terrorism to the Department of Homeland Security. The CIA would be limited to its operational mission only. The FBI would be split into two agencies, creating a Domestic Counter-Terrorism Service reporting to the DCI.[255]

The debate rumbled on. In April 2004, President Bush, perhaps sensing the inevitable, opened the door, at least in theory, to the creation of a much more powerful Director of National Intelligence to oversee what were now said to be 15 intelligence agencies. The FBI had by this date increased the number of counter-terrorism agents from 1344 to 2835; counter-terrorism analysts from 218 to 406; and linguists from 555 to 1204. The failings of the CIA in its slowness to identify bin Laden as a threat to the USA before 1999 had become clearer.[256] The real issue, however, remained: how to penetrate the terrorist networks more effectively; how to place bodies on the ground in the right places. Democratic Representative Jane Harman stated the need forcefully:[257]

> my bottom line is: we have to penetrate these cells. The only ways we will know the plans and intentions of these people is to have somebody in the room, or as close to the room as we can get it. Signals intelligence – what we can hear flying around with very impressive air and satellite power… is not enough. It's not enough. They're smart. You know, they know. It became public knowledge that Osama bin Laden was using a Sirius cell phone. And as soon as that came out, he stopped using it. And we have beautiful pictures of buildings – Colin Powell showed a lot of them at the U.N. last year [2003] – but we didn't have

> enough ground troops to know what was going on in those buildings. So we need a combination. We need more human intelligence, but we need to do it under some set of rules that is as reasonable as it can be.

The present strategy in the 'war on terror' has focused on certain actual (Afghanistan) or presumed ('Irāq: presumed incorrectly) homes for al-Qaeda: but al-Qaeda is inherently a transnational phenomenon. This strategy fails to recognize the imperatives of 'fifth-generation' warfare, which must be to secure the homeland first through drastically improved security while retaining constitutionality and democratic consent. Jane Harman commented on the need to improve 'homeland security' in a meaningful way:[258]

> I think New York City… has enormous vulnerabilities. I've been having a fight with the Mayor because I think we have to raise the possibilities that subways are soft targets… One of the things we predicted in the Bremer Commission was that there would be major terrorist attacks – plural – on U.S. soil. Everybody should get it that al-Qaeda is looking for big opportunities again in America to hit the icons of America. There are some of those on the West Coast. I worry enormously about LAX [Los Angeles International Airport], which my district surrounds, the ports of Los Angeles and Long Beach, which are the largest container ports in America – 43 per cent of our container traffic goes through those – movie studios, [even] the Hollywood sign. All of this kind of stuff is vulnerable. And everyone is working hard to protect it. But we can't – we can't in my view forget that the world has changed and the [terrorists] are here.

Offensive warfare against a hydra-headed cell-like structure is incapable of securing victory over the al-Qaeda phenomenon whether bin Laden and his fellow terrorists are caught or not. For an offensive strategy to succeed it needs a much more broadly-based coalition than current US policy is able to secure without a fundamental change of direction. Unilateralism and pre-emption will have to be renounced and there must be a serious commitment to multilateralism. Without such a change of policy, there is every reason to presuppose failure against 'fifth-generation' warfare. As Ma'sūd Akhtar Shaikh wrote in June 2004:[259]

> America has indirectly helped its enemies by opening for them a worldwide front on which the [terrorists] are free to select the targets of their choosing. Today they strike at a vulnerable point in Saudi Arabia; tomorrow they will hit an important establishment in Pakistan, and the day after, a fully loaded passenger train in France will be their target. While the invisible 'terrorists' (always designated by the Americans as the al-Qaeda men) can be present

everywhere in the world, those fighting against them just cannot… It may be interpreted as rather pessimistic, but the fact remains that unless the basic reasons for the start and subsequent escalation of terrorism in the world are removed, the war against terrorism is bound to be lost, despite immeasurable human and material losses. The earlier the main culprits playing this wasteful game realize this fact, the better for the whole mankind.

Conclusion

'Enlightened Moderation': Towards a Muslim Consensus on the Future Development of Islām and its Relations with the West

M. S. GORBACHEV. As regards Afghanistan, Iran is trying to have a fundamentalist government formed there.
SHEVARDNADZE. And not only there.
G. SHULTZ. In my opinion, the Iranians would not object to fundamentalist governments in the Kremlin and Washington (*laughter*).
M. S. GORBACHEV. All the same, they can scarcely hope for this. Possibly it is true they pray for it…

Record of Meeting of Gorbachev with George Shultz,
US Secretary of State, 22 February 1988[1]

Western intellectuals should bear in mind that this democracy which destroyed the empire of the Soviet Union is capable also of overthrowing these corrupt regimes in the Muslim world. They should bear in mind that Islām can be a friend to the West… moderate and tolerant. But Islām can be hard as well, and angry and seeking revenge. And the West has the power to shape this by its approach to Islām.

Rachid Ghannouchi (Rāshid al-Ghannūshī),
interview with Joyce M. Davis[2]

Two contrasting encounters help us to focus on the issues for the conclusion on the future of Islām, and the relations between the Islamic World and the West. Underlying the Shultz–Gorbachev repartee in 1988 was the belief, shared by the capitalist United States and the Communist Soviet Union, that militant Islamism (they called it 'fundamentalism') sought/seeks supremacy in the world. It was the view of outsiders to the world of Islām, who had lived through the turbulent decade since the Khomeini revolution and had witnessed the Afghan '*jihād*' against the Soviet occupation. This view[3] will not be addressed in the

Conclusion, since it has been dealt with in Chapter 11, particularly in the section on 'fifth-generation' warfare and the approach which states should take towards the phenomenon of terrorism which knows no frontiers and no moral constraints. Although the term *jihād* is used by violent Islamists perpetrating such atrocities, it is clear that there is neither a moral nor an historical justification for them to do so. It is a deliberate confusion of terminology, which uses an ambiguous term that is central to Islām as a spurious justification for actions which cannot be justified morally and as a device for gaining recruits to the cause of violence. This deliberate confusion of terminology is removed from consideration in this Conclusion, which discusses only the views of peaceful Islamists and the Muslim mainstream. Our central concern here is the revival of Islamic political ethics, on which there is much recent debate,[4] and how it should be integrated with such debates in the West. For example, as Muslims and non-Muslims 'continue their recently-begun dialogue on the just international order, they may well find a level of agreement on the ethics of war and peace that will ultimately be reflected in a revised and more universal law of war and peace'.[5] The dialogue is likely to prove fruitful in other areas. In the West, there is perhaps too much emphasis on rights at the expense of responsibilities;[6] the reverse is perhaps true in the Islamic world, where the obligations (*takālīf*)[7] and responsibilities of the group are sometimes emphasized at the expense of the individual. The ideal must be, as Ṭāriq Ramaḍān expresses it, that 'all people should have the means to fulfil their responsibilities and to protect their rights'.[8]

Many mainstream Muslims have condemned the confusion of an extremist political agenda with the terminology, beliefs and ethics of one of the world's great faiths. The exiled Tunisian Islamist Rāshid al-Ghannūshī also did so, which illustrates the point that, however radical their views in the political context in which they operate, peaceful Islamists are by no means outside the pale of legitimate discourse in the Islamic world, or between the Islamic world and the West.[9] He issued a statement after 9/11 which ended with the Qur'ānic text 'help ye one another in righteousness. But help ye not one another in sin and rancour' (Q.5:3). The declaration stated:[10]

> The principles and tenets of Islām can by no means provide any justification whatsoever to what has happened, or may happen in the future, in terms of aggression against innocent people or destruction of institutions and establishments… [Muslims] should never retreat from presenting an authentic and genuine image of Islām as a force advocating peace, tolerance and human brotherhood…

The second encounter, the interview given by Rāshid al-Ghannūshī to Joyce M. Davis, is that of an insider speaking frankly to someone who has gradually gained his trust. It carried with it both a positive and a negative message. The positive message was that the spread of democracy in the Islamic world is not necessarily a challenge to Islām and its values – though no doubt al-Ghannūshī would have added caveats about the need for a degree of flexibility in the democratic system to meet the aspirations and requirements of the people it would be designed to serve and how in practice democracy would be implemented (that is, under whose 'control'?)[11] – but democracy clearly *is* a profound challenge to autocratic and corrupt regimes in the Muslim world. For al-Ghannūshī, democracy 'is simply a way of resolving political and intellectual disputes in a peaceful manner… to call democracy *kufr* is a misguided *ijtihād*'.[12]

The negative message arises from his understanding of the history of the relations between Muslim countries and the West. For, as he made explicit on another occasion, al-Ghannūshī remains convinced that the West withdrew 'only tactically' after decolonization, 'leaving behind agents through whom it continued to control most of the Muslim world. The agents are represented in the Westernized elites, that are cut off from the faith and interests of the masses ruled by them.'[13] Their only legitimacy is derived from 'their suppression of the people and their loyalty to the West'.[14] The response of the Islamic world to the West could be 'hard… and angry and seeking revenge'. Much would depend, in his view, on what approach the West adopts.[15] The phrasing of Rāshid al-Ghannūshī's letter to Edward Djerejian, former US Assistant Secretary of State for Near East and South Asian Affairs, written on 14 June 1992, repays careful reading:[16]

> We want you to know that we, the Muslims, harbour no ill feelings for you or for your superpower status, but we want our freedom in our countries; we want our right to choose the system we feel comfortable with. We want the relationship between you and us to be based on friendship and not subordination. We see a potential for an exchange of ideas, for a flow of information and for cultural exchange in an era governed by the rules of competition[17] and co-operation rather than the rules of hegemony and subordination. We call upon you to halt your aggression against our people and against our religion. We invite you to a[n] historic reconciliation, to rapprochement and to co-operation…

Khurram Murad implies that cooperation rather than conflict between the West and Islām ought to be seen as the norm, and also contends that it is essential for the future of humanity:[18]

> There are over 1.3 billion Muslims in the world. Every fifth person walking on the globe is a Muslim and they inhabit areas which are strategically important. Muslims are not out to deprive the West of the resources that are in their lands. They have to trade with the West. They have to have economic ties with the West. They have to sell their oil as well. Of course, they will guard against the extravagances of their rulers who have been doing the bidding of foreign

> powers, who have been squandering the resources of Muslim countries, but they are not basically hostile [to the West]. Hostility is only a reaction against what has been done to them and what is being done unashamedly.

Reciprocity in the relationship is the key to an improvement in the relations between the Islamic world and the West in the view of an Islamist thinker such as al-Ghannūshī as well as for those who espouse the cause of 'enlightened moderation' in Islām. President Musharraf of Pakistan, who has given this term widespread publicity as well as some thoughtful analysis, has made it clear that there must be reciprocity in the relationship between the Islamic world and the West.[19]

It has not often been stated in this debate that the principle of reciprocity is the underlying 'Golden Rule' to which all the world's main religions subscribe.[20] The problem, however, is that the world's main religions do not have a voice, even at the United Nations – although Dr Rowan Williams, Archbishop of Canterbury, has called for them to have one.[21] Could the world's religions help facilitate the desperately needed peace in the Middle East? It has been argued that they can and should.[22] Whether or not this proves to be case, it is clear that the rejection of religion should no longer continue to be 'inscribed in the genetic code of the discipline' of international relations.[23] John L. Esposito and John O. Voll argue that the

> reconceptualizations of Islām and activist politics of [Mohammad] Khatami, Anwar [Ibrahīm] and [Abdurrahman] Wahid [respectively in Iran, Malaysia and Indonesia] reveal and reflect differing responses to their diverse political and cultural contexts. They challenge those who see the world of the early twenty-first century in polarities, either confrontation-and-conflict or dialogue-and-cooperation, to appreciate the limitations and failures of old paradigms. Ultimately, they demonstrate the need and ability to develop paradigms for governance and policy that are sensitive to the importance of religion and culture in domestic and international affairs.[24]

The practical difficulty is that the relations between international powers are not faith-based[25] or even based on reciprocity but on hegemony: a peaceful world in the future will require international relations to be governed by multilateral initiatives and a perception of what is good for the world, not just the good of the world's greatest power or whichever state has the upper hand in a hegemonic relationship; and also, critically, by equality, reciprocity and justice. In other words, there have to be 'rules to the game'. It has to be recognized in retrospect that the best interests of Chile and the best interests of the United States did *not* coincide in 1973.[26] Sayyid Quṭb argued that the USA had wanted an 'Americanized Islām', but a castrated one: 'the Americans and their allies in the Middle East reject an

Islām that resists imperialism and oppression, and opt for an Islām that resists only Communism'.[27] With the collapse of the Soviet Union, perspectives have changed, but to create a state in its image where Islām is the majority faith might still be said to be an objective of American foreign policy: what else was its role in 'Irāq prior to the restoration of 'Irāqī sovereignty on 30 June 2004?

To understand each *jihād* we must understand its historical context

It will be evident to those who have read the previous eleven chapters of this study that *jihād* is a multi-faceted phenomenon both in theory and practice. There is no, single, all-embracing concept that has been applied within the long, complex and sometimes even tortuous, course of Islamic history. Rather, there have been continual selections of texts and doctrines and the adoption of different practices, in accordance with cultural traditions and the needs and circumstances of the period. Few of the *jihāds* which we have considered in their historical context have conformed to a modern understanding of the theory: they were either preached and launched by individuals, not by the state, and subject to an excess of violence; or, alternatively, they were launched by states acting not in the defence of the faith but in their dynastic or national interests. Fewer still among the *jihāds* of history would have passed the modern test of 'just war'. The best case for *jihāds* which would have passed such tests, those launched against colonial expansion by the Western powers, inevitably ended in failure because of the superior military resources which the colonial power could deploy against them.

Against this complex pattern – the reality of history, which defies simplistic generalization – it is possible nevertheless to draw a number of important conclusions which do not necessarily conform to the accepted stereotype of *jihād*, particularly as interpreted in the West since the events of 11 September 2001. Professor John L. Esposito argues correctly that

> the history of the Muslim community from Muḥammad to the present can be read within the framework of what the Qur'ān teaches about *jihād*. The Qur'ānic teachings have been of essential significance to Muslim self-understanding, piety, mobilization, expansion and defence. *Jihād* as struggle pertains to the difficulty and complexity of living a good life… Depending on the circumstances in which one lives, it can also mean fighting injustice and oppression, spreading and defending Islām, and creating a just society through preaching, teaching and, if necessary, armed struggle or holy war.

Whatever the differences of interpretation in Islām, 'all testify to the centrality of *jihād* for Muslims today. *Jihād* is a defining concept or belief in Islām, a key element in what it means to be a believer and follower of God's Will.'[28]

Defining concept it indeed is; but it is a concept on which there is considerable disagreement. Many Muslims find themselves at a loss to explain quite why the issues are so complex and at the same time intractable: simple explanations are easier for non-Muslims to understand, but are misleading. It would be a useful exercise for Muslim intellectuals to produce their equivalent of John L. Esposito's *What Everyone Needs to Know about Islām*, positing the questions and answers from their own perspective and offering to non-Muslims, in an informed way, their understanding of the issues of their faith, both in theory and in historical practice. More work needs to be done in that most valuable of fields, the comparisons and contrasts between the Abrahamic faiths – a field which Professor F. E. Peters has made his own, both in his two-volume study *The Monotheists* and in his *Islām: A Guide for Jews and Christians*. (It would, incidentally, be highly desirable for Islamic scholars to produce their own impartial guides on Judaism and Christianity for Muslims.) In both his studies, Peters passes the same intriguing remark that the effectiveness of *jihād* has been diminished over time as a practical instrument of policy:[29]

> in the centuries after Muḥammad, the combination of juridically-imposed conditions and political realities has diminished the effectiveness of *jihād* as a practical instrument of policy, though it remains a potent propaganda weapon both for Muslim fundamentalists to brandish and for their Western opponents to decry. Muslim jurists have rarely agreed on the exact fulfilment of the conditions they have laid down for a genuine *jihād* (and Muslim public opinion even less often), while the *ummah* on whose behalf it is to be waged has now been divided, perhaps irretrievably, into nation-states that generally subscribe to a quite different (and decidedly non-Islamic) version of international law.

There is no legitimate offensive *jihād*; nor should Islām be regarded as a 'religion of the sword'

'The religion of Muḥammad was [established] by the sword!'[30] The claim has often been made by the enemies of Islām,[31] and less frequently, by its friends.[32] As was seen in Chapter 8, Gandhi considered this issue carefully and rejected the idea that Islām was a religion of the sword. It is unusual to find Muslims themselves proclaiming that Islām is a religion of the sword, not least because it contravenes the clear Qur'ānic precept that 'there is no coercion in religion' (Q.2:256). One contrary example cited at the beginning of this section is the 'war cry' of Muslims, chiefly the peasantry, involved in anti-immigrant rioting in Palestine in 1920 and 1929. The Muslims feared that the Zionist immigrants were threatening the sanctity of the al-Aqṣā Mosque, the third holiest site in Islām. These concerns had been exploited by the religious and political leaders of the Palestinians, notably the Grand Muftī of Jerusalem, al-Ḥājj Amīn al-Ḥusayni. The situation was different in Palestine from elsewhere, because Jewish immigration meant that there were competing claims to the same territory: Chaim Weizmann claimed, in a speech delivered on 23 April 1936 that further antagonized the Arabs, that there was a conflict between the forces of the desert and destruction on the one side (the Arabs, that is) and the forces of civilization and building on the other (the Jews).[33] Lord Curzon's prediction that the Palestinians would 'not be content either to be expropriated for Jewish immigrants, or to act merely as hewers of wood and drawers of water' to them had become the reality by the end of the 1920s. The virulence of the sentiment expressed in the slogan 'the religion of Muḥammad was [established] by the sword!' is explicable, though not to be condoned, by these circumstances.

For the rest of Islamic history, it was only in the 'revivalist' *jihāds* of the nineteenth century that there was an attempt to impose a single religious viewpoint on Muslims and non-Muslims. The norm was that *jihād* was not begun in order to bring about religious conversions. In that sense, on balance, Islām cannot be viewed as a 'religion of the sword', as has frequently been claimed. In Islamic history, the number of years when there was peace greatly outnumber the number of years of *jihād*. Islām made its greatest progress in terms of conversions in eras of peace rather than of war; war hindered, rather than advanced, the gaining of converts. Rather than facilitate the advance of Islām itself as a religion, historically the *jihād* served the political interests of a Muslim elite which wished to assert, or reassert, its power: for this reason, the *jihād* could be, and frequently was, opposed by other Muslim groups which contested the right of this or that particular elite to claim exclusive authority.

Muḥammad Raheem Bawa Muhaiyaddeen, a Tamil Ṣūfī mystic from Sri Lanka, who died in the USA on 8 December 1986, wrote an important short book called *Islām and World Peace*. In this work, he argued that[34]

> the holy wars that the children of Adam are waging today are not true holy wars. Taking other lives is not true *jihād* We will have to answer for that kind of war when we are questioned in the grave. That *jihād* is fought for the sake of men, for the sake of earth and wealth, for the sake of one's children, one's wife, and one's possessions. Selfish intentions are intermingled within it. True *jihād* is to praise God and cut away the inner satanic enemies… Until we reach that kingdom, we have to wage a holy war within ourselves. To show us how to cut away this enemy within and to teach us how to establish the connection with Him, Allāh sent down 124,000 prophets, twenty-five of whom are described thoroughly in the Qur'ān. These prophets came to teach us how to wage holy war against the inner enemy. This battle within should be fought with faith, certitude, and determination, with the [declarations of faith] (*kalimah*),[35] and with the Qur'ān. No blood is shed in this war. Holding the sword of wisdom, faith, certitude, and justice, we must cut away the evil forces

that keep charging at us in different forms. This is the inner *jihād*… Praising Allāh and then destroying others is not *jihād*. Some groups wage war against the children of Adam and call it holy war. But for man to raise his sword against man, for man to kill man, is not holy war. There is no point in that. There can be no benefit from killing a man in the name of God. Allāh has no thought of killing or going to war. Why would Allāh have sent His prophets if He had such thoughts? It was not to destroy men that Muḥammad came; he was sent down as the wisdom that could show man how to destroy his own evil…

Those who fight for the sake of wife, children, or house follow other rules. If even an atom's worth of such thoughts are present, it is not a true holy war, but rather, a political war. It is fought for the sake of land and country, not for the sake of Allāh. With wisdom, we must understand what the true *jihād* is, and we must think about the answers we will have to give on the Day of Questioning… True holy war means to kill the inner enemy, the enemy to truth. But instead people shout, '*Jihād*!' and go to kill an external enemy. That is not holy war. We should not spread Islām through the sword; we must spread it through the *kalimah*, through truth, faith, and love.

Jihād as the right of defence of the community: the 'just war' argument in new historical circumstances

There are two senses in which a war may be said to be 'just' from the Muslim viewpoint. The first view is that it is *ipso facto* just because it is Muslim. The second view is that only a *jihād* is 'just' – that is, not any war (*qitāl*), but only a war that adheres to Qur'ānic principles and the traditional understanding of the Muslim community as to what constitutes a just cause and just conduct in war. The difference between the two viewpoints is fundamentally important and is often misunderstood.

From the first perspective, any *jihād* is 'just' because the Islamic cause, however it is defined by the group in particular, is of itself deemed just. This is a supremacist viewpoint and must be rejected. Nejatullah Siddiqi repudiates this approach and explains the Qur'ānic basis of his judgement:[36]

> We declare that since we are the people who hold the truth and God has ordained us to work for the domination of this truth, we have to make this truth dominate the globe. We will make it dominant if we have the means… Some of us think that there can be only one dominant system acceptable in the world and that is Islām. We quote the verse: 'and fight them until *fitnah* is no more and the religion is for Allāh' (Q.2:193). Political superiority, the rulership of un-Islām is illegitimate. This is our stand. That is the stand of many in the Islamic movement. Now this is just not acceptable. It is not acceptable because it is not fair. How can 20 per cent of humanity, even if we take all Muslims as belonging

> to the Islamic movement, claim that they have the right to dominate the rest of humanity? Is this what the Divine right of Islām to dominate operationally means: the right of upholders of Islām to dominate?...
>
> What is fair is that everyone is free... Nobody, individual or group, has the right to harm others. So if any group or nation among the collectivity of nations is going to organize itself in a way that it becomes a threat to humanity, then others must come together to stop it. Beyond that, fairness demands that everyone should have equal rights. Personally, I think that this is the formula which gives Islām the best of chances. Whether Muslims are in the majority or in the minority, the best chance which Islām gets is on the basis of equal, individual rights in affairs.... I think that the word *fitnah* [in Q.2:193] does not mean the rule of non-Islām... It means persecution in the name of religion... [if] you cannot practise Islām, you cannot have Islamic faith, then that is *fitnah*, that is persecution. The *jihād* for which that *āyah* calls is *jihād* to end *fitnah* in this sense, not the *jihād* to end *fitnah* in the sense of rule of non-Islām. That is going too far. Very few [Muslims] in the last fourteen hundred years have given this meaning to that verse.

The second view, that only a *jihād* is 'just' that adheres to Qur'ānic principles and the traditional understanding of the Muslim community as to what constitutes a just cause and just conduct in war contains the necessary materials for a fruitful dialogue between the Islamic world and the West.[37] The Christian and the Islamic traditions do not diverge as much on these issues as is sometimes imagined, not least because historically the Christian tradition was influenced by that of Islām.[38] The threat to each tradition no longer seems to come from the other, but from the penetration of the neo-conservative Right ideologues into the argument with a quasi-justification for the doctrine of 'pre-emption' – a quasi-justification that has been found wanting as an intellectual defence for the intervention in 'Irāq in 2003–04.[39] To the extent that the neo-conservative Right ideologues argued that the doctrine of 'just war' had to be brought up to date, the discussion was unexceptionable. Muslims have usefully brought up to date the dichotomy of the abode of Islām/peace (*Dār al-Islām*) and the abode of war (*Dār al-Ḥarb*) on the grounds that it no longer conforms to the reality of the world in which we live. They have proposed instead the terminology of the abode of Covenant/Treaty (*Dār al-'Ahd*)[40] or even the 'abode of testimony'. Muslims in the West, argues Ṭāriq Ramaḍān, are not in 'other societies', desperately hoping for a return to the *Dār al-Islām*; they are, on the contrary, at home in the West; the old terminology thus appears 'completely restrictive and out of context'. At the very least, Muslims in the West owe civic allegiance to their countries of residence, and in some cases asylum.[41]

The danger with the approach of the neo-conservative Right ideologues arguing that the doctrine of 'just war' had to be brought up to date is that it has proved to be not only a justification for pre-emption, but for unilateral (that is, hegemonic) intervention by means of the pre-emptive strike. It therefore overturned the rule of international law and the role of the United Nations, as was evident in the launching of war against 'Irāq in 2003. Ironically, in order for the United States and Britain to disengage from a conflict of their own making, the United Nations had to be brought back into the arena in 2004 to approve the arrangements for the transfer of sovereignty back to the 'Irāqī people. The whole argument for unilateral action has therefore been proven to be fallacious as well as damaging to international cooperation.

The second of the two Muslim arguments about 'just war' – that is, the traditional understanding of the Muslim community as to what constitutes a just cause and just conduct in war – may be, though this is not necessarily the case, pluralist in its understanding of the issues at stake in the war. In terms of group practice rather than individual experience, it may recognize Islām as a world religion like any other[42] and accept that it is subject to an international rule of law as well as cognizant of, and sensitive towards other considerations such as human rights.[43] Waris Maẓharī, the editor of the Urdu monthly *Tarjumān Dār ul-'Ulūm*, the official publication of the Delhi-based Deoband *Madrasah* Old Boys' Association, whose views have been quoted above (Chapter 11), is adamant that Islām is not, and cannot be, a 'religion of the sword'; nor can it engage in offensive war, self-proclaimed war, or war by proxy:[44]

> Islām cannot be imposed on anyone, and in any case the Qur'ān allows for violence only in defence. It is sheer foolishness, in addition to being wholly un-Islamic, to imagine that Islām can be established through violent or offensive means…
>
> The classical scholars of Islamic jurisprudence (*fuqahā'*) have clearly laid down that only the *amīr* or leader of an established state can issue a declaration of *jihād*. Private individuals or groups do not have the right to do so, and so… self-styled *jihād* has no Islamic legitimacy. Furthermore, Islamic law lays down that Muslims must honour the agreements that they enter into with others, and if they wish to disengage from these agreements they must openly declare so. This rules out proxy war on the part of a Muslim state…

The unity of Islām and tolerance within Islām

The imperative for a modern, enlightened Islām is to stress the importance of the 'greater', instead of the lesser, *jihād*. A preoccupation with the 'greater *jihād*' will help the reassertion of an Islām that is 'enlightened' in the sense of being a full participant in the dialogue of civilizations and an equal player among the world's great faiths. It is an objective that has to be nurtured and guided by the political and religious leaders of the Islamic world. One of the problems these leaders face is a sense of division and separation rather than consensus (*ijmā*'). Muslims are enjoined to cooperate to promote virtue and righteousness, not evil and aggression (Q.5:2) and to support universal peace based on justice (Q.2:208; 8:61). On matters of doctrine, Muslims are wise to remember the saying attributed to Imām al-Shāf'ī: 'our view – as we believe it to be – is right, but it could later be proved to be wrong, and the view of others – as we believe it to be – is wrong, but it could probably be later proved to be right.'[45] Muslims may decide to form several Islamic parties if they have differences regarding the concepts, or strategy, or even the structure and the leadership with which they feel comfortable.[46]

Though the Qur'ān warns against religious extremism (Q.4:171; 5:77), Walid Saif accepts that 'the harm which can be done to religion by religious extremism may well surpass any such harm by secular extremism'. In his view, the most dangerous division may not be that of religion versus religion: 'religious freedoms across religious communities or within the same community can be more suppressed by zealous, narrow-minded and exclusivist interpretations of religion.' For belief not to breed exclusion, 'it should never imply that the believer is the truest. Religion and religiosity are not one and the same. No one can claim to be the sole and exclusive representative of God.'[47] In spite of the rhetoric of Muslim unity, Muslims, particularly the '*ulamā*', are fiercely divided on sectarian (*maslakī*) lines.

The tendency is to look at the affiliation of a particular Muslim with greater attention than the views he or she is advocating. Any hint that the Muslim belongs to one school or tradition rather than another, and the risk is that viewpoint is not heeded by Muslims of another school or tradition. Whether or not the Ahmadī community makes positive suggestions,[48] their views are immediately discounted because they are perceived as heretics, whatever their importance historically in the formation and early history of Pakistan. A Wahhābī will not listen to the views of a Ṣūfī, or vice versa, even though such categorization is by no means easy to sustain.[49] More attention sometimes seems to be spent trying to prove that Osama bin Laden is a Wahhābī, or if not a Wahhābī then allegedly a Ṣūfī,[50] rather than in confronting his insidious ideology. It is as if al-Ghazālī had never asked the question as to who could lay claim to 'this monopoly over the truth… Why should one of these parties enjoy a monopoly over the truth to the exclusion of the other?' (see Chapter 6). In particular, the failure of Islām to speak with consensus on the issue of pluralism leaves the political arena to the militant Islamists, who repudiate the concept in entirety.

In an interview with Yoginder Sikand, Waris Maẓharī emphasized the need to try to heal sectarian differences which divide the house of Islām:[51]

> Followers of the different *maslaks* must peacefully coexist with each other despite their differences and cease from condemning other *maslaks*. In place of the sharp polemical exchanges between them, that sometimes take the form

of hurdling *fatāwā* of infidelity against each other, they must learn to relate to each other through peaceful dialogue. For this it is essential that the '*ulamā*' of the different *maslaks* desist from raking controversial and minor issues, and focus instead on the larger issues facing the entire Muslim community as a whole, such as education, poverty, social inequalities, violence, the challenge of the West or *Hindutva* and so on, about which the '*ulamā*', at least in India, know little, if at all, and about which they do next to nothing about in concrete, practical terms. If they were to focus their energies on these larger issues they would have no time for peripheral sectarian matters. They must also put an immediate halt to sectarian polemical literature…

Many '*ulamā*' simply do not possess a universal understanding of Islām that goes beyond the narrow boundaries of their own *maslak*. They also have a vested interest in perpetuating and promoting sectarianism as it gives them the authority to speak for their own flock. In turn, this brings them rich rewards, in concrete financial and political terms… They deliberately ignore the fact that what is common to all the *maslaks* and what unites them – basic beliefs such as faith in God and the prophethood of Muḥammad, which makes them all Muslim in a fundamental sense – far outweighs their differences…

Those who present themselves as the greatest champions of the cause of their *maslak* are the least concerned with the plight of the poor and the needy. It is as if Islām is simply a set of complicated rituals, and that it has nothing to do with social affairs as such. By reducing Islām to a body of external rituals, the spirit of Islām is effectively eclipsed, and what is a universal movement for social emancipation is made to appear as the manifesto of a set of mutually bickering sects. This is, in part, an outcome of the sort of education that the '*ulamā*' receive in the *madrasahs*, where great stress is paid to teaching students about such issues as the length of the beard, the right method of performing their ablutions, and so on, while the Islamic duty of working for the real-world issues of the needy – such as poverty, hunger, war and unemployment – is totally ignored.

Waris Maẓharī makes it clear that in his view, nothing is happening by way of inter-community dialogue in India. Most writers drawn from the *madrasahs* engage only in meaningless rivalry with those from other traditions, seeking thereby to establish a reputation for themselves. All the various Sunnī groups, despite their differences on certain matters, share the same basic beliefs and hence are all Muslims, because the matters on which they differ are not so significant as to be beyond the pale of Islām. With regard to the Shī'a tradition, the situation is more complicated.

Some Shī'a sects believe that the Qur'ān has been distorted or that Imām 'Alī was God. Naturally, we cannot consider them as Muslims. On the other hand,

> many other Shīʻa, particularly the largest Shīʻa group, the Twelver Imāmī or Ithna Ashari Shīʻa, do not hold such views and, in many matters of basic beliefs are much closer to the Sunnīs despite their differences in understanding Islamic history and jurisprudence That is why some leading Deobandī *'ulamā'*... considered them as fellow Muslims. True, some Deobandīs in Pakistan insist that the Ithna Asharis be declared non-Muslims, but not all Deobandīs, even in Pakistan, hold that opinion. As a Sunnī, I don't accept all the beliefs of the Ithna Ashari Shīʻa as valid, but I still consider them to be part of the wider Muslim *ummat*, and hence feel the need for us to work together on issues of common concern.

The plain fact is that Muslims do not have a Council or representative body with which the Christian churches (for example) can do business, and it would be helpful if they did. The Sunnī tradition of Islām is fiercely individualistic, self-assertive and independent. In the absence of Councils such as the early Christian Councils, Islamic thought was developed by various scholars and jurists. In the contemporary era, those who have studied in *madrasahs* (Islamic seminaries) 'have, to a large extent, been out of touch with developments in the field of science, technology and even other areas of thought and society'. An *'ālim* or *muftī* is required to have some basic knowledge and awareness of the 'custom' (*'urf*) and practice (*'ādāt*) of the people where he lives and works. Virtually no *madrasah* trains future *'ulamā'* in even the basic concepts and ideas of the Judaeo-Christian traditions, which are an important foundational study for understanding the West. There is also 'an urgent need to introduce the intellectual and cultural trends of Western society into Muslim seminaries' syllabi'.[52] In contrast, some of the mainstream Christian churches in the West now include necessary courses for appreciating religious diversity and helping future priests to be able to deal sensitively with persons from other faith communities. Doubtless, this is no more than a beginning, but it is nevertheless an encouraging start. Problems remain with the evangelical free Christian churches, many of which stubbornly refuse to acknowledge the equality of other faith traditions even in terms of the practicalities of living in a multi-faith society.

The need for mainstream Islām to embrace positively the existence of pluralist societies in the contemporary world

Most of the jurists whose views have been discussed in earlier chapters of this book would have found it very difficult to conceive of a modern multi-faith and multicultural society. Their preoccupation was with Islām and how best to defend it and propagate it. We do not know what their response would have been towards the key development of the contemporary world, the emergence of pluralist societies, simply because this was a question which was not asked at the time. The indications, however, are that the jurists would not have found it easy to wrestle with the issues of pluralism, posing as they do issues of acculturation and intercultural understanding.

At first sight, we might assume that the inner struggle, the *jihād al-nafs*, would have posed no problems of itself to the contemporary world of pluralist societies. For this was a struggle essentially directed inward, towards holiness, right thinking and right living, and the straight path towards salvation. Yet, as we have seen, in the African *jihād* of the early nineteenth century, a deep-thinking Muslim whose central preoccupation was *jihād al-nafs* (the Ṣūfī Shehu 'Uthmān dan Fodio) turned to a violent '*jihād* of the sword' to impose his vision of a pure Islām on what he perceived to be a faith and society corrupted by acculturation, by concessions to the indigenous African tribal religions. It was as if, in the present day, the notoriously conservative Cardinal Ratzinger, guardian of traditionalism at the Vatican, went to war to impose his viewpoint on those Catholics who, in his view, were making too many concessions to the indigenous faiths of the Indian subcontinent.

In this positive embrace of multi-faith societies, there cannot be half-measures or exclusion clauses as proposed by some of the militant Islamist groups. Two possible scenarios for the relationship between the three Abrahamic faiths have been suggested by the violent Islamist movements. The first is the subordination of Judaism and Christianity to Islamic rule. Thus the 1988 Foundation Charter of the militant Islamist organization Hamas pronounces that[53]

> under the wing of Islām, it is possible for the followers of the three religions – Islām, Christianity and Judaism – to coexist in peace and quiet with each other. Peace and quiet would not be possible except under the wing of Islām. Past and present history are the best witness to that… Islām confers upon everyone his legitimate rights. Islām prevents the incursion on other people's rights. The Zionist Nazi activities against our people will not last for long…

An alternative scenario, described by 'Alī Fayyid of Ḥizb'ullah, is that of 'cold cohabitation' (*al-tasākun al-bārid*) between Muslims and Jews, with no normalization of relations, so severely does he judge the Qur'ānic criticisms of Judaism.[54]

These approaches are ethically incorrect and will not well serve the cause of Islām in the future. Nor are they consistent with the essence of the faith. For, as Walid Saif and Ataullah Siddiqui note, there are several Qur'ānic aids to Muslims in encountering religious diversity in the modern world. 'If thy Lord so willed, He could have made mankind one people' (Q.11:18); 'O mankind, We created you from a single (pair) of a male and a female, and made you into nations and tribes that you may know each other (not that you may despise each other)' (Q.49.13); 'And indeed We have honoured the Children of Adam…' (Q.17:70); and finally, more intriguingly, 'and among His signs is the creation of the heavens and the earth, and the variations in your languages and colour. Verily in that are signs for those who know' (Q.30:22).[55] Every people, we are told in the Qur'ān, 'has been sent its guide (Q.35:24). Differences of belief are thus seen as part of God's plan. The abolition of such differences is not the purpose of the Qur'ān, nor was the Prophet sent for that reason. Human beings have a common spirituality and morality (Q.7:172; Q.91:7–10), but the differences in faith traditions are present because God has given human beings the freedom to choose: 'if it had been the Lord's will, they would all have believed – all who are on earth! Will you then compel people against their wills to believe?' (Q.10:99).[56]

Waris Maẓharī, whose views we have quoted on the subject of divisions with Islām itself, has equally pertinent comments about the need for positive interaction between Islām and other faiths. 'We need considerably to revise our ways of looking at the theological other', he argues, 'and of notions such as *ḥijrat* and *jihād*, because today the world has been radically transformed.' He continues:[57]

> in place of the categories of the abode of Islām and the abode of war, I prefer to speak of the entire world as an abode of agreement. In turn, this points to the pressing need for radically revising how we look at people of other faiths. Some Muslims, including sections of the '*ulamā*', regard all non-Muslims as enemies of Islām. As I see it, this is not at all Islamic. The primary duty of a Muslim is [proselytism, or to reach out, publicize] (*tablīgh*), to convey the truth of Islām to others. Our task is simply to tell others about Islām in our capacity of being witnesses to all humankind. Others are free to choose whether to accept our message or not, for the Qur'ān lays down that there can be no compulsion in religious matters. Now, how can Muslims communicate the message of Islām to others if they consider and treat them all as their enemies?

In arguing that most Muslim clerical figures are unable to engage positively with people of other faiths, Maẓharī places the failure down to the *madrasahs*:[58]

> *Tablīgh* being a principal duty of the Muslims, instead of branding all non-Muslims as enemies of Islām, Muslims should seek dispassionately to understand them and seek to build bridges of friendship and dialogue with them. Unfortunately, few contemporary '*ulamā*' are engaged in this sort of work. A major reason for this is that the *madrasahs* have little or no contact with the wider society, and the '*ulamā*' are cocooned in their own narrow circles, lacking any awareness of the demands of the contemporary world. This is a vicious circle the *madrasahs* train their students in such a way that they are insulated from the world around them, and these students, when they graduate as '*ulamā*', come back to the *madrasahs* to teach and thus perpetuate

the same system. They have little or no knowledge of other faiths, which is a must for the task of *tablīgh* or even simply for building good relations with others. Hardly any *madrasahs* teach their students about other religions. Hence, their graduates are incapable of interacting with others, and some are even scared of doing so. Even if they try to communicate with others, they generally fail, because they cannot communicate with them in an appropriate mode, being ignorant of their traditions, beliefs, cultures, histories, languages and ways of thinking.

For Maẓharī, helping the needy of other communities, as the Prophet did, must receive the attention that it deserves; yet the task is almost entirely neglected. This could today take the form of setting up quality institutions such as schools and hospitals that would serve not just Muslims alone but others as well. In this way, others would be forced to reconsider the ways in which they look at Muslims, seeing them as genuinely concerned about the problems of society as a whole.

Mainstream Muslims engage with the key issues as perceived by others: *ijtihād*, *sharī'ah* modernization, common citizenship, power-sharing in civil society, apostasy

Within the Muslim supremacist position, Mawdūdī defined 'justice' as the removal of constraints on the development of Islām towards its place as the predominant world religion. In his ideology, Islām cannot long accept minority status; it must achieve supremacy over all other faiths (see Chapter 8). No person committed to another faith is likely to accept this concept of 'justice' as fair or reasonable in a multi-faith society with significant religious minorities. Even in Pakistan, where the Muslim population is about 97 per cent of the total population, the theory is open to serious objection in that it contravenes the ideology of the founding father of Pakistan, Jinnah, as defined in his speech to the Constituent Assembly of Pakistan in August 1947:

> you may belong to any religion or caste or creed – that has nothing to do with the business of the state... We are starting in the days when there is no discrimination, no distinction between one community and another, no discrimination between one caste or creed and another. We are starting with this fundamental principle that we are all citizens and equal citizens of one state.[59]

Great words of a great statesman, but alas words that have often been forgotten in periods of communal religious tension in Pakistan's history. The theory of Mawdūdī was diametrically opposed to Jinnah's: the Islamic State he envisaged was 'universal and all-embracing'. Since the definition of the state was ideological, nationhood too was to be derived from ideological (that is, Islamic) convictions. At best non-Muslims might have a system of separate communal electorates from which they would vote for a candidate of their own community. More recent Muslim thinkers have recognized that Mawdūdī's ideology does not resolve the various related problems which they confront. Even if they reject the secularism of Faraj Fūda, who was assassinated in 1992, more recent Muslim thinkers are likely to agree that there is no alternative to his maxim of 'enlightened *ijtihād*, courageous analogy and visionary horizons'.[60]

Any discussion of the role of independent judgement on legal or theological questions based on interpretation and the application of the roots of Islamic law (*ijtihād*) has to begin with the thought and writings of 'Allāma Sir Muḥammad Iqbal (1877–1938), called by President Khamein'i of Iran the 'poet-philosopher of Islamic resurgence'.[61] Nicolas Aghnides had argued in 1916 that *ijtihād* was a mechanical principle that had led Islamic society to fossilization and prevented it from progress ('the Islamic system of law does not possess an evolutionary view of life and the qualifications and limitations for *ijtihād* illustrate the mechanical nature of law'). The urge to refute this statement was the motivation behind Iqbal's lecture on 'The Principle of Movement in the Structure of Islām' (December 1924).[62] 'What then is the principle of movement in the structure of Islām?', Iqbal asked. 'This is known as *ijtihād*' was his reply. 'The word literally means to exert. In the terminology of Islamic law it means to exert with a view to form an independent judgement on a legal question.'[63] Iqbal maintained that 'the closing of the door of *ijtihād*[64] [was] a pure fiction'. Instead of stasis, he focused on the dynamic character of the universe, the Islamic mode of prayer, the self, the Qur'ān and *ijtihād*. Within his definition of *ijtihād* there were a number of dynamic elements: the Qur'ān's anti-classical spirit; a dynamic concept of universe, the nature of Islamic society and culture; the idea of the changeability of life; the realism of juristic reasoning in Islām; and the evolutionary and dynamic concept of the intellect and thought in Islām. Iqbal regarded the conflict between the legists of Ḥijāz and 'Irāq as a source of life and movement in the law of Islām. Iqbal preferred the title *Reconstruction of Religious Thought in Islām* to *Reformation and modernization*, since *Reconstruction* aimed at restoring the original universalism and dynamism of Islām, which could not be achieved by adopting the terms 'reformation' or 'modernization'.[65]

Iqbal cited Q.29:69 ('to those who exert We show Our path') as the origin of the concept of *ijtihād*. He told one of his followers that 'all efforts in the pursuit of sciences and for [the] attainment of perfection and high goals in life which in one way or other are beneficial to humanity are man's exerting in the way of Allāh'.[66] For Iqbal, *sharī'ah* values (*aḥkām*) (for example, rules providing penalties for crimes) are in a sense specific to a particular people; and since their observance is not an end in itself they cannot be strictly enforced in future generations. He argued that there are two spheres of Islām: one is '*ibādāt* which is based on the religious obligations (*arkān-i-dīn*) – these do not require any change; the other sphere is that of *mu'āmalāt* (or social relations) which is subject to the law of change. Not only are social relations subject to the law of change; they are also not necessarily subject to the influence of the *'ulamā'*, at least in Iqbal's view.[67] In the contemporary era, Iqbal's ideas have been taken up by those who, like Rāshid al-Ghannūshī, argue that there must be sufficient 'space' (*farāghat*) in which *ijtihād* can operate.[68]

Muḥammad Asad draws the practical inference with regard to the relationship between *ijtihād* and *sharī'ah*:[69]

> The *sharī'ah* does not attempt the impossible. Being a Divine ordinance, it duly anticipates the fact of historical evolution, and confronts the believer with no more than a very limited number of broad principles; beyond that, it leaves a vast field of constitution-making activity, of governmental methods, of day-to-day legislation to the *ijtihād* of the time concerned…

For Dr Ashgar Ali Engineer, 'it becomes obvious that social dynamism ultimately leads to legal dynamism and the legal philosophy should not be based on outdated medieval concepts. Legal philosophy while based on Islamic and Qur'ānic values should not become stagnant but should remain dynamic and *ijtihād* should be a continuous process'.[70] Ataullah Siddiqui develops the argument further:[71]

> as soon as one utters the word '*sharī'ah*' it conjures up images of severed hands, floggings and the like, but the *sharī'ah* is also the Muslims' prayer (*ṣalāt*), their fasting (*ṣawm*), their charity (*ṣadaqah*) and their love of God, as well as connoting their more mundane activities. The impression created by the word *sharī'ah* is partly the result of the actions of some military leaders (as well as others) keen to demonstrate their Islamic credentials to their people: flogging is an easy option and a visible expression of such desire. In addition, they wish the *sharī'ah* not to be other than what they themselves project. They have no desire to see beyond this image, which suits them well…
>
> *Sharī'ah* means 'the path', 'the way to the water'. As water symbolises the source of life, so the *sharī'ah* represents the source of Muslim existence. The basis of the *sharī'ah* is the Qur'ān and the Prophet Muḥammad's example. But the function of the *sharī'ah* is essentially to outline some basic principles, norms and values. The details of the implementation are left to interpretation. The objectives (*al-maqāṣid*) of the *sharī'ah* are, as al-Ghazālī puts it, 'to promote the welfare of the people, which lies in safeguarding their faith, their life, their intellect, their posterity and their wealth. Whatever ensures the safeguarding of these five serves the public interest and is desirable.' Ibn al-Qayyim states that 'the basis of the *sharī'ah* is wisdom and the welfare of

> the people in this world as well as in the Hereafter. Welfare lies in complete justice, mercy, well-being and wisdom. Anything that departs from welfare to misery, or from wisdom to folly, has nothing to do with the *sharīʻah*.' Ibn Taymīyah further enhanced the objectives, and listed things like the fulfilment of contracts, the preservation of ties of kinship and respect for the rights of one's neighbours. Essentially, he left this list open. Today, Yūsuf al-Qaraḍāwī has further extended the list of the *maqāṣid* to include human dignity, freedom, social welfare and human fraternity among the higher aims of the *sharīʻah*... What we need is to promote the commonly acknowledged good (*ma'ūf*) and discourage the commonly acknowledged evil (*munkar*) in human society...

Furthermore, Ataullah Siddiqui argues elsewhere, 'anything that is commonly known as good (*ma'ūf*) should not wait for any religious approval but should be regarded as part of one's own heritage.[72] Thus *sharīʻah* should no longer be regarded as simply as the imposition of fixed penalties (*ḥudūd*) imposed arbitrarily on minorities without their consent. This does not imply that it has not happened, and may well continue to happen in parts of the Islamic world. It means that it should not. Nejatullah Siddiqi makes this distinction clear:[73]

> When I say that the *sharīʻah* cannot be implemented unless people accept it, I do not mean that the *sharīʻah* is not acceptable. It is acceptable, but... you cannot impose it on an unwilling people, Muslim or non-Muslim. Nobody has the right, not even an Islamic movement gaining whatever percentage of the vote, to impose the *sharīʻah* on an unwilling people...

Arskal Salim, a Muslim academic from Indonesia, argues that Indonesia

> should learn from the Nigerian experience in the application of *sharīʻah*, a process that has triggered bloody riots and created splits in certain provinces of Nigeria. The aftermath of *sharīʻah* implementation in some northern provinces of Nigeria should demonstrate to Indonesians that failure to engage openly with the more sensitive aspects of such a process will only lead to political instability... Every proposal for the implementation of *sharīʻah* must be submitted within... a constitutional framework. This... would seem to be the only realistic option for the future of Indonesian religious pluralism and the only and the best alternative to a complete abandonment of the notion of a legal system based on *sharīʻah*.[74]

What this means is that Muslims have to be committed to the principle of common citizenship on the basis of equality and without discrimination for reasons of faith.[75] 'We sent you not, but as a mercy for all of creation' (Q.21:107): Muslims are enjoined to respect diversity and display inclusiveness.[76] As was seen above

(Chapter 2), one of the few good things about the Khārijīs in the early history of Islām was their tolerant attitude to other faiths: they were prepared to allow a revised *shahādah*, that 'Muḥammad is the Apostle of God to the Arabs but not to us'. Equal citizenship is mandatory on all, because of the prior requirement of reciprocity in our dealings with each other. Fathi Osman argues that[77]

> global pluralism is a solid reality that cannot be escaped, and national pluralism will always be what the majority of the world persistently requires. Besides, Muslims cannot demand justice for their minorities all over the world unless they secure it for non-Muslim minorities living among Muslim majorities… Islām teaches justice, understanding, cooperation, and kindness in dealing with non-Muslims and all 'others' at country and international levels (Q.49:13; 60:7–8).

It is frequently said that the democratic system is not a Qur'ānic concept and that the Islamic equivalent is the *shūrā*. Fathi Osman argues that 'the *shūrā*–democracy polemics have to be settled once and for all. Until we have a really efficient concretization… of the concept of *shūrā*, we have no equal or parallel to the existing developed mechanisms of democracy'.[78] There are two powerful arguments in favour of democracy as against the concept of *shūrā*. The first is that, as has been seen above (Chapter 2), *shūrā* was not formalized by the Prophet as a system for the transfer of power.[79] It existed historically in the early history of Islām as no more than occasional advice given by an extremely narrow group, answerable to no one but themselves, a group moreover which could not be removed from power except through rebellion. The second reason is expressed cogently by Abdelwahab El-Affendi:[80]

> wisdom dictates that we should be pessimistic about the qualities of our rulers, something which should not be too difficult, given our experiences. The institutions of a Muslim polity, and the rules devised to govern it, should therefore be based on expecting the worst. Human experience shows that democracy, broadly defined, offers the best possible method of avoiding such disappointment in rulers, and affords a way of remedying the causes of such disappointment once they occur.

All of which presupposes firstly, the existence of an effective, functioning, civil society and secondly, an unshakeable commitment on the part of Islamist parties to power sharing and the peaceful transfer of power (including the transfer of power at their expense). On the first point, there have been significant developments in recent years in parts of the world-wide Islamic community. Robert W. Hefner argues the case powerfully for the existence of a 'civil Islām' in Indonesia prior to the 'baleful influence' of Islamist paramilitaries after 1998:[81]

> democracy depends not just on the state but on cultures and organizations in society as a whole... In the end, however, democratization involves not one of these phenomena but both in mutual and ever emergent interaction.
>
> The Indonesian example also makes clear, however, that these two developments come to nothing if they are not reinforced by a third above and beyond society: the creation of a civilized and self-limiting state. As in Indonesia, the culture of civility remains vulnerable and incomplete if it is not accompanied by a transformation of [the] state... The state must open itself up to public participation. At the same time, independent courts and watchdog agencies must be ready to intervene when, as inevitably happens, some citizen or official tries to replace democratic proceduralism with netherworld violence. As vigilantes and hate groups regularly remind us, not all organizations in society are *civil*, and the state must act as a guardian of public civility as well as a vehicle of the popular will.
>
> Despite thirty years of authoritarian rule, Indonesia today is witness to a remarkable effort to recover and amplify a Muslim and Indonesian culture of tolerance, equality and civility. The proponents of civil Islām are a key part of this renaissance. Civil Muslims renounce the mythology of an Islamic state.... The majority have learned that they make their understanding of God's commands more relevant when they related them to an ecumenical interpretation of Indonesian history and culture. The majority have also learned from Soeharto's excesses that the critical scepticism of their forebears toward the all-subsuming state is an attitude still relevant for politics in our age.

The creation of a civilized and self-limiting state is not inevitable. In Indonesia, the challenge posed by Laskar Jihād commander Ustadz Ja'far 'Umar Thalib's declaration of *jihād* in early May 2002[82] is one obvious type of challenge; the group closed down the public side of its operations on the eve of the Bali bombings. Seyyed Vali Reza Nasr argues that, instead of a civilized and self-limiting state, the reverse has occurred in Pakistan and Malaysia, where the state has accommodated Islamic ideology but in so doing has 'secured the upper hand in its dealings with... society, entrenched its powers, and expanded its authority, control and reach'.[83]

Apart from Indonesia, another example of 'civil Islām' is provided by the Ahbash in Lebanon. In contrast to Ḥizbu'llah and the Islamic Association, the Ahbash opposes the establishment of an Islamic state on the grounds that this divides Muslims. Instead, it accepts Lebanon's confessional system (which used to give Christians six slots for every five Muslim slots, and now gives them parity). The Ahbash has emerged as a Sunnī middle-class movement that attracts intellectuals, professionals, and businessmen, particularly the traditional Sunnī commercial families of the urban centres. Among these social groups, the Ahbash call for religious moderation, political civility, and peace – specifically in the Lebanese context, inter-sectarian accord and political stability; an enlightened Islamic spiritualism within a modern secularist framework; a Lebanese identity wedded to Arab nationalism; and an accommodating attitude toward the Arab regimes, particularly the Syrian government – has had a powerful resonance. Its foreign policy orientation is equally mild, making no reference to *jihād* and directing no anger toward the West. To achieve a civilized Islamic society, it recommends that members study Western learning. Also, the Ahbash has established branches in Australia, Canada, Denmark, France, Sweden, Switzerland, the Ukraine and the United States (with headquarters in Philadelphia). It enjoys excellent relations with most Arab states, particularly Syria.[84]

Are peaceful Islamists prepared to play by the rules of democracy and accept the peaceful transfer of power even at their own expense? There is an unequivocal answer from Rāshid al-Ghannūshī:[85]

> our movement has worked for a very long time to present an Islām that works together with democracy. This is our main approach. This is our identity as a movement. We believe, and we've said since 1981, that if the people are going to choose Communists, and you know Communists are the enemies of Islām, we should accept the verdict of the people. And we should blame ourselves and try to convince the people that what they have done is wrong. And we should protect minorities, even if they are against us. So our agenda is 100 per cent democratic…
>
> The main point which distinguishes us from the rulers in place is that we want a government which is serving the people and they want a state served by the people.

Rāshid al-Ghannūshī proclaims his credentials as a 'democratic Islamist':[86]

> there is no acceptable alternative other than democracy, one that is not exclusive, recognizing all perspectives. Stability will not occur unless we have a democracy of equality that embodies the people's right to control their civil agendas without mandate; one that adheres stringently to the rotation of power; and one that strives for the fair distribution of wealth and establishment of a free-market economy.

Among the most important of modern human rights is the right of freedom of belief. Freedom of belief implies the freedom to change one's religion. Not so for Mawdūdī. He writes: 'as regards Muslims, none of them will be allowed to change creed. In case any Muslim is inclined to do so, it will be he who will be taken to task for such a conduct, and not the non-Muslim individual or organization whose influence might have brought about this change of mind.' Moreover, for Mawdūdī, the state also had a role in declaring who was and who was not a true Muslim; he was a key figure in the anti-Aḥmadī agitation of 1953 (see Chapter 8).

While the denial of the *shahādah* is the foremost indication of apostasy, it is not the only one.[87] 'Whoever claims prophethood after the completion of Muḥammad's mission, or gives support to such a claimant, becomes an apostate: such claims contradict the idea of the finality of Muḥammad's prophethood which became so central to [the] Islamic creed.'[88] This was the objection to the Aḥmadīs in Pakistan, which led eventually to their being declared heretics. Though the Prophet is said to have accepted the repentance of several persons who had abandoned Islām,[89] the crime of apostasy from Islām is still deemed by many Muslims to be punishable by death. It is, however, a view which has no basis in the Qur'ān. As the leading authority on the subject, Mahmoud Ayoub, argues[90]

> had the Qur'ān considered apostasy a public offence deserving maximum punishment (*hadd*) like theft, adultery or murder, these verses [Q.3:86–91] would have been the proper place for such a ruling. In fact, traditions concerning the occasions for the revelation of the verses do not mention that the persons who had turned away from the faith and later returned penitent were required to make a public confession of their repentance. Nor was apostasy an issue of major concern for classical commentators on these verses.

The early sources of the Islamic tradition reveal an increasingly severe attitude towards apostasy, which was seen not only as a crime against God but as a potential act of treachery against the Islamic state and society. Some evidence for the hard-line view of the jurists was found in a widely-accepted Prophetic *ḥadīth*: 'he who changes religion, kill him'.[91] On the other hand, although there was unanimity among the jurists that apostasy was a crime deserving the death penalty, not all were in agreement that the death penalty should be applied: at least two jurists advocated that an apostate should not be killed, but enjoined to repent until the end of his life.[92]

While accepting that freedom of religion in practice is a recent phenomenon in human history, such interpretations appear to be at variance with the command that there should be no compulsion in religion (Q.2:256). How can it be said that there is no compulsion if a Muslim should lose his faith and find that it now lies elsewhere? He is imprisoned in it, has to dissimulate his change of faith, or has to face, in principle, the death penalty. Some prominent voices now disagree openly with this tradition. The Sudanese Islamist ideologue, Dr Ḥasan al-Turābī, refused to join in the furore concerning Salman Rushdie's *Satanic Verses*. He stated in an interview in June 1996:[93]

> the Prophet… explained that one who abandons his religion and deserts his fellows [in a military sense] should be killed. Regrettably, people of subsequent

> generations have taken the Prophet's saying out of its historic context and generalized it. In so doing they deny one of the basic truths of Islām: the freedom of the faith.

His reasoning was particularly interesting, leading us back to *ijtihād*:

> in matters of faith and day-to-day worship you have to exercise your intellectual faculties... under all circumstances a Muslim should always be spiritually ready to exercise the grace of mind and reason...
>
> *Ijtihād* is a duty which should be carried out by all people, regardless of their individual abilities. If we grasp this concept fully, then it will be easy to consider that knowledge and *ijtihād* are both community activities... *Ijtihād* is a community activity, and to deny the community that right is nothing more than a gross manifestation of religious decadence and ignorance.

The *jihād* for justice and the betterment of the human condition

'Enlightened moderation', as President Musharraf suggested both at the OIC summit in Autumn 2003 and at the Davos Seminar on Promoting Inter-Civilization Dialogue and Action on 23 January 2004, has to be a two-way process.[94] There has to be some give from the West, and that means from the United States, Japan and the enlarged European Union in particular. However, it is *not* opting out of the pressing need for conflict resolution in the world's troubled areas to proclaim loud and clear that the 'greater *jihād*' must be the struggle for the betterment of the human condition in Muslim countries where living standards are far below the human expectations in the West.[95] More than half the *ummah* is illiterate: the illiteracy rate in certain Muslim majority countries sometimes comprises two-thirds of the population. Economic survival itself is imperilled by such high levels of illiteracy; it is evident that the population's capacity for self-betterment is gravely diminished.[96] There is a pressing need to make up on the technological gap between the Islamic world and the West. Waris Maẓharī argues that the answer lies with Muslims themselves:[97]

> To subvert Western dominance, Muslims must learn the secrets of Western strength, which lie in organization, science and technology. It is because of this, rather than because of any supposed anti-Islamic conspiracy, that the West has been able to subjugate the Muslim world. It is only by mastering modern sciences that Muslims can effectively resist Western domination. However, Muslims lag far behind others in this regard, and some '*ulamā'* have only made the situation more precarious with their claims that modern science will lead Muslims away from the path of the faith. This is completely incorrect.

Yūsuf al-Qaraḍāwī's concept of Islām, which we are told is being disseminated by the 'committed, balanced and enlightened trend of Islamic resurgence', is dominated by the characteristics of 'reason, renewal, [capacity for independent judgement] (*ijtihād*), "middle-roadness" (*wasaṭiyyah*), pragmatism, respect for women[98] and family, belief in education and oneness, rejection of priesthood, belief in the right of the *ummah* to elect its leaders, preservation of private property, taking good care of the poor and downtrodden, and encouraging love of the nation and *ummah*'.[99] Jeremy Henzell-Thomas comments further on this concept of 'middle-roadness':[100]

> It is stated in the Qur'ān that Muslims are 'a community of the middle way' (Q.2:143), suggesting, according to Muḥammad Asad, 'a call to moderation on every aspect of life' and 'a denial of the view that there is an inherent conflict between the spirit and the flesh'.[101] A closed, exclusive, puritanical, hostile and inward-looking version of Islām, which regards all non-Muslims as enemies and infidels and refuses to engage with the rest of humankind, corresponds with no period of greatness in Islām and will bring none. Let us remember the words of… [the] Prophet…: 'All God's creatures are his family; and he is the most beloved of God who does most good to God's creatures'.

The full extract of the *ḥadīth* is justly placed on a 9/11 multi-faith commemorative website:[102]

> What actions are most excellent?
> *To gladden the heart of a human being.*
> *To feed the hungry.*
> *To help the afflicted.*
> *To lighten the sorrow of the sorrowful.*
> *To remove the wrongs of the injured.*
> That person is the most beloved of God who does most good to God's creatures.

Now that is the sort of Islām about which we hear very little in the mass media. Lest this universal truth be questioned, a second *ḥadīth qudsī*, or divine saying, is worth quoting because of its close parallel with the *locus classicus* of Christian social responsibility, the Gospel of Matthew (25:41–5). In this text, the Christian is called to recognize God in the hungry, in the thirsty, in the stranger, in the naked, in the sick and in the imprisoned. As the former Roman Catholic Primate in Britain, the late Cardinal Basil Hume remarked,

> loving God must indeed include loving those made in His image and likeness… active concern for our neighbour is an essential element of our spiritual lives

> and our religious growth. The... commandment... must drive me to build the kingdom of God on earth. It must leave me longing for the heavenly Jerusalem.[103]

The *hādīth qudsī* reminds us of Jesus' saying and demonstrates the possibility for peaceful cooperation between Muslims and Christians on the most important issues that affect the human predicament:[104]

> ...the Apostle of God said: 'God says on the Day of Resurrection: "O son of Adam, I was sick and you did not visit me." He [that is, man] says: "O my Lord, how could I visit You, when you are the Lord of all beings?" He [that is, God] says: "Did you not know that My servant so-and-so was sick, and you did not visit him? Did you not know that if you had visited him, you would have found Me with him? – O son of Adam, I sought food from you, and you did not feed me." He [that is, man] says: "O my Lord, how could I feed you, when you are the Lord of all beings?" He [God] says: "Did you not know that my servant so-and-so sought food from you, and you did not feed him? Did you not know that if you had fed him, you would have found that to have been for Me? – O son of Adam, I asked you for drink, and you did not give Me to drink." He [man] says: "O my Lord, how could I give you to drink, when You are the Lord of all beings?" He [God] says: "My servant so-and-so asked you for drink, and you did not give him to drink. [Did you not know that] if you had given him to drink, you would have found that to have been for Me?"'

Those Muslims who follow 'the straight path of those whom [God] ha[s] guided... no[t] of those who are astray', another *ḥadīth qudsī*,[105] will respond when the West shows its commitment to peace with justice in those lands presently torn by long-standing conflict.

The future of the human family. I: six principles for consideration

There must be a *jihād* against militant *jihād*, a struggle against terrorism, the new fanaticism which shows no respect for human life and therefore no respect for the values of civilization. Terrorism must be indeed defeated, but there are more subtle, or better calibrated, ways of doing so than are presently deployed by the West in the so-called 'war on terrorism'. Instead of waging a war without end, with unclear objectives, against an elusive enemy which recognizes no frontiers, clear distinctions should be made between the general interests of the international community in suppressing terrorism and the short-term pursuit of any one state's hegemonic ambitions.

The problem with the wars in Afghanistan and 'Irāq for many in the West, as well as for the majority of Muslims, is that the military strategy adopted appears to have punished the weak and innocent just as much the evil and powerful. The overall economic condition of the population has worsened rather than improved, whether or not they are said to be 'freer' than before. More than two years after the intervention in Afghanistan, there is yet to be any serious effort at reconstructing that war-torn country.

We must never lose sight of the need for the greater *jihād*, a peaceful struggle for the betterment of the human condition, irrespective of religion, colour or nationality. This must include a concern for the betterment of the lot of the Palestinian and Kāshmīrī peoples, among others. They cannot be left as the debris of history, the victims of double standards (*al-izdiwājiyyah*),[106] sacrificed on the altar of alliances of convenience with repressive state structures. It is only through its commitment to that larger struggle for the betterment of the human condition of all peoples – rather than the imposition of a new imperialism, or an alien set of values – that the true interest of the United States and other Western democracies resides. This will not stop the terrorists from perpetrating atrocities, for their objectives are political, not religious or moral; but it will greatly reduce their popular support in Muslim-majority countries and also the facility with which terrorists recruit suicide bombers and other adherents to their cause.

Since 9/11, the world has become a much more dangerous place, with international terrorism manifesting itself against the weak and defenceless, the implementation of the principle of pre-emptive war in international relations, the increased danger of a war between civilizations and a dangerous threat to civil liberties in a number of countries resulting from the imposition of draconian anti-terrorism laws. It is our contention that, within the human family, irrespective of political, religious and cultural divisions, numerous hitherto unheard voices share a commitment to:

- the peaceful resolution of conflicts using the full diplomatic processes of mediation in accordance with the rules of international law and with full justice to all the parties (including the rights of displaced persons/refugee populations)
- resort to war only in the last resort, when the cause is just and of a defensive (and not pre-emptive) nature, and the means deployed are limited so as to restrict the number of casualties, whether civilian or military
- pursue dialogue between cultures and religions (rather than accept the inevitability of 'conflict between civilizations'), adopting the appropriate mode of discourse, searching for mutual understanding and the acceptance of difference, and eschewing hegemonic ambitions
- accept the diversity of traditions within faith communities and the diversity of the world's religions ('religious pluralism')
- respect the human rights of the individual, including the right to gender equality and the right to freedom of religion

- share in the process of democratic political participation on the basis of equal citizenship.

Instead of being known as the era of unilateralism and the pre-emptive strike, it is the hope of this author that politicians, opinion-formers and intellectuals, and the various cultures and faiths which make up the human family will ensure that the twenty-first century of the Common Era is an era of peace with justice on the basis of such principles.

The future of the human family. II: Islām's need to engage in 'public diplomacy'

If these principles find a resonance within the world-wide Muslim community, as the author believes they will, then it would be helpful for the cause of improved understanding if Muslim leaders, opinion-formers, intellectuals and others make their commitment to them loud and clear. There is a parable of Jesus, recounted in the Gospels of Matthew, Mark and Luke[107] which has led to the English saying 'to hide one's light under a bushel', meaning 'to conceal one's talents; to be self-effacing and modest about one's abilities. The bushel as measured in a wooden earthenware container, hence *under a bushel* is to hide something.'[108] For too long the Muslim mainstream has hidden its 'light under a bushel'. The debates within Islām, which have addressed the main issues of modernity, pluralism and human rights, issues which are of concern for the West, are not well known in the West.

Because of their mastery of the internet as a means of diffusion and recruitment, the extremist voice is actually better known in the West than the voice of the mainstream majority. Online databases of *fatāwā* for the faithful are of value for instructing Muslims.[109] They serve little purpose for the broader purpose of engaging with, and informing, non-Muslims. It is time for Muslims to engage more actively in that activity known euphemistically as 'public diplomacy' (a term coined in 1965). Public diplomacy and propaganda are two sides of the same coin. Propaganda based on falsehood is 'disinformation'. If there is a commitment to truth, then that is the best form of public diplomacy/propaganda. 'To be persuasive we must be believable; to be believable we must be credible; to be credible we must be truthful.' The maxim of Edward R. Murrow, the Director of the United States Information Agency (USIA),[110] in May 1963,[111] has not always been followed by the United States, but is a worthy statement of intent. It is for the Muslim mainstream to engage truthfully and authentically both with its own pluralist tradition and to explain that rich and diverse heritage and understanding to others in a relevant, comprehensible and public way.

Without Muslim 'public diplomacy', 'enlightened moderation' in Islām may not in any case succeed; it will certainly not be understood by the West. It is vital for all of us that it does.[112] As Rāshid al-Ghannūshī said in an open debate at a conference in 1996 on the theme of 'Rethinking Islām and Modernity':[113]

> The idea of complementarity between cultures and civilizations is... a fundamental concept of Islām. This meeting presupposes this point of departure: we are all in one boat, and have to worry about how to save it.

Appendix
Extract from a Legal Ruling (*Fatwā*) Pronounced by Ibn Taymīyah on the Mongols, 702/1303

[The *fatwā* is given in reply to a specific question, which provides the framework within which it should be read. The question is stated as follows:]

What is the view of top scholars concerning these Mongols who came [to the Muslim lands] in 699[/1299] and committed well known atrocities? They killed Muslims, enslaved their children, plundered their property and committed offences against what is religiously sacred, such as humiliating Muslims and their mosques, particularly the al-Aqṣā Mosque in Jerusalem, committed corruption there, pilfered large quantities of Muslim property belonging to individuals and the state, took captive a large number of Muslims, driving them from their homeland. Despite all this, they claim to be committed to the declaration of faith (*shahādah*: 'I bear witness that there is no deity other than God, and that Muḥammad is God's messenger'). They maintain that it is forbidden in Islām to fight against them because they follow the basic principles of Islām, and they did not go ahead with their objective of exterminating all Muslims. Is fighting them permissible or a duty? Whichever answer is given, what is the basis of this ruling? Please give us a ruling on this matter; may God reward you.

[In answer Ibn Taymīyah wrote:] All praise is due to God. Any community or group that refuses to abide by any clear and universally accepted Islamic law, whether belonging to these people or to some other group, must be fought until they abide by its laws. This applies even though they make the verbal declaration [that brings a person into the Islamic fold] and abide by some of its laws. Such was the attitude of Abū Bakr [the first caliph] and the Prophet's companions when they fought against those who refused to pay *zakāt*. All scholars in subsequent generations agree to this ruling, even though at first 'Umar questioned Abū Bakr over it. The Prophet's companions were unanimous in their support for fighting to achieve the rights of Islām. This is in line with the Qur'ān and the *Sunnah*.

It must be stressed that all religious requirements in war must be fulfilled. This means that they [that is, the Mongols] should be called upon to abide by and implement Islamic law, if this has not been conveyed to them. In the past, when the first declaration of Islām was still unknown to unbelievers, they were first called upon to make it.

The Prophet says: 'There will be rulers who will be oppressive, treacherous and wicked. Anyone who believes in their lies and helps them does not belong to me; nor do I belong to him. He will not come to me on the Day of Judgement. On the other hand, anyone who rejects their lies and lends them no support in their oppression belongs to me and I belong to him. He will come to me on the Day of Judgement.'

When we understand what the Prophet has ordered of struggle [that is, *jihād*], which has been, and will be undertaken by Muslim rulers until the Day of Judgement, and what he has forbidden towards helping those who are oppressive, we realize that the middle way, which is pure Islām, means to join *jihād* against those who must be targeted with *jihād*, such as the people who are the subject of this enquiry. We should join *jihād* against them with any ruler, commander or group that is closer to Islām than they, if such is the only means of fighting them. We must refrain from helping the side we join in with in *jihād* in committing anything that represents disobedience to God. We obey such rulers in anything conducive to obeying God, and disobey them in anything conducive to disobeying Him. The applicable rule states: 'No creature may be obeyed in what constitutes disobedience of the Creator.'

Such is the line chosen by the best people in the Muslim community, over many generations. It is incumbent on everyone who is required to observe Islamic obligations. It is a middle way between the Khawārij [see Chapter 2] and those like them whose approach is one of strict but faulty piety owing to their lack of thorough knowledge of Islām, and al-Murji'ah [Murji'ites, the extreme opponents of the Khawārij] who advocate obeying rulers in all cases, even though they may be uncommitted to Islām.

We have seen the Mongol army, and ascertained that most of their soldiers do not attend to their prayers. We did not see in their army camps anyone who calls to prayers, *mu'adhdhin*, or any *imām* who leads prayers. They have unlawfully taken away much property belonging to Muslims, and many of their children, and destroyed their land, in numbers and at scales known only to God. In their government they have only people who are among the worst of mankind. These are either unbelievers or hypocrites who do not really believe in Islām, or else some of the more extreme followers of deviant sects, such as the extreme Sh'īa, al-Jahmīyyah [the followers of Jahm ibn Ṣafwān, who denied the divine attributes], etc. Others who follow them are among the most hardened sinners. When they are in their own land, they do not offer the pilgrimage to Mecca, in spite of being able to do so. Although some of them may pray and fast, the majority neither attend to prayers nor pay their *zakāt*.

They fight for Chinghis-Khān (Genghis Khān). Whoever collaborates with them they accept him as belonging to them, even though he may be an unbeliever. They consider anyone who abstains from such collaboration an enemy, even though he may be among the best Muslims. They do not fight to defend Islām. Those who are Muslims among their top leaders and ministers give no Muslim any higher status than they give to unbelievers, Jews or Christians. This is what was stated by their chief emissary in Syria who said to the Muslim delegation, as he tried to gain their good will claiming that they were Muslims. He said: 'Muḥammad and Chinghis-Khān are two great signs of God.' Thus the maximum he can say in order to win favour with Muslims is to equate God's last messenger, who is the dearest to God, with an unbeliever king who ranks as one of the most wicked, corrupt idolaters and aggressors.

To sum up, every type of hypocrisy, unbelief and outright rejection of the faith is found among the Mongol followers. They are among the most ignorant of all people, who least know the faith and are far from following it. They follow their own desires…

Translation by Dr Adil Salahi. Source: *Majmū'a Fatāwā Shaykh al-Islām Aḥmad Ibn Taymiyya*, 37 vols (*Compilation of Legal Opinions of Shaykh al-Islām Aḥmad Ibn Taymiyya*), ed. and comp. 'Abd al-Raḥman ibn Qāsim (n.p., n.d.), xxviii. 501–8.

Notes

Author's preface

1. Khurram Murad, 'Islaām and Terrorism', *Encounters: Journal of Inter-cultural Perspectives*, 4 (1998), 103–14, at 106.
2. M. Nejatullah Siddiqi, 'Future of the Islamic Movement', *Encounters: Journal of Inter-cultural Perspectives*, 4 (1998), 91–101 at 94.
3. Murad, 'Islaām and Terrorism', 103.

Introduction

1. Steve Bird, 'Leicester Algerians accused of funding al-Qaeda holy war', *The Times*, 6 February 2003.
2. Quoted by R. J. Bonney, *Understanding and Celebrating Religious Diversity. The Growth of Diversity in Leicester's Places of Religious Worship since 1970* (University of Leicester, Studies in the History of Religious and Cultural Diversity, 1, 2003), 70. Those convicted launched an appeal on the grounds that the general coverage of terrorism made it difficult for them to have a fair trial: *Leicester Mercury*, 17 October 2003.
3. Four titles purchased on a brief visit to the United States in October 2003: Robert Spencer, *Islam Unveiled. Disturbing Questions about the World's Fastest-Growing Faith* (San Francisco: Encounter Books, 2002); Kenneth R. Timmerman, *Preachers of Hate. Islam and the War on America* (New York: Crown Forum, 2003); Stephen Schwartz, *The Two Faces of Islam. Saudi Fundamentalism and its Role in Terrorism* (New York: Anchor Books, repr. 2003); Robert Spencer, *Onward Muslim Soldiers. How Jihād Still Threatens America and the West* (Washington DC: Regnery Publishing, 2003).
4. Schwartz, *The Two Faces of Islam*, 192. 'Christopher Hitchens was dead on when he described this ideology as "fascism with an Islamic face", and radio host Michael Savage hit the nail on the head when he came up with the term "Islamofascism"': <www.freerepublic.com/focus/news/792626/posts>
5. Elaine Monaghan, 'US "must sell its image" in Muslim World', *The Times*, 2 October 2003.
6. The phrase 'wretched, backward, philosophy of these terrorists' was transcribed from his television address. It does not appear in the official transcript: 'it is a war that strikes at the heart of all that we hold dear, and there is only one response that is possible or rational: to meet their will to inflict terror with a greater will to defeat it; to confront their philosophy of hate with our own of tolerance and freedom; and to challenge their desire to frighten us, divide us, unnerve us with an unshakeable unity of purpose; to stand side by side with the United States of America and with our other allies in the world, to rid our world of this evil once and for all'. <www.pm.gov.uk/output/page4871.asp>
7. Douglas E. Streusand, 'What Does *Jihād* Mean?', *Middle East Quarterly* (September 1997): <www.ict.org.il/articles/jihad.htm>
8. C. Cox and John Marks, *The 'West', Islam and Islamism. Is Ideological Islam Compatible with Liberal Democracy?* (London: Civitas, 2003), 64.
9. Ibn Taymīyah, *Public Duties in Islām. The Institution of the Ḥisba*, trans. Muhtar Holland (Leicester: Islamic Foundation, 1982), 121.

10. <www.youngmuslims.ca/online_library/books/peace_in_islam/index.htm#peace_fighting>
11. H. Hanafī, 'The Preparation of Societies for Life in Peace. An Islamic Perspective', in Hanafī, *Islam in the Modern World. II. Tradition, Revolution and Culture* (Cairo: Dar Kebaa, 2000), 232–3, 244, 247–8, 252–3.
12. H. Hanafī, 'Global Ethics and Human Solidarity. An Islamic Approach', ibid. 285–7.
13. The technical issues concerning abrogation are considered below in Chapter 1.
14. The single most important study for the classical period is that of Alfred Morabia (1931–1986), *Le Ĝihâd dans l'Islam médiéval. Le «combat sacré» des origines au xiie siècle* (Paris: Albin Michel, 1993). For *jihād* and the commandment to do good: ibid. 315, 512–13, n. 197. This maxim is highlighted in Michael Cook, *Commanding Right and Forbidding Wrong in Islamic Thought* (New York: Cambridge University Press, 2000).
15. Cox and Marks, *The 'West', Islam and Islamism*, 79.
16. Cf. Suhas Majumdar, *Jihād: The Islamic Doctrine of Permanent War* (New Delhi: Voice of India, 1994). Few works rival that of Serge Trifkovic in the vilification of Islām: Trifkovic, *The Sword of the Prophet. Islam. History, Theology and Impact on the World* (Boston, MA: Regina Orthodox Press, 2002). He writes: 'the problem of collective historical ignorance – or even deliberately induced amnesia – is the main difficulty in addressing the history of Islam in today's English-speaking world, where claims about far-away lands and cultures are made on the basis of domestic multiculturalist assumptions rather than evidence. The absence of historical memory has taken too many well-meaning Westerners interested in Islam right through the looking glass into the virtual-reality world of superficial reportage, ideological treatises, and agenda-driven academic research that ignores the reality of what Islam actually is and what it does to its adherents.' And again: 'Muslims are *obliged* to wage struggle against unbelievers and may contemplate tactical ceasefires, but never its complete abandonment short of the unbelievers' submission. That struggle is always "defensive", by definition. *This is the real meaning of Jihād.* Far from signifying mere "inner striving" and "spiritual struggle", to generations of Muslims before our time – and to a majority of Muslims today – the meaning of *Jihād* as the obligatory and permanent war against non-Muslims is clear and beyond dispute…' <www.reginaorthodoxpress.com/swordofprophet.html>; <www.chroniclesmagazine.org/News/Trifkovic/NewsST050903.html> For a Muslim response, cf. Habib Siddiqui, 'The Repulsive World of Serge Trifkovic': <www.mediamonitors.net/habibsiddiqui3.html>
17. Quṭb, *Milestones* (Beirut and Damascus: International Islamic Federation of Student Organizations, 1978), 116: 'what kind of man is he, after listening to the commandment of God and the Traditions of the Prophet… and after reading about the events which occurred during the Islamic *jihād*, still thinks that it is a temporary injunction related to transient conditions and that it is concerned only with the defence of the borders?'
18. Cemal Kafadar, *Between Two Worlds. The Construction of the Ottoman State* (Berkeley, CA: University of California Press, 1995), 81.
19. Abdulaziz Sachedina, *The Islamic Roots of Democratic Pluralism* (New York: Oxford University Press, 2001), 23.
20. Hanafī, 'Islam, Religious Dialogue and Liberation Theology', in Hanafī, *Islam in the Modern World. II. Tradition, Revolution and Culture*, 216.
21. Ibid. 222.
22. Hanafī, 'The Preparation of Societies for Life in Peace', 239.
23. Marc Gopin, *Holy War, Holy Peace. How Religion Can Bring Peace to the Middle East* (New York: Oxford University Press, 2002), 54.
24. Although we will make judicious references to the material on the Memri (Middle East Media Research Institute) website, there are criticisms of its selection of material as well as to its independence and objectivity: <www.terrorism101.org/archive/memri_1020371954.html>

An idea of the flavour of this article is given by the following: 'These accursed ones are a catastrophe for the human race. They are the virus of the generation, doomed to a life of humiliation and wretchedness until Judgement Day. They are also accursed because they repeatedly tried to murder the Prophet Muḥammad. They threw a stone at him, but missed. Another time, they tried to mix poison in his food, but providence saved him from their treachery and their crimes. Allāh cursed them when they carried out the criminal massacre of the peaceful Palestinians in Sabra and Shatilla. They are accursed, they, their fathers, and their forefathers… until Judgment Day, because they burst into al-Aqṣā Mosque with their defiled, filthy feet and violated its sanctity. Finally, they are accursed, fundamentally, because they are the plague of the generation and the bacterium of all time. Their history always was and always will be stained with treachery, falseness, and lying. Historical documents prove it…'

25. Memri Special Dispatch 558, 27 August 2003: <www.memri.org/bin/articles.cgi?Page=archives&Area=sd&ID=SP55803>
26. 'Ramadan TV Special: The Protocols of the Elders of Zion' Memri Special Dispatch, 309, 6 December 2001: <www.memri.org/bin/articles.cgi?Page=archives&Area=sd&ID=SP30901> Tsar Nicholas II's *Okhranka*, or secret police, devised this document in 1903. The *Okhranka* fashioned the document as the purported agreement of a group of Jewish elders meeting in Switzerland in 1897 to plot Jewish hegemony through the destruction of Christian civilization. It was first published in Russia on the eve of, and as the instrument of, the vicious 1903 Odessa pogrom. In November 1993 even a Moscow district court accepted that the document was an anti-semitic forgery: <www.nizkor.org/ftp.cgi?documents/protocols/protocols.001>; <www.holocaust-history.org/short-essays/protocols.shtml> It is to be hoped that the Conference on the Protocols, held by the organization Academics against Anti-Semitism in Venice in December 2003, finally laid this myth to rest.
27. It is possible to have anti-semitism against Arabs and there are examples of this in the recent British press. See introduction to Chapter 3.
28. Kenneth R. Timmerman, *Preachers of Hate. Islam and the War on America* (New York: Crown Forum, 2003). Images of children with weapons on the Palestinian Authority's website: <www.ipc.gov.ps/photo%20gallery/photo/pal/a/0017.jpg>; <www.ipc.gov.ps/photo%20gallery/photo/pal/a/0011.jpg>
29. Memri Special Report 22, 14 November 2003: <www.memri.org/bin/opener_latest.cgi?ID=SR2203>
30. Joyce M. Davis, *Martyrs. Innocence, Vengeance and Despair in the Middle East* (New York: Palgrave Macmillan, 2003), 32. The source are extracts from S'ad ibn Muḥammad al-'Āqilī's letter published on 16 April 2001. He used the term 'Jews' rather than 'Israelis'. Memri Special Dispatch 206, 18 April 2001: <www.memri.org/bin/articles.cgi?Page=archives&Area=sd&ID=SP20601>
31. Amru Nasif, a columnist for the Egyptian opposition weekly *Al-Usbū'*, calls for martyrdom operations to defeat Israel and volunteers to join the ranks of martyrs: 'all that it requires is to concentrate on acts of martyrdom, or what is known as "the strategy of the balance of fear"… Let us do some mathematical calculations: 250 Palestinians have signed up for martyrdom operations, and it is not impossible to raise this number to 1,000 throughout the Arab world, i.e., one *fidā'ī* (martyr) out of every 250,000 Arabs. The average harvest of each act of martyrdom is 10 dead and 50 wounded. Thus, 1,000 acts of martyrdom would leave the Zionists with at least 10,000 dead and 50,000 wounded. This is double the number of Israeli casualties in all their wars with the Arab[s] since 1948. They cannot bear this. There is also the added advantage, not noted by many, of negative Jewish emigration, which, as a consequence of the 1,000 martyrdom operations will come to at least 1,000,000 Jews, followed by the return of every Jew to the place from whence he came…' Memri Special Dispatch 224, 4 June 2001: <www.memri.org/bin/articles.cgi?Page=subjects&Area=jihad&ID=SP22401>

32. Memri Special Dispatch 418, September 2002: <www.memri.org/bin/articles.cgi?Page=subjects&Area=jihad&ID=SP41802>
33. Osāma al-Baz counselled 'against conspiracy theorizing. It is all too easy to suggest that Jews or Israelis who criticize Israeli policy are simply playing the role assigned to them as part of a greater scheme to deceive the Arabs and the rest of the world. History cannot be condensed into a series of conspiracies. It is also important, in this regard, that we refrain from succumbing to such myths as the *Protocols of the Elders of Zion* and the use of Christian blood in Jewish rituals. We should not sympathize in any way with Hitler or Nazism. The crimes they committed were abominable, abhorrent to our religion and beliefs…' Memri Special Dispatch Series 454, 3 January 2003: <www.memri.org/bin/articles.cgi?Page=archives&Area=sd&ID=SP45403>
34. Rabbi Abraham Cooper, associate dean of the Simon Wiesenthal Centre in Los Angeles, stated that Mahathir's speech was 'an absolute invitation for more hate crimes and terrorism against Jews'. Complete transcript of Mahathir's speech at: <www.vancouver.indymedia.org/news/2003/10/73779_comment.php>
35. Memri Special Dispatch 95, 23 May 2000: <www.memri.org/bin/articles.cgi?Page=subjects&Area=jihad&ID=SP9500>
36. J. Riley-Smith, 'Islam and the Crusades in History and Imagination, 8 November 1898–11 September 2001', *Crusades*, 2 (2003), 151–67, at 166. Osama bin Laden's comments on uniting in the face of the Christian crusade, quoted ibid., were echoed in Mahathir's speech in October 2003. I am indebted to my colleague Professor Norman Housley, another distinguished historian of the Crusades, for this reference.
37. Ibid. 167.
38. The author is indebted to Professor Norman Housley for this information.
39. Riley-Smith, 'Islam and the Crusades', 153.
40. Ibid. 167.
41. N. J. Housley, *Religious Warfare in Europe, 1400–1536* (Oxford: Oxford University Press, 2002), 205.
42. Ibid. 19.
43. Ibid. 4, 15. In ch. 36, Vitoria argues that women and children of the Turks are 'guiltless' and therefore may be spared death: <www.constitution.org/victoria/victoria_5.htm>
44. *Oxford Dictionary of World Religions*, ed. J. Bowker (Oxford: Oxford University Press, 1997), 258. '*Dār al-Ḥarb*', *Oxford Dictionary of Islam*, ed. John L. Esposito (Oxford: Oxford University Press Inc., 2003). *Oxford Reference Online*. Oxford University Press. <www.oxfordreference.com/views/ENTRY.html?subview=Main&entry=t125.000490> 'Dar al-Islam', Ibid. <www.oxfordreference.com/views/ENTRY.html?subview=Main&entry=t125.000491> 'Dar al-Sulh', Ibid. <www.oxfordreference.com/views/ENTRY.html?subview=Main&entry=t125.000496>

    Shaybānī did not use the two terms *Dār al-Islām* and *Dār al-Ḥarb* consistently in his treatise: *The Islamic Law of Nations. Shaybānī's Siyar*, trans. and ed. Majīd Khaddūrī (Baltimore, MD: Johns Hopkins University Press, 1966), 142 n. 1, 168, 170.
45. Qamaruddin Khan, *The Political Thought of Ibn Taymīyah* (Islamabad: Islamic Research Centre, Pakistan, 1973), 157–8.
46. Ibid. Khan writes: 'the Muslim jurists were not prepared to be convinced by these facts [that is, the division of the world in practice into territorial or dynastic units]. They continued to preach the theory of undiluted *jihād*. It is difficult to read their real motive but it can easily be seen that they certainly erred in their classification of the world.'
47. Morabia, *Le Ĝihâd dans l'Islam médiéval*, 195–6, does not consider the differences between the Sunnī and Shī'a traditions to be fundamental. The author is also grateful to have consulted the PhD thesis of Abdulrahman Muḥammad Alsumaih, 'The Sunnī Concept of *Jihād* in classical *Fiqh* and Modern Islamic Thought' (unpublished PhD, University of Newcastle upon Tyne

1998). However, Dr Alsumaih does not discuss the Shī'a tradition, while his interpretation of the Sunnī tradition is based on the principles that a defensive *jihād* is an innovation in the doctrine, while the greater *jihād* is a military effort in the cause of Allāh. According to Dr Alsumaih (ibid. 6), 'the main change in Sunnī *jihād* doctrine is made by the Islamic modernists and not by contemporary Islamists…' This view would be rejected, for example, by those who follow the Ṣūfī tradition.

48. A similar viewpoint to this has been propounded by Tom Knowlton in 'Moderate Muslims Must Seize the Reins of Islām' (26 February 2003). He writes: 'from the 13th century onward, the Muslim world moved into steady decline and was rapidly eclipsed by empires in Europe and Asia. Now, far from its golden age, the Middle East is a society largely steeped in economic despair, poverty, failed states and oppressive and stagnant regimes. The main catalyst for this decline is the failure of the Muslim world, not to Westernize, but to modernize. Muslims should be bitter about the decline of their empires, but not with the West and certainly not with the United States. No, the betrayal of the *ummah*, the Muslim community, has come not from without but from within. Beginning in the 13th century, the advancement of the Muslim world has been stymied by radical heretics, *mubtadi'ah*, who have hijacked and perverted the tenets of Islām.' <www.hometown.aol.com/ahreemanxi/page41.html>
49. Muḥammad Hamidullah, *Muslim Conduct of the State, Being a Treatise on Siyar* (revised 7th edn, Lahore: Sh. Muḥammad Ashraf, 1977), 170 (para. 323).
50. Ibid. 175–6 (paras 334–5).
51. Quṭb, *Islām: The Religion of the Future* (repr. Beirut: Holy Koran Publishing House, 1978), 106.
52. <www.famulus.msnbc.com/FamulusIntl/ap05-06-010001.asp?reg=ASIA#body>
53. MacArthur, *Terrorism, Jihād and the Bible: A Response to the Terrorist Attacks*, 32, 33. MacArthur observes that Islām 'pretends to regard the Bible as a holy writing, but denies every fundamental doctrine about sin and salvation taught in the Bible' and then proceeds to state that 'Islām is based on lies': ibid. 40–1.
54. M. M. al-A'zamī, *The History of the Qur'ānic Text from Revelation to Compilation. A Comparative Study with the Old and New Testaments* (Leicester: UK Islamic Academy, 2003), 13, 341. Ibid. 342: 'Western scholars feel obliged to instruct Muslims as to how they must interpret their own religion'. To some extent, the author undermines his own argument by making a comparative study with the Old and New Testaments, though the analysis lacks objectivity. We accept that non-Muslims should not claim the right to issue legal opinions (*fatwās*), but we all need to recognize that sometimes outsiders help us to understand ourselves better.
55. Wilfrid Cantwell Smith, 'Comparative Religion: Whither and Why?', *The History of Religions: Essays on Methodology*, ed. M. Eliade and J. Kitagawa (Chicago, IL: University of Chicago Press, 1959), 31–58 at 43. Cited by Clinton Bennett, *In Search of Muḥammad* (London and New York: Cassell, 1998), 3.
56. The late Edward Said commented on 25 July 2000, thus over a year before 9/11, but the remark has become even more pertinent since: 'the search for a post-Soviet foreign devil has come to rest, as it did beginning in the eighth century for European Christendom, on Islām… What matters to "experts" like [Judith] Miller, Samuel Huntington, Martin Kramer, Bernard Lewis, Daniel Pipes, Steven Emerson and Barry Rubin, plus a whole battery of Israeli academics, is to make sure that the "threat" is kept before our eyes, the better to excoriate Islām for terror, despotism and violence, while assuring themselves profitable consultancies, frequent TV appearances and book contracts. The Islamic threat is made to seem disproportionately fearsome, lending support to the thesis (which is an interesting parallel to anti-Semitic paranoia) that there is a worldwide conspiracy behind every explosion.' Said, 'A Devil Theory of Islām': <www.thenation.com/doc.mhtml?i=19960812&c=2&s=said>

On 5 March 2002, Said voiced his concerns about American intellectual life after 9/11: 'when the intellectuals of the most powerful country in the history of the world align themselves so flagrantly with that power, pressing that power's case instead of urging restraint, reflection, genuine communication and understanding, we are back to the bad old days of the intellectual war against communism, which we now know brought far too many compromises, collaborations and fabrications on the part of intellectuals and artists who should have played an altogether different role.' Said, 'Thoughts about America': <www.counterpunch.org/saidamerica.html>

57. R. Peters, *Islam and Colonialism. The Doctrine of Jihād in Modern History* (The Hague: Mouton, 1979), 1, 5. R. Peters, *Jihād in Classical and Modern Islam* (Princeton, NJ Markus Weiner, 1996).
58. H. M. Zawātī, *Is Jihād a Just War? War, Peace and Human Rights under Islamic and Public International Law* (Lewiston, NY: Edwin Mellen Press, 2001), 107–8.
59. Ibid. 109–110.
60. Daniel Pipes, '*Jihād* and the Professors' (November 2002): <www.danielpipes.org/article/498>: 'As for the conditions under which *jihād* might be undertaken – when, by whom, against whom, with what sort of declaration of war, ending how, with what division of spoils, and so on – these are matters that religious scholars worked out in excruciating detail over the centuries. But about the basic meaning of *jihād* – warfare against unbelievers to extend Muslim domains – there was perfect consensus.'
61. Daniel Pipes, 'What is *Jihād*?', *New York Post*, 31 December 2002.
62. Alsumaih, 'The Sunni concept of *Jihād*', 35, notes that only God can determine whether the intention of the *Mujtahid* is truly for the cause of God. The Prophet's tradition, reported by al-Bukhari and Muslim, made it clear that intention determines the worth of a person's actions and that person will attain what he intends.
63. Streusand, 'What Does *Jihād* Mean?'
64. A. Nizar Hamzeh and R. Hrair Dekmejian, 'A Ṣūfī Response to Political Islamism: Al-Ahbash of Lebanon', *International Journal of Middle East Studies*, 28 (1996), 217–29. <www.almashriq.hiof.no/ddc/projects/pspa/al-ahbash.html>
65. <www.speeches.commemoratewtc.com/epilogue/musharraf.php>
66. Fazlur Rahman, *Major Themes of the Qur'ān* (Minneapolis, MN: Bibliotheca Islamica, 1980), 63–4, quoted by Streusand, 'What Does *Jihād* Mean?'
67. Streusand, 'What Does *Jihād* Mean?'
68. <www.muslimindia.com>
69. In his many ways excellent account, Peter Partner talks of 'holy war' in Islām and Christianity. However, he notes: 'one of the virtues most required of the helpers of Muhammad, after their submission to Allāh, was "struggle in the way of Allāh" (*jihād*)': P. Partner, *God of Battles. Holy Wars of Christianity and Islam* (London: HarperCollins, 1997), 32. Cf. Zawātī, *Is Jihād a Just War?*, 13, 111, calls 'the description of… *jihād* as a "holy war"… utterly misleading'.
70. *Speeches and Statements of Quaid-i-Millat Liaquat 'Ali Khan, 1941–51*, ed. M. Rafique Afzal (Research Society of Pakistan, Lahore, 1967), 563. Ibid. 622, Nehru to Liaquat 'Ali Khan, 24 July 1951.
71. Shaykh Abdullah bin Muhammad bin Humaid, '*Jihād* in the Qur'ān and Sunnah': <www.islamworld.net/jihad.html> For an online hypertext of the Holy Qur'ān in several versions: <www.sacred-texts.com/isl/htq/index.htm>
72. <www.geocities.com/al_mo_minah/jihad2.html>
73. *Encyclopedia of Politics and Religion,* ed. Robert Wuthnow (Washington DC: Congressional Quarterly, Inc., 2 vols, 1998), 425–6. <www.cqpress.com/context/articles/epr_jihad.html>
74. Published in *The Islamic World and the West. An Introduction to Political Cultures and International Relations*, ed. Kai Hafez (Leiden: E. J. Brill, 2000), 70–85, at 82.

75. 'Arab Press Reacts to National Security Advisor Condoleezza Rice's Statements on Democracy and Freedom', Memri Special Dispatch 427, 11 Oct. 2002: <www.memri.org/bin/articles.cgi?Page=archives&Area=sd&ID=SP42702>
76. In March 2004, Saudi Arabia's Foreign Minister Prince Saud al-Faisal criticized US-led calls for reform in the Middle East and stated that Arab countries could tackle their problems by themselves. The US proposals 'include clear accusations against the Arab people and their governments that they are ignorant of their own affairs', the official Saudi Press Agency quoted the prince as saying in the Yemeni capital Sanaa. 'Those behind these plans ignore the fact that our Arab people have cultures rooted deep in history and that we are able to handle our own affairs', he was reported as stating. The USA has argued that a lack of democracy in Arab states has helped fuel Islamic militancy: <www.dawn.com/2004/03/22/top14.htm>
77. S. Parvez Manzoor, 'Against the Nihilism of Terror: *Jihād* as Testimony to Transcendence', *Muslim World Book Review*, 22:3 (April–June 2002), 5–14. <www.algonet.se/~pmanzoor/jihad-mwbr.htm>
78. Riley-Smith, 'Islam and the Crusades in History', 166. Riley-Smith comments: 'it is somewhat pedantic to engage in argument whether the policies of the developed world are "crusading" or not. Disputes about the terms we employ will not alter the facts that a very large number of people in the Islamic world, moderates as well as extremists, are attached to a history which satisfies their feelings of both superiority and humiliation and that they perceive themselves to be exploited by westerners, while the religious among them believe themselves to be threatened by values which they loathe. Since it is important for us to understand why they feel as they do, it is worrying that most people in the West are ignorant of the potent historical and moral force [described by the author].' Ibid. 167.
79. Khalid Masud, 'Changing Concepts of *Jihād*', unpublished paper dated 31 October 2003 kindly communicated to the author.
80. <www.fpri.org/fpriwire/1103.200310.kelsay.newjihad.html>
81. Cited in the opening remarks of his sermon for the feast of the sacrifice in 2003: <www.memri.org/bin/articles.cgi?Page=subjects&Area=jihad&ID=SP47603>

Prologue

1. Yohanan Friedmann, *Tolerance and Coercion in Islam. Inter-faith Relations in the Muslim Tradition* (Cambridge: Cambridge University Press, 2003), 22.
2. Philip D. Stern, *The Biblical Ḥerem: A Window on Israel's Religious Experience* (Atlanta, GA: Scholars Press, 1991).
3. Ibid. 105.
4. Ibid. 104.
5. Ibid. 106–7.
6. Ibid. 91. Cf. ibid. 121: it was perceived as 'a weapon against chaos (represented by the autocthonous nations), but in a way that would impede its improper use [by Jereboam II] in a vengeful crusade against Moab'. Ibid. 102: 'there is unanimous agreement that at the time of the war legislation, the six or seven candidates for the [*ḥerem*] were not in the picture, whatever had been the case in the past… the framers of the laws wanted to eliminate the possibility of using the [*ḥerem*] against others…'
7. Ibid. 96.
8. Ibid. 141.
9. Ibid. 153, 160.
10. Ibid. 178.
11. Ibid. 174, 177.
12. Ibid. 92, 173, 175.

13. Ibid. 217.
14. Ibid. 220.
15. Ibid. 222.
16. Philip Stern rejects the term 'genocide' for the war against the Amalekites: 'the *ḥerem*', he writes, 'was not modern genocide, and Amalekites remained: cf. 2 Samuel 1 and 1 Chronicles 4:43. The final elimination of Amalek is more likely to be ideology at work than history…' This might be called special pleading: the genocidal implication is clear, only its effectiveness may be called into question.
17. Ibid. 221.
18. Ibid. 9, 226.
19. Ibid. 96.

Chapter 1

1. Abdulaziz Abdulhussein Sachedina, *The Islamic Roots of Democratic Pluralism* (Oxford: Oxford University Press, 2001), 38–9.
2. Yohanan Friedmann, *Tolerance and Coercion in Islam. Inter-faith Relations in the Muslim Tradition* (Cambridge: Cambridge University Press, 2003), 20.
3. R. Firestone, *Jihād. The Origin of Holy War in Islam* (Oxford, Oxford University Press, 1999), 64, citing Morton Smith, *Palestinian Parties and Politics that Shaped the Old Testament* (repr., London, 1987).
4. Muhammad ibn Idris ibn al-ʻAbbās ibn ʻUthmān ibn Shāfiʻī: *Islamic Jurisprudence. Shāfiʻī's Risāla*, trans. Majīd Khaddūrī (Baltimore, MD: Johns Hopkins University Press, 1961), 88 (para. 55), 91 (para. 61), 94 (para. 65).
5. The original version of the Qur'ān is often thought of as a book preserved in Heaven: Q.85:22. For some commentators, however, this has a metaphorical meaning, that is an allusion to the imperishable quality of the divine writ.
6. Thus the Prophet was not handed a scroll on which the writing was already placed as was Ezekiel in Ezekiel 2:10.
7. According to Abū Bakr al-Kalābādhī (d. 385/995), Ṣūfīs were in agreement that 'the Qur'ān is the real word of God, and that it is neither created, nor originated in time, nor an innovation; that it is recited by our tongues, written in our books and preserved in our breasts, but not dwelling therein. They are also agreed that it is neither body, nor element, nor accident… The word Qur'ān… is… understood in its general connotation to mean the speech of God, and in that case it is uncreated': *The Doctrine of the Ṣūfīs*, ed. and trans. A. J. Arberry (Lahore: Sh. Muhammad Ashraf, repr. 1966), 23, 26.
8. J. Burton, 'The Collection of the Qur'ān', reprinted in I. Edge (ed.), *Islamic Law and Legal Theory* (Aldershot: Dartmouth, c. 1996), ch. 5, 113. J. Burton, *The Collection of the* Qur'ān (Cambridge: Cambridge University Press, 1977). Al-Bukhārī 9/89/301. The first caliph, Abū Bakr, collected the Qur'ān soon after the battle of Yamama[h] which led to the death of at least 70 of the memorizers of the Qur'ān: al-Bukhārī 6/60/201 and 6/61/509. Mohammed Hashim Kamālī, *Principles of Islamic Jurisprudence*, rev. edn (Cambridge: Islam Texts Society, 1991), 17.
9. Muḥammad Hamidullah, *Muslim Conduct of the State, Being a Treatise on Siyar* (rev. 7th edn, Lahore: Sh. Muḥammad Ashraf, 1977), 20 (para. 30).
10. Ibid. Burton, 'The Collection of the Qur'ān', ch. 7. 'Qurān', *Oxford Dictionary of Islam*. ed. John L. Esposito, (Oxford: Oxford University Press Inc., 2003). *Oxford Reference Online*. <www.oxfordreference.com/views/ENTRY.html?subview=Main&entry=t125.001945>

    M. M. al-Aʻzamī, *The History of the Qur'ānic Text from Revelation to Compilation. A Comparative Study with the Old and New Testaments* (Leicester: UK Islamic Academy, 2003),

69: 'based on the total number of scribes, the Prophet's custom of summoning them to record all new verses, we can safely assume that in his own lifetime the entire Qur'ān was available in written form'. However, full compilation 'in one master volume', came later. No copy was sent out without its reciter, the main purpose being to eliminate all occasion for disputes in recitation: ibid. 77, 93–5.

11. '…write it in the dialect of Quraish, for the Qur'ān was revealed in this dialect': al-Bukhārī 6/61/507. Al-A'zamī, *The History of the Qur'ānic Text from Revelation to Compilation*, 86. For reflections on revelation from a Muslim perspective: William A. Graham, *Divine Word and Prophetic Word in Early Islām. A Reconsideration of the Sources, with Special Reference to the Divine Saying or Ḥadīth Qudsī* (The Hague: Mouton, 1977). Ditto from a Christian perspective: R. Swinburne, *Revelation. From Metaphor to Analogy* (Oxford: Clarendon Press, 1992) and N. Wolterstorff, *Divine Discourse. Philosophical Reflections that Claim that God Speaks* (Cambridge: Cambridge University Press, 1995).
12. Kamālī, *Principles of Islamic Jurisprudence*, 16.
13. Shaykh 'Abdul Rahman 'Abdul Khaliq cites as the first of his twelve proofs that Muḥammad was a true prophet the fact that he was illiterate: <www.islaam.com/Article.aspx?id=61> For Al-A'zamī, 'his complete illiteracy preclude[ed] any knowledge of Jewish or Christian practices': al-A'zamī, *The History of the Qur'ānic Text from Revelation to Compilation*, 25. In contrast, the Shī'a community cannot accept that the 'fountainhead of knowledge, Sayyid al-anbiya [leader of all the prophets]' could be illiterate: <www.islamoriginal.co.uk/Muhammad.htm>
14. Zaid bin Thabit al-Ansari 'was one of those who used to write the Divine Revelation': al-Bukhārī 6/60/201. Defection of a Christian convert who acted as a scribe: ibid. 4/56/814.
15. The Prophet himself amended a peace treaty with the people of Hudaibiya: al-Bukhārī 3/49/862. He called for 'a bone of scapula, so that I may write something for you after which you will never go astray': ibid. 4/53/393. Same words, but writing materials unspecified: ibid. 5/59/716, 717.
16. J. P. Berkey, *The Formation of Islam. Religion and Society in the Near East, 600–1800* (Cambridge: Cambridge University Press, 2003), 67, favours 'unscriptured'. <www.bismikaallahuma.org/Polemics/umiyy.htm>; <www.debate.org.uk/topics/theo/muhammad.htm#B1ii>
17. For reasons of familiarity to readers, the spellings Mecca and Medina have been preferred to Makka and Madīna.
18. Kamālī, *Principles of Islamic Jurisprudence*, 18. How *naskh* is used depends on the law school and the goal of the interpreter – some believe it should be used chronologically, that is, the latest revelation is the most authoritative, while others, like Muḥammad Maḥmud Ṭāhā of Sudan, have argued that the chronology should be reversed so that the universal prescriptions are considered more authoritative than those that are more limited to specific historical conditions. Others believe that verses should be placed either in their historical context or in the broader context of the Qur'ān. The biggest 'problem' with *naskh* is that it suggests that there is an error in the revelation itself, a real conundrum if one believes that the Qur'ān is the exact and direct revelation of God. How can God make an error? Natana DeLong Bas notes that Muḥammad ibn 'Abd Al-Wahhāb raised this issue with respect to *naskh* and tended therefore to use it in as limited a manner as possible. He preferred to distinguish between absolute prescriptions and those that were limited to a specific situation: Natana DeLong Bas, *Wahhābī Islām. From Revival and Reform to Global Jihād* (New York: Oxford University Press, 2004), 103–5. The author is particularly grateful to Dr DeLong Bas for sharing her findings prior to publication and for other comments on the draft text of this book.
19. Kamālī, *Principles of Islamic Jurisprudence*, 149. Aḥmad Ḥasan, *The Early Development of Islamic Jurisprudence* (Islamabad: Islamic Research Institute, 1970), 60–84.

20. Al-Bukhārī 6/60/53.
21. Sunan Abū-Dāwūd 14/2526 (narrated by Anas ibn Mālik): '*jihād* will be performed continuously since the day Allāh sent me as a prophet until the day the last member of my community will fight with the Dajjāl [Antichrist]. The tyranny of any tyrant and the justice of any just [ruler] will not invalidate it. One must have faith in Divine decree.' <www.masmn.org/Hadith/Sunan_Abu_Dawud/014.htm>
22. Kamālī, *Principles of Islamic Jurisprudence*, 153.
23. <www.usc.edu/dept/MSA/law/alalwani_usulalfiqh/ch4.html>
24. Kamālī, *Principles of Islamic Jurisprudence*, 158–9. Ibid. 160, Kamālī concludes that 'notwithstanding the strong case that al-Shāfi'ī has made in support of his doctrine, the majority opinion, which admits abrogation of the Qur'ān and *Sunnah* by one another is preferable, as it is based on the factual evidence of having actually taken place'. He cites al-Ghāzāli's evidence in this respect.
25. Fathi Osman, *Concepts of the Qur'ān. A Topical Reading* (Los Angeles, CA: MVI Publishers, 1997), 974 (except for verses 128–9).
26. Ibid. 165. The Basrah-based scholar and Qur'ānic exegete Qatāda, known as Abū al-Khaṭṭāb (60–117/679–735), gave a lower figure. He contended that Q.8:61 had been abrogated by the verse of the sword (Q.9:5) and that a total of 113 or so verses had been abrogated. Hibatullah considered the total to be 124 verses: J. Burton, *The Sources of Islamic Law: Islamic Theories of Abrogation* (Edinburgh: Edinburgh University Press, 1990), 184. Others stated 114: J. J. G. Jansen, *The Neglected Duty. The Creed of Sadat's Assassins and Islamic Resurgence in the Middle East* (New York: Macmillan, 1986), 195–6. Hasan, *The Early Development of Islamic Jurisprudence*, 67–8.
27. Kamālī, *Principles of Islamic Jurisprudence*, 165.
28. Ḥasan, *The Early Development of Islamic Jurisprudence*, 68, for those who denied the theory of *naskh* entirely, or severely reduced the number of Qur'ānic verses to which the theory might apply.
29. Quoted by M. S. bin Jani, 'Sayyid Quṭb's View of *Jihād*: An Analytical Study of his Major Works' (University of Birmingham, unpublished PhD. thesis, 1998), 117. The author acknowledges his debt to this thesis, valuable both for its analysis of Quṭb's writings as for the classical doctrine of *jihād*. The thesis was supervised by Professor Jørgen Nielsen. The depiction of al-Sarakhsī's work is that of Chibli Mallat, 'On Islām and Democracy': <www.www.soas.ac.uk/Centres/IslamicLaw/PublicIntro.html>
30. The Qur'ān confirmed a pre-Islamic custom here, according to which four months (Muḥarram, Rajab, Dhū 'l-Qa'dah and Dhū 'l-Ḥijjah) were considered 'sacred' in the sense that all tribal warfare had to cease. This custom was preserved with the intention of promoting peace among warring tribes.
31. Ayatullah Murtazá Muṭahharī, *Jihād: The Holy War of Islām and its Legitimacy in the Qur'ān* (Tehran: Islamic Propogation Society, 1998), 74–5. <www.al-islam.org/short/jihad/>
32. Firestone, *Jihād*.
33. Contrast the holistic interpretation of Alec Motyer, *The Prophecy of Isaiah* (Leicester: Inter-Varsity Press, 1993) with the first, second and third Isaiahs analysed in *The New Jerome Biblical Commentary*, ed. R. E. Brown, J. A. Fitzmeyer and R. E. Murphy (London: Geoffrey Chapman, 1990), chs 15 and 21.
34. <www.geocities.com/al_mo_minah/jihad3.html>
35. Sayyid Quṭb, *Milestones* (Beirut and Damascus: International Islamic Federation of Student Organizations, 1978), 78. Earlier on in his analysis, Quṭb cited the account of the pupil of Ibn Taymīyah, Ibn Qayyim al-Jawziyah, entitled *Zad al-Ma'ād* for the various stages: ibid. 73–7.

36. Thus Reuven Firestone argues that the conflicting Qur'ānic verses fail to prove 'an evolution of the concept or sanction for religiously authorized warring in Islām from a non-aggressive to a militant stance'. Firestone, *Jihād.*
37. Burton, *The Sources of Islamic Law*, 5, 98.
38. Ibid. 182, 198, 208.
39. Ibid. 31.
40. Firestone, *Jihād*, 69.
41. Sachedina, *The Islamic Roots of Democratic Pluralism*, 36.
42. Friedmann, *Tolerance and Coercion in Islam*, 56, 88.
43. Firestone, *Jihād*, 75. Firestone states these as the Sacred Months, in which fighting was forbidden; the sacred city, in which fighting was forbidden; and the state of ritual consecration of the pilgrim, who was forbidden to bear arms.
44. <web.umr.edu/~msaumr/Quran/> Hanna E. Kassis, *A Concordance of the Qur'ān* (Berkeley, CA: University of California Press, 1983), 587–8, s.v. *jāhada. An Exhaustive Concordance of the Meaning of the Qur'ān*, ed. J. Cason, K. El-Fad and F. Walker (n.p.: 2000), 751, s.v. 'strive'. Alfred Morabia also accepts that there are 35 verses and contends that 22 refer to general effort, ten to warlike activity, and three are of a spiritual tone: Alfred Morabia, *Le Ĝihâd dans l'Islam médiéval. Le «combat sacré» des origines au xiie siècle* (Paris: Albin Michel, 1993), 141, 417 n. 204, 205, 106. It should be noted that the enumeration of the verses differs somewhat between the authorities. Abdulrahman Muḥammad Alsumaih, 'The Sunnī Concept of *Jihād* in Classical Fiqh and Modern islamic Thought' (University of Newcastle upon Tyne, unpublished PhD thesis, 1998), 14, states that there are more than 32 occurrences, but he does not analyse these systematically. He contends (ibid. 15) that there is no significant difference between the words *Jihād* and *Qitāl* (fighting), which are used with the same meaning in the Qur'ān. This inference is rejected here as unwarranted. Even Mawdūdī himself rejected it: 'in the terminology of the *Sharī'ah*, *qitāl* and *jihād* were two different things. *Qitāl* is applied to the military venture undertaken against the armies of the enemy. *Jihād* is applied to the total effort mounted by the whole nation for the success of the objective for which the war began. During this struggle, *qitāl* may stop at times, and may also be suspended. But *jihād* continues till the time when that aim is achieved for which it began.' Mawdūdī in the newspaper *Mashriq*, Lahore, 12 October 1965.
45. Moulavi Cherāgh 'Alī, *A critical exposition of the popular 'jihad': showing that all the wars of Mohammad were defensive, and that aggressive war, or compulsory conversion, is not allowed in the Koran, with appendices providing that the word 'jihad' does not exegetically mean 'warfare', and that slavery is not sanctioned by the Prophet of Islam* (Delhi, 1885; repr. Karachi, 1977). The numerical order in the following notes follows that of Cherāgh 'Alī.
46. Among the Medina *sūrahs*: 30) Q.9:82; 31) Q.9:86; 32) Q.9:88; 33) Q.5:36.
47. Among the Mecca *sūrahs*: 1) Q.21:14–15; 2) Q.25:53–4; 3) Q.12:77–8; 4) Q.16:110; 5) Q.29:6; 6) Q.29:8; 8) Q.16:38; 9) Q.35:42. Among the Medina *sūrahs*: 15) Q.6:109; 19) Q.24:53; 34) Q.5:53.
48. Among the Mecca *sūrahs*: 7) Q.29:69. Among the Medina *sūrahs*: 10) Q.2:218; 11) Q.3:142; 12) Q.8:72; 13) Q.8:74; 14) Q.8:75; 16) Q.47:31; 17) Q.61:11; 18) Q.4:95; 20) Q.66:9; 21) Q.9:73; 22) Q.60:1; 23) Q.49:15; 24) Q.9:16; 25) Q.9:19; 26) Q.9:20; 27) Q.9:24; 28) Q.9:41; 29) Q.9:44; 35) Q.5:54.
49. <www.islamistwatch.org/main.html> The website carries the following note: 'The sole reason for reproducing these *sūrahs* is to note those to which the Islamists refer repeatedly in order to establish the rationale for their goals and activities. These are not all the *sūrahs* referred to by the Islamists, but they are central to their arguments.'

50. Burton, *The Sources of Islamic Law*, 1, calls Q.9:5 the sword verse: 'So kill the unbelievers (*mushriks*) wherever you find them immediately after the end of the Forbidden Month.' Q.9:5 and Q.9:36 are clearly related verses.
51. Muḥammad Iqbal, the spiritual father of the idea of a separate Pakistan, persuaded Asad to abandon plans to travel to eastern Turkestan, China and Indonesia and instead 'to help elucidate the intellectual premises of the future Islamic state'. When Pakistan was born in 1947, Asad was appointed its undersecretary of state for Near Eastern Affairs and became its permanent representative to the United Nations in 1952.
52. Citing Tabari and Ibn Kathir. Muḥammad Asad, *The Message of the Qur'ān* (Gibraltar: Dar Al-Andalus, 1980), 41 n. 167 (comment on Q.2:190), 93 n. 128 (comment on Q.3:165).
53. Ibid. 41 n. 168. Also ibid. 242 n. 25 ('temptation to evil').
54. Ibid. 42 n. 172.
55. Ibid. 256, n. 7 and n. 9.
56. Shaykh Muḥammad al-Ghazālī, *A Thematic Commentary on the Qur'ān* (Herndon, VA: International Institute of Islamic Thought, 2000), 105.
57. Ibid. 183.
58. Ibid. 19. He calls this 'an objectionable undermining of eternal Islamic principles, inviting noxious charges against Islām, for which we have only ourselves to blame'. The index of his commentary distinguishes between *jihād* in peace-time and in war-time: ibid. 787.
59. Ibid. 263.
60. Ibid. 746.
61. Ibid. 429–30.
62. Ibid. 576.
63. Ibid. 584, 642, 760–1. Controversially, in June 1993, he issued a *fatwā* stipulating that any Muslim who argued for the suspension of *sharī'ah* law was an apostate (*murtadd*) and should be killed with impunity: *The Tablet*, 3 October 1998, 1275.
64. Faruq Sherif, *A Guide to the Contents of the Qur'ān* (London: Ithaca Press, 1985), 3–4.
65. Ibid. 113.
66. Ibid. 114.
67. Osman, *Concepts of the Qur'ān*, 927.
68. Ibid. 944.
69. Ibid. 948. An important caveat.
70. Ibid. 945. Ibid. 951, where Q.2:190 is cited.
71. Ibid. 948.
72. Ibid. 953.
73. See below, Chapter 2.
74. See below, Chapter 4.
75. Qamaruddin Khan, *Political Concepts in the Qur'ān* (Karachi: Institute of Islamic Studies, 1973), 70.
76. Ibid. 69.
77. Ibid. 73–6.
78. Ibid. 68.
79. Qamaruddin Khan, *The Political Thought of Ibn Taymīyah* (Islamabad: Islamic Research Centre, Pakistan, 1973), 157.
80. Friedmann, *Tolerance and Coercion in Islam*, 11–12. Sachedina, *The Islamic Roots of Democratic Pluralism*, 38–9.
81. '...it may be noted that *ḥadīth* means any report that records the word, deed, or approval of the Prophet. And *Sunnah* is the justice ruling derived from such a report. The *ḥadīth* is, therefore, the vehicle of the *Sunnah*...': Qamaruddin Khan, *Political Concepts in [the] Sunnah. A Treatise on the Political Concepts of the Holy Prophet*, ed. H. M. Arshad Qureshi (Lahore,

Islamabad and Washington DC: Islamic Book Foundation, 1988), 23. It should be noted that this important book was published posthumously and regrettably contains numerous editorial and typesetting errors. Its sense is, however, clear. The scholar Wensinck similarly stated that *ḥadīth* was the form, *sunnah* the matter: Khadūrī, *Islamic Jurisprudence*, 30.

82. Al-Bukhārī 9/88/208.
83. <www.usc.edu/dept/MSA/fundamentals/hadithsunnah/hadithqudsi.html>
84. <www.hadith.org.za/qudsi.htm> The sixth *ḥadīth qudsī* warns against false claims of martyrdom: 'The first of people against whom judgement will be pronounced on the Day of Resurrection will be a man who died a martyr. He will be brought and Allāh will make known to him His favours and he will recognize them. [The Almighty] will say: And what did you do about them? He will say: I fought for you until I died a martyr. He will say: You have lied – you did but fight that it might be said [of you]: He is courageous. And so it was said. Then he will be ordered to be dragged along on his face until he is cast into Hell-fire.'
85. Graham, *Divine Word and Prophetic Word in Early Islām*, 169–70. Graham comments: 'the references to the spoils of war and to being killed indicate in particular that it is the willingness to die in the cause of God rather than a more spiritual striving that is being encouraged. This is further substantiated by the specific mention of *Ghāzā*, "to go out on raids, to raid" in version ii and other variants. Similarly, the apparent meaning of Muḥammad's final statement is that he would gladly be killed again and again by his enemies to gain fulfilment of God's promises.' Ibid. 200, Saying 75, where those 'killed in the cause of God' are not reckoned as dead, but are fed a heavenly sustenance with their Lord (Q.3:169). This is the definitive answer to the false *ḥadīth* of the 72 black-eyed virgins as the reward for the martyr. Ibid. 193, Saying 67: 'they [could] not be given anything that they would love more than the vision of their Lord'.
86. Jamilah Kolocotronis, *Islamic Jihād: An Historical Perspective* (Indianapolis, IN: American Trust Publications, 1990), 28. The author applies this categorization to both the Qur'ān and the *ḥadīth*.
87. Al-Bukhārī 4/52/73: <www.usc.edu/dept/MSA/fundamentals/hadithsunnah/bukhari/052.sbt.html#004.052.073> Also Muslim 19/4314: <www.iiu.edu.my/deed/hadith/muslim/019_smt.html>
88. Al-Bukhārī 4/52/112; 4/56/839. Mālik has the term 'a man who takes the rein of his horse to do *jihād* in the way of Allāh': 21/1/4. Good as a permanent quality on the foreheads of horses for *jihād*: al-Bukhārī 4/52/104. Similarly, *jihād* was a legitimate charge on the land (that is, for the payment of the costs of war): al-Bukhārī 4/51/33. If armour was kept for *jihād*, then no obligatory charity (*zakāt*) was payable: 2/24/547.
89. Al-Bukhārī 3/31/121; 5/57/18. Mālik 21/19/49. Muslim 5/2239.
90. Al-Bukhārī 4/53/352.
91. Al-Bukhārī 9/93/519. Mālik states that the Prophet 'stimulated people for *jihād* and mentioned the Garden': 21/18/42.
92. Mālik 21/14/32.
93. Al-Bukhārī 9/93/549; 9/93/555. '…if it is solely *jihād* and trust in his promise that brings him out of his house': Mālik 21/1/2.
94. Mālik 9/18/56: one who goes to the mosque and goes nowhere else 'either to learn good or teach it' is like someone who does *jihād* 'and returns with booty'.
95. Al-Bukhārī 4/53/352.
96. Abū-Dāwūd 14/2510.
97. Abū-Dāwūd 14/2513.
98. Al-Bukhārī 1/2/35. Muslim 1/210.
99. Al-Bukhārī 2/15/86.
100. Muslim 17/4198; 17/4199.

101. Muslim 20/4626. Ditto, but without multiple deaths: 20/4630. Ditto, but fighting and dying twice: 20/4631. Al-Bukhārī 1/2/35.
102. Muslim 20/4638; 20/4639; 20/4641; better than anything on which the sun rises or sets: 20/4643; 20/4644.
103. Muslim 20/4646, on the testimony of Abū Qatada.
104. Muslim 20/4645, on the testimony of Abū Sa'id Khudri.
105. Al-Bukhārī 5/58/254: Abū Musa said that 'we took part in *jihād* after Allāh's Apostle, prayed and did plenty of good deeds, and many people have embraced Islām at our hands, and no doubt, we expect rewards from Allāh for these good deeds'.
106. Muslim 37/6670. 'The three who were left behind' did not mean that they remained back from *jihād*: ibid.
107. Al-Bukhārī 4/52/41; 1/10/505; 8/73/1; 9/93/625.
108. Muslim 1/151; 1/152; 1/153.
109. Al-Bukhārī 1/2/25; 2/26/594. Muslim 1/148; 20/4597; 20/4599.
110. Al-Bukhārī 4/52/87; 4/52/208; 4/53/412.
111. Al-Bukhārī 4/52/79; 4/52/311; 5/58/240; 5/59/602.
112. Muslim 7/3116.
113. Al-Bukhārī 4/52/311; 4/53/412.
114. Al-Bukhārī 4/52/85. Muslim 20/4676; 20/4677.
115. Al-Bukhārī 4/52/248; 8/73/3. Muslim 32/6184; 32/6185; 32/6186.
116. Muslim 1/151; 1/152; 1/153.
117. The Ismāʻīlī jurist al-Nuʻmān added two more 'pillars', devotion to the *imām* (*walāyah*) and *jihād*: *Mediaeval Ismāʻīlī History and Thought*, ed. F. Daftary (Cambridge: Cambridge University Press, 1996), 127.
118. Al-Bukhārī 6/60/40.
119. Al-Bukhārī 5/59/285.
120. Abū-Dāwūd 40/4631.
121. Mālik 21/15/35.
122. Muslim 1/209.
123. Mālik 21/14/28. Muslim 20/4658; 20/4660.
124. Muslim 20/4688.
125. Al-Bukhārī 8/77/604.
126. Abū-Dāwūd 14/2533.
127. Al-Bukhārī 2/23/381. Muslim 31/6042.
128. Abū-Dāwūd 20/3131.
129. Mālik 21/16/37.
130. Al-Bukhārī 2/23/434.
131. Al-Bukhārī 5/59/325.
132. Muslim 20/4634; 20/4635.
133. Abū-Dāwūd 14/2516.
134. Muslim 20/4706.
135. Al-Bukhārī 2/57/35; 4/56/680; 6/77/616; 7/71/630; 8/77/616.
136. Muslim 20/4707; 20/4708.
137. Muslim 20/4706.
138. Al-Bukhārī 7/71/630.
139. Abū-Dāwūd 20/3105. Mālik 16/12/36. Elsewhere Mālik gives five: plague, disease of the belly, drowning, collapsing building and martyr in the path of Allāh: Mālik 8/2/6.
140. Abū-Dāwūd 14/2493.
141. Al-Bukhārī 3/43/660. Abū-Dāwūd 40/4753.
142. Abū-Dāwūd 40/4753.

143. M. Z. Ṣiddiīqī, *Ḥadīth Literature. Its Origin, Development and Special Features* (Cambridge: Cambridge University Press, 1993), 76–84.
144. J. Burton, *An Introduction to the Ḥadīth* (Edinburgh: Edinburgh University Press, 1994), 125.
145. Ṣiddiīqī, *Ḥadīth Literature*, 18. Muslim was al-Bukhārī's close student and probably followed his *madhhab*, and was a *Mujtahid Murajjiḥ* (senior Mujtahid).
146. Burton, *Introduction to the Ḥadīth*, 125–6.
147. The traditional number of the canonical collections is said to be six, but Graham argues that there are 'at least nine' which have been 'so widely relied upon by later Islām that they can be called in some sense "classical"': Graham, *Divine Word and Prophetic Word in Early Islām*, 83.
148. *Nahj al-Balāgha. Peak of Eloquence. Sermons, Letters and Sayings of Imām 'Alī ibn Abī Ṭālib*, ed. Sayed Ali Reza (New York: Tahrike Tarsile Qur'ān, Inc., 3rd rev. edn, 1984), 423 (sermon 209). 'Alī distinguished between the lying hypocrites; those who were mistaken; those who were ignorant; and those who memorized truthfully.
149. M. M. al-a'zamī, *Studies in Ḥadīth Methodology and Literature* (Plainfield, IN: American Trust Publications, 1978), 33.
150. Burton, *Introduction to the Ḥadīth*, 146.
151. Muslim critiques of the Orientalists' views, including Juynboll, in Ṣiddīqī, *Ḥadīth Literature*, 124–35.
152. Kamālī, *Principles of Islamic Jurisprudence*, 66.
153. G. H. A. Juynboll, *Muslim Tradition. Studies in Chronology, Provenance and Authorship of Early Hadīth*, Cambridge Studies in Islamic Civilization (Cambridge: Cambridge University Press, 1983), 71, 74.
154. Ibid. 144.
155. Al-a'zamī, *Studies in Ḥadīth Methodology*, 26, who notes that the actual number of *aḥadīth* may be no more than 1236. Juynboll, *Muslim Tradition*, 204, asks the reader to consider why Abū Hurayra 'should transmit traditions of a certain tenor to a pupil hailing from a certain city, and transmit fundamentally different traditions [to] a pupil hailing from another city…'
156. Ṣiddīqī, *Ḥadīth Literature*, 18. Al-a'zamī, *Studies in Ḥadīth Methodology*, 26.
157. Ṣiddīqī, *Ḥadīth Literature*, 35.
158. Juynboll, *Muslim Tradition*, 161.
159. Kamālī, *Principles of Islamic Jurisprudence*, 81. Al-a'zamī, *Studies in Ḥadīth Methodology*, 59–60.
160. Ibid. 64–7.
161. Ḥasan, *The Early Development of Islamic Jurisprudence*, 178–216, especially 187: 'al-Shāfi'ī believes that no authentic *ḥadīth* goes against the Qur'ān.' Khaddūrī, *Islamic Jurisprudence*, 123 (para. 101), 125 (para. 101).
162. Khaddūrī, *Islamic Jurisprudence*, 121 (para. 98): 'Among the things with which [the Prophet] was inspired is his *sunnah*. This [*sunnah*] is the Wisdom which God mentioned…'
163. Burton, *Introduction to the Ḥadīth*, 179.
164. Kamālī, *Principles of Islamic Jurisprudence*, 47.
165. Ibid. 82.
166. R. Peters, *Jihād in Classical and Modern Islam* (Princeton, NJ: Markus Weiner, 1996). Of the four principal schools of jurisprudence, the Mālikī school defines it as the fighting of unbelievers by Muslims for the raising of the word of Allāh; the Shāfi'ī school defines it as fighting in the cause of Allāh; the Ḥanafī school states that it means the call to Islām and fighting those who do not accept it; the Ḥanbalī school explains that *jihād* means the fighting of non-believers: Alsumaih, 'The Sunnī concept of *Jihād*', 14. There were, however, divergent tendencies within each tradition.

167. Authorities cited at: <www.israinternational.com/Scholars.html>
168. Ibn Kathīr in his *Tafsīr* of Surah al-Raḥman (55), verse 72: 'It was mentioned by Daraj Ibn Abī Ḥatim that Abū al-Haytham 'Abdullah Ibn Wahb narrated from Abū Sa'īd Al-Khudrī, who heard the Prophet Muḥammad saying: "The smallest reward for the people of Paradise is an abode where there are 80,000 servants and 72 wives, over which stands a dome decorated with pearls, aquamarine, and ruby, as wide as the distance from al-Jābiyya [a Damascus suburb] to Ṣana'ā'."'
169. Al-Tirmidhī 4/21/2687. This work contains 3956 *aḥādīth.* He stated that 'nobody agrees with anyone' (in assessing transmitters): Juynboll, *Muslim Tradition*, 177–8. Al-Tirmidhī was al-Bukhārī's close student and a *Mujtahid Murajjiḥ* and comparatist of the first rank. The fact that neither al-Bukhārī nor Muslim include this *ḥadīth* is particularly suspicious given the relationship between the three *ḥadīth* collectors.
170. A. K. S. Lambton counted six: those of al-Bukhārī, Muslim, Abū-Dāwūd, al-Tirmidhī, Ibn Māja and al-Nasā'ī. Lambton, *State and Government in Medieval Islam. An Introduction to the Study of Islamic Political Theory: The Jurists* (New York: Oxford University Press, 1981), 6. <www.answering-islam.org/Gilchrist/Vol1/6b.html> The IHSAN Network *ḥadīth* database counts seven by including the *Al-Muwaṭṭa'* of Mālik ibn Anas: <www.cmeis.cam.ac.uk/ihsan/>
171. Commentary on Q.55:72.
172. Al-A'zamī, *Studies in Ḥadīth Methodology*, 105. Al-A'zamī contends that all those contained in al-Bukhārī and Muslim are true.
173. See above, note 85.
174. Though it must be said that Professor Reuven Firestone considers that the *ḥadīth* on 72 black-eyed virgins is probably genuine. He cites 'a *ḥadīth* in an authoritative collection called *Sunan* al-Tirmidhī, which would be on the shelves of any Muslim scholar. In my edition, published in Beirut, it can be found in a section called "The Book of Description of the Garden", chapter 23, titled "The least reward for the people of Heaven", *ḥadīth* number 2562. The *ḥadīth* reads literally as follows: "Sawda (Tirmidhī's grandfather) reported that he heard from 'Abdullah, who received from Rishdīn bin Sa'd, who in turn learned from 'Amr b. al-Ḥārith, from Darrāj, from Abu'l-Haytham, from Abū Sa'īd al-Khudrī, who received it from the Apostle of God [Muḥammad]: the least [reward] for the people of Heaven is 80,000 servants and 72 wives, over which stands a dome of pearls, aquamarine and ruby, as [wide as the distance] between al-Jābiyya and Ṣana'ā'." That these 72 wives are virgin[s] is confirmed by the Qurān (55:74) and commentaries on that verse. Al-Jābiyya was a suburb of Damascus, according to the famous 14th century commentator, Ismā'īl Ibn Kathīr, so one personal jewelled dome would stretch the distance from Syria to Yemen, some 1,600 miles.' Reuven Firestone, 'Islām hijacked', *The Jewish Journal of Greater Los Angeles*, 28 September 2001. <www.islamfortoday.com/firestone01.htm>
175. Kamālī, *Principles of Islamic Jurisprudence*, 67. The terminology, but not the attribution to this particular false *Ḥadīth*, is that of Professor Kamālī.
176. Friday (17 August 2001) sermon broadcast live on Palestinian TV from the Shaykh 'Ijlin mosque in Gaza. The preacher was Shaykh Isma'il Aal Ghadwan: '… The martyr, if he meets Allāh, is forgiven with the first drop of blood; he is saved from the torments of the grave; he sees his place in Paradise; he is saved from the Great Horror [of the day of judgment]; he is given 72 black-eyed women; he vouches for 70 of his family to be accepted to Paradise; he is crowned with the Crown of glory, whose precious stone is better than all of this world and what is in it…' Memri Special Dispatch 261, 23 August 2001: <www.memri.org/bin/articles.cgi?Page=subjects&Area=jihad&ID=SP26101>
177. 'Are "the black-eyed" available for sex? Some evidently think they are. The Israeli media reported on a suicide bomber caught before he managed to carry out his mission; he was

wearing a towel as a loincloth to protect his genitals for use in Paradise. In Islām, marriage is the only legal location for sexual relations. Because marriage requires the man to provide his wife with a dowry and maintenance, many Palestinian youth are unable to marry – they simply cannot fulfil the financial obligations. They thus have no legal outlet for sexual fulfilment, which would add to the attraction of the promise of sexual relations in Paradise. The appeal of the virgins may be taken as a sign of the desperation of the conditions in which the Palestinian people live, rather than as evidence of sexual obsession. The question of sexual relations was also brought up in an interview that Shaykh of al-Azhar, Muḥammad Sayyed Ṭanṭāwī gave to the Egyptian weekly *Aakher Sa'a*. To the question "What is the meaning of the Koranic verse 'And we will marry them to the "black-eyed?"'" Ṭanṭāwī replied, "This verse heralds to faithful believers that in the world to come, Allāh will set 'the black-eyed' to serve them, so that they will have wives, along with the righteous women from this world." Getting straight to the point, the interviewer asked, "Do people in Paradise have sexual relations?" "This issue is known only to Allāh," said Ṭanṭāwī. "It is enough that we know that Paradise offers [everything] to satisfy the soul and gladden the eye. Regarding other, private matters, only Allāh knows. It is enough for us that the Koran says, 'It has [everything] to satisfy the soul and gladden the eye, and in it you have life everlasting.'" In a review of the Egyptian press in the London daily *Al-Quds Al-'Arabī*, the veteran Egyptian journalist Ḥasanayn Karrūm explained that Shaykh Ṭanṭāwī knowingly gave a vague answer to the question, so as to avoid a scandal like the one created a few years earlier by the late author and journalist Muḥammad Jalāl al-Kushk. Al-Kushk wrote, "the men in Paradise have sexual relations not only with the women [who come from this world] and with 'the black-eyed', but also with the serving boys." According to Karrūm, al-Kushk also stated, "in Paradise, a believer's penis is eternally erect".' Memri Inquiry and Analysis Series No. 74, 30 October 2001. <www.memri.org/bin/articles.cgi?Page=archives&Area=ia&ID=IA7401> There is need for the same note of caution about the source material here as stated above in the introduction (note 24).

178. Mark Werlin, 'An interview with Prof. Mark Jurgensmeyer': <www.pariswerlin.com/articles/juergens.html>
179. Though the author does not fully align himself with the remarks of Alan Dershowitz, 'Suicide Bombing is advocated by privileged elites', *Guardian* (4 June 2004): <www.guardian.co.uk/comment/story/0,3604,1231148,00.html>
180. See note 125 above.
181. Khan, *Political Concepts in [the] Sunnah*, 39, uses volume six of the *Kanz al-'Ummal* of 'Alī al-Muttaqī (d. 974/1567) on the grounds that this classifies the *ḥadīth* by subject headings.
182. Ibid. 64, 74–5, 107.
183. Ibid. 124.
184. Khan, *Political Concepts in the Qur'ān*, 70. Khan, *Political Concepts in [the] Sunnah*, 155: 'the states of the Prophet and the Orthodox Caliphs were purely secular states, and administered by entirely secular laws'.
185. Khan, *Political Concepts in the Qur'ān*, 69.
186. Khan, *The Political Thought of Ibn Taymīyah*, 157.
187. F. E. Peters, *Muḥammad and the Origins of Islām* (New York: State University of New York Press, 1994), 173–6. Hamidullah, *Muslim Conduct of the State*, 121–2 (para. 220). <www.sharjahfm.com/english-sharjah/biography/MUSLIMS_MIGRATE_TO_ABYSSINIA.htm>
188. *The First Written Constitution in the World. An Important Document of the Time of the Holy Prophet*, trans Muḥammad Hamidullah (Lahore: Sh. Muhammad Ashraf, 2nd rev. edn, 1968).
189. Fahmi Huweidi, 'Non-Muslims in Muslim Society', in *Rethinking Islām and Modernity. Essays in Honour of Fathi Osman*, ed. Abdelwahab El-Affendi (Leicester: Islamic Foundation, 2001), 84–91 at 89.

190. Quoted by Peters, *Muḥammad and the Origins of Islām*, 200.
191. Bahtiar Effendy, *Islām and the State in Indonesia* (Singapore: Institute of Southeast Asian Studies, 2003), 108.
192. Quoted by Peters, *Muḥammad and the Origins of Islām*, 151.
193. Khan, *Political Concepts in [the] Sunnah*, 158–9.
194. As evidenced, for example, in the 1600 pages in *The Historical Jesus. Critical Concepts in Religious Studies*, ed. Craig Evans (London: Routledge, 2004). Geza Vermes, *The Changing Faces of Jesus* (London: Allen Lane, 2000), 267 n. 3, gives some of the titles of works on the 'historical' Jesus. One of the more notable of these works is E. P. Sanders, *The Historical Figure of Jesus* (London: Allen Lane, 1993). Most recently: Vermes, *The Authentic Gospel of Jesus the Jew* (London: Allen Lane, 2003).
195. *Judaisms and their Messiahs at the Turn of the Christian Era*, ed. J. Neusner, W. S. Green and E. Frerichs (Cambridge: Cambridge University Press, 1987).
196. Vermes, *The Changing Faces of Jesus*, 260. Cf. Vermes, *The Religion of Jesus the Jew* (London: SCM Press, 1993).
197. The *Muslim Jesus. Sayings and Stories in Islamic Literature*, ed. Tarif Khālidī (Cambridge, MA: Harvard University Press, 2001), introduction.
198. Ibid. 71–2, on the saying of Jesus 'place your treasures in heaven, for the heart of man is where his treasure is'.
199. Ibid. 42.
200. Mawlānā Amīn Aḥsan Iṣlāḥī, 'Self-Development in the Context of Man's Relationship with Allāh', in *Tazkiyah. The Islamic Path to Self-Development*, ed. Abdur Rashid Siddiqui (Leicester: The Islamic Foundation, 2004), 133–214, at 201–2. For an older study of Islamic references to Jesus and Christianity: Rev. James Robson, *Christ in Islām* (London: John Murray, 1929). Web version at: <www.sacred-texts.com/isl/cii/cii.htm>
201. Matthew 16:24; Mark 8:34; Luke 9:23. Luke 9:23 has the command to take up one's cross as a daily endeavour or struggle.
202. A higher figure of 26 or 27 expeditions commanded by the Prophet (*ghazwas*) is cited by Majīd Khaddūrī, *War and Peace in the Law of Islām* (Baltimore, MD, and London: Johns Hopkins University Press, 1955; orig. edn, 1940), 87 n. 23. When campaigns were commanded by Companions, the term used was *sariyyas*. Peters, *Muḥammad and the Origins of Islām*, ch. 9, provides a brief history of the campaigns.
203. Solail H. Hashmī, 'Interpreting the Islamic Ethics of War and Peace', in *Islamic Political Ethics. Civil Society, Pluralism and Conflict*, ed. Solail Hashmī (Princeton, NJ: Princeton University Press, 2002), 203.
204. Peters, *Muḥammad and the Origins of Islām*, 224. *The Pillars of Islam. Da'ā'im al-Islām of al-Qāḍī al-Nu'mān. I. Acts of Devotion and Religious Observances*, trans. Asaf A. A. Fyzee, revised by Ismail Kurban Husein Poonawala (New Delhi: Oxford University Press, 2002), 466. Akram Diyā' al 'Umarī, *Madīnan Society at the Time of the Prophet. I. Its Characteristics and Organization*, trans. H. Khattāb (Herndon, VA: International Islamic Publishing House and International Institute of Islamic Thought, 1992), 137. The atrocity is only partly explained by two previous expulsions, that of the Banū Qaynuqā and the Banū al Nadīr.
205. Robert Spencer, *Onward Muslim Soldiers. How Jihād Still Threatens America and the West* (Washington DC: Regnery, 2003), 160. The evidence is compiled in <www.answering-islam.org.uk/Muhammad/Jews/BQurayza/banu3.html> and preceding pages.
206. Falwell, a prominent supporter of President Bush, explained: 'I said I have read both Muslim and non-Muslim biographers on Muhammad – of course he lived hundreds of years ago – but they all seem to uniformly agree that he was a man of war and man of violence. And today he would probably be associated with Arafat and Saddam Hussein as a terrorist. Killing people didn't bother him.' Falwell argued that Jesus and Moses stood in contrast to that. 'Their's was

a model of love', he said. 'So I would say that you cannot equate Islam with Christianity and Judaism.' <www.worldnetdaily.com/news/article.asp?ARTICLE_ID=29175>; <www.cnsnews.com/ViewForeignBureaus.asp?Page=\\ForeignBureaus\\archive\\200210\\FOR20021015f.html>

207. Norbert Brox, *A History of the Early Church*, trans. John Bowden (London: SCM, 1994), 9.
208. Muḥammad Hamidullah, *Muslim Conduct of the State*, 54–5.
209. M. J. Kister, 'Land Property and *Jihād*: A Study of Some Early Traditions', *Journal of the Economic and Social History of the Orient*, 34 (1991), 281, repr. in Kister, *Concepts and Ideas at the Dawn of Islam* (Aldershot: Ashgate, Variorum, 1997), ch. 4.
210. *The Islamic Law of Nations. Shaybānī's Siyar*, trans and ed. Majīd Khaddūrī (Baltimore, MD: Johns Hopkins University Press, 1966), 76 para. 1 and 92 para. 47. Khaddūrī suggests the insertion of 'only'. Hamidullah, *Muslim Conduct of the State*, 299–300 (para. 646).
211. Khaddūrī, *Islamic Jurisprudence*, 82 (para. 40), 84–6 (paras 42–51).
212. Ibid. 195 (para. 228), 204–6 (paras 451–70).
213. Clinton Bennett, *In Search of Muhammad* (London and New York: Cassell, 1998), 61. <www.sufism.org/society/articles/MomentofTruth.htm>; <www.sufism.org/society/articles/PeaceHadith.htm>
214. B. G. Weiss and A. H. Green, *A Survey of Arab History* (rev. edn, Cairo: American University in Cairo Press, 1987), 43.
215. Reza, *Nahj al-Balāgha*, 326 (sermon 160).
216. Ibid. 370 (sermon 184).
217. Sherman A. Jackson, *Islamic Law and the State: the Constitutional Jurisprudence of Shihāb al-Dīn al-Qarāfī* (Leiden and New York: E. J. Brill, 1996), 219.
218. Ibid. 220: 'what he says in these capacities constitutes universal law which remains binding until the Day of Judgement'.
219. Morabia, *Le Ĝihâd dans l'Islam médiéval*, 79.
220. Weiss and Green, *A Survey of Arab History*, 55.
221. For a categorical assertion that he did: Taqiuddin an-Nabhani, *The Islamic State* (New Delhi: Milli Publications, 2001), 148: 'the Messenger of Allāh had designed the plan of the conquests before his death.' For the categorical assertion that he did not, and that at most he thought of Arabs under Byzantine or Persian rule: *Concise Encyclopedia of Islām*, ed. H. A. R. Gibb and J. H. Kramers (London: Stacey International, 1966), 402: 'it cannot be proved that he ever went beyond this in his schemes.' Ibid. 401: 'it is very doubtful if Muḥammad ever thought at all of his religion as a universal religion of the world'.
222. Al 'Umarī, *Madīnan Society at the Time of the Prophet. I*, 99–120. H. M. Zawātī, *Is Jihād a Just War? War, Peace and Human Rights under Islamic and Public International Law* (Lewiston, NY: Edwin Mellen, 2001), 115–20.
223. Michael Cook, *Commanding Right and Forbidding Wrong in Islamic Thought* (New York: Cambridge University Press, 2000), 13.
224. Cook, ibid. 15, gives the Qur'ānic references.
225. Ibid. 27.
226. Ibid. 30.
227. Ibid. 562, 568–9.
228. P. Crone and M. Hinds, *God's Caliph. Religious Authority in the First Centuries of Islam* (Cambridge: Cambridge University Press; repr. 2003), 26–8, 33, 116–26.
229. Ibid. 39. Cf. Q.3:98: 'hold you fast to God's rope together and do not scatter'.
230. Annemarie Schimmel, *Mystical Dimensions of Islam* (Chapel Hill, NC: University of North Carolina Press, 1975, repr. 1976), 94, 216–17. Ibid. 27: 'in a comparatively short time, Muḥammad's personality gained great importance for the spiritual life of his community: He was the ideal leader, and the duty of every Muslim was to imitate him'.

231. <www.hozien.com/pdf/Bk-XX.rtf>
232. *Nahj al-Balāgha*, 325 (sermon 159).
233. John Alden Williams (ed.) *Themes of Islamic Civilization* (Berkeley and London: University of California Press, 1971), 260–61.
234. Gibb and Kramers, *Concise Encyclopedia of Islām*, 405.
235. Ibid.
236. Friedmann, *Tolerance and Coercion in Islam*, 118, n. 168.
237. Ibid. 23.
238. Gibb and Kramers, *Concise Encyclopedia of Islām*, 404. Friedmann, *Tolerance and Coercion in Islam*, 14, 26. For Abdulaziz Sachedina, the *dictum* 'only my religion possesses the intrinsic religious value for attaining religious perfection' denies the basis of pluralism: Sachedina, *The Islamic Roots of Democratic Pluralism*, 38.
239. *A Muslim Theologian's Response to Christianity. Ibn Taymiyya's al-Jawāb al-ṣaḥīḥ*, edited and translated by Thomas F. Michel (Delmar, NY: Caravan Books, 1984), 155–6.

Chapter 2

1. Majīd Khaddūrī, *War and Peace in the Law of Islām* (Baltimore and London: Johns Hopkins University Press, 1955), 74 [original edn, 1940]. Alfred Morabia, *Le Ĝihâd dans l'Islam médiéval. Le «combat sacré» des origines au xiie siècle* (Paris: Albin Michel, 1993), 303, 502 n. 66.
2. Patricia Crone, *Meccan Trade and the Rise of Islam* (Oxford: Blackwell, 1987), 243, cited by J. P. Berkey, *The Formation of Islam. Religion and Society in the Near East, 600–1800* (Cambridge: Cambridge University Press, 2003), 67.
3. Hugh Kennedy, *The Prophet and the Age of the Caliphates. The Islamic Near East from the Sixth to the Eleventh Century* (London and New York: Longman, 1986), 59.
4. Ibid. Berkey, *The Formation of Islam*, 72.
5. Muḥammad Hamidullah, *Muslim Conduct of the State, Being a Treatise on Siyar* (rev. 7th edn, Lahore: Sh. Muḥammad Ashraf, 1977), 52–6 (paras 102–6).
6. Ira M. Lapidus, *A History of Islamic Societies* (2nd edn Cambridge, Cambridge University Press, 2002), 30.
7. Patricia Crone, *Medieval Islamic Political Thought* (Edinburgh: Edinburgh University Press, 2004), 367, 372.
8. Lapidus, *A History of Islamic Societies*, 198: 'earlier generations of European scholars believed that conversion to Islām were made at the point of the sword, and that conquered peoples were given the choice of conversion or death. It is now apparent that conversion by force, while not unknown in Muslim countries, was, in fact, rare. Muslim conquerors ordinarily wished to dominate rather than convert, and most conversions to Islām were voluntary.' Ibid. 201: 'the conversion of North Africa… began with the Arab conquests, but… primarily involved the adoption of Islam, notably in sectarian form, by the chiefs of Berber societies as the basis of tribal coalitions and state formation. Khariji states in Algeria and Morocco adopted Islam to help regulate tribal relations and long-distance trade…' At first, before the conversions took place on any significant scale, some of the Christian *dhimmīs* subjected to the new Muslim empire provided a quasi-resistance culture of opposition to Islām: John V. Tolan, *Saracens. Islam in the Medieval European Imagination* (New York: Columbia University Press, 2002), 276.
9. R. W. Bulliet, *Conversions to Islam in the Medieval Period: an Essay in Quantitative History* (Cambridge, MA: Harvard University Press, 1979), 82, 97, 109. These figures, not surprisingly, are contested.
10. Tolan, *Saracens*, 205–6.

11. 'This is the protection which the servant of Allāh, 'Umar, the Ruler of the Believers, has granted to the people of Eiliya [Jerusalem]. The protection is for their lives and properties, their churches and crosses, their sick and healthy and for all their coreligionists. Their churches shall not be used for habitation, nor shall they be demolished, nor shall any injury be done to them or to their compounds, or to their crosses, nor shall their properties be injured in any way. There shall be no compulsion for these people in the matter of religion, nor shall any of them suffer any injury on account of religion... Whatever is written herein is under the covenant of Allāh and the responsibility of His Messenger, of the Caliphs and of the believers, and shall hold good as long as they pay [the] *jizya* [the tax for their defence] imposed on them.' <www.islamknowledge.faithweb.com/umar_bin_khattab.htm>
12. Tolan, *Saracens*, 41.
13. Tastan Osman, 'The Jurisprudence of Sarakhsī with Particular Reference to War and Peace: A Comparative Study in Islamic Law' (unpublished PhD thesis, University of Exeter, 1993), 139–40, 143–4, esp. 147, notes that al-Sarakhsī used Abū Bakr's decision as the legitimate precedent for such wars in the absence of prior guidance from the Prophet. For al-Qarāfi's unconventional view of Abū Bakr's decisions in the wars of *riddah* as no more than *fatwās*, not binding on 'Umar, his successor: Sherman A. Jackson, *Islamic Law and the State: the Constitutional Jurisprudence of Shihāb al-Dīn al-Qarāfī* (Leiden and New York: E. J. Brill, 1996), 217. Wilferd Madelung notes that the notion of 'rebellion' had in reality no basis in the Qur'an [the proof text was Q.49:9, but this 'could not be applied to the 'rebel' tribes] or the practice of the Prophet but arose out of the caliphate as conceived by Abū Bakr... The caliph was to be not so much the religious leader of the *ummah*, the community of Islām, as Muḥammad had been, but the ruler of all Arabs, commanding their obedience in the name of Islām': Wilferd Madelung, *The Succession to Muḥammad. A Study of the Early Caliphate* (Cambridge: Cambridge University Press, 1997; repr. 2001), 48–9.
14. Kennedy, *The Prophet and the Age of the Caliphates*, 55.
15. Ibid. 56.
16. S. Abdullah Schleifer, '*Jihād* and the Traditional Islamic Consciousness', *Islamic Quarterly*, 4th quarter (1983), 182–3. Web version at: <webdev.webstar.co.uk/salaam/knowledge/schleifer_2.php>
17. Ibid.
18. Elie Adib Salem, *Political Theory and Institutions of the Khawārij* (Baltimore, MD: Johns Hopkins University Press, 1956), 26. They refused to be called *al-māriq* (dissenters) for this reason.
19. Michael Cook, *Commanding Right and Forbidding Wrong in Islamic Thought* (New York: Cambridge University Press, 2000), 393–6.
20. H. A. R. Gibb and J. H. Kramer (eds) *Concise Encyclopedia of Islām* (London: Stacey International, 1966), 248. P. Crone and M. Hinds, *God's Caliph. Religious Authority in the First Centuries of Islam* (Cambridge: Cambridge University Press; repr. 2003), 63.
21. Salem, *Political Theory and Institutions of the Khawārij*, 84: 'I take an oath to declare *jihād* against those, who though they profess Islām, do in fact deviate from the Book and follow their fancies.' Even Mawdūdī, not known for his tolerant views of others (see Chapter 8), described this as an abuse: 'calling others wrong-doers is not merely the violation of the rights of an individual, rather it is also a crime against society. It is an act of injustice against the entire Islamic society, and it does immense harm to the Muslims as a community...' <www.muslim.org/light/96-6.htm>
22. Ibid. 85, correcting Khaddūrī, *War and Peace in the Law of Islām*, 141. and H. M. Zawātī, *Is Jihād a Just War?, War, Peace and Human Rights under Islamic and Public International Law* (Lewiston, NY: Edwin Mellen, 2001), 15.

23. Salem, *Political Theory and Institutions of the Khawārij*, 85. To abstain from this duty constituted *kufr*: ibid. 87.
24. Berkey, *The Formation of Islam*, 87.
25. Lapidus, *A History of Islamic Societies*, 49. Though Kennedy adds that 'not all Khawārij were violent, nor did all embrace the desert life': Kennedy, *The Prophet and the Age of the Caliphates*, 80. The concept of force, however, was fundamental in their doctrine of *jihād*: Salem, *Political Theory and Institutions of the Khawārij*, 82.
26. Salem, *Political Theory and Institutions of the Khawārij*, 60–1.
27. Ibid. 55. For Patricia Crone, their *imām* was 'the most meritorious scholar, just as he was the most meritorious statesman… There was no separation of powers. The Khārijites continued to see the *imām* as a multi-purpose leader presiding over a multi-purpose community. They merely cut him down to size, religious authority, political power and all': Crone, *Medieval Islamic Political Thought*, 59.
28. Salem, *Political Theory and Institutions of the Khawārij*, 84, 90. They rejected any real peace with those whom they considered unbelievers: ibid. 93.
29. Ibid. 27. Ibid. 29, where Salem notes that only two of them survive today, the Ṣufriyya in Oran and the Ibāḍīyya in Morocco. Gibb and Kramers, *Concise Encyclopedia of Islām*, 248. Schleifer calls them 'more in the nature of an explosive chain of sub-sects than a unified movement': Schleifer, '*Jihād* and the Traditional Islamic Consciousness', 180. 'Alī faced the opposition of five Khārijī splinter-groups until 38/February 659: Madelung, *The Succession to Muḥammad*, 295–7.
30. They were prepared to allow a revised *shahādah*, that 'Muḥammad is the Apostle of God to the Arabs but not to us'.
31. Crone comments that 'some suspected the Khārijites of meaning "no government" with their slogan "no judgement except God's"': Crone, *Medieval Islamic Political Thought*, 61.
32. *The Islamic Law of Nations. Shaybānī's Siyar*, trans and ed. Majīd Khaddūrī (Baltimore, MD: Johns Hopkins University Press, 1966), 231 (para. 1372). For 'Alī's sermons against the Khawārij: Sayed Ali Reza (ed.), *Nahj al-Balāgha. Peak of Eloquence. Sermons, Letters and Sayings of Imām 'Alī ibn Abī Ṭālib* (New York: Tahrike Tarsile Qur'ān, Inc., 3rd rev. edn, 1984), 187 (sermon 57), 188 (sermon 59), 235–7 (sermon 92), 369 (sermon 183). Madelung, *The Succession to Muḥammad*, 150, argues that 'Alī's sermons 'tended to alienate many of his lukewarm supporters, but also to arouse the enthusiastic backing and fervour of a minority of pious followers'.
33. Subsequently, this was the call in all the main revolts up to the end of the Umayyad period: Crone and Hinds, *God's Caliph*, 60.
34. Alsumaih, 'The Sunnī Concept of *Jihād* in Classical *Fiqh* and Modern Islamic Thought' (unpublished PhD thesis, University of Newcastle upon Tyne, 1998), 195. Asma Afsaruddin, *Excellence and Precedence. Medieval Islamic Discourse on Legitimate Leadership* (Leiden: E. J. Brill, 2002), 60.
35. Al-Bukhārī 6/61/577. Al-Bukhārī 9/84/67 (another narration of 'Alī); 4/56/807; 4/59/638; 6/61/578; 8/73/184; 9/93/651; 9/93/527 (narrations of Abū Sa'īd al-Khudrī); 9/84/65 (narration of 'Abdullah bin 'Amr bin Yasīr recalling Abu Said al-Khudri); 9/84/66 (narration of 'Abdullah bin 'Umar); 9/84/67 (narration of Abu Sa'id); 9/84/68 (narration of Yusair bin 'Amr). Among both the Sunnīs and the Shī'a, the frequency with which a *ḥadīth* is transmitted serves as a gauge of its reliability: Afsaruddin, *Excellence and Precedence*, 213. For the chain of narrators concerning 'Alī as the first to become a believer after Khadīja: ibid. 209. Asaf A. A. Fyzee (trans), *The Pillars of Islām. Da'ā'im al-Islām of al-Qāḍī al-Nu'mān. I. Acts of Devotion and Religious Observances*, revised Ismail Kurban Husein Poonawala (New Delhi: Oxford University Press, 2002), 481–2.

36. Chase F. Robinson, *Empire and Elites after the Muslim Conquest. The Transformation of Northern Mesopotamia* (Cambridge: Cambridge University Press, 2000), 109–26.
37. Ibid. 124: 'To Khārijite eyes, it was the Umayyads who were innovating, and the innovation lay in the state's shutting down of *hijrah* and *jihād*, a *ḥadīth*-driven programme that took institutional form in the professionalization of caliphal armies, armies in which they no longer had a place. To historians' eyes, these ideological and institutional changes mark nothing less than the state's attempt to monopolize legitimate violence, while the Khārijites' showy raids represent an attempt to demonstrate the tribesmen's continuing right to commit the sacral violence that God had made incumbent upon all Muslims.'
38. Crone, *Medieval Islamic Political Thought*, 23, 55, 58.
39. Robinson, *Empire and Elites after the Muslim Conquest*, 115. The citations are from Q.4:95, 9:46, 9:83, 9:20 and 4:100.
40. Ibid. 116.
41. Schleifer, '*Jihād* and the Traditional Islamic Consciousness', 180.
42. Crone and Hinds, *God's Caliph*, 41.
43. Ibid. 42, 97. The significance of the date is the abolition of the Mu'tazilite inquisition (*miḥna*).
44. There was heavy criticism of his nepotism, which required him to make a public statement of repentance: Madelung, *The Succession to Muḥammad*, 122.
45. Crone and Hinds, *God's Caliph*, 57, 129–32. Though Abū Ḥamza allowed for the 'good intentions' of 'Umar II: ibid. 74.
46. Berkey, *The Formation of Islam*, 87–8. Hamidullah, *Muslim Conduct of the State*, 81 (para. 138). The Prophet's son Ibrāhīm did not survive, according to statements ascribed to several Companions, because he would have been seen as a Prophet, and thus Muḥammad would not have been the 'last' or 'Seal' of the Prophets: Madelung, *The Succession to Muḥammad*, 17.
47. For Crone and Hinds, 'the classical view that 'Alī was the fourth ['rightly guided'] caliph reflects doctrinal developments of the ninth century [CE], not contemporary opinion: in contemporary perspective, 'Alī was a pretender, on a par with the other protagonists of the first civil war': Crone and Hinds, *God's Caliph*, 32.
48. Though the motivation for the assassination of 'Umar does not seem to have been political: Kennedy, *The Prophet and the Age of the Caliphates*, 69.
49. Berkey, *The Formation of Islam*, 70. 'Umar prevented the Prophet in his last illness from writing a document, perhaps naming 'Alī as his successor: Madelung, *The Succession to Muḥammad*, 24. 'Umar, allegedly, was worried about the Banū Hāshim arrogating the caliphate to themselves and depriving the 'people' (that is, the Quraysh), of their collective right to it: ibid. 29. Abū Bakr stressed that the Quraysh were 'the most central [= noble] of the Arabs in lineage and abode': ibid. 31.
50. Afsaruddin, *Excellence and Precedence*, 168, to which he had added, 'however, let Abū Bakr lead you in prayer'.
51. Ibid. 160–1.
52. Ibid. 172, 185, 222. In contrast, Wilferd Madelung asserts: 'in the eyes of Muḥammad, the leadership of the prayer had no significance for the succession. He did not care whether Abu Bākr or 'Umar performed the task. When Abū Bakr still hesitated, the Prophet rudely grasped him by his clothes, pushing him into his place…': Madelung, *The Succession to Muḥammad*, 25.
53. Afsaruddin, *Excellence and Precedence*, 54.
54. Ibid. 57, 66.
55. Ibid. 152–3, 166–7. Contrary view: ibid. 159.
56. Ibid. 219.
57. Ibid. 131.

58. Ibid. 131–2.
59. Ibid. 115. The Shī'a contention is that this was a coup, while 'Alī was concerned with the funeral arrangements for the Prophet, and that 'Alī initially refused the oath of allegiance to Abū Bakr: <www.al-islam.org/imamate/3.htm>

    It was only later that 'Alī publicly submitted to Abū Bakr: Madelung, *The Succession to Muḥammad*, 53.
60. Ibid. 31–7.
61. M. A. Shaban, *Islamic History, A.D. 600–750 (A.H. 132). A New Interpretation* (Cambridge: Cambridge University Press, 1971), 19. Abū Bakr stated that he was 'not the caliph of God, but caliph of the Prophet of God': A. K. S. Lambton, *State and Government in Medieval Islam. An Introduction to the Study of Islamic Political Theory: The Jurists* (New York:Oxford University Press, 1981), 87. After the first two 'rightly-guided' caliphs, the title *Khalīfat Allāh*, 'deputy of God' (not 'God's successor', since God is still alive), was adopted after 23/644: Crone and Hinds, *God's Caliph*, 4, 11, 21–2. Crone, *Medieval Islamic Political Thought*, 195 n. 113.
62. Madelung, *The Succession to Muḥammad*, 55: 'since Abū Bakr did not view the caliphate as an elective office, it was only natural that he appointed, without prior consultation, his successor, 'Umar bin al-Khaṭṭāb'. Ibid. 56: 'Abū Bakr owed him a considerable debt. 'Umar had made the coup at the Saqīfa in his favour possible…'
63. John Alden Williams (ed.), *Themes of Islamic Civilization* (Berkeley, CA, and London: University of California Press, 1971), 262.
64. Hamidullah, *Muslim Conduct of the State*, 300–1 (para. 649).
65. Account of al-Balādhurī (d. *c.* 279/892) in Williams, *Themes of Islamic Civilization*, 263–6.
66. Shaban, *Islamic History, A.D. 600–750*, 29.
67. Khaddūrī, *War and Peace in the Law of Islām*, 134.
68. Shaban, *Islamic History, A.D. 600–750*, 166.
69. Ibid. 80. Kennedy, *The Prophet and the Age of the Caliphates*, 80.
70. Khaddūrī, *War and Peace in the Law of Islām*, 62, for whom 'the importance of the *jihād* in Islām lay in shifting the focus of attention of the tribes from their inter-tribal warfare to the outside world…'
71. Ibid. 89.
72. Khaddūrī, *The Islamic Law of Nations. Shaybānī's Siyar*, 107 (para. 157). Hugh Kennedy, *The Armies of the Caliphs. Military and Society in the Early Islamic State* (London and New York: Routledge, 2001), 10, 51.
73. Hamidullah, *Muslim Conduct of the State*, 302 (para. 652).
74. Kennedy, *The Prophet and the Age of the Caliphates*, 68–9.
75. For differences between Mālik and al-Qarāfī as to whether Egypt could be considered 'conquered' and forcibly despoiled: Jackson, *Islamic Law and the State*, 126.
76. Kennedy, *The Armies of the Caliphs*, 2, 6.
77. Hamidullah, *Muslim Conduct of the State*, 36 (para. 71).
78. Lapidus, *A History of Islamic Societies*, 34.
79. 'Alī, the fourth rightly guided caliph, regarded them as people of the book, although their record had disappeared: Fyzee, *The Pillars of Islām*, 469–70. M. J. Kister, 'Social and Religious Concepts of Authority in Islam', *Jerusalem Studies in Arabic and Islam*, 18 (1994), 88–9, repr. in M. J. Kister, *Concepts and Ideas at the Dawn of Islam* (Aldershot: Ashgate, Variorum, 1997), ch. 5. For the complexity of early Muslim attitudes to Zoroastrians: Yohanan Friedmann, *Tolerance and Coercion in Islam. Inter-faith Relations in the Muslim Tradition* (Cambridge: Cambridge University Press, 2003), 72–6, esp. 75: 'the Zoroastrians are People of a Book other than the Tawrāt [the Torah] and the Injīl [the New Testament]. They forgot their book and corrupted it. [Nevertheless,] the Messenger of God allowed to take *jizya* from them.'

80. Berkey, *The Formation of Islam*, 100–1.
81. Kennedy, *The Prophet and the Age of the Caliphates*, 63.
82. Robinson, *Empire and Elites after the Muslim Conquest*, 166: 'the exclusivity and insularity of the ruling élite determined a great deal of first-century history. The state apparatus remained in Arabia and Arabized Syria, while outside it social boundaries were reinforced and institutionalized…'
83. Shaban, *Islamic History, A.D. 600–750*, 169.
84. Morabia, *Le Ĝihâd dans l'Islam médiéval*, 171. B. Lewis, *The Crisis of Islam. Holy War and Unholy Terror* (London: Weidenfeld and Nicolson, 2003), xxv. Khaddūrī, *War and Peace in the Law of Islām*, 160. Friedmann, *Tolerance and Coercion in Islam*, 91. Ibid. 93: 'non-Muslim communities living under Islam experienced far less expulsions and persecutions than Jews or "deviant" Christians, living under medieval Christendom'.
85. Fyzee, *The Pillars of Islām*, 460. Both prayers are on the testimony of the fourth rightly-guided caliph, 'Alī.
86. Berkey, *The Formation of Islam*, 73.
87. Ibid.
88. Fyzee, *The Pillars of Islām*, 458 and 458 n. 17.
89. Kennedy, *The Armies of the Caliphs*, 9.
90. Fyzee, *The Pillars of Islām*, 457.
91. Morabia, *Le Ĝihâd dans l'Islam médiéval*, 185. Ibid. 184: 'La théorie n'a pas determiné l'action des conquérants. C'est plutôt celle-ci qui a imprimé sa marque à la théorie, à tout le moins pour le premier siècle de l'Islam.' Similar conclusion on the first treatise in Alsumaih, 'The Sunnī Concept of *Jihād*', 3–4.
92. Kister, 'Social and Religious Concepts of Authority in Islam', ch. 5.
93. Muḥammad Qāsim Zamān, *Religion and Politics under the Early 'Abbāsids: The Emergence of the Proto-Sunni Elite* (Leiden: E. J. Brill, 1997), 164.
94. Sohail H. Hāshmī, 'Interpreting the Islamic Ethics of War and Peace', in S. H. Hāshmī (ed.) *Islamic Political Ethics. Civil Society, Pluralism and Conflict* (Princeton, NJ: Princeton University Press, 2002), 205.
95. H. A. R. Gibb, *Studies on the Civilization of Islam*, ed. S. J. Shaw and W. R. Polk (London: Routledge, 1962), 162. Quoted by Lambton, *State and Government in Medieval Islam*, 84. Crone, *Medieval Islamic Political Thought*, 126, 219: 'the Sunnīs have their roots in, and derive their name from, the partisans of *ḥadīth* who came to prominence in the ninth century under the name of *ahl al-sunnah wa'l-jamā'a* [roughly, "adherents of right practice and communal solidarity"]'.
96. Cf. Taqiuddin An-Nabhani, *The Islamic State* (New Delhi: Milli Publications, 2001), 150: '…the carrying of the Message of Islām was the basis on which the Islamic State was founded and for which the Muslim army had been prepared. *Jihād* was decreed and this was the method followed in the conquering of other countries…' Hamidullah, *Muslim Conduct of the State*, 191 (para. 378): '*Jihād* is to be waged solely for the purpose that "the Word of God shall alone prevail".' Crone, *Medieval Islamic Political Thought*, 364–8, distinguishes between 'the earliest concept of holy war' and 'the classical concept of *jihād*'.
97. Reza, *Nahj al-Balāgha*, 302 (sermon 145).
98. The text of Abū Yūsuf on 'Umar's rationale of the division of the lands is reproduced by Bat Ye'or, *The Dhimmī Jews and Christians under Islam* (Cranbury, NJ: Associated University Press, rev. edn 1985), 165: 'do you not think that these vast countries… do not have to be covered with troops who must be well paid? Where can one obtain their pay if the land is divided up, as well as its inhabitants?'

99. Aḥmad Ḥasan, *The Early Development of Islamic Jurisprudence* (Islamabad: Islamic Research Institute, 1970), 119–20. Khaddūrī, *The Islamic Law of Nations. Shaybānī's Siyar*, 99–100 (para. 91).
100. Khaddūrī, *The Islamic Law of Nations. Shaybānī's Siyar*, 269 (para. 1684).
101. Ḥasan, *The Early Development of Islamic Jurisprudence,* 146. Hamidullah, *Muslim Conduct of the State*, 26 (para. 45). *The Islamic Law of Nations. Shaybānī's Siyar*, 25, who notes that the text of Shaybānī 'is essentially an exposition of Abū Ḥanīfa's system of the *siyar*…' Khaddūrī states that Abū Ḥanīfa 'was perhaps the first to develop a set of principles governing Islām's external relations with other communities as well as a coherent system of relationship between the Islamic and non-Islamic communities'.
102. Hossein Modarressi Tabātabā'ī, *Kharāj in Islamic Law* (London: Anchor Press, 1983), 45–6, 78–9.
103. Madelung, *The Succession to Muḥammad*, 73–4: 'the great conquests outside Arabia had turned the mass of the Arabs, deprived of their former freedom and reduced to tax-paying subjects by [the] Qurayshīs during the *riddah*, into a military caste sustained by a numerically much larger non-Arab and non-Muslim subject population. It may be questioned whether the caliphate of [the] Qurayshīs would have lasted very long without this imperial expansion… The successful diversion of all energy into vast military conquests, in the name of Islām, kept any longing for a restoration of the past at bay… The Arab warriors (*muqātilah*) were subject to strict, sometimes brutal, military discipline. But in return they were provided with generous stipends and pensions apart from their share in the booty gained in battle. They thus had a stake in the imperial policies of [the] Qurayshīs…' Ibid. 77: 'the domination of Arabs over non-Arabs on an ethnic basis was also in essential conflict with the universal call of Islām. This, however, became patent only in the later Umayyad age when masses of non-Arabs converted to Islām and loudly demanded equality in its name.'
104. Ibid. 68.
105. Ibid. 80.
106. Though this policy was not fully realized until the caliphate of Mu'āwiya: ibid. 85.
107. Khaled Abou El Fadlh, *Islam and the Challenge of Democracy. A Boston Review Book*, ed. Joshua Cohen and Deborah Chasman (Princeton, NJ, and Oxford: Princeton University Press, 2004), 17.
108. Madelung, *The Succession to Muḥammad*, 78.
109. Ibid. 113.
110. Ibid. 134.
111. '"Ever Since the Murder of 'Uthmān [the third caliph]" – Arab Literary Scholar on the Evil Spirit of Murder and Violence in Early Islam Re-Appearing Today': Memri Special Dispatch 704, 30 April 2004: 'they murdered him, and their murder of him was tantamount to the crushing of the symbol of consensus in the [Islamic] nation, and the violation of its sanctity, and the tearing to shreds of the garment of awe and reverence without which the ruled cannot be pleased with the ruler. This garment is woven spontaneously, by free nations, of their own choice, and they bestow it upon an individual whom they choose from amongst them, so that this individual will, despite his shortcomings, become a symbol of their collective will. [This individual] becomes an idea greater than his limited personal capabilities. When the nation is pleased with him, it is in fact pleased with itself.' <www.memri.org/bin/opener_latest.cgi?ID=SD70404>
112. Madelung, *The Succession to Muḥammad*, 141. His irregular election left the community divided into three factions: ibid. 146–7.
113. Kennedy, *The Prophet and the Age of the Caliphates*, 70, 77. Kennedy, *The Armies of the Caliphs*, 74–5. Shaban, *Islamic History, A.D. 600–750*, 72. Reza, *Nahj al-Balāgha* (sermon 204). The jurist Ibn Ḥanbal was instrumental in the rehabilitaton of 'Alī and Sunnī recognition

of him as the fourth 'rightly-guided' caliph: Zamān, *Religion and Politics under the Early 'Abbāsids*, 169.

114. Madelung, *The Succession to Muḥammad*, 150–1.
115. Ibid. 276.
116. Ibid. 313.
117. Reza, *Nahj al-Balāgha*, 153–4 (sermon 27).
118. Ibid. 156–7 (sermon 29).
119. Ibid. 166–7 (sermon 34).
120. Ibid. 174 (sermon 39).
121. Ibid. 185 (sermon 55).
122. Ibid. 266–7 (sermon 118).
123. Ibid. 271–2 (sermon 123).
124. Ibid. 359–60 (sermon 181).
125. Madelung, *The Succession to Muḥammad*, 309.
126. Ibid. 335. Ibid. 321: it was said that he 'entered Islām under duress, stayed in it out of fear, and left it voluntarily without faith preceding…'
127. Ibid. 326.
128. Ibid. 334. Marwān, the architect of Umayyad dynastic rule, stated 'our reign would not be sound without that', that is, the cursing of 'Alī from the pulpits.
129. Kennedy, *The Prophet and the Age of the Caliphates*, 81, 84.
130. Shaban, *Islamic History, A.D. 600–750*, 80.
131. Kennedy, *The Armies of the Caliphs*, 18, 30–1, 49.
132. Shaban, *Islamic History, A.D. 600–750*, 155.
133. Kennedy, *The Armies of the Caliphs*, 78.
134. D. J. Wasserstein, *The Caliphate in the West. An Islamic Political Institution in the Iberian Peninsula* (Oxford: Clarendon Press, 1993), 80–1.
135. Ibid. 111, 162–3. In 483/1091 the Almoravids took Cordoba, Almería, Badajoz and Seville, sending the Sevillan king al-Mutamid into exile. Their advance along the east coast was only impeded by El Cid in Valencia.
136. Kennedy, *The Armies of the Caliphs*, 87–8.
137. R. Firestone, *Jihād. The Origin of Holy War in Islam* (Oxford: Oxford University Press, 1999), 61.
138. Majīd Khaddūrī (trans.) *Islamic Jurisprudence. Shāfi'ī's Risāla* (Baltimore, MD: Johns Hopkins University Press, 1961), 84. Hasan, *The Early Development of Islamic Jurisprudence*, 39.
139. Khaddūrī, *Islamic Jurisprudence. Shāfi'ī's Risāla*, 86–7. Hasan, *The Early Development of Islamic Jurisprudence*, 57 n. 36.
140. A. Ben Shamesh (ed.), *Taxation in Islām. Yaḥyā Ben Adam's Kitāb al Kharāj* (3 vols, Leiden: E. J. Brill, 1958–69), i. 17.
141. Tabātabā'ī, *Kharāj in Islamic Law*, 82–3.
142. Lambton, *State and Government in Medieval Islam*, 55.
143. Ibid. 56–7. Zamān, *Religion and Politics under the Early 'Abbāsids*, 95–9.
144. Hasan, *The Early Development of Islamic Jurisprudence*, 104.
145. Zamān, *Religion and Politics under the Early 'Abbāsids*, 213.
146. Ibid. 148.
147. Hasan, *The Early Development of Islamic Jurisprudence*, 98–100.
148. Kister, 'Social and Religious Concepts of Authority in Islam', 99.
149. A. A. at-Tarjumana and Y. Johnson (trans), *Al-Muwaṭṭa'* (Norwich: Diwan Press, 1982), 197–206. <www.iiu.edu.my/deed/hadith/malik/021_mmt.html>
150. Ibid. 203. Mālik 21/14/29.
151. Hamidullah, *Muslim Conduct of the State*, 26 (para. 46).

152. For the following quotation: Fyzee, *The Pillars of Islām*, 457. For the fundamental similarity, in spite of special Shī'a features: Morabia, *Le Ĝihâd dans l'Islam médiéval*, 195–6. For the general principle: Hamidullah, *Muslim Conduct of the State*, 192 (para. 380).
153. Ibid. 426: Bedouins who do not engage in *jihād* do not have a share in the booty.
154. M. S. bin Jani, 'Sayyid Quṭb's View of *Jihād:* An Analytical Study of his Major Works' (unpublished PhD thesis, University of Birmingham, 1998), 338.
155. Ibid. 109–11.
156. Ibid. 128 n. 76.
157. For Khaddūrī, 'the Ḥijāzī jurists, somewhat remote from the areas in which Muslims and non-Muslims came into direct contact, paid little or no attention to the questions arising from the encounters between Islām and other communities'. Khaddūrī, *The Islamic Law of Nations. Shaybānī's Siyar*, 23. He also asserts (ibid. 16–17) that 'there was no essential difference among leading jurists [about *jihād*]... whether in orthodox or heterodox doctrine'. However, AbūSulaymān notes that Khaddūrī is 'overly selective in [his] choice of interpretations of some jurists while neglecting others': 'AbdulḤamīd A. AbūSulaymān, *Towards an Islamic Theory of International Relations: New Directions for Methodology and Thought* (Herndon, VA: International Institute of Islamic Thought, 1993), 20. Ibid. 22: Khaddūrī relied 'basically on one juristic opinion, that of al-Shāfi'ī'. [We would add al-Shaybānī.] Alsumaih, 'The Sunnī Concept of *Jihād*', 4, relies chiefly on al-Shaybānī and al-Mawārdī. The full classical legal sources for *jihād* are described by Morabia, *Le Ĝihâd dans l'Islam médiéval*, 185–94.
158. Bin Jani, 'Sayyid Quṭb's View of *Jihād*', 111.
159. Ibid. 116. Friedmann, *Tolerance and Coercion in Islam*, 102–3: 'both verses that are said to have abrogated [Q.2:256] speak about *jihād*. It can be inferred from this that the commentators who consider [Q.2:256] as abrogated perceive *jihād* as contradicting the idea of religious freedom. While it is true that religious differences are mentioned in both [Q.9:29 and Q.9:73] as the reason because of which the Muslims were commanded to wage war, none of them envisages the forcible conversion of the vanquished enemy. [Q.9:29] defines the purpose of the war as the imposition of the *jizya* on the People of the Book and their humiliation, while [Q.9:73] speaks only about the punishment awaiting the infidels and the hypocrites in the hereafter, and leaves the earthly purpose of the war undefined. *Jihād* and religious freedom are not mutually exclusive by necessity: religious freedom could be granted to the non-Muslims after their defeat, and commentators who maintain that [Q.2:256] was not abrogated freely avail themselves of this exegetical possibility with regard to the Jews, the Christians and the Zoroastrians.'
160. Bin Jani, 'Sayyid Quṭb's View of *Jihād*', 114.
161. Ibid. 121–2.
162. AbūSulaymān, *Towards an Islamic Theory of International Relations*, 20 and 20 n. 6.
163. Ibid. 20–3.
164. Fyzee, *The Pillars of Islām*, v.
165. Ibid. 424–5.
166. Ibid. 423.
167. Ibid. 426.
168. Ibid. 427. Also ibid. 459, 'for those who are engaged in *jihād*, the best course is never to be without arms under any circumstances'.
169. Ibid. 482.
170. Ibid. 477.
171. Osman, 'The Jurisprudence of Sarakhsī'.
172. Ibid. 108. Caliph 'Umar Ibn 'Abd-al 'Aziz (d. 101/720) was, exceptionally, also regarded as *de facto* 'rightly guided': ibid. 248.
173. Ibid. 121.

174. Ibid. 129.
175. Ibid. 234.
176. Ibid. 203.
177. Ibid. 204.
178. Ibid. 247.
179. Alsumaih, 'The Sunnī Concept of *Jihād*', 50–1. Zawātī, *Is Jihād a Just War?*, 14.
180. Zawātī, *Is Jihād a Just War?*, 36. Cf. Crone, *Medieval Islamic Political Thought*, 109: 'the unpalatable truth was that the price of civilization was submission to tyrants. The alternative was tribalism, with or without religious beautification.'
181. Cook, *Commanding Right and Forbidding Wrong in Islamic Thought*, 346. Lambton, *State and Government in Medieval Islam*, 104–6. Bin Jani, 'Sayyid Quṭb's View of *Jihād*', 100.
182. Cook, *Commanding Right and Forbidding Wrong in Islamic Thought*, 390, 478, 496, 511. Bin Jani, 'Sayyid Quṭb's View of *Jihād*', 101–4.
183. Cook, *Commanding Right and Forbidding Wrong in Islamic Thought*, 478.
184. Ibid. 226.
185. Ibn Taymīyah, *Public Duties in Islām. The Institution of the Ḥisba*, trans. Muhtar Holland (Leicester: Islamic Foundation, 1982), 80.
186. Al-Māwardī is considered the author/early advocate of the doctrine of 'necessity' in political science and was in favour of a strong caliphate with only limited powers delegated to the regional governors. He laid down clear principles for the election of the caliph and qualities of the voters, chief among which are attainment of a degree of intellectual level and purity of character: Abū al-Ḥaṣan al- Māwardī, *Oxford Dictionary of Islam*, ed. John L. Esposito (Oxford: Oxford University Press 2003). <www.oxfordreference.com/views/ENTRY.html?subview=Main&entry=t125.001474> Qamaruddin Khan, 'Al-Māwardī', in *A History of Muslim Philosophy*, ed. M. M. Sharif (Karachi: Royal Book Company, 1963–66; repr. 1983) ch. 36, 727. <www.muslimphilosophy.com/hmp/default.htm> Qamaruddin Khan, *Al-Māwardī's Theory of the State* (New Delhi: Idarah-i-Adabiyat-i-Delli, repr. 1979). E. I. J. Rosenthal, *Political Thought in Medieval Islam. An Introductory Outline* (Cambridge: Cambridge University Press, 1958), 27–37.
187. Lambton, *State and Government in Medieval Islam*, 92.
188. Hamidullah, *Muslim Conduct of the State*, 129 (para. 239): 'we have seen above [110–11, para. 199] that all Muslims belong to one and the same nation. We have also seen that the division of Islām into several states, hostile at times, had to be admitted by jurists by force of facts [86–8, paras 148–150]'.
189. Lambton, *State and Government in Medieval Islam*, 78. Al-Baghdādi had stressed the requirement of outward probity of character and conformity with the *sharī'ah*. He did not discuss the deposition of an unjust *imām*: ibid. 80. Gibb, *Studies on the Civilization of Islam*, 157.
190. M. A. Shaban, *Islamic History. A New Interpretation, Part 2: AD 750–1055 (A.H. 132–448)* (Cambridge: Cambridge University Press), 159–87 ('the Būyid Confederacy'), especially 160. Lambton, *State and Government in Medieval Islam*, 92. Gibb, *Studies on the Civilization of Islam*, 159. Hamidullah, *Muslim Conduct of the State*, 88 (para. 151): 'during the decadence of the 'Abbāsid Empire, its provincial governors became hereditary and virtually independent. They could wage war, make peace or conclude other treaties, without reference to the Caliph, and administer all their internal as well as external affairs at their own will…'
191. Farouk Mitha, *Al-Ghazālī and the Ismailis. A Debate on Reason and Authority in Medieval Islam* (London: I. B. Tauris, 2001), 79–80. *Ghazālī's Book of Counsel for Kings (Naṣīḥat al-Mulūk)*, trans. F. R. C. Bagley (London: Oxford University Press, 1964), lv–lvi (since this latter work may not have been written by al-Ghazālī, it does not receive separate attention).
192. Lambton, *State and Government in Medieval Islam*, 98–102.

193. Gibb, *Studies on the Civilization of Islam*, 162. For the compact theory of concordats: <www.newadvent.org/cathen/04196a.htm>
194. This work was called *The Book of Perfecting the Distinction between Legal Responsa, Judicial Decisions and the Discretionary Actions of Judges and Caliphs*: Jackson, *Islamic Law and the State*, xix.
195. Ibid. 215–16. Cf. ibid. 198 n. 32. Ibid. 133: 'this is a novel use of the term' *fatwā*, in Jackson's view 'unique to al-Qarāfī'.
196. *Jihād in Mediaeval and Modern Islam. The Chapter on Jihād from Averroës' Legal Handbook 'Bidāyat al-Mudjtahid' and the Treatise 'Koran and Fighting' by the late Shaykh al-Azhar, Maḥmūd Shaltūt*, trans. and ed. R. Peters (Leiden: E. J. Brill, 1977), 9–25, 80–4. Another translation is *The Distinguished Jurist's Primer. I. Bidāyat al-Mudjtahid. Ibn Rushd*, trans. Imran Ahsan Khan Nyazee (Reading: Centre for Muslim Contribution to Civilization: Reading, 1994, repr. 2000), 454–87, which includes material which Peters omits.
197. Peters, *Jihād in Mediaeval and Modern Islam*, 17.
198. AbūSulaymān, *Towards an Islamic Theory of International Relations*, 23. Nyazee, *The Distinguished Jurist's Primer. I. Bidāyat al-Mudjtahid. Ibn Rushd*, 463–4.
199. Nyazee, *The Distinguished Jurist's Primer. I. Bidāyat al-Mudjtahid. Ibn Rushd*, 475.
200. Ibid. 479.
201. Ibid. 480–1.
202. Jackson, *Islamic Law and the State*, xxvi.
203. Morabia, *Le Ĝihâd dans l'Islam médiéval*, 192–3, 195. Rosenthal, *Political Thought in Medieval Islam*, 203.
204. Rosenthal, *Political Thought in Medieval Islam*, 196.
205. The most recent study of his philosophy, which does not deal with his views of war, is that of Miriam Galston, *Politics and Excellence. The Political Philosophy of Alfarabi* (Princeton, NJ: Princeton University Press, 1990).
206. Sayf al-Dawla's (r. 332/944–356/967) preoccupation with *jihād* is well attested: Rosenthal, *Political Thought in Medieval Islam*, 134, 139.
207. Ibid. 131.
208. Majīd Khaddūrī, *The Islamic Conception of Justice* (Baltimore, MD, and London: Johns Hopkins University Press, 1984), 172.
209. Ibid. 173. Khaddūrī, *War and Peace in the Law of Islām*, 70–2. Zawātī, *Is Jihād a Just War?*, 107.
210. Friedmann, *Tolerance and Coercion in Islam*, 85–6. Friedmann's second chapter, on the classification of unbelievers, is of fundamental importance, as is the discussion in ch. 3 on whether or not there is compulsion in religion.
211. The *locus classicus* is Ye'or, *The Dhimmī*, and the author's website: <www.dhimmi.org> An idea of the author's ideological viewpoint is gained from her comment that 'human rights and the concept of *jihād* are two incompatible ideas'. '*Jihād* and Human Rights Today' (National Review Online, 1 July 2002): <www.nationalreview.com/comment/comment-yeor070102.asp>
212. Khaddūrī, *The Islamic Law of Nations. Shaybānī's Siyar*, 275 (para. 1702, thought to be the levy in 'Irāq). Ye'or, *The Dhimmī*, 181, 183 (Ghāzī b al-Wāsiṭi and Ibn Naqqāsh citing 'Umar's ruling). D. Goitein, 'Evidence on the Muslim Poll Tax, from Non-Muslim Sources: A Geniza Study', *Journal of the Economic and Social History of the Orient*, 6 (1963), 278–95 at 286–8, finds evidence of rates of 4 and $^{1}/_{6}$ *dīnārs* as the highest rate, 2 and $^{1}/_{12}$ as the medium rate and 1 and $^{5}/_{8}$ as the lowest rate.
213. Nyazee, *The Distinguished Jurist's Primer. I. Bidāyat al-Mudjtahid. Ibn Rushd*, 484–5. Khaddūrī, *War and Peace in the Law of Islām*, 193. Morabia, *Le Ĝihâd dans l'Islam médiéval*,

273. Cf. the comment that one and a quarter *dirhams* a day was 'really not a salary' (*c.* 1140–59): Goitein, 'Evidence on the Muslim Poll Tax', 286.

214. Khaddūrī, *The Islamic Law of Nations. Shaybānī's Siyar*, 124 (para. 316).

215. During the blockade of Damascus in 643/1246, a house worth 10,000 *dirhams* sold for only 1500 *dirhams*, but this was 'enough to buy precisely one sack of wheat': R. Stephen Humphreys, *From Saladin to the Mongols. The Ayyubids of Damascus, 1193–1260* (Albany, NY: State University of New York Press, 1977), 285.

216. Goitein, 'Evidence on the Muslim Poll Tax', 295, suggests that, because of abuses in the levy, the burden on the lower classes was 'intolerable' and may have prompted mass conversions to Islām. An equivalent to 6 per cent of capital as an annual tax payment, a high rate cited by one critic of the system, is not attested in the sources: <www.hindunet.org/alt_hindu/1994/msg00746.html>

217. Aurangzab's *fatwā* is reproduced at: <www.sscnet.ucla.edu/southasia/History/Mughals/Aurnag_fatwa.html>

218. The efficiency of Fāṭimid minting and the debasement of gold coins struck at Baghdād during the Būyid period are among the points made by Andrew S. Ehrenkreutz, 'Studies in the Monetary History of the Near East in the Middle Ages. II. The Standard of Fineness of Western and Eastern *Dīnārs* Before the Crusades', *Journal of the Economic and Social History of the Orient*, 6 (1963), 243–77.

219. Baron Charles de Secondat Montesquieu, *The Spirit of the Laws*, trans. Anne M. Cohler, Basia C. Miller and Harold Stone, Cambridge Texts in the History of Political Thought (Cambridge: Cambridge University Press, 1989) bk 13 ch. 16: 'It was this excess of taxes that occasioned the prodigious facility with which the Mahometans carried on their conquests. Instead of a continual series of extortions devised by the subtle avarice of the Greek emperors, the people were subjected to a simple tribute which was paid and collected with ease. Thus they were far happier in obeying a barbarous nation than a corrupt government, in which they suffered every inconvenience of lost liberty, with all the horror of present slavery.' Contrast Ibn Naqqāsh (d. 763/1362), cited by Ye'or, *The Dhimmī*, 185, that 'it is proper for the *imām* to show his zeal for the faith by increasing the sum of the *jizya*…'

220. However, in *The Message of the Qur'ān* (Gibraltar: Dar Al-Andralus, 1980), 262 n. 43, the very reliable Muḥammad Asad comments: 'from all available Traditions it is evident that it is considerably lower than the tax called *zakāh* ("the purifying dues") to which Muslims are liable…'

221. Raymond W. Goldsmith, *Premodern Financial Systems. A Historical Comparative Study* (Cambridge: Cambridge University Press, 1987), 78. Ibid. table 5–2 at 77, suggests variations from 2 *dirhams* (Libya) to 4 (Arabia), 5 (Sind and Tunisia including part of Algeria), 8 (Syria), 11 (Egypt), 15 (Afghanistan and Turkestan), 17 (caliphate = Iraq), 34 (Iran) and 37 (Mesopotamia). Whether any real reliance can be placed on these figures is a matter of conjecture.

222. After the initial conquest, revenue from the *jizya* amounted to 12 million *dirhams* from Egypt alone. During the caliphate of Mu'āwiya, ten years later or ao, the amount fell to 5 million; and in the caliphate of 'Umar ibn 'Abdul 'Azīz (62/682) the amount dwindled to nothing, causing the governor to seek authority to raise new taxes to cover the expenses of his administration: <www.masnet.org/history.asp?id=422>

223. *Al-Muwaṭṭa'* 17/24/46. Livestock was accepted as payment: ibid. 17/24/45.

224. Abū Yūsuf commented: 'no-one of the people of the *dhimma* should be beaten in order to extract payment of the *jizya*, nor made to stand in the hot sun, nor should hateful things be inflicted upon their bodies, or anything of that sort. Rather, they should be treated with leniency…' Khadduri, *War and Peace in the Law of Islām*, 196. Ye'or, *The Dhimmi*, 168.

225. Khaddūrī, *War and Peace in the Law of Islām*, 189, 191.

226. Williams, *Themes of Islamic Civilization*, 265–6.
227. <www.ummah.org.uk/what-is-islam/war/war6.htm> Cf. Abū Yūsuf's dictum that 'their lives and possessions are guaranteed [in] safety only upon payment of the *jizya*, which is comparable to tribute money'.
228. Khadduri̅, *War and Peace in the Law of Islām*, 197–8. Ye'or, *The Dhimmī*, 169, quoting Abū Yūsuf.
229. Ye'or, *The Dhimmī*, 201–2, for later traditions regarding the degrading conditions for the payment, which were designed to make the *dhimmī* consider the option of conversion.
230. Kemal H. Karpat, *An Inquiry into the Social Foundations of Nationalism in the Ottoman State: From Social Estates to Classes, From Millets to Nations* (Princeton, NJ: Princeton University Press, 1973), 32, quoting Sir Harry Luke, writing in 1936.
231. Ibid. 32–3. Kemal H. Karpat, 'Ottoman Migration: Ethnopolitics and the Formation of Nation-States', in *The Great Ottoman Turkish Civilization. I. Politics*, ed. Kemal Çiçek (Ankara: Yeni Türkiye, 2000), 382–98 at 384.
232. Thomas Ambrosio, 'Ottoman "Hegemonic Control" in the Balkans…' (1997): <www.ndsu.nodak.edu/ndsu/ambrosio/hegemony.html>
233. Khaddūrī, *War and Peace in the Law of Islām*, 198.
234. A Jewish author writing shortly afterwards found that 'the most generally accepted estimate is 50,000 families, or, as others say, 53,000', that is, about 250,000 persons: <www.fordham.edu/halsall/jewish/1492-jews-spain1.html> Henry Kamen revises the figure downwards substantially to 100,000 for the population, of whom about a half went abroad; but his views have not found universal acceptance: H. Kamen, 'The Mediterranean and the Expulsion of Spanish Jews in 1492', *Past and Present*, 119 (1988), 30–55.
235. N. J. Housley, *The Later Crusades, 1274–1580: From Lyons to Alcazar* (Oxford: Oxford University Press, 1992).
236. The Turkish governor of Baṣra in 519/1126 and Mosul in 521/1127 and tutor to the sultan's two sons. For his campaign: P. Partner, *God of Battles. Holy Wars of Christianity and Islam* (London: HarperColins, 1997), 92. Morabia, *Le Ĝihâd dans l'Islam médiéval*, 106, calls the sermons of Ibn Nubāta the Syrian, court preacher to Sayf al-Dawla's '*jihāds* of the sword', a precedent for this.
237. Gibb, *Studies on the Civilization of Islam*, 96.
238. Malcolm C. Lyons and David E. P. Jackson, *Saladin. The Politics of the Holy War* (Cambridge: Cambridge University Press, 1982, repr. 1984), 45, 47, 228.
239. Humphreys, *From Saladin to the Mongols*, 27.
240. Gibb, *Studies on the Civilization of Islam*, 96. P. M. Holt, A. K. S. Lambton and B. Lewis (eds) *Cambridge History of Islām* (Cambridge: Cambridge University Press, 1970), i. 203. Lyons and Jackson, *Saladin*, 48–9.
241. Humphreys, *From Saladin to the Mongols*, 23.
242. Ibid. 32, 39, 53.
243. Lyons and Jackson, *Saladin*, 112.
244. Humphreys, *From Saladin to the Mongols*, 125. *Cambridge History of Islām*, i. 204.
245. Jonathan Riley-Smith, 'Islam and the Crusades in History and Imagination, 8 November 1898–11 September 2001', *Crusades*, 2 (2003), 152.
246. The expression is that of Humphreys, *From Saladin to the Mongols*, 28.
247. Ibid. 85. Saladin regarded Egypt as a special case: ibid. 74.
248. Ibid. 36. Lyons and Jackson, *Saladin*, 371.
249. Partner, *God of Battles*, 94.
250. Lyons and Jackson, *Saladin*, 156, 370–1.
251. Chase F. Robinson, *Islamic Historiography* (Cambridge: Cambridge University Press, 2003), 122.

252. Lyons and Jackson, *Saladin*, 280–1.
253. Ibid. 155, 163, 194, 370.
254. Ibid. 360. Cf. the delay before Aleppo: ibid. 201. Twelve years 'chiefly against Muslim adversaries, with only an occasional skirmish with the Franks': *Cambridge History of Islām*, i. 204.
255. Humphreys, *From Saladin to the Mongols*, 21.
256. Ibid. 365.
257. Ibid. 368.
258. Lyons and Jackson, *Saladin*, 368–9.
259. Ibid. 369. Gibb, *Studies on the Civilization of Islam*, 104.
260. <www.fordham.edu/halsall/source/1192peace.html>
261. Humphreys, *From Saladin to the Mongols*, 132–3.
262. Ibid. 267.
263. Lyons and Jackson, *Saladin*, 227.
264. Humphreys, *From Saladin to the Mongols*, 8, 45.
265. Ibid. 338.
266. Rosenthal, *Political Thought in Medieval Islam*, 43–51, especially 49. Lambton, *State and Government in Medieval Islam*, 138–43, especially 142. Jackson, *Islamic Law and the State*, xxxix.

Chapter 3

1. From the legal rulings section of the website of the Islamic Supreme Council of America: <www.islamicsupremecouncil.org/bin/site/wrappers/default.asp?pane_2=content-legal-jihad_dhikr>
2. 'Robert Kilroy-Silk's anti-Arab diatribe is not only offensive and stupid; it also speaks of a startling degree of ignorance': Derek Brown in the *Guardian*, 15 January 2004.
3. Annemarie Schimmel, *Mystical Dimensions of Islam* (Chapel Hill, NC: University of North Carolina Press, 1975, repr. 1976), 29.
4. The threefold relationship comes out most clearly in al-Bukhārī 1/2/47: 'one day while the Prophet was sitting in the company of some people, [the angel] Gabriel came and asked, "What is faith?" Allāh's Apostle replied: "faith is to believe in Allāh, His angels, [the] meeting with Him, His Apostles, and to believe in Resurrection." Then he further asked, "What is Islām?" Allāh's Apostle replied: "to worship Allāh Alone and none else, to offer prayers perfectly, to pay the compulsory charity (*zakāt*) and to observe fasts during the month of Ramadan." Then he further asked, "What is *iḥṣān*?" Allāh's Apostle replied: "to worship Allāh as if you see Him, and if you cannot achieve this state of devotion then you must consider that He is looking at you."' Cf. Muslim 1/1: 'that you worship Allāh as if you are seeing Him, for though you don't see Him, He, verily, sees you.'
5. Cf. Rabia Harris' comments in Abū'l-Qāsim 'Abd al-Karīm bin Hawāzin al-Qushayrī, *The Risālah: Principles of Ṣūfism*, trans. Rabia Harris (Chicago, IL: Great Books of the Islamic World, Kazi Publications, 2002), xv–xvii, where the criticism is directed at W. M. Watt, *The Formative Period of Islamic Thought* (Edinburgh: Edinburgh University Press, 1973), 238. W. A. Graham, *Divine Word and Prophetic Word in Early Islam* (The Hague: Mouton, 1977), 95 is also quoted.
6. George Makdisi, 'Ibn Taymīya: A Ṣūfī of the Qādirīya Order', *American Journal of Arabic Studies*, 1 (1974), 118–129, especially 129.
7. J. Spencer Trimingham, *The Sufi Orders in Islam* (Oxford: Clarendon Press, 1971), 1. However, there are doubts about this: 'the Ṣūfīs are not distinguished by the wearing of wool!' Al-Qushayrī, *The Risālah*, 337. According to Abū Bakr al-Kalābādhī, wool is the dress of the

Prophets and the garb of saints. The wool explanation of the term 'Ṣūfī' had all the necessary meanings, such as 'withdrawal from the world, inclining the soul away from it, leaving all settled abodes, keeping constantly to travel, denying the carnal soul its pleasures, purifying the conduct, cleansing the conscience, dilation of the breast and the quality of leadership': *The Doctrine of the Ṣūfīs*, ed. and trans. A. J. Arberry (Cambridge: Cambridge University Press, 1935, repr. 1977), 7, 10.

8. Al-Qushayrī, *The Risālah*, 18.
9. Michael Cook, *Commanding Right and Forbidding Wrong in Islamic Thought* (New York: Cambridge University Press, 2000), 459–60.
10. Schimmel, *Mystical Dimensions of Islam*, 99. Schimmel's ch. 3 provides a guide to the Ṣūfī path. Alfred Morabia, *Le Ĝihâd dans l'Islam médiéval. Le combat sacré des origines au xiie siècle* (Paris: Albin Michel, 1993), 328.
11. Schimmel, *Mystical Dimensions of Islam*, 227. There were originally four main *ṭarīqah*: the Naqshabandiyyah, the Qādiriyyah, the Chishtiyyah, and Suhrawardiyyah, named after their four founders. Rembrandt copied a Mughal miniature showing the four founders drinking tea under a tree: ibid. 233. Trimingham, *The Sufi Orders in Islam*, ch. 2, 'the chief *ṭarīqah* lines'. There are now numerous orders and sub-orders: <www.haqq.com.au/~salam/sufilinks/>
12. Schimmel, *Mystical Dimensions of Islam*, 112.
13. *The Faith and Practice of Al-Ghazālī*, trans. W. Montgomery Watt (London: Allen and Unwin, repr. 1967), 131.
14. Ibid. 14–15. Cf. Schimmel, *Mystical Dimensions of Islam*, 95: 'this teaching – a marriage between mysticism and law – has made Ghazālī the most influential theologian of medieval Islām'.
15. For al-Makkī: Schimmel, *Mystical Dimensions of Islam*, 85.
16. *Ghazālī's Book of Counsel for Kings (Naṣīḥat al-Mulūk)*, trans. F. R. C. Bagley (London: Oxford University Press, 1964), xxxvi.
17. Cf. Schimmel, *Mystical Dimensions of Islam*, 59: Abū'l-Qāsim Junayd (d. 298/910) 'refined the art of speaking in *ishārāt*, subtle allusion to the truth – a trend, attributed first to Kharrāz [Abū Sa'īd Aḥmad al-Kharraz (d. 285/899 or 276/890)], that became characteristic of later Ṣūfī writings'.
18. Arberry, *The Doctrine of the Ṣūfīs*, 82–5.
19. Al-Qushayrī, *The Risālah*, 37. Later date in Trimingham, *The Sufi Orders in Islam*, 265 n. 2.
20. Arberry, *The Doctrine of the Ṣūfīs*, 45–6.
21. Schimmel, *Mystical Dimensions of Islam*, 65.
22. Arberry, *The Doctrine of the Ṣūfīs*, xiv.
23. Morabia, *Le Ĝihâd dans l'Islam médiéval*, 329.
24. Schimmel, *Mystical Dimensions of Islam*, 36–7, 109.
25. Al-Qushayrī, *The Risālah*, 18–19.
26. Arberry, *The Doctrine of the Ṣūfīs*, 13.
27. Schimmel, *Mystical Dimensions of Islam*, 156.
28. *Concise Encyclopedia of Islām*, ed. H. A. R. Gibb and J. H. Kramers (Leiden: E. J. Brill, repr. 2001), 287.
29. Al-Qushayrī, *The Risālah*, xlvii–xlviii. Ibid. 1, for the date of publication.
30. Ibid. 14.
31. Ibid. 93.
32. Ibid. 107–8.
33. Ibid. 108.
34. Ibid. 111.

35. Ibid. 119–23. The author is indebted to Professor Khalid Masud, who first drew his attention to the importance of al-Qushayrī's *Risālah* and this section in particular, and kindly supplied translations from the Urdu version of the text.
36. Muḥammad Asad, *The Message of the Qur'ān* (Gibraltar: Dar Al-Andalus, 1980), 616 and n. 61, has 'but as for those who strive hard in Our cause – We shall most certainly guide them onto paths that lead unto Us: for, behold, God is indeed with the doers of good.' Asad adds the comment that the plural 'our paths' is meant to stress the fact that 'there are many paths which lead to a cognizance… of God'.
37. The *ḥadīth* is reported in two of the seven canonical books (Abū-Dāwūd and al-Tirmidhī).
38. Al-Qushayrī, *The Risālah*, 30. Ibid. 31: 'for thirty years I prayed, though in each prayer the conviction within me was no better than it would have been had I been a fire-worshipper…' For Bisṭāmī: Schimmel, *Mystical Dimensions of Islam*, 47–50.
39. Al-Qushayrī, *The Risālah*, 120.
40. Ibid.
41. Ibid. 122 for this and the previous quotation.
42. Ibid. 129.
43. Ibid. 169.
44. Ibid. 205. Though he added, 'laugh little, for much laughter kills the heart'.
45. Ibid. 246–7.
46. Ibid. 250.
47. Ibid. 257.
48. In a later manual of *futuuwah* in 688/1290, al-Naṣīrī effectively defines it as a stage that could be attained by those who were unable to reach the ultimate goal of the mystic, the absolute truth which the Ṣūfī acquires when he sees God in everything created. Ethel Sara Wolper, *Cities and Saints. Sufism and the Transformation of Urban Space in Medieval Anatolia* (Pennsylvania: Pennsylvania State University Press, 2003), 77.
49. Al-Qushayrī, *The Risālah*, 275.
50. Ibid. 276.
51. Ibid. 317.
52. Ibid. 475.
53. Cf. Trimingham, *The Sufi Orders in Islam*, 6: 'the Karrāmiyya was relatively short-lived (two centuries) whereas the Ṣūfī movement went on from an individualistic discipline to change the whole devotional outlook of Muslims'.
54. For early (ninth-century) Khurāsān Ṣūfīs: Schimmel, *Mystical Dimensions of Islam*, 35, 38.
55. Al-Qushayrī, *The Risālah*, li.
56. Abū'l-Futūḥ Aḥmad ibn Muḥammad al-Ghazālī (d. 520/1126): Trimingham, *The Sufi Orders in Islam*, 32–3. Cf. Abū Bakr al-Kalābādhī: 'the real essence of the spiritual states of the Ṣūfīs is such that expressions are not adequate to describe it'. Arberry, *The Doctrine of the Ṣūfīs*, 114.
57. Cf. 'Ghazzālī was, among other things, a Ṣūfī': Cook, *Commanding Right and Forbidding Wrong in Islamic Thought*, 459. Yet when writing his long account of forbidding wrong, 'there is little or nothing to indicate' that al-Ghazālī was writing 'as a Ṣūfī': ibid. 460.
58. Trimingham, *The Sufi Orders in Islam,* 3. Schimmel, *Mystical Dimensions of Islam*, 94.
59. *The Book of Invocation* (*Iḥyā' 'ūlūm al-Dīn*), translated by Kojiro Nakamura as *Ghazālī on Prayer* (Tokyo: University of Tokyo Press, 1975), 167.
60. Trimingham, *The Sufi Orders in Islam,* 152, 155, 156. <www.sunnah.org/tasawwuf/jihad002.html>
61. Schimmel, *Mystical Dimensions of Islam*, 94.

62. This section draws on the magisterial study of Cook, *Commanding Right and Forbidding Wrong in Islamic Thought*, 427–59. M. Cook, *Forbidding Wrong in Islam: An Introduction* (Cambridge: Cambridge University Press, 2003), 27–35, 73–7, offers a summary.
63. Cook, *Commanding Right and Forbidding Wrong in Islamic Thought*, 401–2, 458.
64. Wolper, *Cities and Saints*, 100.
65. Ibid. 31.
66. Ibid. 64.
67. Ibid. 9, 11.
68. Ibid. 80, where the author notes the many references to Christians which 'stress Christianity as a necessary stage in true enlightenment'.
69. Ibid. 24–5.
70. Ibid. 78.
71. Ibid. 79. Gibb and Kramers, *Concise Encyclopedia of Islām*, 61–2.
72. Literally, 'tethering of horses' (*ribāṭ al-khayl*), an expression which signifies 'holding in readiness mounted troops at all points open to enemy invasion'. Thus, the place where the mounts were assembled and kept in readiness for an expedition; possibly also a relay for horses. Gibb and Kramers, *Concise Encyclopedia of Islām*, 473.
73. <www.muslimheritage.com/topics/default.cfm?ArticleID=336>
74. Gibb and Kramers, *Concise Encyclopedia of Islam*, 474.
75. <www.bewley.virtualave.net/ibnyasin.html#preaching>
76. Muḥammad al-Ghazālī, *The Socio-Political Thought of Shāh Walī Allāh* (Islamabad: International Institute of Islamic Thought and Islamic Research Institute, 2001), 37.
77. Ibid. 42, 76. Ibid. 278 (paras 335, 336), where he cites the precedents of Abū Bakr and 'Umar.
78. Ibid. 85.
79. Ibid. xiii. Definitive English edition by Marcia K. Hermansen, *The Conclusive Argument from God. Shāh Walī Allāh of Delhi's Ḥujjat Allāh al-Bāligha* (Leiden: E. J. Brill, 1996). Ibid. 360, for the assertion that once the caliphate appeared (namely, that the Prophet assumed a political role), '*jihād* with the enemies of God became possible'.
80. Al-Ghazālī, *The Socio-Political Thought of Shāh Walī Allāh*, 86.
81. Ibid. 93, 96–7.
82. Ibid. 260.
83. Ibid. 102.
84. Ibid. 98.
85. Ibid. 103, 261 (para. 272).
86. Ibid. 103, 261 (para. 273).
87. Ibid. 103, 260 (para. 272).
88. Ibid. 263 (para. 283).
89. Hermansen, *The Conclusive Argument from God*, 369.
90. Ibid. 13.
91. Ibid. 36.
92. Ibid. 230. Ibid. 359, for the requirement that *jihād* is not limited in time.
93. Al-Ghazālī, *The Socio-Political Thought of Shāh Walī Allāh*, 265, 267–78 (paras 289, 291, 296).
94. Ibn Taymīyah wrote: Abū Bakr 'and others from the Companions began with the *jihād* against the apostates before the *jihād* against the Unbelievers from the *Ahl al-Kitāb*. The *jihād* against [the apostates] is to secure the conquered lands of the Muslims, and to bring into Islām those who intend to disengage from it. The *jihād* against those who do not fight against us from the Polytheists and *Ahl al-Kitāb* is an extra contribution to the manifestation of Religion. To preserve the capital is prior to making the profit.' Quoted by Tastan Osman, 'The Jurisprudence

of Sarakhsī with Particular Reference to War and Peace: A Comparative Study in Islamic Law' (unpublished PhD thesis, University of Exeter, 1993), 145.

95. Saiyid Athar Abbas Rizvī, *Shāh Walī Allāh and His Times. A Study of Eighteenth-Century Islām, Politics and Society in India* (Canberra: Ma'rifat, 1980), 293, 397.
96. Ibid. 313.
97. Ibid. 305. Satish Chandra, *Parties and Politics and the Mughal Court, 1707–1740* (New Delhi: Oxford University Press, 1959, repr. 2002), 290–2.
98. Rizvī, *Shāh Walī Allāh and His Times*, 305.
99. Ibid. 399: 'more as the successor of the great Sunnī revivalist Ibn Taymīyah... than... of Ibn Khaldūn'.
100. Ibid. 296. Ibid. 159: 'the cry of Sunnī *jihād* against the heretic Shī'a dominance was a significant feature of the trial of strength between the rival parties'.
101. Ibid. 273, 314.
102. In this respect, he cited the words of the Prophet: 'the one who fights so that the word of God is exalted, that is reckoned in the way of God'. Al-Ghazālī, *The Socio-Political Thought of Shāh Walī Allāh*, 263 (para. 280).
103. Ibid. 264 (paras 286, 287).
104. Ibid. 103–4, 262 (paras 276, 279).
105. Ibid. 266–7 (paras 291, 292, 293, 294).
106. Ibid. 75 contrasted with <www.saag.org/papers7/paper629.html>
107. Al-Ghazālī, *The Socio-Political Thought of Shāh Walī Allāh*, 19–21.
108. Ibid. 4. Qeyamuddin Ahmad, *The Wahhābi Movement in India* (New Delhi: Manohar, 1994), 14, 43–4, 66.
109. F. E. Peters, *Islām. A Guide for Jews and Christians* (Princeton, NJ and Oxford: Princeton University Press, 2003), 249.
110. Mark J. Sedgwick, *Sufism: the Essentials* (Cairo and New York: American University in Cairo Press, 2003), 81–2.
111. Azzam Karam (ed.), *Transnational Political Islām. Religion, Ideology and Power* (London and Sterling, VA: Pluto Press, 2004).
112. The authors provide a significant bibliography and set of questions, so interested scholars should consult the website for this international conference held in September 2003: <www.let.uu.nl/~martin.vanbruinessen/personal/conferences/sufism_and_the_modern.html>

Chapter 4

1. 'Faraj and the *Neglected Duty*'. Interview with Professor Johannes J. G. Jansen (8 December 2001): <www.religioscope.com/info/dossiers/textislamism/faraj_jansen.htm>

   The author is grateful to Dr Ataullah Siddiqui for drawing his attention to the importance of the following work: *A Muslim Theologian's Response to Christianity. Ibn Taymiyya's al-Jawāb al-ṣaḥīḥ*, ed. and trans. Thomas F. Michel (Delmar, NY: Caravan Books, 1984).
2. Full name: Taqī ad-Dīn Abū l-'Abbās Aḥmad ibn 'Abd al-Ḥalīm ibn 'Abd as-Salām Ibn Taymīyah al-Harranī al-Ḥanbalī. Biographical notice by M. ben Cheneb in *Shorter Encyclopedia of Islām*, ed. H. A. R. Gibb and J. H. Kramers (Ithaca, NY: Cornell University Press, 1953; reprinted as *Concise Encyclopedia of Islām*: Boston, MA, and Leiden: E. J. Brill, 2001), 151–2, which notes the historic as well as contemporary divergence of views on his orthodoxy, and his influence on Wahhābism. Also 'Ibn Taimiy(y)a', *The Concise Oxford Dictionary of World Religions*, ed. John Bowker (Oxford: Oxford University Press, 2000). *Oxford Reference Online*. <www.oxfordreference.com/views/ENTRY.html?subview=Main&entry=t101.003335> 'Ibn Taymīyah, Taqī al-Dīn Aḥmad', *The Oxford Encyclopedia of the Modern Islamic World*, ed. John Esposito (4 vols New York: Oxford University Press, 1995), ii. 165–6. 'Abd al-Salām

Faraj pronounced that 'Ibn Taymīyah's collection of *fatwās* is useful in the present age': J. J. G. Jansen, *The Neglected Duty. The Creed of Sadat's Assassins and Islamic Resurgence in the Middle East* (New York: Macmillan, 1986), 175. For Ibn Taymīyah's view of *jihād*: A. Morabia, 'Ibn Taymiyya, dernier grand théoricien du *Ĝihâd* médiéval', *Bulletin d'études orientales*, 30 (1978), 85–99. There is a useful account of his philosophical thought by Serajul Haque, 'Ibn Taymīyah', in *A History of Muslim Philosophy*, ed. M. M. Sharif (Pakistan Philosophical Congress, n.d.), ch. 41, 796–819. <www.muslimphilosophy.com/hmp/default.htm>

3. The title was denied by his detractors: J. H. Kramers in Gibbs and Kramers, *Concise Encyclopedia of Islām*, 519. The Ḥanafī scholar 'Alā' al-Dīn al-Bukhārī issued a *fatwā* whereby anyone who called Ibn Taymīyah Shaykh al-Islām committed disbelief.
4. *Majmū'a Fatāwā Shaykh al-Islām Aḥmad ibn Taymiyya*, 37 vols [*Compilation of Legal Opinions of Shaykh al-Islām Aḥmad Ibn Taymiyya*], ed. and comp. 'Abd al-Raḥman ibn Qāsim (n.p., n.d.). Full list of his writings in Qamaruddin Khan, *The Political Thought of Ibn Taymīyah* (Islamabad: Islamic Research Centre, Pakistan, 1973), 186–200, but he also cites by title (201–5) 115 titles that appear to be lost.
5. For his work as a legist in this school: A. H. I. Matroudi, 'The Role of Ibn Taymīyah in the Ḥanbalī School of Law' (unpublished PhD thesis, University of Leeds, 1999).
6. E. I. J. Rosenthal, *Political Thought in Medieval Islam. An Introductory Outline* (Cambridge: Cambridge University Press, 1958), 60. S. A. Jackson, *Islamic Law and the State: The Constitutional Jurisprudence of Shihāb al-Dīn al-Qarāfī* (Leiden: E. J. Brill, 1996), xxii–xxiii.
7. The Ḥanbali scholar Najm al-Dīn Sulaymān ibn 'Abd al-Qawī al-Ṭufī stated: 'a time came when his companions took to over-praising him and this drove him to be satisfied with himself until he became conceited before his fellow human beings. He became convinced that he was a scholar capable of independent reasoning (*mujtahid*). Henceforth he began to answer each and every scholar great and small, past and recent, until he went all the way back to 'Umar… and faulted him in some matter… Others considered him a dissimulator (*munāfiq*) because of what he said about 'Alī:… namely, that he had been forsaken everywhere he went, had repeatedly tried to acquire the caliphate and never attained it, fought out of lust for power rather than religion, and said that "he loved authority while 'Utmān loved money". He would say that 'Abū Bakr had declared Islām in his old age, fully aware of what he said, while 'Alī had declared Islām as a boy, and the boy's Islām is not considered sound upon his mere word….' <www.www.sunnah.org/history/Innovators/ibn_taymiyya.htm>

   'Uthmān refuted at the time the charge that he loved money: 'I do not think it right to spend anything on my kinsmen out of public funds. In fact, I do not get anything out of these funds for my own expenses either. The revenue of each province is spent on the people of that province. The public treasury at Medina receives nothing but the fifth part of booty. This money is spent by the people themselves in times of need.' <www.hannityandcolmes.com/uthman.htm>
8. Morabia, 'Ibn Taymiyya, dernier grand théoricien du *Ĝihâd* médiéval', 92.
9. Michel, *A Muslim Theologian's Response to Christianity*, 80.
10. <www.www.sunnah.org/history/Innovators/ibn_taymiyya.htm> The quotation is from Ibn Hadjar al-Haitamī (909/1504–974/1567), a jurist of the Shāf'ite school.
11. A. Nizar Hamzeh and R. Hrair Dekmejian, 'A Ṣūfī Response to Political Islamism: Al-Ahbash of Lebanon', *International Journal of Middle East Studies*, 28 (1996), 217–29. <www.almashriq.hiof.no/ddc/projects/pspa/al-ahbash.html>
12. Jackson, *Islamic Law and the State*, 206.
13. For the position of Baghdād: Janet L. Abu-Lughod, *Before European Hegemony. The World System AD 1250–1350* (New York: Oxford University Press, 1989), 193–7.

14. R. S. Humphreys, *From Saladin to the Mongols. The Ayyubids of Damascus, 1193–1260* (Albany, NY: State University of New York Press, 1977), 11. P. M. Holt, A. K. S. Lambton and B. Lewis (eds) *Cambridge History of Islām* (Cambridge: Cambridge University Press), i. 212.
15. Majīd Khaddūrī, *War and Peace in the Law of Islām* (Baltimore, MD, and London: Johns Hopkins University Press, 1955), 269. Yahya Armajani, *Middle East: Past and Present* (Englewood Cliffs, NJ: Prentice-Hall, 1970), 132. In 703/1313, Ghāzān visited Karbalā' and made lavish gifts to the sanctuary: Gibb and Kramers, *Concise Encyclopedia of Islām*, 360. The Shī'a leanings of Ghāzān would have simplified Ibn Taymīyah's task had he been aware of them. 'The '*ulamā*' of Syria and Egypt were faced with an excruciating dilemma', writes Emmanuel Sivan. 'Was war against the Mongols still a *jihād* or had it become a mere clash between two Sunnī states? If the latter, it created a deplorable conflict but not one that called for recourse to Holy War (which would have been the case had the Mongols become Shī'a)': Emmanuel Sivan, *Radical Islām: Medieval Theology and Modern Politics* (New Haven, CT, and London: Yale University Press, 1985), 96.
16. Michel, *A Muslim Theologian's Response to Christianity*, 71–3.
17. J. Riley-Smith, 'Islam and the Crusades in History and Imagination, 8 November 1898–11 September 2001', 165. Morabia, 'Ibn Taymiyya, dernier grand théoricien du *Ĝihâd* médiéval', 90.
18. Ibid. 85, 99.
19. Ibid. 134.
20. Quoted by 'Abd al-Salām Faraj in Jansen, *The Neglected Duty*, 177.
21. Hülagü Khān (1215–1265), a grandson of the notorious Chinghis-Khān. After his defeat in 658/1260, he withdrew to Azerbaijan, adopted Islām, and founded the Ilkhān dynasty.
22. Khan, *Political Thought*, 114: 'it is well known that Qazān Khān, the Mongol conqueror, had given a pledge to Ibn Taymīyah that the city of Damascus would not be stormed if Muslims ceased to resist'.
23. Though a popular work, John Man, *Genghis Khan. Life, Death and Resurrection* (London and New York: Bantam Press, 2004) usefully portrays Mongol culture. Chinghis-Khān himself seems to have converted to Taoism by the time of his death: ibid. 204–6.
24. <www.coldsiberia.org/webdoc9.htm#Its%20message>
25. Humphreys, *From Saladin to the Mongols*, 356.
26. For a translation of Ibn Taymīyah's *fatwā*, a key text in modern violent Islamist discussion, see Appendix. The version included here is derived from Jansen's translation of 'Abd al-Salām Faraj's *al-Farīḍa al-Ghā'iba*: Jansen, *The Neglected Duty*, 166, 168. Ibid. 175: 'it is neither a House of Peace which is ruled by the laws of Islām because its soldiers are Muslims, nor a House of War, the inhabitants of which are infidels'. Instead it was in a third category, one in which a Muslim 'should be treated accorded to what is due to him and someone who rebels against the Law of Islām should [in his turn] be treated in accordance to what is due to him'.
27. Ibid. 173–4.
28. Rudolph Peters, '*Jihād*', in Esposito, *The Oxford Encyclopedia of the Modern Islamic World*, ii. 372, argues that there was a second ruling of its kind. Professor Hassan Hanafī, in a private communication to the author, has refuted this suggestion.
29. Sivan, *Radical Islām*, 98.
30. Shaykh ul-Islām Taqi-ud-Deen Aḥmad Ibn Taymīyah, *The Religious and Moral Doctrine of Jihād* (Birmingham: Maktabah Al Ansaar Publications, 2001), 24–5. Also at <www.islamistwatch.org/main.html>; <www.allaahuakbar.net/scholars/ibn_taymiyyah/ibn_taymiyyah_on_jihaad.htm>

A different translation, with the complete text of the work from which the chapter on *jihād* is drawn, is provided by Omar A. Farrukh, *Ibn Taymīyah on Private and Public Law in Islām or Public Policy in Islamic Jurisprudence* (Beirut: Khayats, 1966), 135–61, with the quotation at 138. Farrukh translates the last phrase as 'the tip of its hump is *jihād*'. [However, Farrukh's chapter title, '*Jihād* (Holy War) and decisive fight' is unacceptable.] Full publication of Ibn Taymīyah's *The Criterion between the Allies of the Merciful and the Allies of the Devil*, trans. Salim Abdallah Ibn Morgan: <www.java-man.com/Pages/Books/criterion.html>

31. Ibn Taymīyah, *The Religious and Moral Doctrine of Jihād*, 25–6.
32. Ibid. 26–7.
33. Ibid. 27–8.
34. Ibn Taymīyah, *The Criterion between the Allies of the Merciful and the Allies of the Devil*. 'Al 'Iraqy… states: "the mentioned *Hadīth* is related by Imām Bayhaqī with a *Sanad ḍa'īf* [weak chain of narrators] from Jābir". Apart from the *ḥadīth* related by Imām Baihaqi there is also a *ḥadīth* related by al-Khaṭib Al-Baghadadī from Jābir. It turns out that this *ḥadīth* is weak because within its *Sanad* there is a narrator by the name of Khalaf bin Muḥammad bin Ismā'il al-Khiyam [whose *aḥādīth*] according to al-Ḥakim… "are unreliable". And Abū Ya'lā al-Khalīlī says: "he often adulterates, is very weak and narrates unknown *ḥadīth*." Al-Ḥakim and Ibn Abī Zur'a state: "we often write statements from Khalaf bin Muḥammad bin Isma'il only as an example, and we remove ourselves of responsibility from him." And even more doubtful than that, there is within the *Sanad* of this *ḥadīth* a narrator by the name of Yaḥya bin Al-'Ulā al Bajīli who according to Imām Aḥmad is a known *Kadhdhaab* – liar – and forger of *ḥadīth*. Also, 'Amr bin 'Alī, An Nasā'i and Darāquṭnī state: "his *ḥadīth* are renounced." Ibn 'Adī states: "his *ḥadīth* are false"… Furthermore, besides the two stated weak *ḥadīths*, there is the statement of a Tabi'i by the name of Ibrāhīm bin Abī 'Abla to people who had returned from battle, which states: "you have returned from *Jihād Asghar* so is the *Jihād Akbar* you intend to do *Jihād qalbī* (*Jihād* of the heart)?" Darāquṭnī states that Ibrāhīm bin Abī 'Abla himself is believable but the chain of transmission is broken… As a result of that, the statement above cannot be attributed to Ibrāhīm bin Abī 'Abla unless the chain of transmission is authentic…' On the basis of the above statements, a modern commentator concludes that "the evidence used as proof or the basis for establishing that *Jihād* against disbelievers on the battlefield is *Jihād Asghar* and *Jihād* against the desires and *Shyṭān* is *Jihād Akbar*, are weak if not false *ḥadīth*".' <www.islam.org.au/articles/26/jihad.htm>

    Khan, *Political Thought*, 169, notes that Ibn Taymīyah is inconsistent in his citation of *aḥādīth*, quoting 'many… which are certainly not genuine'. Ibid. 83: 'this *ḥadīth* on which Ibn Taymīyah has built a whole political theory is of a spurious origin'.
35. Morabia, 'Ibn Taymiyya, dernier grand théoricien du *Ĝihâd* médiéval', 96. Ibn Taymīyah, *Enjoining Right and Forbidding Wrong*, trans. Salim Abdallah Ibn Morgan: <www.java-man.com/Pages/Books/alhisba.html>
36. Khan, *Political Thought*, seems erroneously to imply that Ibn Taymīyah did not cite this *ḥadīth*.
37. He wrote: 'Allāh does not like chaos and corruption. All that which Allāh has enjoined is beneficial, and the epitome of benefit. Allāh has praised *ṣalāḥ* (righteousness) and the *muṣliḥīn* (reformers, or those who bring about *ṣalāḥ*). And He has praised those who believe and do good works (*sāliḥāt*), while condemning corruption (*fasād*) and those who cause it in many places in the Qur'ān. Thus whenever the adverse effects (*mafsadah*) of any act of enjoining or forbidding are greater than its benefit (*maṣlaḥah*), it is no longer part of what Allāh has enjoined upon us, even if it be a case of neglecting obligations or committing the forbidden.'
38. Cook, *Commanding Right and Forbidding Wrong*, 152–3, 155. Cook, *Forbidding Wrong in Islam*, 69. Ibn Taymīyah wrote: 'as for the previous nations, none of them enjoined all people with all that is right, nor did they prohibit all that is wrong to all people. Furthermore, they

did not make *jihād* in this cause. Some of them did not take up armed struggle at all, and those who did, such as the Jews, their struggle was generally for the purpose of driving their enemy from their land, or as any oppressed people struggles against their oppressor, and not for sake of calling the people of the world to guidance and right, nor to enjoin on them right and to prohibit to them wrong.'

39. 'All of what we are saying comes under the general principle which says that when benefit and harm are mixed up together, and one must choose between doing good with some bad side effects, or leaving that good to avoid its bad side effects, it is obligatory to choose the course having the greater overall or net benefit. This is because enjoining and forbidding though they entail the attainment of some benefit, and the prevention of some harm, its opposite must also be considered. If, in carrying out this enjoining or forbidding, there is benefit lost greater than the benefit gained, or harm is brought about greater than the harm which was avoided, then this is not part of that which Allāh has ordered us to do, rather it is *ḥaram*, because of the fact that its net harm is greater than its net benefit.'
40. <www.memri.org/bin/articles.cgi?Page=countries&Area=saudiarabia&ID=SP53503>
41. Riley-Smith, 'Islam and the Crusades in History', 165.
42. Muslim 20/4638.
43. Khan, *Political Thought*, 151.
44. Morabia, 'Ibn Taymiyya, dernier grand théoricien du *Ĝihâd* médiéval', 91–2.
45. Khan, *Political Thought*, 155.
46. Ibid. 54.
47. Al-Bukhārī 4/52/198: 'Whenever Allāh's Apostle intended to carry out a *Ghazwah*, he would use an equivocation to conceal his real destination till it was the *Ghazwah* of Tabūk which Allāh's Apostle carried out in very hot weather. As he was going to face a very long journey through a wasteland and was to meet and attack a large number of enemies. So, he made the situation clear to the Muslims so that they might prepare themselves accordingly and get ready to conquer their enemy.'
48. R. Peters, *Jihād in Classical and Modern Islam* (Princeton, NJ: Marks and Weiner, 1996), 49–50. H. A. Haleem, O. Ramsbotham, S. Risaluddin and B. Wicker (eds) *The Crescent and the Cross. Muslim and Christian Approaches to War and Peace* (Basingstoke: Macmillan, 1998), 81.
49. Khan, *Political Thought*, 160.
50. G. Makdisi, 'Hanbalite Islam', in *Studies on Islam*, ed. and trans. Merlin L. Swartz (New York: Oxford University Press, 1981), 249. G. Makdisi, 'Ibn Taymīya: A Ṣūfi of the Qādirīya Order', *American Journal of Arabic Studies*, 1 (1974), 118–29.
51. In contrast, modern technology allows this now to be achieved: while *ijmā* is now possible technically, it may yet be impossible for political or doctrinal reasons. I am grateful to Dr Adil Salahi for this point.
52. Morabia, 'Ibn Taymiyya, dernier grand théoricien du *Ĝihâd* médiéval', 93. Ibn Taymīyah dismissed the Shī'a belief that 'Alī was the rightful successor of the Prophet; if God had designated him thus, he must have been willing to appoint to the caliphate a man who was not going to enjoy the total allegiance of the community and whose rule would lead to civil war: Hamid Enayat, *Modern Islamic Thought. The Response of the Shī'ī and Sunnī Muslims to the Twentieth Century* (London and Basingstoke: Macmillan, 1982), 35.
53. *The Concise Encyclopedia of Islām*, ed. C. Glassé (London: Stacey International, 1966), 176.
54. Hamzeh and Dekmejian, 'Ṣūfī Response to Political Islamism'.
55. These are listed as his ordeal because of his treatise *Al- Ḥamawiyyah* in the year 698/1298; his ordeal and debates because of his treatise *Al-Wāṣiṭiyya* in the year 705/1305; his ordeal, summons to Egypt and imprisonment there in the year 705/1305 for 18 months; his ordeal

with the [Ṣūfīs] in Egypt after his release; his deportation to Alexandria in the year 709/1309 and imprisonment there for eight months; his ordeal because of specific verdicts related to divorce and resultant imprisonment in the year 720/1320, for five months; finally his ordeal because of his legal verdict banning the undertaking of journeys specifically to visit graves and resultant imprisonment in the year 726/1325 until his death in the year 728/1328: <www.sunnahonline.com/ilm/seerah/0047.htm>

56. H. M. Zawati, *Is Jihad a Just War? War, Peace and Human Rights under Islamic and Public International Law* (Lewiston, NY: Edwin Mellen, 2001), 12.
57. Majīd Khaddūrī, *The Islamic Conception of Justice* (Baltimore, Md: Johns Hopkins University Press, 1984), 169–70.
58. M. S. Bin Jani, 'Sayyid Quṭb's View of *Jihād*: An Analytical Study of his Major Works' (unpublished PhD thesis, University of Birmingham, 1998), 94. Ibn Taymīyah, *Enjoining Right and Forbidding Wrong*, s.v. 'Methodology of enjoining right and forbidding wrong'. Cook, *Commanding Right and Forbidding Wrong*, 33: 'this tradition is referred to, quoted and commented upon with great frequency in subsequent literature'.
59. Morabia, 'Ibn Taymiyya, dernier grand théoricien du *Ĝihâd* médiéval', 95. Haleem et al., *The Crescent and the Cross*, 81.
60. John L. Esposito, *Unholy War. Terror in the Name of Islām* (Oxford: Oxford University Press, 2002), 46.
61. Natana DeLong Bas, *Wahhābī Islām. From Revival and Reform to Global Jihād* (New York: Oxford University Press, 2004), 256.
62. *The Tablet*, 22 August 1998. <www.artsweb.bham.ac.uk/bmms/1998/09September98.html>
63. Morabia, 'Ibn Taymiyya, dernier grand théoricien du *Ĝihâd* médiéval', 95.
64. Posted on the Memri website on 5 March 2003: <www.memri.org/bin/articles.cgi?Page=subjects&Area=jihad&ID=SP47603>
65. Memri Special Dispatch series 539, 18 July 2003: <www.memri.org/bin/opener_latest.cgi?ID=SD53903>
66. R. Spencer, *Onward Muslim Soldiers. How Jihād Still Threatens America and the West* (Washington DC: Regnery Publishing, 2003), 174.
67. Dr Muhammad Sayyed Ṭanṭāwī, who served as Muftī of Egypt from 1986 until 1996 when he was appointed Shaykh of Al-Azhar, called for *jihād* against US forces in 'Irāq at a press conference on 5 April 2003. Shaykh Ṭanṭāwī called on the 'Irāqī people to 'continue its *jihād* in defence of religion, faith, honour, and property, because *jihād* is a religious ruling of Islām aimed at opposing aggressors. It is the right of the 'Irāqīs to carry out any operation in defence of their homeland, whether martyrdom operations [i.e. suicide operations] or [by] any other means.' Shaykh Ṭanṭāwī encouraged volunteers from Arab and Islamic countries to go to 'Irāq 'to support the *jihād* of their oppressed brethren there, because resistance to oppression is an Islamic obligation, whether the oppressor is Muslim or not'. Memri Special Dispatch 145, 14 August 2003: <memri.org/bin/articles.cgi?Page=archives&Area=ia&ID=IA14503>
68. Khan, *Political Thought*, 183.
69. It is true that the text is not that of bin Laden himself, but of a prominent supporter. However, it appeared on a website linked, according to Memri, to al-Qaeda: <www.memri.org/bin/articles.cgi?Page=subjects&Area=jihad&ID=SP41802>
70. Farrukh, *Ibn Taymīyah on Public and Private Law in Islām*, 165.
71. Ibid. 164.
72. Ibn Taymīyah, *Public Duties in Islām. The Institution of the Ḥisba*, trans. Muhtar Holland (Leicester: Islamic Foundation, 1982), 109.
73. Ibid. 80. He added: 'if the right is preponderant it should be commanded, even if it entails a lesser wrong. But a wrong should not be forbidden if to do so entails the loss of a greater right…'

74. Khan, *Political Thought,* 142.
75. Ibid. 144.
76. Ibid. 166–7. Far from condoning deposition, Khan states that Ibn Taymīyah observed 'a judicious silence on this matter' because of the serious international military situation.
77. Ibid. 170.
78. Memri Special Dispatch 418, 4 September 2002: <www.memri.org/bin/articles.cgi?Page=subjects&Area=jihad&ID=SP41802>
79. Ibn Taymīyah, *Public Duties in Islām*, 106.

Chapter 5

1. Metin Kunt and Christine Woodhead (eds), *Süleyman the Magnificent and his Age. The Ottoman Empire in the Early Modern World* (London: Longman, 1995), 139.
2. Gábor Ágoston, 'Ottoman Warfare in Europe, 1453–1826', in *European Warfare, 1453–1815*, ed. Jeremy Black (Basingstoke: Macmillan, 1999), 119. The author is grateful to Professor Ágoston for making available to him some of his other articles and also the conclusion of his forthcoming book, *Guns for the Sultan: Military Power and the Weapons Industry in the Early Modern Ottoman Empire* (Cambridge: Cambridge University Press, 2005).
3. Ágoston, 'Ottoman Warfare in Europe', 124.
4. Nurten Kiliç-Schubel, 'Unity and Diversity in Political Culture. Muslim Empires of Sixteenth-Century Eurasia: Ottomans, Mughals, Safavids and Uzbeks', in *The Great Ottoman Turkish Civilization. I. Politics*, ed. Kemal Çiçek (Ankara: Yeni Türkiye, 2000), 275–84 at 277, 281–82.
5. P. M. Holt, A. K. S. Lambton and B. Lewis (eds), *The Cambridge History of Islām* (Cambridge: Cambridge University Press, 1970), i. 323.
6. Kemal H. Karpat, *The Politicization of Islam: Reconstructing Identity, State, Faith and Community in the Late Ottoman State* (New York: Oxford University Press, 2001), 48.
7. Ibid. 4.
8. Holt et al., *The Cambridge History of Islām*, i. 320.
9. Peter Sugar, *South-eastern Europe Under Ottoman Rule, 1354–1804* (Seattle, WA: University of Washington Press, 1977), ch. 1.
10. Cemal Kafadar, *Between Two Worlds. The Construction of the Ottoman State* (Berkeley, CA: University of California Press, 1995), 80: 'gaza was a lesser category than jihad. Canonical works describe it as a lesser *farż* [religious duty]; that is, contributing to it was not incumbent upon everyone in the Muslim community, as was the case with jihad'.
11. For a description of the siege: John Alden Williams (ed.), *Themes of Islamic Civilization* (Berkeley, CA and London: University of California Press, 1971), 289–93. Ibid. 293: from being part of the *dār al-ḥarb*, hostile territory, it became the *dar al-zarb*, the city of the mint, hence the capital. Holt et al., *The Cambridge History of Islām*, i. 295.
12. Ibid. i. 291.
13. Ibid. i. 300.
14. Gábor Ágoston, 'Early Modern Ottoman and European Gunpowder Technology', in *Multicultural Science in the Ottoman Empire*, ed. Ekmeleddin Ihsanoglu, Kostas Chatzis, Efthymios Nicolaidis (Turnhout, Belgium: Brepols, 2003), 18.
15. Gábor Ágoston, 'Ottoman Artillery and European Military Technology in the Fifteenth and Seventeenth Centuries', *Acta Orientalia Academia Scientiarum Hungaricae*, 47 (1994), 29–30.
16. Ibid. 18. The caveat was as follows: 'they cannot, however, be induced as yet to use printing, or to establish public clocks, because they think that the scriptures – that is, their sacred books

– would no longer be scriptures if they were printed, and that, if public clocks were introduced, the authority of their muezzins and their ancient rites would be thereby impaired'.

17. Ebru Turan, 'Some Reflections on the Ottoman Grand Vizierate in the Classical Age, 1300–1600', at <www.humanities.uchicago.edu/orgs/institute/sawyer/archive/islam/ebru.html>
18. Kafadar, *Between Two Worlds*, 136–7.
19. In 907/1501.
20. Holt et al., *The Cambridge History of Islām*, i. 398–9.
21. A. K. S. Lambton, *State and Government in Medieval Islam. An Introduction to the Study of Islamic Political Theory: The Jurists* (New York: Oxford University Press, 1981), 264.
22. Kunt and Woodhead, *Süleyman the Magnificent and his Age*, 22.
23. Holt et al., *The Cambridge History of Islām*, i. 315.
24. Kunt and Woodhead, *Süleyman the Magnificent and his Age*, 147–8.
25. It was not until almost the end of the dynasty, the reign of Shāh Sulṭān Ḥusayn (1105/1694–1135/1722), that the Shī'a '*ulamā*' reappropriated this responsibility, the long delay being a possible reason for the 'decline' of the Ṣafavid State. Ibid. 425, for decline in this reign. For the loss of '*ijtihād*: Colin Paul Mitchell, *Sir Thomas Roe and the Mughal Empire* (Karachi: Area Study Centre for Europe, 2000), 12, 43 n. 43.
26. Kiliç-Schubel, 'Unity and Diversity in Political Culture', 278.
27. Ágoston, 'Early Modern Ottoman and European Gunpowder Technology', 19.
28. Holt et al., *The Cambridge History of Islām*, i. 400. Cf. ibid. i. 315.
29. Richard Knolles, *The generall historie of the Turkes, from the first beginning of that nation to the rising of the Othoman familie: with all the notable expeditions of the Christian princes against them: together with The lives and conqvests of the Othoman kings and emperours* (5th edn, London: Adam Islip, 1638), 517.
30. Ibid. 512–13.
31. Holt et al., *The Cambridge History of Islām*, i. 321.
32. In 1305/1888, following the questioning of the legitimacy of the Ottoman caliph in the *Punjab Times*, the Ottoman government claimed that the renunciation of 922/1517 was both public knowledge and confirmed by the existence of a written document that was available for inspection (though there is no evidence that such a document did, in fact, exist): Karpat, *The Politicization of Islam*, 249.
33. Holt et al., *The Cambridge History of Islām*, i. 319–21.
34. Karpat, *The Politicization of Islam*, 244.
35. Kunt and Woodhead, *Süleyman the Magnificent and his Age*, 23.
36. By 1533, Barbarossa's fame was sufficient for a medal to be issued in Germany by L. Neufarer in his honour: National Maritime Museum London, MEC0367. Ercüment Kuran, 'Maghreb History During the Ottoman Period', in Çiçek *The Great Ottoman Turkish Civilization. I. Politics*, 245.
37. Miguel Angel de Bunes Ibarra, 'Kanuni Sultan Süleyman, Barbaros Pasha and Charles V: The Mediterranean World', in Çiçek *The Great Ottoman Turkish Civilization. I. Politics*, 240.
38. Cornell H. Fleischer, '*Mahdī* and Millennium: Messianic Dimensions in the Development of Ottoman Imperial Ideology', *The Great Ottoman Turkish Civilization. III. Philosophy, Science and Institutions*, ed. Kemal Çiçek (Ankara: Yeni Türkiye, 2000), 44–5.
39. Kiliç-Schubel, 'Unity and Diversity in Political Culture', 275.
40. Ágoston, 'Ottoman Warfare in Europe', 118.
41. Francis I was born on 12 September in 899/1494 and Süleyman on 6 November in 900/1494. Süleyman's 46-year rule was considerably longer that of both Francis I and Charles V. Süleyman outlived his rival monarchs (Francis died in 1547, Charles V abdicated in 1555–56 and died in 1558).

42. Esin Atil, *Süleymanname. The Illustrated History of Süleyman the Magnificent* (National Gallery of Art, Washington DC/New York: Harry N. Abrams, 1986), 16–17. The lands cited were Rumelia, Anatolia, Karaman, Dulkadir, Diyarbakir, Azerbaijan, Iran, Syria, Egypt, Mecca, Medina and 'all the Arab lands'.
43. Knolles, *The generall historie of the Turkes*, 570: 'in managing of warres, the opportunity of the time is especially to be followed'. F. A. Fischer-Galati, *Ottoman Imperialism and German Protestantism, 1521–1555* (2nd edn nd, New York: Octagon Books, 1972; 1st edn. 1959).
44. Atil, *Süleymanname*, 136–7. For the importance of this miniature as a depiction of Ottoman artillery techniques: Ágoston, 'Ottoman Warfare in Europe', 126.
45. Ágoston, 'Early Modern Ottoman and European Gunpowder Technology', 19.
46. Ibid. 20.
47. Gábor Ágoston, 'The Ottoman–Habsburg Frontier in Hungary, 1541–1699', in Çiçek *The Great Ottoman Turkish Civilization. I. Politics*, 284.
48. Kunt and Woodhead, *Süleyman the Magnificent and his Age*, 42, for the Venetian treaty. For the French treaty: Majīd Khaddūrī, *War and Peace in the Law of Islām* (Baltimore, MD, and London: Johns Hopkins University Press, 1955), 272–5 [original edn, 1940]. R. J. Knecht, *Renaissance Warrior and Patron: the Reign of Francis I* (Cambridge: Cambridge University Press, 1994), 329–30, concludes that 'no formal agreement... seems to have been concluded, and a commercial treaty, if discussed at all, seems to have been shelved'. This comment was based on the fact that the only surviving document in the West is a draft treaty. Holt et al., *The Cambridge History of Islām*, i. 327, accepts that the treaty was signed. It should be noted that the *bonne et sure paix* was to last during the lives of the two rulers: technically, the treaty lapsed on the death of Francis I in 954/1547, but there is no evidence that it did so in practice. Subsequently, in 987/1580, to the dismay of the French, a treaty was signed with the English. The *sulṭān* 'sought to establish a principle which would apply to other Christian princes as well': Khaddūrī, *War and Peace in the Law of Islām*, 273–4. A treaty with the Dutch Republic followed in 1021/1612: Holt et al., *The Cambridge History of Islām*, i. 330. This treaty facilitated Anglo–Ottoman trade in goods prohibited by Spain and Venice: Gábor Ágoston, '*Merces Prohibitae*: The Anglo–Ottoman Trade in War Materials and the Dependence Theory', *Oriente Moderno*, 20 (2001), 184.
49. Knecht, *Renaissance Warrior and Patron*, 489.
50. Holt et al., *The Cambridge History of Islām*, i. 329. R. J. Bonney, '"For God, Fatherland and Freedom": Rethinking Pluralism in Hungary in the Era of Partition and Rebellion, 1526–1711', in *The First Millennium of Hungary in Europe*, ed. K. Papp, J. Barta et al. (Debrecen, 2002), 377–96.
51. Fleischer, '*Mahdī* and Millennium', 48.
52. Ibid. 48, 51. Kunt and Woodhead, *Süleyman the Magnificent and his Age*, 172.
53. Fleischer, '*Mahdī* and Millennium', 46–9.
54. Ibid. 45–6.
55. Kunt and Woodhead, *Süleyman the Magnificent and his Age*, 78.
56. Fleischer, '*Mahdī* and Millennium', 50–1.
57. Kunt and Woodhead, *Süleyman the Magnificent and his Age*, 152–3.
58. Ágoston, 'Ottoman Warfare in Europe', 137.
59. Ibid. 135, shows that the number of Janissaries increased from 12,798 in 1567–68 to 54,222 in 1680, with a trebling of the number already by 1609.
60. Gábor Ágoston, 'Ottoman Conquest and the Ottoman Military Frontier in Hungary', in *A Millennium of Hungarian Military History*, ed. László Veszprémy and Béla K. Király (Boulder, CO: Social Science Monographs, distr. New York: Columbia University Press, 2002), 95.
61. Karen Barkey, *Bandits and Bureaucrats: The Ottoman Route to State Centralization* (Ithaca, NY: Cornell University Press, 1994).

62. Sir Thomas Roe, *A continuation of the Turkish history, from the beginning of the yeare… 1620, vntil the ending of the yeare .. 1628. Collected ovt of the papers and dispatches of Sr. Thomas Rowe… And since by him re-viewed and corrected* [in Knolles, *The general historie of the Turkes*], 1409–11.
63. Ibid. 1424.
64. Turan, 'Some Reflections on the Ottoman Grand Vizierate'.
65. Roe, *A continuation of the Turkish history*, 1398.
66. Ágoston, 'Ottoman Warfare in Europe', 138.
67. Kunt and Woodhead, *Süleyman the Magnificent and his Age*, 117.
68. Ibid. 184–5. E. I. J. Rosenthal, *Political Thought in Medieval Islam. An Introductory Outline* (Cambridge: Cambridge University Press, 1958), 226–33.
69. Kunt and Woodhead, *Süleyman the Magnificent and his Age*, 14.
70. Holt et al., *The Cambridge History of Islām*, i. 676–7.
71. Ágoston, 'Early Modern Ottoman and European Gunpowder Technology', 27.
72. Raymond W. Goldsmith, *Premodern Financial Systems. A Historical Comparative Study* (Cambridge: Cambridge University Press, 1987), 95. John F. Richards, *The New Cambridge History of India. 1.5 The Mughal Empire* (Cambridge: Cambridge University Press, repr. 2001), 1.
73. Goldsmith, *Premodern Financial Systems*, 81.
74. John F. Richards, *Power, Administration and Finance in Mughal India* (Ashgate: Variorum, 1993), ch. 5, 292, 296, 299.
75. Ibid., ch. 1.
76. Williams, *Themes of Islamic Civilization*, 282–4 for a description of a battle in 392/1001.
77. Richards, *Power, Administration and Finance in Mughal India*, ch. 2, 186, 191.
78. R. M. Eaton, *Essays on Islam and Indian History* (New Delhi: Oxford University Press, 2000), 99.
79. Ibid. 116. Romila Thapar writes: 'in [416/]1026, Maḥmud of Ghaznī raided the temple of Somanatha and broke the idol. Reference is made to this in various sources, or reference is omitted where one expects to find it. Some of the references contradict each other. Some lead to our asking questions which do not conform to what we have accepted so far in terms of the meaning and the aftermath of the event. An event can get encrusted with interpretations from century to century and this changes the perception of the event.' Romila Thapar, 'Somanatha and Mahmud', *Frontline*, 16:8 (1999): <www.flonnet.com/fl1608/16081210.htm>
80. Saiyid Athar Abbas Rizvī, *Religious and Intellectual History of the Muslims in Akbar's Reign: With Special Reference to Abū'l Fazl* (New Delhi: Munshiram Manoharlal, 1975), 276–67.
81. Ibid. 19.
82. Ibid. 399.
83. Eaton, *Essays on Islam and Indian History*, 116.
84. Muzaffar Alam and Sanjay Subrahmanyam (eds), *The Mughal State, 1526–1750*, (New Delhi: Oxford University Press, 2002), 145.
85. Ibid. 117.
86. Ibid. 118, 146.
87. Eaton, *Essays on Islam and Indian History*, 109 n. 37, who notes the 'absence of firm evidence of temple desecration by any of the early Mughals, in Ayodhya or elsewhere'. Jan-Peter Hartung, Gillian Hawkes, Anuradha Bhattacharjee, *Ayodhya, 1992–2003: The Assertion of Cultural and Religious Hegemony* (Leicester, 2003; Delhi: Media House, 2004).
88. Ram Puniyani, *Communal Politics. Facts versus Myths* (New Delhi: Sage, 2003), 54 (box 2.4). Annette Beveridge (who translated Bābur's chronicle, the *Bābur-nama*, into English) dismissed Bābur's Will as a forged document citing 15 different problems, including the conflict of dates (though both are in Bābur's lifetime). The evidence is inconclusive, although

the *Bābur-nama* has Bābur extolling his life of violence: 'For Islām's sake, I wondered in the wilds, / Prepared for war with pagans and Hindus, / Resolved myself to meet the martyr's death. / Thanks be to God! a *ġhāzī* I became'. Different translation in *Memoirs of Zehīr-ed-Dīn Muhammed Bābur…*, ed. William Erskine and Lucas King (2 vols) (Oxford: Oxford University Press, 1921; repr. New Delhi: Vintage Books, 1993), ii. 307 (at the date 933/1526). However, the apparent contradiction between the two texts may not be as great as appears, being the difference between the early ideology of raiding (the father) and a later philosophy of settled rule (the father's suggestions to his son).

89. Alam and Subrahmanyam, *The Mughal State, 1526–1750*, 120–2.
90. Ibid. 123.
91. Ibid. 98.
92. Mitchell, *Sir Thomas Roe and the Mughal Empire*, 10.
93. Eaton, *Essays on Islam and Indian History*, 255.
94. Alam and Subrahmanyam, *The Mughal State, 1526–1750*, 166.
95. Ibid. 132–5.
96. Rizvī, *Religious and Intellectual History of the Muslims in Akbar's Reign*, 69–70.
97. Ibid. 361.
98. Ibid. 362, 381, 382.
99. Ibid. 146–8, 150, 156.
100. *The Embassy of Sir Thomas Roe to the Court of the Great Mogul, 1615–1619, as Narrated in his Journal and Correspondence* (2 vols) (London: Hakluyt Society, 1899; repr. Leichtenstein: Klaus reprints, 1967), ii. 313. The 'new law' was mistakenly attributed to Akbar's successor: ibid. ii. 314.
101. Rizvī, *Religious and Intellectual History of the Muslims in Akbar's Reign*, 393–394, lists 18 followers.
102. Ibid. 164, for a different version of events. Ibid. 309, for his holding the governorship of Kabul under Akbar until his death in 993/1585, when it was annexed.
103. Mitchell, *Sir Thomas Roe and the Mughal Empire*, 11–12, 41 n. 21.
104. <www.vohuman.org/Article/Persia%20Past%20and%20Present.htm>
105. Saiyid Athar Abbas Rizvī, *Shāh Walī Allāh and His Times. A Study of Eighteenth-Century Islām, Politics and Society in India* (Canberra: Ma'rifat, 1980), 395. Ibid. 393: there was also a strong reason for its remission in that Hindus had often fought on the side of the Mughals. Ibid. 394, for rival groups and policies on the issue of Akbar's abolition of the *jizya* and whether the Hindus' duty was to remain subservient to Islām.
106. Satish Chandra, *Parties and Politics and the Mughal Court, 1707–1740* (New Delhi: Oxford University Press, 1959, repr. 2002), 297.
107. M. Athar Ali, *The Mughal Nobility Under Aurangzeb* (London: Asia Publishing House for Department of History, Aligarh Muslim University, 1966).
108. Ibid. 107. Rizvī, *Shāh Walī Allāh and His Times*, 101.
109. Ali, *The Mughal Nobility Under Aurangzeb*, 99.
110. Ibid. 174.
111. Ibid. 79–80.
112. Ibid. 89. Bernier called them 'timariots'.
113. Jagadish Narayan Sarkar, *Mughal Polity* (Delhi: IAD, 1984), 322, who concludes 'the creature killed its creator'.
114. Ali, *The Mughal Nobility Under Aurangzeb*, 92–4. Chandra, *Parties and Politics*, 35–9, 304.
115. Ali, *The Mughal Nobility Under Aurangzeb*, 94.
116. Chandra, *Parties and Politics*, 302.
117. Ibid. 198. Rizvī, *Shāh Walī Allāh and His Times*, 303.

118. Chandra, *Parties and Politics*, 304.
119. Daniel Goffman, *The Ottoman Empire and Early Modern Europe* (Cambridge: Cambridge University Press, 2002), 91–2.
120. Karpat, *The Politicization of Islam*, 50, 71.
121. Ibid. 51. Alan Palmer, *The Decline and Fall of the Ottoman Empire* (London: John Murray, repr. 1993), 58.
122. Palmer, *The Decline and Fall of the Ottoman Empire*, 70–2.
123. Karpat, *The Politicization of Islam*, 92–3.
124. Ibid. 111.
125. Palmer, *The Decline and Fall of the Ottoman Empire*, 92–3.
126. Kemal H. Karpat, *An Inquiry into the Social Foundations of Nationalism in the Ottoman State: From Social Estates to Classes, from Millets to Nations* (Princeton, NJ: Princeton University Press, 1973), 77–8.
127. Palmer, *The Decline and Fall of the Ottoman Empire*, 84–5.
128. Karpat, *The Politicization of Islam*, 95.
129. Karpat, *An Inquiry into the Social Foundations of Nationalism*, 85.
130. Erik Jan Zürcher, 'Ottoman Labour Battalions in World War I': <www.hist.net/kieser/aghet/Essays/EssayZurcher.html>
131. Karpat, *The Politicization of Islam*, 336.
132. Ibid. 341.
133. Karpat, *An Inquiry into the Social Foundations of Nationalism*, 87–93, 334.
134. Palmer, *The Decline and Fall of the Ottoman Empire*, 149.
135. Karpat, *The Politicization of Islam*, 125, 148. Ibid. 413, where Karpat asserts that the war was launched 'without any provocation'. Ibid. 147: 'this was an imperialist war, in fact an anti-Ottoman crusade'.
136. Ibid. 174. (Ibid. 256: 'Abdülhamid never made use of the call to *cihād* except in the war with Greece in 1897' appears to be made in error, a small slip in a remarkable study.) See Palmer, *The Decline and Fall of the Ottoman Empire*, 150–1.
137. Palmer, *The Decline and Fall of the Ottoman Empire*, 162. Karpat, *The Politicization of Islam*, 134, 149, 151, 186–8, 413.
138. Karpat, *The Politicization of Islam*, 165–6.
139. Palmer, *The Decline and Fall of the Ottoman Empire*, 184. Karpat, *The Politicization of Islam*, 171–2.
140. Palmer, *The Decline and Fall of the Ottoman Empire*, 177. Janet Klein, 'Power in the Periphery: The Hamidiye Light Cavalry and the Struggle over Ottoman Kurdistan, 1890–1914' (unpublished PhD thesis, Princeton University, 2002).
141. Karpat, *The Politicization of Islam*, 210.
142. Palmer, *The Decline and Fall of the Ottoman Empire*, 177.
143. At an otherwise tendentious website: <www.atmg.org/ArmenianProblem.html>

    A similar assertion of a revolutionary made to Dr Hamlin: Hunchak bands would 'watch their opportunity to kill Turks and Kurds, set fire to their villages, and then make their escape into the mountains. The enraged Moslems will then rise, and fall upon the defenceless Armenians and slaughter them with such barbarity that Russia will enter in the name of humanity and Christian civilisation and take possession': <www.atmg.org/ImperialDiplomacy.html>
144. Palmer, *The Decline and Fall of the Ottoman Empire*, 179, though the figures are disputed.
145. Donald Bloxham, 'The Armenian Genocide of 1915–1916: Cumulative Radicalization and the Development of a Destruction Policy', *Past and Present*, 181 (2003), 141–91, at 149.
146. Karpat, *The Politicization of Islam*, 176.
147. Ibid. 248.
148. Ibid. 246.

149. Jonathan Riley-Smith, '*Jihād* Crusaders. What Osama Bin Laden means by "crusade"' (5 January 2004): <www.nationalreview.com/comment/riley-smith200401050839.asp>
150. Karpat, *The Politicization of Islam*, 349–50. Karpat, *An Inquiry into the Social Foundations of Nationalism*, 112.
151. The military tribunal investigating the Armenian genocide affirmed: 'the Executive of the Ittihad and Terakki Party had taken decisive and audacious steps involving the fate of the nation and the country, that it declared war on its own without even consulting the Council of Ministers and obtaining that body's consent, something which it found to be unnecessary – although even the kings cannot arbitrarily declare war.' Verdict (*Kararname*) of the Turkish Military Tribunal, 1337/5 July 1919: <www.armenian-genocide.org/Affirmation.237/current_category.50/affirmation_detail.html>
152. Karpat, *The Politicization of Islam*, 256, 370. R. Peters, *Jihād in Classical and Modern Islam* (Princeton, NJ: Markus Weiner, 1996), 55–7. Bloxham, 'The Armenian Genocide', 157, notes that 'measures against Armenians developed initially in tandem with a general anti-Christian chauvinism, encouraged by a declaration of *cihād*... in November 1914, and illustrating the lengths to which the government was prepared to go to protect its territory. Christians and Entente nationals were cast as collective targets when Talat and Cemal threatened reprisals against them respectively for any Muslims that died in bombardments of coastal settlements; there was of course no mention of reprisals for Ottoman Christian deaths.' Ibid. 158: 'the *cihād* was announced with German encouragement to smooth the path for invasion of the Caucasus and Persian Azerbaijan, and to appeal to Muslims subjects of Britain and Russia. It was one of a broader set of strategies used by both sides in the conflict to undermine the other by stimulating anti-imperial insurgency on ethnic and/or national grounds, conceptually comparable to the British sponsorship of the Arab revolt or the German appeals to Ukrainian nationalists.' Ibid. 161: '...the *cihād* declaration was also largely unsuccessful, though the (false) religious imperative probably encouraged some Muslims to participate in the coming Armenian genocide'.
153. M. Naeem Qureshi, *Pan-Islam in British Indian Politics. A Study of the Khilafat Movement, 1918–1924* (Leiden, Boston and Cologne: Brill, 1999), 73. Ibid. 72: 'nowhere in India is there any sign that a Turkish *jihād* would induce Indian Mahomedans to give really serious trouble...' (15 September 1914).
154. Palmer, *The Decline and Fall of the Ottoman Empire*, 226.
155. Morgenthau added that on a subsequent occasion, Talat asked him to obtain from 'the American life insurance companies to send us a complete list of their Armenian policy holders. They are practically all dead now and have left no heirs to collect the money. It of course all escheats to the State. The Government is the beneficiary now. Will you do so?' Morgenthau records that he lost his temper at this suggestion, calling it 'the most astonishing request I had ever heard'. Henry Morgenthau, *Ambassador Morgenthau's Story* [dedicated to President Woodrow Wilson] (New York: Doubleday: 1918): <www.cilicia.com/morgenthau/Morgen25.htm>
156. Nubar commented that 'the [Armenian] volunteers were a danger in pushing the Turks to revenge', though he declared he 'could never have imagined that revenge would reach such a degree of cruelty and savagery': quoted by Bloxham, 'The Armenian Genocide', 185–6. Ibid. 184: 'the Armenian political leaderships were not simply dupes, fooled into collaboration by Russian lies about future autonomy, though the Russian role in fostering an explosive situation does need to be highlighted'.
157. The Entente issued this declaration at Russian instigation: Bloxham, 'The Armenian Genocide', 179–80. Bloxham comments that the Armenians 'actually received least in the way of help from the Entente when they most needed it': ibid. 187. Report Prepared by the United Nations War Crimes Commission, 28 May 1948: <www.armenian-genocide.org/Affirmation.168/current_category.6/affirmation_detail.html>

158. Bloxham, 'The Armenian Genocide', 180–1. Bloxham adds: 'the circumstances surrounding these developments were truly poisonous. New elements of vengeance for the events of the War, the fall of Van and attacks on Muslims had been amalgamated with established suspicions of Russian–Armenian collaboration, and with a sense of the Committee of Union and Progress having nothing left to lose since criminal culpability had already been invoked on the international stage in this latest chapter of great power intervention in Ottoman affairs.' Talat claimed that he feared the international condemnation general deportation would bring: ibid. 179 and n. 198. The auctioning off of Armenian property authorized by Talat's Ministry of the Interior on 9 June 1915 shows that the Armenians were clearly not expected to return: ibid. 182.
159. Cf. ibid. 190: 'if we think more along the lines of a "policy of annihilation", we get the idea of a general consensus of destruction of the Armenian national community, a consensus which developed and was augmented over time… thus phases of acceleration and radicalization become more appropriate terms of reference than discernible, discrete shifts in intent'.
160. Palmer, *The Decline and Fall of the Ottoman Empire*, 234. Bloxham, 'The Armenian Genocide', 141, suggests that 'at least one million Armenians died, more than two-thirds of those deported'. There were also forcible conversions to Islām. The official Turkish estimate is in the region of 300,000, which seems improbably low. Many of the mass graves have not been found. Robert Fisk records how, 'in the spring of 1993, with my car keys, I slowly unearthed a set of skulls from the clay wall of a hill in northern Syria. I had been looking for the evidence of a mass murder – the world's first genocide – for the previous two days but it took a 101-year-old Armenian woman to locate the river bed where her family were murdered in the First World War. The more I dug into the hillside next to the Habur river, the more skulls slid from the earth, bright white at first then, gradually, collapsing into paste as the cold, wet air reached the calcium for the first time since their mass murder. The teeth were unblemished – these were mostly young people – and the bones I later found stretched behind them were strong. Backbones, femurs, joints, a few of them laced with the remains of some kind of cord. There were dozens of skeletons here. The more I dug away with my car keys, the more eye sockets peered at me out of the clay. It was a place of horror.' Robert Fisk in the *Independent* (5 August 2000): <www.hr-action.org/archive3/090800ind.html>
161. There were eight or nine different cabinets between 1918 and 1923, but several of them were headed by Damad Ferit Pasha.
162. Verdict (*Kararname*) of the Turkish Military Tribunal, 1337/5 July 1919: 'they created an even greater atmosphere of harassment of the non-Islamic elements of the land, the Armenians in particular, who had hoped, from our precious Constitution, for justice and peace. These people now understood that they had been victimised by hypocrisy, and they assumed the posture of awaiting that opportune moment when they would be able to realize their former national aspirations. And the cause of all this were the Ittihadists themselves. They even raised national and racial issues among the Moslems of the land, they promoted divisiveness and conflict and jeopardized Ottoman unity.' <www.armenian-genocide.org/Affirmation.237/current_category.50/affirmation_detail.html>
163. The Grand National Assembly conferred this title on Kemal on 19 September 1921: Williams, *Themes of Islamic Civilization*, 299.
164. Palmer, *The Decline and Fall of the Ottoman Empire*, 259.
165. Ibid. 265.

Chapter 6

1. Even his dates of birth and death are a matter of controversy. Birth dates range from 1702 to 1707 and death dates from 1791 to 1797. Natana DeLong Bas, the most recent scholar studying his thought has taken 1702/3 for the birth date and 1791/2 for the death date. Natana

DeLong Bas, *Wahhābī Islām. From Revival and Reform to Global Jihād* (New York: Oxford University Press, 2004). The author is particularly grateful to Dr DeLong Bas for sharing her findings prior to publication and for her comments on the draft text of this book. Part of the dating difference is due to the conversion of *hijrah* years to common era, but there are still variations in the sources. In contrast to the balanced reassessment by DeLong Bas, Jalal Abualrub, *Muḥammad Ibn Abdul Wahhāb. His Life-Story and Mission* (Orlando, FL: Madinah, 2003) is an anti-Ṣūfī work (cf. ibid. 280 n. 757, where Ṣūfīsm is described as 'an ideology [*sic*] completely alien to Islām').

2. Stephen Schwartz, *The Two Faces of Islam. The House of Sa'ud from Tradition to Terror* (New York: Doubleday, 2002), 107, 192. This was reprinted as Schwartz, *The Two Faces of Islam. Saudi Fundamentalism and its Role in Terrorism* (New York: Anchor Books, repr. 2003), which reveals the agenda more clearly.
3. Hamid Algar, *Wahhabism: A Critical Essay* (Oneonta, NY: Islamic Publications International, 2002), 2, 4, 5.
4. This view diverges completely from the later argument of Sayyid Quṭb, for whom conversion has never been a goal of *jihād*: see Chapter 8.
5. DeLong Bas, *Wahhābī Islām*, conclusion, at 287–90.
6. Though it is possible to argue that followers of Muḥammad Ibn 'Abd al-Wahhāb who espoused violence may have been doing no more than pursuing the internal logic of the master's arguments, but this is a semantic distinction. The important point is the chronological divide between what happened in his lifetime and after his death.
7. Michael Cook, *Commanding Right and Forbidding Wrong in Islamic Thought* (New York: Cambridge University Press, 2000), 170. Ibid. 169, where Cook notes that he did not write a separate treatise on the subject.
8. Cf. Algar, *Wahhabism: A Critical Essay*, 21: 'it is in the Khārijite movement that some historical antecedent for the Wahhābīs can perhaps be discovered…, with respect not to the details of doctrine but to their mode of interaction with others.' Cf. the criticism of contemporaries: 'Sometimes his opponents argued that he belonged to the Khawārij. Sometimes they said that he tore apart the Consensus (of the scholars) and claimed the authority for absolute *Ijtihad* [independent judgement in a legal or theological question], without considering any scholar or jurisprudent before him': <www.alinaam.org.za/library/hist_bio/ibnwahhaab.htm>

   For a modern detailed refutation of any similarities to the Khārijīs: Abualrub, *Muḥammad Ibn Abdul Wahhāb*, 287–95.
9. DeLong Bas, *Wahhābī Islām*, 195–9. For the doctrines of the Mu'tazilites: Cook, *Commanding Right and Forbidding Wrong in Islamic Thought*, 195–226.
10. <www.sunnah.org/tasawwuf/scholr25.htm>
11. At this point the two *shaykhs* laughed, though the seriousness of the objection was recognized by Ibn Taymīyah: 'you have spoken well. If only your master were as you say, for he would then be as far as possible from unbelief. But what he has said cannot sustain the meanings that you have given in my view.'
12. H. A. R. Gibb and J. H. Kramer (eds), *Concise Encyclopedia of Islām* (Leiden: E. J. Brill, 2001), 618.
13. Algar, *Wahhabism: A Critical Essay*, 11.
14. Ibid. 34.
15. Eamon Duffy, *The Stripping of the Altars. Traditional Religion in England, 1400–1580* (New Haven, CT, and London: Yale University Press, 1992), ch. 5, 155–205.
16. Abdul Hamid Siddiqi, 'Renaissance in Arabia, Yemen, Iraq, Syria and Lebanon: Muḥammad Bin 'Abd al-Wahhāb and His Movement', ch. 72 of *A History of Muslim Philosophy*, ed. M. M. Sharif (Lahore: Pakistan Philosophical Congress, n.d.), 1447. <www.muslimphilosophy.com/hmp/title.htm>

17. Schwartz, *The Two Faces of Islam*, 73. A *ḥadīth* cited by Schwartz has the Prophet prophesying about Najd: 'from that place will come only earthquakes, conflicts, and the horns of Satan'. Among Najd's sins were those of having produced not only a false prophet, rather succinctly known to history as 'Musaylima the Liar', but the rebel sect of the Khawārij, who were responsible for the assassination of Muḥammad's nephew the fourth caliph, 'Alī – an event pivotal to the rise of Shī'īsm. The same powerful tribe that produced the Khawārij, the Banu Tamim, was to produce al-Wahhāb over a thousand years later. For an Islamic website which quotes the Prophet's prophesy about Najd: <www.naqshbandi.org/ottomans/wahhabis.htm>
18. Madawi Al-Rasheed, *A History of Saudi Arabia* (New York and Cambridge: Cambridge University Press, 2002), 17.
19. Ameen Rihani, *Ibn Sa'oud of Arabia. His People and His Land* (London: Constable, 1928), 242, states specifically 'in the same year the alliance was contracted… the war against the *mushrikūn* and for unitarianism was declared'.
20. Algar, *Wahhabism: A Critical Essay*, 20, gives the date as 1159/1746. A still later date of 1160/1747 is given in Gibb and Kramers, *Concise Encyclopedia of Islām*, 619.
21. Rihani, *Ibn Sa'oud of Arabia*, 243, who placed the number of casualties as 4000 in total, 1700 Wahhābis and 2300 on the part of their opponents, 'that is, only 133 every year died by the sword'. However, this does suggest that the *jihād* was not a purely pacific struggle.
22. Cook, *Commanding Right and Forbidding Wrong in Islamic Thought*, 174–5.
23. DeLong Bas, *Wahhābī Islām*, 202.
24. Ibid. 207.
25. Ibid. 209, 211.
26. Abualrub, *Muḥammad Ibn Abdul Wahhāb*, 544, 550–7. The letter begins ibid. 509.
27. DeLong Bas, *Wahhābī Islām*, 206, 214. For a modern view (that of Ibn Āshūr) that *jihād* was only about conversion in the earliest period of Islām (a view which the neo-Ṣūfī *jihāds* of the nineteenth century seem to contradict): Yohanan Friedmann, *Tolerance and Coercion in Islam. Inter-faith Relations in the Muslim Tradition* (Cambridge: Cambridge University Press, 2003), 103.
28. DeLong Bas, *Wahhābī Islām*, 215.
29. Ibid. 224.
30. Ibid. 205.
31. Ibid. 200–1, 221, 223–4. For earlier discussions of apostasy: Friedmann, *Tolerance and Coercion in Islam*, ch. 4.
32. Mark J. Sedgwick, *Sufism. The Essentials* (Cairo and New York: the American University in Cairo Press, rev. edn. 2003), 95.
33. DeLong Bas, *Wahhābī Islām*, 225.
34. Kuwait was not recovered by the Ottomans (via Ibrāhīm Pasha of Egypt) until 1231/1816. Schwartz contends that Ibn 'Abd al-Wahhāb issued a *fatwā* of *jihād* against the Ottoman caliphate, which he considered to have deviated from the original teachings of the Prophet, but this is strongly contested by Natana DeLong Bas in a communication to the author, and there is no mention of such a *fatwā* in Kemal H. Karpat, *The Politicization of Islam: Reconstructing Identity, State, Faith and Community in the Late Ottoman State* (New York: Oxford University Press, 2001), 23. Cf. Schwartz, *The Two Faces of Islām*, 79. Mohajir Al-Saeed, 'Al-Saud: Past and Present': <www.khyber.demon.co.uk/history/saudi-arabia/past.htm>; <www.hizb-ut-tahrir.org/english/books/howthekwasdestroyed/chapter_03.html>
35. Jerzy Zdanowski, 'Some Comments on Islamic Welfare: The Case of the Wahhābī State': <www.valt.helsinki.fi/kmi/Julkais/WPt/1999/wp799.htm>
36. <www.alinaam.org.za/library/hist_bio/ibnwahhaab.htm> This incident is also recorded briefly in Gibb and Kramers, *Concise Encyclopedia of Islām*, 619.
37. Quoted in <www.english.russ.ru/politics/20011228-pr.html>

38. '*Fatwā* on the Heresy of Wahhābism': <www.hizmetbooks.org/Advice_for_the_Muslim/wah-31.htm>
39. Rihani, *Ibn Sa'oud of Arabia*, 248, 250.
40. Hamza Hendawi, 'Some Shiites blame Wahhābīs for Atrocities' (6 March 2004): 'Shiites have long been a favourite target of Wahhābī warriors in Iraq, who sacked Karbalā' in 1801 [*sic*], and twice laid siege to the nearby holy city of Najaf early in the 19th century.' <www.wtopnews.com/index.php?nid=255&sid=176467>
41. 'Against Wahhābism': <www.village.flashnet.it/users/fn034463/warning.html>
42. Karpat, *The Politicization of Islam*, 25–6. The pilgrimage was declared no longer compulsory for Muslims while Mecca remained under Wahhābī occupation. Reconstruction took place on a significant scale after the ending of the occupation: <www.hizmetbooks.org/Advice_for_the_Muslim/wah-42.htm>
43. <www.au.af.mil/au/awc/awcgate/loc/sa/wahhabi.htm>
44. Cook, *Commanding Right and Forbidding Wrong in Islamic Thought*, 189: 'if the Sa'ūdī state was not to lose its religious identity, it had to turn its righteousness inwards'. Clearly more work is required on the nineteenth century. Professor David Commins of the Clarke Center, Dickinson College, Carlisle, Philadelphia, has undertaken research on the religious scholars of the nineteenth century: <www.dickinson.edu/~commins/>
45. Schwartz, *The Two Faces of Islam*, 90.
46. Al-Rasheed, *A History of Saudi Arabia*, 23.
47. Schwartz, *The Two Faces of Islam*, 97.
48. Ibid. 104.
49. Cook, *Commanding Right and Forbidding Wrong in Islamic Thought*, 181. Rihani, *Ibn Sa'oud of Arabia*, 193, 203.
50. Rihani, *Ibn Sa'oud of Arabia*, 192.
51. Ibid. 202.
52. Ibid. 208.
53. Ibid. 210.
54. John S. Habib, *Ibn Sa'ud's Warriors of Islam. The Ikhwān of Najd and Their Role in the Creation of the Sa'udi Kingdom, 1910–1930* (Leiden: E. J. Brill, 1978).
55. Schwartz, *The Two Faces of Islam*, 103.
56. Apart from his invaluable publication, not a great deal seems to be known about Rutter. He did, however, produce an account of slavery in Arabia in 1933 and therefore seems to have formed part of the anti-slavery international movement: Anti-Slavery International, Part 2: Publications and Reports of Anti-Slavery International and predecessors, 1880–1979 (Adam Matthew Publications Ltd), reel 17 of microfilm sources: <www.adam-matthew-publications.co.uk/collect/p612.htm>
57. Gibb and Kramers, *Concise Encyclopedia of Islām*, 619. Eldon Rutter, *The Holy Cities of Arabia* (2 vols) (London: Putnam, 1928), i. 79.
58. Rutter, *The Holy Cities of Arabia*, i. 200.
59. Ibid. i. 270–1.
60. Ibid. i. 272–3.
61. Ibid. i. 282.
62. Ibid. ii. 256–7.
63. Ibid. i. 175. The allusion is to the iconoclasm in England during the Puritan Revolution, particularly the period of the Commonwealth and Protectorate (1649–60).
64. Ibid. i. 190.
65. Ibid.
66. Ibid. i. 189.

67. M. Naeem Qureshi, *Pan-Islam in British Indian Politics. A Study of the Khilafat Movement, 1918–1924* (Leiden, Boston and Cologne: E. J. Brill, 1999), 397. Ibid. 396: the Khilafatists had initially welcomed the *jihād* of ibn Sa'ūd in 1924.
68. Cook, *Commanding Right and Forbidding Wrong in Islamic Thought*, 182–7.
69. Al-Rasheed, *A History of Saudi Arabia*, 3. The author is a descendant of the al-Rashīd family that was displaced by ibn Sa'ūd.
70. Rihani, *Ibn Sa'oud of Arabia*, xv.
71. H. St J. B. Philby, *Arabia of the Wahhabis* (London: Constable, 1928), 66–7.
72. H. St J. B. Philby, *The Heart of Arabia. A Record of Travel and Exploration. I* (London: Constable, 1922), 306.
73. Schwartz, *The Two Faces of Islām*, 110–11: 'oil, more than control of Mecca and Medina, made Sa'ūdī Arabia a global power and its theopolitical [*sic*] ideology an issue for the world.' Elsewhere he writes: 'the Sa'ūdī regime poses as an ally of the democracies in the anti-terrorist coalition, while continuing to spend vast sums of its oil revenues to promote Wahhābī radicalism throughout the Islamic world and the Muslim communities in the West, including America.' <www.weeklystandard.com/Content/Public/Articles/000/000/000/277gkmhl.asp?pg=2>
74. <www.saudiaramco.com>
75. Loretta Napoleoni, *Modern Jihād. Tracing the Dollars behind the Terror Networks* (London: Pluto Press, 2003), 119.
76. <www.eia.doe.gov/emeu/cabs/saudi.html>; <www.saudinf.com/main/z002.htm>
77. Schwartz, *The Two Faces of Islam*, 273: 'the question is therefore not whether Sa'ūdī Arabia is a friend or a foe, but whether the Sa'ūdī regime can survive, and whether we [*sic*: the USA] should conspire with the Wahhābī–Sa'ūdī establishment to continue propping it up'. Schwartz concludes (ibid. 281) that the regime is part of the 'axis of evil – and possibly the most dangerous part'. Ibid. 286, where Schwartz argues that 'the fall of Wahhābīsm could help foster new relations between Jews, Christians and Muslims', while the liberation of Mecca and Medina from the dictatorship of the Wahhābīs 'would doubtless be greeted as a positive event' by other Muslims.
78. Algar, *Wahhabism: A Critical Essay*, 69–70: 'what has inspired this essay… is… a concern that the Wahhābīs have seriously distorted fundamental teachings of Islām; functioned for many decades as the ideological mainstay of a regime that has squandered the wealth of the Arabian peninsula; vilified Muslims, both Sunnī and Shī'a, as non-Muslim and shed blood; introduced or exacerbated division and strife wherever they have gone; destroyed a significant part of the cultural patrimony of all Muslims, first in the Ḥijāz and then in places such as Chechnya, Bosnia and Kosovo; and signally failed to contribute anything either to the intellectual elaboration of Islām or the advancement of its political and civilizational agenda in the present age.' It has been observed that Hamid Algar was Khomeini's official biographer. Relations between Iran and Sa'ūdī Arabia during Khomeini's period of rule were particularly strained. There is also, as has been observed, a long history of hatred between Wahhābīs and Shī'a. For these various reasons, although there is truth in Hamid Algar's observations they are not entirely objective.
79. For example, the Roman Catholic Archbishop of Liverpool, Derek Worlock, and the Anglican Bishop of Liverpool, David Sheppard, wrote: 'Thank God, the Lord's Prayer does not belong to any one Christian denomination. It is his gift to all of us and shared by all of us. To pray "Our Father" helps to remind us that we belong to the whole people of God. It is our experience that we are enriched when we bring ourselves to receive from others the insights which they are able to bring.' Worlock and Sheppard, *With Christ in the Wilderness. Following Lent Together* (London: Bible Reading Fellowship, 1990), 35.

80. Yokinder Sikand, 'How many more Karbalas and Quettas?', circulated on 4 March 2004 on the Indiadebates email distribution list. Sikand correctly adds with regard to Christian ecumenism: 'Christian theologians active in the movement remain committed to their own different interpretations of their faith. And yet that has not deterred them from reaching out in a spirit of positive appreciation to other Christian groups who have traditionally been considered their rivals. It is not the fear or hatred of a religious "other" that drives them to promote Christian unity. Rather, it is a spirit of openness and love and commitment to their common (although divergently understood) faith that impels many involved in the ecumenical movement.' See also Conclusion for further discussion of this issue.
81. Dr Natana DeLong Bas is undertaking new research on this issue.
82. <www.63.175.194.25/index.php?ln=eng&ds=qa&lv=browse&QR=783&dgn=4> The website is 'maintained by a famous Wahhābi scholar': <www.islamicweb.com/beliefs/creed/wahhab.htm>
83. Tariq Rahman, 'The *Madrassa* and the State of Pakistan', *Himāl South Asian* (February 2004): <www.himalmag.com/2004/february/essay.htm>
84. Ibid. The author comments: 'according to [P. W.] Singer, about 10–15 percent of the *madrassas* are "radical", including anti-American rhetoric in their instruction and even imparting military training. No proof for these claims is offered, but they are credible given the fact that *madrassa* teachers often repeat the line that the United States is at war with Islām.'
85. *Unfulfilled Promises: Pakistan's Failure to Tackle Extremism* (ICG Asia Report 73, 16 January 2004), 11. <www.crisisweb.org//library/documents/asia/south_asia/073_unfulfil_promises_pakistan_extr.pdf>
86. Al-Ghazālī, *On the Boundaries of Theological Tolerance*, translated by Sherman A. Jackson (Karachi: Oxford University Press, 2002). The author's attention was drawn to this work by two important articles. Asma Barlas, 'Reviving Islamic Universalism: East/s, West/s, and Coexistence Conference on Contemporary Islamic Synthesis', Alexandria, Egypt, 4–5 October 2003: <www.ithaca.edu/faculty/abarlas/papers/barlas_20031004.pdf>

    Asma Barlas, 'Al-Ghazālī on Tolerance', *Daily Times* [Pakistan], 29 September 2003: <www.dailytimes.com.pk/?page=story_29-7-2003_pg3_4>
87. Al-Ghazālī, *On the Boundaries of Theological Tolerance*, 119–20.
88. Ibid. 85, 89.
89. Ibid. 106.
90. Ibid. 88.
91. Ibid. 120.
92. G. Makdisi, 'Hanbalite Islam', in *Studies on Islam*, ed. and trans. Merlin L. Swartz (New York: Oxford University Press, 1981), 252.
93. Ibid. 253, quoting the nineteenth-century Orientalist Goldziher. Goldziher argued that 'consensus' (*ijmā'*) was the distinctive mark of Sunnī Islām, although what one group regards as consensus another may reject. In contrast, Shī'īsm 'is based on authority'.
94. Ibid. 256–64.
95. <www.hazara.net/hazara/geography/Buddha/buddha.html>
96. <www.alhewar.com/statement_on_taliban_destruction.htm>; <www.alhewar.com/american_muslims_condemn_taliban.htm>
97. <www.portal.telegraph.co.uk/news/main.jhtml?xml=/news/2001/11/18/wbud18.xml&sSheet=/news/2001/11/18/ixhomer.html>; <www.hss.fullerton.edu/comparative/b%C3%A2mi%C3%A2n_buddhas_exart.htm>
98. DeLong Bas, *Wahhābī Islām*, is a careful re-evaluation of the distinctions between Ibn 'Abd Al-Wahhāb's views and post-Quṭb *jihādī* viewpoints. The main point at issue is Ibn 'Abd Al-Wahhāb's failure fully to distance himself from Ibn Sa'ūd's violence in his lifetime. He

cannot be blamed for violence after his death, or for more violent interpretations than his own. For present-day intolerance of others, see note 1.

Chapter 7

1. Usman Muḥammad Bugaje, 'A Comparative Study of the Movements of 'Uthman Dan Fodio in early nineteenth-century Hausaland and Muḥammad Aḥmad al-Mahdī in late nineteenth-century Sudan', 193. <www.webstar.co.uk/~ubugaje/comparativestudy1.pdf>
2. John O. Voll, 'Renewal and Reform in Islamic History: *Tajdīd* and *Iṣlaḥ*', in *Voices of Resurgent Islam*, ed. John L. Esposito (New York, Oxford University Press, 1983), 32–3.
3. Kemal H. Karpat, *The Politicization of Islam: Reconstructing Identity, State, Faith and Community in the Late Ottoman State* (New York: Oxford University Press, 2001), 21, 23, 44.
4. The era of colonialism is the focus of Rudolph Peters, *Islām and Colonialism. The Doctrine of Jihād in Modern History* (The Hague: Mouton, 1979).
5. I. M. Lewis (ed.) *Islam in Tropical Africa* (London: Hutchinson, 1969, repr. 1980), 135.
6. Nehemia Levtzion, *Muslims and Chiefs in West Africa. A Study of Islam in the Middle Volta Basin in the Pre-Colonial Period* (Oxford: Clarendon Press, 1968), 164.
7. P. M. Holt, Ann K. S. Lambton and Bernard Lewis (eds), *The Cambridge History of Islām* (Cambridge: Cambridge University Press, 1970), ii. 359.
8. Lewis, *Islam in Tropical Africa*, 142.
9. The modern translation of 'Arab Faqīh, *Futūḥ al-Ḥabaša. The Conquest of Abyssinia (16th Century)*, trans. Paul Lester Stenhouse (Hollywood, CA: Tsehai Publishers, 2003) has proved extremely difficult to locate.
10. Holt et al., *The Cambridge History of Islām*, i. 332. Ibid. ii. 385–6.
11. David Vô Vân, 'A propos du jihâd dans le *Futūḥ al-Ḥabasha*. De la lecture d'Alfred Morabia à la relecture d'"Arab-Faqīh"'. The author talks of 'la description minutieuse des faits et gestes d'Aḥmad Grañ, paragon du "combat dans la voie d'Allâh"': <www.cfee-fces.org/code/vovan.htm>
12. Lewis, *Islam in Tropical Africa*, 8–9.
13. A term which first appears in French travel accounts in 1617 and 1637: James Kritzeck and William H. Lewis (eds), *Islam in Africa* (New York: Van Nostrand–Reinhold, 1969), 88. J. Spencer Trimingham, *Islam in West Africa* (Oxford: Clarendon Press, 1959), 68. The term comes from *murābiṭ*, the inhabitant of a *ribāṭ*, a Muslim fortress, fortified monastery or advanced communications post: S. Abdullah Schleifer, '*Jihād* and the Traditional Islamic Consciousness', *Islamic Quarterly*, 4th quarter (1983), 191. H. A. R. Gibb and J. H. Kramers (eds), *Concise Encyclopedia of Islām* (Leiden: E. J. Brill, 2001), 475.
14. The crucial points here are that his title to rule (*elimān*) was different from that of Nāṣir al-Dīn (*Amīr al-Mu'minīn)*, and that Bundu was not governed by the rules of the *sharī'ah*. For a drastic scaling down of his *jihādī* credentials and any linkage with the later Futa Jallon *jihād*: Michael A. Gomez, 'The Problem with Malik Si and the Foundation of Bundu', *Cahiers d'Études africaines*, 100 (1985).
15. For Bundu specificity: Michael A. Gomez, *Pragmatism in the Age of Jihād: The Precolonial State of Bundu* (New York: Cambridge University Press, 1992). For Nāṣir al-Dīn's *jihād* to the south: ibid. 49.
16. Lewis, *Islam in Tropical Africa*, 42.
17. David Robinson, *The Holy War of Umar Tal: The Western Sudan in the Mid-Nineteenth Century* (Oxford: Clarendon Press, 1985). Web publication at: <www.pulaaku.net/defte/dRobinson>
18. Holt et al., *The Cambridge History of Islām*, ii. 366–7.

19. Ibraheem Sulaiman, *A Revolution in History. The Jihād of Usman Dan Fodio* (London and New York: Mansell, 1986), especially 122–48.
20. Mervyn Hiskett, *The Sword of Truth. The Life and Times of the Shehu Usuman Dan Fodio* (New York: Oxford University Press, 1973), 74. There is a second edition of this work (Evanston, IL: Northwestern University Press, 1994), but while there is a new introduction the pagination of the main text is the same.
21. Usman Muḥammad Bugaje distinguishes three periods (1) the early development of his thought, 1774–89; (2) the beginning of involvement with the Gobir rulers to the point of conflict period 1789–95; (3) the *Hijrah*, and the preparation for the armed clash, 1795–1804. Bugaje, 'The Contents, Methods and Impact of Shehu Usman Dan Fodio's Teachings, 1774–1804', 35: <www.webstar.co.uk/~ubugaje/udfcontmethimpact.pdf>
22. Hiskett, *The Sword of Truth*, 7–8.
23. Ibid. 66.
24. Ibid. 122.
25. Ibid. 119.
26. Bugaje, 'A Comparative Study of the Movements of 'Uthman Dan Fodio', appendix one, 245–50.
27. Ibid. 95, the prayer of Muḥammad Bello, the Shehu's son and chief commander, *c.* 1220/1806: 'O Lord, thou seest what our enemies are working against us. They are plotting against us by fraud. O God, turn their plotting on their own heads. O God, give the victory and the conquest to our people.'
28. Ibid. 100.
29. Ibid. 93–94, 101.
30. M. G. Smith, 'The *Jihād* of Shehu Dan Fodio: Some Problems', in Lewis, *Islam in Tropical Africa*, 218, 221. Hiskett, *The Sword of Truth*, 89, 100, 102.
31. Hiskett, *The Sword of Truth*, 97–8.
32. Thus Trimingham, *Islam in West Africa*, 82–3: 'from the nature of their training and reading it is not surprising that their religious compositions are stilted, unnatural, couched in legalistic language, and entirely unoriginal in thought… Thus 'Uthmān dan Fodio's tract on the conditions under which the waging of war upon bad Muslims is permissible is based upon the work of al-Majhīlī, which in turn is a compilation of citations from earlier authorities.'
33. Hiskett, *The Sword of Truth*, 63, 69, 132–3, 161. Ibid. 131: 'the reformers of Hausaland were certainly not Wahhābīs'. Holt et al., *Cambridge History of Islām*, ii. 367.
34. Hiskett, *The Sword of Truth*, 136–49.
35. Ibid. 115.
36. Ibid. 64–5, 151.
37. Ibid. 151–2.
38. Ibid. 153.
39. Holt et al., *Cambridge History of Islām*, ii. 372.
40. Smith, 'The *Jihād* of Shehu Dan Fodio', 224. Ibid. 223, where he remarks: 'when Shehu and Yunfa came to blows… latent hostilities and cleavages between Muslim and heathen, pastoralist and farmer, immigrant and native people, Fulāni and Hausa, all poured themselves into this conflict, with the result that the critical principles for which the Shehu stood were often obscured'.
41. Trimingham, *Islam in West Africa*, 83, noting a commentary on at-Tijāni's work, written in 1261/1845 and another work based on Fazāzi. John Ralph Willis, *In the Path of Allāh. The Passion of Al-Hajj 'Umar. An Essay into the Nature of Charisma in Islām* (London: Frank Cass, 1989), 127: 'Allāh informed me that I was authorized to undertake the *jihād*, and He repeated this three times.'
42. David Robinson, *The Holy War of Umar Tal*.

43. Robinson, ibid. uses the term ‘crusade’.
44. Ibid.
45. Cf. Gomez, *Pragmatism in the Age of Jihād*, 131–3.
46. Lewis, *Islam in Tropical Africa*, 237.
47. Peters, *Islām and Colonialism*, 57–9.
48. Ibid. 60–1.
49. 1300 ME = 1882–83 in the Common Era.
50. Usman Muḥammad Bugaje, ‘A Comparative Study of the Movements of ‘Uthman Dan Fodio’, 166–7.
51. ‘I am of the family of the Prophet of God, on whom be blessing and peace. My father is a Husni on his father’s side; and his mother, and my mother likewise, on the side of both father and mother [are] ‘Abbasids, and God knows that I am of the blood of Hussain, and these excellent indications will suffice for him who has been touched by His grace and [by] the faith.’ Ibid., appendix two, 250–6.
52. The various traditions are surveyed in Gibb and Kramers, *Concise Encyclopedia of Islām*, 310–13. An attempt to demonstrate their orthodoxy for Sunnīs is made at: <www.al-islam.org/encyclopedia/chapter2/2.html>
53. Ibid. 408.
54. Peters, *Islām and Colonialism*, 70–4.
55. Gibb and Kramers, *Concise Encyclopedia of Islām*, 311.
56. Lewis, *Islam in Tropical Africa*, 229.
57. Ibid. 230.
58. Ibid. 235.
59. Ibid. 232–3. P. M. Holt, *The Mahdist State in the Sudan, 1881–1898: A Study of its Origins, Development and Overthrow* (Oxford: Clarendon, 1958; 2nd edn Nairobi, 1979), 43–4.
60. Bugaje, ‘A Comparative Study’, 203, cites no authority for this assertion, which seems improbable since (ibid. 202) Gordon had been summoned to accept Islām but had refused; the alternative to conversion was death, as had happened to earlier opponents of the Mahdī.
61. Gibb and Kramers, *Concise Encyclopedia of Islām*, 408. The relevant passage was: ‘thou shall pray in Khartoum, then thou shall pray in the Mosque at Berber, then thou shall pray at the sacred House of God [the Ka‘ba of Mecca] then thou shall pray in the mosque of Yatrib [Medina], then thou shall pray in the mosque of Egypt, then thou shall pray at the Holy House [Jerusalem], then thou shall pray in the mosque of al-Iraq, then thou shall pray in the mosque of al-Kūfa…’ Bugaje, ‘A Comparative Study’, 217.
62. Holt et al., *The Cambridge History of Islām*, ii. 340.
63. Gibb and Kramers, *Concise Encyclopedia of Islām*, 173–4. Bugaje, ‘A Comparative Study’, 212–13.
64. Peters, *Islām and Colonialism*, 67.
65. Bugaje, ‘A Comparative Study’, 199.
66. Yohanan Friedmann, *Tolerance and Coercion in Islam. Inter-faith Relations in the Muslim Tradition* (Cambridge: Cambridge University Press, 2003), 119.
67. Gibb and Kramers, *Concise Encyclopedia of Islām*, 408: ‘Wahhābī influences are very probable in a number of regulations…’
68. Bugaje, ‘A Comparative Study’, 175, seems to justify this decision, while acknowledging that most people in the Sudan belonged to one of the Ṣūfī orders: ‘the Mahdī had to use his authority to submerge or at least enervate the *ṭuruq*, to allow for the emergence of a non-partisan Muslim solidarity. It is in this context that Mahdī’s abrogation of all Ṣūfī *ṭuruq* should be seen, even though the Mahdī justified his action by arguing that the mere appearance of a Mahdī obviates the need for the *ṭuruq*… the Mahdī extended this logic to the *madhāhib* (schools thought in *fiqh*) which he also abrogated…’

69. Ibid.
70. Ibid. 208. Contrast Shehu dan Fodio, for whom 'the role of the scholars in maintaining… solidarity [with the caliph] was vital'. Hiskett, *The Sword of Truth*, 145. Ibid. 123–5, for Shehu dan Fodio's belief in the advent of the Mahdī.
71. Uwe Halbach, '"Holy War" against Czarism: The links between Ṣūfīsm and *Jihād* in the Nineteenth-Century Anti-Colonial resistance against Russia', in *Muslim Communities Re-emerge: Historical Perspectives on Nationality, Politics, and Opposition in the Former Soviet Union and Yugoslavia*, ed. A. Kappeler, G. Simon and G. Brunner (Durham, NC: Duke University Press, 1994), 251–73.
72. Karpat, *The Politicization of Islam*, 34–5.
73. Halbach, '"Holy War" against Czarism', 253.
74. John Frederick Baddeley, *The Russian Conquest of the Caucasus* (London: Longmans, 1908): 'such were the people, who, without any external help, without artillery except what they could capture from the enemy, without trust in anyone but God and His Prophet, their own right hands and flashing swords, defied the Russian might for more than half a century; defeating her troops, attacking her colonies, and laughing with scorn at her wealth, pride, and numbers. And the story of their heroic struggle has its specific justification for the sympathy of its English readers. It is true that they fought for themselves alone – for their faith, freedom, and land. But they also stood, albeit not knowingly, as the protection of the British rule in India.' Quoted at: <www.cc.jyu.fi/~aphamala/pe/2003/tsets-2.htm>
75. Halbach, '"Holy War" against Czarism', 264.
76. Ibid. 266, 268.
77. Karpat, *The Politicization of Islam*, 37.
78. This paragraph draws on David Damrel, 'Ṣūfī Warriors. The Religious Roots of Conflict: Russia and Chechnya': <www.mubai.cc/articles/art34.htm>; Yavus Z. Akhmadov, 'Kunta Hadji and the Kunta Hadjists: The Kunta-Hadji Chechen Religious Movement': <www.jmu.edu/orgs/wrni/cs-part7.html>
79. <www.cc.jyu.fi/~aphamala/pe/2003/tsets-3.htm>
80. The expression is that of Karpat, *The Politicization of Islam*, 41. However, Hasan Israilov, a former journalist, and a member of the Communist Party, was the leader of the movement.
81. <www.utoledo.edu/~nlight/dissch1.htm>
82. Hodong Kim, *Holy War in China. The Muslim Rebellion and State in Chinese Central Asia, 1864–1877* (Stanford, CA: Stanford University Press, 2004), 129.
83. See Chapter 5 above. Kim, *Holy War in China*, 8–9, 39.
84. Ibid. 14.
85. Ibid. 73. Kim states that alien domination commenced in the 1680s.
86. Ibid. 17.
87. Ibid. 18, 179.
88. Ibid. 5, 47, 67, 181.
89. Ibid. 89–93, 182.
90. Ibid. 92–3. The previous regime lasted just three years from June 1864 to June 1867.
91. A 'kingdom of Islām' had been proclaimed there on 26 June 1864: ibid. 43. The revolt had also been led by Naqshbandī Ṣūfīs: ibid. 63.
92. Ibid. 71, 75.
93. Ibid. 130.
94. Ibid. 131. The Russian mission concluded that the population had reached 1.5 million in 1825 but had declined to 1.2 million by 1876: ibid. 124. The British estimated just over 1 million in 1873: ibid. 123. An army of 40,000 represented quite high levels of militarization against the background of a traditional economy, where barter was still the norm: ibid. 125.
95. Ibid. 99, 126–7, 151–2, 154. Karpat, *The Politicization of Islam*, 61.

96. Kim, *Holy War in China*, 187–93.
97. Ibid. 119.
98. Ibid. 160–1.
99. Ibid. 163.
100. Ibid. 169–71.
101. Ibid. 120.
102. Irfan Habib (ed.), *Confronting Colonialism: Resistance and Modernization under Haidar 'Alī and Tīpū Sultan* (orig. edn 1999; repr. London: Anthem Press, 2002).
103. Karpat, *The Politicization of Islam*, 49–51.
104. Ibid. 51.
105. Ibid. 74, 414–15. The Ottoman caliph's attitude was determined by British support in the Crimean War: 'in 1857 when the independent minded Muslims and Hindus of India joined forces to launch a war of independence against British rule, the *khalīfah* wrote and gave to the British a *fatwā* to the effect that the Muslims of India ought not to fight the British because the latter had proved to be supporters and well-wishers of the Islamic *Khilafat*': <www.tariq.bitshop.com/misconceptions/jihad/jbg.htm>
106. Qeyammudin Aḥmad, *The Wahhābī Movement in India* (2nd rev. edn Delhi: Manohar, 1994), 41. In spite of his ranting against 'innovation (*bid'ah*) of the "Ṣūfīstic polytheists"', 'heretics in Ṣūfīstic garb' and 'polytheists in Ṣūfīstic garb', as Aḥmad points out, apart from semantic convenience it is uncertain whether Sayyid Aḥmad should be regarded as a Wahhābī: he was considerably influenced by Ṣūfī ideas and practices: ibid. 66. While he went to Mecca, this was not during the period of Wahhābī domination. Wahhābī was the pejorative description used by the British, who sought to depict the movement as 'rebels' and as extremists and desecrators of shrines. The term *mujāhidīn* was considered by Aḥmad as generic and unspecific to the movement. Perhaps 'Indian proto-*salafīs*' might suffice as a descriptor.
107. Ibid. 50, 67.
108. Ibid. 61.
109. Ibid. 164, 169, 290–1.
110. Ibid. 153.
111. Ibid. 163.
112. Ibid. 169.
113. Ibid. 307.
114. F. W. Buckler, 'The Political Theory of the Indian Mutiny', *Transactions of the Royal Historical Society*, 4th ser. 5 (1922), 71–100, at 76, 99. F. W. Buckler, 'A Rejoinder to "The Political Theory of the Indian Mutiny" by Douglas Dewar and H. L. Garrett', ibid. 7 (1924), 160–5. Douglas Dewar and H. L. Garrett, 'A Reply to Mr. F. W. Buckler's "The Political Theory of the Indian Mutiny"', ibid. 7 (1924), 131–59.
115. The Indian Mahomedans. (Letters to the Editor) AN ANGLO-INDIAN OFFICER. *The Times*, 9 January 1880, 4; Issue 29773; Start column: E. Elec. Coll.: CS68075049.
116. W. W. Hunter, *The Indian Musalmans* (London: Trubner and Co., 1872, 2nd edn, 1872 [the second edition dropped the subtitle *Are They Bound in Conscience to Rebel against the Queen?*]), appendices 2 and 3. Ibid. 218–19, *fatwā* of 18 July 1870: 'the Musalmans here are protected by Christians, and there is no *jihād* in a country where protection is afforded, as the absence of protection and liberty between Musalmans and Infidels is essential in a religious war, and that condition does not exist here. Besides, it is necessary that there should be a probability of victory to Musalmans and glory to the Indians. If there be no such probability, the *jihād* is unlawful.' Ibid. 219, second *fatwā*, given by Maulavi Karamat Ali of the Calcutta Muhammadan Society: '...*jihād* can by no means be lawfully made in *Dār al-Islām*. This is so evident that it requires no argument or authority to support it. Now, if any misguided wretch, owing to his perverse fortune, were to wage war against the Ruling Powers of this

Country, British India, such war would be rightly pronounced rebellion; and rebellion is strictly forbidden by the Islamic Law. Therefore such war will likewise be unlawful; and in case any one would wage such war, the Muslim subjects would be bound to assist their Rulers, and, in conjunction with their Rulers, to fight with such rebels…'

117. Ahmad, *The Wahhābī Movement in India*, 307.
118. For the tribes: S. Iftikhar Hussain, *Some Major Pukhtoon Tribes along the Pak–Afghan Border* (Peshawar: Area Studies Centre and Hans Seidl Foundation, 2000).
119. Ahmad, *The Wahhābī Movement in India*, 189.
120. Ibid. 190–1.
121. Ibid. 192, 194, 267.
122. Peters, *Islām and Colonialism*, 79.
123. Ibid. 86.
124. Ibid. 74.
125. Ibid. 127.
126. Ibid. 93.
127. In so doing, he ran into difficulties with al-Afghānī, who was virulently anti-British: Christian W. Troll, *Sayyid Ahmad Khan: A Reinterpretation of Muslim Theology* (New Delhi: Vikas, 1978), 21–2.
128. Peters, *Islām and Colonialism*, 125.
129. Ibid.

Chapter 8

1. Farhat Haq, '*Jihād* over Human Rights, Human Rights as *Jihād*. Clash of Universals', in *Negotiating Culture and Human Rights*, ed. Lynda S. Bell, Andrew J. Nathan, Ilan Peleg (New York: Columbia University Press, 2001), 242–57, 247.
2. Ibrahīm M. Abū-Rabī', *Intellectual Resurgence in the Modern Arab World* (Albany, NY: State University of New York Press, 1996), 129.
3. K. R. Pruthi (ed.), *Encyclopaedia of Jihād* (5 vols) (New Delhi: Anmol Publications Pvt., 2002), i. 99. <www.islam.org.au/articles/15/PARLIAMT.HTM> Abū Bakr Başīr, a radical Indonesian cleric, is said to be the spiritual leader of Jemaah Islamiah (JI), a shadowy group accused of the 2002 Bali bombings. He has stated that he supports Osama bin Laden's struggle 'because his is the true struggle to uphold Islām, not terror – the terrorists are America and Israel': <www.news.bbc.co.uk/2/hi/asia-pacific/2339693.stm>
4. Sayyid Abul A'la Mawdūdī [or Maudūdī], *Towards Understanding the Qur'ān* [English version of *Tafhīm al-Qur'ān*] (New Delhi: Markazi Maktaba Islami Publishers, in progress: six volumes by 1998). Sayyid Quṭb, *In the Shade of the Qur'ān* [English version of *Fī Ẓilāl al-Qur'ān*], trans. and ed. Adil Salahi (Leicester: Islamic Foundation and Islamonline.net, in progress: nine volumes by 2004).
5. Reported to the author by an eyewitness, M. Risaluddin MBE, in September 2003.
6. Fathi Osman, 'Mawdūdī's Contribution to the Development of Modern Islamic Thinking in the Arabic-Speaking World', *The Muslim World*, 93 (July–October 2003), 465–85.
7. Seyyed Vali Reza Nasr, *Mawdūdī and the Making of Islamic Revivalism* (New York and Oxford: Oxford University Press, 1996), 138.
8. Sayyid Abul A'la Mawdūdī, '*Jihād fi Sabillah*: *Jihād* in Islām', trans. Khurshid Ahmad, ed. Huda Khattab (UK Islamic Mission Dawah Centre, Birmingham, 1995). Sayyid Abul A'la Mawdūdī, *Jihād in Islām* (International Islamic Federation of Student Organizations, Damascus, 1977). References are given to the Damascus edition first.
9. <www.islamistwatch.org/main.html>
10. *Jihād in Islām*, 22. '*Jihād* in Islām', 12.

11. Nasr, *Mawdūdī and the Making of Islamic Revivalism*, 27. Nasr gives a list of Mawdūdī's writings. Another list is to be found at <www.jamaat.org/overview/writings.html>
12. They seem to remain unknown to Muḥammad Qāsim Zamān, *The 'Ulamā' in Contemporary Islām: Custodians of Change* (Princeton, NJ, and Oxford: Princeton University Press, 2002), 47, 103.
13. Nasr, *Mawdūdī and the Making of Islamic Revivalism*, 20.
14. Azād had stated: 'there are serious misconceptions regarding what is *jihād*. Many people think that *jihād* means only to fight. The critics of Islām too labour under this misunderstanding, whereas to think thus is utterly to narrow the practical scope of this sacred commandment. *Jihād* means to strive to the utmost. In the Qur'ān and *Sunnah* terminology, this utmost exertion, which is undertaken for the sake of truth rather than personal ends, is indicated by the word *jihād*. This effort could be with one's life, or property, or expenditure of time, or by bearing labour and hardship, or fighting the enemy and shedding blood.'
15. Nasr, *Mawdūdī and the Making of Islamic Revivalism*, 20.
16. G. R. Thursby, *Hindu–Muslim Relations in British India. A study of the Controversy, Conflict and Communal Movements in Northern India, 1923–1928* (Leiden: E. J. Brill, 1975), 164. Ibid. 3. In *The Discovery of India* (Calcutta, 1946), 398–9, Nehru stated: 'The Arya Samaj was a reaction to the influence of Islam and Christianity, more especially the former. It was a crusading and reforming movement from within, as well as a defensive organization for protection against external attacks. It introduced proselytization into Hinduism and thus tended to come into conflict with other proselytizing religions. The Arya Samaj, which had been a close approach to Islām, tended to become a defender of everything Hindu, against what it considered as the encroachments of other faiths.'
17. Thursby, *Hindu–Muslim Relations in British India*, 40. An earlier Arya Samaj leader, Swami Dayanand (d. 1883), had stated of Islām (ibid. 13): 'such teachings deserve to be utterly discarded. Such a book, such a prophet and such a religion do nothing but harm. The world would be better off without them. Wise men would do well to discard a religion so absurd and accept the Vedic faith which is absolutely free from error.' What of the enshrining of the caste system within the laws of Manu? R. J. Bonney, 'Reflections on the Differences between Religion and Culture', *Clinical Cornerstone,* 6:1 (2004).
18. Gandhi's immediate response on 24 December to the assassination was that, 'in my opinion, if a Mussalman thinks than Abdul Rashid did well he will be disgracing his religion. For that is not his religion. His religion is something else. Now is the opportunity for the Mussalmans to show the real teachings of Islām.' Gandhi, xxxvii. 437 (reference is to the electronic edition of his writings: *The Collected Works of Mahatma Gandhi* (New Delhi: Ministry of Information and Broadcasting, 1999)). On 9 January 1927 he pronounced: 'if any Mussalman considers Swami Shraddhanand to have been an enemy, he is in error. Swamiji died as a hero. He pursued his mission according to truth and *dharma* with great courage.' Gandhi, xxxviii. 37. At various times in his life he denied that Islām was simply a religion of the sword. In his second lecture on Hinduism, delivered at Johannesburg on 11 March 1905 he had stated: 'to this inherent strength was also added the power of the sword. The fanatical raiders who, from time to time, found their way into India, did not hesitate to convert by the sword if they could not do so by persuasion. They more or less overran all parts of India, breaking idols after idols, and although the Rajput valour was at the disposal of Hinduism, it was powerless to afford protection against the Mahomedan inroad…' Gandhi, iv. 209. At a meeting of Christians on 4 August 1925, he stated: 'in my opinion, it is not true to say that Islām is a religion of the sword. History does not bear that out.' Gandhi, xxxii. 246. In a letter dated 22 May 1927 he commented: 'I adhere to the conclusion that Islām is not a religion of the sword; but it is like all other great religions a religion of peace. I say so for this reason. I have met numerous Mussalmans who no more think of slaying men of different faiths than you or I do. And these Mussalmans are

by no means scoffers. They are devout followers of their faith. The long line of Sufīs whose veneration for the Koran cannot be questioned owe their illuminating philosophy of peace and love to the Koran. I have read Mawlānā Shiblī's *Life of the Prophet* as also portions of his *Al-Kalām*. I have read also his *Leaves from the Lives of the Companions of the Prophet*. The sum-total of the impression left on my mind by these writings is of an elevating character. I hope you do not wish to suggest that Maulana Shibli and such other writers on Islām have written what they did not believe and in order to throw dust in the eyes of others. All this does not however mean that I regard the life of the Prophet to have been perfect or that the Koran itself is a perfect book. Like all other religious books including our own, there are passages which cause difficulty. But the difficulties caused in reading the Koran are not greater than those caused by reading books of other faiths.' Gandhi, xxxviii. 424. Swami Vivekananda had earlier stated that Islām could not be regarded as a 'religion of the sword' since conversions had occurred among the depressed classes who sought enfranchisement: *The Complete Works of Swami Vivekananda*, iii. 294–5, 298 (though other passages are much more critical of Islām).

19. Gandhi, xxxvii. 457.
20. Nasr, *Mawdūdī and the Making of Islamic Revivalism*, 22.
21. Sayyed Vali Reza Nasr, *The Vanguard of the Islamic Revolution: The Jama'at-i Islami of Pakistan* (Berkeley, CA: University of California Press, 1994), ch. 1. <www.ark.cdlib.org/ark:/13030/ft9j49p32d/>
22. Hadhrat Mirza Ghulam Aḥmad of Qadian published a statement on *jihād* on 22 May 1900 in which he stated: 'with the advent of the Promised Messiah it is incumbent upon every Muslim to give up *jihād*. If I had not come there could have been some excuse for this misconception. But with my advent you have become witnesses of the appointed hour. Now you have no excuse for using swords for religious battles before God.' <www.alislam.org/library/articles/new/British_Government_and_Jihad.html>
23. Nasr, *Mawdūdī and the Making of Islamic Revivalism*, 21–2. Aḥmadīs naturally consider that theirs is the correct view of *jihad*: <www.ahmadiya.com/jihad/> Mawdūdī was consistent in his opposition to the Aḥmadīs, supporting the anti-Aḥmadī agitation in Pakistan in 1373/1953–54.
24. J. Slomp, 'The "Political Equation" in *Al-Jihād fi al-Islām* of Abul a'la Mawdūdī (1903–1979)', in *A Faithful Presence: Essays for Kenneth Cragg*, ed. David Thomas with Clare Amos (London: Melisende, 2003), 237–55 at 244. This summary draws upon Slomp's chapter-by-chapter discussion of Mawdūdī's work.
25. Michael Cook, *Commanding Right and Forbidding Wrong in Islamic Thought* (New York: Cambridge University Press, 2000), 512 n. 50, cites Mawdūdī only once; but, as with Ibn Taymīyah, the discussion of *jihād* is predicated on this precept.
26. Slomp, 'The "Political Equation"', 248. The passage is translated by C. J. Adams, 'Mawdūdī on "The Necessity of Divine Government for the Elimination of Oppression and Injustice"', in *Muslim Self-Statement in India and Pakistan, 1857–1968*, ed. A. Ahmad and G. E. von Grunebaum (Wiesbaden, 1970), 156–7, at 156.
27. Nasr, *Mawdūdī and the Making of Islamic Revivalism*, 83.
28. Ahmed, *The Concept of an Islamic State. An Analysis of the Ideological Controversy in Pakistan* (London: Pinter, 1987), 104.
29. Ibid. 167 n. 27. Cf. *Jihād in Islām*, 17. '*Jihād* in Islām', 10.
30. Ishtiaq Ahmed, *The Concept of an Islamic State*, 105.
31. 'To alter people's outlook and spark a mental and intellectual revolution through the medium of speech and the written word is a form of *jihād*. To change the old tyrannical system and establish a just new order by the power of the sword is also *jihād*, as is spending wealth and

undergoing physical exertion for this cause.' It must at all times be 'in the cause of Allāh' (*jihād fī sabīl Allāh*). *Jihād in Islām*, 5. '*Jihād* in Islām', 4.

32. Nasr, *Mawdūdī and the Making of Islamic Revivalism*, 88: 'the state as Mawdūdī conceptualised it remained fundamentally antithetical to pluralism…' Ibid. 99: 'Mawdūdī's Islamic state, although an ideal, was intended for India and only later for Pakistan. At first, therefore, it had to confront cultural pluralism, communalism and minorities.' Mawdūdī commented: 'Islamic *jihād* does not seek to interfere with the faith, ideology, rituals of worship and social customs of the people. It allows them complete freedom of religious belief, and permits them to act according to their creed. However, Islamic *jihād* does not recognize their right to administer affairs of state according to a system which, in the view of Islām, is evil. Furthermore, Islamic *jihād* also forbids them to continue with such practices under an Islamic government if those practices are detrimental to the public interest according to Islām… With a view to ensuring the general welfare of the public and for reasons of self-defence, the Islamic government will not permit such cultural activities as may be permissible in non-Muslim systems but which Islām regards as detrimental and even fatal to [its] moral fibre.' *Jihād in Islām*, 27. '*Jihād* in Islām', 14. In other words, minorities remain minorities and cannot have other than second-class status. They do not have rights so much as concessions.
33. *Jihād in Islām*, 19. '*Jihād* in Islām', 11.
34. 'Islamic *jihād* is both offensive and defensive at one and the same time. It is offensive because the Muslim party attacks the rule of an opposing ideology, and it is defensive because the Muslim Party is constrained to capture state power in order to protect the principles of Islām in space-time forces. As a party, it has no home to defend: it upholds certain principles which it must protect. Similarly, this party does not attack the home of the opposing party, but launches an assault on the principles of the opponent. The objective of this attack is not to coerce the opponent to relinquish his principles, but to abolish the government which sustains them.' *Jihād in Islām*, 26. '*Jihād* in Islām', 14.
35. *Jihād in Islām*, 32–3. '*Jihād* in Islām', 16.
36. Sayyid Abul A'la Mawdūdī, *Fundamentals of Islām* [*Khutabat*] (Lahore: Islamic Publications, 1975), 263: 'their work is just to sit comfortably doing service to their *Nafs*…'
37. Nasr, *Mawdūdī and the Making of Islamic Revivalism*, 40.
38. Nasr, *The Vanguard of the Islamic Revolution*, ch. 1.
39. Mawdūdī, *Fundamentals of Islām*, 263. Ibid. 258 n. 1, where the editor provides a reminder that the sermons were given in 1938.
40. Khurram Jah Murad (trans.), *Let Us Be Muslims* (Leicester: Islamic Foundation, 1985). <www.youngmuslims.ca/online_library/books/let_us_be_muslims/ch7top28.html>
41. Sayyid Abul A'la Mawdūdī, *The Islamic Law and the Constitution*, trans. and ed. Khurshid Ahmad (Lahore: Islamic Publications. 1st edn., 1955; 2nd edn 1960, which included speeches of Mawdūdī and additional documents). Quotations are from the 1960 edition. This work is discussed in detail by C. J. Adams, 'Mawdūdī and the Islamic State', in *Voices of Resurgent Islam*, ed. J. L. Esposito (New York and Oxford: Oxford University Press, 1983), 99–133, but there is no discussion of *jihād*, a key concept in Mawdūdī's system of how to arrive at the Islamic state.
42. Nasr, *Mawdūdī and the Making of Islamic Revivalism*, 93.
43. Zaman, *The 'Ulamā' in Contemporary Islām*, 104–5.
44. A. G. Noorani, *Islam and Jihād. Prejudice versus Reality* (London: Zed Books, 2002), 78.
45. Qamaruddin Khan, *Political Concepts in the Qur'ān* (Karachi: Institute of Islamic Studies, 1973), 74.
46. Mawdūdī, *The Islamic Law and the Constitution*, 177. Quoted by Ahmed, *The Concept of an Islamic State*, 94.

47. Mawdūdī, *The Islamic Law and the Constitution*, 160: 'the powers which the dictators of Russia, Germany and Italy have appropriated or which Ataturk has exercised in Turkey have not been granted by Islām to its *Amīr* (leader) '.
48. Mawdūdī, *The Islamic Law and the Constitution*, 154–5. Quoted by Ahmed, *The Concept of an Islamic State*, 94.
49. Mawdūdī, *The Islamic Law and the Constitution*, 148, 235. Ahmed, *The Concept of an Islamic State*, 98.
50. Mawdūdī, *The Islamic Law and the Constitution*, 334: '...joint electorate is totally opposed to the needs of our country and our people, and the system of separate electorate alone is in consonance with the ideology and needs of Pakistan'.
51. Ibid., 318.
52. In a talk given at Lahore on 16 November 1975, he underlined the importance of freedom of conscience and conviction and the protection of religious sentiments: Sayyid Abul A'la Mawdūdī, *Human Rights in Islām*, trans. and ed. Khurshid Ahmad (Leicester: Islamic Foundation, 1976, repr. 1993), 30. Ahmed, *The Concept of an Islamic State*, 103.
53. Ahmed, *The Concept of an Islamic State*. Mawdūdī's essay on the subject was published originally in 1942–1943. It was translated in 1994 by Syed Silas Husain and Ernest Hahn as 'The Punishment of the Apostate according to Islamic Law': <www.answeringislam.org/Hahn/Mawdudi/index.htm#preface>
54. Ahmed, *The Concept of an Islamic State*, 118. For Mawdūdī's conception of political principles in the Qur'ān: Nasr, *Mawdūdī and the Making of Islamic Revivalism*, 81. Qamaruddin Khan, *Political Concepts in the Qur'ān*, 73: '...the Qur'ān does not provide any principle of constitutional law or political theory'; 'the political theory in Islām has developed not out of the Qur'ān, but out of historical circumstances, and hence carries no religious sanctity itself'.
55. Mawdūdī, *Towards Understanding the Qur'ān*, i. 169.
56. Ignaz Goldziher, *Introduction to Islamic Theology and Law*, trans A. and R. Hamori (Princeton, NJ: Princeton University Press, 1981), 102, cites al-Māwardī for the division between lands of war and lands of Islām.
57. Mawdūdī, *Towards Understanding the Qur'ān,* ii. 74.
58. Ibid. iii. 210.
59. Ibid. iii. 233. Later on, the word *qitāl* was used in this context: ibid. iii. 275 (Q.9:123).
60. Ibid. iii. 240.
61. Ibid. iii. 260.
62. Ibid. vi. 70.
63. Nasr, *Mawdūdī and the Making of Islamic Revivalism*, 42–3.
64. M. Rafique Afzal (ed.), *Speeches and Statements of Quaid-i-Millat Liaquat 'Ali Khan, 1941–51* (Research Society of Pakistan, Lahore, 1967), 563. Ibid. 622, Nehru to Liaquat 'Ali Khan, 24 July 1951.
65. Nasr, *Mawdūdī and the Making of Islamic Revivalism*, 45.
66. Sayyid Abul A'la Mawdūdī, *Towards Understanding Islam* (Leicester: The Islamic Foundation, 1980), 94. Mawdūdī's preface is dated 11 September 1960.
67. Nasr, *Mawdūdī and the Making of Islamic Revivalism*, 44.
68. The author is indebted to Dr Ataullah Siddiqui for this reference. *The Muslim*, vol. 6 (February 1969), 104–7, especially at 106–7 in answer the question 'do you think that the Islamic State can be established by an armed revolt?' His answer was that he did not think this 'the right road to pursue and such a policy may, instead of producing anything good, prove to be highly harmful'.
69. This paragraph is capitalized in the original.

70. C. Wendell (ed.), *Five Tracts of Ḥasan al-Bannā', 1906–1949. A selection from the Majmū'at Rasā'il al-Imām al-Shahīd Ḥasan al-Bannā* (Santa Barbara, CA: University of California Press, 1978), 133–61. Online version at: <www.youngmuslims.ca/online_library/books/jihad/>
71. It is thus difficult to portray it as a Wahhābī movement, as does one of the anti-terrorist websites: <www.fas.org/irp/world/para/mb.htm>
72. <www.maschicago.org/library/misc_articles/hassan_banna.htm>
73. Cf. L. Carl Brown, *Religion and the State. The Muslim Approach to Politics* (New York: Columbia University Press, 2000), 146.
74. A. S. Moussalli, *Moderate and Radical Islamic Fundamentalism. The Quest for Modernity, Legitimacy and the Islamic State* (Gainesville, FL: University of Florida Press, 1999), 55.
75. *Five Tracts of Ḥasan al-Bannā'*, 44.
76. Ibid. 82.
77. Ibid. 56–7.
78. A. S. Moussalli, *The Islamic Quest for Democracy, Pluralism and Human Rights* (Gainesville, FL: University of Florida Press, 2001), 123.
79. Kennedy, *The Prophet and the Age of the Caliphates*, 70.
80. Moussalli, *Moderate and Radical Islamic Fundamentalism,* 86–7, 124–5.
81. In contrast, L. Carl Brown terms the Brotherhood 'an Islamist totalitarian movement': Brown, *Religion and the State*, 148.
82. Wendell, *Five Tracts of Ḥasan al-Bannā'*, 113. He also rejected the Nazi and fascist racial policies: ibid. 4.
83. Basheer M. Nafi, 'The Arabs and the Axis: 1933–1940', *Arab Studies Quarterly*, 19 (1997), n. 14 citing Public Record Office, Foreign Office Papers, Lampson to Eden, 24 December 1942, FO 371/35578/J245 (Enclosure, Appendix A).
84. Moussalli, *The Islamic Quest for Democracy*, 67. Moussalli, *Moderate and Radical Islamic Fundamentalism,* 120.
85. M. K. al-Sayyid, 'The Other Face of the Islamist Movement', Carnegie Endowment, Democracy and Rule of Law Project. Global Policy Program, 33, January 2003. <www.ceip.org/files/pdf/wp33.pdf>
86. Moussalli, *Moderate and Radical Islamic Fundamentalism,* 117–18, 122.
87. Ibid. 122. Definition of three types of *shīrk* at: <www.usc.edu/dept/MSA/fundamentals/tawheed/abutaw/abutaw_7.html#HEADING6>
88. Wendell, *Five Tracts of Ḥasan al-Bannā'*, 59. Moussalli, *Moderate and Radical Islamic Fundamentalism,* 116–17.
89. Moussalli, *Moderate and Radical Islamic Fundamentalism,* 126, 128. Cook, *Commanding Right and Forbidding Wrong*, 523.
90. Moussalli, *Moderate and Radical Islamic Fundamentalism,* 129.
91. Ibid. 131.
92. Wendell, *Five Tracts of Ḥasan Al-Bannā'*, 155.
93. Ibid. 143.
94. Ibid. 151.
95. This might be taken to imply that the text was drafted during World War II.
96. S. M. Ḥasan al-Bannā', *Imām Shahīd Ḥasan al-Bannā'. From Birth to Martyrdom* (Swansea: Awakening Publications, 2002), 58.
97. M. S. bin Jani, 'Sayyid Quṭb's view of *Jihād*: An Analytical Study of his Major Works' (unpublished PhD thesis, University of Birmingham, 1998), 257–8.
98. Al-Bannā', *Imām Shahīd Ḥasan al-Bannā'*, 59.
99. A. S. Moussalli, *Radical Islamic Fundamentalism: The Ideological and Political Discourse of Sayyid Quṭb* (Beirut: American University of Beirut, 1992), 30.
100. Ibid. 43–4, 46.

101. Adil Salahi, 'Violence in Sayyid Quṭb's perspective', Quṭb, *In the Shade of the Qur'ān*, ix. Introduction at xviii–xix.
102. The author is grateful to Dr Salahi for this information, which arises from the testimony of Quṭb's sister; and also for his comments on this section.
103. Ibid. x.
104. Moussalli, *Radical Islamic Fundamentalism*, 46. The English translation suggests a date of 'about 1945', which must be in error: S. Kotb, *Social Justice in Islam*, trans. J. B. Hardie (American Council of Learned Societies, 1953; repr. New York: Octagon Books; 1980), frontispiece.
105. Quoted by I. M. Abū-Rabī', *Intellectual Origins of Islamic Resurgence in the Modern Arab World* (Albany, NY: State University of New York Press, 1996), 130.
106. Ibid. 116.
107. Kotb, *Social Justice in Islam*, trans. Hardie, 91.
108. Bin Jani, 'Sayyid Quṭb's View of *Jihād*', 344.
109. Ibid. 337–8.
110. Goldziher, *Introduction to Islamic Theology and Law*, 13. Muḥammad Asad uses the term 'pagan ignorance': Muḥammad Asad, *The Message of the Qur'ān* (Gibraltar: Dar Al-Andalus, 1980), 154 n. 71.
111. Bin Jani, 'Sayyid Quṭb's View of *Jihād*', 346.
112. Zamān, *The 'Ulamā' in Contemporary Islām*, 172, quoting Roxanne L. Euben, *Enemy in the Mirror: Islamic Fundamentalism and the Limits of Modern Rationalism: A Work of Comparative Political Theory* (Princeton, NJ: Princeton University Press, 1999), 85.
113. Moussalli, *Moderate and Radical Islamic Fundamentalism,* 150.
114. This section draws on Moussalli, *Radical Islamic Fundamentalism*, 70–125; Moussalli, *Moderate and Radical Islamic Fundamentalism,* 135–9. Haddad's interpretation substitutes 'lordship' for oneness of God, but these are related concepts: Y. Y. Haddad, 'Sayyid Quṭb: Ideologue of Islamic Revival', in *Voices of Resurgent Islam*, ed. J. L. Esposito (New York and Oxford: Oxford University Press, 1983), 74.
115. Haddad, 'Sayyid Quṭb', 76.
116. '...in a teleological view of the world and of history, the meaning and value of all historical events derives from their ends or purposes, that is, all events in history are *future-directed*': <www.wsu.edu:8080/~dee/GLOSSARY/TELE.HTM>
117. Moussalli, *Moderate and Radical Islamic Fundamentalism,* 154. Moussalli, *Radical Islamic Fundamentalism*, 162–3.
118. There is an error in the designation of translator in Emmanuel Sivan, *Radical Islām: Medieval Theology and Modern Politics* (New Haven, CT and London: Yale University Press, 1985), 23.
119. Quṭb, *In the Shade of the Qur'ān*, vii. 4–6, 179–180. Ibid. viii. 20–22. Ibid. vii. 13 for the division of mankind into three classes after the revelation of *Sūrah* 9: 'Muslims who believed in the Prophet's message; those at peace with him who enjoyed security; and those who were hostile and feared him.'
120. Ibid. vii. 25. Mawdūdī is quoted at length: ibid. vii. 25–37.
121. Bin Jani, 'Sayyid Quṭb's View of *Jihād*', 261–70.
122. Quṭb, *In the Shade of the Qur'ān*, vii. 133–4. Cf. ibid. xii, where Dr Salahi writes: 'it may be said, with some justification, that Sayyid Quṭb was a bit too strong in his argument, providing a platform for extremism to stand on. Here we may find ourselves trying to answer the question: to what extent may a writer be blamed for being misunderstood by his readers? In the case of Sayyid Quṭb, the overwhelming majority of his readers maintain that he reflects the middle path Islām adopts... [An objective, unbiased reading of Sayyid Quṭb's ideas] is bound to support the conclusion that powerful as his exposition of the concept of *jihād* is, it never advocates

extremism. It only defends the right of Islām to act in defence of mankind generally, of each person's right to have free access to God's message, and their freedom to choose their beliefs without pressure or coercion.' It should be noted that if this is indeed the correct interpretation of his writings, it is one that has been lost on many Western commentators.

123. Quṭb, *In the Shade of the Qur'ān*, viii. 41–50.
124. Ibid. vii. 190.
125. Quṭb, *In the Shade of the Qur'ān*, viii. 23–8. Moussalli, *Radical Islamic Fundamentalism*, 205–10.
126. Quṭb, *In the Shade of the Qur'ān*, viii. 77–83.
127. Ibid. viii. 124.
128. Quṭb, *In the Shade of the Qur'ān*, viii.150. The reference is to Wilfrid Cantwell Smith, *Islam in Modern History*.
129. Ibid. viii. 162.
130. Ibid. viii. 266.
131. Ibid. viii. 302.
132. Ibid. viii. 306.
133. Haddad, 'Sayyid Quṭb', 86, who argues that 'Quṭb projects a Manichean view of the world'.
134. <www.youngmuslims.ca/online_library/books/milestones/remember.asp>
135. Bin Jani, 'Sayyid Quṭb's View of *Jihād*', 339.
136. Ibid. 383–4.
137. Sayyid Quṭb, *Milestones* (Beirut and Damascus: International Islamic Federation of Student Organizations; 1978), 18 n. 1, 103. <www.youngmuslims.ca/online_library/books/milestones/chapter_4.asp>
138. Ibid. 139.
139. Bin Jani, 'Sayyid Quṭb's View of *Jihād*', 357.
140. Ibid. 356.
141. Haddad, 'Sayyid Quṭb', 88–9 ('reified by Quṭb to designate a necessary phase in the process of Islamizing society'), 90, 94.
142. Quṭb, *Milestones*, 31–3.
143. Quoted by Natana DeLong Bas, *Wahhābī Islām. From Revival and Reform to Global Jihād* (New York: Oxford University Press, 2004), 264, who calls it an 'absolutist, global vision of *jihād*'. Ibid. 259–61, where Dr DeLong Bas contrasts Muḥammad Ibn 'Abd al-Wahhāb's 'limited, geographically localized activity that could only occur under specific circumstances and conditions', with Quṭb's permanent *jihād* against *dār al-jāhilīyyah.*
144. <memri.org/bin/articles.cgi?Page=archives&Area=sd&ID=SP53903>
145. John L. Esposito, *Unholy War. Terror in the Name of Islām* (Oxford: Oxford University Press, 2002), 50–1, 61. For similarities in the world view of al-Bannā' and Mawdūdī: ibid. 52–3.

Chapter 9

1. Arzina R. Lalani, *Early Shī'ī Thought. The Teachings of Imām Muḥammad al-Bāqir* (London: I. B. Tauris and Institute of Ismaili Studies, 2000), 30.
2. Ibid. 70–1. Patricia Crone, *Medieval Islamic Political Thought* (Edinburgh: Edinburgh University Press, 2001), 73. The *ḥadīth* was reported by many Sunnī scholars, too, with the difference that the term *mawlā* was interpreted to mean the person in whom proximity exists, not necessary the master (hence *imām*) as for the Shī'a. According to Ayatollah Ja'far Sobhani [Subḥānī], 'the *ḥadīth* of Ghadir is accounted *mutawātir* (most authenticated) being related by companions, the followers of the companions, and countless narrators of *ḥadīth* down through the ages. A total of 110 companions, 89 of those in the succeeding generation,

and 3,500 scholars of *ḥadīth* have transmitted this *ḥadīth*, so there can be no question of disputing its authenticity': Sobhani, *Doctrines of Shī'ī Islām: A Compendium of Imāmī Beliefs and Practices*, trans. and ed. Reza Shah-Kazemi (London: I. B. Tauris and Institute of Ismaili Studies, 2001), 105. According to Wilferd Madelung, the Prophet did not settle the succession in his lifetime because he had expected a Qur'ānic revelation but did not receive one. He found it necessary to make a strong public statement of support of his cousin, following criticism of 'Alī's conduct as his representative in the Yemen, but 'it was evidently not a suitable occasion to appoint him successor': Wilferd Madelung, *The Succession to Muḥammad. A Study of the Early Caliphate* (Cambridge: Cambridge University Press, 1997; repr. 2001), 18. 'Alī had the *ḥadīth* of Ghadīr Khumm proclaimed in public and thus 'unequivocally claimed a religious authority superior to that of Abū Bakr and 'Umar': ibid. 253.

3. Lalani, *Early Shī'ī Thought*, 2, 5. The Sunnīs argue that since Aaron died before Moses, he could not be seen as his successor. Ibid. 74. An Ismā'īlī author, Abū al-Fawāris, argued that there was no evidence that 'Alī was deposed on the Prophet's return: ibid. 74.
4. Ibid. 27.
5. Madelung, *The Succession to Muḥammad*, 194.
6. Lalani, *Early Shī'ī Thought*, 76. The other addition to the number of the pillars was to include *jihād*.
7. Ibid. 69.
8. Ibid. 63. Usually interpreted by Sunnīs as either rulers or scholars.
9. 'Rafīḍa was a general abusive name for people considered as Shī'ites': H. A. R. Gibb and J. H. Kramers (eds), *Concise Encyclopedia of Islām* (Leiden: E. J. Brill, 2001), 466. However, the term could also be used for any 'rejectionist', Shī'ī or not: Madelung, *The Succession to Muḥammad*, 196.
10. Wilferd Madelung, noting that the battle of al-Nahrawān 'sealed the division between the Shī'a and Kharijites', and that the 'massacre' was 'the most problematic event in 'Alī's reign', argues that the Khārijīs 'would have been among his most vigorous allies in a war against the distorters of the rule of the Qur'ān… 'Alī could not agree to either of their demands, that he attest his own infidelity in accepting the arbitration, or that he treat his Muslim opponents as infidels. He could not simply condone the murders that some of them had committed. Patient argument with them, however, might have gradually won over most, if not all, of them. His first task would have been to restore a consensus among the Qur'ān readers, or at least the activists among them, who were his natural allies. It was the haste with which he sought to resume the war with Mu'āwiya that forced him to resort to counter-productive threats and violence against his former followers': ibid. 261–2. 'Alī faced the opposition of five Khārijī splinter-groups until 38/Feb. 659: ibid. 295–7.
11. Rasul Ja'fariyan, 'Shī'ism and its Types During the Early Centuries': <www.al-islam1.org/al-tawhid/types/shiism.htm>

    The Khārijīs objected to the personal allegiance to the *imām* expressed in this formula and accused the Basrans of emulating the Syrians in their secessionist pledge to Mu'āwiya: Madelung, *The Succession to Muḥammad*, 253.
12. S. Husain and M. Jafri, *Origins and Early Development of Shī'a Islām* (London and New York: Longman, 1979), 92, 126.

    Electronic publication at: <www.karbala-najaf.org/shiaism/shiaism.html>
13. Madelung, *The Succession to Muḥammad*, 141, 169.
14. *Islām and Revolution. Writings and Declarations of Imām Khomeini*, trans. and ed. Hamid Algar (Berkeley, CA: Mizan Publications, 1981), 200. Ibid. 202: '…Islām came in order to destroy these palaces of tyranny. Monarchy is one of the most shameful and disgraceful reactionary manifestations…'
15. Madelung, *The Succession to Muḥammad*, 317.

16. Ibid. 319.
17. Ibid. 327: 'in spite of his resignation... he was still considered the chief of the Prophet's house by the Banū Hāshim and the partisans of 'Alī, who pinned their hopes on his eventual succession to the Umayyad.'
18. Ibid. 384.
19. It was also the Jewish Day of Atonement.
20. Ayatullah Murtazá Muṭahharī (d. 1979), '*Shahīd*', in *Jihād and Shahādat. Struggle and Martyrdom in Islām*, ed. Mehdi Abedi and Gary Legenhausen (Houston, TX: Institute for Research and Islamic Studies, 1986), 129. <www.ummah.net/khoei/imam.htm#3>
21. In 65/November 684: Husain and Jafri, *Origins and Early Development of Shi'a Islam*, 231. Crone, *Medieval Islamic Political Thought*, 124.
22. Ibid. 249–52.
23. The issue was that Mūsā al-Kāẓim was only the third son of Ja'far al-Sādiq but outlived his father, whereas Ismā'īl was his eldest son, but had predeceased Ja'far. The beginnings of the Twelver movement were evident at the death of the fifth *imām* in 114/732 according to Lambton: A. K. S. Lambton, *State and Government in Medieval Islam. An Introduction to the Study of Islamic Political Theory: The Jurists* (New York: Oxford University Press, 1981), 222 n. 13. For the names and dates of the twelve *imāms*: Sobhani, *Doctrines of Shī'ī Islām*, 113–14.
24. According to Lambton, drawing on Madelung's research, this corpus of writing was 'largely the work of Hishām bin al-Ḥakam'. Lambton, *State and Government in Medieval Islam*, 228 n. 34.
25. Husain and Jafri, *Origins and Early Development of Shī'a Islām*, 290.
26. Ibid. 295.
27. Ibid. 291.
28. Lambton, *State and Government in Medieval Islam*, 229–30.
29. Ibid. 236.
30. Husain and Jafri, *Origins and Early Development of Shī'a Islām*, 298–9. Sobhani, *Doctrines of Shī'ī Islām*, 153–4, emphasizes the 'sacrifices' made by the Shī'a in earlier historical periods, under the Umayyad, 'Abbāsid and Ottoman dynasties: 'the Shī'a underwent such a fate, despite having recourse to dissimulation; what, one wonders, would have been their lot if they [had not had] recourse to this principle? One might well also ask: in such a case, that is, without the principle of dissimulation, would there be anything left of Shī'ism today?'
31. Sobhani, *Doctrines of Shī'a Islām*, 114–15. The '*Imām* of the Age' was born in 255/869 and thus 'is now over eleven centuries old': ibid. 119. Crone, *Medieval Islamic Political Thought*, 118.
32. Lambton, *State and Government in Medieval Islam*, 238.
33. Ibid. 245, 260.
34. Ibid. 261–3. Crone, *Medieval Islamic Political Thought*, 124.
35. Crone, *Medieval Islamic Political Thought*, 197–218.
36. Asaf A. A. Fyzee (trans), *The Pillars of Islam. Da'ā'im al-Islām of al-Qāḍī al-Nu'mān. I. Acts of Devotion and Religious Observances*, revised Ismail Kurban Husein Poonawala (New Delhi: Oxford University Press, 2002).
37. Ismail Kurban Husein Poonawala, 'Al-Qāḍī al-Nu'mān and Ismā'īlī Jurisprudence', in *Medieval Ismā'īlī History and Thought*, ed. Farhad Daftary (Cambridge: Cambridge University Press, 1996), 117–43 at 132.
38. F. Daftary, *A Short History of the Ismā'īlīs. Traditions of a Muslim Community* (Edinburgh: Edinburgh University Press, 1998), 77. The *Ẓāhiriyya* were those who derived the law governing Muslim life from the direct text (*ẓāhir*) of the Qur'ān and the *sunnah*. Daftary adds that al-Nu'mān 'recognized the minoritarian status of the Ismā'īlīs in North Africa and

attempted a legalistic rapprochement with Sunnī Islām, presumably aiming to reduce the precarious isolation of the Ismā'īlīs as a ruling class in the midst of predominantly non-Ismā'īlī surroundings'. Ibid.

39. Wilferd Madelung, 'The Fāṭimids and the Qarmaṭīs of Baḥrayn', in Daftary, *Medieval Ismā'īlī History and Thought*, 22.
40. Heinz Halm, 'The Ismā'īlī Oath of Allegiance and the "Sessions of Wisdom"', in Daftary, *Medieval Ismā'īlī History and Thought*, 94, 98.
41. Madelung, 'The Fāṭimids and the Qarmaṭīs of Baḥrayn', 33, 49.
42. Ibid. 21.
43. Ibid. 38–9.
44. Ibid. 46–7, 51.
45. Ibid. 23.
46. Daftary, *A Short History of the Ismā'īlīs*, 106–7.
47. No source given, but summarized by H. H. Aga Khan at the opening session of an international colloquium entitled 'Word of God, Art of Man: The Qur'ān and its Creative Expressions', organized by the Institute of Ismā'īlī Studies in London: <www.iis.ac.uk/learning/speeches_ak4/2003b.pdf>. Cf. 'Aga Khan aims to educate West about Islām': *The Times*, 20 October 2003.
48. Farouk Mitha, *Al-Ghāzalī and the Ismailis: A Debate on Reason and Authority in Medieval Islām*. Ismā'īlī Heritage Series, 5 (London: I. B. Tauris in association with the Institute of Ismaili Studies, 2001).
49. Ibid. 25.
50. Ibid. 54.
51. Ibid. 45.
52. Ibid. 49.
53. Ibid. 73. For al-Ghāzalī, the innate attributes were maturity or attainment of puberty; intelligence; freedom; that he be of the male gender; descent from the Qurayshīs; that he possess soundness of hearing and sight. The acquired qualities included bravery or military prowess; political competence; piety and knowledge. Ibid. 78–9.
54. Ibid. 77–8.
55. Ibid. 81.
56. Ibid. 91.
57. Ibid. 46, 89.
58. Ibid. 23.
59. Farhad Daftary, 'Ḥasan-i Ṣabbāḥ and the Origins of the Nizārī Movement', in Daftary, *Medieval Ismā'īlī History and Thought*, 193.
60. Ibid. 196.
61. Daftary, *A Short History of the Ismā'īlīs*, 138–41. Bernard Lewis, *The Assassins. A Radical Sect in Islām* (London: Weidenfeld and Nicolson, 1967), 74–5.
62. Daftary, *A Short History of the Ismā'īlīs*, 145–6.
63. Charles Melville 'The Role of the Ismā'īlīs in Mamlūk–Mongol relations', in Daftary, *Medieval Ismā'īlī History and Thought*, 258.
64. For Iran's contemporary aim of preventing the emergence of an independent religious and spiritual leadership emerging there, which might rival that of Qum in Iran: Nimrod Raphaeli, 'Iran's Stirrings in Iraq', Memri Inquiry and Analysis No. 173, 5 May 2004: <www.memri.org/bin/latestnews.cgi?ID=IA17304>
65. Juan Cole, *Sacred Space and Holy War. The Politics, Culture and History of Shī'ite Islām* (London: I. B. Tauris, 2002), 25.
66. Ibid. 5, 7.
67. Ibid. 107.

68. Ibid. 4, 20.
69. Ibid. 107–8, 233 n. 25.
70. Ibid. 118.
71. Ibid. 120.
72. The others were *al-Jamiyyah al-Wataniyya al-Islāmīyyah* (The Muslim National League) and *Haras al-Istiqlal* (The Guardians of Independence): <www.historymedren.about.com/library/text/bltxtiraq9.htm>
73. Askary Hussein, 'Lessons to be Learned: 'Irāqī Resistance to British Occupation 80 Years Ago', *Executive Intelligence Review* (14 November 2003): <www.larouchepub.com/other/2003/3044iraq_history.html> Apart from the intrinsic importance of this article, on which this section draws heavily, the author's grandfather on his mother's side and great-grandfather on his father's side participated in the revolt in 1920.
74. Thus John Darwin concludes that in 'Irāq, 'the problem confronting [British] imperial policy was… not a question of bending an existing system to the imperial design, but of constructing a new regime in a region which lacked almost all the attributes of political, social, economic or religious unity. Indeed, the internal coherence of the territorial unit which became the 'Irāq state derived not from any nation-building tendencies at work in its diverse communities but from the conviction of British policy-makers that the three *vilayets* of Mosul, Baghdād and Basra should fall within Britain's sphere of influence, and that only if they were combined within a single political framework could that influence be effectively exercised.' John Darwin, *Britain, Egypt and the Middle East: Imperial Policy in the Aftermath of War, 1918–1922* (London: Macmillan, 1981), 273.
75. Ali M. Ansari, *Modern Iran since 1921. The Pahlavis and After* (London: Pearson, 2003), 27 n. 10. Nikki R. Keddie, *Modern Iran. Roots and Results of Revolution* (New Haven, CT, and London: Yale University Press, repr. 2003), 80.
76. Algar, *Islām and Revolution. Writings and Declarations of Imām Khomeini*, 136.
77. Ibid. 255, where Khomeini argues that the rule of the son was illegal because of the illegality of the rule of the father: 'the Constituent Assembly was convened in his time at bayonet-point… even if we suppose the rule of Rizā *Shāh* to have been legal, those who put him in power had no right to determine our destiny…. In any event… our people are declaring today that they do not want this ruler.' 'The Pahlavī monarchy', he concluded, 'was void from the outset', and in any case a monarchical regime was 'contrary to all rational laws and precepts as well as human rights'. Speech of 2 February 1979.
78. Ibid. 49, where Khomeini talks of the partition of the Ottoman state by the imperialists 'creating in its territories about ten or fifteen petty states. Then each of these was entrusted to one of their servants or a group of servants, although certain countries were later able to escape the grasp of the agents of imperialism'.
79. For rival versions of the rising, contrast Hussein, 'Lessons to be Learned' with John Darwin, *Britain, Egypt and the Middle East*, 200: 'the lack of any concerted protest in the towns and the absence of any figurehead capable of weaving diffuse local grievances into a rudimentary political programme gave the revolt the character not so much of an outburst of national feeling as of a rural backlash against the age-old oppressions of government and town, a tribal *jacquerie* innocent of any larger political ideas.' Without wishing to criticize John Darwin's superb study on other than a point of detail, the joint letter of al-Shīrāzī and Shaykh al-Shar'īa al-Aṣfahānī to President Wilson in 1337/February 1919, cited above, suggests that the assertion of the right of self-determination was indeed the issue which made the revolt more than a tribal *jacquerie*. Moreover, the signs of Shī'a–Sunnī cooperation also place the movement in a different category.

80. Darwin, *Britain, Egypt and the Middle East*, 200. David Fromkin, *A Peace to End All Peace: The Fall of the Ottoman Empire and the Creation of the Modern Middle East* (London: Phoenix, repr. 2003), 453.
81. Fromkin, *A Peace to End All Peace*, 97.
82. Khomeini was particularly devastating in his critique: 'six hundred of the '*ulamā*' of Najaf and Iran were on the payroll of the British. Shaykh Murtazā [Ansari] took the money for only two years before he realized where it was coming from. The proof may be found in documents preserved in the Indian Office archives.' Algar, *Islām and Revolution. Writings and Declarations of Imām Khomeini*, 140.
83. Fromkin's summary of the viewpoint of Kitchener and his colleagues: Fromkin, *A Peace to End All Peace*, 96–7.
84. Ibid. 104.
85. On 31 August 1920, Churchill recorded his innermost fears: 'it is an extraordinary thing that the British civil administration should have succeeded in such a short time in alienating the whole country to such an extent that the Arabs have laid aside the blood feuds they have nursed for centuries and that the Sunni and Shia tribes are working together'. Quoted by A. G. Noorani, 'Iraq's revolt in 1920', *The Hindu*, 9–16 April 2004: <www.freerepublic.com/focus/f-news/1117519/posts>
86. Though Lawrence thought the Arabs 'even less stable than the Turks. If properly handled they would remain in a state of political mosaic, a tissue of small jealous principalities incapable of political cohesion': cited by Algar in *Islām and Revolution. Writings and Declarations of Imām Khomeini*, 154 n. 44.
87. *Sunday Times*, 22 August 1920.
88. Fromkin, *A Peace to End All Peace*, 497.
89. Algar, *Islām and Revolution. Writings and Declarations of Imām Khomeini*, 214: 'the imperialist states, like America and Britain, brought Israel into existence, and we have seen what misery they have inflicted on the Muslim peoples by means of Israel, and what crimes they are now committing against the Muslims, particularly the Shī'a [of south Lebanon as a result of Israeli attacks]'. Speech of 19 February 1978.
90. The most detailed account is now *Mohammad Mosaddeq and the 1953 Coup in Iran*, ed. Mark J. Gasiorowski and Malcolm Byrne (Syracuse, NY: Syracuse University Press, 2004). CIA Clandestine Service History, 'Overthrow of Premier Mossadeq of Iran, November 1952–August 1953', March 1954, by Dr Donald Wilber: <www.gwu.edu/~nsarchiv/NSAEBB/NSAEBB28/4-Orig.pdf>

    The assumptions underlying American intervention were: '1) the *shāh* could be persuaded to take desired action if all-out pressure were applied; 2) assurance that the *shāh* was behind him would both cause [General] Zahedi to act and would win him the support of many officers in key positions; 3) faced with a choice between following the orders of the *shāh* and those of Mossadeq, the rank and file of the army and its officers would obey the *shāh*'. The report concluded that the coordination of US and UK 'interests and activities' was critical and that it was the conviction that both the US and the UK were ready to support him that convinced the *shāh* to act. 89 people had known of 'the project and its purpose', though the number who knew that there was direct US–UK collaboration was 'kept to a more restricted number': <www.gwu.edu/~nsarchiv/NSAEBB/NSAEBB28/10-Orig.pdf>
91. Algar, *Islām and Revolution. Writings and Declarations of Imām Khomeini*, 215: 'as for America, a signatory to the Declaration of Human Rights, it imposed this *shāh* upon us, a worthy successor to his father. During the period he has ruled, this creature has transformed Iran into an official colony of the U.S. What crimes he has committed in service to his masters… All they [Pahlavī father and son] have to offer humanity is repression; we have witnessed part

of it, and we have heard part of it...' Speech of 19 February 1978. Later he talked of the *shāh*'s 'own tyrannical rule and his parasitic masters': ibid. 242. Speech of 23 November 1978.

92. Mark J. Gasiorowski, 'The 1953 Coup d'État against Mossadeq', in Gasiorowski and Byrne, *Mohammad Mosaddeq and the 1953 Coup in Iran*, 227–60, at 253.

93. Ayatullah Murtazá Muṭahharī, *Fundamentals of Islamic Thought. God, Man and the Universe*, trans. R. Campbell (Berkeley, CA: Mizan Press, 1985), 18, 20 (introduction by Hamid Algar). For an overview of his writings: Vanessa Martin, *Creating an Islamic state. Khoemeini and the Making of a New Iran* (London and New York: I. B. Tauris, repr. 2003), ch. 4, 75–99. Ibid. 75 for his leadership role during Khomeini's exile; ibid. 79 for the role of Furqān in the assassination. Iranian websites call it a 'terrorist' group. According to Abedi and Legenhausen, Furqān claimed to follow the teachings of the late Dr 'Alī Shari'ati, with whom Muṭahharī had been in ideological conflict: *Jihād and Shahādat. Struggle and Martyrdom in Islām...*, ed. Abedi and Legenhausen, 38.

94. Ayatullah Murtazá Muṭahharī, *Master and Mastership* (Islamic Seminary Publications, n.d.), ch. 2: <www.al-islam.org/mastership/>

    Elsewhere, he writes in *Jihād*: 'the Qur'ān states the beliefs of the People of the Book to be confused and misconceived. A Christian, even if he is a learned Christian scholar, recognizes God and even recognizes the Oneness of God, but at the same time, he may have some idea about Jesus or the angel Gabriel that pollutes his belief in the Oneness of God...' On the other hand, he adds: 'if God did not check the aggression of some people by means of others, all the houses of worship of all the different sects and religions would be destroyed. The churches of Christians, the synagogues of Jews, the monasteries, the *masjids*, places of prostration of Muslims, all would exist no longer. Some people would commit such aggression that no one would find the freedom in which to worship God.' Ayatullah Murtazá Muṭahharī, *Jihād: The Holy War of Islām and its Legitimacy in the Qur'ān*, trans. Muḥammad Salman Tawhidi (Tehran: Islamic Propogation Society, 1998). Web version at: <www.al-islam.org/jihad/index.html> All quotations are from this version of the text. There is an alternative translation in M. Abedi and G. Legenhausen (eds) *Jihād and Shahādat. Struggle and Martyrdom in Islam* (Houston, TX: Institute for Research and Islamic Studies, 1986), 81–124.

95. Ibid. ch. 6.

96. Ayatullah Muḥammad Bāqir al-Ṣadr and Ayatullah Murtazá Muṭahharī, *The Awaited Saviour* (Karachi: Islamic Seminary Publications, n.d.), ch. 9: <www.al-islam.org/awaited/>

97. <www.al-islam.org/al-tawhid/1-human-evol.htm>

98. Muṭahharī, *Jihād*.

99. Ibid. 13.

100. Ibid. 25.

101. Ibid. 26: 'whether a war is launched to take possession of land, to seize ownership of national wealth, or due to contempt of others and out of sentiment of racial superiority, i.e. "those people are inferior to us who are superior, and the superior must govern over the inferior", it is a war of aggression.' Or, later (ibid. 78): 'when a group wants to take away the independence of a nation and place that nation under its own mandate, if the people of that nation decide to defend themselves and pick up the gun, this action is lawful, in fact laudable and worthy of admiration. So, defence of life, defence of wealth, property and lands, defence of independence, defence of chastity, all these are lawful defences. No one doubts the fact that in these cases, defence is permissible...'

102. Ibid. 31–2, 41.

103. Ibid. 126–7.

104. Ibid. 80.

105. Ibid. 82, 84.

106. Abedi and Legenhausen, *Jihād and Shahādat*, 129–31.

107. Ibid. 141, citing the words of 'Alī following his wounding by his assassin: 'I have succeeded'; 'by God, nothing unexpected and undesirable has occurred. What has occurred is what I wanted. I have achieved *shahādah*, which I desired. I am like a man who was in search of water, and has suddenly struck upon a well or spring. I am like a man who was actively looking for something, and got it.'
108. Ibid. 132. 'Alī preferred to die from 'a thousand strokes of the sword' rather than in bed: ibid. 134.
109. Ibid. 128–9. His logic before Karbalā' was 'that of a *shahīd*, which is beyond the comprehension of ordinary people': ibid. 135. Ibid. 144: 'the earth of Karbalā'... emits the smell of the *shuhadā'*.'
110. Ibid. 126–7, 148.
111. Ibid. 136–7.
112. Ibid. 139.
113. Ibid. 130.
114. Asghar Schirazi, *The Constitution of Iran. Politics and the State in the Islamic Republic*, trans. John O'Kane (London: I. B. Tauris, 1997, repr. 1998), 269–70.
115. Ibid. 145.
116. Algar, *Islām and Revolution. Writings and Declarations of Imām Khomeini*, 326: 'a nation of thirty million or more has stood up and defied tanks and machine guns. If the Soviets or the Americans come marching in, they will be met with the same defiance…'
117. Ibid. 349–62 (extracts only). <www.al-islam1.org/al-tawhid/greater_jihad.htm> There is a brief biography of Khomeini by Hamid Algar at: <www.khomeini.com/gatewaytoheaven/Information/imamsbiography.htm>
118. <www.ummah.net/khoei/khomeini.htm> The comment of Martin, *Creating an Islamic State*, 43, is pertinent: 'all forms of *jihād* that may be waged in the world depend on this greater *jihād*. Without the inner *jihād* the outer *jihād* is impossible.'
119. <www.muslimedia.com/archives/features98/khomeini.htm>
120. <www.math.nyu.edu/phd_students/amirishs/Html/imam.html#will>
121. Iran acknowledged that nearly 300,000 people died in the war; more than 500,000 were injured, out of a total population which by the war's end was nearly 60 million. It is worth pointing out Germany's losses in World War I, relative to total national population, were at least five times higher than Iran's: <www.globalsecurity.org/military/world/war/iran-iraq.htm>
122. Efraim Karsh, *The Iran–Iraq War, 1980–1988* (Wellingborough: Osprey, 2002), 85–6.
123. Ibid. 29.
124. <www.math.nyu.edu/phd_students/amirishs/Html/imam.html#msg01>
125. In his [Islamic] New Year's message on 21 March 1980, Khomeini proclaimed: 'we must strive to export our Revolution throughout the world, and must abandon all idea of not doing so, for not only does Islām refuse to recognize any difference between Muslim countries, it is the champion of all oppressed people. Moreover, all the powers are intent on destroying us, and if we remain surrounded in a closed circle, we shall certainly be defeated…': Algar, *Islām and Revolution. Writings and Declarations of Imām Khomeini*, 286. He added that 'satanic counter-revolutionary conspiracies' were on the increase: ibid. 288.
126. In a subsequent message, on 12 September 1980, Khomeini mentioned not only the 'extensive propaganda campaign being waged apparently against Iran, but in reality against Islām…' but also 'the difficulties that America and its satellites have created for us – economic boycott, military attack and the planning of extensive coups d'état…': ibid. 303.
127. 'Here I officially apologize to the mothers, fathers, sisters, brothers, the spouses and the children of the martyrs and the self-sacrificing devotees because of some of the wrong analyses of these days. I beg God to accept me alongside the martyrs of the imposed war. We do not repent, nor are sorry even for one moment for our performance during the war.'

128. Christoph Reuter, *My Life is A Weapon. A Modern History of Suicide Bombing*, trans. Helena Ragg-Kirkby (Princeton, NJ, and Oxford: Princeton University Press, 2004; orig. German edn 2002), 38.
129. Ibid. 34.
130. Ibid. 35.
131. 'Our war was the war of the right against the wrong, and this war can never come to a conclusion. Our war was the war of poverty against affluence. Our war was the war of faith against deceit and this war has been going on from the time of Adam (AS) to the end of life.'
132. Schirazi, *The Constitution of Iran*, 71.
133. He also called the USA 'the world-devourer'. For the virulence of Iranian anti-American propaganda: Adam Parfrey (ed.) *Extreme Islām. Anti-American Propaganda of Muslim Fundamentalism* (Los Angeles, CA: Feral, 2001), 192–207.
134. Algar, *Islām and Revolution. Writings and Declarations of Imām Khomeini*, 305: 'let the Muslim nations be aware that Iran is a country effectively at war with America and that our martyrs – the brave young men of our army and the Revolutionary Guards – are defending Iran and the Islām we hold dear against America.' Moreover the clashes along the Iranian border with 'Irāq were 'caused by America' because of the Islamic content of the Revolution. Written before the end of the war, Asaf Hussain's study noted this reality: Asaf Hussain, *Islamic Iran. Revolution and Counter-Revolution* (London: Pinter, 1985), 196–7.
135. Algar, *Islām and Revolution. Writings and Declarations of Imām Khomeini*, 306. In his speech on 12 Sept. 1980, Khomeini stated his demands for the release of the hostages: 'I have said repeatedly that the taking of hostages by our militant, committed Muslim students was a natural reaction to the blows our nation suffered at the hands of America. They can be set free if the property of the dead *shāh* is returned, all claims of America against Iran are annulled, a guarantee of political and military non-interference in Iran is given by America, and all our capital is released. Of course, I have turned the affair over to the Islamic Assembly for it to settle in whatever way it deems best…' Asghar Schirazi calls this an example of a case where Khomeini was not the initiator of the faction, but where his actual response was perfectly clear to the actual initiators. Khomeini had no hesitation in giving the action his approval, though it was carried out against the will of the provisional government: Schirazi, *The Constitution of Iran*, 62.
136. Parfrey, *Extreme Islām*, 192–3.
137. On 12 February 1997, the Iranian charitable foundation, 15 Khordad, reportedly announced an increase in the reward for the murder of Salman Rushdie to $2.5 million. The head of the 15 Khordad Foundation, Ayatollah Shayh Hassan Sanei, a senior member of the religious establishment and personal representative of the Leader of the Islamic Republic, was reported in the newspaper *Jomhuri-ye Islamī* as saying that anyone who killed the 'apostate' writer could claim the reward, including non-Muslims and his bodyguards. President Hashemi Rafsanjani was quick to distance himself from this position and was reported to have said that day at a press conference: 'This is a non-governmental organization and its decisions have nothing to do with the government's policies. I don't know what their motive was, but the government's policy towards the (Rushdie affair) is the same as before, and one which we have repeatedly announced.' However, the failure of the Government of the Islamic Republic of Iran publicly to condemn and put an end to such threats indicates official acquiescence in a threat of an extrajudicial execution. On 13 February 1997, a statement by the Revolutionary Guards printed in the *Jomhuri-ye Islamī* was reported to have said, among other things, 'Muslims of the world have always considered Salman Rushdie an apostate and shall not rest until *Imām* Khomeini's order is implemented'. Since the Revolutionary Guards are an official part of the

state apparatus, this carries an alarming implication that the *fatwā* might be implemented by state forces: <www.web.amnesty.org/library/Index/engMDE130171997>

However, the British government accepted President Khatami's assurance in September 1999 that the case was 'completely finished' and restored diplomatic relations: Keddie, *Modern Iran*, 262, 272.

138. M. M. Ahsan and A. R. Kidwai (eds), *Sacrilege versus Civility: Muslim Perspectives on the Satanic Verses Affair* (Leicester: Islamic Foundation, rev. and enlarged edn, 1993).

139. Hussain, *Islamic Iran*, 67. Ibid. 68–78, where Hussain distinguishes between a) the [belief in one God] *tawḥīdī* dimension ('pivotal to his interpretation of Islām'); b) the *jihādic* dimension ('a revolutionary one, encouraging both the Muslim masses and the religious leadership to engage in political action'), c) the [unity of believers] *ummah* dimension; and (d) the [needy and the oppressed] *mostazafeen* dimension.

140. Kazem Ghazi Zadeh, 'General Principles of *Imām* Khumaynī's Political Thought', trans. A. N. Baqirshahi, in *Message of Thaqalayn. A Quarterly Journal of Islamic Studies,* 2, Nos 2 and 3. Web version at: <www.aalulbayt.org/html/eng2/books/massage-of-thagalain/17generl-tl.html>; <www.al-shia.com/html/eng/books/message-of-thaqalayn/17generl-tl.html>; <www.al-islam.org/mot/default.asp?url=17Generl.HTM>

141. Quotations from Zadeh, 'General Principles of *Imām* Khumaynī's Political Thought', unless otherwise indicated.

142. 'Unlike other atheist schools of thought, Islām take into account all aspects of human life, that is, individual, social, physical, spiritual, cultural, political, economic and military affairs of life. Islām does not leave any point helpful for the material and spiritual development of man and society untouched. Islām elaborates on the obstacles of such development in society and teaches how to remove them.'

143. 'Belief in the necessity of forming a government and establishing an executive system is a part of *wilāyat*. Likewise any attempt in this direction is also part of the belief in *wilāyat*. We believe in *wilāyat* and hold that the Prophet appointed a caliph in obedience to God. Therefore, we have to believe that it is necessary for Muslims to form a government… The struggle for forming an Islamic state is one of the foundations of the belief in *wilāyat*'.

144. '*Ijtihad* in… religious centres is insufficient for the holders of authority. That is, if a person is the most learned in religious studies but is unable to recognize the interests of his society or to distinguish between useful and useless persons for the society, i.e., he does not have a proper social insight, he is not in fact [a qualified Islamic scholar (*Mujtahid*)] and cannot pass decrees on socio-political affairs. Hence, he is not eligible to hold the authority.'

145. Algar, *Islām and Revolution. Writings and Declarations of Imām Khomeini*, 326–7.

146. Ibid. 47: 'then a handful of wretched Jews (the agents of America, Britain and other foreign powers) would never have been able to accomplish what they have, no matter how much support they enjoyed from America and Britain. All this has happened because of the incompetence of those who rule over the Muslims.' David Zeidan, 'The Islamic Fundamentalist View of Life as a Perennial Battle', *Meria (Middle East Review of International Affairs)*, 5 (December 2001), 26–53, at 47. Web version at: <www.meria.idc.ac.il/journal/2001/issue4/zeidan.pdf>

147. Algar, *Islām and Revolution. Writings and Declarations of Imām Khomeini*, 127: '…the imperialists, the oppressive and treacherous rulers, the Jews, Christians and materialists are all attempting to distort the truths of Islam and lead the Muslims astray… We see today that the Jews (may God curse them) have meddled with the text of the Qur'ān and have made certain changes in the Qur'āns they have had printed in the occupied territories.' Ibid. 89, where he seems to approve of the massacre of the Banū al Qurayẓah, which resulted in between 400 and 700 executions (see Chapter 1).

148. Ibid. 175 (3 April 1963), 187 (27 October 1964), 197 (6 February 1971). Amir M. Haji-Yousefi, 'Foreign Policy of the Islamic Republic of Iran towards Israel, 1979–2002', *Strategic Studies*, Quarterly Journal of the Institute of Strategic Studies, Islamabad, 23 (2003), 55–75.
149. Algar, *Islām and Revolution*, 177–8.
150. Executive Summary of the investigation of the Office of Independent Counsel: 'contrary to their testimony to the presidentially-appointed Tower Commission and the Select Iran/Contra Committees of Congress, Independent Counsel determined that Secretary Weinberger and his closest aides were consistently informed of proposed and actual arms shipments to Iran during 1985 and 1986. The key evidence was handwritten notes of Weinberger, which he deliberately withheld from Congress and the OIC until they were discovered by Independent Counsel in late 1991. The Weinberger daily diary notes and notes of significant White House and other meetings contained highly relevant, contemporaneous information that resolved many questions left unanswered in early investigations. The notes demonstrated that Weinberger's early testimony that he had only vague and generalized information about Iran arms sales in 1985 was false, and that he in fact had detailed information on the proposed arms sales and the actual deliveries. The notes also revealed that Gen. Colin Powell, Weinberger's senior military aide, and Richard L. Armitage, assistant secretary of defence for international security affairs, also had detailed knowledge of the 1985 shipments from Israeli stocks. Armitage and Powell had testified that they did not learn of the November 1985 HAWK missile shipment until 1986. Weinberger's notes provided detailed accounts of high-level Administration meetings in November 1986 in which the President's senior advisers were provided with false accounts of the Iran arms sales to protect the President and themselves from the consequences of the possibly illegal 1985 shipments from Israeli stocks. Weinberger's notes provided key evidence supporting the charges against him, including perjury and false statements in connection with his testimony regarding the arms sales, his denial of the existence of notes and his denial of knowledge of Saudi Arabia's multi-million dollar contribution to the Contras. He was pardoned less than two weeks before trial by President Bush on December 24, 1992. There was little evidence that Powell's early testimony regarding the 1985 shipments and Weinberger's notes were wilfully false. Powell cooperated with the various Iran/Contra investigations and, when his recollection was refreshed by Weinberger's notes, he readily conceded their accuracy. Independent Counsel declined to prosecute Armitage because the OIC's limited resources were focused on the case against Weinberger and because the evidence against Armitage, while substantial, did not reach the threshold of proof beyond a reasonable doubt': <www.webcom.com/pinknoiz/covert/icsummary.html>
151. Schirazi, *The Constitution of Iran*, 70.
152. Ansari, *Modern Iran since 1921*, 238.
153. Karsh, *The Iran–Iraq War, 1980–1988*, 51. Note: for consistency, the calendar used throughout is the Sunnī one, according to which the year 1407 began on 6 September 1986.
154. Schirazi, *The Constitution of Iran*, 71.
155. Algar, *Islām and Revolution. Writings and Declarations of Imām Khomeini*, 114.
156. Michael Cook, *Commanding Right and Forbidding Wrong in Islamic Thought* (New York: Cambridge University Press, 2000), 534.
157. Ibid. 536.
158. Ibid. 541–2.
159. Algar, *Islām and Revolution. Writings and Declarations of Imām Khomeini*, 115–16.
160. Ibid. 132.
161. Schirazi, *The Constitution of Iran*, 52, according to whom, the comparison gives 'at least an approximate idea… of how many people Khomeini was no longer able to mobilize after only nine months'. The same writer comments on the first referendum, 'the turn-out for

the referendum was very high and 98.2 per cent of the participants said "yes" to a form of government about which they were sadly misinformed': ibid. 27.

162. Ibid. chs 8, 9, 10.
163. Ibid. 8.
164. Ibid. 11. Cook, *Commanding Right and Forbidding Wrong*, 545–6: 'according to the Constitution, the duty is one that must be fulfilled "by the people with respect to one another, by the government with respect to the people, and by the people with respect to the government". In practice, the first and third have been relatively muted by the din of the second. Iran, like Saudi Arabia, has become a society in which forbidding wrong is overwhelmingly a function of the state apparatus, in this case involving a plurality of organs which do not always act in concert.'
165. Schirazi, *The Constitution of Iran*, 9, 13.
166. Ibid. 10, 69.
167. Ibid. 294.
168. Ibid. 297.
169. Philip Shenon, 'President says U.S. to examine Iran–[al-]Qaeda Tie', *New York Times*, 20 July 2004: <www.nytimes.com/2004/07/20/politics/20panel.html?th>
170. Though there is little organized opposition inside the country and financing it directly or through front organizations would probably play into the hands of the mullahs: Michael Binyon and Bronwen Maddox, 'U.S. sets sights on toppling Iran regime', *The Times*, 17 July 2004: <www.timesonline.co.uk/newspaper/0,,175-1181969,00.html>
171. Karsh, *The Iran–Iraq War*, 89–92.
172. 'Secretary of Defence Dick Cheney said on [5 March 1991] that "it would be very difficult for us to hold the coalition together for any particular course of action dealing with internal Iraqi politics, and I don't think, at this point, our writ extends to trying to move inside Iraq"': *Endless Torment. The 1991 Uprising in Iraq and Its Aftermath* (Human Rights Watch, June 1992): <www.hrw.org/reports/1992/Iraq926.htm>

    Cheney stated additionally: 'when the President offered to send forces, King Fahd agreed, and he did so on the basis that he knew he could trust the United States of America; that we would come, we would keep our word, we would bring enough force to be able to roll back Saddam Hussein's aggression, and that when we were no longer needed or no longer wanted, we would leave…' UN Security Council Resolution 660, adopted on 2 August 1990, defined the objective toward which the Security Council subsequently authorized military action in the Gulf. It demanded that Iraqi forces withdraw 'to the positions in which they were located on 1 August 1990'. No resolution prior to, or during the war, authorized the US-led coalition to invade 'Irāq. It therefore followed that it was irresponsible to encourage internal rebellion when there was no immediate prospect of assisting it from abroad.
173. 'The strongest signal of U.S. support for a popular rebellion came toward the end of the air war, when President Bush declared on [15 February 1991]: "[t]here's another way for the bloodshed to stop, and that is for the Iraqi military and the Iraqi people to take matters into their own hands to force Saddam Hussein, the dictator, to step aside." This remark was heard by Iraqis on the Voice of America [radio transmission]. Soon after the uprising began, however, fears of a disintegrating Iraq led the Administration to distance itself from the insurgents…' There had been earlier remarks from President Bush as early as 11 August 1990 and 30 August 1990. In addition, 'there [we]re allegations that the U.S. further encouraged the rebellion by launching in January 1991 the Voice of Free Iraq, a clandestine radio station that preached sedition against Saddam in clear terms. While administration officials, including spokespersons for the CIA, State Department and Pentagon all denied or declined to confirm U.S. involvement in the station, its programming and language bore the marks of CIA sponsorship.' Ibid.

174. 'On [5 March 1991], Rear Admiral Mike McConnell, director of intelligence for the [US] Joint Chiefs of Staff, acknowledged that "chaotic and spontaneous" uprisings were under way in thirteen Iraqi cities, but stated the Pentagon's view that Saddam would prevail because of the rebels' "lack of organization and leadership"'. Ibid.
175. The US International Religious Freedom Report for 2002 noted: 'while a precise statistical breakdown is impossible to ascertain because of likely inaccuracies in the latest census (conducted in 1997), according to best estimates, 97 percent of the population of 22 million persons are Muslim. Shī'a Muslims – predominantly Arab, but also including Turkomen, Faili Kurds, and other groups – constitute a 60 to 65 per cent majority. Sunnī Muslims make up 32 to 37 per cent of the population (approximately 18 to 20 per cent are Sunni Kurds, 12 to 15 per cent Sunnī Arabs, and the remainder Sunnī Turkomen). The remaining approximately 3 per cent of the overall population consist of Christians (Assyrians, Chaldeans, Roman Catholics, and Armenians), Yazidis, Mandaeans, and a small number of Jews.' <www.state.gov/g/drl/rls/irf/2002/13996.htm>
176. 'When asked by M[iddle] E[ast] W[atch] to describe the uprising, Kurdish refugees often began by recounting past government persecution: arbitrary arrest and torture, disappearances, eviction from the countryside, the destruction of villages, and the use of chemical weapons against Kurdish civilians in 1987 and 1988. The Shī'a of the south also spoke of arbitrary arrests, torture and disappearances, and about the expulsion of thousands of Shī'a to Iran in the early years of the Iran–'Irāq war.' Human Rights Watch, *Endless Torment*.
177. During the *intifāḍah*, he had issued no call for *jihād*, but two communiqués. The first, dated 5 March 1991, urged Muslims to 'guard the territory of Islām', to 'look after the holy places', and to guard the honour and the property of the people and preserve the public institutions of 'Irāq. As the rebels appeared increasingly likely to consolidate their victory, he issued a statement establishing a 'Supreme Committee' under whose leadership the Shī'a would preserve 'Irāq's security and stabilize public, religious and social affairs. The second communiqué, with its implication of a rival governing body, may have contributed to the decision by Saddam to force him to make a statement on national television denouncing the violence that was taking place.
178. Rami El-Amine, 'The Shī'a Rise Up' (14 May 2004): <www.zmag.org/content/showarticle.cfm?SectionID=15&ItemID=5529>, 'Shī'a Backgrounder: A History of Oppression': <www.leftturn.org/Articles/Viewer.aspx?id=520&type=M>
179. Peter Hennessy, 'A devastating indictment', *The Tablet* (17 July 2004), 5. Cf. the same author's earlier article, 'Ground rules for war', *The Tablet* (29 May 2004).
180. 'The current American peace will be short-lived if the United States becomes vulnerable to rogue powers with small, inexpensive arsenals of ballistic missiles and nuclear warheads or other weapons of mass destruction. We cannot allow North Korea, Iran, Iraq or similar states to undermine American leadership, intimidate American allies or threaten the American homeland itself. The blessings of the American peace, purchased at fearful cost and a century of effort, should not be so trivially squandered.' *Rebuilding America's Defenses: Strategies, Forces and Resources for a New Century*, written in September 2000 by the neo-conservative think-tank, Project for the New American Century (PNAC), 87. 'The United States has for decades sought to play a more permanent role in Gulf regional security. While the unresolved conflict with Iraq provides the immediate justification, the need for a substantial American force presence in the Gulf transcends the issue of the regime of Saddam Hussein.' The plan referred to 'key allies' such as the UK as 'the most effective and efficient means of exercising American global leadership'.
181. Cf. the Pax Christi International statement (Brussels, 26 March 2003): 'The war in Iraq demonstrates a tragic failure of international diplomacy. U.S.-led military intervention began without the consent of the UN Security Council, ignoring the warning of a big part of the

diplomatic world, civil society, churches and other faith communities worldwide. The Holy See has condemned the policy of "pre-emptive war". Non-violent means to solve the conflict were far from exhausted. The disarmament of Iraq could have been achieved without war. Weapon inspectors have not been given enough time to finish their work. This war is politically dangerous, culturally unwise and discounts the growing importance of religion and culture for the political identification of many people…' <www.paxchristi.net/PDF/ME50E03.pdf>

182. This is documented in R. J. Bonney, 'Impossible to Reconcile? Christian Just War Theory and the Second Iraq War', *Encounters. Journal of Inter-cultural Perspectives*, 9 (March 2003), 69–91. The Archbishop of Canterbury, Dr Rowan Williams, subsequently pronounced that the attack on 'Irāq 'cannot be defended as [Christian] just war': *The Times*, 15 October 2003. Earlier he had taken a more moderately critical stance: Rowan Williams, 'Weaknesses and Moral Inconsistency led us to war', *The Times*, 25 March 2003: <www.timesonline.co.uk/newspaper/0,,2741-622772,00.html>

183. Richard A. Clarke, formerly President Bush's counter-terrorism coordinator, stated that on the evening of 12 September 2001, Bush had required him 'in a very intimidating way' to investigate whether Saddam Hussein was involved: Tim Reid, 'Mr President, al-Qaeda did this. "I know, I know, but see if Saddam was involved. Just look. I want to know any shred"', *The Times* (23 March 2004), following the publication of Clarke's book, *Against All Enemies*.

184. Bob Woodward, *Plan of Attack* (New York: Simon and Schuster, 2004), 21: 'the intellectual godfather and fiercest advocate for toppling Saddam was Paul Wolfowitz…' Ibid. 9: 'Cheney had been secretary of defence during George H. W. Bush's residence… and he harboured a deep sense of unfinished business about Iraq.' In March 2003, Deputy Defense Secretary Paul Wolfowitz admitted to the magazine *Vanity Fair* that the weapons of mass destruction issue was never the United States' prime reason for invading 'Irāq. 'For bureaucratic reasons, we settled on one issue, weapons of mass destruction, because it was the one reason everyone could agree on', Wolfowitz stated in the interview. Wolfowitz was one of the authors of *Rebuilding America's Defences: Strategies, Forces and Resources for a New Century*: <www.dawn.com/2003/08/28/top14.htm>

185. Reuters, cited by Security Watch: Monday, 24 February 2003.

186. Jeremy Bransten, 'Iraq Invasion Fuels Terrorism, Say Analysts' (15 September 2003): 'the U.S. administration based its assertion on, among other things, the presence of camps containing several hundred fighters from the Ansar al-Islām terrorist group in northern 'Irāq. Most of those fighters are believed to have either fled or to have been killed during the US-led invasion of 'Irāq. Many are believed to have since returned to 'Irāq, but their links to Hussein's former regime and al-Qaeda remain unproven. At the time of the war, experts had not detected any other international terrorist group with a major presence in 'Irāq and none outside the country with a demonstrable link to Hussein's regime. But many experts now fear a connection may be forming, paradoxically, thanks to the US-led invasion and the chaos that has existed in 'Irāq since Hussein's ouster.' <www.isn.ethz.ch/infoservice/secwatch/index.cfm?service=cwn&parent=detail&menu=8&sNewsID=7441>

The US Commission investigating the terror attacks on 11 September 2001 subsequently confirmed that there was no evidence linking Saddam's regime to the al-Qaeda network, in contradiction to previous statements from the White House. There was 'no credible evidence that 'Irāq and al-Qaeda cooperated on attacks against the United States'. Although the report stated that bin Laden had sought support from 'Irāq in the early 1990s, it concluded that the bid had been unsuccessful.

187. Bonney, 'Impossible to Reconcile?', 75.

188. In a statement aired by Qatar's al-Jazeera channel on 21 May 2003, Ayman al-Ẓawāhirī, number two in the al-Qaeda network, lashed out at Arab countries which had helped the US during the invasion. Muslim rulers had been 'unmasked' by granting the United States facilities to attack

'Irāq and hosting troops of the United States and the United Kingdom, he said, naming Sa'ūdī Arabia, Kuwait, Qatar, Bahrain, Egypt, Yemen and Jordan. 'Here is Sa'ūdī Arabia [allowing] planes to take off from its airfields, here is Kuwait [hosting] huge armies marching [on 'Irāq] from its territory, while the campaign's command is based in Qatar, Bahrain hosts the Fifth Fleet and warships sail through Egypt's [Suez] Canal', he stated. As to Yemen, 'Crusader ships are refuelling from its ports'; moreover 'Crusader armies have deployed in Jordan, where batteries of Patriot missiles have also been set up, to protect Israel': <www.dawn.com/2003/05/22/top17.htm>; <www.news.bbc.co.uk/2/low/middle_east/3048275.stm>

189. Statement read out on 'Irāqī television by Information Minister Mohammed Saeed al-Sahaf, on behalf of Saddam, at the end of March 2003: '…for ages and ages, religious scholars could not reach such a consensus as they have reached today – that this aggression against the fortress of faith is an aggression on religion, wealth, honour and life and is an aggression against the homes of Islām. Therefore, *jihād*… is a duty in facing them and whoever dies on its fields is rewarded by heaven. Seize it [*jihād*], O brothers, for within it are one of two good deeds for the sake of God and great principles… Answer the call of *jihād* and long live the *mujāhidīn* of our nation. God is great and may the criminals fail.' <www.news.bbc.co.uk/2/hi/middle_east/2908147.stm>

190. Shayk Abd al-Ghafour al-Qaysi, quoted by Memri Special Dispatch series, 487, 28 March 2003: <www.memri.org/bin/articles.cgi?Page=subjects&Area=jihad&ID=SP48703>

191. Audio message purported to be by al-Qaeda leader Osama bin Laden, broadcast on Arab television station al-Jazeera, 11 February 2003: 'the enemy fears city and street wars most, a war in which the enemy expects grave human losses. We stress the importance of the martyrdom operations against the enemy – operations that inflicted harm on the United States and Israel that have been unprecedented in their history, thanks to Almighty God…. We also point out that whoever supported the United States, including the hypocrites of 'Irāq or the rulers of Arab countries, those who approved their actions and followed them in this crusade war by fighting with them or providing bases and administrative support, or any form of support, even by words, to kill the Muslims in 'Irāq, should know that they are apostates and outside the community of Muslims. It is permissible to spill their blood and take their property… Needless to say, this crusade war is primarily targeted against the people of Islām. Regardless of the removal or the survival of the socialist party or Saddam, Muslims in general and the 'Irāqīs in particular must brace themselves for *jihād* against this unjust campaign and acquire ammunition and weapons. This is a prescribed duty… Fighting in support of the non-Islamic banners is forbidden.' <www.news.bbc.co.uk/2/hi/middle_east/2751019.stm>; <www.news.bbc.co.uk/1/hi/not_in_website/syndication/monitoring/media_reports/2768873.stm>

192. They called for a legitimate, defensive, *jihād* on 11 March 2003: 'according to Islamic law, if the enemy steps on Muslims' land, *jihād* becomes a duty on every male and female Muslim'. The statement by the Islamic Research Academy called 'upon Arabs and Muslims throughout the world to be ready to defend themselves and their faith'. <www.smh.com.au/articles/2003/03/11/1047144977495.html>

For contradictory views at al-Azhar: <www.memri.org/bin/articles.cgi?Page=subjects&Area=jihad&ID=SP48003>

193. Memri Special Dispatch 747, 20 July 2004: <www.memri.org/bin/opener_latest.cgi?ID=SD74704>

194. Matthew Levinger, 'U.S. invasion of 'Irāq would play into bin Laden's hands', *Seattle Post-Intelligencer*, 4 October 2002: <www.seattlepi.nwsource.com/opinion/89688_matt04.shtml>

195. Saddam's regime claimed more than 4000 had joined the *jihād* from the Arab world: Justin Huggler, 'Iraq claims thousands of foreigners are joining *Jihād*', *Independent* (UK, 31 March 2003).

196. <www.news.bbc.co.uk/2/hi/middle_east/3204230.stm>
197. <www.defenselink.mil/news/Oct2003/n10202003_200310201.html>
198. A later statement was issued on 5 January 2004, which claimed: 'we have to liberate the Islamic world from the military occupation of the crusaders by raising up the banner of *jihād* for God's sake'. Bin Laden criticized Gulf countries for receiving members of the 'Irāqī Governing Council. He called for the establishment of a council to replace the Arab rulers and take on the role of unifying Arab positions and raising the banner of *jihād*. 'The occupation of 'Irāq is the beginning of the full occupation of the other Gulf states. The Gulf is the key for control of the world in the point of view of the big powers because of the presence of the biggest deposits of oil': <www.iht.com/articles/123712.html>; <www.uk.news.yahoo.com/040105/140/eibd4.html>
199. <www.money.cnn.com/2003/03/25/news/companies/war_contracts/> Jeffrey Donovan, 'Bush Backers Reap US$8bn in Iraq Contracts' (4 November 2003): <www.isn.ethz.ch/infoservice/secwatch/index.cfm?service=cwn&parent=detail&menu=8&sNewsID=7763>
200. '…The threat of the Shī'a to the [*ummah*] nation is equal to the threat posed by the Jews and the Christians. They harbour the same ill-will against the nation, which needs to protect itself from them and from being deceived by them… They pose a danger not only to 'Irāq, but to the whole region. If the Shī'a have influence over 'Irāq, or if they obtain some kind of autonomy in southern 'Irāq, they will be so much closer to extending their influence. After all, they exist in considerable numbers in Saudi Arabia, Kuwait, and Bahrain. If those Shī'a get organized and if their initiatives get support from countries that sponsor them – Iran, Syria, and Lebanon – it will mean that they have reached advanced stages in their 50-year plan.' Memri Special Dispatch series 498, 2 May 2003: <www.memri.org/bin/articles.cgi?Page=subjects&Area=jihad&ID=SP49803>

    Al-Qaeda denied any role in the Karbalā' bombing in March 2004, but did not wish to associate itself with the Shī'a cause, claiming that it was 'infidel': <www.shianews.com/hi/europe/news_id/0000494.php>
201. Neo-conservatives and their supporters argue that 'moral equivalence' is unacceptable and that a figure such as Osama bin Laden, with blood on his hands, has no moral case. This fails to address the extent to which the moral argument projected by bin Laden may have resonance within his constituency in parts of the Islamic world: cf. Garry Kasparov, 'Stop the Moral Equivalence: suicide-bombing and hostage-taking vs. democracy' (EDT *Wall Street Journal*, 19 May 2004) <www.opinionjournal.com/editorial/feature.html?id=110005100>
202. Woodward, *Plan of Attack*, 434–7.
203. Warren Hoge, 'Ex-UN Inspector has harsh words for Bush', *New York Times* (16 March 2004): <www.nytimes.com/2004/03/16/international/middleeast/16BLIX.html?th> James Risen, 'How pair's finding on terror led to clash on shaping intelligence', *New York Times* (28 April 2004): 'the Senate Select Committee on Intelligence is investigating whether the unit – named the Counter-Terrorism Evaluation Group by its creator, Douglas J. Feith, the under secretary of defence for policy – exaggerated the threat posed by Iraq to justify the war'. <www.nytimes.com/2004/04/28/politics/28INTE.html?th>
204. Reuters, quoted by Security Watch: Monday, 29 September 2003.
205. Reuters, quoted by Security Watch: Tuesday, 30 September 2003. Woodward, *Plan of Attack*, 433: '[Richard] Armitage believed that Chalabi had provided hyped WMD intelligence that had made its way to Bush and Cheney before the war.'
206. Richard A. Oppel Jr, 'U.S. to halt payments to Iraqi group headed by a one-time favourite', *New York Times*, 18 May 2004: <www.nytimes.com/2004/05/18/politics/18CHAL.html?th>
207. His deputy press secretary, Emily Miller, had tried to prevent him making the statement: <www.state.gov/secretary/rm/32505.htm>
208. Woodward, *Plan of Attack*, 425–6.

209. Simon Jenkins, 'I saw our failure through the bars of Abu Ghraib', *The Times* (5 May 2004).
210. A 'report of investigation' was sent to then Secretary of Defense Richard Cheney in March 1992, 'nine months after the Defense Department began an internal investigation into how seven counter-intelligence and interrogation manuals used for years by the Southern Command throughout Latin America had come to contain "objectionable" and prohibited material. Army investigators traced the origins of the instructions on use of beatings, false imprisonment, executions and truth serums back to "Project X" – a program run by the Army Foreign Intelligence unit in the 1960s. The report to Cheney found that the "offensive and objectionable material in the manuals" contradicted the Southern Command's priority of teaching respect for human rights, and therefore "undermines U.S. credibility, and could result in significant embarrassment". Cheney concurred with the recommendations for "corrective action" and [ordered the] recall and destruction of as many of the offending manuals as possible': <www.gwu.edu/~nsarchiv/NSAEBB/NSAEBB122/index.htm>
211. Tom Reid, 'Rumsfeld authorised "tactics that led to jail abuse"', *The Times* (17 May 2004).
212. The documents released by the US administration were made available on the BBC website on 23 June 2004, together with a video from the BBC's Clive Myrie arguing that 'Donald Rumsfeld approved a number of aggressive techniques': <www.news.bbc.co.uk/1/hi/world/americas/3831399.stm>

    The memorandum from Rumsfeld to the Commander, US Southern Command is at: <www.news.bbc.co.uk/nol/shared/bsp/hi/pdfs/23_06_04_apr03rums.pdf>

    The Rumsfeld memorandum contained an appendix 'to be declassified on 2 April 2013' but now indicated 'unclassified'. Table A listed the following interrogation techniques. 'A. Direct: asking straightforward questions. B. Incentive/Removal of Incentive: provide a reward or removing a privilege, above and beyond those that are required by the Geneva Convention from detainees. [Caution: other nations that believe that detainees are entitled to POW protections may consider that provision and retention of religious items (e.g. the Koran) are protected under international law (see Geneva III, Article 34). Although the provisions of the Geneva Convention are not applicable to the interrogation of unlawful combatants, consideration should be given to these views prior to application of the technique.] C. Emotional Love: playing on the love a detainee has for an individual or group. D. Emotional Hate: playing on the hatred a detainee has for an individual or group. E. Fear Up Harsh: significantly increasing the fear level in a detainee. F. Fear Up Mild: moderately increasing the fear level in a detainee. G. Reduced Fear: reducing the fear level in a detainee. H. Pride and Ego Up: boosting the ego of a detainee. I. Pride and Ego Down: attacking or insulting the ego of a detainee, not beyond the limits that would apply to a POW. [Caution: Article 17 of Geneva III provides, 'prisoners of war who refuse to answer may not be threatened, insulted, or exposed to any pleasant or disadvantageous treatment of any kind.' Other nations that believe that detainees are entitled to POW protections may consider this technique inconsistent with the provisions of Geneva. Although the provisions are not applicable to the interrogation of unlawful combatants, consideration should be given to these views prior to application of the technique.] J. Futility: invoking the feeling of futility of a detainee. K. We Know All: convincing the detainee that the interrogator knows the answer to questions he has asked the detainee. L. Establish Your Identity: convincing the detainee that the interrogator knows the answer to questions he asks the detainee. M. Repetition Approach: Continuously repeating the same question to the detainee within interrogation periods of normal duration. N. File and Dossier: Convincing detainee that the interrogator has a damning and inaccurate file, which must be fixed. O. Mutt and Jeff: A team consisting of a friendly and a harsh interrogator. The harsh interrogator must employ the Pride and EgoDown technique. [Caution: other nations that believe that POW protections apply to detainees may view this technique as inconsistent with Geneva III, Article 13 which provides that POWs must be protected against acts of intimidation. Although the provisions of Geneva are not applicable to the interrogation of unlawful combatants, consideration should be given to these views prior to application of the technique.] P. Rapid Fire: questioning in rapid succession without allowing the detainee to answer. Q. Silence: staring at the detainee to encourage discomfort. R. Change of Scenery Up: removing the detainee from the standard interrogation setting (generally to a location more pleasant, but no worse). S. Change of Scenery Down: removing the detainee from the standard interrogation setting and placing him in a setting that may be less comfortable; would not constitute a substantial change in environmental quality. T. Dietary manipulation: changing the diet of a detainee; no intended deprivation of food or water; no adverse medical or cultural effect and without intent to deprive subject of food or water, e.g. hot rations to MREs. U. Environmental Manipulation: altering the environment to create moderate discomfort (e.g. adjusting temperature or introducing an unpleasant smell). Conditions would not be such that they would injure the detainee. Detainee would be accompanied by interrogator at all times. [Caution: based on court cases in other countries, some nations may view application of this technique in certain circumstances to be inhumane. Consideration of these views should be given prior to use of this technique.] U. Sleep Adjustment: adjusting the sleeping times of the detainee (e.g. reversing sleep cycles from night to day). This technique is NOT sleep deprivation. W. False Flag: convincing the detainee that individuals from a country other than the United States are interrogating him. X. Isolation: isolating the detainee from other detainees while still complying with basic standards of treatment. [Caution: the use of isolation as an interrogation technique requires detailed implementation instructions, including specific guidelines regarding the length of isolation, medical and psychological review, and approval for extensions of the length of isolation by the appropriate level in the chain of command. This technique is not known to have been generally used for interrogation purposes for longer than 30 days. Those nations that believe detainees are subject to POW protections may view use of this technique as inconsistent with the requirements of Geneva III, Article 13 which provides that POWs must be protected against acts of intimidation; Article 14 which provides that POWs are entitled to respect for their person; Article 34 which prohibits coercion and Article 126 which ensures access and basic standards of treatment. Although the provisions of Geneva are not applicable to the interrogation of unlawful combatants, consideration should be given to these views prior to application of the technique.] Table B, General Safeguards, states that A) application of these interrogation techniques is subject to the following general safeguards: (i) limited to use only at strategic interrogation facilities; (ii) there is a good basis to believe that the detainee possesses critical intelligence; (iii) the detainee is medically and operationally evaluated as suitable (considering all techniques to be used in combination); (iv) interrogators are specifically trained for the technique(s); (v) a specific interrogation plan… has been developed; (vi) there is appropriate supervision; and (vii) there is appropriate specified senior approval for use with any specific detainee (after considering the foregoing and receiving legal advice). B) The purpose of all interviews and interrogations is to get the most information from a detainee with the least intrusive method, always applied in a humane and lawful manner with sufficient oversight by trained investigators and interrogators… C) Interrogations must always be planned, deliberate actions… D) Interrogation approaches are designed to manipulate the detainee's emotions and weaknesses to gain his willing cooperation…'

213. Christopher Findlay, 'Democratic Values Eroded in Debate over Legal Torture' (15 May 2004): <www.isn.ethz.ch/infoservice/secwatch/index.cfm?service=cwn&parent=detail&menu=8&sNewsID=8833>

214. 'Rumsfeld in the dock', *The Times* (8 May 2004).

215. Valentinas Mite, 'U.S. abuse of 'Irāqī Prisoners Inflames Arab World' (4 May 2004): <www.isn.ethz.ch/infoservice/secwatch/index.cfm?service=cwn&parent=detail&menu=8&sNewsID=8766>
216. Valentinas Mite, 'Coalition Losing Initiative in 'Irāq' (6 May 2004): <www.isn.ethz.ch/infoservice/secwatch/index.cfm?service=cwn&parent=detail&menu=8&sNewsID=8777>
217. <www.usinfo.state.gov/usinfo/Archive/washfile_feature2.html>
218. Philip Webster, 'Britain and U.S. plan for quick exit from Iraq', *The Times* (17 May 2004).
219. <www.iht.com/articles/94522.html> Reuters quoted by Security Watch: Wednesday, 28 May 2003.
220. In an intercepted letter to bin Laden, Abū Musab al-Zarqawī had stated: 'the only solution is to strike the religious, military and other cadres of the Shī'a so they will strike against the Sunnīs'. 'Souls will perish and blood will be spilt. That is exactly what we want.' Quoted by Richard Beeston and James Hider, 'Al-Qaeda linked with attempt to provoke civil war', *The Times* (3 March 2004): <www.timesonline.co.uk/newspaper/0,,172-1023894,00.html>
221. See the obituary by Michael Wood, 'Wise and moderate Shī'a cleric murdered before he could contribute to the rebuilding of Iraq', *Guardian* (12 April 2003): <www.guardian.co.uk/Iraq/Story/0%2C2763%2C935242%2C00.html> Cf. Nimrod Raphaeli, 'Muqtada al-Sadr Not Supported by Other Iraqi Leaders', Memri Inquiry and Analysis series, 170 (9 April 2004): 'not unlike the Iranian Ayatollahs who prepared him as a cleric, al-Sadr is interested, first and foremost, in achieving an Islamic state in 'Irāq, and he will not avoid confrontation if it enables him to achieve his ultimate objective…' <memri.org/bin/articles.cgi?Page=archives&Area=ia&ID=IA17004>

    The arrest warrant was signed by an 'Irāqī judge in August 2003, but it was not enforced for fear of sparking unrest among his Shī'a supporters.
222. <www.timesonline.co.uk/article/0,,1-1076885,00.html>
223. <www.timesonline.co.uk/newspaper/0,,175-1110728,00.html>
224. Anthony Loyd, 'U.S. tanks fire at will in valley of peace', *The Times* (15 May 2004): <www.timesonline.co.uk/newspaper/0,,175-1110726,00.html>
225. <www.dawn.com/2004/08/31/top13.htm>
226. <www.news.bbc.co.uk/1/hi/world/middle_east/3720161.stm> 'Al-Zarqawi's fingerprints are all over this murder', stated Hamid al-Bayati, the Deputy Foreign Minister and a friend of Mr Salim. 'Ezzedine Salim was a humble and faithful servant to his country and people. By killing him, when he served as president of the governing council, al-Zarqawi hoped to jeopardise the transfer of authority.' Al-Zarqawi is also suspected of the murder of Nick Berg: Richad Beeston, 'Car bomb blasts Iraqi council's hopes for peace', *The Times*, 18 May 2004: <www.timesonline.co.uk/newspaper/0,,171-1112687,00.html>
227. Nimrod Raphaeli, 'Moqtada al-Sadr: The Young Rebel of the 'Irāqī Shī'a Muslims' (Irạqi Leadership Biographical Series, 11 Feb. 2004) has argued that he initially sought to distance himself from a commitment to *jihād*. This was subsequently reversed, as is evidenced above. What is at issue is the permanence of the commitment to *jihād*: <www.memri.de/uebersetzungen_analysen/laender/persischer_golf/irak_sadr_11_02_04.pdf>
228. Charles Recknagel, 'Iraqi radical Shī'ite cleric, from outlaw to politician' (17 June 2004), argued that there were increasing signs that al-Ṣadr was emerging as a politician from being an outlaw/resistance leader: <www.isn.ethz.ch/infoservice/secwatch/index.cfm?service=cwn&parent=detail&menu=8&sNewsID=9034> Subsequently, the 'Irāqī prime minister lifted the US-imposed ban on the newspaper belonging to al-Ṣadr in what appeared to be an attempt to broaden the government's support as it coped with growing violence. Jeffrey Donovan, 'Allawi Lifts U.S. Ban on Radical Cleric's Newspaper' (20 July 2004): <www.isn.ethz.ch/infoservice/secwatch/index.cfm?service=cwn&parent=detail&menu=8&sNewsID=9265>

Chapter 10

1. <www.fas.org/irp/world/para/ayman_bk.html>
2. Possibly as a result of Lord Curzon's strenuous objections to earlier drafts of the declaration: 'what is to become of the people of this country…? They and their forefathers have occupied the country for the best part of 1,500 years… they will not be content either to be expropriated for Jewish immigrants, or to act merely as hewers of wood and drawers of water to the latter.' Zafarul-Islam Khan (ed.), *Palestine Documents* (New Delhi: Pharos Media, 1998), 62–3. His statement to the Cabinet was dated 26 October 1917, just a few days before the Balfour Declaration of 2 November 1917. Curzon's reference to 'hewers of wood and drawers of water' drew on the Hebrew Scriptures: Joshua 9:21.
3. Khan, *Palestine Documents*, 64–5. The telegram of Sir Reginald Wingate to Amīr Faisal ibn Husain of 8 February 1918 had stated that the British were 'determined to stand by the Arab people in their struggle for the establishment of an Arab world in which law shall replace Ottoman injustice, and in which unity shall prevail over rivalries artificially provoked by the policy of Turkish officials. His Majesty's Government re-affirm their former pledge in regard to the liberation of the Arab peoples…' Ibid. 71. A subsequent statement of the British government, dated 16 June 1918, known as the Declaration to the Seven [anonymous Arabs] which distinguished four categories of lands. In the first two categories (territories which were free and independent before the outbreak of World War I and territories liberated from Turkish rule by the action of the Arabs themselves) 'the complete and sovereign independence of the Arabs inhabiting those territories' was envisaged; in the third category (territories liberated from Turkish rule by the action of the Allied armies), 'the future government of those territories should be based upon the principle of the consent of the governed'. Ibid. 73. Amīr Faisal ibn Husain's memorandum to the supreme council at the Paris Peace Conference (1 January 1919), while noting that 'the Jews are very close to the Arabs in blood' and that 'there is no conflict of character between the two races. In principle, we are absolutely at one', nevertheless stressed that 'the Arabs cannot risk irresponsibility of holding level the scales in the clash or races and religions that have, in this one province [Palestine] so often involved the world in difficulties'. Ibid. 77. Faisal's alleged agreement with Chaim Weizmann, the Zionist leader, on 3 [?] January 1919, was not utilized by the Zionists until after Faisal's death in 1936, because it had a clause in Arabic inserted which declared it null and void if there was 'the slightest modification or departure' from the demands in his memorandum to the British. Ibid. 80–1. The Peel Commission report of June 1937 noted that the Arabs of Palestine 'put their trust in the Proclamation which Lord Allenby issued in 1917 in the name of the Governments of Great Britain and France that it was the solemn purpose of the Allies to further the cause of Arab self-determination and to establish Arab national governments': ibid. 159.
4. In this account, the beginning of the revolt is taken to be the murder of two Jews on 15 April 1936: Baruch Kimmerling and Joel S. Migdal, *The Palestinian People: A History* (Cambridge, MA: Harvard University Press, 2003), 102.
5. In the first years of Islām, the Prophet instructed his followers to direct their prayers towards Jerusalem, in the manner of the Jews.
6. Madawi al-Rasheed, *A History of Saudi Arabia* (New York and Cambridge: Cambridge University Press, 2002), 102–3.
7. This is a moot point, given the relatively restricted Jewish immigration in the pre-war years, though the 'Arabs of Palestine saw in the issue of immigration the matter of their political and cultural survival': Mark Tessler, *A History of the Israeli–Palestinian Conflict* (Bloomington and Indianapolis, IN: Indiana University Press, 1994), 170 (who gives the annual figures for immigration between 1919 and 1939). The Peel Commission report of June 1937 talked of 'about a million Arabs in strife, open or latent, with some 400,000 Jews': ibid. 242.

8. Khan, *Palestine Documents*, 197 (no source cited).
9. Ibid. 198.
10. This included clauses to the effect that Britain and the United States '…desire to see no territorial changes that do not accord with the freely expressed wishes of the peoples concerned;…they respect the right of all peoples to choose the form of government under which they will live; and they wish to see sovereign rights and self-government restored to those who have been forcibly deprived of them…'
11. Ibid. 197, 199, February and 5 April 1945.
12. Ritchie Ovendale, *The Origins of the Arab–Israeli Wars* (4th edn, Harlow: Pearson, 2004), 90–8, 109–10, 127–35, 330–1, 333–4.
13. For example, the American–Israel Public Affairs Committee (AIPAC), which calls itself 'America's pro-Israel lobby' and seeks to encourage 'Jewish political activism and student activism in campaigns and elections and by lobbying Congress…': <www.aipac.org>
14. Tessler, *A History of the Israeli–Palestinian Conflict*, 237.
15. Khan, *Palestine Documents*, 155–6.
16. Yehoshua Porath, *The Palestinian Arab National Movement. From Riots to Rebellion. II. 1929–1939* (London: Frank Cass, 1977), ii. 138–9. Publicly at least up to 1936, the Muftī had urged the Arabs to target the Jews, not the British: Kimmerling and Migdal, *The Palestinian People*, 108. The British thought him 'not an outstanding personality nor a great leader', and one who 'has allowed himself to be pushed into extreme courses against his better judgement': Yehuda Taggar, *The Muftī of Jerusalem and Palestine Arab Politics, 1930–1937* (New York and London: Garland Publishing Inc., 1986), 359. Ibid. 375, 'the Muftī and the Supreme Committee did very little leading, and in fact were dragged most of the way'.
17. Nels Johnson, *Islām and the Politics of Meaning in Palestinian Nationalism* (London: Kegan Paul International, 1982), 41.
18. Ibid. 42.
19. Porath, *The Palestinian Arab National Movement*, ii. 136.
20. Ibid. ii. 183.
21. Shaul Mishal and Avraham Sela, *The Palestinian Hamas. Vision, Violence and Coexistence* (New York: Columbia University Press, 2000), 64. <www.us-israel.org/jsource/biography/alqassam.html>
22. Martin Kolinsky, *Law, Order and Riots in Mandatory Palestine, 1928–35* (Basingstoke: Macmillan, 1993), 38. The Muftī claimed that the Jews wanted 'the thing behind it [the Temple] – the mosque of 'Umar'. Some of the British officials attributed to the Muftī a greater share in the responsibility for the 1929 disturbances than in the majority Shaw Commission Report: '…the Muftī must bear the blame for his failure to make any effort to control the character of the agitation conducted in the name of a religion of which he was the head.' Ibid. 75. The Muftī was seen by the British as an astute politician capable of raising a religious cry which 'in time of unrest might sweep the country and compel his present opponents to follow his banner'. Ibid. 165.
23. Recommendations of the King–Crane Commission, 28 August 1919: Khan, *Palestine Documents*, 98. 'With the best possible intentions, it may be doubted whether the Jews could possibly seem to either Christians or Muslims proper guardians of the holy places, or custodians of the Holy Land as a whole.'
24. Porath, *The Palestinian Arab National Movement*, ii. 168–9. Kimmerling and Migdal, *The Palestinian People*, 111.
25. Philip Mattar, *The Mufti of Jerusalem. Al-Hajj Amin al-Husayni and the Palestine National Movement* (New York: Columbia University Press, 1988), 82.

26. Porath, *The Palestinian Arab National Movement*, ii. 233–6 (who shows that the government intended to deport the Muftī before Andrews' assassination). Kimmerling and Migdal, *The Palestinian People*, 113–14.
27. Rudolph Peters, *Islām and Colonialism. The Doctrine of Jihād in Modern History* (The Hague: Manton, 1979), 99.
28. Porath, *The Palestinian Arab National Movement*, ii. 242, 248. Kimmerling and Migdal, *The Palestinian People*, 114, 117–19.
29. Porath, *The Palestinian Arab National Movement*, ii. 238.
30. Ibid. ii. 264, 269 (lack of Christian support and the revolt as an anti-Christian movement). For the rural character of the revolt: Ylana N. Miller, *Government and Society in Rural Palestine, 1920–1948* (Austin, TX: University of Texas Press, 1985), 121–38. Tessler, *A History of the Israeli–Palestinian Conflict*, 240. The Peel Commission Report of July 1937 argued that the cause of the disturbances was to be found in 'the desire of the Arabs for national independence' and 'their hatred and fear of the establishment of the Jewish National Home'. Ibid. 241. Khan, *Palestine Documents*, 158 (where the wording is not quite the same).
31. Tessler, *A History of the Israeli–Palestinian Conflict*, 245.
32. Kimmerling and Migdal, *The Palestinian People*, 129–31.
33. The failure to reach an Anglo-Arab Accord with the Arab states in April 1939 may have been over a matter of wording ('consultation with' Arab states in the British proposal, 'consent of' the Arab states in their version): Khan, *Palestine Documents*, 178.
34. Ibid. 118.
35. Ibid. 189.
36. Ibid.
37. Francis R. Nicosia, *The Third Reich and the Palestine Question* (London: I. B. Tauris, 1985), 187–91.
38. His letter to Hitler, dated 20 January 1941, is reproduced by Zvi Elpeleg, *The Grand Muftī of Jerusalem, Haj[j] Amin al-Hussaīni, Founder of the Palestinian National Movement* (London: Frank Cass, 1993), 202–5. Ibid. 64–73 for his period in Nazi Germany.
39. Mattar, *The Mufti of Jerusalem*, 95: 'the *fatwā* was the most anti-British statement he had ever made, pointing out how the British had promised to free the Arabs from the Ottomans but instead had divided the Arab world and imposed British rule, and committed "unheard of barbarism" in Palestine'.
40. The British Air Vice-Marshal (Commanding) in 'Irāq wrote: 'the rest of the Arabic speaking countries watched the rebellion with feigned indifference. Fresh in their memories, however, were our recent defeats in Libya and Greece and they were well aware of our obvious embarrassment in Crete. A further defeat in 'Irāq, which would have been made much of by the enemy propagandists, would surely have shaken Arab confidence in our invincibility. And Syria in particular – would not the Syrian attitude have stiffened in consequence? How would Turkey and Iran, two of the few remaining neutrals anxiously watching every move in this world struggle, have reacted to the presence of an Axis-controlled Arab State on their frontiers?' <www.raf.mod.uk/history/opsrep.html>
41. <www.eretzyisroel.org/~jkatz/fatwa.html>
42. Mattar, *The Mufti of Jerusalem*, 102–3.
43. Named after the sword carried by Turkish policemen: George Lepré, *Himmler's Bosnian Division: the Waffen–SS Handschar Division, 1943–1945* (Atglen, PA: Schiffer Military History, 1997), 47. Ibid. 34, for the regiment having the 'appearance of a pan-Islamic fighting unit'; ibid. 261, for criticism of it as 'not even a military formation but merely a political instrument'. Although the Muftī claimed that the entire Muslim world was united with Germany against Britain (ibid. 33), in reality an estimated 10,000 Muslims served with Tito's partisans (ibid. 135), while nearly half the *Hanzar* Division comprised conscripts: ibid. 315.

44. Mattar, *The Mufti of Jerusalem*, 104–5, is summary on this period but notes (ibid. 107) 'only a thorough and non-partisan study, based on captured German documents, could elucidate the role of the Muftī in Germany'.
45. Lepré, *Himmler's Bosnian Division*, 75. Himmler is reported to have stated: 'I have nothing against Islām because it educates the men in this division for me and promises them heaven if they fight and are killed in action. A very practical and attractive religion for soldiers.' On 6 August 1943 Himmler ordered that 'all Muslim members of the Waffen SS and police are to be afforded the undeniable right of their religious demands never to touch pork, pork sausages nor to drink alcohol…' These details, and the quotations from the Muftī, are given in a somewhat tendentious account by Carl K. Savich, 'Islām under the Swastika: The Grand Muftī and the Nazi Protectorate of Bosnia–Hercegovina, 1941–1945': <www.rastko.org.yu/rastko-bl/istorija/kcsavic/csavich-islam_e.html>
46. He was surprisingly well informed about the size of the US population, which was estimated at 136.7 million in July 1943: Historical National Population Estimates, 1 July 1900 to 1 July 1999. Source: Population Estimates Program, Population Division, US Census Bureau Internet Release Date: 11 April 2000. Revised date: 28 June 2000. <www.census.gov/population/estimates/nation/popclockest.txt> There were about 4.6 million Jews in the USA at this date, whereas there were only about 100,000 Arabs.
47. Seven Arab states had objected to resolutions in the House of Representatives and the Senate in late January and early February in favour of the establishment of an independent Jewish state in Palestine: Ovendale, *The Origins of the Arab–Israeli Wars*, 87.
48. John Darwin, *Britain and Decolonisation. The Retreat from Empire in the Post-War World* (Basingstoke and London: Macmillan, 1988, repr. 1992), 118. Lepré, *Himmler's Bosnian Division*, 316: 'Muslim autonomists' were 'the big losers'.
49. Satadru Sen, 'Subhas Chandra Bose': <www.andaman.org/book/app-m/textm.htm> Also Ranjan Borra, 'Subhas Chandra Bose, l'armée nationale indienne et la guerre de libération de l'Inde': <www.angelfire.com/folk/library/bose2_fr.htm> There is, however, a claim that Bose lived on until 16 September 1985 as a hermit: <www.hindustantimes.com/news/specials/Netaji/netajihomepage.shtml>
50. Avi Shlaim, *The Iron Wall. Israel and the Arab World* (London: Penguin, repr. 2001), 18.
51. Ibid. 41.
52. Ibid. 51–2. Shlaim comments that the leadership of the Muftī 'remained as uncompromising in its opposition to Zionism in the late 1940s as it had been over the preceding quarter of a century': ibid. 29.
53. Ibid. 55–6, 78.
54. Khan, *Palestine Documents*, 63.
55. Tessler, *A History of the Israeli–Palestinian Conflict*, 285.
56. Khan, *Palestine Documents*, 270–2. Ibid. 37, for the Basel Protocol of August 1897; ibid. 63–5 for the Balfour Declaration; ibid. 242–68 for Resolution 181 (II).
57. Tessler, *A History of the Israeli–Palestinian Conflict*, 287. Ernest Bevin, British Foreign Secretary, talked of 'more than twenty centuries' of Arab occupation, i.e. referring to pre-Islamic times, in the debate in the House of Commons on 26 January 1949. Churchill talked of a perspective of 2000 or 3000 years: Ovendale, *The Origins of the Arab–Israeli Wars*, 141.
58. Ovendale, *The Origins of the Arab–Israeli Wars*, 135. The Arab League Secretary, General Azzam Pasha, declared *jihād*, stating 'this will be a war of extermination and a momentous massacre which will be spoken of like the Mongolian massacres and the Crusades'. The Muftī of Jerusalem stated: 'I declare *jihād*, my Muslim brothers! Murder the Jews! Murder them all!'
59. <www.hashd.org/english/readinbook/arablegue.htm>
60. Ovendale, *The Origins of the Arab–Israeli Wars*, 135.

61. <www.domino.un.org/UNISPAL.NSF/561c6ee353d740fb8525607d00581829/5fbced3943293bbd0525656900654aa6!OpenDocument>
62. Tessler, *A History of the Israeli–Palestinian Conflict*, 279–80.
63. <www.un.org/unrwa/overview/index.html>
64. Sami Hadawi, *Palestinian Rights and Losses in 1948. A Comprehensive Study. V. An Economic Assessment of Total Palestinian Losses. Written by Dr Atef Kubursi* (London: Saqi Books, 1988), 183.
65. Ibid. 308–14.
66. Shlaim, *The Iron Wall*, 59.
67. Ibid. 101.
68. Ibid. 77.
69. Ibid. 67. Avi Shlaim, *The Politics of Partition. King Abdullah, the Zionists and Palestine, 1921–1951* (Oxford: Oxford University Press, 1988, repr. 1990), 395–6, 417. The cousin of the former Grand Muftī was the only one to plead 'guilty' to the charges as read out in the trial.
70. Shlaim, *The Iron Wall*, 128. Speech of Golda Meir to the UN General Assembly, 17 January 1957: '...amongst the destitute elements of the local population and refugee camps, the Egyptian High Command organized *fedayeen* units as military formations of the Egyptian army. In the past eighteen months, these units carried out an intensified campaign of attack upon Israel. They ambushed road traffic, killed men, women and children, blew up wells and water installations, mined roads at night, demolished houses in which farmers and their families were peacefully asleep. These outrages culminated in major outbreaks during August and September 1955, April 1956 and October 1956. In the ominous build-up of Egyptian forces, with offensive weapons obtained during the first half of 1956, the Gaza strip had an essential role both as a centre for *fedayeen* groups, and as the forward base of an Egyptian Army division which was stationed there within an hour's drive from Tel Aviv. Since the expulsion of Egyptian forces from Gaza *fedayeen* have ceased to infest the countryside...' <www.gos.sbc.edu/m/meir2.html>
71. <www.us-israel.org/jsource/Terrorism/Fedayeen.html>
72. Khan, *Palestine Documents*, 290–1.
73. Ovendale, *The Origins of the Arab–Israeli Wars*, 193.
74. Yezid Sayigh, *Armed Struggle and the Search for State. The Palestinian National Movement, 1949–1993* (Institute for Palestine Studies, Washington DC: Oxford University Press, 1997; repr. 1999), 78–9, 83–4, 89. Nasser argued, according to Habash, that the 'enemy is not Israel but the U.S.A.': ibid. 109.
75. Tessler, *A History of the Israeli–Palestinian Conflict*, 374.
76. 'Zionism is a colonialist movement in its inception, aggressive and expansionist in its goal, racist in its configurations, and fascist in its means and aims. Israel, in its capacity as the spearhead of this destructive movement and as the pillar of colonialism, is a permanent source of tension and turmoil in the Middle East, in particular, and to the international community in general.' The Palestine National Charter (May 1964): <www.palestine-un.org/plo/pna_two.html>
77. Sayigh, *Armed Struggle and the Search for State*, 84. '1) Palestine is part of the Arab World, and the Palestinian people are part of the Arab Nation, and their struggle is part of its struggle. 2) The Palestinian people have an independent identity. They are the sole authority that decides their own destiny, and they have complete sovereignty [over] all their lands. 3) The Palestinian Revolution plays a leading role in liberating Palestine. 4) The Palestinian struggle is part and parcel of the world-wide struggle against Zionism, colonialism and international imperialism. 5) Liberating Palestine is a national obligation which necessities the materialistic and human support of the Arab Nation.' The precise dating is unclear: <www.fateh.net/e_public/constitution.htm#Introduction%20to%20the>

78. Sayigh, *Armed Struggle and the Search for State*, 103, 107–8, 123–5. Shlaim, *The Iron Wall*, 230, 232.
79. Sayigh, *Armed Struggle and the Search for State*, 139, 141.
80. Shlaim, *The Iron Wall*, 245.
81. Khan, *Palestine Documents*, 294–5. <www.hashd.org/english/readinbook/hartom67.htm>
82. Sayigh, *Armed Struggle and the Search for State*, 147, 192, 196. Under the terms of the Bistami–Arafat agreement of 3 November 1969, Palestinians resident in Lebanon were to be permitted to 'participate in the Palestinian revolution through the Armed Struggle and in accordance with the principles of the sovereignty and security of Lebanon. It was agreed to facilitate commando activity by means of: 1. facilitating the passage of commandos and specifying points of passage and reconnaissance in the border areas;…4. establishing a joint command control of the Armed Struggle and the Lebanese Army; 5. ending the propaganda campaigns by both sides; 6. conducting a census of Armed Struggle personnel in Lebanon by their command; 7. appointing Armed Struggle representatives at Lebanese Army headquarters to participate in the resolution of all emergency matters…' The two delegations affirmed that 'the Palestinian armed struggle is in the interest of Lebanon as well as in that of the Palestinian revolution and all Arabs…': <www.lebanese-forces.org/lebanon/agreements/cairo.htm> <www.radiobergen.org/plo/cairo.html>
83. Sayigh, *Armed Struggle and the Search for State*, 158. Ibid. 157, where Sayigh comments that 'Arafat…, characteristically, exaggerated the extent of Fateh organizational and military preparedness.'
84. Ibid. 162–3.
85. Ibid. 172–3.
86. Ibid. 178–9. Tessler, *A History of the Israeli–Palestinian Conflict*, 425–6. Joseph Nevo, 'The Jordanian, Palestinian and the Jordanian-Palestinian Identities', The Fourth Nordic conference on Middle Eastern Studies. 'The Middle East in [a] Globalizing World'. Oslo, 13–16 August 1998: <www.hf.uib.no/smi/pao/nevo.html>
87. Sayigh, *Armed Struggle and the Search for State*, 217.
88. Ibid. 209–10.
89. Ibid. 210–15. Ibid. 219, for the Fateh self-analysis that 'Fateh is the Palestinian revolution… the history of Fateh is the history of the modern Palestinian revolution.'
90. Ibid. 226, 236. These are listed, at the date of 1970, in Tessler, *A History of the Israeli–Palestinian Conflict*, 431.
91. Johnson, *Islām and the Politics of Meaning in Palestinian Nationalism*, 75.
92. Ibid. 76.
93. Ibid.
94. Tessler, *A History of the Israeli–Palestinian Conflict*, 434. Khan, *Palestine Documents*, 297–303. <www.iap.org/charter.htm>; <www.electronicintifada.net/bytopic/historicaldocuments/44.shtml>
95. Sayigh, *Armed Struggle and the Search for State*, 280.
96. Ibid. 281, 307.
97. Shlaim, *The Iron Wall*, 312.
98. Ibid. 313.
99. Ibid. 316. Khan, *Palestine Documents*, 307.
100. Shlaim, *The Iron Wall*, 319, 330. Sayigh, *Armed Struggle and the Search for State*, 316–17.
101. Khan, *Palestine Documents*, 309.
102. Ibid. 313–14. Web version at: <www.palestine-un.org/plo/doc_one.html>
103. Khan, *Palestine Documents*, 314–15.
104. Ibid. 315.
105. <www.weltpolitik.net/texte/policy/israel/Speecharafat_1974.pdf>

106. Khan, *Palestine Documents*, 317.
107. Ibid. 318–19.
108. Ibid. 332–5.
109. <www.us-israel.org/jsource/Peace/sadat_speech.html>
110. Shlaim, *The Iron Wall*, 369–70.
111. Khan, *Palestine Documents*, 337–42.
112. Shlaim, *The Iron Wall*, 376–7.
113. <www.nobel.se/peace/laureates/1978/>
114. Khan, *Palestine Documents*, 352–3.
115. Sayigh, *Armed Struggle and the Search for State*, 501.
116. Khan, *Palestine Documents*, 346–52. Significantly, the USA signed a separate memorandum with Israel on the same day, confirming its support in the eventuality that Egypt did not honour the peace: <www.israel-mfa.gov.il/mfa/peace%20process/guide%20to%20the%20peace%20process/us-israel%20memorandum%20of%20agreement>
117. Johnson, *Islām and the Politics of Meaning in Palestinian Nationalism*, 75.
118. Sayigh, *Armed Struggle and the Search for State*, 503.
119. <www.cfrterrorism.org/groups/jamaat2.html> 'Following the release of most of the Islamic prisoners from the Egyptian jails by President Sādāt after 1971, several groups of militants began to organize themselves. These militant groups or cells took names such as the Islamic Liberation Party, al-Takfīr wal-Hijra (Excommunication and Emigration), al-Najāt min al-nār (Saved from the Inferno), and Jihād (Holy War), as well as many others, including al-Gama'a al-Islāmiyya (The Islamic Group). Each cell operated separately and was self-contained, a fact that allowed the organization to be structured, but at the same time loosely organized. It seems that there was some kind of organized contact between the leaders of the different groups, but whether there was ever an effective overall direction of all the groups is not clear': <www.ict.org.il/inter_ter/orgdet.cfm?orgid=12>
120. The treaty was signed on 26 October 1994: <www.kinghussein.gov.jo/peacetreaty.html>; <www.us-israel.org/jsource/Peace/isrjor.html>; <www.mfa.gov.il/MFA/Peace%20Process/Guide%20to%20the%20Peace%20Process/Israel-Jordan%20Peace%20Treaty> Draft Israel–Syria peace (January 2000): <www.us-israel.org/jsource/Peace/syrdraft.html>
121. Shlaim, *The Iron Wall*, 390–1.
122. The siege lasted from 20 November until 3 December 1979: al-Rasheed, *A History of Saudi Arabia*, 144–5. Gwenn Okruhlik, 'Networks of Dissent: Islamism and Reformism in Saudi Arabia': <www.ssrc.org/sept11/essays/okruhlik.htm>
123. K. R. Pruthi (ed.), *Encyclopaedia of Jihad* (5 vols, New Delhi: Anmol Publications Pvt., 2002), i. 204.
124. Ibid. 205: Khalid al-Islambuli, the actual assailant, acted because of the arrest of his brother in Sādāt's purge. Hani Mustafa and Khaled El-Fiqi photographed the picture of Khālid al-Islāmbulī, 'the most famous picture in Iran', in the street named after him in Tehran in the course of April 2002. The naming of this street has caused diplomatic difficulties between Egypt and Iran: <www.weekly.ahram.org.eg/2002/582/feature.htm>

    An extraordinary celebration of the act of assassination was penned by the self-proclaimed 'freedom fighter' and Moroccan-born Islamist, Aḥmed Rāmī: 'when treason reigns, and mischief overwhelms, and mouths are muzzled, there is nothing one can do except what Khālid al-Islāmbulī and his brothers did. Khālid al-Islāmbulī, the people's spear in the heart of treason, the thunder of anger in an environment of submission and lack of championship. He is a torch in the gutter of silence, and the cry of truth in the face of the despots. Peace be upon thee the day the first bullet hit the chest of the despot and overthrew the throne: He was soaking in his blood, and you took revenge for Egypt and all the Muslims. Peace be upon thee when you fell, a martyr, reverberating from the deep earth your last Adieus…' <www.abbc.net/rami/>

125. <www.sptimes.ru/archive/times/906/rest/r_10476.htm>

126. For the organization of al-Jihād, David Zeidan, 'Radical Islām in Egypt: A Comparison of Two Groups', *Meria* (*Middle East Review of International Affairs*), 3 (September 1999): <www.meria.idc.ac.il/journal/1999/issue3/jv3n3a1.html>; David Zeidan, 'The Islamic Fundamentalist View of Life as a Perennial Battle', *Meria* [*Middle East Review of International Affairs*], 5 (December 2001): <www.meria.idc.ac.il/journal/2001/issue4/zeidan.pdf>

127. J. J. G. Jansen, *The Neglected Duty. The Creed of Sadat's Assassins and Islamic Resurgence in the Middle East* (New York: Macmillan, 1986). Another edition, ed. Abu Umamah, *Jihād. The Absent Obligation* (Birmingham: Maktabah Al Ansaar Publications, 2000). Another title is *Forgotten Obligation*, Pruthi, *Encyclopaedia of Jihad*, i. 264–9: <www.omislam.com/engelska-artiklar/jihad_the_forgotten_obligation_2.htm>

128. The Shaykh did not become Grand Imām until March 1982: <www.islamophile.org/spip/article40.html> Jansen, *The Neglected Duty*, ch. 2, 'the response from Al-Azhar', 3, 54–60. Abdulrahman Muḥammad Alsumaih, 'The Sunnī Concept of *Jihād* in Classical *Fiqh* and Modern Islamic Thought' (unpublished PhD, University of Newcastle upon Tyne, 1998), 347–53.

129. Jansen, *The Neglected Duty*, 160–1.

130. Ibid. 193, 197.

131. Ibid. 22, 201. Alsumaih, 'The Sunnī Concept of *Jihād*', 348.

132. Jansen, *The Neglected Duty*, 167. In Asad's translation, it is 'they who do not judge in accordance with what God has bestowed on them on high are, indeed, deniers of the truth!'

133. Ibid. 54. Alsumaih, 'The Sunnī Concept of *Jihād*', 349–50.

134. Jansen, *The Neglected Duty*, 8, 169, 193.

135. Ibid. 55.

136. Jansen, *The Neglected Duty*, 8, 10, 169, 183, 191–2.

137. Alsumaih, 'The Sunnī Concept of *Jihād*', 351.

138. Jansen, *The Neglected Duty*, 55.

139. Ibid. 55, 195.

140. Ibid. 56. Alsumaih, 'The Sunnī Concept of *Jihād*', 351–2.

141. Jansen, *The Neglected Duty*, 56–7. Alsumaih, 'The Sunnī Concept of *Jihād*', 352.

142. Jansen, *The Neglected Duty*, 165.

143. Ibid. 11, 186.

144. Ibid. 18, 192.

145. Ibid. 172. Ibid. 136: 'we must concentrate on our own Islamic situation; we have to establish the rule of God's religion in our own country first, and to make the Word of God supreme… There is no doubt that the first battlefield for *jihād* is the extermination of these infidel leaders and to replace them by a complete Islamic order.'

146. Quṭb was cited, but in the context of those who preferred 'cheap comfort' over 'noble toil', 'base safety' over the 'sweet danger' of *jihād*: ibid. 30, 226.

147. Ibid. 21, 200.

148. However, the term is problematic. 'Every year the Islamic conference calls for upholding the principle of the *shūrā* [council] – which no Arab government has honoured since the dawn of Islām. No one knows what the *shūrā* is, or how to implement it in the world of today, because we have had no actual experience in implementing it.' Memri Special Dispatch Series, 721, 25 May 2004: <www.memri.org/bin/opener_latest.cgi?ID=SD72104>

149. The Qur'ānic source is Q.42:36. James Piscatori, 'Islām, Islamists and the Electoral Principle in the Middle East', *ISIM Papers* (Leiden, 2000): <www.isim.nl/files/paper_piscatori.pdf>

150. Jansen, *The Neglected Duty*, 57–60.

151. Ibid. 31. Zeidan, 'Radical Islam in Egypt'.
152. Pruthi, *Encyclopaedia of Jihād*, iv. 1076.
153. Ibid. iv. 1098–102. Ibid. iv. 1099, where Faraj's *Neglected Obligation* is cited as reflecting 'the opinions of the Islamic Jihād'.
154. <www.web.amnesty.org/report2003/lby-summary-eng>
155. Sayigh, *Armed Struggle and the Search for State*, 497, 499.
156. Shlaim, *The Iron Wall*, 411: 'the text of the telegram [of Begin to Reagan], which was published in the *Jerusalem Post*, shocked many Israelis, who felt that the memory of the Holocaust should not be invoked to justify the Lebanon War or the siege of Beirut'.
157. Ibid. 405: Sharon spoke in cabinet of a limit of 40 kilometres for the incursion into Lebanon. Ibid. 410: the southern outskirts of Beirut, where the Israeli army ended up, marked 'a distance considerably longer than forty kilometres'.
158. Ibid. 401.
159. <www.rense.com/general24/nil.htm> <www.econ161.berkeley.edu/movable_type/2003_archives/000352.html>; William N. Dale, 'Cursed is the U.S. Envoy who tries to bring peace to the Middle East', in *American Diplomacy*: <www.unc.edu/depts/diplomat/archives_roll/2002_07-09/book_sept02/book_dale_peace.html>

John Boykin's study won the Douglas Dillon Award for a Book of Distinction on the Practice of American Diplomacy (2002). Boykin wrote (*Cursed is the Peacemaker. The American Diplomat Versus the Israeli General. Beirut 1982* [Belmont, CA: Applegate Press, 2002], 271): 'As Sharon tells the story, the problem was not hundreds of people got killed. It was that too many of the wrong people got killed. The Phalangists just "went too far", he says, killing… civilians when they were supposed to be killing only terrorists. To Phil Habib and most of the rest of the world, the problem was that no such operation should have happened at all. In the first place, the Israelis had no right to take over West Beirut at all. As Dillon puts it, "the Israelis, who had promised to stay out of Beirut, immediately invaded to 'restore order'. That was just a pretext; there was no disorder." In the second place, neither they nor any surrogates had any right to be in the camps killing anybody at all. [Maurice] Draper put it this way to Sharon on Saturday morning: "you must stop the acts of slaughter. They are horrifying. I have a representative in the camp counting the bodies. You should be ashamed. The situation is absolutely appalling. They're killing children! You have the field completely under you control and are therefore responsible for that area."…[Habib] was the one who had promised the civilians safety. "I had signed this paper which guaranteed that these people [the Palestinians] in west Beirut would not be harmed. I got specific guarantees on this from Bashir [Gemayel] and from the Israelis – from Sharon… On the basis of those assurances we had given our word. We had been deceived… I had given Arafat an undertaking that his people would not be harmed, but this was totally disregarded by Sharon whose word was worth nothing."' At p. 442, Boykin notes from Sharon's own account [Ariel Sharon with David Chanoff, *Warrior: An Autobiography* (New York: Simon and Schuster, 1989), 505] that civilian collateral casualties were inevitable. At p. 443, Boykin notes that Sharon explains 'his lack of concern on the fact that everyone knew Bashir [Gemayel] had not been killed by a Palestinian. Therefore, he reasoned, the Phalange would have no reason to wreak vengeance on the Palestinians and "no-one had batted an eye at the idea of sending in the Phalangists": ibid. 507; also Kahan Report, 22, 27. But, Boykin adds, 'it was not at all certain in the hours and days following the assassination whom the bomber was working for. Indeed, it is still not… Regardless of who thought who had killed Bashir, the Phalange had wanted revenge against Palestinians for various wrongs for years before Bashir was assassinated… Of the IDF's instructions to the Phalange to conduct themselves honourably in the camps Schiff and Ya'ri write [Ze'ev Schiff and Ehud Ya'ari, *Israel's Lebanon War* (New York: Simon and Schuster, 1984), 257] that "such repeated warnings would seem to indicate, especially in light of the

Phalange's known record of atrocities, that the senior military men in the field were wary of their intentions from the start…"'

160. Shlaim, *The Iron Wall*, 422.
161. Sayigh, *Armed Struggle and the Search for State*, 508.
162. Ibid. 506.
163. Shlaim, *The Iron Wall*, 403. This attack was almost certainly carried out by Arafat's sworn enemy, Abu Nidal. Shlaim comments that 'Mossad sources had intelligence to suggest that the attempt on Argov's life was intended to provoke an Israeli assault on Arafat's stronghold in Lebanon in order to break his power'.
164. Khan, *Palestine Documents*, 380–1.
165. The figure given in Sayigh, *Armed Struggle and the Search for State*, 539.
166. <www.nilemedia.com/Columnists/Ahmed/2001/Feb/Sharon_Knew_0.html> The McBride Commission commented on the Israeli argument that its army was not occupying the country and thus had no responsibility for law and order and 'had nothing to do with the Lebanese settling their accounts': 'in the view of the Commission, this Israeli argument is legally unacceptable. By making use of the militias which it controlled, and by leaving them free to do what they liked, or by permitting the activities of smaller groups which were not under its control, the Israelis as the occupying power bear responsibility for the acts they have committed. The Commission has established that the Israeli policy was to make the Lebanese auxiliaries carry out the tasks which Israel did not wish its own army to execute…' Khan, *Palestine Documents*, 391. Ibid. 398: 'the denial of nationality to Palestinians has resulted in all Palestinian social institutions being considered to be part of the apparatus of the "terrorists of the PLO". The borderline between Mr Begin's claim to "eliminate the PLO" and the total destruction of the social organization of the Palestinian peoples in Lebanon is a very narrow one, and the constant reference to the need to "purify" the territory of the Lebanon of PLO elements has been conducive to attacks on the autonomy of the Palestinian people.'
167. The Kahan Commission defined the 'indirect responsibility' of Israel thus: 'we assert that the atrocities in the refugee camps were perpetrated by members of the Phalangists, and that absolutely no direct responsibility devolves upon Israel or upon those who acted in its behalf. At the same time, it is clear from what we have said above that the decision on the entry of the Phalangists into the refugee camps was taken without consideration of the danger – which the makers and executors of the decision were obligated to foresee as probable – that the Phalangists would commit massacres and pogroms against the inhabitants of the camps, and without an examination of the means for preventing this danger. Similarly, it is clear from the course of events that when the reports began to arrive about the actions of the Phalangists in the camps, no proper heed was taken of these reports, the correct conclusions were not drawn from them, and no energetic and immediate actions were taken to restrain the Phalangists and put a stop to their actions. This both reflects and exhausts Israel's indirect responsibility for what occurred in the refugee camps': <www.caabu.org/press/documents/kahan-commission-part6.html>

    For Sharon's role: <www.caabu.org/press/documents/kahan-commission-part9.html>

    Finally, the Commission recommended: 'in our opinion, it is fitting that the Minister of Defence draw the appropriate personal conclusions arising out of the defects revealed with regard to the manner in which he discharged the duties of his office – and if necessary, that the Prime Minister consider whether he should exercise his authority under Section 21-A(a) of the Basic Law: the Government, according to which "the Prime Minister may, after informing the Cabinet of his intention to do so, remove a minister from office".' <www.caabu.org/press/documents/kahan-commission-part13.html>
168. Shlaim, *The Iron Wall*, 417.

169. Judith Palmer Harik, *Hezbollah. The Changing Face of Terrorism* (London and New York: I. B. Tauris, 2004), 65.
170. Sayigh, *Armed Struggle and the Search for State*, 541, 545, 569.
171. Helen Cobban, 'The Growth of Shīʿī Power in Lebanon', in *Shīʿīsm and Social Protest*, edited by Juan R. I. Cole and Nikki R. Keddie (New Haven, CT, and London: Yale University Press, 1986), 137–55, at 147. Quoted by Amal Saad-Ghorayeb, *Hizbu'llah: Politics and Religion* (London and Sterling, VA: Pluto Press, 2002), 11.
172. The founding of Hamas may be seen as an indirect consequence because the expulsion of the PLO from Lebanon appeared to have rendered it militarily and politically bankrupt. A new body was needed, therefore, to fill the void. <www.wrmea.com/archives/november02/0211020.html>
173. Saad-Ghorayeb, *Hizbu'llah*, 14.
174. R. Hrair Dekmejian, *Islām in Revolution. Fundamentalism in the Arab World* (Syracuse, NY: Syracuse University Press, 1985), fig. 5 at 98.
175. Saad-Ghorayeb, *Hizbu'llah*, 11. The Ḥizbu'llah website's statement of 'identity and goals' comments: 'Hezbollah went through various decisive moments in its history, with the most important moment being in 1982, the year of the Zionist invasion of Lebanon. This invasion led to the occupation of the capital Beirut making it the second Arab capital to be occupied during the Arab–"Israeli" conflict, with Jerusalem being the first. This crossroad speeded up the presence of Hezbollah as a struggle movement that is totally affiliated in the long complicated and complex fight against the Zionist enemy. The starting point of that struggle being the Zionist occupation of Palestine, and then to many of the Arab lands in Egypt, Syria and Jordan, leading up to Lebanon. All that led to the establishment of the identity of Hezbollah as a struggle movement against the Zionists.' <www.hizbollah.tv/english/info.htm>
176. Saad-Ghorayeb, *Hizbu'llah*, 45.
177. Ibid. 51.
178. Ibid. 36–7.
179. Ibid. 25.
180. <www.almashriq.hiof.no/lebanon/300/320/324/324.2/hizballah/warn/hizballah.html#5.2>
181. Saad-Ghorayeb, *Hizbu'llah*, 25.
182. Though a Shīʿa-led movement, Ḥizbu'llah tries to deny that it is a closed group or sectarian in outlook: 'every Muslim is automatically a member of Ḥizbu'llah' is the claim: Saad-Ghorayeb, *Hizbu'llah*, 69–70. It seeks to realize the interests of 'all Lebanese citizens': ibid. 84.
183. Ibid. 114–15.
184. Ibid. 73, 78, 84, 112.
185. Ibid. 116–17, 119.
186. Ibid. 123.
187. Ibid. 124.
188. Ibid. 125.
189. Ibid. 127.
190. Ibid. 128.
191. Ibid. 129.
192. Martin Kramer, 'Hizbu'llah: The Calculus of Jihād', in *Fundamentalisms and the State: Remaking Polities, Economies, and Militance* (The Fundamentalism Project, vol. 3), ed. M. Marty and R. S. Appleby (Chicago, IL: University of Chicago Press, 1993), 539–56: <www.geocities.com/martinkramerorg/Calculus.htm>
193. Saad-Ghorayeb, *Hizbu'llah*, 95–6.
194. Terry Waite, *Taken on Trust* (London: Hodder and Stoughton, 1993); Brian Keenan, *An Evil Cradling* (London: Hutchinson, 1992); John McCarthy and Jill Morrell, *Some Other Rainbow* (London: Bantam, 1993). Keenan and McCarthy were incarcerated together and five years later

travelled to Chile to revisit their imagination and past experiences: Keenan and McCarthy, *Between Extremes* (London: Bantam, 1999).

195. <www.news.bbc.co.uk/2/hi/middle_east/300995.stm>
196. Upon his release, Terry Waite's hostage-takers told him, 'we don't believe we have achieved much by keeping you': <www.livingbetter.org/livingbetter/articles/terrywaite.html>
197. <www.ghazi.de/security.html>
198. <www.timpritchard.com/hostage.htm>
199. <www.fortunatepipedream.org/Services/seanresumefolders/specialforces/williambuckley1.html>
200. <www.ghazi.de/security.html>
201. <www.livingbetter.org/livingbetter/articles/terrywaite.html>; <www.news.bbc.co.uk/onthisday/hi/dates/stories/november/18/newsid_2520000/2520055.stm>
202. A. Nizar Hamzeh, 'Islamism in Lebanon: A Guide to the Groups', *Middle East Quarterly*, 4 (1997), who identifies a number of 'affiliates and offshoots' of Ḥizbu'llah: <www.meforum.org/article/362>
203. And Augustus Richard Norton, who writes: 'The hostage seizures were fully consistent with Hizballah's declared goal of expunging both the American diplomatic presence and Americans from Lebanon, and the hostages' fate was often manipulated in order to serve the interests of Hizballah's sponsor, Iran… Hizballah's close links to Iran, from which it has received generous financial and matériel support since 1982, suggest that it is less a phenomenon of Lebanese politics than a geopolitical foothold for Tehran. Hizballah also maintains a close working relationship with Syria, with which it has willingly cooperated, at least in recent years. Hizballah's relentless attacks on the Israeli occupation zone in southern Lebanon have served Syria's purposes by violently underlining the insistence of Damascus that Israel withdraw completely from both the Golan Heights and southern Lebanon.' Norton, 'Hizballah: From Radicalism to Pragmatism?', *Middle East Policy Council Journal*, 5 (1998): <www.mepc.org/public_asp/journal_vol5/9801_norton.asp>
204. Martin Kramer writes: 'consensus eluded Hizballah regarding the extraordinary means of hostage-taking. Fadlallah's preaching created a moral dilemma for Hizballah and necessitated a more careful reformulation of Hizballah's own position. Husayn al-Musawi continued to support the kidnapping of "spies or military personnel", actions that were "undoubtedly useful" to the cause. But hostage taking had got… out of hand after "some excited Muslims in Beirut" began to take "anyone off the streets". No good had come of these ill-conceived operations, and Muslims were now widely regarded as kidnappers. Hostage-taking had become "chaotic", overshadowing and tarnishing "the major acts of hostage taking which were done to serve the nation of Hizballah". Musawi's was a plea for discriminate rather than indiscriminate hostage taking, in accord with what he called "Islamic decision-making" – a euphemism in Hizballah's lexicon for Iran. Musawi even reached the conclusion that if hostages were innocent, then "I am against hostage-taking, even if the captives are American or French". Subhi al-Tufayli also concluded that the hostage situation "harms the Islamic cause". The growing unease in Hizballah over the method of hostage taking had its origins in the moral logic of Fadlallah, who sought to serve as the movement's unacknowledged conscience. But if hostage deals should ever begin to provide substantial benefits to Hizballah, that may force a change in his moral logic by altering perceptions of cost and benefit in hostage taking': Martin Kramer, 'The Moral Logic of Hizballah', in *Origins of Terrorism: Psychologies, Ideologies, Theologies, States of Mind*, ed. Walter Reich (Cambridge: Cambridge University Press, 1990), 131–57: <www.geocities.com/martinkramerorg/MoralLogic.htm>
205. Saad-Ghorayeb, *Hizbu'llah*, 97, 100–2.
206. Walid Phares argues: 'those 350 men who died that morning were deployed for a humanitarian mission. But the attack was aimed at destroying peace, exactly as Osama bin Laden's last

speech explained to the world: "We do not believe in peaceful and democratic solutions."… When Hizbollah unilaterally attacked US peacekeepers in 1983, Jihādists around the world were watching carefully. Washington chose to pull out in silence and shame. That dictated the further bombings and hostage taking. The US abandoned the Eastern Mediterranean all together, including weak Lebanon to Syria. A message travelled into Jihādist minds: that the US was a sinner that was caught, not a victim of its own mistakes. Bin Laden took it from there… No doubt about it, the Marines attacks in 1983 paved the Jihād route to September 11.' <www.frontpagemag.com/Articles/ReadArticle.asp?ID=10470>

207. Saad-Ghorayeb, *Hizbu'llah*, 100. George Schultz, the US Secretary of State, had been involved in a ten-day 'shuttle' diplomatic mission prior to the treaty. Text of the Israel–Lebanese Peace Treaty of 17 May 1983, 'witnessed by Morris Draper for the Government of the United States of America': <www.lebanese-forces.org/lebanon/agreements/may17.htm> <www.almashriq.hiof.no/israel/300/320/327/israel-lebanon.html>; <www.clhrf.com/unresagreements/may17.agreement.htm>

208. In the explosion at the US embassy in Beirut in April 1983, 63 Americans died and the entire top rank of the CIA's Middle East-Persian Gulf department, 29 officers in all, were massacred, including Middle East desk chief Robert Ames. The US embassy, it is alleged, 'was not ravaged by a truck-bomb as reported then and since, but by a bomb planted on the floor above the conference room in which the CIA officers assembled… Intelligence sources [also] disclose for the first time, 19 years after the event, that two minutes after the CIA chiefs took their seats in that room, the explosive charge was detonated by remote control': <www.cuttingedge.org/na/na010.html>

209. Christoph Reuter, *My Life is A Weapon. A Modern History of Suicide Bombing*, trans. Helena Ragg-Kirkby (Princeton, NJ, and Oxford: Princeton University Press, 2004; orig. German edn 2002), 53.

210. Saad-Ghorayeb, *Hizbu'llah*, 67; Martin Kramer, 'Sacrifice and "Self-Martyrdom" in Shi'ite Lebanon', *Terrorism and Political Violence*, vol. 3, no. 3 (Autumn 1991), 30–47; revised in Martin Kramer, *Arab Awakening and Islamic Revival* (New Brunswick, NJ: Transaction Publishers, 1996), 231–43. Web version at: <www.martinkramer.org/pages/899526/index.htm>.

211. The most reliable account of his views is to be found in Ibrāhīm M. Abū-Rabī', *Intellectual Origins of Islamic Resurgence in the Modern Arab World* (Albany, New York: State University of New York Press, 1996), ch. 8: 'Toward an Islamic Liberation Theology: Muḥammad Ḥusayn Faḍlallāh and the Principles of Shī'ī Resurgence', 220–47. Ibid. 237: 'Faḍlallāh defines martyrdom as a mechanism that helps to alleviate suffering, draws Muslims closer to each other, and alerts the world to the plight of the downtrodden'.

212. Saad-Ghorayeb, *Hizbu'llah*, 6, denies Faḍlallāh was linked to Ḥizbu'llah 'in any organizational sense'. Faḍlallāh repeatedly that denied any formal connection to Ḥizbu'llah: 'the claim that I am the leader of Ḥizbu'llah is baseless and untrue. I am not the leader of any organization or party. It seems that when they could not find any prominent figure to pin this label on, and when they observed that I was active in the Islamic field, they decided to settle on me. It could be that many of those who are considered to be part of Ḥizbu'llah live with us in the mosque and they might have confidence in me. Who is the leader of Ḥizbu'llah? Obviously, he is the one who has influence. So, when they cannot see anybody on the scene, no spokesman, no prominent political figure speaking out for Ḥizbu'llah, they try to nail it on a specific person, whose name is then linked to every incident.' Cited by Martin Kramer, 'The Oracle of Hizbu'llah: Sayyid Muhammad Husayn Fadlallah', in *Spokesmen for the Despised. Fundamentalist Leaders of the Middle East*, ed. R. Scott Appleby (Chicago: University of Chicago Press, 1997), 83–181: <www.geocities.com/martinkramerorg/Oracle1.htm>

213. Harik, *Hezbollah*, 65, 70.

214. Abū-Rabī', *Intellectual Origins of Islamic Resurgence in the Modern Arab World*, 242.
215. Kramer, 'The Moral Logic of Hizballah', 131–57: <www.geocities.com/martinkramerorg/MoralLogic.htm>
216. Ibid.
217. Kramer, 'The Oracle of Hizbu'llah': <www.geocities.com/martinkramerorg/Oracle2.htm>
218. Ibid.
219. <www.bayynat.org/bayynatsite/www/english/islamicinsights/martyr.htm>
220. Reuter, *My Life is A Weapon*, 79–80.
221. Cobban, 'The growth of Shī'ī Power in Lebanon', 138. The call for them to resign from the army came from Nabih Berri of Amal.
222. Hassan Krayem, 'The Lebanese Civil War and the Tai'f Agreement' (American University of Beirut, n.d.): <www.ddc.aub.edu.lb/projects/pspa/conflict-resolution.html>
223. Shlaim, *The Iron Wall*, 454. It was an image used by Arafat in his speech to the United Nations in December 1988: 'greetings to you from the stone-throwing children, who are challenging the occupation and its aircraft, tanks and weaponry, recalling the new image of the defenceless Palestinian David opposing the heavily-armed Israeli Goliath'.
224. Lt. Col. Thomas X. Hammes, 'The Evolution of War: The Fourth Generation', *Marine Corps Gazette* (September 1994): <www.d-n-i.net/fcs/hammes.htm>
225. Khan, *Palestine Documents*, 408–10.
226. Danny Rabinowitz, 'Recognizing the Original Sin', *Ha'aretz* (17 October 2000). Rabinowitz wrote: 'The Jewish public in Israel, self-centred and insensitive, consistently forgets that its state is built on the destruction of another nation… An original sin which is not dealt with, and which is not exposed, is like an internal wound which is not tended to. When one tries to cover and suppress it, it festers…' <www.malaysia.net/lists/sangkancil/2000-10/frm00683.html>
227. Shlaim, *The Iron Wall*, 462, 464, 501.
228. Mishal and Sela, *The Palestinian Hamas*, 42.
229. Khan, *Palestine Documents*, 412–14. Web version at: <www.al-bab.com/arab/docs/pal/pal3.htm>
230. Not in the excerpt published in ibid., 415–17. Web version at: <www.al-bab.com/arab/docs/pal/pal4.htm>
231. <www.al-bab.com/arab/docs/pal/pal5.htm>
232. Khan, *Palestine Documents*, 423–4. Arafat stated in his speech that 'the PLO will work to reach a comprehensive peaceful settlement between the parties involved in the Arab-Israeli struggle, including the state of Palestine and Israel, as well as the other neighbouring states, within the framework of an international conference for peace in the Middle East in order to realise equality and a balance of interests, particularly the right of our people to freedom and national independence, and the respect of the right to life and the right of peace and security for everyone, namely, all the parties involved in the struggle in the area, in accordance with Resolutions 242 and 338'.
233. Mishal and Sela, *The Palestinian Hamas*, 43, 49.
234. Ibid. 44 and appendix two. Web version of Hamas Charter at: <www.yale.edu/lawweb/avalon/mideast/hamas.htm>
235. Mishal and Sela, *The Palestinian Hamas*, 53, 71, 108.
236. Islamic Jihād was established in the Gaza Strip in 1981 by Fatḥī al-Shiqāqī: Meir Hatin, *Islam and Salvation in Palestine. The Islamic Jihad Movement* (Tel Aviv: Tel Aviv University, Moshe Dayan Center for Middle Eastern and African Studies, 127, 2001). The charter of this movement (ibid. 161–8) cited armed *jihād* against Israel as the primary goal; unusually, though a Palestinian Sunnī grouping, it placed itself firmly under the ideological leadership of Iran (ibid. 112).

237. Mishal and Sela, *The Palestinian Hamas*, 84, 99.
238. Ibid. 85. Ibid. 45, 196: Hamas subscribed to some of the rabid anti-semitic conspiracy theories such as the Protocols of the Elders of Zion (article 32 of its Charter).
239. Ibid. 87.
240. Allegra Pacheco, 'Flouting Convention: The Oslo Agreements', in *The New Intifada. Resisting Israel's Apartheid*, ed. Roane Carey (London and New York: Verso, 2001), 181–206, at 188. Hanan Ashrawi commented: 'it's clear that the ones who initialled this agreement have not lived under occupation'.
241. Mishal and Sela, *The Palestinian Hamas*, 95–6.
242. Ibid. 97–9.
243. Ibid. 65–6.
244. Reuter, *My Life is A Weapon*, 100.
245. Mishal and Sela, *The Palestinian Hamas*, 104.
246. Declaration of 5 May 1994: Khan, *Palestine Documents*, 534–5.
247. Laetitia Bucaille, *Growing Up Palestinian. Israeli Occupation and the Intifada Generation*, trans. Anthony Roberts (Princeton, NJ, and Oxford: Princeton University Press, 2004), 138.
248. Mishal and Sela, *The Palestinian Hamas*, 107.
249. Shlaim, *The Iron Wall*, 609.
250. Ibid. 605.
251. Mouin Rabbani, 'A Smorgasord of Failure: Oslo and the al-Aqṣā *Intifada*', in *The New Intifada. Resisting Israel's Apartheid*, ed. Roane Carey (London and New York: Verso, 2001), 69–89 at 78.
252. Glenn E. Robinson, 'The Peace of the Powerful', in Carey, *The New Intifada*, 111–23 at 114.
253. Ovendale, *The Origins of the Arab–Israeli Wars*, 309.
254. <www.eretzyisroel.org/~jkatz/planned.html>
255. Robinson, 'The Peace of the Powerful', 122.
256. <www.palestinefacts.org/pf_1991to_now_alaqsa_start.php>
257. Ghassan Andoni, 'A Comparative Study of *Intifada* 1987 and *Intifada* 2000', in Carey, *The New Intifada*, 209–18.
258. Ovendale, *The Origins of the Arab–Israeli Wars*, 322, 335.
259. Ibid. 316.
260. Ibid. 320. Bucaille, *Growing Up Palestinian*, 151. For the greater difficulty for the Palestinians in mounting a resistance movement in the second *intifāḍah*: ibid. 124. Also Joyce M. Davis, *Martyrs. Innocence, Vengeance and Despair in the Middle East* (New York: Palgrave Macmillan, 2003).
261. They had already exchanged *fatāwā* at the time of the Oslo Accords, when Yūsuf al-Qaraḍawī had sided with Hamas: Mishal and Sela, *The Palestinian Hamas*, 109.
262. Gary R. Bunt, *Islām in the Digital Age. E-Jihād, Online Fatwās and Cyber Islamic Environments* (London and Sterling, VA: Pluto Press, 2003), 108. 'Debating the Religious, Political and Moral Legitimacy of Suicide Bombings. I. The Debate Over Religious Legitimacy', May 2001: <www.memri.org/bin/articles.cgi?Page=archives&Area=ia&ID=IA5301>; <www.mediareviewnet.com/SHEIKH%20QARADAWIs%20lecture.htm>
263. Ibid.
264. Ibid.
265. Memri Inquiry and Analysis Series, 100, 4 July 2002: <www.memri.org.bin.articles.cgi?Page=subjects&Area=conflict&ID=IA10002>
266. Memri Special Dispatch Series, 655, 4 February 2002: <www.memri.org/bin/articles.cgi?Page=archives&Area=sd&ID=SP65504>
267. <www.dawn.com/2004/06/07/top16.htm>
268. Edward W. Said, 'Palestinians under Siege' (14 December 2000), reprinted in Carey, *The New Intifada*, 27–42 at 41.

269. Apart from the excellent studies of Avi Shlaim, cited above, one might add, for example, Bernard Wasserstein, *Divided Jerusalem. The Struggle for the Holy City* (London: Profile Books, 2001). Bernard Wasserstein, *Israel and Palestine. Why They Fight and Can They Stop?* (London: Profile Books, 2003), 137: 'under the impact of terrorist assault, a large body of opinion in Israel today favours [an] iron wall between the two peoples. But is such a "separation" feasible, given other mounting pressures on the two populations cohabiting a narrow strip of territory with limited natural resources?'
270. Avi Shlaim, 'The United States and the Israeli–Palestinian Conflict', *Worlds in Collision. Terror and the Future of Global Order*, ed. Ken Booth and Tim Dunne (Basingstoke: Palgrave Macmillan, 2002), 172–83 at 182.
271. Robert Fisk, 'This is a place of filth and blood which will forever be associated with Ariel Sharon', *Independent* (6 February 2001), reprinted in Carey, *The New Intifada*, 293–6.
272. The Human Rights Watch report 'Erased in a Moment: Suicide Bombing Attacks Against Israeli Civilians' (October 2002) commented (p. 3): 'The greatest failure of President Arafat and the PA leadership – a failure for which they must bear heavy responsibility – is their unwillingness to deploy the criminal justice system decisively to stop the suicide bombings, particularly in 2001, when the PA was most capable of doing so. President Arafat and the PA also failed to take aggressive measures to ensure that the intensely polarized political atmosphere [did] not serve as a justification for such attacks. Certain Israeli actions, such as the destruction of PA police and security installations, gradually undermined the PA's capacity to act. But even when their capacity to act was largely intact, Arafat and the PA took no effective action to bring to justice those in Hamas, Islamic Jihad, the PFLP, and the al-Aqsa Martyrs' Brigades who incited, planned or assisted in carrying out bombings and other attacks on Israeli civilians. Instead, Arafat and the PA pursued a policy whereby suspects, when they were detained, were not investigated or prosecuted, but typically were soon let out onto the street again. Indeed, the PA leadership appeared to treat its duty to prosecute murderers as something that was negotiable and contingent on Israel's compliance with its undertakings in the Oslo Accords, not as the unconditional obligation that it was.' The report (p. 4) refused to acknowledge any merit in the Palestinian arguments: 'Palestinian armed groups have sought to justify suicide bombing attacks on civilians by pointing to Israeli military actions that have killed numerous Palestinian civilians during current clashes, as well as the continuing Israeli occupation of the West Bank and much of the Gaza Strip. Such excuses are completely without merit. International humanitarian law leaves absolutely no doubt that attacks targeting civilians constitute war crimes when committed in situations of armed conflict, and cross the threshold to become crimes against humanity when conducted systematically, whether in peace or war. As the latter term denotes, these are among the worst crimes that can be committed, crimes of universal jurisdiction that the international community as a whole has an obligation to punish and prevent.'<www.hrw.org/reports/2002/isrl-pa/> Palestinian resistance groups claimed that Israel's continuing military occupation, and its vastly superior means of combat, made such attacks their only option and also asserted that their targets were not really civilians because 'all Israelis are reservists' and because Israeli-imposed residents of sprawling illegal settlements had forfeited their civilian status. However, the HRW report stated that while civilian Israeli settlements in the West Bank and Gaza were illegal under international humanitarian law, persons residing there were entitled to protection as civilians except when they were directly participating in hostilities: <www.islam-online.net/english/news/2002-11/01/article01.shtml>

Chapter 11

1. K. R. Pruthi (ed.), *Encyclopaedia of Jihād* (5 vols) (New Delhi; Anmol Publications pvt., 2002), iii. 804. <www.islam.org.au/articles/older/INT-KSHM.HTM>

2. <www.fas.org/irp/world/para/ayman_bk.html>
3. Because of the profusion of dates in this chapter, and the fact that they are all relatively recent, the alternative dating in the Muslim calendar is not given.
4. I. Malik, F. Noshab and S. Abdullah, '*Jihād* in the Modern Era: Image and Reality', *Islamabad Papers* 18 (Institute of Strategic Studies, Islamabad, 2001), 38, 40.
5. Pakistan had supported a UN General Assembly motion for the previous twelve years reaffirming the principle of self-determination: *Dawn*, 22 November 2003: 'India [on this occasion] opposed the resolution, which for the last 12 years was adopted with consensus, with India voting for it. Some 30 countries sponsored the resolution. By adopting the resolution, the U.N. General Assembly declared its firm opposition to the acts of foreign military intervention, aggression and occupation, since those acts have resulted in the suppression of the right of self-determination': <www.dawn.com/2003/11/22/top17.htm>
6. John Gray, *Al-Qaeda and What it Means to be Modern* (London: Faber, 2003), 82.
7. <www.pbs.org/newshour/terrorism/international/fatwa_1996.html>: <www.daveross.com/binladen.html>; Yossef Bodansky, *Bin Laden: The Man Who Declared War on America* (New York: Random House, 2001), 186: 'the fact that no previous terrorist leader of any ideology dared to confront the United States so directly testifies to bin Laden's resolve and dedication.' The significance of bin Laden's remarks was also emphasized by Air Marshal (Retd) Ayaz Ahmed Khan, 'Terrorism and Asymmetrical Warfare: International and Regional Implications': <www.defencejournal.com/2002/february/terrorism.htm>
8. 'Bin Laden Lieutenant Admits to September 11 and Explains al-Qaeda's Combat Doctrine'. Memri Special Dispatch 344, 10 February 2002: <www.memri.org/bin/articles.cgi?Page=archives&Area=sd&ID=SP34402>; <www.metimes.com/2K2/issue2002–7/reg/al_qaeda_theoretician.htm>
9. Quoted above, Chapter 10. Lt. Col. Thomas X. Hammes, 'The Evolution of War: The Fourth Generation', *Marine Corps Gazette* (September 1994): <www.d-n-i.net/fcs/hammes.htm>
10. Gray, *Al-Qaeda and What it Means to be Modern*, 81–2.
11. 'America's Asymmetrical Wars: Following a Failed Israeli Military Doctrine'. Report from a Palestine Center briefing by Marwan Bishara: <www.palestinecenter.org/cpap/pubs/20040112ftr.html>
12. 'The Americanization of Israel and the Israelization of American Policy'. Report from a Palestine Center briefing by Marwan Bishara: <www.palestinecenter.org/cpap/pubs/20030903ftr.html>
13. By 'fabrication' here, the author does not imply that the original Afghan *jihād* was not a 'genuine' phenomenon, that is, an indigenous response of Afghans to foreign invasion; instead, the term describes the process of undercover assistance and money laundering chiefly by the USA (assisted by Sa'ūdī Arabia) to conceal its role in supporting the *jihād*, which was really part of the Cold War with the Soviet Union. The allusion to 'sowing the wind' here is to Hosea 8:7, quoted in the next section (p. 335).
14. Larry P. Goodson, *Afghanistan's Endless War. State Failure, Regional Politics and the Rise of the Taliban* (Seattle and London: University of Washington Press, 2001), 58.
15. Shoshana Keller, *To Moscow, not Mecca. The Soviet Campaign against Islām in Central Asia, 1917–1941* (Westport, CT, and London: Praeger, 2001), conclusion: 'damaged but not destroyed'.
16. David B. Edwards, *Before Taliban: Genealogies of the Afghan Jihād* (Berkeley, CA: University of California Press, *c.* 2002). Web edition at: <www.ark.cdlib.org/ark:/13030/ft3p30056w/>
17. <www.wwics.si.edu/index.cfm?topic_id=1409&fuseaction=library.document&id=526> A telephone call between Kosygin and Noor Muḥammad Tarakī indicated that there was little government support at Herat, which was 'almost wholly under the influence of Shī'a slogans – follow not the heathens, but follow us'. 'Kosygin: Ḥundreds of Afghan officers were trained

in the Soviet Union. Where are they all now? Tarakī: Most of them are Muslim reactionaries. We are unable to rely on them, we have no confidence in them.' <www.wwics.si.edu/index.cfm?topic_id=1409&fuseaction=library.document&id=39>

18. <www.wwics.si.edu/index.cfm?topic_id=1409&fuseaction=library.document&id=181>
19. Question: 'The former director of the CIA, Robert Gates, stated in his memoirs [*From the Shadows*], that American intelligence services began to aid the *mujāhidīn* in Afghanistan 6 months before the Soviet intervention. In this period you were the national security adviser to President Carter. You therefore played a role in this affair. Is that correct?' Brzezinski: 'Yes. According to the official version of history, CIA aid to the *mujāhidīn* began during 1980, that is to say, after the Soviet army invaded Afghanistan, 24 Dec 1979. But the reality, secretly guarded until now, is completely otherwise. Indeed, it was July 3, 1979 that President Carter signed the first directive for secret aid to the opponents of the pro-Soviet regime in Kābul. And that very day, I wrote a note to the president in which I explained to him that in my opinion this aid was going to induce a Soviet military intervention.' Interview with *Le Nouvel Observateur*, 15–21 January 1998: <www.globalresearch.ca/articles/BRZ110A.html>

    For US–Pakistan relations in these years: Dennis Kux, *The United States and Pakistan, 1947–2000. Disenchanted Allies* (Baltimore, MD, and London: Johns Hopkins University Press and Woodrow Wilson Center Press, Washington DC, 2001), 245–75.
20. Edwards, *Before Taliban.*
21. Ibid.
22. Goodson, *Afghanistan's Endless War*, 59.
23. Ibid. 63.
24. Ibid. 65.
25. 'The agreement with the Americans was for an annual allocation of 250 grip-stocks, together with 1,000–1,200 missiles': Muḥammad Yousaf and Mark Adkin, *Afghanistan: The Bear Trap. The Defeat of a Superpower* (Havertown: Casemate, orig. edn. 1992, repr. 2001). Web edition at: <www.sovietsdefeatinafghanistan.com/beartrap/english/15.htm>
26. 'New Evidence on the War in Afghanistan', *Cold War International History Project Bulletin*, 14/15 (Winter 2003–Spring 2004), 161. Web version at: <www.wwics.si.edu/topics/pubs/c-afghanistan.pdf>
27. Yezid Sayigh, *Armed Struggle and the Search for State. The Palestinian National Movement, 1949–1993* (Institute for Palestine Studies, Washington DC: Oxford University Press, 1997, repr. 1999), 640.
28. Aḥmed Rashīd's estimate (that the *mujāhidīn* received 'over $10 million') is too low: Aḥmed Rashīd, *Ṭālibān. The Story of the Afghan Warlords* (London: Pan Books, repr. 2001), 18. Dilip Hiro, 'The cost of an Afghan "victory"', *The Nation*, 28 January/15 February 1999: <www.thenation.com/doc.mhtml?i=19990215&c=2&s=hiro>
29. Loretta Napoleoni, *Modern Jihād. Tracing the Dollars behind the Terror Networks* (London: Pluto Press, 2003), 80–3.
30. <www.sovietsdefeatinafghanistan.com/beartrap/english/08.htm>
31. <www.rjsmith.com/kia_tbl.html>
32. As quoted by Yousaf and Adkin, *Afghanistan: The Bear Trap*, from the *Daily Telegraph* (14 January 1985).
33. <www.sovietsdefeatinafghanistan.com/beartrap/english/07.htm>
34. KHAD, or Khedamat-e Etelea'at-e Dawlati, was Soviet-run Afghanistan's secret police, sometimes known euphemistically as the State Information Agency.
35. In an interview in June 1997, Zbigniew Brzezinski, President Carter's National Security Adviser, confirms this view: '…I have to pay tribute to the guts of the Pakistanis: they acted with remarkable courage, and they just weren't intimidated and they did things which one would have thought a vulnerable country might not have the courage to undertake. We, I am

pleased to say, supported them very actively and they had our backing, but they were there, they were the ones who were endangered, not we.' <www.gwu.edu/~nsarchiv/coldwar/interviews/episode-17/brzezinski2.html>

36. An explosion at the camp in early April 1988, which was followed by secondary blasts over the next two days, destroyed the ISI depot for the Afghan war and severely hindered *mujāhidīn* operations the following year.
37. <www.sovietsdefeatinafghanistan.com/beartrap/english/06.htm>
38. 'It should be concluded so [that] Afghanistan becomes a neutral country. Apparently, on our part there was an underestimation of difficulties, when we agreed with the Afghan government to give them our military support. The social conditions in Afghanistan made the resolution of the problem in a short amount of time impossible. We did not receive domestic support there. In the Afghan army the number of conscripts equals the number of deserters… Concerning the Americans, they are not interested in the settlement of the situation in Afghanistan. On the contrary, it is to their advantage for the war to drag out.' <www.wwics.si.edu/index.cfm?topic_id=1409&fuseaction=library.document&id=342>
39. 'New Evidence on the War in Afghanistan', *Cold War International History Project Bulletin*, 14/15 (Winter 2003–Spring 2004), 148.
40. Ibid. 150: 'in the period after the proclamation of the policy of reconciliation, of a total of around 164,000 [rebels], 15,000 armed rebels openly came over to the side of the government. More than 600 groups with a total strength of 53,000 men are holding talks with the government. Part of the counter-revolutionary formations, about 50,000 men, are taking a wait-and-see position. However, as before, there is an active nucleus of the irreconcilable opposition numbering 46,000 men.'
41. Ibid. 150–1.
42. Ibid. 157–8.
43. Ibid. 159.
44. Ibid. 163, 165.
45. Ibid. 176. On this occasion, Najibullah affirmed: 'not one government in Afghanistan has yet recognized this "Durand Line" as the border [with Pakistan]. And if we do this now, an explosive situation would arise in society. Therefore we have tried to select a formula such that an Afghan–Pakistani agreement about non-interference would not signify official recognition of the "Durand Line" by us or cause any concern among the Pashtuns. We found such a formulation in the end.'
46. Ibid. 178.
47. Ibid. 184–5.
48. Ibid. 188.
49. Ibid. 191: 'Pakistan can be compared to a boiling kettle which is full of various contradictions and antagonisms – religious, national, and ethnic. In order to keep this "kettle" from exploding Pakistani leaders are trying to let off the "steam" of public dissatisfaction, diverting the attention of their people to problems of an external nature. At one time it seized upon the Afghan problem eagerly and actively heated it up. At the present time the Kashmir issue has become a safety valve. For decades the military has decided and dictated the policy of Pakistan. And even after B[enazir] Bhutto came to power the policy of the Pakistani administration regarding Afghanistan remained unchanged: it was only… dressed "in civilian clothes"…'
50. Ibid. 190.
51. On the night Najibullah was to leave Kābul, opposition forces took control of the airport, preventing him from leaving the country. He took refuge in the UN compound in Kābul, where he remained for four years until he was captured and murdered by the Ṭālibān, with no pretence of legal process, in September 1996. Rashīd, *Ṭālibān*, 49.

52. M. E. Yapp, 'Lines in the Sand', review of Neamatollah Nojumi, *The Rise of the Ṭālibān in Afghanistan* (London: Palgrave, 2003) in *Times Literary Supplement* (18 April 2003), 11.
53. Yvonne Yazbeck Haddad, 'Islamism: A Designer Ideology for Resistance, Change and Empowerment', in *Muslims and the West. Encounter and Dialogue*, edited by Zafar Ishaq Ansari and John L. Esposito (Islamabad: Islamic Research Institute, Islamabad and Center for Christian–Muslim Understanding, Washington DC, 2002), 274–95 at 278–9, 287–8.
54. Hiro, 'The cost of an Afghan "victory"'.
55. Interview with *Le Nouvel Observateur*, 15–21 January 1998: <www.globalresearch.ca/articles/BRZ110A.html> The interview is also quoted by Tariq Ali, *The Clash of Fundamentalisms. Crusades, Jihads and Modernity* (London: Verso, 2002), 207–8.
56. Christianity is still estimated to have more followers in the world than Islām, so Brzezinski's wording has been changed from 'the' to 'a'. He would have been correct to have called it the world's fastest growing religion.
57. Carl Conetta, 'Strange Victory: A Critical Appraisal of Operation Enduring Freedom and the Afghanistan War', *Project on Defense Alternatives*, Research Monograph 6 (30 January 2002). This section draws particularly on Appendix 3. 'The Rise and Fall of the Ṭālibān: A Note on Their Strategy and Power': <www.comw.org/pda/0201strangevic.html>
58. Rohan Gunaratna, *Inside Al-Qaeda. Gobal Network of Terror* (London: C. Hurst and Co., 2002), 30–1.
59. 'Interpreting Islām', *Indian Currents* (6 June 2004), 22–3. Web version at: <www.islaminterfaith.org/june2004/interview-06-04.htm>
60. Joshua Rey, 'However cruel the regime, the Ṭālibān created law and peace', *The Times*, 20 January 2004 (the author is an aid consultant who worked in Afghanistan before and after the Ṭālibān's fall; he contends that the movement was cultural rather than religious).
61. Rashīd, *Ṭālibān*, 42.
62. Norimitsu Onishi, 'A tale of the Mullāh and Muḥammad's amazing cloak', *New York Times*, 19 December 2001: <www.faughnan.com/scans/011219_MuhammedCloak.pdf>
63. Rashīd, *Ṭālibān*, 58–9.
64. Ibid. 73. Some 400 Hazārā women were allegedly taken as concubines: ibid. 75. This account follows that of the Human Rights Watch report of November 1998: <www.hrw.org/reports98/afghan/Afrepor0.htm>
65. <www.shianews.com/hi/europe/news_id/0000189.php>
66. Rashīd, *Ṭālibān*, 87–8. Rashīd proceeds to describe them as Deobandīs, but notes that the Ṭālibān's 'interpretation of the creed has no parallel anywhere in the Muslim world'.
67. For a hostile Indian comment on 'Afghanistan: Pakistan's Black Hole' (17 April 2001): <www.saag.org/papers3/paper228.html#top>
68. Gunaratna, *Inside Al-Qaeda*, 43.
69. In an address on 3 February 2002, Prince Turkī recalled his second visit to Mullāh Omar in 1998. 'The man turned abusive to the Kingdom and was insulting and totally out of order. He said things like, "The Kingdom should be ashamed of itself for wanting to try this upright, fantastic human being, bin Laden. You are doing this at the behest of the United States, the enemy of Islām". So, I just cut the meeting short and I said, "I am not going to take any more abuse". As I was leaving, I turned to him and I said, "Mullāh Omar, you are going to regret this act. It is going to bring harm not just to you, but to Afghanistan"': <www.ccasonline.org/publicaffairs/turki_02032002.html> Another account of the meeting is provided by Steve Coll, *Ghost Wars. The Secret History of the CIA, Afghanistan and Bin Laden from the Soviet Invasion to September 10, 2001* (New York: Penguin Press, 2004), 400–2. Steve Coll has been Managing Editor of the *Washington Post* since 1998. His account of the Afghanistan years of bin Laden is one of the most detailed.
70. <www.washingtoninstitute.org/media/henderson/henderson-saudis.htm>

71. Gunaratna, *Inside Al-Qaeda*, 50.
72. Conetta, 'Strange Victory'.
73. Ibid.
74. <www.dawn.com/2004/06/14/top8.htm>
75. <www.islam.org.au/articles/older/INT-KSHM.HTM>
76. <www.sikhspectrum.com/062003/hizb.htm>
77. Sumantra Bose, *Kashmir: Roots of Conflict, Paths to Peace* (Cambridge, MA.: Harvard University Press, 2003), 108–9. The precise numbers of those killed is uncertain. Others give lower figures, say 100 killed. More partisan accounts do not mention the shootings at all: Šumit Ganguly, *The Crisis in Kashmir. Portents of War, Hopes of Peace* (Cambridge: Cambridge University Press and Woodrow Wilson Center Series, 1997), 106, talks only of the shooting of three protesters crossing the Line of Control (LoC), while emphasizing 'the insurgency' and 'the level of violence in the state'.
78. Truth and Reconciliation Commission Verdict: <www.africanhistory.about.com/library/bl/blTRCFindings-Sharpeville.htm>
79. Bose, *Kashmir*, 117.
80. Ibid. 112. The JKLF was founded in 1964 and was committed to 'one fully independent and truly democratic state' of Jāmmū and Kāshmīr with the borders as prior to 1947. It advocated 'equal political, economic, religious and social rights' for all citizens of the proposed state, 'irrespective of race, religion, region, culture and sex'. Yoginder Sikand, *Muslims in India since 1947. Islamic Perspectives on Inter-Faith Relations* (London and New York: RoutledgeCurzon, 2004), 195.
81. Sikand, *Muslims in India since 1947*, 197–9.
82. Ibid. 212.
83. Ibid. 208–12, especially 209.
84. <www.islam.org.au/articles/older/INT-KSHM.HTM>
85. Bose, *Kashmir*, 112.
86. Ibid. 120, 124.
87. Ibid. 128.
88. <www.islam.org.au/articles/older/INT-KSHM.HTM>
89. Bose, *Kashmir*, 126–7, 130. It is impossible to ascertain support for the ideological stance of guerrilla movements retrospectively.
90. Ibid. 141.
91. 'The statistics available with the [Ministry of Home Affairs] showed that out of 2,215 terrorists killed in the year 2002, 1,702 were foreign mercenaries. The foreign mercenaries killed in the years 2001, 2000, 1999 and 1998 were 2,028, 1,520, 1,082 and 999 out of total 2,655, 1,956, 1,387 and 1,318 terrorists respectively.' <www.dailyexcelsior.com/web1/03sep27/edit.htm#5>; <www.ribt.org/nuke/html/modules.php?op=modload&name=News&file=article&sid=97>; Bose's figures are significantly lower: Bose, *Kashmir*, 136.
92. 'The Jaish-e-Moḥammad… was formed by Masood Azhar upon his release from prison in India in early 2000. The group's aim is to unite Kashmir with Pakistan. It is politically aligned with the radical political party, Jamiat 'Ulamā'-i-Islām Fazlur Rehman faction (JUI-F). The United States announced the addition of JEM to the US Treasury Department's Office of Foreign Asset Control (OFAC) list – which includes organizations that are believed to support terrorist groups and have assets in U.S. jurisdiction that can be frozen or controlled – in October 2001 and the Foreign Terrorist Organization list in December 2001. By 2003, JEM had splintered into Khuddam ul-Islām (KUI) and Jamaat ul-Furqan (JUF). Pakistan banned KUI and JUF in November 2003': <www.fas.org/irp/world/para/jem.htm>
93. 'The LT is the armed wing of the Pakistan-based religious organization, Markaz-ud-Dawa-wal-Irshad (MDI) – a Sunnī anti-US missionary organization formed in 1989. The LT is led

by Hafiz Muhammad Saeed and is one of the three largest and best trained groups fighting in Kashmir against India; it is not connected to a political party. The United States in October 2001 announced the addition of the LT to the US Treasury Department's Office of Foreign Asset Control (OFAC) list – which includes organizations that are believed to support terrorist groups and have assets in U.S. jurisdiction that can be frozen or controlled. The group was banned, and the Pakistani Government froze its assets in January 2002. The LT is also known by the name of its associated organization, Jamaat ud-Dawa (JUD). Musharraf placed JUD on a watchlist in November 2003': <www.fas.org/irp/world/para/lashkar.htm>

94. <www.saag.org/papers8/paper768.html>
95. 'The U.S. envoy had said in her speech: "The government of Pakistan must ensure its pledges are implemented to prevent infiltration across the Line of Control and end the use of Pakistan as a platform for terrorism." Pakistan's position regarding the LoC was reiterated, namely, that there was no infiltration on the LoC. Pakistan had taken all measures not to allow any infiltration, said the Foreign Office spokesman. The U.S. ambassador was informed during the meeting that Pakistan had repeatedly called for the deployment of UN observers on both sides of the LoC to verify Indian allegations of infiltration. According to the statement, it was made clear to Ms Powell that Pakistan stood by its commitment to the international community and in the same spirit Pakistan expected the international community to fulfil its commitments for the peaceful resolution of the Kashmir dispute.' <www.dawn.com/2003/01/25/top2.htm>
96. Bose, *Kashmir*, 146.
97. <www.kashmirwatch.com/iptalkscr.htm>
98. Aḥmed Rashīd, *Jihād. The Rise of Militant Islām in Central Asia* (New Haven, CT, and London: Yale University Press, 2002), 108.
99. Ibid 109.
100. Ibid.
101. Nargis Zokirova, 'Tajikistan: Clock Ticking on Corruption. One of the World's Most Corrupt Countries Struggles to Create a Fairer Business Environment': <www.iwpr.net/index.pl?archive/rca/rca_200406_293_3_eng.txt>
102. Rashīd, *Jihād*, 135.
103. Aḥmed Rashīd notes that this is an area of potential recruitment for militant groups because of very high levels of unemployment: 'the Central Asian regimes do not take the Ferghana Valley seriously; that has been half the problem. They do not believe that there is an economic crisis or an unemployment crisis in Ferghana whereas, in fact, there is something like 90 per cent unemployment there.'
104. Ibid. 133–4.
105. <www.209.52.189.2/discussion.cfm/investing/72641/619611> That the American military might be beginning to wake up to the implications of anti-Americanism resulting from their presence is suggested by the National Defense University paper by Lt. Col. Daniel S. Rogerts on 'Tajikistan: Pol[itical]/Mil[itary] Informed Questions': <www.ndu.edu/NWC/writing/AY03/5604/5604K.pdf>
106. Kambiz Arman, 'Tājīkistān Shuns U.S., Tilts Towards Russia' (15 June 2004): <www.isn.ethz.ch/infoservice/secwatch/index.cfm?service=cwn&parent=detail&menu=8&sNewsID=9011>
107. Rashīd, *Jihād*, 148–9.
108. Ibid. 249. Somewhat at variance with its title, this declaration implies that the *jihād* in Ūzbekistān was already underway, but explains the 'reason for the start of the *jihād* in Kyrgyzstan'.
109. Ibid. 133.
110. Ron Synovitz, 'Pakistan arrests al-Qaida suspects with links to Uzbek militants' (15 June 2004): <www.isn.ethz.ch/infoservice/secwatch/index.cfm?service=cwn&parent=detail&menu=8&sNewsID=9019>; David Rhode and Mohammed Khan, 'Ex-fighter for Taliban dies in strike

in Pakistan', *New York Times* (19 June 2004): <www.nytimes.com/2004/06/19/international/asia/19STAN.html?th>

111. Brian Glyn Williams, 'Shattering the al-Qaeda–Chechen Myth' (October 2003): <www.peaceinchechnya.org/news/200310-11%20-%20BGW%20Article.htm>
112. Ibid.
113. Human Rights Watch report, 'In the Name of Counter-Terrorism: Human Rights Abuses Worldwide. A Human Rights Watch Briefing Paper for the 59th Session of the United Nations Commission on Human Rights' (25 March 2003): <www.hrw.org/un/chr59/counter-terrorism-bck4.htm#P286_64797>
114. Both quoted ibid.
115. Andrei Piontkovsky, 'Putin's blind alley in Chechnya', *Washington Post* (30 March 2004): <www.washingtonpost.com/wp-dyn/articles/A34685-2004Mar29.html>
116. <www.diacritica.com/sobaka/dossier/khattab.html>. *Pravda* carried a story on 29 April 2002 that 'Khattab did not die like a hero in a battle', with a photograph of him with his eyes closed: <www.english.pravda.ru/main/2002/04/29/28081.html>
117. Andrew McGregor, 'Chechnya: Amīr Abū al-Walīd and the Islamic Component of the Chechen War' (26 February 2003): <www.religioscope.info/article_88.shtml>
118. <www.interpol.int/public/wanted/notices/data/1999/43/1999_14843.asp>
119. Andrew McGregor, '"Operation Boomerang": Shāmil Basaev's Justification for Terrorism' (26 February 2004): <www.jamestown.org/publications_details.php?volume_id=400&issue_id=2914&article_id=23566>
120. <www.kavkazcenter.com/eng/article.php?id=2028>
121. 'Foundation of Chechen Society challenged' (16 June 2004): 'According to unwritten laws, a Chechen can only lay his hand on his fellow Chechen to defend his honour. According to the unofficial Chechen code, political persecution of fellow countrymen (and especially relatives of opponents), or especially murder is viewed as the gravest crime before the people and as vile treason, and is subject to overall ostracism and condemnation. Members of the Chechen society, who resort to such crimes, are struck out of this society and are subjected to implacable prosecution with no statute of limitations. No protracted wars, total extermination, or even Stalin's deportation have ever been able to strip Chechens of these principles. Kadyrov's gangs and other groups of collaborators, which are playing the key role in Russia's strategy of "Chechenization" of this war, are thus challenging the foundations of the Chechen society'. <www.kavkaz.org.uk/eng/article.php?id=2883>
122. 'President Maskhadov, a resolute advocate of the application of international law in the conflict, nevertheless displays some understanding of Basaev's rage against Russia: "Basaev is a warrior. He is somebody who is exerting revenge. He employs the same methods as the enemy, who uses them against the Chechens, civilians. It is an eye for an eye… If it were possible to subordinate Basaev and to funnel all his energy against the enemy, employing acceptable methods, he would achieve much more."' Quoted by McGregor, '"Operation Boomerang"'.
123. Pruthi, *Encyclopaedia of Jihād*, v. 1463.
124. Graham E. Fuller and S. Frederick Starr, 'The Xinjiang Problem', Central Asia-Caucasus Institute, Paul H. Nitze School of Advanced International Studies, Johns Hopkins University (n.d.), 62–3: <www.cornellcaspian.com/pub2/xinjiang_final.pdf>
125. Pruthi, *Encyclopaedia of Jihād*, v. 1464–65.
126. <www.uygur.org/wunn03/2003_12_22a.htm>
127. <www.time.com/time/europe/timetrails/algeria/al881017.html>
128. Pruthi, *Encyclopaedia of Jihād*, v. 1311.
129. Ibid. Gilles Kepel, *Jihād. The Trail of Political Islām* (London and New York: I. B. Tauris, 2002), 167.

130. Graham E. Fuller, 'Algeria: The Next Fundamentalist State?' (1996): <www.rand.org/publications/MR/MR733/>
131. Kepel, *Jihād*, 174. There are some electoral statistics that suggest less than overwhelming support (the vote for the FIS had declined by over a million since the local elections, while 5 million people, or more than 40 per cent of the 13.2 million registered voters, did not cast their ballots): Pruthi, *Encyclopaedia of Jihād*, v. 1321. Such shortcomings are common to democratic systems, however, and the high percentage of votes cast must be seen as the decisive evidence that the FIS represented the popular will in December 1991.
132. Pruthi, *Encyclopaedia of Jihād*, v. 1313.
133. Kepel, *Jihād*, 175.
134. Cf. Martin Stone, *The Agony of Algeria* (New York: Columbia University Press, 1997), 178, who talks of the GIA as having emerged in the summer of 1993.
135. Pruthi, *Encyclopaedia of Jihād*, iii. 889–91.
136. <www.isnet.org/archive-milis/archive96/may96/0273.html>
137. Reuters report, Algiers, 13 June 2004.
138. <www.fas.org/irp/world/para/ayman_bk.html>
139. Montasser al-Zayyāt, *The Road to Al-Qaeda. The Story of bin Lāden's Right-Hand Man* (London and Ann Arbor, MI: Pluto Press, 2004), 69. Al-Ẓawāhirī stated: 'it is clear from a careful reading of Muntasir al-Zayyāt's statements above that he does not belong to the jihādist movement and does not agree with its choice of *jihād* as a method. Indeed he describes *jihād* in the cause of God as violence, which is exactly the term that the government uses. He alleges that he tried to promote an end to violence, which is the way the government describes *jihād* in the cause of God.'
140. Pruthi, *Encyclopaedia of Jihād*, iii. 852.
141. Variously quoted, e.g. by Peter L. Bergen, *Holy War Inc.: Inside the Secret World of Osama bin Laden* (London: Phoenix, repr. 2003), 56.
142. Pruthi, *Encyclopaedia of Jihād*, i. 19–31. Dr 'Abdullah 'Azzam, *Defense of the Muslim Lands. The First Obligation after Iman*, trans. Brothers in Ribatt, n.d., 23: <www.geocities.com/johnathanrgalt/Defence_of_the_Muslim_Lands.pdf>
143. Quoted by Jason Burke, *Al-Qaeda: Casting a Shadow of Terror* (London: I. B. Tauris, 2003), 8. <www.guardian.co.uk/alqaida/story/0,12469,997063,00.html> Burke contends: 'modern radical Islamic thought is heavily influenced by Western radical political thought, on the right and the left, and the concept of the vanguard is only one of a number of concepts, and tactics, borrowed from thinkers ranging from Trotsky and Mao to Hitler and Heidegger'. Also Yoginder Sikand excerpting Burke, *Al-Qaeda*, and reviewing Gunaratna, *Inside Al-Qaeda* in *The Milli Gazette* (Delhi), 1–15 August 2003, 28.
144. <www.iacsp.com/itobli3.html>
145. Pruthi, *Encyclopaedia of Jihad*, iii. 933.
146. A largely credible reading of bin Laden's statements is provided by the former CIA analyst, Anonymous, *Through Our Enemies' Eyes. Osama bin Laden, Radical Islam and the Future of America* (Washington DC: Brassey's Inc., 2002). Of interest, but less credible, is Jean E. Rosenfeld, 'The Religion of Usamah bin Ladin: Terror as the Hand of God': <www.publiceye.org/frontpage/911/Islam/rosenfeld2001.html>
147. Bodansky, *Bin Laden*, 225–6. The other signatories were Ayman al-Ẓawāhirī; Rifai Ahmad Taha (Abu-Yassir); Shaykh Mir Hamzah, secretary of the *Jamiat-ul-Ulema-e-Pakistan*; Fazlul Rahman Khalil, leader of the Ansar Movement in Pakistan; and Shaykh Abdul Salam Muhammad, *amīr* of the *Jihād* Movement in Bangladesh. The *fatwā* was published in *Al-Quds al-'Arabi* on 23 Febuary 1998.
148. Reported by Arnaud de Borchgrave, United Press International Editor at Large on 13 July 2001 from Kandahar: <www.metimes.com/2K1/issue2001-28/reg/sheikh_omar_says.htm>

149. Al-Zayyat, *The Road to al-Qaeda*, 68–70.
150. <www.ict.org.il/articles/fatwah.htm> Text with moderate Muslim commentary: <www.understanding-islam.com/related/text.asp?type=question&qid=1007>
151. Al-Zayyat, *The Road to al-Qaeda*, 72.
152. Jason Burke, 'The making of the world's most wanted man', *Sunday Observer*, 28 October 2001: <www.observer.guardian.co.uk/waronterrorism/story/0,1373,582274,00.html> This figure, it seems, was arrived at by the US State Department by taking the value of the bin Laden family net worth – estimated at US $5 billion – by the number of bin Laden senior's sons (20 sons). A fact rarely mentioned is that in 1994 the bin Laden family disowned Osama and took control of his share: <www.greenleft.org.au/back/2001/465/465p15.htm>
153. Bergen, *Holy War Inc.*, 104.
154. Nick Fielding, 'Sa'ūdīs paid bin Laden £200 million', *The Times*, 25 August 2002: <www.timesonline.co.uk/article/0,,2089-393584,00.html>
155. Napoleoni, *Modern Jihād*, 164.
156. Ibid. 199.
157. Al-Zayyāt, *The Road to al-Qaeda*, 96–8.
158. John Kelsay, 'The New *Jihād* and Islamic Tradition' on the Foreign Policy Research Institute website: <www.fpri.org/fpriwire/1103.200310.kelsay.newjihad.html>
159. Natana DeLong Bas, *Wahhābī Islām. From Revival and Reform to Global Jihād* (New York: Oxford University Press, 2004), 266. In her opinion, Ibn Taymīyah and Sayyid Quṭb 'figure more prominently in bin Laden's world view' than does Muḥammad Ibn 'Abd al-Wahhāb. Ibid. 250–65.
160. Ṭāriq Ramaḍān, *Western Muslims and the Future of Islām* (New York: Oxford University Press, 2004), 27. Ṭāriq Ramaḍān, *To be a European Muslim. A Study of the Islamic Sources in European Context* (Leicester: Islamic Foundation, 1999), 243, has slightly different wording.
161. <www.frontpagemag.com/Articles/Printable.asp?ID=8978>
162. One reason for the invasion of 'Irāq, which slipped by 'almost unnoticed, but [which was of] huge [importance]', according to Paul Wolfowitz, was that the attack would allow a withdrawal of US troops from Sa'ūdī Arabia. The presence of US troops there has been one of the main bones of contention for the al-Qaeda network: 'just lifting that burden from the Sa'ūdīs is itself going to open the door' to a more peaceful Middle East, Wolfowitz claimed. The Americans had withdrawn from the bases by the end of August 2003, thus decreasing 'to almost zero its military profile', since 'the presence of American troops ha[d] generated resentment because of their proximity to Islām's holiest sites': <www.dawn.com/2003/08/28/top14.htm>
163. <www.fas.org/irp/world/para/ayman_bk.html>
164. Samuel P. Huntington, 'The Clash of Civilizations?', *Foreign Affairs*, 72 (1993), 22–49, reprinted in his *The Clash of Civilizations? The Debate* (New York: Foreign Affairs, 1996), with other contributions. Samuel P. Huntington, *The Clash of Civilizations and the Remaking of the World Order* (repr. London: The Free Press, 2002), 258: 'no single statement in my *Foreign Affairs* article attracted more critical comment than "Islam has bloody borders". I made the judgement on the basis of a casual survey of inter-civilizational conflicts. Quantitative evidence from every disinterested source (*sic*) conclusively demonstrates its validity.'
165. <www.yale.edu/lawweb/avalon/mideast/hamas.htm>
166. David Zeidan, 'The Islamic Fundamentalist View of Life as a Perennial Battle', *Meria (Middle East Review of International Affairs)*, 5 (2001), 26–53, at 47–8. Web version at: <www.meria.idc.ac.il/journal/2001/issue4/jv5n4a2.htm>

    The author has changed Zeidan's terminology from 'fundamentalist' to 'militant Islamist' to clarify the issue. We should be wary of confusing peaceful 'fundamentalists' with 'militant Islamists' though it may be in the political interest of some to do so. See also Martin Kramer,

'Coming to Terms: Fundamentalists or Islamists?', *Middle East Quarterly*, 10 (Spring 2003). Web version at: <www.meforum.org/article/541>

167. Kramer, 'Coming to Terms: Fundamentalists or Islamists?'.
168. Quoted by Masoud Kazemzadeh, 'Teaching the Politics of Islamic Fundamentalism': <www.apsanet.org/PS/march98/kazemzadeh.cfm>
169. Ibid.
170. Sayyid Abu'l-A'la Mawdūdī, *Human Rights in Islām*, trans. and ed. Khurshid Aḥmad (Leicester: Islamic Foundation, 1976, repr. 1993), 15.
171. Ebrahim Afsah, 'Islamic Exceptionalism. How Valid is the Concept of "Islamic Human Rights"?': <www.ksg.harvard.edu/ksr/article_EA.htm> The quotation is from Fouad Zacharia, 'Human Rights in the Arab World: The Islamic Context', in *Philosophical Foundations of Human Rights* (Paris: UNESCO, 1986), 237.
172. Ibid.
173. Gary R. Bunt, *Islām in the Digital Age. E-Jihād, Online Fatwās and Cyber Islamic Environments* (London and Sterling, VA: Pluto, 2003), 68–9. For the taking out of Jehad.net: <www.encyclopedia4u.com/j/jehad-net.html>
174. Memri Special Report 31 (16 July 2004): <www.memri.org/bin/latestnews.cgi?ID=SR3104>
175. Jeffrey Donovan, 'Islamic Militants take *Jihād* to the Internet' (17 June 2004): <www.isn.ethz.ch/infoservice/secwatch/index.cfm?service=cwn&parent=detail&menu=8&sNewsID=9035>
176. '"Job Application" Online for Suicide Bombers' (10 June 2004), with facsimile of the original: <www.worldnetdaily.com/news/printer-friendly.asp?ARTICLE_ID=38895>
177. Reuters report, 7 June 2004.
178. Ayelet Savyon, 'The Internal Debate in Iran: How to Respond to Western Pressure Regarding its Nuclear Program', Memri Inquiry and Analysis Series 181 (17 June 2004): <www.memri.org/bin/opener_latest.cgi?ID=IA18104>
179. <www.theage.com.au/articles/2003/11/07/1068013395809.html?from=storyrhs&oneclick=true>
180. Shirl McArthur estimated that 'the roughly $3.3 billion in annual aid [to Israel] compares with some $2 billion for Egypt, $225 million for Jordan, and $35 million for Lebanon. Aid for the Palestinian Authority (PA) is not earmarked, but has been running at about $100 million. Furthermore, aid to the PA is strictly controlled by the U.S. Agency for International Development, and goes for specific projects, mostly civil infrastructure projects such as water and sewers': Shirl McArthur, 'A Conservative Total for U.S. Aid to Israel: $91 Billion – and Counting', *Congress Watch*, January/February 2001, 15–16: <www.washington-report.org/backissues/010201/0101015.html>
181. Quoted by Ali, *The Clash of Fundamentalisms*, ix.
182. Ibid. 260. Smedley Butler's speech is at: <www.fas.org/man/smedley.htm> His book has a web version at: <www.ratical.com/ratville/CAH/warisaracket.html#c1>
183. Henry C. K. Liu, 'War and the military–industrial complex', *Asia Times* (31 January 2003): <www.atimes.com/atimes/Front_Page/EA31Aa03.html>
184. Ali, *The Clash of Fundamentalisms*, 144–5.
185. Ira M. Leonard,'Violence is the American Way' (22 April 2003): <www.alternet.org/story/15665> Professor Leonard provides figures for murders in the USA as: 596,984 (1900–71); 592,616 (1971–97).
186. Peter J. Spiro, 'The New Sovereigntists: American Exceptionalism and its False Prophets', *Foreign Affairs*, 79 (November/December 2000), 915. Web version at: <www.globalpolicy.org/globaliz/law/intllaw/newamsov.htm>
187. Ibid. 13, 15.
188. <www.un.org/apps/news/story.asp?NewsID=11081&Cr=ICC&Cr1=>

189. Andrew Tully, 'Ashcroft Denies Torture, but Holds Back Memos' (10 June 2004): <www.isn.ethz.ch/infoservice/secwatch/index.cfm?service=cwn&parent=detail&menu=8&sNewsID=8986>
190. Michael A. Weinstein, 'Readjustments to U.S. Weakness Indicate Power Vacuum' (15 July 2004): <www.isn.ethz.ch/infoservice/secwatch/index.cfm?service=cwn&parent=detail&menu=8&sNewsID=9233>
191. Amal Saad-Ghorayeb, *Hizbu'llah: Politics and Religion* (London and Sterling, VA: Pluto Press, 2002), 108.
192. Ibid. 81.
193. Lewis Wright, 'The Man behind bin Laden. How an Egyptian doctor became a master of terror', *The New Yorker* (16 September 2002): <www.newyorker.com/fact/content/?020916fa_fact2>
194. Azzam S. Tamimi, *Rachid Ghannouchi. A Democrat within Islām* (New York: Oxford University Press, 2001), 143.
195. Liu, 'War and the military–industrial complex'.
196. Hammes, 'The Evolution of War'.
197. Cf. 'What the Senate Intelligence Committee Report Missed' (week of 27 July 2004): 'The Senate Select Committee in Intelligence report on the CIA's failure in Iraq is exhaustive and makes for fascinating reading. One major problem, intelligence analysts say, is that the report misses the mark. The report focuses on CIA analysis and does not properly address collection. In other words, insufficient or improper collection always leads to faulty analysis. One problem is a lack of understanding of how the CIA mines and processes intelligence and shares it with the rest of the U.S. intelligence community. Indeed, the CIA has sometimes only grudgingly shared intelligence with such agencies as the Defence Intelligence Agency, the State Department or National Geospatial Intelligence Agency. The result is that the CIA can't hope to obtain sufficient information to fill in blanks or develop analysis that could alter the bureaucratic momentum of policy. Robert David Steele Vivas, a veteran U.S. intelligence operative, said the report fails to address the CIA's dismissal of open-source information. Vivas said that since 1989, the CIA knew more about Osama bin Laden and terrorism worldwide from open sources of information than from classified sources, but that CIA and other agencies largely ignored open source intelligence. Indeed, Vivas said, Congress and the White House sanctioned the U.S. intelligence attitude toward open-source intelligence since 1992. That, in itself, demands an investigation of the lack of congressional oversight.' Subscription newsletter: <www.geostrategy-direct.com/geostrategy-direct/secure/2004/7_27/me.asp?>
198. Quite the reverse, it has made them less safe. This view is supported by the study of a senior CIA operative 'Anonymous', *Imperial Hubris: Why the West is Losing the War on Terror* (Dulles, VA: Brasseys, 2004), which was announced as forthcoming when this book was at press: 'a growing segment of the Islamic world strenuously disapproves of specific U.S. policies and their attendant military, political and economic implications. Capitalising on growing anti-U.S. animosity, Osama bin Laden's genius lies not simply in calling for *jihād*, but in articulating a consistent and convincing case that Islam is under attack by America. Al Qaeda's public statements condemn America's protection of corrupt Muslim regimes, unqualified support for Israel, the occupation of Iraq and Afghanistan, and a further litany of real-world grievances. Bin Laden's supporters thus identify their problem and believe their solution lies in war. "Anonymous" contends they will go to any length, not to destroy [Western] secular, democratic way of life, but to deter what they view as specific attacks on their lands, their communities and their religion. Unless U.S. leaders recognise this fact and adjust their policies abroad accordingly, even moderate Muslims will join the bin Laden camp.'
199. 'Top officials, including Bush, Rumsfeld, National Security Adviser Condoleezza Rice, and U.S. Vice President Dick Cheney have unequivocally linked Iraq to the al-Qaida network.

Cheney repeated the insinuations this week, claiming that Iraqi President Saddam Hussein had had "long-established ties with al-Qaida". But the 9/11 commission report said there was "no credible evidence that Iraq and al-Qaida cooperated on attacks against the United States". Although the report said bin Laden had sought support from Iraq in the early 1990s, it said that bid had been inconclusive. The commission also rubbished rumours of a meeting between 9/11 hijacker Mohammed Atta and Iraqi intelligence agents in Prague. "We do not believe that such a meeting occurred", the report said. The White House has pointed to Abu Mussab al-Zarqawi, a Jordanian who is said to have links to al-Qaida and may be operating in Baghdad, as evidence of collusion between the former regime and extremists. However, former CIA chief George Tenet has testified that al-Zarqawi was not under Hussein's control and may be operating independently of the al-Qaida network, too. Bush has cited al-Zarqawi as the "best evidence of [the Iraqi regime's] connection to al-Qaida"' (17 June 2004): <www.isn.ethz.ch/infoservice/secwatch/index.cfm?service=cwn&parent=detail&sNewsID=9041&menu=1>

200. Simon Jenkins, 'Case Proven – War does not eradicate terrorism', *The Times*, 14 May 2003: <www.timesonline.co.uk/article/0,,1059-679557,00.html>
201. Robert Worth, 'Bin Laden wants U.S. to strike back disproportionately – the Deep Islamic Roots of Islamic Terror', *New York Times*, 13 October 2001: <www.nytimes.com/2001/10/13/arts/13ROOT.html?todaysheadlines>
202. 'After decades of authoritarian rule and nearly a quarter of a century of conflict, the election is seen as an opportunity for Afghans to finally curb ethnic rivalries and Islamic extremism and embrace peace and democracy. The elections are also important to the West. A successful vote would be seen as vindicating the decision to crush al-Qaeda and topple the Taleban after the attacks of September 11, 2001. And with violence and unrest continuing in Iraq, President Bush is hoping that the election of the pro-Western Hamed Karzai will help his own prospects for re-election in November. But so far, only 2.5 million people have registered. The UN had set a target of registering 10.5 million eligible voters. Large parts of the south and east of the country, the so-called "Pashtun belt", are no-go areas for UN staffers because of growing Taleban and al-Qaeda violence'. Nahim Qaderi, 'Voter Registration Lags in North' (4 June 2004): <www.iwpr.net/index.pl?archive/arr/arr_200406_121_1_eng.txt>; <www.isn.ethz.ch/infoservice/secwatch/index.cfm?parent=news&menu=1#9051>
203. The Spanish government was also punished by the electorate in the spring of 2004 for claiming at first that the atrocities were the work of ETA (when they were on a scale never before contemplated by ETA), and (allegedly) for covering up evidence of an al-Qaeda link. Subsequently, well after the elections, two letters were published by the newspaper *El Mundo* which suggested that there was at least an embryonic link between al-Qaeda and ETA. There was some suggestion that there was a 'double pressure' strategy emerging, in which the Spanish government would not be able to cope with an ETA campaign in northern Spain simultaneously with a violent Islamist campaign in the south: David Sharrock, 'Letters suggest ETA link with Islamic terror', *The Times*, 2 June 2004: <www.timesonline.co.uk/newspaper/0,,172-1130851,00.html>
204. Anatole Kaletsky, 'Terrorism is more than a match for democracy', *The Times*, 25 March 2004: <www.timesonline.co.uk/newspaper/0,,2728-1050272,00.html>
205. Simon Jenkins, 'Now is the time for furious common sense. That bin Laden has not yet been neutralized is the pre-eminent scandal of the new world order', *The Times* (17 March 2004): <www.timesonline.co.uk/newspaper/0,,172-1040736,00.html>
206. 'Al-Qaida Describes Cell Strategy to Topple Saudis, Undermine Western Economy.' Subscription service: <www.geostrategy-direct.com/geostrategy%2Ddirect/secure/2004/6_08/1.asp>
207. 'Oil Experts Gauge Impact of Saudi Terror Attack' (1 June 2004): <www.isn.ethz.ch/infoservice/secwatch/index.cfm?service=cwn&parent=detail&menu=8&sNewsID=8918>

208. Memri Special Dispatch series 726, 3 June 2004: <www.memri.org/bin/latestnews.cgi?ID=SD72604>
209. 'Al-Qaida Describes Cell Strategy to Topple Saudis'.
210. Martin Bright, Nick Pelham and Paul Harris, 'Britons left in jail amid fears that Saudi Arabia could fall to al-Qaeda', *Observer* (London), 28 July 2002: <www.guardian.co.uk/saudi/story/0%2C11599%2C764617%2C00.html>
211. Bronwen Maddox, 'Shouldn't oil prices rise even higher?', *The Times*, 2 June 2004: <www.timesonline.co.uk/newspaper/0,,172-1130917,00.html>
212. Michael Binyon, 'The House of Saud is doomed by its contradictions', *The Times*, 1 June 2004: <www.timesonline.co.uk/newspaper/0,,171-1129928,00.html>
213. 'Even Saudi Princes are Pulling their Money out of Kingdom'. Subscription service: <www.geostrategy-direct.com/geostrategy%2Ddirect/secure/2004/6_08/me.asp>
214. 'U.S. Intelligence Sees Danger that Saudi Monarchy could Suffer Shah's Fate' (week of 27 July 2004). Subscription service: <www.geostrategy-direct.com/geostrategy-direct/secure/2004/7_27/me.asp> 'U.S.–Saudi Ties Seen Changing Whatever the Outcome of 2004 Election' (week of 27 July 2004). Subscription service: <www.geostrategy-direct.com/geostrategy-direct/secure/2004/7_27/2.asp>
215. Michael Gove, 'The upsurge in terror is a sign of the West's success', *The Times*, 26 August 2003: <www.timesonline.co.uk/newspaper/0,,171-792582,00.html>
216. 'The presidential motorcade has special jamming equipment, which blocks all remote-controlled devices in a 200-metre radius', a senior security official investigating the blast commented. 'That is why the bomb exploded after Musharraf's motorcade had crossed the bridge', the official said, requesting anonymity: <www.dawn.com/2003/12/17/top7.htm>; <www.dawn.com/2003/12/18/top5.htm>; <www.news.bbc.co.uk/1/hi/world/south_asia/3319497.stm>
217. '"Network-centric Warfare" called $84 Billion Industry in Next Decade'. Subscription service: <www.geostrategy-direct.com/geostrategy%2Ddirect/secure/2004/6_22/mi.asp>
218. Liu, 'War and the military–industrial complex'.
219. The letter was undated but was post-November 2001: Tariq Ali, *The Clash of Fundamentalisms*, 305–6.
220. 'Reactions in the Arab Press to President Bush's Address on Democracy in the Middle East'. Memri Special Dispatch 615 (25 November 2003): <www.memri.org/bin/articles.cgi?Page=archives&Area=sd&ID=SP61503>
221. Katrina vanden Heuvel, 'Diplomats and Soldiers vs. Bush' (17 June 2004): <www.alternet.org/election04/18979/>
222. <www.comw.org/pda/0201oef.html>
223. <www.cursor.org/stories/civilian_deaths.htm> There is an incident-by-incident online database: <www.cursor.org/stories/casualty_count.htm>
224. There is an incident-by-incident online database: <www.iraqbodycount.net/>
225. <www.cursor.org/stories/civilian_deaths.htm>
226. 'Citing an American academic, a BBC report on [3 January 2002 stated] the number of Afghan civilians killed by U.S. bombs had surpassed the death toll of the 11 September attacks. Nearly 3,800 Afghans had died between 7 October and 7 December, Prof. Marc Herold of the University of New Hampshire said in a research report. Basing his findings on data collected from news agencies, major newspapers and first-hand accounts since the attacks began, Herold placed the civilian death toll conservatively at 3,767. "I think that a much more realistic figure would be around 5,000", he reportedly said. This figure was well in excess of the estimated 2,998 people killed in the 11 September attacks on New York and Washington, the BBC report added.' UN Integrated Regional Information Network, 7 January 2002: <www.globalpolicy.org/wtc/analysis/2002/0107civil.htm>; <www.news.bbc.co.uk/1/hi/world/south_asia/1740538.stm>

227. Carl Conetta, 'Strange Victory'.
228. According to the website dedicated to 11 September 2001 victims, the number recorded as at 7 July 2004 was 2996 victims, comprising WTC Victims: 2626; Flight 11 (America Airlines) Victims: 87; Flight 77 (American Airlines) Victims: 59; Flight 93 (United) Victims: 40; Flight 175 (United) Victims: 5; Pentagon Victims: 125. <www.september11victims.com/september11victims/STATISTIC.asp>
229. In order to advertise a conference in the UK on 11 September 2003, the al-Muhajiroun group courted deliberate controversy by placing a poster on the internet entitled 'the magnificent 19 that divided the world on Sept 11th', with images of the 9/11 terrorists and a smiling bin Laden, together with the Qur'ānic passage 'they were youths who believed in their Lord and we increased them in guidance' (Q.18:13). The image could not fail, as the newspaper caption put it, to give the impression of an 'extremist poster' which 'celebrates [the] 9/11 killers': *The Times*, 25 August 2003. The author downloaded the poster from the website in October 2003, but as at 9 July 2004, the offending website (www.almuhajiroun.com) was no longer active. The message of the poster in any case seemed to be at variance with the position defined by Shaykh Omar Bakrī Muḥammad on the same website the previous year, in which he rejected three different but, in his view, equally mistaken interpretations of the term *jihād*. These interpretations were: 1) that the purpose of *jihād* was the forcible conversion of non-Muslims; 2) that its purpose was the establishment of an Islamic state; 3) that *jihād* referred to the personal efforts of the individual to become 'a model citizen in whatever society one finds oneself in'. Having rejected all of these interpretations, he continued: 'rather, *jihād* is the method adopted by Islām to protect land, honour and life and to save humanity from slavery to man-made regimes'. Shaykh Bakrī stressed that when fighting to liberate occupied Muslim land, 'the divine rules of *jihād* must be observed, i.e. Muslims are forbidden from killing women, children, the elderly… unless [they are] killed accidentally and unavoidably because, for example, they are located amongst the enemy. But the military institutions and governments of any country occupying Muslim land are legitimate targets and if its liberation cannot be achieved without their destruction, then their destruction will become obligatory.' This position was at variance with the al-Qaeda strategy e.g. of targeting 'impious' Muslim regimes. Memri Special Dispatch 435, 30 October 2002: <www.memri.org/bin/articles.cgi?Page=subjects&Area=jihad&ID=SP43502>

    The BBC reported as early as 19 September 2001 that the activities of the al-Muhajiroun group were being monitored: <www.news.bbc.co.uk/1/hi/uk/1552682.stm>

    Bakrī was reported on 18 November 2002 as stating that cyber-attacks were a likely strategy of al-Qaeda: 'in a matter of time you will see attacks on the stock market[s]', he said, referring specifically to the markets in New York, London and Tokyo: <www.computerworld.com/securitytopics/security/story/0,10801,76000,00.html>
230. <www.jihadunspun.com/articles/08212002-Casualty.Report/casualty03.html>
231. <www.antiwar.com/casualties/>
232. Yevgeny Bai, 'Lavrov: American casualties in 'Irāq proportional to Soviet casualties in Afghanistan', *Izvestia*, 10 September 2003: <www.cdi.org/russia/273-9.cfm>
233. <www.guardian.co.uk/usa/story/0,12271,1077361,00.html>
234. <www.costofwar.com/index-aids.html>
235. <www.geostrategy-direct.com/geostrategy%2Ddirect/>
236. Israeli Prime Minister Ariel Sharon personally supervised the attack on Yāsīn, Israeli public radio reported. Sharon had given the green light to Yāsīn's assassination and supervised the operation, the radio stated. After the attack, Sharon congratulated security forces and said 'the war on terror' would continue. 'The state of Israel this morning hit the first and foremost leader of the Palestinian terrorist murderers', Sharon said in his first public reaction to the strike.

'I want to make clear the war on terrorism is not over and will continue daily everywhere.' <www.defencetalk.com/news/publish/printer_1458.shtml>

237. On 18 April 2004, UN Secretary-General Kofi Annan condemned Israel's assassination of Hamas leader 'Abd al-'Azīz al-Rantīsī, calling on the Israeli government to 'immediately end' the practice of 'extrajudicial killings'. Such killings, Annan said in a statement issued by his spokesman, 'are violations of international law'. The spokesman said Annan was 'apprehensive that such an action would lead to further deterioration of an already distressing and fragile situation' in the Middle East. 'The only way to halt an escalation in the violence is for Israelis and Palestinians to work towards a viable negotiating process aimed at a just, lasting and comprehensive settlement, based on the Quartet's Road Map', said Annan. He referred to the 'road map' for Middle East peace drawn by the United States, the United Nations, the European Union and Russia. Rantīsī was killed by an Israeli helicopter rocket attack on his car, less than a month after he succeeded as leader of Hamas. <www.theage.com.au/articles/2004/04/18/1082226620678.html?oneclick=true>
238. In a leaflet issued in Hebron, the closest Palestinian city to Beersheba, Hamas said it was avenging the assassinations of its two leaders earlier in 2004. Addressing Prime Minister Sharon and Defence Minister Shaul Mofaz, it stated: 'You are wrong if you think the assassination of our leaders is going to damage our determination to fight.' <www.news.bbc.co.uk/1/hi/world/middle_east/3614614.stm>
239. Anatole Kaletsky, 'Terrorism is more than a match for democracy'.
240. Sir Max Hastings writes: 'charges of anti-Semitism are not infrequently levelled against the growing number of Jews who express dismay about the behaviour of the Israeli government; they are "self-hating Jews", who betray their own kin. Yet surely it is those who make such cruel allegations who bring shame upon themselves. Jewish genius through the centuries has been reflected in the highest intellectual standards. Attempts to equate anti-Zionism, or even criticism of Israeli policy, with anti-Semitism reflect a pitiful intellectual sloth, an abandonment of reasoned attempts to justify Israeli actions in favour of moral blackmail. In the short run, such intimidation is not unsuccessful, especially in America. Yet in the long term, grave consequences may ensue. In much of the world, including Europe, a huge head of steam is building against Israeli behaviour. More than a few governments are cooperating less than wholeheartedly with America's war on terror because they are unwilling to be associated with what they see as an unholy alliance of the Sharon and Bush governments… It is ironic that Israel's domestic critics – former intelligence chiefs and serving fighter pilots – have shown themselves much braver than overseas Jews. If Israel persists with its current policies, and Jewish lobbies around the world continue to express solidarity with repression of the Palestinians, then genuine anti-Semitism is bound to increase. Herein lies the lobbyists' recklessness. By insisting that those who denounce the Israeli state's behaviour are enemies of the Jewish people, they seek to impose a grotesque choice…' Max Hastings, 'A grotesque choice. Israel's repression of the Palestine people is fuelling a resurgence of anti-Semitism.' *Guardian*, 11 March 2004: <www.guardian.co.uk/comment/story/0,3604,1166637,00.html>
241. <www.fbi.gov/mostwant/topten/fugitives/laden.htm> The bounty was just $5 million on 12 September 2001: <www.archives.tcm.ie/breakingnews/2001/09/12/story23432.asp>
242. <www.fbi.gov/mostwant/terrorists/teralzawahiri.htm>
243. <www.fbi.gov/mostwant/terrorists/fugitives.htm>
244. A (possibly apocryphal) story is that one bounty hunter claimed that he had beheaded bin Laden aide Ayman al-Ẓawāhirī and asked the Pentagon for the $25 million reward. An official reported that the Pentagon asked for proof and received the head, which the FBI found wasn't Ẓawāhirī's. 'There are a lot of con men over there [in Afghanistan]', said the official: <www.freerepublic.com/focus/news/719547/posts>; <www.afio.com/sections/wins/2002/2002-29.html#Bounty>

245. <www.classbrain.com/artfree/publish/article_153.shtml>
246. Nek was eating dinner with four other men, all of whom were killed. They were, naturally, assumed to be 'terrorists' since they had been killed in proximity to him. Quite how the remote-controlled missile knew this is unclear! Ismail Khan and Dilawar Khan Wazir, 'Night raid kills Nek, four other militants: Wana operation', *Dawn* (19 June 2004): <www.dawn.com/2004/06/19/top1.htm>. David Rhode and Mohammed Khan, 'Ex-fighter for Taliban dies in Strike in Pakistan', *New York Times* (19 June 2004): <www.nytimes.com/2004/06/19/international/asia/19STAN.html?th>
247. <www.usatoday.com/news/washington/2003-09-11-gitmo-detainees_x.htm>
248. <www.uncommonknowledge.org/700/717.html>
249. 'Personal liberty should not be a casualty of the campaign against terrorism', said Kenneth Roth, Executive Director of Human Rights Watch on 22 October 2001. 'We believe Congress can develop anti-terrorism measures that protect the nation without sacrificing important rights.' According to Human Rights Watch, the breadth and vagueness of the criteria for the certification and detention of non-citizens in the USA raised the possibility of arbitrary or abusive application: <www.hrw.org/press/2001/10/terrorism1022.htm>
250. Human Rights Watch, World Report 2004. Kenneth Roth, 'Drawing the Line: War Rules and Law Enforcement Rules in the Fight against Terrorism': <www.hrw.org/wr2k4/9.htm>
251. Although not as complex as the US structure, there are several different intelligence sources in the UK. Cf. the BBC website's article 'The UK's Intelligence Agencies' (13 July 2004): <www.news.bbc.co.uk/1/hi/uk/3460275.stm> Robert Fox, 'Pricking the Balloon', *The Tablet* (17 July 2004) notes: 'The intelligence available on Iraq was huge, but [the] Butler Report reveals it as fool's gold, shaped by what politicians on both sides of the Atlantic wanted to hear. Pre-emptive wars will now be off the agenda… In his masterpiece *On the Psychology of Military Incompetence*, Professor Norman F. Dixon reflects on the way leaders justify mistaken decisions and intelligence assessments. "In short, an inability to admit one has been in the wrong will be greater the more wrong one has been," he says, "and the more wrong one has been the more bizarre will be subsequent attempts to justify the unjustifiable."' <www.thetablet.co.uk/cgi-bin/archive_db.cgi/tablet-00918>
252. ISN Security Watch Newsletter, 21 June 2004.
253. 'The Intelligence Community was not created, and does not operate, as a single, tightly knit organization. Rather, it has evolved over nearly 50 years and now amounts to a confederation of separate agencies and activities with distinctly different histories, missions and lines of command. Some were created to centralize the management of key intelligence disciplines. Others were set up to meet new requirements or take advantage of technological advances. Not surprisingly, the *ad hoc* nature of their growth resulted in some duplication of activities and functions. All but the CIA reside in policy departments and serve departmental as well as national interests. Except for the CIA, which for reasons of security is funded in the Defence budget, they are funded by their parent department's appropriation. Their directors are selected by the Secretaries of the departments they serve, although in some cases consultation with the DCI is required': <www.access.gpo.gov/int/int009.html>
254. 'Intelligence Agencies to Link Databases' (27 June 2002). '"We are examining how best to create and share a multi-agency, government-wide database that captures all information relevant to any of the many watch lists that are currently managed by a variety of agencies", CIA Director George Tenet told the Senate Governmental Affairs Committee. "The new department must connect electronically with members of the intelligence community", he said': <www.govexec.com/dailyfed/0602/062702td1.htm>
255. Richard A. Stubbing and Melvin A. Goodman, 'How to fix U.S. Intelligence', *Christian Science Monitor*, 26 June 2002: <www.csmonitor.com/2002/0626/p11s02-coop.html> They also proposed that George Tenet should be sacked as DCI. His resignation in June 2004 met

their condition. 'Few surprised at CIA Chief's resignation' (7 June 2004): <www.isn.ethz.ch/infoservice/secwatch/index.cfm?service=cwn&parent=detail&menu=8&sNewsID=8956>

256. Richard K. Betts, 'The New Politics of Intelligence: Will Reforms Work this Time?', *Foreign Affairs* (May–June 2004): <www.foreignaffairs.org/20040501facomment83301-p10/richard-k-betts/the-new-politics-of-intelligence-will-reforms-work-this-time.html> Michael Duffy, 'How to fix our intelligence' (19 April 2004): <www.edition.cnn.com/2004/ALLPOLITICS/04/19/intelligence.tm/> Duffy states: 'though al-Qaeda was formed in 1988, the CIA "did not describe" the organization comprehensively on paper until 1999. For years the agency believed that bin Laden was a financier rather than an engineer of terrorism – even after it received what a commission report called "new information revealing that bin Laden headed his own terrorist organization, with its own targeting agenda and operational commanders". And though the CIA drafted "thousands" of reports on aspects of al-Qaeda's operation beginning in June 1998 – some of them for the "highest officials in the government", the panel said – the agency never produced an "authoritative portrait of [bin Laden's] strategy and the extent of his organization… or the scale of the threat his organization posed to the United States".'

257. Council on Foreign Relations. Transcript. 'After 'Irāq: New Direction for U.S. Intelligence and Foreign Policy': <www.cfr.org/publication.php?id=6862>

258. Ibid.

259. Ma'sūd Akhtar Shaikh, 'Terrorism: the real culprit', *The News International*, Islamabad, 4 June 2004.

Conclusion

1. 'New Evidence on the War in Afghanistan', *Cold War International History Project Bulletin*, 14/15 (Winter 2003–Spring 2004), 172. Web version at: <www.wwics.si.edu/topics/pubs/c-afghanistan.pdf>
2. Joyce M. Davis, *Between Jihād and Salaam. Profiles in Islām* (Basingstoke: Macmillan, repr. 1999), 105.
3. It is a view of *jihād* that is the only one discussed in an otherwise valuable study by A. J. Coates, *The Ethics of War* (Manchester and New York: Manchester University Press, 1997), 46.
4. Solail H. Hashmi (ed.), *Islamic Political Ethics. Civil Society, Pluralism and Conflict* (Princeton, NJ: Princeton University Press, 2002).
5. Ibid. 215.
6. An emphasis which David Selbourne tried single-handedly to correct: David Selbourne, *The Principle of Duty. An Essay on the Foundations of the Civic Order* (London: Sinclair-Stevenson, 1994).
7. Fathi Osman, 'Islām and Human Rights: The Challenge to Muslims and the World', in *Rethinking Islām and Modernity. Essays in Honour of Fathi Osman*, ed. Abdelwahab El-Affendi (Leicester: Islamic Foundation, 2001), 27–65 at 35.
8. Ṭāriq Ramaḍān, *Western Muslims and the Future of Islām* (New York: Oxford University Press, 2004), 149. In fact, his list of rights is a long one (ibid. 149–52), and implemented in few societies, Western or Muslim: the right to life and the minimum necessary to sustain it; the right to family; the right to housing [frequently neglected in the West as well as in the Islamic world]; the right to education; the right to work; the right to justice; and the right to [social] solidarity.
9. How peaceful Islamists might enter the pale, instead of being excluded, is thoughtfully discussed by Graham E. Fuller, *The Future of Political Islām* (Basingstoke: Palgrave Macmillan, 2003). Also Azzam Karam (ed.), *Transnational Political Islām. Religion, Ideology and Power* (London and Sterling, VA: Pluto Press, 2004).

10. Davis, *Between Jihād and Salaam*, 91.
11. Thus, in an address on 29 February 1992, al-Ghannūshī remarked: 'we want modernity… but only insofar as it means absolute intellectual freedom; scientific and technological progress; and promotion of democratic ideals. However, we will accept modernity only when we dictate the pace with which it penetrates our society and not when the French, British or American interpretations impose it upon us. It is our right to adopt modernity through methods equitable to our people and their heritage.' Ibid. 85–6.
12. François Burgat, *Face to Face with Political Islām* (London and New York: I. B. Tauris, 2003), 130. Also Khaled Abou El Fadlh, *Islam and the Challenge of Democracy. A Boston Review Book*, ed. Joshua Cohen and Deborah Chasman (Princeton, NJ, and Oxford: Princeton University Press, 2004). John L. Esposito and John O. Voll, *Islām and Democracy* (New York: Oxford University Press, 1996).
13. The introduction of genuine democracy would seem to pose a severe threat to the regimes which Fu'ād Zakariyyā calls those of 'Petro-Islām', which preserve social relations where the few at the top of the ladder 'possess the lion's share of this wealth'. Quoted by Ibrāhīm M. Abū-Rabī', *Intellectual Resurgence in the Modern Arab World* (Albany, NY: State University of New York Press, 1996), 253.
14. Davis, *Between Jihād and Salaam*, 84.
15. Cf. also Amin Saikal, *Islam and the West. Conflict or Cooperation* (Basingstoke: Palgrave Macmillan, 2003).
16. Azzam S. Tamimi, *Rachid Ghannouchi. A Democrat within Islām* (New York: Oxford University Press, 2001), 181. Djerejian had stated on 2 June 1992 that the US intent was honourable and affirmed that its foreign policy would include 'support for human rights, pluralism, women's and minority rights and popular participation in government'. He also affirmed America's 'rejection of extremism, oppression and terrorism': Yvonne Yazbeck Haddad, 'Islamism: A Designer Ideology for Resistance, Change and Empowerment', in *Muslims and the West. Encounter and Dialogue*, ed. Zafar Ishaq Ansari and John L. Esposito (Islamabad: Islamic Research Institute, Islamabad and Center for Christian–Muslim Understanding, Washington DC, repr. 2002), 274–95, at 292.
17. This seems to imply acceptance of the economic aspects of globalization. On this: Benjamin R. Barber, *Jihād vs. McWorld. Terrorism's Challenge to Democracy* (London: Corgi Books, 2003); Benjamin Barber, 'Democracy and Terror in the Era of *Jihād* vs. McWorld', in *Worlds in Collision. Terror and the Future of Global Order*, ed. Ken Booth and Tim Dunne (Basingstoke: Palgrave Macmillan, 2002), 245–62.
18. Khurram Murad, 'Islām and Terrorism', *Encounters: Journal of Inter-cultural Perspectives*, 4 (1998), 103–14 at 112–13. Web version at: <www.robert-fisk.com/islam_and_terrorism_khurram_murad.htm>
19. 'With the Muslim world executing one prong of the strategy of rejecting extremism in favour of self-emancipation through human resource development, it is in the wider interest of the international community simultaneously to deliver the second pincer in the Strategy of Enlightened Moderation for global peace and harmony. It can do so in two principal ways: i) by helping to secure just solutions for the political disputes where Muslim peoples are being unjustly oppressed; ii) by assisting the Muslim world in its internal strategy of socio-economic development within the Strategy of Enlightened Moderation. Quite clearly this strategy of "Enlightened Moderation" cannot be one-sided, that the Muslim world responds positively while the West shows inaction in its prong. Both the prongs have to be launched simultaneously and both must succeed.' Speech of President Musharraf to the Summit of the Organization of Islamic Countries, Putrajaya, Malaysia, 2003, para. 21: <www.infopak.gov.pk/President_Addresses/OIC_2003.htm>

20. Islām: 'not one of you truly believes until you wish for others what you wish for yourself' (the Prophet in the Hadith); Christianity: 'in everything, do to others as you would have them do to you; for this is the law and the prophets' (Jesus, in Matthew 7:12); Judaism: 'what is hateful to you, do not do to your neighbour. This is the whole Torah; all the rest is commentary' (Hillel, Talmud, Shabbath 31a). The Golden Rule is thus: 'treat others only in ways that you are willing to be treated in the same situation'. Web version with quotations from other faiths: <www.reconnecting.com/docs/goldenrule.html>
21. Dr Williams commented on 18 June 2004: 'every religious tradition concentrates upon what is good for human beings as such, not upon what is good exclusively for a nation state or even an empire. We all know how this has been distorted by self-interest in the past, but we all know equally how religious traditions renew themselves self-critically, so that they become agents of constructive critique in their social and national settings. It is at best an open question whether secularism can deliver a robust sense of general accountability for the common human good. Despite the divisive potential of many kinds of religious thought and practice, the positive element of focus upon a good that is not local and merely short-term, the sense of being answerable for all and for the whole of a limited material environment is not easily to be found where the religious perspective is systematically ruled out (think of the ecological record of the twentieth century's most thoroughly anti-religious regimes, for example). Hence the importance of religious representation at the UN – and specially, if we can presume to put it in this way, representation that guarantees a voice which can draw on long and sophisticated traditions of moral and political reflection…' <www.archbishopofcanterbury.org/sermons_speeches/040618.html>
22. Marc Gopin, *Holy War, Holy Peace. How Religion can bring Peace to the Middle East* (New York: Oxford University Press, 2002).
23. Fabio Petito and Pavlos Hatzopoulos (ed.), *Religion in International Relations. The Return from Exile* (New York: Palgrave Macmillan, 2003), 1.
24. John L. Esposito and John O. Voll, 'Islām and the West. Muslim Voices of Dialogue', in Petito and Hatzopoulos, *Religion in International Relations*, 236–69 at 265.
25. 'Abdulḥamīd A. AbūSulaymān, *Towards an Islamic Theory of International Relations* (Herndon, VA: International Institute of Islamic Thought, 1993), provides a modern faith-based approach.
26. Murad, 'Islām and Terrorism', 109–10, quotes the Report of Brigadier General James H. Doolittle to President Eisenhower in 1954 ('there are no rules in such a game. Hitherto acceptable norms of human conduct do not apply. If the United States is to survive [against Communism] the long-standing American concept of fair play must be reconsidered. We must develop effective espionage and counter-espionage services and must learn to subvert, sabotage and destroy our enemies by more clever, more sophisticated, and more effective methods than those used against us. It may become necessary that the American people be made acquainted with, understand and support this fundamentally repugnant philosophy') and the comment of President Gerald Ford on the military *coup* against Allende ('I think this was in the best interests of the people of Chile. And certainly in our best interests'). Robert Parry argues that 'while Eisenhower and later presidents did implement the first part of Doolittle's recommendation – ordering covert actions around the world – they finessed the latter. Rather than explain the choices to the American people, U.S. leaders dropped a cloak of state secrecy around "this fundamentally repugnant philosophy"'. Robert Parry, 'Is Media a Danger to Democracy?' (21 March 2000): <www.consortiumnews.com/2000/032000a.html>

    Henry Kissinger famously remarked that he saw 'no reason' for the US to stand by and let a nation [namely Chile] 'go Marxist' because 'its people are irresponsible'.
27. Quoted by Abū-Rabī', *Intellectual Resurgence in the Modern Arab World*, 131.

28. John L. Esposito, *Unholy War. Terror in the Name of Islām* (New York: Oxford University Press, 2002), 26–8. John L. Esposito, *What Everyone Needs to Know about Islām* (New York: Oxford University Press, 2002), 117–18.
29. F. E. Peters, *Islām. A Guide for Jews and Christians* (Princeton, NJ, and Oxford: Princeton University Press, 2003), 209. F. E. Peters, *The Monotheists. Jews, Christians and Muslims in Conflict and Competition. I. The Peoples of God* (Princeton, NJ, and Oxford: Princeton University Press, 2003), 271. The second volume of Peters' study is subtitled *The Words and Will of God.*
30. Rudolph Peters, *Islām and Colonialism. The Doctrine of Jihād in Modern History* (The Hague: Mouton, 1979), 99.
31. A sole example among many will suffice: '…there is no basis for the idea that the beliefs espoused by Islamic terrorists are unrepresentative of the true spirit of Islām. In fact, they are the natural outgrowth of a religion that is in its essential character brutal and violent.' The website is a Christian evangelical one ['The Apostle Paul spoke and inerrantly wrote (Romans to Philemon) on behalf of the Lord Jesus Christ, who is currently seated in Glory. Our material is not presented from the perspective of academic scholars with copious footnotes, but rather that of folks who enjoy passionate thinking' (*sic*)]. Chuck Sligh, 'The Bloody Legacy of Islām': <www.withchrist.org/csligh.htm>
32. Esposito, *Unholy War*, 75: 'for Westerners, Islām is a religion of the sword, of holy war or *jihād*. For Muslims, Christianity is the religion of the Crusades and hegemonic ambitions.'
33. Ritchie Ovendale, *The Origins of the Arab–Israeli Wars* (4th edn) (Harlow: Pearson, 2004), 74.
34. Muḥammad Raheem Bawa Muhaiyaddeen, *Islām and World Peace. Explanations of a Ṣūfī* (Philadelphia, PA: Fellowship Press, 1987). Web edition at: <www.bmf.org/iswp/index.html>
35. These are short and are learnt in Arabic: 'I seek protection in Allāh from the Shaytan, the cursed one'; 'in the name of Allāh, the Beneficent, the Merciful…' Arabic text online at: <www.islam.tc/kalimah/>
36. M. Nejatullah Siddiqi, 'Future of the Islamic Movement', *Encounters: Journal of Inter-cultural Perspectives*, 4 (1998), 91–101 at 96–7.
37. Louay M. Safi, *Peace and the Limits of War. Transcending the Classical Conception of Jihād* (London and Washington: International Institute of Islamic Thought, repr. 2003).
38. James Turner Johnson, *The Holy War Idea in Western and Islamic Traditions* (Philadelphia, PA: Pennsylvania State University Press, 1997; repr. 2002). Harfiyah Abdel Haleem, Oliver Ramsbotham, Saba Risaluddin and Brian Wicker (eds), *The Crescent and the Cross. Muslim and Christian Approaches to War and Peace* (Basingstoke: Macmillan, 1998). Unfortunately, Jalal Abualrub, *Holy Wars, Crusades, Jihād. I. Jihād* (Orlando, FL: Madinah Publishers, 2002) is tendentious (see comment on another work by the same author, Chapter 6, note 1).
39. R. J. Bonney, 'Impossible to reconcile? Christian Just War Theory and the Second Iraq War', *Encounters. Journal of Inter-cultural Perspectives*, 9 (2003), 69–91.
40. The outcome of a meeting of '*ulamā*' at Château-Chinon in July 1992: Ataullah Siddiqui, 'Ethics in Islām: Key Concepts and Contemporary Challenges', *Journal of Moral Education*, 26 (1997), 423–31 at 427–8.
41. Ramaḍān, *Western Muslims and the Future of Islām*, 53, 63, 93, 159.
42. This is not to suggest that Muslims subordinate their truth claim to any one else's. Some theorists of pluralism, a minority, argue that it is only by some degree of subordination of the truth claims of each religion / all religions that inter-faith dialogue can really make progress. Nicholas Rescher makes the fundamental distinction between the standpoint of the individual and the standpoint of the group: 'pluralism is a feature of the collective group: it turns on the fact that different experiences engender different views. But from the standpoint of the

individual this cuts no ice. We have no alternative to proceeding as best we can on the basis of what is available to us.' Nicholas Rescher, *Pluralism: Against the Demand of Consensus* (Oxford: Clarendon Press, 1993), 88–9, cited by Osman, 'Islām and Human Rights', 52.

43. H. M. Zawātī, *Is Jihād a Just War? War, Peace and Human Rights under Islamic and Public International Law* (Lewiston, NY, 2001). Ann Elizabeth Mayer, *Islam and Human Rights. Tradition and Politics* (London; Boulder, CO and San Francisco: Pinter and Westview Press, 1991). Abdullah A. An-Na'im, 'Islamic Foundations of Religious Human Rights', in *Religious Human Rights in Global Perspective. Religious Perspectives*, ed. John Witte Jr and Johan D. van der Vyver (The Hague: Martinus Nijhoff, 1996), 337–59. Farhat Haq, '*Jihād* over Human Rights, Human Rights as *Jihād*. Clash of Universals', in *Negotiating Culture and Human Rights*, ed. Lynda S. Bell, Andrew J. Nathan Ilan Peleg (New York: Columbia University Press, 2001), 242–57.
44. 'Interpreting Islām', *Indian Currents* (6 June 2004), 22–3. Web version at: <www.islaminterfaith.org/june2004/interview-06-04.htm>
45. Osman, 'Islām and Human Rights', 52.
46. Ibid. 50.
47. Walid Saif, 'Reflections on Muslim–Christian Dialogue: Core Values and Common Responsibilities', *Encounters: Journal of Intercultural Perspectives*, 7 (2001), 91–9; quotations at 95, 96, 97.
48. 'The kind of peace we should be looking for today is not only mere lack of terrorism, weapons of mass destruction and armed conflict. Real peace is a realisation of individual human potential for peace in the community with oneself, one's neighbours and the natural world but above all peace with God Almighty. The world is looking for peace with justice, [an] equitable sharing of world resources and freedom for every people to develop their own gifts and follow their own vision without racial or religious oppression or colonial domination…' *Eid* Message of Hadhrat Mirzā Masroor Aḥmad, November 2003. It is reasonable to argue that, if one makes a comparison with Christianity, the Ahmadīs are the equivalent of the Jehovah's Witnesses: they think of themselves as Muslims/Christians but are not so regarded by mainstream Muslims/Christians. However, sensible ideas (as against doctrines) are sensible ideas, irrespective of their origins.
49. Seyyed Hossein Nasr, *The Heart of Islām. Enduring Values for Humanity* (San Francisco: HarperSanFrancisco, 2002), 63.
50. Haneef James Oliver, 'Dispelling the "Wahhābī" Myth: Dispelling Prevalent Fallacies and the Fictitious Link with Bin Laden' (n.p., 2002), 9, 15, 24. Electronic publication available at: <www.thewahhabimyth.colm/khawarij.htm>
51. 'Interpreting Islām', *Indian Currents* (6 June 2004), 23–6. Web version at: <www.islaminterfaith.org/june2004/interview-06-04.htm>
52. Siddiqui, 'Ethics in Islām', 426. Ataullah Siddiqui, 'Fifty Years of Christian–Muslim Relations: Exploring and Engaging in a New Relationship', *IslamoChristiana*, 20 (2000), 51–77 at 73.
53. Shaul Mishal and Avraham Sela, *The Palestinian Hamas. Vision, Violence and Coexistence* (New York: Columbia University Press, 2000), appendix two. Web version of Hamas Charter at: <www.yale.edu/lawweb/avalon/mideast/hamas.htm>
54. Quoted by Amal Saad-Ghorayeb, *Hizbu'llah: Politics and Religion* (London and Sterling, VA: Pluto Press, 2002), 185–6.
55. Saif, 'Reflections on Muslim–Christian Dialogue', 91. Ataullah Siddiqui, 'Believing and Belonging in a Pluralist Society: Exploring Resources in Islamic Traditions', in *Multi-Faith Britain*, ed. David Hart (London: O'Books, 2002), 23–33 at 23.
56. Siddiqui, 'Believing and Belonging in a Pluralist Society', 24.
57. 'Interpreting Islām', *Indian Currents* (6 June 2004), 23–6. Web version at: <www.islaminterfaith.org/june2004/interview-06-04.htm>

58. Ibid.
59. Ian Talbot, *Jinnah: Role Model for Future Generations of Pakistanis* (Leicester: INPAREL South Asian History Academic Papers, 1; 2001), 20. Richard Bonney, *Three Giants of South Asia: Gandhi, Ambedkar and Jinnah on Self-Determination* (Leicester, 2002: INPAREL South Asian History Academic Papers, 5). Reprinted New Delhi, Media House Publications, with a new introduction for India: 2004 imprint.
60. Cited by Abū-Rabī‘, *Intellectual Resurgence in the Modern Arab World*, 257.
61. Richard Bonney, introduction to Fateh Muḥammed Malik, *Iqbal's Reconstruction of Political Thought in Islām* (Leicester, 2002: INPAREL South Asian History Academic Papers, 6; reprinted New Delhi: Media House Publications, 2004) and document eight. References to the New Delhi edition.
62. M. K. Masud, 'Iqbal's lecture on *Ijtihād*', *Selections from the Iqbal Review*, ed. W. Qureshi (Lahore: Iqbal Academy, Pakistan, 1983), 109–17. This article dates from 1978. Masud dates the delivery of the lecture in Lahore as 13 December 1924: ibid. 116.
63. Malik, *Iqbal's Reconstruction of Political Thought in Islām*, 97.
64. B. Weiss, 'Interpretation in Islamic Law: The Theory of *Ijtihād*', *American Journal of Comparative Law*, 26 (1978), 208. Weiss comments on 'closing the door' (ibid. 209): 'yesterday's rules, transformed by the Consensus into timeless principles, become material out of which today's rules may be derived'. Ṭāriq Ramaḍān, *Western Muslims and the Future of Islām*, 48, also asserts that the doors of *ijtihād* were never closed.
65. Without citing Iqbal's name, Weiss endorses this quest: '…it is by virtue of the theory that Islamic law is Islamic. Obviously, the theory must eventually be related to actual practice, and for this a renewed *ijtihād*, resembling in its vigour and zeal the *ijtihād* of the earliest centuries of Islām, clearly must be undertaken.' Weiss, 'Interpretation in Islamic Law', 212.
66. Cited at n. 6 of the web edition of this section of Iqbal's *Restoration*: <www.allamaiqbal.com/works/prose/english/reconstruction/notes.htm#Lecture%20VI:>
67. Bonney, introduction to Malik, *Iqbal's Reconstruction of Political Thought in Islām*, 11. The view of the legislative assembly should, in Iqbal's view, prevail over that of the '*ulamā*'.
68. Tamimi, *Rachid Ghannouchi*, 187.
69. Muḥammad Asad, *The Principles of State and Government in Islām* (Gibraltar: Dar-al-Andalus, repr. 1985), 23. The same author (ibid. 16) states that the need for a free enquiry, a rediscovery of the 'open road' of Islam 'is urgently needed'.
70. Ashgar Ali Engineer, 'Evolution of *Sharī'ah* Law and its Potentiality for Change', *Islam and the Modern Age* (June 2004): <www.csss-isla.com/IIS/index.php>
71. Ataullah Siddiqui, 'People of Faith in Britain Today and Tomorrow', Centre for the Study of Islām and Christian–Muslim Relations, University of Birmingham, Occasional Papers 4 (1999).
72. Siddiqui, 'Believing and Belonging in a Pluralist Society', 26–7.
73. Siddiqi, 'Future of the Islamic Movement', 99.
74. Arskal Salim and Azyumardi Azra (ed.), *Sharī'ah and Politics in Modern Indonesia* (Singapore: Institute of Southasian Studies, 2003), 230.
75. Donna E. Artz, 'The Treatment of Religious Dissidents under Classical and Contemporary Islamic Law', in *Religious Human Rights in Global Perspective. Religious Perspectives*, ed. John Witte Jr and Johan D. van der Vyver (The Hague: Martinus Nijhoff, 1996), 387–453.
76. Saif, 'Reflections on Muslim–Christian Dialogue', 97.
77. Osman, 'Islām and Human Rights', 47.
78. Ibid. 54.
79. A modern Indonesian Muslim intellectual, Abdurrahman Wahid, argues: 'if it is true that the Prophet aspired for the formation of an "Islamic State" it is impossible that the issues of leadership succession and transfer of power were not formally formulated. [In this case] the

Prophet simply ordered [the Muslim community] to "consult in matters". It was amazing that issues of such great significance were not concretely institutionalized, rather it sufficed for him to regulate those issues in a single dictum: "their affairs should be consulted among them." Is there a state in such a form?' Quoted by Bahtiar Effendy, *Islām and the State in Indonesia* (Singapore: Institute of Southeast Asian Studies, 2003), 107.

80. Abdelwahab El-Affendi, *Who Needs an Islamic State?* (London: Grey Seal Books, 1991), 93.
81. Robert W. Hefner, *Civil Islām. Muslims and Democratization in Indonesia* (Princeton, NJ, and Oxford: Princeton University Press, 2000), 215, 218. Cf. also Robert W. Hefner, 'Islamic Orders' [review of John R. Bowen, *Islam, Law and Equality in Indonesia: an Anthropology of Public Reasoning* (New York: Cambridge University Press, 2003) in *Times Literary Supplement*, 28 November 2003].
82. 'Listen, you accomplices of the United States. Listen, you accomplices of the World Church Council. Listen, you accomplices of Zionist evangelists. Listen, you Jews and Christians: we Muslims are inviting the U.S. military to prove its power in Maluku. Let us fight to the finish. Let us prove for the umpteenth time that the Muslim faithful cannot be conquered by over-exaggerated physical power. The second Afghanistan war will take place in Maluku if you are determined to carry out the threat…' Text of the 'Declaration of War' by Laskar Jihād Commander Ustadz Ja'far 'Umar Thalib, broadcast on Radio SPMM (Voice of the Maluku Muslim Struggle) on 1–3 May 2002; as published by Indonesian newspaper *Berdarah* on its website on 8 May: <www.websitesrcg.com/ambon/documents/laskar-jihad-010502.htm>
83. Seyyed Vali Reza Nasr, 'States and Islamization', in Ansari and Esposito, *Muslims and the West. Encounter and Dialogue*, 296–310 at 309.
84. A. Nizar Hamzeh and R. Hrair Dekmejian, 'Al-Ahbash: A Ṣūfī Response to Political Islamism', *International Journal of Middle East Studies,* 28 (1996), 217–29: <www.almashriq.hiof.no/ddc/projects/pspa/al-ahbash.html> However, the authors note: 'the fact is that both al-Ahbash and al-Jamā'ā are engaged in mutual *takfīr*, refusing to recognize each other's Islamic legitimacy'.
85. Davis, *Between Jihād and Salaam*, 95–6.
86. Ibid. 89.
87. Mahmoud Ayoub, 'Religious Freedom and the Law of Apostasy in Islām', *IslamoChristiana*, 20 (1994), 75–91 at 76.
88. Yohanan Friedmann, *Tolerance and Coercion in Islam. Inter-faith Relations in the Muslim Tradition* (Cambridge: Cambridge University Press, 2003), 121–2.
89. Ibid. 131.
90. Ayoub, 'Religious Freedom and the Law of Apostasy in Islām', 79.
91. Ibid. 83, 88.
92. Ibid. 89.
93. 'Opinion on Apostasy stirs a Heated Debate in Islamic Juristic Circles', *The Diplomat*, 2 (June 1996), 38–9. The author is indebted to Dr Ataullah Siddiqui for this reference, and for suggesting materials for this section.
94. <www.infopak.gov.pk/President_Addresses/OIC_2003.htm>; <www.infopak.gov.pk/President_Addresses/Seminar_Davos.htm>
95. At the OIC summit, President Musharraf stated: 'the first prong of this strategy has to be executed by… ourselves. We have to address and overcome [the] internal weaknesses and vulnerabilities of the Islamic world, while simultaneously rejecting recourse to militancy and extremism. Our shortcomings are visible. Our human development indicators are among the lowest in the world; poverty is pervasive; literacy is less than 50%; institutions of higher learning are insignificant. Poverty and illiteracy breed extremism and orthodoxy. Our economic underdevelopment consigns us to the margins of international power structure. Our intellectual

impoverishment diminishes our ability to defend our just causes. Our shortage of scientific skills erodes our ability to energize our economies, to compete commercially and to cater for the defence of our countries. To promote dynamic development, prosperity and peace within our nations and societies, we must focus on poverty reduction, employment generation, expansion of production, science and technology, higher education, health and human resource development. This will require considerable and focused investment of resources. These are limited but can be generated, domestically and externally, by policies that place the interests of our peoples at the centre of our political agendas. We can also help each other. Collectively, we can, and must, assist the poorest amongst our members. Socio-economic progress and growing prosperity will also provide the best antidote to extremism and violent proclivities which accompany it.'

96. Siddiqi, 'Future of the Islamic Movement', 98.
97. 'Interpreting Islām', *Indian Currents* (6 June 2004), 23–6. Web version at: <www.islaminterfaith.org/june2004/interview-06-04.htm>
98. On this theme: Riffat Hassan, 'Rights of Women within Islamic Communities', in Witte Jr and van der Vyver, *Religious Human Rights in Global Perspective*, 361–85.
99. Abū-Rabī', *Intellectual Resurgence in the Modern Arab World*, 263.
100. Jeremy Henzell-Thomas, *The Challenge of Pluralism and the Middle Way of Islām*, Association of Muslim Social Scientists (UK), Occasional Paper Series 1 (AMSS UK, 2002).
101. Muḥammad Asad, *The Message of the Qur'ān* (Gibraltar: Dar Al-Andalus, 1980), 30 n. 118: 'a community that keeps an equitable balance between extremes…' However, in *The Bounteous Korān. A Translation of Meaning and Commentary* by M. M. Khatib (authorized by al-Azhar, 1984: London: Macmillan, 1986), 27 n. 47 (at 28), a rather different gloss is given: 'a just nation, with virtuous qualities, able to teach people the correct path to righteousness'. *The Holy Qur'ān. Text, Translation and Commentary*, ed. A. Yusuf Ali (1st edn, 1934; Brentwood: Maryland: Amana Corp., 1983), 57 n. 143 has 'justly balanced': 'the essence of Islām is to avoid all extravagances on either side'. A. J. Arberry's translation has 'the midmost nation': *The Koran*, trans. A. J. Arberry (Oxford: Oxford University Press, 1983), 18.
102. <www.uahc.org/sept11/interfaith.shtml> Cf. Sayings of the Prophet, s.v. 'kindness': <www.twf.org/Sayings/Sayings3.html#Kindness>
103. Cardinal Basil Hume, O.S.B., *To be a Pilgrim. A Spiritual Notebook* (Slough: St Paul Publications, 1968), 168.
104. William A. Graham, *Divine Word and Prophetic Word in Early Islām. A Reconsideration of the Sources, with Special Reference to the Divine Saying or Ḥadīth Qudsī* (The Hague: Mouton, 1977), 179 (Saying 54).
105. Ibid. 183 (Saying 57).
106. Haddad, 'Islamism: A Designer Ideology for Resistance, Change and Empowerment', 291.
107. Matthew 5:15–16: 'neither do men light a candle, and put it under a bushel, but on a candlestick; and it giveth light unto all that are in the house. Let your light so shine before men, that they may see your good works, and glorify your Father which is in heaven.' Mark 4:21–2: '…is a candle to be put under a bushel, or under a bed? And not to be set on a candlestick? For there is nothing hid, which shall not be manifested; neither was anything kept secret, but that it should come abroad.' Luke 11:33: 'no man, when he hath lighted a candle, putteth it in a secret place, neither under a bushel, but on a candlestick, that they which come in may see the light'.
108. *Brewer's Dictionary of Phrase and Fable*, ed. Ivor H. Evans (revised edn London: Guild Publishing, 1985), 174.
109. There are useful discussions of the various aspects of *jihād* for Muslims in such databases. If they had been read by some of the authors of vitriolic anti-Muslim literature noted in the Introduction, then there might have been better informed works on *jihād* available to

the public. They were not read by so-called specialists. They are not likely to be read by the general public. For a very interesting set of online *fatawā* on *jihād*: <www.islamonline.net/fatwa/english/FatwaDisplay.asp?hFatwaID=96325> and others to be found there. Nor are the published *fatawā* likely to be read: *Fatawā Islamiya. Islamic Verdicts. Volume 8. Jihād, Da'wah, and Commanding Good and Forbidding Evil*, collected by Muhammad bin 'Abdul-'Aziz al-Musnad (Riyadh, London, etc.: Darussalam, 2002).

110. The USIA merged with the Department of State on 1 October 1999: <www.dosfan.lib.uic.edu/usia/usiahome/pdforum/homepage.htm>

The history of this institution is analysed by Professor Nicholas J. Cull of the University of Leicester in a forthcoming study to be published by Cambridge University Press. In a private communication to the author, Professor Cull writes: 'there is certainly a need for greater Islamic public diplomacy. The world would be better for more dialogue – but we also must be open to Muslim public diplomacy – reciprocal partners in their equivalents of the Fulbright or Rhodes-type scholarship. We have to feel that we have something to learn from them and be ready to see ourselves through their eyes. Many in U.S. public diplomacy circles do not really want to be open to "their" ideas and I suspect use talk of dialogue as just another way to open the door to U.S. ideological transmission.' When a Planning Group reported on the integration of USIA into the Department of State (20 June 1997), it defined public diplomacy as an activity which 'seeks to promote the national interest of the United States through understanding, informing and influencing foreign audiences'. The Planning Group distinguished Public Affairs from Public Diplomacy as follows: 'Public Affairs is the provision of information to the public, press and other institutions concerning the goals, policies and activities of the U.S. Government. Public affairs seeks to foster understanding of these goals through dialogue with individual citizens and other groups and institutions, and domestic and international media. However, the thrust of public affairs is to inform the domestic audience': <www.publicdiplomacy.org/1.htm>

111. <www.publicdiplomacy.org/1.htm#propaganda>

112. Hence the University of Leicester Centre for the History of Religious and Political Pluralism's website on the development of mainstream Islām's relations with the West: '*Nur*: Islām and Enlightenment': <www.le.ac.uk/pluralism/nur>

113. *Rethinking Islām and Modernity. Essays in Honour of Fathi Osman*, ed. Abdelwahab El-Affendi (Leicester: Islamic Foundation, 2001), 174.

Select Bibliography

Abedi, Mehdi, and Legenhausen, Gary (eds), *Jihād and Shahādat. Struggle and Martyrdom in Islām...* (Houston: Institute for Research and Islamic Studies, 1986).

Abou El Fadlh, Khaled, *Islam and the Challenge of Democracy. A Boston Review Book*, ed. Joshua Cohen and Deborah Chasman (Princeton, NJ, and Oxford: Princeton University Press, 2004).

Abualrub, Jalal, *Holy Wars, Crusades, Jihād. I. Jihād* (Orlando, FL: Madinah Publishers, 2002).

Abualrub, Jalal, *Muḥammad Ibn Abdul Wahhāb. His Life-Story and Mission* (Orlando, FL: Madinah Publishers, 2003).

Abu-Lughod, Janet L., *Before European Hegemony. The World System AD 1250–1350* (New York: Oxford University Press, 1989).

Abū-Rabī', Ibrāhīm M., *Intellectual Resurgence in the Modern Arab World* (Albany, NY: State University of New York Press, 1996).

AbūSulaymān, 'AbdulḤamīd A., *Towards an Islamic Theory of International Relations: New Directions for Methodology and Thought* (Herndon, VA: International Institute of Islamic Thought, 1993).

Adams, C. J., 'Mawdūdī on "The Necessity of Divine Government for the Elimination of Oppression and Injustice"', in *Muslim Self-Statement in India and Pakistan, 1857–1968*, ed. A. Ahmad and G. E. von Grunebaum (Wiesbaden, 1970).

Adams, C. J., 'Mawdūdī and the Islamic State', in *Voices of Resurgent Islam*, ed. J. L. Esposito (New York and Oxford: Oxford University Press, 1983), 99–133.

Afsaruddin, Asma, *Excellence and Precedence. Medieval Islamic Discourse on Legitimate Leadership* (Leiden: E. J. Brill, 2002).

Afzal, M. Rafique (ed.), *Speeches and Statements of Quaid-i-Millat Liaquat 'Ali Khan, 1941–51* (Lahore: Research Society of Pakistan, 1967).

Ágoston, Gábor, 'Ottoman Artillery and European Military Technology in the Fifteenth and Seventeenth Centuries', *Acta Orientalia Academia Scientiarum Hungaricae*, 47 (1994).

Ágoston, Gábor, 'Ottoman Warfare in Europe, 1453–1826', in *European Warfare, 1453–1815*, ed. Jeremy Black (Basingstoke: Macmillan, 1999).

Ágoston, Gábor, 'The Ottoman–Habsburg Frontier in Hungary, 1541–1699', in *The Great Ottoman Turkish Civilization. I. Politics*, ed. Kemal Çiçek (Ankara: Yeni Türkiye, 2000).

Ágoston, Gábor, '*Merces Prohibitae*: The Anglo–Ottoman Trade in War Materials and the Dependence Theory', *Oriente Moderno*, 20 (2001).

Ágoston, Gábor, 'Ottoman Conquest and the Ottoman Military Frontier in Hungary', in *A Millennium of Hungarian Military History*, ed. László Veszprémy and Béla K. Király (Boulder, CO: Social Science Monographs, distr. New York: Columbia University Press, 2002).

Ágoston, Gábor, 'Early Modern Ottoman and European Gunpowder Technology', in *Multicultural Science in the Ottoman Empire*, ed. Ekmeleddin Ihsanoglu, Kostas Chatzis, Efthymios Nicolaidis (Turnhout, Belgium: Brepols, 2003).

Ágoston, Gábor, *Guns for the Sultan: Military Power and the Weapons Industry in the Early Modern Ottoman Empire* (Cambridge: Cambridge University Press, 2005).

Ahmad, Qeyammudin, *The Wahhābī Movement in India* (New Delhi: Manohar, 1994).

Ahmed, Ishtiaq, *The Concept of an Islamic State. An Analysis of the Ideological Controversy in Pakistan* (London: Pinter, 1987).

Ahsan, M. M., and Kidwai, A. R. (eds), *Sacrilege versus Civility: Muslim Perspectives on the Satanic Verses Affair* (Leicester: The Islamic Foundation, 1993).

Akhmadov, Yavus Z., 'Kunta Hadji and the Kunta Hadjists: The Kunta-Hadji Chechen Religious Movement' at: <www.jmu.edu/orgs/wrni/cs-part7.html>

Alam, Muzaffar, and Subrahmanyam, Sanjay (eds), *The Mughal State, 1526–1750* (New Delhi: Oxford University Press, 2002).

Al-A'zamī, M. M., *The History of the Qur'ānic Text from Revelation to Compilation. A Comparative Study with the Old and New Testaments* (Leicester: UK Islamic Academy, 2003).

Al-Bannā', S. M. Ḥasan, *Imām Shahīd Ḥasan al-Bannā'. From Birth to Martyrdom* (Swansea: Awakening Publications, 2002).

Algar, Hamid (ed. and trans.), *Islām and Revolution. Writings and Declarations of Imām Khomeini* (Berkeley, CA: Mizan Publications, 1981).

Algar, Hamid, *Wahhabism: A Critical Essay* (Oneonta, NY: Islamic Publications International, 2002).

Al-Ghazālī, Muḥammad, *The Socio-Political Thought of Shāh Walī Allāh* (Islamabad: International Institute of Islamic Thought and Islamic Research Institute, 2001).

Al-Ghazālī, Muḥammad, *On the Boundaries of Theological Tolerance*, trans. Sherman A. Jackson (Karachi: Oxford University Press, 2002).

Ali, A. Yusuf, *The Holy Qur'ān. Text, Translation and Commentary* (1st edn 1934; Brentwood: MD: Amana Corp., 1983).

Ali, M. Athar, *The Mughal Nobility Under Aurangzeb* (London: Asia Publishing House for Department of History, Aligarh Muslim University, 1966).

Ali, Tariq, *The Clash of Fundamentalisms. Crusades, Jihads and Modernity* (London: Verso, 2002).

Al-Qushayrī, Abū-l-Qāsim 'Abd-al-Karīm bin Hawāzin, *The Risāla: Principles of Ṣūfism*, trans. Rabia Harris (Chicago, IL: Great Books of the Islamic World, Kazi Publications, 2002).

Al-Rasheed, Madawi, *A History of Saudi Arabia* (New York and Cambridge: Cambridge University Press, 2002).

Al-Sadr, Ayatullah Muḥammad Baqir and Muṭahharī, Ayatullah Murtazá, *The Awaited Saviour* (Karachi: Islamic Seminary Publications, n.d.), at: <www.al-islam.org/awaited/>

Al-Saeed, Mohajir, 'Al-Saud: Past and Present' at: <www.khyber.demon.co.uk/history/saudi-arabia/past.htm>

Al-Sayyid, M. K., 'The Other Face of the Islamist Movement' (Carnegie Endowment, Democracy and Rule of Law Project. Global Policy Program, 33, January 2003) at: <www.ceip.org/files/pdf/wp33.pdf>

Alsumaih, Abdulrahman Muḥammad, 'The Sunnī Concept of *Jihād* in Classical Fiqh and Modern Islamic Thought' (unpublished PhD thesis, University of Newcastle upon Tyne, 1998).

Al-Zayyāt, Montasser, *The Road to Al-Qaeda. The Story of bin Lāden's Right-Hand Man* (London and Ann Arbor, MI: Pluto Press, 2004).

Ambrosio, Thomas, 1997, 'Ottoman "Hegemonic Control" in the Balkans' at: <www.ndsu.nodak.edu/ndsu/ambrosio/hegemony.html>

Andoni, Ghassan, 'A Comparative Study of *Intifada* 1987 and *Intifada* 2000', in *The New Intifada. Resisting Israel's Apartheid*, ed. Roane Carey (London and New York: Verso, 2001), 209–18.

An-Nabhani, Taqiuddin, *The Islamic State* (New Delhi: Milli Publications, 2001).

An-Na'im, Abdullah A., 'Islamic Foundations of Religious Human Rights', *Religious Human Rights in Global Perspective. Religious Perspectives*, ed. John Witte Jr and Johan D. van der Vyver (The Hague: Martinus Nijhoff, 1996), 337–59.

Anonymous, *Through Our Enemies' Eyes. Osama bin Laden, Radical Islam and the Future of America* (Washington DC: Brassey's Inc., 2002).

Anonymous, *Imperial Hubris. Why the West is Losing the War on Terrorism* (Washington DC: Brassey's Inc., 2004).

Ansari, Ali M., *Modern Iran since 1921. The Pahlavis and After* (London: Pearson, 2003).

Arberry, Arthur J. (ed. and trans.), *The Doctrine of the Ṣūfīs* [*Kitāb al-Ta'arruf li-madhhab ahl al-tasawwuf* translated from the Arabic of Abū Bakr al-Kalābādhī] (AMS Press, 1935; Cambridge, Cambridge University Press: 1935; Lahore: Sh. Muhammad Ashraf, 1966; New York: Cambridge University Press, 1977).

Arberry, Arthur J. (trans.), *The Koran* (Oxford: Oxford University Press, 1983).

Armajani, Yahya, *Middle East: Past and Present* (Englewood Cliffs, NJ: Prentice-Hall, 1970).

Artz, Donna E., 'The Treatment of Religious Dissidents under Classical and Contemporary Islamic Law', in *Religious Human Rights in Global Perspective. Religious Perspectives*, ed. John Witte Jr and Johan D. van der Vyver (The Hague: Martinus Nijhoff, 1996), 387–453.

Asad, Muḥammad, *The Message of the Qur'ān* (Gibraltar: Dar Al-Andalus, 1980).

Asad, Muḥammad, *The Principles of State and Government in Islām* (Gibraltar: Dar-al-Andalus, repr. 1985).

Atil, Esin, *Süleymanname. The Illustrated History of Süleyman the Magnificent* (National Gallery of Art, Washington DC/New York: Harry N. Abrams, 1986).

At-Tarjumana, A. A., and Johnson, Y. (trans.), *Al-Muwaṭṭa'* (Norwich: Diwan Press, 1982).

Ayoub, Mahmoud, 'Religious Freedom and the Law of Apostasy in Islām', *IslamoChristiana*, 20 (1994), 75–91.

Azzam, 'Abdullah, *Defense of the Muslim Lands. The First Obligation after Iman*, trans. Brothers in Ribatt, n.d., at: <www.geocities.com/johnathanrgalt/Defence_of_the_Muslim_Lands.pdf>

Baddeley, John Frederick, *The Russian Conquest of the Caucasus* (London: Longman, 1908).

Bagley, F. R. C. (trans.), *Ghazālī's Book of Counsel for Kings (Nasīḥaṭ al-Mulūk)* (London: Oxford University Press, 1964).

Barber, Benjamin R., 'Democracy and Terror in the Era of *Jihād* vs. McWorld', in *Worlds in Collision. Terror and the Future of Global Order*, ed. Ken Booth and Tim Dunne (Basingstoke: Palgrave Macmillan, 2002), 245–62.

Barber, Benjamin R., *Jihād vs. McWorld. Terrorism's Challenge to Democracy* (London: Corgi Books, 2003).

Barkey, Karen, *Bandits and Bureaucrats: The Ottoman Route to State Centralization* (Ithaca, NY: Cornell University Press, 1994).

Barlas, Asma, 'Reviving Islamic Universalism: East/s, West/s, and Coexistence' Conference on Contemporary Islamic Synthesis, Alexandria, Egypt, 4–5 October 2003 at: <www.ithaca.edu/faculty/abarlas/papers/barlas_20031004.pdf>

Bennett, Clinton, *In Search of Muḥammad* (London and New York: Cassell, 1998).

Bergen, *Holy War Inc.: Inside the Secret World of Osama bin Laden* (London: Phoenix, repr. 2003).

Betts, Richard K., 'The New Politics of Intelligence: Will Reforms Work this Time?', *Foreign Affairs* (May–June 2004) at: <www.foreignaffairs.org/20040501facomment83301-p10/richard-k-betts/the-new-politics-of-intelligence-will-reforms-work-this-time.html>

Bin Jani, M. S., 'Sayyid Quṭb's View of *Jihād*: An Analytical Study of his Major Works' (unpublished PhD thesis, University of Birmingham, 1998.)

Bishara, Marwan, 'The Americanization of Israel and the Israelization of American Policy'. Report from a Palestine Center briefing at: <www.palestinecenter.org/cpap/pubs/20030903ftr.html>

Bishara, Marwan, 'America's Asymmetrical Wars: Following a Failed Israeli Military Doctrine'. Report from a Palestine Center briefing at: <www.palestinecenter.org/cpap/pubs/20040112ftr.html>

Bloxham, Donald, 'The Armenian Genocide of 1915–1916: Cumulative Radicalization and the Development of a Destruction Policy', *Past and Present*, 181 (2003), 141–91.

Bodansky, Yossef, *Bin Laden: The Man Who Declared War on America* (New York: Random House, 2001).

Bonney, Richard, Introduction to Fateh Muḥammed Malik, *Iqbal's Reconstruction of Political Thought in Islām* (Leicester: INPAREL South Asian History Academic Papers, 6, 2002; repr. New Delhi: Media House, 2004).

Bonney, R. J., '"For God, Fatherland and Freedom": Rethinking Pluralism in Hungary in the Era of Partition and Rebellion, 1526–1711', in *The First Millennium of Hungary in Europe*, ed. K. Papp, J. Barta et al. (Debrecen: Debrecen University Press, 2002), 377–96.

Bonney, R. J., 'Impossible to Reconcile? Christian Just War Theory and the Second Iraq War', *Encounters. Journal of Inter-cultural Perspectives*, 9 (March 2003), 69–91.

Bonney, R. J., *Understanding and Celebrating Religious Diversity. The Growth of Diversity in Leicester's Places of Religious Worship since 1970* (University of Leicester, Studies in the History of Religious and Cultural Diversity, 1, 2003).

Bonney, R. J., 'Reflections on the Differences between Religion and Culture', *Clinical Cornerstone*, 6:1 (2004), 25–33.

Bonney, Richard, *Three Giants of South Asia: Gandhi, Ambedkar and Jinnah on Self-Determination* (Leicester: INPAREL South Asian History Academic Papers, 5, 2002; New Delhi: Media House, 2004).

Borra, Ranjan, 'Subhas Chandra Bose, l'armée nationale indienne et la guerre de libération de l'Inde' at: <www.angelfire.com/folk/library/bose2_fr.htm>

Bose, Sumantra, *Kashmir: Roots of Conflict, Paths to Peace* (Cambridge, MA: Harvard University Press, 2003).

Bowker, J. (ed.), *Oxford Dictionary of World Religions* (Oxford: Oxford University Press, 1997).

Boykin, John, *Cursed is the Peacemaker: The American Diplomat Versus the Israeli General, Beirut 1982* (Belmont, CA: Applegate Press, 2002).

Brown, L. Carl, *Religion and the State. The Muslim Approach to Politics* (New York: Columbia University Press, 2000).

Bucaille, Laetitia, *Growing Up Palestinian. Israeli Occupation and the Intifada Generation*, trans. Anthony Roberts, (Princeton, NJ, and Oxford: Princeton University Press, 2004).

Buckler, F. W., 'The Political Theory of the Indian Mutiny', *Transactions of the Royal Historical Society*, 4th ser. 5 (1922), 71–100.

Buckler, F. W., 'A Rejoinder to "The Political Theory of the Indian Mutiny" by Douglas Dewar and H. L. Garrett', *Transactions of the Royal Historical Society*, 7 (1924), 160–5.

Bugaje, Usman Muḥammad, 'A Comparative Study of the Movements of 'Uthman Dan Fodio in Early Nineteenth-Century Hausaland and Muḥammad Aḥmad al-Mahdī in Late Nineteenth-Century Sudan' at: <www.webstar.co.uk/~ubugaje/comparativestudy1.pdf>

Bugaje, Usman Muḥammad, 'The Contents, Methods and Impact of Shehu Usman Dan Fodio's Teachings, 1774–1804' at: <www.webstar.co.uk/~ubugaje/udfcontmethimpact.pdf>

Bulliet, R. W., *Conversions to Islam in the Medieval Period: An Essay in Quantitative History* (Cambridge, MA: Harvard University Press, 1979).

Bunt, Gary R., *Islām in the Digital Age. E-Jihād, Online Fatwās and Cyber Islamic Environments* (London and Sterling, VA: Pluto Press, 2003).

Burgat, François, *Face to Face with Political Islām* (London and New York: I. B. Tauris, 2003).

Burke, Jason, *Al-Qaeda: Casting a Shadow of Terror* (London and New York: I. B. Tauris, 2003).

Chandra, Satish, *Parties and Politics and the Mughal Court, 1707–1740* (New Delhi: Oxford University Press, 1959, repr. 2002).

Coates, A. J., *The Ethics of War* (Manchester and New York: Manchester University Press, 1997).

Cobban, Helen, 'The Growth of Shī'ī Power in Lebanon', in *Shī'īsm and Social Protest*, ed. Juan R. I. Cole and Nikki R. Keddie (New Haven, CT and London: Yale University Press, 1986), 137–55.

Cole, Juan, *Sacred Space and Holy War. The Politics, Culture and History of Shī'ite Islām* (London: I. B. Tauris, 2002).

Coll, Steve, *Ghost Wars. The Secret History of the CIA, Afghanistan and Bin Laden from the Soviet Invasion to September 10, 2001* (New York: Penguin Press, 2004).

Conetta, Carl, 'Strange Victory: A Critical Appraisal of Operation Enduring Freedom and the Afghanistan War', *Project on Defense Alternatives*, Research Monograph 6 (30 January 2002) at: <www.comw.org/pda/0201strangevic.html>

Cook, Michael, *Commanding Right and Forbidding Wrong in Islamic Thought* (New York: Cambridge University Press, 2000).

Cook, Michael, *Forbidding Wrong in Islam: An Introduction* (Cambridge: Cambridge University Press, 2003).

Cox, C. and Marks, John, *The 'West', Islam and Islamism. Is Ideological Islam Compatible with Liberal Democracy?* (London: Civitas, 2003).

Crone, Patricia, *Meccan Trade and the Rise of Islām* (Oxford: Blackwell, 1987).

Crone, Patricia, *Medieval Islamic Political Thought* (Edinburgh: Edinburgh University Press, 2004).

Daftary, Farhad, *A Short History of the Ismā'īlīs. Traditions of a Muslim Community* (Edinburgh: Edinburgh University Press, 1998).

Daftary, Farhad, 'Ḥasan-i Ṣabbāḥ and the Origins of the Nizārī Movement', in *Mediaeval Ismā'īlī History and Thought*, ed. Farhad Daftary (Cambridge: Cambridge University Press, 1998).

Dale, William N., 'Cursed is the U.S. Envoy who Tries to Bring Peace to the Middle East', in *American Diplomacy* at: <www.unc.edu/depts/diplomat/archives_roll/2002_07-09/book_sept02/book_dale_peace.html>

Damrel, David, 'Ṣūfī Warriors. The Religious Roots of Conflict: Russia and Chechnya' at: <www.mubai.cc/articles/art34.htm>

Darwin, John, *Britain, Egypt and the Middle East: Imperial Policy in the Aftermath of War, 1918–1922* (London: Macmillan, 1981).

Darwin, John, *Britain and Decolonisation. The Retreat from Empire in the Post-War World* (Basingstoke and London: Macmillan, 1988).

Davis, Joyce M., *Between Jihād and Salaam. Profiles in Islām* (Basingstoke: Macmillan, repr. 1999).

Davis, Joyce M., *Martyrs. Innocence, Vengeance and Despair in the Middle East* (New York: Palgrave Macmillan, 2003).

Dekmejian, R. Hrair, *Islām in Revolution. Fundamentalism in the Arab World* (Syracuse, NY: Syracuse University Press, 1985).

DeLong Bas, Natana, *Wahhābī Islām. From Revival and Reform to Global Jihād* (New York: Oxford University Press, 2004).

Dewar, Douglas and Garrett, H. L., 'A Reply to Mr. F. W. Buckler's "The Political Theory of the Indian Mutiny"', *Transactions of the Royal Historical Society*, 7 (1924), 131–59.

Duffy, Eamon, *The Stripping of the Altars. Traditional Religion in England, 1400–1580* (New Haven, CT and London: Yale University Press, 1992).

Eaton, Richard M., *Essays on Islam and Indian History* (New Delhi: Oxford University Press, 2000).

Edwards, David B., *Before Taliban: Genealogies of the Afghan Jihād* (Berkeley, CA: University of California Press, *c.* 2002) at: <www.ark.cdlib.org/ark:/13030/ft3p30056w/>

Ehrenkreutz, Andrew S., 'Studies in the Monetary History of the Near East in the Middle Ages. II. The Standard of Fineness of Western and Eastern *Dīnārs* Before the Crusades', *Journal of the Economic and Social History of the Orient*, 6 (1963), 243–77.

El-Affendi, Abedlwahab, *Who Needs an Islamic State?* (London: Grey Seal Books, 1991).

El-Affendi, Abdelwahab (ed.), *Rethinking Islām and Modernity. Essays in Honour of Fathi Osman* (Leicester: Islamic Foundation, 2001).

Elpeleg, Zvi, *The Grand Muftī of Jerusalem, Haj[j] Amin al-Hussaīni, Founder of the Palestinian National Movement* (London: Frank Cass, 1993).

Enayat, Hamid, *Modern Islamic Thought. The Response of the Shī'ī and Sunnī Muslims to the Twentieth Century* (London and Basingstoke: Macmillan, 1982).

Engineer, Ashgar Ali, 'Evolution of *Sharī'a* Law and its Potentiality for Change', *Islam and the Modern Age* (June 2004) at: <www.csss-isla.com/IIS/index.php>

Erskine, William, and King, Lucas, (eds), *Memoirs of Zehīr-ed-Dīn Muhammed Bābur* (2 vols) (Oxford: Oxford University Press, 1921; repr. New Delhi: Vintage Books, 1993).

Esposito, John L. (ed.), *The Oxford Encyclopaedia of the Islamic World* (4 vols) (New York: Oxford University Press, 1995).

Esposito, John L., *Unholy War. Terror in the Name of Islām* (Oxford: Oxford University Press, 2002).

Esposito, John L., *What Everyone Needs to Know about Islām* (New York: Oxford University Press, 2002).

Esposito John L. and Voll, John O., *Islām and Democracy* (New York: Oxford University Press, 1996).

Esposito John L. and Voll, John O., 'Islām and the West. Muslim Voices of Dialogue', in *Religion in International Relations. The Return from Exile*, ed. Fabio Petito and Pavlos Hatzopoulos (New York: Palgrave Macmillan, 2003), 236–69.

Euben, Roxanne L., *Enemy in the Mirror: Islamic Fundamentalism and the Limits of Modern Rationalism: A Work of Comparative Political Theory* (Princeton, NJ: Princeton University Press, 1999).

Evans-Pritchard, E. E., *The Sanusi of Cyrenaica* (Oxford: Oxford University Press, 1949).

Faqīh, 'Arab, *Futūḥ al-Ḥabaša. The Conquest of Abyssinia (16th Century)*, trans. Paul Lester Stenhouse (Hollywood, CA: Tsehai Publishers, 2003).

Farouk Mitha, *Al-Ghazālī and the Ismailis. A Debate on Reason and Authority in Medieval Islam* (London: I. B. Tauris, 2001).

Farrukh, Omar A., *Ibn Taymīyah on Private and Public Law in Islam or Public Policy in Islamic Jurisprudence* (Beirut: Khayats, 1966).

'*Fatwā* on the Heresy of Wahhābism' at: <www.hizmetbooks.org/Advice_for_the_Muslim/wah-31.htm>

Firestone, R., *Jihād. The Origin of Holy War in Islam* (Oxford: Oxford University Press, 1999).

Fischer-Galati, F. A., *Ottoman Imperialism and German Protestantism, 1521–1555* (2nd edn) (New York: Octagon Books, 1972).

Fleischer, Cornell H., '*Mahdī* and Millennium: Messianic Dimensions in the Development of Ottoman Imperial Ideology', in *The Great Ottoman Turkish Civilization. III. Philosophy, Science and Institutions*, ed. Kemal Çiçek (Ankara: Yeni Türkiye, 2000).

Foster, William (ed.), *The Embassy of Sir Thomas Roe to the Court of the Great Mogul, 1615–1619, as Narrated in his Journal and Correspondence* (2 vols) (London: Hakluyt Society, 1899; repr. Leichtenstein: Klaus reprints, 1967).

Friedmann, Yohanan, *Tolerance and Coercion in Islam. Inter-faith Relations in the Muslim Tradition* (Cambridge: Cambridge University Press, 2003).

Fromkin, David, *A Peace to End All Peace: The Fall of the Ottoman Empire and the Creation of the Modern Middle East* (London: Phoenix, 2003).

Fuller, Graham E., *The Future of Political Islām* (Basingstoke: Palgrave Macmillan, 2003).

Fuller, Graham E. and Starr, S. Frederick, 'The Xinjiang Problem', Central Asia-Caucasus Institute, Paul H. Nitze School of Advanced International Studies, Johns Hopkins University (n.d.) at: <www.cornellcaspian.com/pub2/xinjiang_final.pdf>

Fyzee, Asaf A. A. (trans.), *The Pillars of Islam. Da'ā'im al-Islām of al-Qāḍī al-Nu'mān. I. Acts of Devotion and Religious Observances*, revised by Ismail Kurban Husein Poonawala (New Delhi: Oxford University Press, 2002).

Galston, Miriam, *Politics and Excellence. The Political Philosophy of Alfarabi* (Princeton, NJ: Princeton University Press, 1990).

Ganguly, Šumit, *The Crisis in Kashmir. Portents of War, Hopes of Peace* (Cambridge: Cambridge University Press and Woodrow Wilson Center Series, 1997).

Gasiorowski, Mark J. and Byrne, Malcolm, *Mohammad Mosaddeq and the 1953 Coup in Iran* (Syracuse, NY: Syracuse University Press, 2004).

Gibb, H. A. R., *Studies on the Civilization of Islam*, ed. S. J. Shaw and W. R. Polk (London: Routledge, 1962).

Gibb, H. A. R., and Kramers, J. H. (eds), *Concise Encyclopedia of Islām* (Leiden: E. J. Brill, 2001).

Glassé, C., *The Concise Encyclopedia of Islām* (London: Stacey International, 1966).

Goffman, Daniel, *The Ottoman Empire and Early Modern Europe* (Cambridge: Cambridge University Press, 2002).

Goitein, D., 'Evidence on the Muslim Poll Tax, from Non-Muslim Sources: A Geniza Study', *Journal of the Economic and Social History of the Orient*, 6 (1963), 278–95.

Goldsmith, Raymond W., *Premodern Financial Systems. A Historical Comparative Study* (Cambridge: Cambridge University Press, 1987).

Goldziher, Ignaz, *Introduction to Islamic Theology and Law*, trans. A. Hamori and R. Hamori (Princeton, NJ: Princeton University Press, 1981).

Gomez, Michael A., 'The Problem with Malik Si and the Foundation of Bundu', *Cahiers d'Études africaines*, 100 (1985).

Gomez, Michael A., *Pragmatism in the Age of Jihād: The Precolonial State of Bundu* (New York: Cambridge University Press, 1992).

Goodson, Larry P., *Afghanistan's Endless War. State Failure, Regional Politics and the Rise of the Taliban* (Seattle and London: University of Washington Press, 2001).

Gopin, Marc, *Holy War, Holy Peace. How Religion Can Bring Peace to the Middle East* (New York: Oxford University Press, 2002).

Graham, William A., *Divine Word and Prophetic Word in Early Islām. A Reconsideration of the Sources, with Special Reference to the Divine Saying or Ḥadīth Qudsī* (The Hague: Mouton, 1977).

Gray, John, *Al-Qaeda and What it Means to be Modern* (London: Faber, 2003).

Gunaratna, Rohan, *Inside Al-Qaeda: Global Network of Terror* (London: Hurst and Co., 2002).

Habib, Irfan (ed.), *Confronting Colonialism: Resistance and Modernization under Haidar 'Alī and Tīpū Sultan* (orig. edn 1999; repr. London: Anthem Press, 2002).

Habib, John S., *Ibn Sa'ud's Warriors of Islam. The Ikhwān of Najd and their Role in the Creation of the Sa'udi Kingdom, 1910–1930* (Leiden: E. J. Brill, 1978).

Hadawi, Sami, *Palestinian Rights and Losses in 1948. A Comprehensive Study. V. An Economic Assessment of Total Palestinian Losses. Written by Dr Atef Kubursi* (London: Saqi Books, 1988).

Haddad, Yvonne Yazbeck, 'Sayyid Quṭb: Ideologue of Islamic Revival', in *Voices of Resurgent Islam*, ed. J. L. Esposito (New York and Oxford: Oxford University Press, 1983).

Haddad, Yvonne Yazbeck, 'Islamism: A Designer Ideology for Resistance, Change and Empowerment', in *Muslims and the West. Encounter and Dialogue*, ed. Zafar Ishaq Ansari and John L. Esposito (Islamabad: Islamic Research Institute, Islamabad and Center for Christian–Muslim Understanding, Washington DC, repr. 2002), 274–95.

Hafez, Kai (ed.), *The Islamic World and the West. An Introduction to Political Cultures and International Relations* (Leiden: E. J. Brill, 2000).

Haji-Yousefi, Amir M., 'Foreign Policy of the Islamic Republic of Iran Towards Israel, 1979–2002', *Strategic Studies,* Quarterly Journal of the Institute of Strategic Studies, Islamabad, 23 (2003), 55–75.

Halbach, Uwe, '"Holy War" against Czarism: The links between Ṣūfīsm and *Jihād* in the Nineteenth-Century Anti-Colonial Resistance against Russia', in *Muslim Communities Re-emerge: Historical Perspectives on Nationality, Politics, and Opposition in the Former Soviet Union and Yugoslavia*, ed. A. Kappeler, G. Simon and G. Brunner (Durham, NC: Duke University Press, 1994), 251–73.

Haleem, H. A., Ramsbotham, O., Risaluddin, S. and Wicker, B. (eds), *The Crescent and the Cross. Muslim and Christian Approaches to War and Peace* (Basingstoke: Macmillan, 1998).

Halm, Heinz, 'The Ismā'īlī Oath of Allegiance and the 'Sessions of Wisdom', in *Medieval Ismā'īlī History and Thought*, ed. Farhad Daftary (Edinburgh: Edinburgh University Press, 1998).

Hamidullah, Muḥammad (trans.), *The First Written Constitution in the World. An Important Document of the Time of the Holy Prophet* (Lahore: Sh. Muḥammad Ashraf, 2nd rev. edn, 1968).

Hamidullah, Muḥammad, *Muslim Conduct of the State, Being a Treatise on Siyar* (rev. 7th edn, Lahore: Sh. Muḥammad Ashraf, 1977).

Hammes, Thomas X., 'The Evolution of War: The Fourth Generation', *Marine Corps Gazette* (September 1994) at: <www.d-n-i.net/fcs/hammes.htm>

Hamzeh, A. Nizar, 'Islamism in Lebanon: A Guide to the Groups', *Middle East Quarterly*, 4 (1997) at: <www.meforum.org/article/362>

Hamzeh, A. Nizar, and Dekmejian, R. Hrair, 'A Ṣūfī Response to Political Islamism: Al-Ahbash of Lebanon', *International Journal of Middle East Studies,* 28 (1996), 217–29 at: <www.almashriq.hiof.no/ddc/projects/pspa/al-ahbash.html>

Hanafī, H., 'Global Ethics and Human Solidarity. An Islamic Approach', in H. Hanafī, *Islam in the Modern World. II. Tradition, Revolution and Culture* (Cairo: Dar Kebaa, 2000).

Hanafī, H., 'Islam, Religious Dialogue and Liberation Theology', in H. Hanafī, *Islam in the Modern World. II. Tradition, Revolution and Culture* (Cairo: Dar Kebaa, 2000).

Hanafī, H., 'The Preparation of Societies for Life in Peace. An Islamic Perspective', in H. Hanafī, *Islam in the Modern World. II. Tradition, Revolution and Culture* (Cairo: Dar Kebaa, 2000).

Haq, Farhat, '*Jihād* over Human Rights, Human Rights as *Jihād*. Clash of Universals', in *Negotiating Culture and Human Rights*, ed. Lynda S. Bell, Andrew J. Nathan, Ilan Peleg (New York: Columbia University Press, 2001), 242–57.

Haque, Serajul, 'Ibn Taymīyah', in *A History of Muslim Philosophy*, ed. M. M. Sharif. Pakistan Philosophical Congress, nd. ch. 41, 796–819 at: <www.muslimphilosophy.com/hmp/default.htm>

Harik, Judith Palmer, *Hezbollah. The Changing Face of Terrorism* (London and New York: I. B. Tauris, 2004).

Hartung, Jan-Peter, Hawkes, Gillian, and Bhattacharjee, Anuradha, *Ayodhya, 1992–2003: The Assertion of Cultural and Religious Hegemony* (Leicester, 2003; Delhi: Media House, 2004).

Harvey, Robert, *Global Disorder. How to Avoid a Fourth World War* (London: Robinson, 2003).

Ḥasan, Aḥmad, *The Early Development of Islamic Jurisprudence* (Islamabad: Islamic Research Institute, 1970).

Hāshmī, Sohail H. (ed.), 'Interpreting the Islamic Ethics of War and Peace', in *Islamic Political Ethics. Civil Society, Pluralism and Conflict*, ed. Sohail Hāshmī (Princeton, NJ: Princeton University Press, 2002).

Hashmi, Solail H. (ed.), *Islamic Political Ethics. Civil Society, Pluralism and Conflict* (Princeton, NJ: Princeton University Press, 2002).

Hassan, Riffat, 'Rights of Women within Islamic Communities', in *Religious Human Rights in Global Perspective. Religious Perspectives*, ed. John Witte Jr and Johan D. van der Vyver (The Hague: Martinus Nijhoff, 1996), 361–85.

Hatin, Meir, *Islam and Salvation in Palestine. The Islamic Jihad Movement* (Tel Aviv: Tel Aviv University, Moshe Dayan Center for Middle Eastern and African Studies, 127, 2001).

Hefner, Robert W., *Civil Islām. Muslims and Democratization in Indonesia* (Princeton, NJ, and Oxford: Princeton University Press, 2000).

Henzell-Thomas, Jeremy, *The Challenge of Pluralism and the Middle Way of Islām*. Association of Muslim Social Scientists (UK). Occasional Paper Series 1 (AMSS UK, 2002).

Hermansen, Marcia K., *The Conclusive Argument from God. Shāh Walī Allāh of Delhi's Ḥujjat Allāh al-Bāligha* (Leiden: E. J. Brill, 1996).

Hinds, M., 2003, *God's Caliph. Religious Authority in the First Centuries of Islam* (Cambridge: Cambridge University Press, 2003).

Hiskett, Mervyn, *The Sword of Truth. The Life and Times of the Shehu Usuman Dan Fodio* (New York: Oxford University Press, 1973).

Holt, P. M., *The Mahdist State in the Sudan, 1881–1898: A Study of its Origins, Development and Overthrow* (Oxford: Clarendon, 1958; 2nd edn Nairobi, 1979).

Holt, P. M., Lambton, Ann K. S., and Lewis, Bernard (eds), *The Cambridge History of Islām* (Cambridge: Cambridge University Press, 1970).

Housley, N. J., *The Later Crusades, 1274–1580: From Lyons to Alcazar* (Oxford: Oxford University Press, 1992).

Housley, N. J., *Religious Warfare in Europe, 1400–1536* (Oxford: Oxford University Press, 2002).

Humaid, Sheikh Abdullah bin Muḥammad bin, '*Jihād* in the Qur'ān and Sunnah' at: <www.islamworld.net/jihad.html>

Human Rights Watch, 'Endless Torment. The 1991 Uprising in Iraq and Its Aftermath' (1992) at: <www.hrw.org/reports/1992/Iraq926.htm>

Human Rights Watch, 'In the Name of Counter-Terrorism: Human Rights Abuses Worldwide. A Human Rights Watch Briefing Paper for the 59th Session of the United Nations Commission on Human Rights' (25 March 2003) at: <www.hrw.org/un/chr59/counter-terrorism-bck4.htm#P286_64797>

Hume, Basil, O.S.B., *To be a Pilgrim. A Spiritual Notebook* (Slough: St Paul Publications, 1968).

Humphreys, R. Stephen, *From Saladin to the Mongols. The Ayyubids of Damascus, 1193–1260* (Albany, NY: State University of New York Press, 1977).

Hunter, W. W., *The Indian Musalmans* (London: Trubner and Co., 2nd edn, 1872).

Huntington, Samuel P., 'The Clash of Civilizations?', *Foreign Affairs*, 72 (1993), 22–49.

Huntington, Samuel P., *The Clash of Civilizations? The Debate* (New York: Foreign Affairs, 1996).

Huntington, Samuel P., *The Clash of Civilizations and the Remaking of the World Order* (London: The Free Press, 2002).

Husain, S. and Jafri, M., *Origins and Early Development of Shī'a Islām* (London and New York: Longman, 1979).

Hussain, Asaf, *Islamic Iran. Revolution and Counter-Revolution* (London: Pinter, 1985).

Hussain, S. Iftikhar, *Some Major Pukhtoon Tribes along the Pak–Afghan Border* (Peshawar: Area Studies Centre and Hans Seidl Foundation, 2000).

Hussein, Askary, 'Lessons to be Learned: 'Irāqī Resistance to British Occupation 80 Years Ago', *Executive Intelligence Review* (14 November 2003) at: <www.larouchepub.com/other/2003/3044iraq_history.html>

Huweidi, Fahmi, 'Non-Muslims in Muslim Society', in *Rethinking Islām and Modernity. Essays in Honour of Fathi Osman*, ed. Abdelwahab El-Affendi (Leicester: Islamic Foundation, 2001), 84–91.

Ibarra, Miguel Angel de Bunes, 'Kanuni Sultan Süleyman, Barbaros Pasha and Charles V: The Mediterranean World', in *The Great Ottoman Turkish Civilization. I. Politics*, ed. Kemal Çiçek (Ankara: Yeni Türkiye, 2000).

Ibn Morgan, Salim Abdallah (trans. and ed.), 'The Criterion between the Allies of the Merciful and the Allies of the Devil' at: <www.java-man.com/Pages/Books/criterion.html>

Ibn Taymīyah, Taqī al-Dīn Aḥmad, *Public Duties in Islām. The Institution of the Ḥisba*, trans. Muhtar Holland (Leicester: Islamic Foundation, 1982).

Ibn Taymīyah, Taqī al-Dīn Aḥmad, *Enjoining Right and Forbidding Wrong*, trans. Salim Abdallah Ibn Morgan: <www.java-man.com/Pages/Books/alhisba.html>

Ibn Taymīyah, Taqī al-Dīn Aḥmad, *The Religious and Moral Doctrine of Jihād* (Birmingham: Maktabah Al Ansaar Publications, 2001).

ICG Asia Report, 'Unfulfilled Promises: Pakistan's Failure to Tackle Extremism'. Report No. 73 (16 January 2004).

Iṣlāḥī, Mawlānā Amīn Aḥsan, 'Self-Development in the Context of Man's Relationship with Allāh', in *Tazkiyah. The Islamic Path to Self-Development*, ed. Abdur Rashid Siddiqui (Leicester: The Islamic Foundation, 2004), 133–214.

Ja'fariyan, Rasul, 'Shī'ism and its Types During the Early Centuries' at: <www.al-islam1.org/al-tawhid/types/shiism.htm>

Jackson, Sherman A., *Islamic Law and the State: the Constitutional Jurisprudence of Shihāb al-Dīn al-Qarāfī* (Leiden and New York: E. J. Brill, 1996).

Jansen, Johannes J. G., *The Neglected Duty. The Creed of Sadat's Assassins and Islamic Resurgence in the Middle East* (New York: Macmillan, 1986).

Johnson, James Turner, *The Holy War Idea in Western and Islamic Traditions* (Philadelphia, PA: Pennsylvania State University Press, 1997; repr. 2002).

Johnson, Nels, *Islām and the Politics of Meaning in Palestinian Nationalism* (London: Kegan Paul International, 1982).

Kafadar, Cemal, *Between Two Worlds. The Construction of the Ottoman State* (Berkeley, CA: University of California Press, 1995).

Kamen, H., 'The Mediterranean and the Expulsion of Spanish Jews in 1492', *Past and Present*, 119 (1988), 30–55.

Karam, Azzam (ed.), *Transnational Political Islām. Religion, Ideology and Power* (London and Sterling, VA: Pluto Press, 2004).

Karpat, Kemal H., *An Inquiry into the Social Foundations of Nationalism in the Ottoman State: From Social Estates to Classes, from Millets to Nations* (Princeton, NJ: Princeton University Press, 1973).

Karpat, Kemal H., 'Ottoman Migration: Ethnopolitics and the Formation of Nation-States', in *The Great Ottoman Turkish Civilization. I. Politics*, ed. Kemal Çiçek (Ankara: Yeni Türkiye, 2000), 382–98.

Karpat, Kemal H., *The Politicization of Islam: Reconstructing Identity, State, Faith and Community in the Late Ottoman State* (New York: Oxford University Press, 2001).

Karsh, Efraim, *The Iran–Iraq War, 1980–1988* (Wellingborough: Osprey, 2002).

Keddie, Nikki R., *Modern Iran. Roots and Results of Revolution* (New Haven, CT, and London: Yale University Press, 2003).

Keller, Shoshana, *To Moscow, not Mecca. The Soviet Campaign against Islām in Central Asia, 1917–1941* (Westport, CT, and London: Praeger, 2001).

Kennedy, Hugh, *The Prophet and the Age of the Caliphates. The Islamic Near East from the Sixth to the Eleventh Century* (London and New York: Longman, 1986).

Kennedy, Hugh, *The Armies of the Caliphs. Military and Society in the Early Islamic State* (London and New York: Routledge, 2001).

Kepel, Gilles, *Jihād. The Trail of Political Islām* (London and New York: I. B. Tauris, 2002).

Khaddūrī, Majīd, *War and Peace in the Law of Islām* (Baltimore, MD, and London: Johns Hopkins University Press, 1955).

Khaddūrī, Majīd (ed.), *Islamic Jurisprudence. Shāfi'ī's Risāla* (Baltimore, MD: Johns Hopkins University Press, 1961).

Khaddūrī, Majīd (trans. and ed.), *The Islamic Law of Nations. Shaybānī's Siyar* (Baltimore, MD: Johns Hopkins University Press, 1966).

Khaddūrī, Majīd, *The Islamic Conception of Justice* (Baltimore, MD: Johns Hopkins University Press, 1984).

Khan, Ayaz Ahmed, 'Terrorism and Asymmetrical Warfare: International and Regional Implications': at: <www.defencejournal.com/2002/february/terrorism.htm>

Khan, Qamaruddin, *Political Concepts in the Qur'ān* (Karachi: Institute of Islamic Studies, 1973).

Khan, Qamaruddin, *Al-Māwardī's Theory of the State* (New Delhi: Idarah-i-Adabiyat-i-Delli, repr. 1979).

Khan, Qamarrudin, *Political Concepts in [the] Sunnah. A Treatise on the Political Concepts of the Holy Prophet*, ed. H. M. Arshad Qureshi (Lahore, Islamabad and Washington DC: Islamic Book Foundation, 1988).

Khan, Zafarul-Islam (ed.), *Palestine Documents* (New Delhi: Pharos Media, 1998).

Khatib, M. M., *The Bounteous Korān. A Translation of Meaning and Commentary,* authorized by al-Azhar (London: Macmillan, 1986).

Kiliç-Schubel, Nurten, 'Unity and Diversity in Political Culture. Muslim Empires of Sixteenth-Century Eurasia: Ottomans, Mughals, Safavids and Uzbeks', in *The Great Ottoman Turkish Civilization. I. Politics*, ed. Kemal Çiçek (Ankara: Yeni Türkiye, 2000), 275–84.

Kim, Hodong, *Holy War in China. The Muslim Rebellion and State in Chinese Central Asia, 1864–1877* (Stanford, CA: Stanford University Press, 2004).

Kimmerling, Baruch, and Migdal, Joel S., *The Palestinian People: A History* (Cambridge, MA: Harvard University Press, 2003).

Kister, M. J., *Concepts and Ideas at the Dawn of Islam* (Aldershot: Ashgate, Variorum, 1997)

Klein, Janet, 'Power in the Periphery: The Hamidiye Light Cavalry and the Struggle over Ottoman Kurdistan, 1890–1914' (unpublished PhD thesis, Princeton University, 2002). Abstract at: <www.princeton.edu/~klein/dissabstract.html>

Knecht, R. J., *Renaissance Warrior and Patron: The Reign of Francis I* (Cambridge: Cambridge University Press, 1994).

Knolles, Richard, *The generall historie of the Turkes, from the first beginning of that nation to the rising of the Othoman familie: with all the notable expeditions of the Christian princes against them: together with The lives and conqvests of the Othoman kings and emperours* (5th edn) (London: Adam Islip, 1638).

Kolinsky, Martin, *Law, Order and Riots in Mandatory Palestine, 1928–35* (Basingstoke: Macmillan, 1993).

Kotb, Sayyid [Quṭb], *Social Justice in Islam*, trans. J. B. Hardie (American Council of Learned Societies, 1953).

Kramer, Martin, 'The Moral Logic of Hizballah', in *Origins of Terrorism: Psychologies, Ideologies, Theologies, States of Mind*, ed. Walter Reich (Cambridge: Cambridge University Press, 1990), 131–57 at: <www.geocities.com/martinkramerorg/MoralLogic.htm>

Kramer, Martin, 'Sacrifice and "Self-Martrydom" in Shi'ite Lebanon', *Terrorism and Political Violence*, vol. 3, no. 3 (Autumn 1991), 30–47; revised in Martin Kramer, *Arab Awakening and Islamic Revival* (New Brunswick, NJ: Transaction Publishers, 1996), 231–43. Web version at <www.martinkramer.org/pages/899526/index.htm>

Kramer, Martin, 'Hizbu'llah: The Calculus of *Jihād*', in *Fundamentalisms and the State: Remaking Polities, Economies, and Militance* (The Fundamentalism Project, vol. 3), ed. M. Marty and R. S. Appleby (Chicago, IL: University of Chicago Press, 1993), 539–56 at: <www.geocities.com/martinkramerorg/Calculus.htm>

Kramer, Martin, 'The Oracle of Hizbu'llah: Sayyid Muhammad Husayn Fadlallah', in *Spokesmen for the Despised. Fundamentalist Leaders of the Middle East*, ed. R. Scott Appleby (Chicago, IL: University of Chicago Press, 1997), 83–181 at: <www.geocities.com/martinkramerorg/Oracle1.htm>

Kramer, Martin, 'Coming to Terms: Fundamentalists or Islamists?', *Middle East Quarterly*, 10 (Spring 2003) at: <www.meforum.org/article/541>

Krayem, Hassan, 'The Lebanese Civil War and the Tai'f Agreement' (American University of Beirut, n.d.) at: <www.ddc.aub.edu.lb/projects/pspa/conflict-resolution.html>

Kritzeck, James and Lewis, William H., *Islam in Africa* (New York: Van Nostrand–Reinhold, 1969).

Kunt, Metin, and Woodhead, Christine (eds), *Süleyman the Magnificent and his Age. The Ottoman Empire in the Early Modern World* (London: Longman, 1995).

Kuran, Ercüment, 'Maghreb History During the Ottoman period', in *The Great Ottoman Turkish Civilization. I. Politics*, ed. Kemal Çiçek (Ankara: Yeni Türkiye, 2000).

Kux, Dennis, *The United States and Pakistan, 1947–2000. Disenchanted Allies* (Baltimore, MD, and London: Johns Hopkins University Press and Woodrow Wilson Center Press, Washington DC, 2001).

Lalani, Arzina R., *Early Shī'ī Thought. The Teachings of Imām Muḥammad al-Bāqir* (London: I. B. Tauris and Institute of Ismaili Studies, 2000).

Lambton, A. K. S., *State and Government in Medieval Islam. An Introduction to the Study of Islamic Political Theory: The Jurists* (New York: Oxford University Press, 1981).

Lapidus, Ira M., A *History of Islamic Societies* (2nd edn) (Cambridge: Cambridge University Press, 2002).

Lepré, George, *Himmler's Bosnian Division: The Waffen–SS Handschar Division, 1943–1945* (Atglen, PA: Schiffer Military History, 1997).

Levtzion, Nehemia, *Muslims and Chiefs in West Africa. A Study of Islam in the Middle Volta Basin in the Pre-Colonial Period* (Oxford: Clarendon Press, 1968).

Lewis, Bernard, *The Assassins. A Radical Sect in Islām* (London: Weidenfeld and Nicolson, 1967).

Lewis, Bernard, *The Crisis of Islam. Holy War and Unholy Terror* (London: Weidenfeld and Nicolson, 2003).

Lewis, I. M., *Islam in Tropical Africa* (London: Hutchinson, 1969, repr. 1980).

Lyons, Malcolm C., and Jackson, David E. P., *Saladin. The politics of the Holy War* (Cambridge: Cambridge University Press, 1982, repr. 1984).

Madelung, Wilferd, 'The Fāṭimids and the Qarmaṭīs of Baḥrayn', in *Medieval Ismā'īlī History and Thought*, ed. Farhad Daftary (Edinburgh: Edinburgh University Press, 1998).

Madelung, Wilferd, *The Succession to Muḥammad. A Study of the Early Caliphate* (Cambridge: Cambridge University Press, 1997, repr. 2001).

Majumdar, Suhas, *Jihād: The Islamic Doctrine of Permanent War* (New Delhi: Voice of India, 1994).

Makdisi, George, 'Ibn Taymīya: A Ṣūfī of the Qādirīya Order', *American Journal of Arabic Studies*, 1 (1974), 118–29.

Makdisi, G., 'Hanbalite Islam', in *Studies on Islam*, ed. and trans. Merlin L. Swartz (New York: Oxford University Press, 1981).

Malik, I., Noshab, F. and Abdullah, S., '*Jihād* in the Modern Era: Image and Reality', *Islamabad Papers* 18 (Institute of Strategic Studies, Islamabad, 2001).

Man, John, *Genghis Khan. Life, Death and Resurrection* (London and New York: Bantam Press, 2004).

Manzoor, S. Parvez, 'Against the Nihilism of Terror: *jihād* as Testimony to Transcendence', *Muslim World Book Review*, 22:3 (April–June 2002), 5–14.

Martin, Vanessa, *Creating an Islamic state. Khoemeini and the Making of a New Iran* (London and New York: I. B. Tauris, 2003).

Masud, Khalid, 'Iqbal's Lecture on *Ijtihād*', in *Selections from the Iqbal Review*, ed. W. Qureshi (Lahore: Iqbal Academy, Pakistan, 1983), 109–117.

Masud, Khalid, 'Changing Concepts of *Jihād*', unpublished paper dated 31 October 2003.

Matroudi, A. H. I., 'The Role of Ibn Taymīyah in the Ḥanbalī School of Law' (unpublished PhD University of Leeds, 1999).

Mattar, Philip, *The Mufti of Jerusalem. Al-Hajj Amin al-Husayni and the Palestine National Movement* (New York: Columbia University Press, 1988).

Mawdūdī, Sayyid Abu'l-A'la, *The Islamic Law and the Constitution*, trans. and ed. Khurshid Ahmad (Lahore: Islamic Publications, 1st edn, 1955; 2nd. edn, 1960).

Mawdūdī, Sayyid Abu'l-A'la, *Fundamentals of Islām [Khutabat*] (Lahore: Islamic Publications, 1975).

Mawdūdī, Sayyid Abu'l-A'la, *Human Rights in Islām*, trans. and ed. Khurshid Aḥmad (Leicester: Islamic Foundation, 1976; repr. 1993).

Mawdūdī, Sayyid Abu'l-A'la, '*Jihād fī Sabillah*: *Jihād* in Islām', trans. Khurshid Ahmad, ed. Huda Khattab (Birmingham: UK Islamic Mission Dawah Centre, 1995).

Mawdūdī, Sayyid Abul-A'la, *Towards Understanding the Qur'ān* (*Tafhīm al-Qur'ān*) (New Delhi: Markazi Maktaba Islami Publishers, 1998).

Mayer, Ann Elizabeth, *Islam and Human Rights. Tradition and Politics* (London; Boulder, CO, and San Francisco: Pinter and Westview Press, 1991).

Melville, Charles, 'The Role of the Ismā'īlīs in Mamlūk–Mongol Relations', in *Medieval Ismā'īlī History and Thought*, ed. Farhad Daftary (Edinburgh: Edinburgh University Press, 1998).

Michel, Thomas F. (trans. and ed.), *A Muslim Theologian's Response to Christianity. Ibn Taymiyya's al-Jawāb al-Ṣaḥīḥ* (Delmar, NY: Caravan Books, 1984).

Miller, Ylana N., *Government and Society in Rural Palestine, 1920–1948* (Austin, TX: University of Texas Press, 1985).

Mishal, Shaul and Sela, Avraham, *The Palestinian Hamas. Vision, Violence and Coexistence* (New York: Columbia University Press, 2000).

Mitchell, Colin Paul, *Sir Thomas Roe and the Mughal Empire* (Karachi: Area Study Centre for Europe, 2000).

Mitha, Farouk, *Al-Ghāzalī and the Ismailis: A Debate on Reason and Authority in Medieval Islām.* Ismaili Heritage Series, 5 (London: I. B. Tauris in association with the Institute of Ismaili Studies, 2001).

Morabia, A., 'Ibn Taymiyya, dernier grand théoricien du *Ĝihâd* médiéval', *Bulletin d'études orientales*, 30 (1978), 85–99.

Morabia, Alfred, *Le Ĝihâd dans l'Islam médiéval. Le «combat sacré» des origins au xiie siècle* (Paris: Albin Michel, 1993).

Morgenthau, Henry, *Ambassador Morgenthau's Story* (New York: Doubleday, 1918) at: <www.cilicia.com/morgenthau/Morgen25.htm>

Moussalli, A. S., *Radical Islamic Fundamentalism: The Ideological and Political Discourse of Sayyid Quṭb* (Beirut: American University of Beirut, 1992).

Moussalli, A. S., *Moderate and Radical Islamic Fundamentalism. The Quest for Modernity, Legitimacy and the Islamic State* (Gainesville, FL: University of Florida Press, 1999).

Moussalli, A. S., *The Islamic Quest for Democracy, Pluralism and Human Rights* (Gainesville: University of Florida Press, 2001).

Muhaiyaddeen, Muḥammad Raheem Bawa, *Islām and World Peace. Explanations of a Ṣūfī* (Philadelphia, PA: Fellowship Press, 1987) at: <www.bmf.org/iswp/index.html>

Murad, Khurram Jah (trans.), *Let Us Be Muslims* (Leicester: Islamic Foundation, 1985).

Murad, Khurram, 'Islām and Terrorism', *Encounters: Journal of Inter-cultural Perspectives*, 4 (1998), 103–14 at: <www.robert-fisk.com/islam_and_terrorism_khurram_murad.htm>

Muṭahharī, Ayatullah Murtazá, *Fundamentals of Islamic Thought. God, Man and the Universe*, trans. R. Campbell (Berkeley, CA: Mizan Press, 1985).

Muṭahharī, Ayatullah Murtazá, '*Shahīd*', in *Jihād and Shahādat. Struggle and Martyrdom in Islām,* ed. Mehdi Abedi and Gary Legenhausen (Houston, TX: Institute for Research and Islamic Studies, 1986).

Muṭahharī, Ayatullah Murtazá, *Jihād: The Holy War of Islām and its Legitimacy in the Qur'ān*, trans. Muḥammad Salman Tawhidi (Tehran: Islamic Propagation Society, 1998) at: <www.al-islam.org/short/jihad/index.html>

Muṭahharī, Ayatullah Murtazá, *Master and Mastership* (Islamic Seminary Publications, n.d.) at: <www.al-islam.org/mastership/>

Nafi, Basheer M., 'The Arabs and the Axis: 1933–1940', *Arab Studies Quarterly*, 19 (1997).

Nakamura, Kojiro (trans.), *The Book of Invocation* (*Ihya ulum al-Din*) (translated as *Ghazālī on Prayer*). (Tokyo: University of Tokyo Press, 1975).

Napoleoni, Loretta, *Modern Jihād. Tracing the Dollars behind the Terror Networks* (London: Pluto Press, 2003).

Nasr, Seyyed Hossein, *The Heart of Islām. Enduring Values for Humanity* (San Francisco: HarperSanFrancisco, 2002).

Nasr, Seyyed Vali Reza, *The Vanguard of the Islamic Revolution: The Jama'at-i Islami of Pakistan* (Berkeley, CA: University of California Press, 1994) at: <www.ark.cdlib.org/ark:/13030/ft9j49p32d/>

Nasr, Seyyed Vali Reza, *Mawdūdī and the Making of Islamic Revivalism* (New York and Oxford: Oxford University Press, 1996).

Nasr, Seyyed Vali Reza, 'States and Islamization', in *Muslims and the West. Encounter and Dialogue*, ed. Zafar Ishaq Ansari and John L. Esposito (Islamabad: Islamic Research Institute, Islamabad and Center for Christian–Muslim Understanding, Washington DC, 2002), 296–310.

Nevo, Joseph, 'The Jordanian, Palestinian and the Jordanian-Palestinian Identities', The Fourth Nordic conference on Middle Eastern Studies. 'The Middle East in [a] Globalizing World'. Oslo, 13–16 August 1998 at: <www.hf.uib.no/smi/pao/nevo.html>

'New Evidence on the War in Afghanistan', *Cold War International History Project Bulletin*, 14/15 (Winter 2003–Spring 2004) at: <www.wwics.si.edu/topics/pubs/c-afghanistan.pdf>

Nicosia, Francis R., *The Third Reich and the Palestine Question* (London: I. B. Tauris, 1985).

Noorani, A. G., *Islam and Jihād. Prejudice versus Reality* (London: Zed Books, 2002).

Norton, Augustus Richard, 'Hizballah: From Radicalism to Pragmatism?', *Middle East Policy Council Journal*, 5 (1998) at: <www.mepc.org/public_asp/journal_vol5/9801_norton.asp>

Nyazee, Imran Ahsan Khan (trans.), *AbūSulaymān, Towards an Islamic Theory of International Relations*, 23. *The Distinguished Jurist's Primer. I. Bidāyat al-Mudjtahid. Ibn Rushd* (Reading: Centre for Muslim Contribution to Civilization, 1994, repr. 2000).

Okruhlik, Gwenn, 'Networks of Dissent: Islamism and Reformism in Saudi Arabia' at: <www.ssrc.org/sept11/essays/okruhlik.htm>

Oliver, Haneef James, 'Dispelling the "Wahhabi" Myth: Dispelling Prevalent Fallacies and the Fictitious Link with Bin Laden' (n.p., 2002) at: <www.thewahhabimyth.com/khawarij.htm>

Osman, Fathi, 'Islām and Human Rights: The Challenge to Muslims and the World', in *Rethinking Islām and Modernity. Essays in Honour of Fathi Osman*, ed. Abdelwahab El-Affendi (Leicester: Islamic Foundation, 2001), 27–65.

Osman, Fathi, 'Mawdūdī's Contribution to the Development of Modern Islamic Thinking in the Arabic-Speaking World', *The Muslim World*, 93 (July–October 2003), 465–85.

Osman, Tastan, 'The Jurisprudence of Sarakhsī with Particular Reference to War and Peace: A Comparative Study in Islamic Law' (unpublished PhD thesis, University of Exeter, 1993).

Ovendale, Ritchie, *The Origins of the Arab–Israeli Wars* (4th edn) (Harlow: Pearson, 2004).
Pacheco, Allegra, 'Flouting Convention: The Oslo Agreements', in *The New Intifada. Resisting Israel's Apartheid*, ed. Roane Carey (London and New York: Verso, 2001), 181–206.
Palmer, Alan, *The Decline and Fall of the Ottoman Empire* (London: John Murray, repr. 1993).
Parfrey, Adam (ed.), *Extreme Islām. Anti-American Propaganda of Muslim Fundamentalism* (Los Angeles, CA: Feral, 2001).
Partner, P., *God of Battles. Holy Wars of Christianity and Islam* (London: HarperCollins, 1997).
Peters, F. E., *Muḥammad and the Origins of Islām* (New York: State University of New York Press, 1994).
Peters, F. E., *Islām. A Guide for Jews and Christians* (Princeton, NJ, and Oxford: Princeton University Press, 2003).
Peters, F. E., *The Monotheists. Jews, Christians and Muslims in Conflict and Competition. I. The Peoples of God; II. The Words and Will of God* (Princeton, NJ, and Oxford: Princeton University Press, 2003).
Peters, R. (ed. and trans.), *Jihād in Mediaeval and Modern Islam. The Chapter on Jihād from Averroës' Legal Handbook 'Bidāyat al-Mudjtahid' and the Treatise 'Koran and Fighting' by the Late Shaykh Al-Azhar, Maḥmūd Shaltūt* (Leiden: E. J. Brill, 1977).
Peters, Rudolph, *Islām and Colonialism. The Doctrine of Jihād in Modern History* (The Hague: Mouton, 1979).
Peters, Rudolph, 'Jihād', *The Oxford Encyclopedia of the Modern Islamic World*, ed. J. L. Esposito (4 vols) (New York: Oxford University Press, 1995).
Peters, R., *Jihād in Classical and Modern Islam* (Princeton, NJ: Markus Weiner, 1996).
Petito, Fabio and Hatzopoulos, Pavlos (eds), *Religion in International Relations. The Return from Exile* (New York: Palgrave Macmillan, 2003).
Philby, H. St J. B., *The Heart of Arabia. A Record of Travel and Exploration. I.* (London: Constable, 1922).
Philby, H. St J. B., *Arabia of the Wahhabis* (London: Constable, 1928).
Piscatori, James, 'Islām, Islamists and the Electoral Principle in the Middle East', *ISIM Papers* (Leiden, 2000) at: <www.isim.nl/files/paper_piscatori.pdf>
Poonawala, Ismail Kurban Husein, 'Al-Qāḍī al-Nu'mān and Ismā'īlī Jurisprudence', in *Medieval Ismā'īlī History and Thought*, ed. Farhad Daftary (Cambridge: Cambridge University Press, 1996), 117–43.
Poonawala, Ismail Kurban Husein (ed.), *The Pillars of Islam. Da'ā'im Al-Islām of Al-Qāḍī Al-Nu'mān. I. Acts of Devotion and Religious Observances*, trans. Asaf A. A. Fyzee (New Delhi: Oxford University Press, 2002).
Porath, Yehoshua, *The Palestinian Arab National Movement. From Riots to Rebellion. II. 1929–1939* (London: Frank Cass, 1977).
Pruthi, K. R. (ed.), *Encyclopaedia of Jihād* (5 vols) (New Delhi: Anmol Publications Pvt., 2002).
Puniyani, Ram, *Communal Politics. Facts versus Myths* (New Delhi: Sage, 2003).
Qureshi, M. Naeem, *Pan-Islam in British Indian Politics. A Study of the Khilafat Movement, 1918–1924* (Leiden, Boston and Cologne: E. J. Brill, 1999).
Quṭb, Sayyid, *Islām: The Religion of the Future* (repr. Beirut: Holy Koran Publishing House, 1978).
Quṭb, Sayyid, *Milestones* (Beirut and Damascus: International Islamic Federation of Student Organizations, 1978).
Quṭb, Sayyid, *In the Shade of the Qur'ān* [*Fī Ẓilāl al-Qur'ān*], trans. and ed. Adil Salahi (Leicester: Islamic Foundation and Islamonline.net, in progress: nine volumes by 2004).
Rabbani, Mouin, 'A Smorgasord of Failure: Oslo and the al-Aqṣā *Intifada*', in *The New Intifada. Resisting Israel's Apartheid*, ed. Roane Carey (London and New York: Verso, 2001), 69–89.
Rahman, Fazlur, *Major Themes of the Qur'ān* (Minneapolis, MN: Bibliotheca Islamica, 1980).

Rahman, Ṭāriq, 'The *Madrassa* and the State of Pakistan', *Himāl South Asian* (February 2004) at: <www.himalmag.com/2004/february/essay.htm>

Ramaḍān, Ṭāriq, *To be a European Muslim. A Study of the Islamic Sources in European Context* (Leicester: Islamic Foundation, 1999).

Ramaḍān, Ṭāriq, *Western Muslims and the Future of Islām* (New York: Oxford University Press, 2004).

Rashīd, Aḥmed, *Ṭālibān. The Story of the Afghan Warlords* (London: Pan Books, repr. 2001).

Rashīd, Aḥmed, *Jihād. The Rise of Militant Islām in Central Asia* (New Haven, CT, and London: Yale University Press, 2002).

Rescher, Nicholas, *Pluralism: Against the Demand of Consensus* (Oxford: Clarendon Press, 1993).

Reuter, Christoph, *My Life is A Weapon. A Modern History of Suicide Bombing*, trans. Helena Ragg-Kirkby, Princeton, NJ, and Oxford: Princeton University Press, 2004; orig. German edn 2002).

Reza, Sayed Ali (ed.), *Nahj al-Balāgha. Peak of Eloquence. Sermons, Letters and Sayings of Imām 'Alī ibn Abī Ṭālib* (New York: Tahrike Tarsile Qur'ān, Inc., 3rd rev. edn, 1984).

Richards, John F., *Power, Administration and Finance in Mughal India* (Ashgate: Variorum, 1993).

Richards, John F., *The New Cambridge History of India. 1.5 The Mughal Empire* (Cambridge: Cambridge University Press, repr. 2001).

Rihani, Ameen, *Ibn Sa'oud of Arabia. His People and His Land* (London: Constable, 1928).

Riley-Smith, Jonathan, 'Islam and the Crusades in History and Imagination, 8 November 1898–11 September 2001', *Crusades*, 2 (2003), 151–67.

Riley-Smith, Jonathan, '*Jihād* Crusaders. What Osama Bin Laden Means by "Crusade"' (5 January 2004) at: <www.nationalreview.com/comment/riley-smith200401050839.asp>

Rizvī, Saiyid Athar Abbas, *Religious and Intellectual History of the Muslims in Akbar's Reign: With Special Reference to Abu'l Fazl* (New Delhi: Munshiram Manoharlal, 1975).

Rizvī, Saiyid Athar Abbas, *Shāh Walī Allāh and His Times. A Study of Eighteenth-Century Islām, Politics and Society in India* (Canberra: Ma'rifat, 1980).

Robinson, Chase F., *Empire and Elites after the Muslim Conquest. The Transformation of Northern Mesopotamia* (Cambridge: Cambridge University Press, 2000).

Robinson, Chase F., *Islamic Historiography* (Cambridge: Cambridge University Press, 2003).

Robinson, David, *The Holy War of Umar Tal: The Western Sudan in the Mid-Nineteenth Century* (Oxford: Clarendon Press, 1985) at: <www.pulaaku.net/defte/dRobinson>

Robinson, Glenn E., 'The Peace of the Powerful', in *The New Intifada. Resisting Israel's Apartheid*, ed. Roane Carey (London and New York: Verso, 2001), 111–23.

Roe, Sir Thomas, *A continuation of the Turkish history, from the beginning of the yeare … 1620, vntil the ending of the yeare … 1628. Collected ovt of the papers and dispatches of Sr. Thomas Rowe… And since by him re-viewed and corrected* [in Richard Knolles, *The general historie of the Turkes…* (5th edn) (London: Adam Islip, 1638)], 1409–11.

Rosenthal, E. I. J., *Political Thought in Medieval Islam. An Introductory Outline* (Cambridge: Cambridge University Press, 1958).

Rutter, Eldon, *The Holy Cities of Arabia* (2 vols) (London: Putnam, 1928).

Saad-Ghorayeb, Amal, *Hizbu'llah: Politics and Religion* (London and Sterling, VA: Pluto Press, 2002).

Sachedina, Abdulaziz, *The Islamic Roots of Democratic Pluralism* (New York: Oxford University Press, 2001).

Safi, Louay M., *Peace and the Limits of War. Transcending the Classical Conception of Jihād* (London and Washington: International Institute of Islamic Thought, repr. 2003).

Said, Edward W., 'Palestinians under Siege' (14 December 2000), reprinted in *The New Intifada. Resisting Israel's Apartheid*, ed. Roane Carey (London and New York: Verso, 2001), 27–42.

Saif, Walid, 'Reflections on Muslim–Christian Dialogue: Core Values and Common Responsibilities', *Encounters: Journal of Intercultural Perspectives*, 7 (2001), 91–9.

Saikal, Amin, *Islam and the West. Conflict or Cooperation* (Basingstoke: Palgrave Macmillan, 2003).

Salem, Elie Adib, *Political Theory and Institutions of the Khawārij* (Baltimore, MD: Johns Hopkins University Press, 1956).

Salim, Arskal, and Azra, Azyumardi (eds), *Sharī'a and Politics in Modern Indonesia* (Singapore: Institute of Southasian Studies, 2003).

Sarkar, Jagadish Narayan, *Mughal Polity* (Delhi: IAD, 1984).

Savich, Carl K., 'Islām under the Swastika: The Grand Muftī and the Nazi Protectorate of Bosnia–Hercegovina, 1941–1945': <www.rastko.org.yu/rastko-bl/istorija/kcsavic/csavich-islam_e.html>

Sayigh, Yezid, *Armed Struggle and the Search for State. The Palestinian National Movement, 1949–1993* (Institute for Palestine Studies, Washington DC: Oxford University Press, 1997; repr. 1999).

Schimmel, Annemarie, *Mystical Dimensions of Islam* (Chapel Hill, NC: University of North Carolina Press, 1975, repr. 1976).

Schirazi, Asghar, *The Constitution of Iran. Politics and the State in the Islamic Republic*, trans. John O'Kane (London: I. B. Tauris, 1997).

Schleifer, S. Abdullah, '*Jihād* and the Traditional Islamic Consciousness', *Islamic Quarterly*, 4th quarter (1983) at: <webdev.webstar.co.uk/salaam/knowledge/schleifer.php>

Schwartz, Stephen, *The Two Faces of Islam. The House of Sa'ud from Tradition to Terror* (New York: Doubleday, 2002).

Schwartz, Stephen, *The Two Faces of Islam. Saudi Fundamentalism and its Role in Terrorism* (New York: Anchor Books, repr. 2003).

Sedgwick, Mark J., *Sufism: The Essentials* (Cairo and New York: American University in Cairo Press, rev. edn, 2003).

Selbourne, David, *The Principle of Duty. An Essay on the Foundations of the Civic Order* (London: Sinclair-Stevenson, 1994).

Sen, Satadru, 'Subhas Chandra Bose' at: <www.andaman.org/book/app-m/textm.htm>

Shaban, M. A., *Islamic History, A.D. 600–750 (A.H. 132). A New Interpretation* (Cambridge: Cambridge University Press, 1971).

Shaban, M. A. *Islamic History: A New Interpretation, Part 2: A.D. 750–1055 (A.H. 132–448)* (Cambridge: Cambridge University Press, 1976).

Shamesh, A. Ben (ed.), *Taxation in Islām. Yaḥyā Ben Adam's Kitāb al Kharāj* (3 vols) (Leiden: E. J. Brill, 1958–69).

Shlaim, Avi, *The Politics of Partition. King Abdullah, the Zionists and Palestine, 1921–1951* (Oxford: Oxford University Press, 1988).

Shlaim, Avi, *The Iron Wall. Israel and the Arab World* (London: Penguin, repr. 2001).

Shlaim, Avi, 'The United States and the Israeli–Palestinian Conflict', in *Worlds in Collision. Terror and the Future of Global Order*, ed. Ken Booth and Tim Dunne (Basingstoke: Palgrave Macmillan, 2002), 172–83.

Siddiqi, Abdul Hamid, 'Renaissance in Arabia, Yemen, Iraq, Syria and Lebanon: Muḥammad Bin 'Abd al-Wahhāb and His Movement', in *A History of Muslim Philosophy*, ed. M. M. Sharif (Lahore: Pakistan Philosophical Congress, n.d.) at: <www.muslimphilosophy.com/hmp/default.htm>

Siddiqi, M. Nejatullah, 'Future of the Islamic Movement', *Encounters: Journal of Inter-cultural Perspectives*, 4 (1998), 91–101.

Siddiqui, Ataullah, 'Ethics in Islām: Key Concepts and Contemporary Challenges', *Journal of Moral Education*, 26 (1997), 423–31.

Siddiqui, Ataullah, 'People of Faith in Britain Today and Tomorrow', Centre for the Study of Islām and Christian–Muslim Relations, University of Birmingham, Occasional Papers 4 (1999).

Siddiqui, Ataullah, 'Fifty Years of Christian–Muslim Relations: Exploring and Engaging in a New Relationship', *IslamoChristiana*, 20 (2000), 51–77.

Siddiqui, Ataullah, 'Believing and Belonging in a Pluralist Society: Exploring Resources in Islamic Traditions', in *Multi-Faith Britain*, ed. David Hart (London: O'Books, 2002), 23–33.

Siddiqui, Habib, 'The Repulsive World of Serge Trifkovic', at: <www.mediamonitors.net/habibsiddiqui3.html>

Sikand, Yoginder, *Muslims in India since 1947. Islamic Perspectives on Inter-Faith Relations* (London and New York: RoutledgeCurzon, 2004).

Sivan, Emmanuel, *Radical Islām: Medieval Theology and Modern Politics* (New Haven, CT, and London: Yale University Press, 1985).

Slomp, J., 'The "Political Equation" in *Al-Jihād fi al-Islām* of Abul a'la Mawdūdī (1903–1979)', in *A Faithful Presence: Essays for Kenneth Cragg*, ed. David Thomas with Clare Amos (London: Melisende, 2003), 237–55.

Smith, M. G., 'The *Jihād* of Shehu Dan Fodio: Some Problems', in *Islam in Tropical Africa*, ed. I. M. Lewis (London: Hutchinson, 1969, repr. 1980).

Smith, Wilfrid Cantwell, 'Comparative Religion: Whither and Why?', in *The History of Religions: Essays on Methodology*, ed. M. Eliade and J. Kitagawa (Chicago, IL: University of Chicago Press, 1959), 31–58.

Sobhani, Ayatollah Ja'far, *Doctrines of Shī'ī Islām: A Compendium of Imāmī Beliefs and Practices*, trans. and ed. Reza Shah-Kazemi (London: I. B. Tauris and Institute of Ismaili Studies, 2001).

Spencer, Robert, *Islām Unveiled. Disturbing Questions about the World's Fastest-Growing Faith* (San Francisco: Encounter Books, 2002).

Spencer, Robert, *Onward Muslim Soldiers. How Jihād Still Threatens America and the West* (Washington DC: Regnery Publishing, 2003).

Stern, Jessica, *Terror in the Name of God. Why Religious Militants Kill* (New York: HarperCollins, 2003).

Stone, Martin, *The Agony of Algeria* (New York: Columbia University Press, 1997).

Streusand, Douglas E., 'What Does *Jihād* Mean?', *Middle East Quarterly* (September 1997) at: <www.ict.org.il/articles/jihad.htm>

Sugar, Peter, *South-eastern Europe Under Ottoman Rule, 1354–1804* (Seattle, WA: University of Washington Press, 1977).

Sulaiman, Ibraheem, *A Revolution in History. The Jihād of Usman Dan Fodio* (London and New York: Mansell, 1986).

Tabātabā'ī, Hossein Modarressi, *Kharāj in Islamic Law* (London: Anchor Press, 1983).

Taggar, Yehuda, *The Muftī of Jerusalem and Palestine Arab Politics, 1930–1937* (New York and London: Garland Publishing Inc., 1986).

Talbot, Ian, *Jinnah: Role Model for Future Generations of Pakistanis* (Leicester: INPAREL South Asian History Academic Papers, 1, 2001).

Tamimi, Azzam S., *Rachid Ghannouchi. A Democrat within Islām* (New York: Oxford University Press, 2001).

Tessler, Mark, *A History of the Israeli–Palestinian Conflict* (Bloomington and Indianapolis, IN: Indiana University Press, 1994).

Thapar, Romila, 'Somanatha and Mahmud', *Frontline*, 16:8, 1999 at: <www.flonnet.com/fl1608/16081210.htm>

Thursby, G. R., *Hindu–Muslim Relations in British India. A Study of the Controversy, Conflict and Communal Movements in Northern India, 1923–1928* (Leiden: E. J. Brill, 1975).

Timmerman, Kenneth R., *Preachers of Hate. Islam and the War on America* (New York: Crown Forum, 2003).

Tolan, John V., *Saracens. Islam in the Medieval European Imagination* (New York: Columbia University Press, 2002).

Trifkovic, S., *The Sword of the Prophet. Islam. History, Theology and Impact on the World* (Boston, MA: Regina Orthodox Press, 2002).

Trimingham, J. Spencer, *Islam in West Africa* (Oxford: Clarendon Press, 1959).

Trimingham, J. Spencer, *The Sufi Orders in Islām* (Oxford: Clarendon Press, 1971).

Troll, Christian W., *Sayyid Ahmad Khan: A Reinterpretation of Muslim Theology* (New Delhi: Vikas, 1978).

Turan, Ebru, 'Some Reflections on the Ottoman Grand Vizierate in the Classical Age, 1300–1600', at: <www.humanities.uchicago.edu/orgs/institute/sawyer/archive/islam/ebru.html>

Umamah, Abu (ed.), *Jihād. The Absent Obligation* (Birmingham: Maktabah Al Ansaar Publications, 2000).

United Nations War Crimes Commission (28 May 1948), at: <www.armenian-genocide.org/Affirmation.168/current_category.6/affirmation_detail.html>

Voll, John O., 'Renewal and Reform in Islamic History: *Tajdīd* and *Iṣlaḥ*', in *Voices of Resurgent Islam*, ed. John L. Esposito (New York: Oxford University Press, 1983).

Wasserstein, Bernard, *Divided Jerusalem. The Struggle for the Holy City* (London: Profile Books, 2001).

Wasserstein, Bernard, *Israel and Palestine. Why They Fight and Can They Stop?* (London: Profile Books, 2003).

Wasserstein, D. J., *The Caliphate in the West. An Islamic Political Institution in the Iberian Peninsula* (Oxford: Clarendon Press, 1993).

Watt, W. Montgomery (trans.), *The Faith and Practice of Al-Ghazālī* (London: Allen and Unwin, repr. 1967).

Watt, W. M., *The Formative Period of Islamic Thought* (Edinburgh: Edinburgh University Press, 1973).

Weiss, B., 'Interpretation in Islamic Law: The Theory of *Ijtihād*', *American Journal of Comparative Law*, 26 (1978).

Wendell, C., *Five Tracts of Ḥasan al-Bannā', 1906–1949. A Selection from the Majmū'at Rasā'il al-Imām al-Shahīd Ḥasan al-Bannā* (Santa Barbara, CA: University of California Press, 1978).

Wilber, Donald, 'Overthrow of Premier Mossadeq of Iran, November 1952–August 1953', March 1954, CIA Clandestine Service History, at: <www.gwu.edu/~nsarchiv/NSAEBB/NSAEBB28/4-Orig.pdf>

Williams, Brian Glyn, 'Shattering the al-Qaeda–Chechen Myth' (October 2003) at: <www.peaceinchechnya.org/news/200310-11%20-%20BGW%20Article.htm>

Williams, John Alden, *Themes of Islamic Civilization* (Berkeley, CA, and London: University of California Press, 1971).

Willis, John Ralph, *In the Path of Allāh. The Passion of Al-Hajj 'Umar. An Essay into the Nature of Charisma in Islām* (London: Frank Cass, 1989).

Wolper, Ethel Sara, *Cities and Saints. Sufism and the Transformation of Urban Space in Medieval Anatolia* (Pennsylvania: Pennsylvania State University Press, 2003).

Woodward, Bob, *Plan of Attack* (New York: Simon and Schuster, 2004).

Yapp, M. E., 'Lines in the Sand', review of Neamatollah Nojumi, *The Rise of the Ṭālibān in Afghanistan* (London: Palgrave, 2003) in *Times Literary Supplement,* 18 April 2003, 11.

Ye'or, Bat, *The Dhimmī. Jews and Christians under Islam* (Cranbury, NJ: Associated University Press, rev. edn 1985).

Yousaf, Muḥammad, and Adkin, Mark, *Afghanistan: The Bear Trap. The Defeat of a Superpower* (Havertown: Casemate, 1992; repr. 2001) at: <www.sovietsdefeatinafghanistan.com/beartrap/english/01.htm>

Zacharia, Fouad, 'Human Rights in the Arab World: The Islamic Context', in *Philosophical Foundations of Human Rights* (Paris: UNESCO, 1986).

Zadeh, Kazem Ghazi, 'General Principles of *Imām* Khumaynī's Political Thought', trans. A. N. Baqirshahi, in *Message of Thaqalayn. A Quarterly Journal of Islamic Studies*, 2, Nos 2 and 3, at: <www.aalulbayt.org/html/eng2/books/massage-of-thagalain/17generl-tl.html>; <www.al-shia.com/html/eng/books/message-of-thaqalayn/17generl-tl.html>; <www.al-islam.org/mot/default.asp?url=17Generl.HTM>

Zamān, Muḥammad Qāsim, *Religion and Politics under the Early 'Abbāsids: The Emergence of the Proto-Sunni Elite* (Leiden: E. J. Brill, 1997).

Zamān, Muḥammad Qāsim, *The 'Ulamā' in Contemporary Islām: Custodians of Change* (Princeton, NJ, and Oxford: Princeton University Press, 2002).

Zawātī, H. M., *Is Jihād a Just War? War, Peace and Human Rights under Islamic and Public International Law* (Lewiston, NY: Edwin Mellen Press, 2001).

Zdanowski, Jerzy, 'Some Comments on Islamic Welfare: The Case of the Wahhābī State', at: <www.valt.helsinki.fi/kmi/Julkais/WPt/1999/wp799.htm>

Zeidan, David, 'Radical Islām in Egypt: A Comparison of Two Groups', *Meria (Middle East Review of International Affairs)*, 3 (Sept. 1999) at: <www.meria.idc.ac.il/journal/1999/issue3/jv3n3a1.html>

Zeidan, David, 'The Islamic Fundamentalist View of Life as a Perennial Battle', *Meria (Middle East Review of International Affairs)*, 5 (December 2001) at: <www.meria.idc.ac.il/journal/2001/issue4/zeidan.pdf>

Zürcher, Erik Jan, 'Ottoman Labour Battalions in World War I' at: <www.hist.net/kieser/aghet/Essays/EssayZurcher.html>

Index

Compiled by Sue Carlton